Denitrogenative Transformation of Nitrogen Heterocycles

Denitrogenative Transformation of Nitrogen Heterocycles

Synthesis, Reactions and Applications

Edited by Pazhamalai Anbarasan

WILEY-VCH

Editor

Prof. Pazhamalai Anbarasan
Department of Chemistry
Indian Institute of Technology Madras
Chennai 600036
Tamil Nadu
India

Cover Images:
© aiperfectportraits/pixabay,
Courtesy of Prof. Pazhamalai Anbarasan

Library of Congress Card No.: applied for

British Library Cataloguing-in-Publication Data
A catalogue record for this book is available from the British Library.

Bibliographic information published by the Deutsche Nationalbibliothek
The Deutsche Nationalbibliothek lists this publication in the Deutsche Nationalbibliografie; detailed bibliographic data are available on the Internet at <http://dnb.d-nb.de>.

Print ISBN: 978-3-527-35316-3
ePDF ISBN: 978-3-527-84483-8
ePub ISBN: 978-3-527-84484-5
oBook ISBN: 978-3-527-84485-2

Typesetting Straive, Chennai, India
Printing and Binding CPI Group (UK) Ltd, Croydon, CR0 4YY

C9783527353163_211124

Dedicated to all the contributors in this field

Contents

Preface

Development of elegant synthetic methodologies for the construction of diverse structural motifs in pharmaceuticals, materials, and total synthesis of biologically important natural products is highly desirable in modern organic synthesis. Nitrogen heterocycles are unique structural moieties living in the core of various drug molecules and possessing the ability to act as ligands as well as directing groups in transition metal catalysis. More importantly, polynitrogen heterocycles such as *N*-sulfonyl-1,2,3-triazoles have been found to be shelf-stable and safe-to-handle diazo surrogate in various denitrogenative transformations to access structurally complex molecules of biological importance. Hence, development of methodologies exploiting these *N*-heterocycles including 5-iodotriazoles, F-containing triazoles, tetrazoles, aminoindazoles, benzotriazinones, and so on as synthetic tools has caught significant attention of researchers in past years. Therefore, a book that comprehends all type of nitrogen heterocycles in denitrogenative transformation got our attention.

The ready accessibility of a diverse range of these *N*-heterocycles is the key attribute for high reliance of chemists on related methodologies. Hence, the book begins with Chapter 1 highlighting the remarkable modifications encountered in the advancement of synthesis of these *N*-heterocycles with better yields and selectivity. Interestingly, these modifications yielded metal-free, cost-economical, and greener conditions which can even sustain in water.

The core of the book consists of 13 chapters (Chapters 2–14), which discuss the extensively studied chemistry of *N*-sulfonyl-1,2,3-triazoles and its analogs such as pyridotriazoles, triazoloindoles, benzotriazoles, and thiadiazoles. These diazo surrogates can afford access to unique α-diazoimine without the requirement of any additional coupling partner for generation of metal carbenoids. The beginning of a long story was crafted with simple trapping of the chain isomer of pyridotriazole and *N*-sulfonyl-1,2,3-triazole with Rh(II) catalyst to generate corresponding rhodium azavinyl carbenoid producing nitrogen as green by-product. Generally, this strategy revolves around the donor–acceptor metal carbenoids, which are compatible in plenty of denitrogenative transformations such as transannulations, ylide formations and rearrangements, and insertion with diverse coupling partners including alkyne, nitrile, carbonyl, carbo/heterocycles, etc. The beauty of this carbenoid system lies in the presence of an imino group next to it which facilitates the construction of various complex *N*-heterocycles in atom and step-economical, and waste-free manner. Of note, denitrogenative transformations of benzotriazoles involve generation of diazonium salt instead of diazoimine.

Just like the cherry on the cake, Chapters 15–17 are the beautiful addition in this book showcasing the diversity present in these *N*-heterocycles while keeping the key reactivity intact. These chapters introduce the denitrogenative transformations of few unique heterocycles such as benzotriazinones, 3-aminoindazoles, tetrazoles, and pyridotetrazoles, which can facilitate access into distinguished *N*-containing building blocks. Next, Chapter 18 flashes a light on the aspect of radical denitrogenative transformations exclusively. Eventually, the book ends at where it began as Chapter 19 emphasizes the importance of *N*-heterocycles in synthesis of some natural products exploiting denitrogenative transformation as key step.

Last but not the least, *N*-heterocycles as diazo surrogates have omitted the requirement of slow addition of carbene source in reaction mixture in contrary to traditional diazo compounds and created ample scope to tune the reactivity of the metal carbenoids by introducing electronically different substituents around. Still the chemistry of these heterocycles offers a plenty of opportunities for further development of these synthetic tools and their applications as the book depicts three major limitations manifested in the system: (i) synthesis of triazoloindoles, pyridotriazoles, thiadiazoles, and benzotriazinones are still quite underdeveloped; (ii) a diverse range of transition metal catalysts have been screened over years, but mostly, rhodium(II) carboxylates are found to be productive in denitrogenative transformation of *N*-heterocycles; (iii) application of these diazo surrogates in asymmetric transformations based on denitrogenative transformations is still in its infancy. Hence, further development of synthetic methodologies to attain better accessibility and applicability is desirable due to the growing demand of these heterocycles as synthetic tool. Furthermore, development of methodologies involving the photo and electrochemical denitrogenative transformations could pave the way toward a completely new synthetic arena of these nitrogen heterocycles in near future.

Pazhamalai Anbarasan
Indian Institute of Technology Madras
Chennai, Tamil Nadu, India

1

Synthesis of Diverse Nitrogen Heterocycles Explored in Denitrogenative Transformations

Monalisa Akter and Pazhamalai Anbarasan

Department of Chemistry, Indian Institute of Technology Madras, Chennai, Tamil Nadu, India

1.1 Introduction

Nitrogen heterocycles are extensively present in numerous drug molecules which manifest potent therapeutic activities such as antibacterial, anticancer, antiallergic, potassium-channel activator, antiplatelet, glucosidase, and HIV-1 reverse transcriptase inhibitory activities (Figure 1.1) [1].

Beside living in the core of biologically active molecules, N-heterocycles can act as ligands as well as directing groups in various transition metal catalysis. Importantly, polynitrogen heterocycles can participate in various fruitful transformations that pave the way to access structurally complex molecules of biological importance (Figure 1.2). In this context, *N*-sulfonyl-1,2,3-triazole and its analogs have been exclusively exploited as safe-to-handle diazo surrogates in various denitrogenative transformations. These methods yield a diverse range of structural motifs to facilitate structural modification and total synthesis of natural products and drug molecules [2–4]. Over a period of time, denitrogenative transformations of some related heterocycles such as 5-iodotriazoles [5], F-containing triazoles [6–8], tetrazoles [9, 10], pyridotetrazoles [11], and aminoindazoles [12] have also been explored and established as efficient synthetic tools.

Consequently, easily accessible, atom-economical, and widely compatible synthetic methodologies to access these diverse N-heterocycles are highly desirable. This chapter briefly showcases the development of such primitive to advanced synthetic methodologies to serve the aforementioned purposes.

1.2 Synthesis of 1,2,3-Triazoles

1.2.1 Synthesis of *NH*-Triazoles

NH-triazole is one of the polynitrogen heterocycles that undergoes denitrogenative transformation and their recent development emphasized its importance as a

Denitrogenative Transformation of Nitrogen Heterocycles: Synthesis, Reactions and Applications, First Edition. Edited by Pazhamalai Anbarasan.

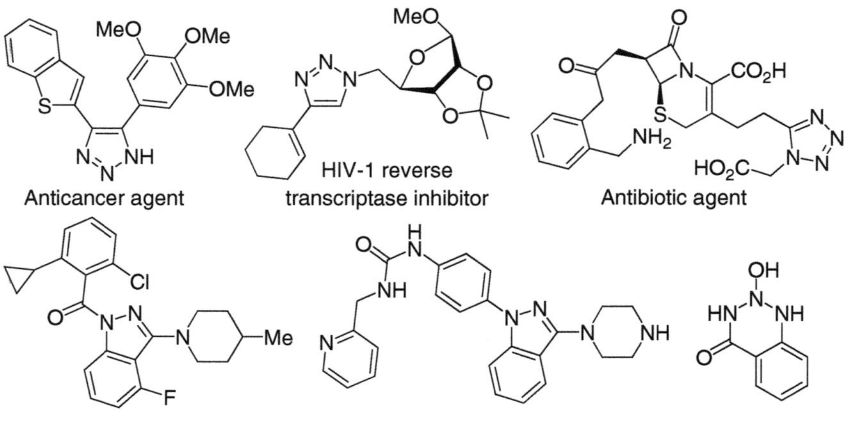

Figure 1.1 Representative drug molecules containing N-heterocycles.

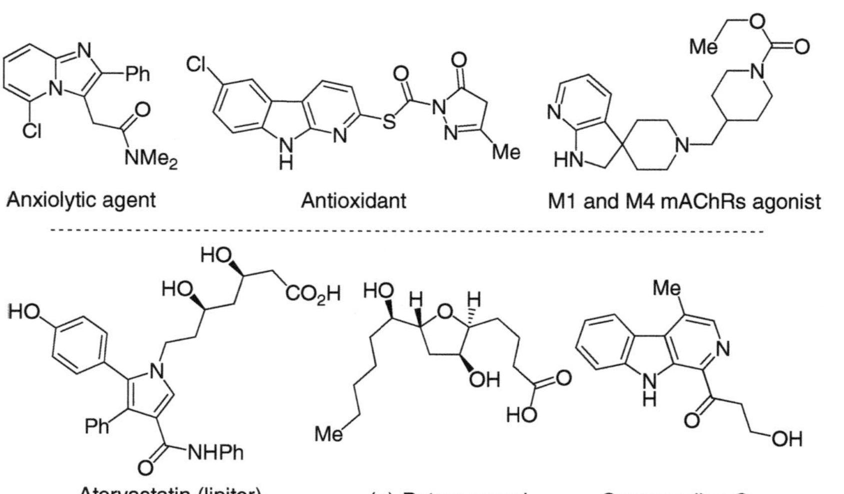

Figure 1.2 Representative drug molecules and natural products synthesized through denitrogenative transformations.

building block and their elegant synthesis. Initially, *NH*-triazoles were prepared *via* the deprotection of various N-protected triazoles. In this context, various organic azides with removable protecting groups such as benzyl [13], tropylium [14], trimethylsilyl [15, 16], tosyl [17, 18], (trimethylsilyl)ethoxymethyl (SEM) [19], and *p*-methoxybenzyl [20] azides have been explored along with sodium azide [21, 22]. Later, Sharpless and co-workers [23] introduced three more organic azides, azidomethyl pivalate, azidomethyl morpholine-4-carboxylate, and azidomethyl *N,N*-diethylcarbamate, which deliver a base-labile N-protected triazole (Scheme 1.1).

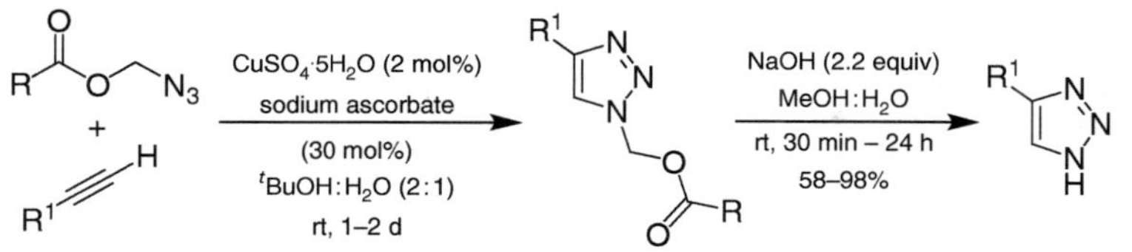

Scheme 1.1 Synthesis of *NH*-triazoles *via* N-protected triazoles. Source: Adapted from Sharpless [23].

In 1989, Banert [24] demonstrated an efficient strategy to synthesize *NH*-triazoles from propargyl azides under mild conditions (Scheme 1.2). Mechanistically, propargyl azide **1.2b** is obtained by treatment of propargyl halide **1.2a** with sodium azide, which undergoes a [3, 3]-sigmatropic rearrangement to generate the reactive allenyl azides **1.2c** that readily cyclizes to triazafulvene intermediate **1.2d**. Finally, the intermediate **1.2d** is trapped by various nucleophiles to afford corresponding triazole **1.2e**.

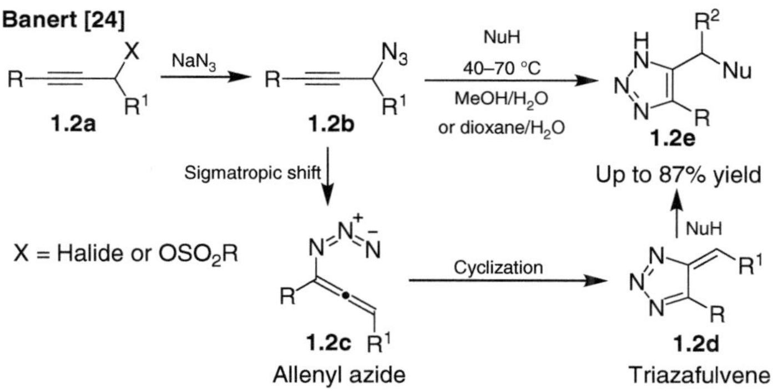

Scheme 1.2 Synthesis of *NH*-triazoles by using Banert cascade. Source: Adapted from Banert [24].

Despite its high reliability and wide substrate scope, its synthetic utility was hardly explored. In 2005, Sharpless and co-workers [25] exclusively studied this pathway to access diverse *NH*-triazoles and expanded the scope of nucleophiles involved in the process (Scheme 1.3). Recently, Topczewski and co-workers [26] exploited silver(I) fluoride as nucleophile, which facilitated access to α-fluorinated *NH*-1,2,3-triazoles **1.3e** in excellent yields.

In 2016, Dehaen and co-workers [27] disclosed the synthesis of various mono-, di-, and tri-substituted triazoles **1.4c** from the reaction of enolizable ketones **1.4a**, NH_4OAc, and nitrophenyl azide **1.4b** under mild acidic condition (Scheme 1.4). Simultaneously, a β-cyclodextrin-mediated multicomponent synthesis of *NH*-triazoles **1.4e** from propynals **1.4d**, trimethylsilyl azide, and malononitrile in water was reported by Medvedeva and co-workers [28]. Besides, the use of amine in place of malononitrile under microwave irradiation furnished the imine-substituted

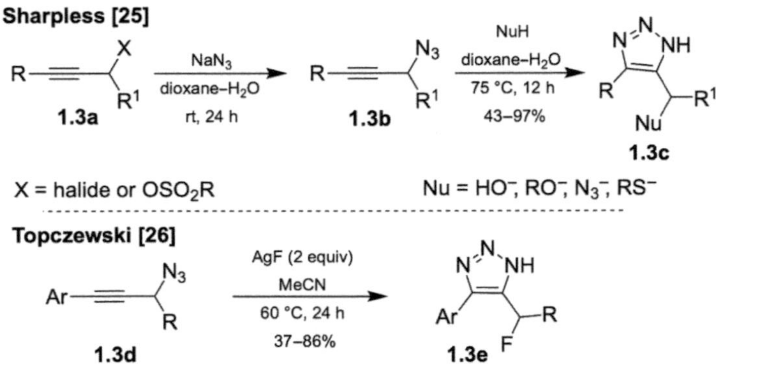

Scheme 1.3 Synthesis of diverse *NH*-triazoles.

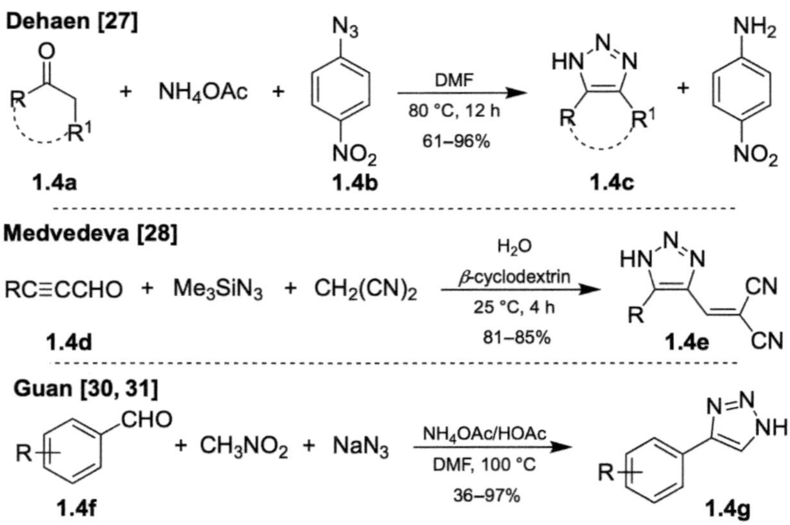

Scheme 1.4 Synthesis of *NH*-triazoles through three-component reactions.

triazole with a shorter reaction time [29]. Subsequently, Guan and co-workers [30–33] accomplished the synthesis of various 4-aryl-*NH*-1,2,3-triazoles **1.4g** through three-component reaction of aldehydes **1.4f**, nitromethane, and NaN_3. Later, Negrón-Silva and co-workers [34] developed a heterogeneous catalytic system consisting of Al-MCM-41 and sulfated zirconia to accomplish the same synthesis.

In 2019, Shu and Wu reported a molecular iodine-mediated cascade [4 + 1] cyclization of *N*-tosylhydrazones **1.5a** and sodium azide in presence of MsOH to access 4-aryl-*NH*-1,2,3-triazoles **1.5b** (Scheme 1.5) [35]. Subsequently, the group of Gao and Shu achieved the synthesis of 4-aryl-*NH*-1,2,3-triazoles **1.5d** *via* an iodine-mediated condensation-cyclization of α-azido ketones **1.5c** with *p*-toluenesulfonyl hydrazide

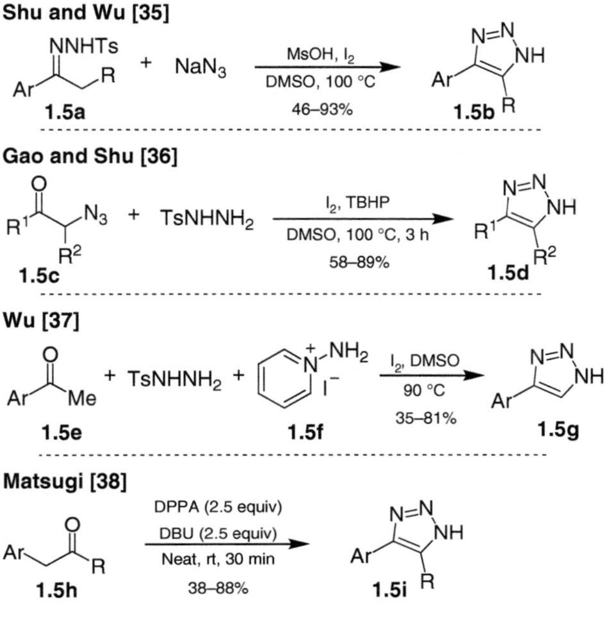

Scheme 1.5 Synthesis of 4-aryl-*NH*-1,2,3-triazoles.

[36]. Recently, Wu and co-workers demonstrated the synthesis of *NH*-triazole under azide-free conditions *via* an iodine-mediated [2 + 2 + 1] cyclization of methyl ketones **1.5e**, *p*-toluenesulfonyl hydrazide, and 1-aminopyridinium iodide **1.5f** [37]. Solvent-free synthesis of 4-aryl-*NH*-1,2,3-triazoles **1.5i** has been demonstrated by Matsugi and co-workers [38] from benzyl ketones **1.5h** exploiting diphenyl phosphorazidate in the presence of 1,8-diazabicyclo[5.4.0]undec-7-ene (DBU).

Catalyst-free synthesis of 4-acyl-*NH*-1,2,3-triazoles **1.6b** was reported by Wen and Wan, which involves water-mediated cycloaddition reactions of enaminones **1.6a** and tosyl azide (Scheme 1.6) [39, 40]. Instead of enaminones, Gribanov et al. [41] employed alkylnitriles **1.6c** and azide **1.6d** in the presence of KO^{t}Bu

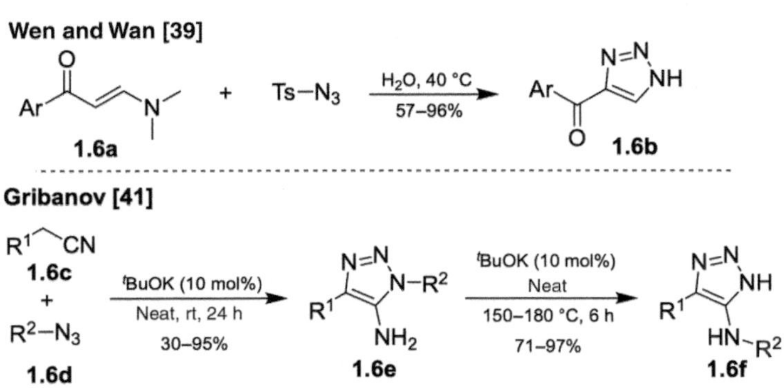

Scheme 1.6 Synthesis of various *NH*-triazoles.

for the synthesis of 5-amino-1,2,3-triazoles **1.6e**, which on subsequent Dimroth rearrangement affords **1.6f** at elevated temperature under solvent-free conditions in one pot.

1.2.2 Synthesis of *N*-Sulfonyl-1,2,3-triazoles

For years, a large number of *N*-sulfonyl-1,2,3-triazoles have been extensively exploited as diazo surrogate in numerous denitrogenative transformations. In general, sulfonylation of *NH*-1,2,3-triazoles **1.7a** with sulfonyl chlorides could furnish the corresponding *N*-sulfonyl-1,2,3-triazoles **1.7b** (Scheme 1.7). But the major drawback of this strategy is the formation of a mixture of regioisomeric products **7.2** and **1.7c**, which significantly reduces its efficiency and applicability [42].

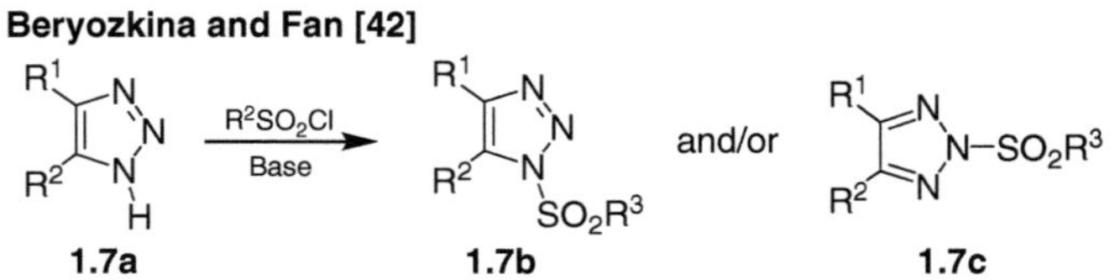

Scheme 1.7 Synthesis of *N*-sulfonyl-1,2,3-triazoles from *NH*-triazoles. Source: Adapted from Beryozkina and Fan [42].

On the other hand, 1,2,3-triazoles **1.8d** were readily achieved through the copper-catalyzed azide-alkyne cycloaddition (CuAAC) as reported by Sharpless and co-workers in 2002 (Scheme 1.8) [43–45]. This reaction appeared to be the most effective click reaction over the traditional Huisgen cycloaddition due to its remarkably high regioselectivity and yields. Various 1,4-disubstituted triazoles **1.8d** could be synthesized from terminal alkynes **1.8a** and azides **1.8b** (Scheme 1.8). However, the use of sulfonyl azides led to the formation of various secondary products **1.8g** instead of the desired triazoles **1.8d** *via* the generation of ketenimine

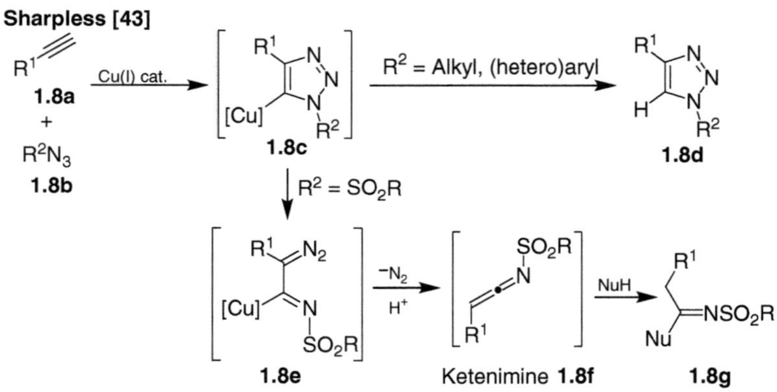

Scheme 1.8 Synthesis of *N*-sulfonyl-1,2,3-triazoles through CuAAC. Source: Adapted from Sharpless [43].

intermediate **1.8f** [46, 47]. The formation of ketenimine was due to the poor stability of the copper-triazole species **1.8c**.

To increase the stability of sulfonyl substituted **1.8c** and for the synthesis of sulfonyltriazoles, in 2007, for the first time, Chang, Fokin and co-workers utilized stoichiometric amount of base, 2,6-dimethylpyridine in the CuI-catalyzed selective synthesis of 4-substituted *N*-sulfonyl-1,2,3-triazoles from the corresponding terminal alkynes and sulfonyl azides (Scheme 1.9) [48, 49]. Subsequently, Fu and co-workers [50] reported an inexpensive catalytic system by exploiting the thioanisole as ligand in combination with CuBr in water medium at room temperature. It was emphasized that the addition of sulfur-containing ligands could inhibit cleavage of N1—N2 bond and stabilize the 5-cuprated triazole species. Two years later, Pérez and co-workers [51] demonstrated the synthesis of *N*-sulfonyl-1,2,3-triazoles using well-defined copper complex, [Tpm*,^{Br}Cu(NCCH$_3$)]BF$_4$. On the other hand, Raushel and Fokin [52] employed Cu(I)-thiophene-2-carboxylate (CuTC) complex, in the absence of external ligand, under mild conditions for more efficient and general synthesis of *N*-sulfonyltriazoles (Scheme 1.9).

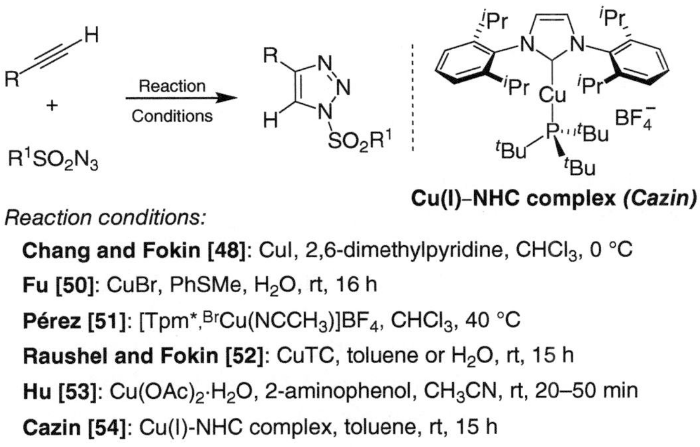

Reaction conditions:

Chang and Fokin [48]: CuI, 2,6-dimethylpyridine, $CHCl_3$, 0 °C

Fu [50]: CuBr, PhSMe, H_2O, rt, 16 h

Pérez [51]: [Tpm*,^{Br}Cu(NCCH$_3$)]BF$_4$, $CHCl_3$, 40 °C

Raushel and Fokin [52]: CuTC, toluene or H_2O, rt, 15 h

Hu [53]: Cu(OAc)$_2$·H$_2$O, 2-aminophenol, CH_3CN, rt, 20–50 min

Cazin [54]: Cu(I)-NHC complex, toluene, rt, 15 h

Scheme 1.9 Cu-catalyzed synthesis of *N*-sulfonyl-1,2,3-triazoles.

In 2011, Hu and co-workers [53] reported 2-aminophenol as a suitable ligand in copper-catalyzed highly selective synthesis of *N*-sulfonyltriazoles. Herein, it was proposed that 2-aminophenol plays a dual role as a reductant and ligand. Importantly, the first triazole possessing two electron-withdrawing groups was synthesized successfully by using this strategy. In 2018, Cazin and co-workers [54] employed Cu(I)-NHC complexes for the selective synthesis of 4-substituted 1,2,3-triazoles through click reaction.

But the limitation of all these above-mentioned methods remains in the selective synthesis of 1,4-disubstituted triazoles. Croatt and co-workers [55] documented the synthesis of 5-substituted *N*-sulfonyl-1,2,3-triazoles **1.10c** from corresponding terminal alkynes **1.10a** and sulfonyl azides **1.10b** using a strong base, such as

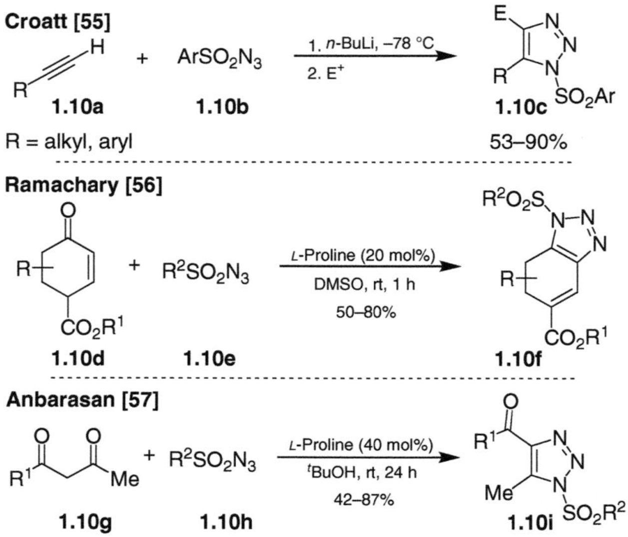

Scheme 1.10 Synthesis of multisubstituted *N*-sulfonyl-1,2,3-triazoles.

n-BuLi (Scheme 1.10). Interestingly, the synthesis of various 1,4,5-trisubstituted triazoles was also achieved by trapping the 4-lithio-1,5-disubstituted triazole in the reaction mixture with suitable electrophiles. In 2008, Ramachary and co-workers reported the first proline-catalyzed synthesis of 4,5-disubstituted triazoles **1.10f** from Hagemann's ester **1.10d** and tosyl azides **1.10e** [56]. Later, Anbarasan and co-workers [57] accomplished a general proline-mediated synthesis of a diverse range of 4,5-disubstituted 1,2,3-triazoles **1.10i** with excellent regioselectivity from substituted 1,3-dicarbonyl compounds **1.10g** and sulfonyl azides **1.10h** through enamine-azide cycloaddition.

In 2016, Zhang and co-worker [58] achieved the synthesis of 4-substituted 1,2,3-triazoles **1.11c** from (*Z*)-arylvinyl bromides **1.11a** and sulfonyl azides through the generation of arylacetylene (Scheme 1.11). This strategy involves KOH-promoted HBr elimination from the vinylbromide **1.11a** followed by a copper-catalyzed [3 + 2] cycloaddition of the resulted alkyne **1.11b** with azide. Subsequently, Ma and co-workers [59] reported a regioselective synthesis of 5-sulfamido-*N*-sulfonyl-1,2,3-triazoles **1.11f** in high yields through a Cu-catalyzed cycloaddition of terminal alkynes **1.11d** and sulfonyl azides **1.11e** using 1.2 equivalents of LiO^{t}Bu. Most recently, Kim and co-workers [60] reported the first continuous flow synthesis of *N*-sulfonyl-1,2,3-triazoles.

1.2.3 Synthesis of *N*-Trifluoromethyl-1,2,3-triazoles

N-trifluoromethyl-1,2,3-triazoles also have been explored in the denitrogenative transformations owing to the unique physicochemical properties of F-containing organic moieties. But, unlike the synthesis of other *N*-sulfonyl-1,2,3-triazoles, the

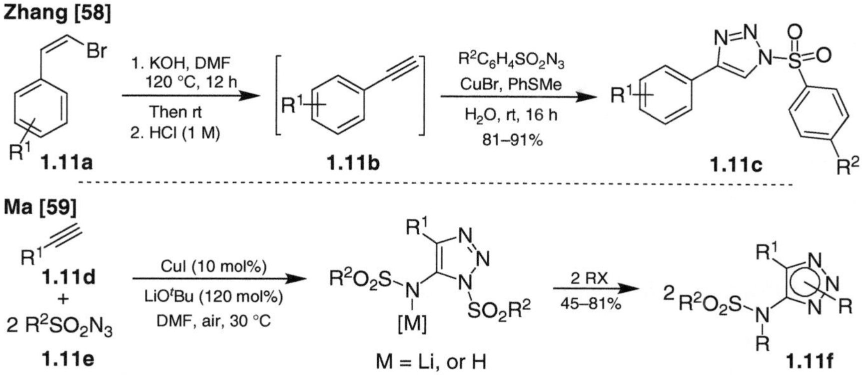

Scheme 1.11 Synthesis of diverse *N*-sulfonyl-1,2,3-triazoles.

synthesis of *N*-fluoroalkyl-1,2,3-triazoles was not achieved following conventional alkyne-azide cycloaddition strategy. In 1977, Holton and co-workers [61] studied various 1,3-dipolar cycloaddition across the C=N bond of polyfluoroazaolefins with diazomethane (Scheme 1.12). A mixture of products was obtained in the reaction of perfluoro-*N*-(trifluoromethyl)butanimidoyl fluoride **1.12a** with excess diazomethane, in which, the 1-(trifluoromethyl)-5-(heptafluoropropyl)-1,2,3-triazole **1.12c** was present as a major product. Recently, Xu, Guan, and Wang demonstrated the synthesis of *N*-trifluoromethyl-1,2,3-triazoles **1.12f** *via* a similar [3 + 2] cyclization of DMAP-stabilized *N*-CF_3 nitrilium salts **1.12d** with diazo compounds **1.12e** [62].

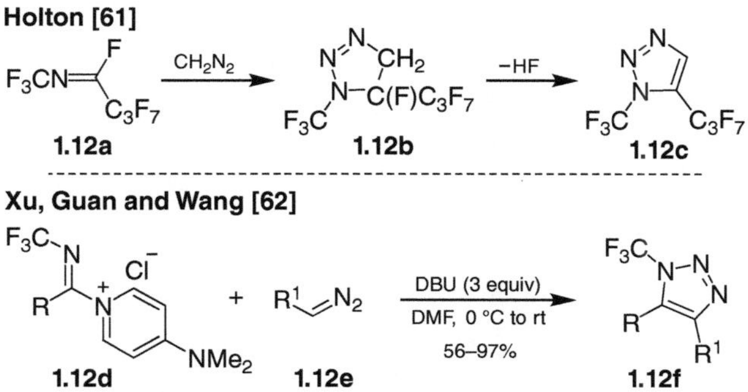

Scheme 1.12 Synthesis of *N*-trifluoromethyl-1,2,3-triazoles.

In 2017, Beier and co-workers [63] developed an efficient strategy to synthesize perfluoroalkyl azides (Scheme 1.13). Subsequently, the reactivity of perfluoroalkyl azide was explored in azide-alkyne cycloaddition reaction to access *N*-perfluoroalkyl triazoles.

Additionally, a one-pot synthesis of such triazoles has also been demonstrated directly starting from trifluoromethylsilane in good yield and regioselectivity.

Scheme 1.13 Synthesis of *N*-perfluoroalkyl-1,2,3-triazoles. Source: Adapted from Beier [63].

1.2.4 Synthesis of 5-Iodo-1,2,3-triazoles

Similar to electron withdrawing group-substituted triazoles, 5-iodo-1,2,3-triazoles having a tethered nucleophile are shown to undergo denitrogenative transformations. These triazoles were also synthesized through azide-alkyne cycloaddition. In this context, in 2005, Wu et al. [64] developed a strategy to access 5-iodo-1,2,3-triazoles **1.14c** in one-pot *via* Cu(I)-catalyzed reaction of organo-azides **1.14a** with terminal alkynes **1.14b** in presence of iodine as an electrophilic trapping reagent (Scheme 1.14). It was proposed that the reaction involves the formation of an intermediate of Cu(I) salt of triazole. Interestingly, yield was increased drastically when ICl was used instead of I_2. Subsequently, Zhang and Zhang [65] reported a CuI-NBS-mediated multicomponent reaction of azides **1.14d** and alkynes **1.14e**, which also facilitated access to 5-iodo-1,4-disubstituted-1,2,3-triazoles **1.14f** [66, 67]. To improve the reactivity and chemoselectivity, ICl was *in situ* generated from the combination of CuI and NCS and used as an effective iodonium source. Recently, these strategies were further modified to accomplish the synthesis in aqueous medium [68, 69].

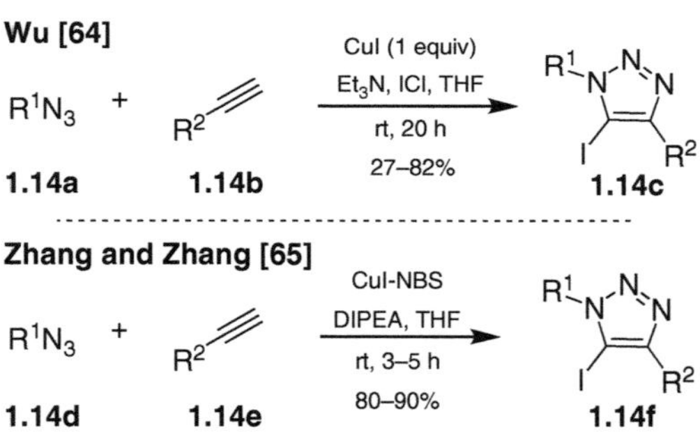

Scheme 1.14 Synthesis of 5-iodo-1,2,3-triazoles.

In 2009, Hein, Fokin, and co-workers [70] reported the synthesis of 5-iodotriazoles from the CuI-catalyzed cycloaddition of 1-iodoalkynes **1.15b** and organic azides **1.15a** (Scheme 1.15). Various ligands such as TBTA or TTTA were screened to furnish the desired products in good yields with low catalyst loadings and shorter reaction times. Zhu and co-workers [71] reported an assisting ligand-free strategy to access various 5-iodo-triazoles **1.15f** (Scheme 1.15). Subsequently, the effect of ligand [72] as well as mechanism [73] involved in the reaction were

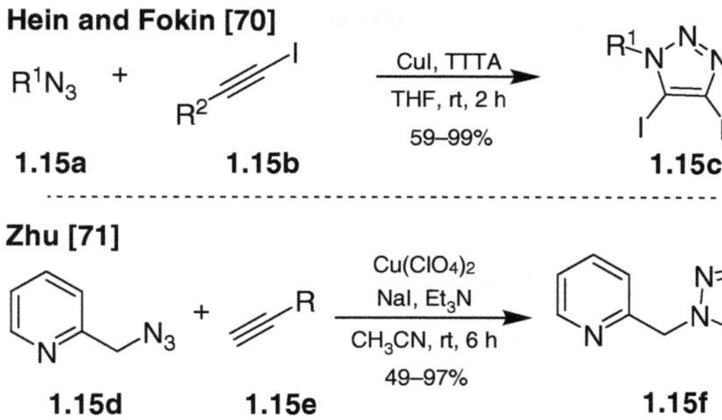

Scheme 1.15 Cu-catalyzed synthesis of 5-iodo-1,2,3-triazoles.

established. Besides, several other groups have also accomplished the synthesis of 5-iodotriazoles [74–80].

1.2.5 Synthesis of Pyridotriazoles

Pyridotriazoles being analogous to triazoles are highly compatible in denitrogenative transformations, but only selected strategies are known for their synthesis. Traditionally, pyridotriazoles **1.16b** have been synthesized from the corresponding hydrazones **1.16a** of pyridine-2-carbaldehydes or 2-pyridylketones *via* oxidative cyclization (Scheme 1.16) [81, 82]. Although some of the hydrazones are just refluxed in methanol under air to afford the corresponding pyridotriazoles, suitable oxidizing agents such as nickel peroxide, air and a cupric salt, potassium ferrocyanide and bicarbonate, manganese dioxide or (diacetoxyiodo)benzene are

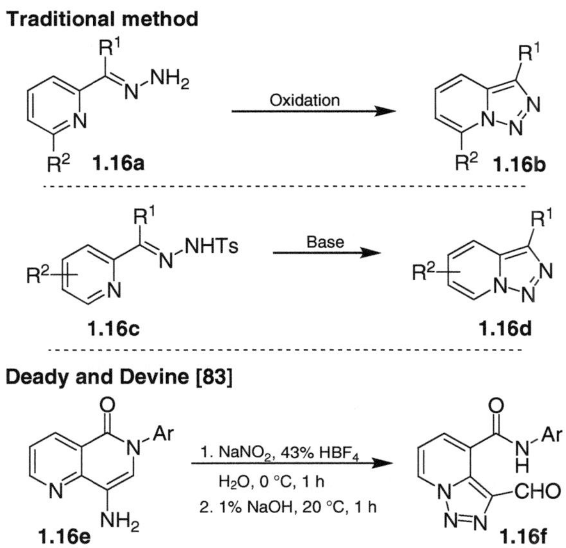

Scheme 1.16 Synthesis of pyridotriazoles. Source: Deady and Devine [83]/with permission of Elsevier.

necessary for most of the hydrazones. In the case of sensitive substrates, pyridotriazoles **1.16d** could be obtained from the reaction of tosyl hydrazones **1.16c** of pyridine-2-carbaldehydes or 2-pyridylketones with base, usually morpholine (Scheme 1.16). In 2006, Deady and Devine demonstrated an alternative strategy to access pyridotriazoles **1.16f** from an aminonaphthyridinone **1.16e** [83]. Compound **1.16e** could form a diazonium salt upon treatment with aqueous sodium nitrate and fluoroboric acid. Then triazolopyrido aldehyde **1.16f** is obtained as a white solid upon treatment of the salt with sodium hydroxide in water.

1.2.6 Synthesis of Triazoloindoles

Due to the growing interest in the denitrogenative transformation of triazoles, in 2014, Lu and Wang accomplished the synthesis of diverse triazoloindoles **1.17b** from alkynes **1.17a** and sulfonylazides *via* a copper-catalyzed tandem process (Scheme 1.17), and demonstrated their reactivity [84]. This strategy involves a sequence of amine group-stabilized CuAAC and C—N coupling reactions. Further study on the triazoloindoles revealed its predominant existence as open-chain form, 3-diazoindolin-2-imines **1.17c**. Subsequently, the same group developed an efficient catalyst-free synthesis of 3-diazoindolin-2-imines **1.17f** from indole and sulfonyl azide [85]. Indoles bearing alkyl, allyl, and benzyl substituents at N1-position and electron-donating substituents in the aryl ring underwent the reaction smoothly and effectively. However, a significant loss in yields was witnessed in the presence of electron-withdrawing groups in aryl ring.

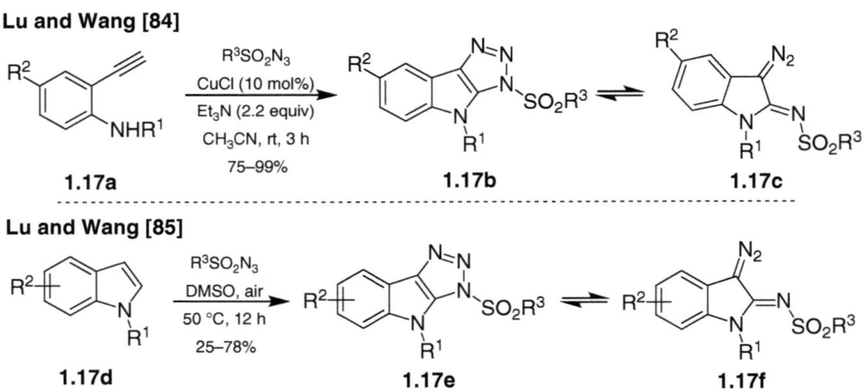

Scheme 1.17 Synthesis of triazoloindoles. Source: Adapted from Lu and Wang [84, 85].

1.2.7 Synthesis of Benzotriazoles

Denitrogenative transformations of benzotriazoles have also witnessed significant development in past years. Particularly, benzotriazoles containing aryl, aroyl, alkenyl, and alkyl substituents at N1-position have been explored well. In general, benzotriazole derivatives are obtained through the selective protection of readily available benzotriazoles (BtH) as shown in Scheme 1.18 [86].

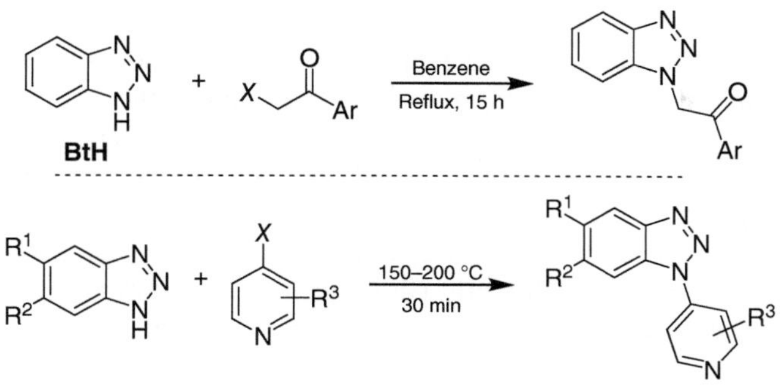

Scheme 1.18 Synthesis of benzotriazole derivatives.

1.3 Synthesis of 1,2,3-Thiadiazoles

1,2,3-Thiadiazoles were also demonstrated to undergo denitrogenative transformation through the generation of thiavinyl carbenes. Their synthesis involves an interesting cyclization with the introduction of sulfur. In 1955, Hard and Mori reported an effective synthesis of 1,2,3-thiadiazoles from acylhydrazones and thionyl chloride (Scheme 1.19) [87]. This method is widely applicable for the synthesis of a variety of substituted thiadiazoles. Alternatively, thiadiazoles could be accessed *via* the reaction of diazo compound with Lawesson's reagent [88].

Hard and Mori [87]

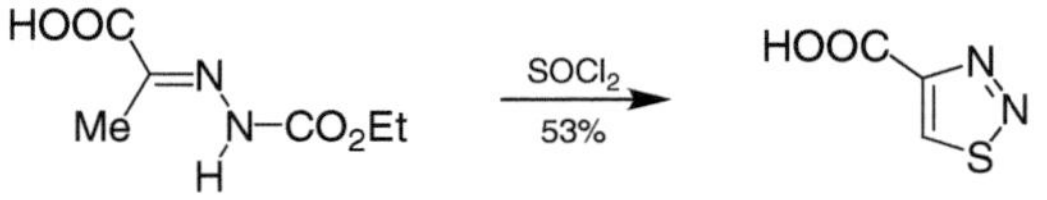

1.19a 1.19b

By using Lawesson's reagent

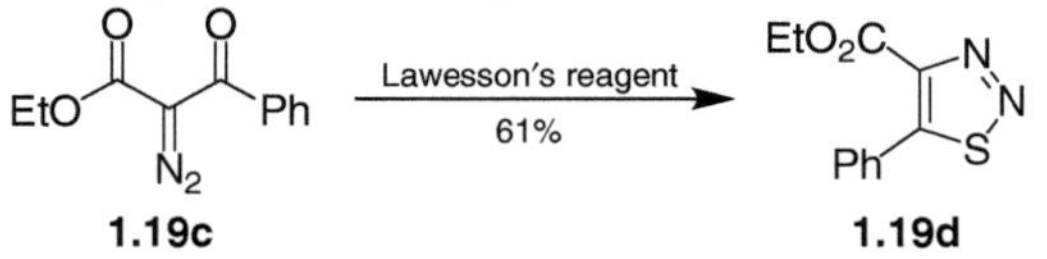

1.19c 1.19d

Scheme 1.19 Synthesis of 1,2,3-thiadiazoles. Source: Hurd and Mori [87]/American Chemical Society.

1.4 Synthesis of Tetrazoles

1.4.1 Synthesis of 1*H*-Tetrazoles

1*H*-tetrazole as a privileged scaffold has received significant attention from researchers in development of drugs as well as in the denitrogenative transformations. Initially, 1*H*-tetrazoles have been exclusively synthesized *via* the

[3 + 2]-cycloaddition reaction of nitriles with various azides such as alkylsilyl azides, alkyltin azides, or sodium azide [89–93]. Major drawbacks of these primitive strategies were use of expensive and toxic metals, water sensitivity, or the possible generation of volatile and explosive hydrazoic acid [94–101]. Consequently, an efficient catalytic synthetic route was highly desired. In 2008, Yamamoto and co-workers [102] developed an efficient Cu(I)-catalyzed synthesis of 5-substituted 1*H*-tetrazoles **1.20b** in good yields *via* the [3 + 2] cycloaddition of nitriles **1.20a** and trimethylsilyl azide (Scheme 1.20). Recently, Gholizadeh and co-workers [103] achieved the synthesis of tetrazole from nitrile using ionic liquid as a catalyst.

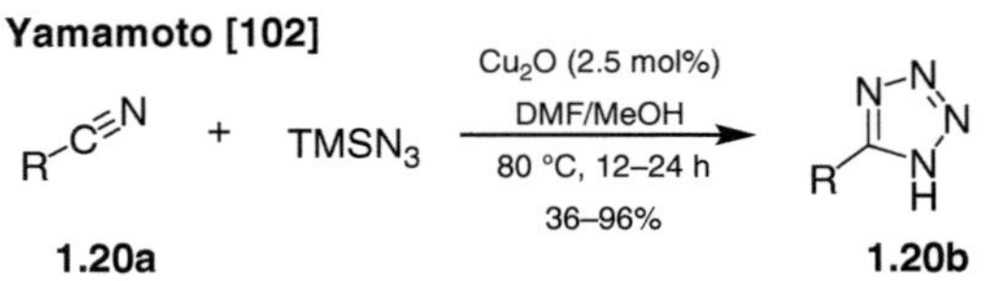

Scheme 1.20 Synthesis of 1*H*-tetrazoles. Source: Yamamoto [102]/with permission of Elsevier.

In 2015, Abdollahi-Alibeik et al. [104] reported a Cu-MCM-41 nanoparticles-catalyzed novel three-component reaction of aldehydes **1.21a**, hydroxylamine, and sodium azide for the synthesis of various tetrazoles **1.21b** (Scheme 1.21). Subsequently, polymer-supported palladium nanoparticles were exploited for the synthesis of tetrazole by Abadi and co-workers [105]. Roy, Islam, and Sen [106] developed a $Ni(OH)_2$ nanoparticles-catalyzed synthesis of 5-substituted 1*H*-tetrazoles **1.21f** from aldoximes **1.21e** and sodium azide in water under mild reaction conditions. Following this development, a remarkable advancement in the synthesis of 1*H*-tetrazole has recently been observed [9, 10, 107–112].

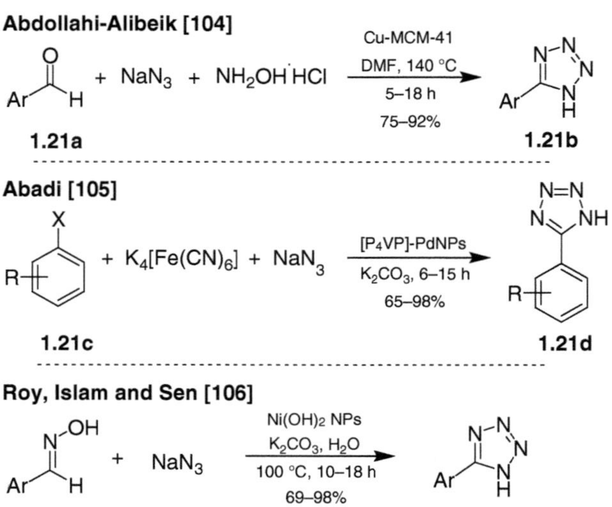

Scheme 1.21 Metal nanoparticles-catalyzed synthesis of 1*H*-tetrazoles.

1.4.2 Synthesis of Pyridotetrazoles

Pyridotetrazoles analogous to tetrazoles also undergo various denitrogenative transformations facilitating access to structurally complex molecules. In 1983, Iyengar and co-workers [113] observed the unprecedented formation of tetrazolo[1,5-*a*]pyridine [114] upon reaction of pyridine-*N*-oxide with arenesulfonyl azide (Scheme 1.22). Later, Keith [115, 116] optimized the reaction conditions and explored the substrate scope extensively. Further, the reaction was modified by Tzschucke and co-workers [117] to achieve the synthesis from pyridine-*N*-oxides, *p*-toluenesulfonyl chloride, and sodium azide.

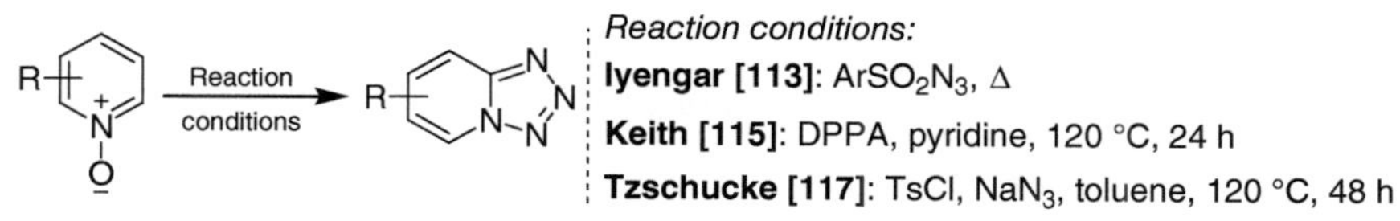

Scheme 1.22 Synthesis of pyridotetrazoles.

In 1997, Noravyan and co-workers [118] synthesized tetrazolo[1,5-*a*]pyrano[3,4-*c*] pyridines **1.23b** from chloropyridines **1.23a** and sodium azide (Scheme 1.23). Later, Cuny and co-workers [119] employed trimethylsilyl azide and tetrabutylammonium fluoride hydrate for the synthesis of various pyridotetrazoles **1.23d** from 2-halopyridines **1.23c**.

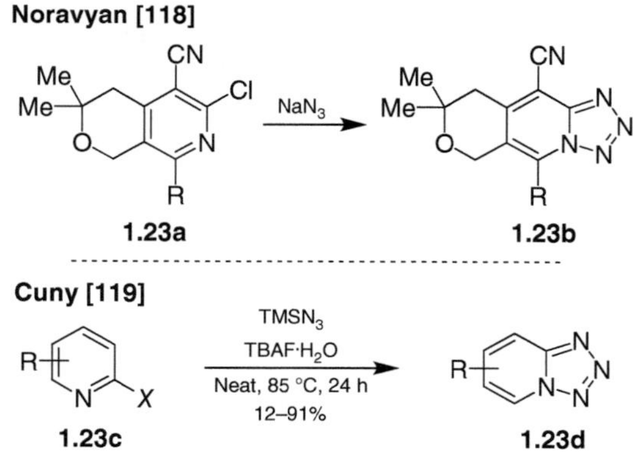

Scheme 1.23 Synthesis of pyridotetrazoles from 2-halopyridines.

Recently, Chattopadhyay and co-workers [120, 121] synthesized various pyridotetrazoles **1.24c** in two steps starting from corresponding 2-halopyridines **1.24a** (Scheme 1.24), and explored their reactivity in denitrogenative transformations. Subsequently, Ghasemzadeh and co-workers [122] reported a three-component reaction of 1*H*-tetrazole-5-amine **1.24d**, benzaldehydes **1.24e**, and 3-cyanoacetyl indole **1.24f** in presence of hexamethylenetetramine-based ionic liquid/MIL-101(Cr) metal-organic framework as a recyclable catalyst, which facilitated access to diverse tetrazolo[1,5-*a*]pyrimidine-6-carbonitriles in excellent yields.

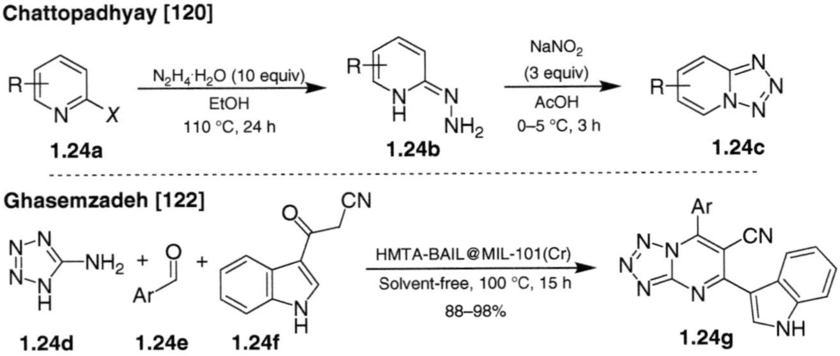

Scheme 1.24 Synthesis of diverse pyridotetrazoles.

1.4.3 Synthesis of Tetrazolo[1,5-*a*]quinolines

Tetrazolo[1,5-*a*]quinolines are largely found in natural products exhibiting biological activities. In 1997, Smalley and co-workers synthesized tetrazolo[1,5-*a*]quinolines **1.25c** by the reaction of 2-azidocarboxaldehyde **1.25a** with acetonitrile in the presence of piperidine or sodium ethoxide as base in ethanol (Scheme 1.25) [123]. The product formation could be rationalized through a Knoevenagel condensation of the acetonitrile with aldehyde **1.25a** followed by an intramolecular 1,3-dipolar cycloaddition reaction of azide with pendant cyano moiety present in **1.25b** intermediate. Later, Bekhit and co-workers reported the synthesis of tetrazolo[1,5-*a*]quinoline-4-carboxaldehyde **1.25e** from corresponding 2-chloroquinoline-3-carboxaldehyde **1.25d** by treating with sodium azide in DMSO/AcOH. [124–126] Recently, Chakroborty et al. modified the reaction conditions to attain a green and more efficient synthesis [127].

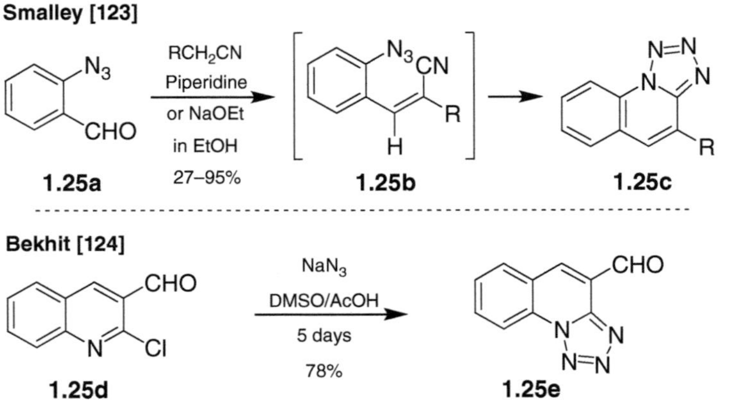

Scheme 1.25 Synthesis of tetrazolo[1,5-*a*]quinolines.

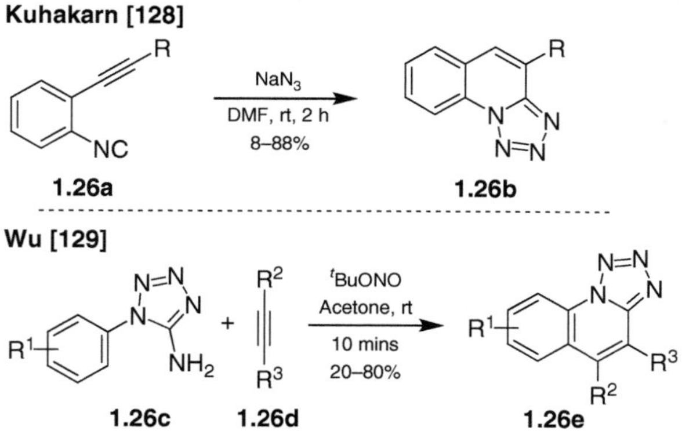

Scheme 1.26 Synthesis of various tetrazolo[1,5-*a*]quinolines.

In 2019, Kuhakarn and co-workers [128] accomplished the synthesis of tetrazolo[1,5-*a*]quinolines **1.26b** from the reaction of *o*-alkynylisocyanobenzenes **1.26a** with sodium azide under metal- and base-free conditions (Scheme 1.26). This strategy involves nucleophilic addition of azide to isocyanide followed by 6-endo cyclization. On the other hand, Wu and co-workers [129] synthesized various tetrazolo[1,5-*a*]quinolines **1.26e** through *tert*-butylnitrite-mediated radical cyclization of 1*H*-tetrazol-5-amines **1.26c** with alkynes **1.26d**.

1.5 Synthesis of 3-Aminoindazoles

Interestingly, 3-aminoindazoles also has emerged as an efficient synthetic tool compatible with denitrogenative transformations. In 2002, Semple and co-workers [130–132] synthesized various 3-aminoindazoles from halobenzonitriles to access important non-covalent thrombin inhibitors. Later, the groups of Ma [133], Devkate [134], and Zhang and Hu [135] accomplished the synthesis of diverse 3-aminoindazoles **1.27c** from reaction of halobenzonitriles **1.27a** with hydrazines **1.27b** under the modified reaction conditions (Scheme 1.27). In 2010, Fabis and co-workers [136] reported a two-step synthesis of substituted 3-aminoindazoles **1.27g** from 2-bromobenzonitriles **1.27d** and hydrazones **1.27e**. This strategy involves *N*-arylation of hydrazone **1.27e** with *o*-bromobenzonitriles **1.27d** followed by deprotection of the hydrazone intermediate **1.27f** in methanol.

In 2011, Callot and co-workers [137] demonstrated a Buchwald–Hartwig C—N coupling reaction of various 3-haloindazoles **1.28a** to access 3-aminoindazoles **1.28b** in good to moderate yields (Scheme 1.28). Subsequently, Charrette and co-workers [138, 139] prepared 3-aminoindazoles from tertiary amides *via* the generation of aminohydrazones **1.28c** followed by intramolecular C—H amination.

Recently, Olmos and co-workers [140] reported the synthesis of 3-aminoindazoles **1.28g** from 3-(2-bromoaryl)-1,2,4-oxadiazol-5(4*H*)-ones **1.28e** and amines **1.28f** which involves an intramolecular N—N bond formation.

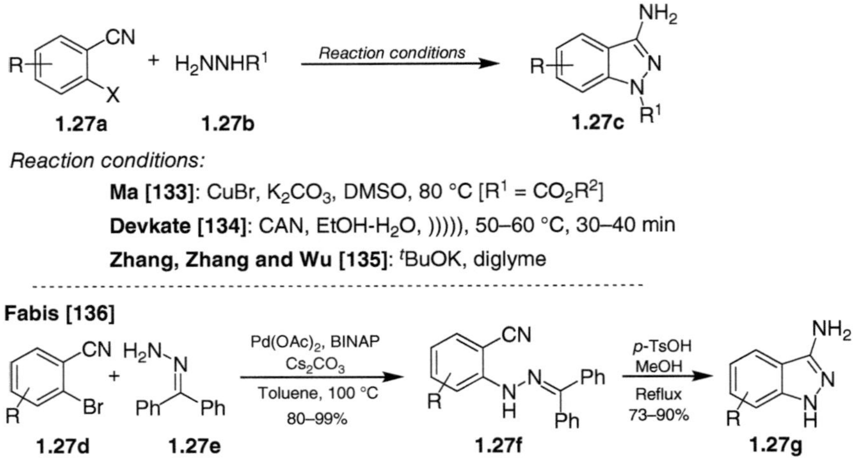

Scheme 1.27 Synthesis of 3-aminoindazoles from 2-halobenzonitriles.

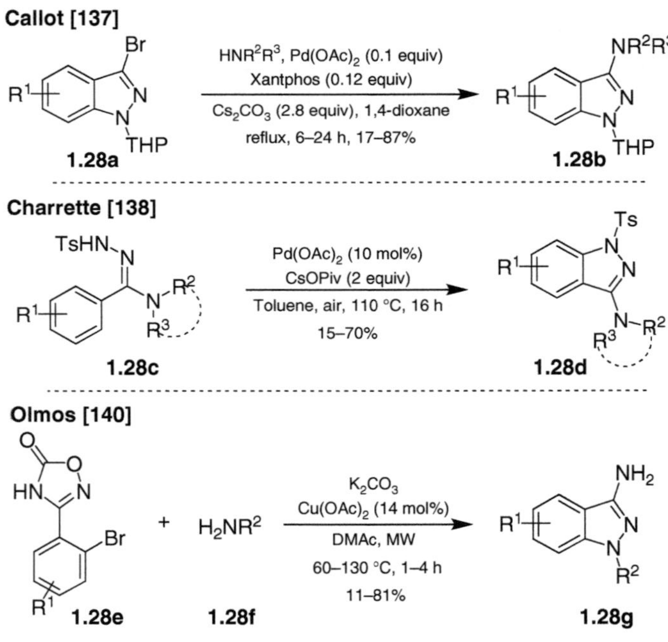

Scheme 1.28 Synthesis of various substituted 3-aminoindazoles.

1.6 Synthesis of Benzotriazinones

Another unique N-heterocycle, benzotriazinone, has been explored in several denitrogenative transformations. In 2015, Xu and co-workers [141–146] prepared several benzotriazinones **1.29b** from anthranilamides **1.29a** *via* diazotization, and cyclization reactions in one-pot (Scheme 1.29). They further derivatized these benzotriazinones by attaching spirocyclic indoline-2-one moieties, and studied

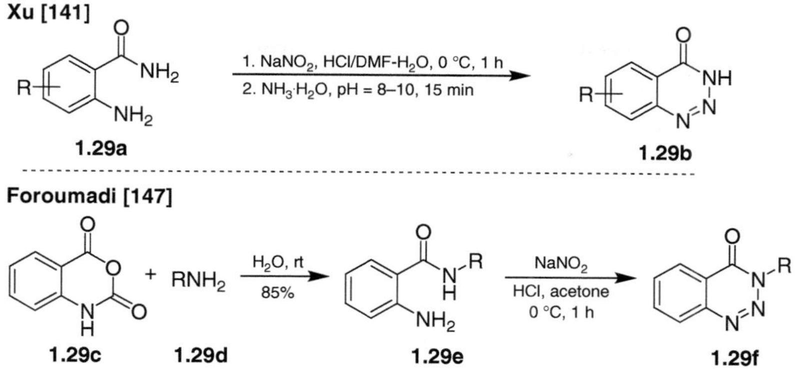

Scheme 1.29 Synthesis of benzotriazinones.

their nematicidal activities. Alternatively, the groups of Foroumadi [147], and Zhang and Li [148] prepared the benzotriazinones **1.29f** from isatoic anhydride **1.29c** and alkyl amines **1.29d** following the reaction sequence shown below. At first, isatoic anhydride **1.29c** and alkyl amines **1.29d** were stirred in water at room temperature to furnish 2-amino-*N*-(prop-2-yn-1-yl)benzamide **1.29e** as a white solid. Next, exposure of **1.29e** to acidic solution of sodium nitrite resulted in the formation of benzotriazinones **1.29f** *via* intramolecular nitrogen–nitrogen bond formation.

In 2017, Rocha-Alonzo and co-workers [149] developed a novel strategy to access 1,2,3-benzotriazinones **1.30b** from the reaction of 2-(*o*-aminophenyl)oxazolines **1.30a** with isoamyl nitrite in methanol (Scheme 1.30). Recently, Rayes and co-workers [150] reported the synthesis from anthranilhydrazide as well as methyl anthranilate.

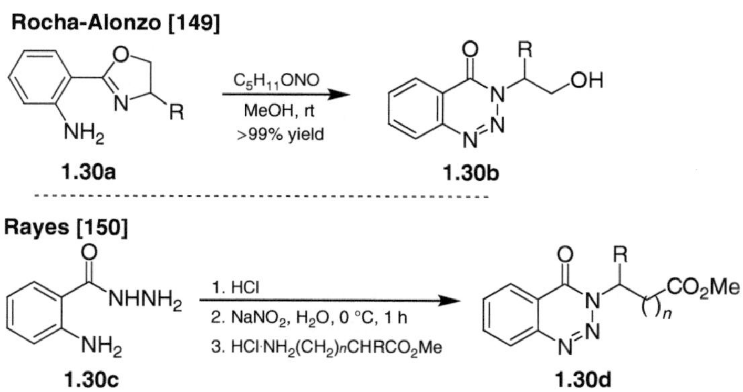

Scheme 1.30 Synthesis of various N-substituted benzotriazinones.

In 2010, Murakami and co-workers reported a Ni-catalyzed annulation of 1,2,3,4-benzothiatriazine-1,1(2*H*)-dioxides **1.31d** with allenes (Scheme 1.31) [151]. Herein, benzothiazinone **1.31d** was synthesized from *ortho*-nitrobenzenesulfonyl chloride **1.31a**. Coupling of **1.31a** with methylamine followed by reduction of

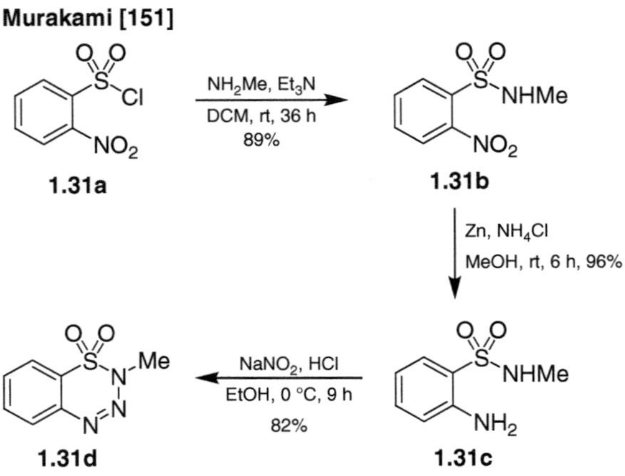

Scheme 1.31 Synthesis of benzothiazinones. Source: Murakami [151]/John Wiley & Sons.

1.31b using Zn gives *ortho*-amino-*N*-methylbenzenesulfonamide **1.31c**. Finally, HONO-mediated ring closing of **1.31c** provides **1.31d** in 82% yield.

1.7 Summary and Outlook

Owing to the remarkable footprints left by various nitrogen heterocycles in the advancement of drugs, and total synthesis of natural products of biological importance, the development of methodologies to access such N-heterocycles has caught significant attention of synthetic chemists in past years. Wherein, atom and cost economy, green approach, and sustainability of the strategy were utmost priorities among other parameters. This chapter highlights the important modifications encountered over the years to accomplish the synthesis of diverse N-heterocycles, for example, triazoles, tetrazoles, 3-aminoindazoles, benzotriazinones, etc. with better yields and selectivity, specifically, which undergo various denitrogenative transformations. Delightfully, many synthetic routes have been developed in a metal-free manner and using milder conditions, some of which are even sustained in a water medium. In recent times, various metal nanoparticles have also been exploited in the advancement of synthesis of these N-heterocycles. Nonetheless, the synthesis of triazoloindoles, pyridotriazoles, thiadiazoles, and benzotriazinones has not received much attention. Hence, further development of synthetic methodologies to attain better accessibility is desirable due to the growing demand for these heterocycles as synthetic tools.

References

1 Anbarasan, P., Yadagiri, D., and Rajasekar, S. (2014). Recent advances in transition-metal-catalyzed denitrogenative transformations of 1,2,3-triazoles and related compounds. *Synthesis* 46: 3004–3023.

2 Akter, M., Rupa, K., and Anbarasan, P. (2022). 1,2,3-Triazole and its analogues: new surrogates for diazo compounds. *Chem. Rev.* 122: 13108–13205.

3 Kadhim, M.M., Mahmood, E.A., Heravi, M.R.P. et al. (2022). The synthesis of biologically active 1-sulfonyl-1,2,3-triazoles from sulfonyl azides and alkynes: a focus review. *J. Sulfur Chem.* 44: 377–391.

4 Bahadorikhalili, S., Divar, M., Damghani, T. et al. (2021). *N*-Sulfonyl ketenimine as a versatile intermediate for the synthesis of heteroatom containing compounds. *J. Org. Chem.* 939: 121773–121812.

5 Danilkina, N.A., Govdi, A.I., and Balova, I.A. (2020). 5-Iodo-1*H*-1,2,3-triazoles as versatile building blocks. *Synthesis* 52: 1874–1896.

6 Kumar, A.S., Kommu, N., Ghule, V.D., and Sahoo, A.K. (2014). Synthesis of trifluoromethyl-substituted *N*-aryl-poly-1,2,3-triazole derivatives. *J. Mater. Chem. A* 2: 7917–7926.

7 Niedermann, K., Früh, N., Senn, R. et al. (2012). Direct electrophilic *N*-trifluoromethylation of azoles by a hypervalent iodine reagent. *Angew. Chem. Int. Ed.* 51: 6511–6515.

8 Usachev, B.I. (2018). Chemistry of fluoroalkyl-substituted 1,2,3-triazoles. *J. Fluor. Chem.* 210: 6–45.

9 Neochoritis, C.G., Zhao, T., and Dömling, A. (2019). Tetrazoles via multicomponent reactions. *Chem. Rev.* 119: 1970–2042.

10 Verma, A., Kaur, B., Venugopal, S. et al. (2022). Tetrazole: a privileged scaffold for the discovery of anticancer agents. *Chem. Biol. Drug Des.* 100: 419–442.

11 Das, S.K., Roy, S., and Chattopadhyay, B. (2023). Transition-metal-catalyzed denitrogenative annulation to access high-valued N-heterocycles. *Angew. Chem. Int. Ed.* 62: e202210912.

12 Guo, Y. and Gao, Q. (2022). Recent advances in 3-aminoindazoles as versatile synthons for the synthesis of nitrogen heterocycles. *Org. Biomol. Chem.* 20: 7138–7150.

13 Wiley, R.H., Hussung, K.F., and Moffat, J. (1956). The preparation of 1,2,3-triazole. *J. Org. Chem.* 21: 190–192.

14 Looker, J.J. (1965). Preparation of 1,2,3-triazoles from 7-azido-l,3, 5-cycloheptatriene. A displacement from nitrogen. *J. Org. Chem.* 30: 638–639.

15 Birkofer, L., Ritter, A., and Richter, P. (1963). Thermolyse silylierter tetrazole. *Chem. Ber.* 96: 2750–2757.

16 Birkofer, L. and Wegner, P. (1966). 1.3-Cycloadditionen mit trimethylsilyl-azid; uber isomere *N*-acetyl-1.2.3-triazole. *Chem. Ber.* 99: 2512–2517.

17 Huisgen, R., Knorr, R., Möbius, L., and Szeimies, G. (1965). Einige Beobachtungen zur Addition organischer Azide an CC-Dreifachbindungen. *Chem. Ber.* 98: 4014–4021.

18 Harvey, G.R. (1966). A new synthesis of 1,2,3-triazoles and diazo esters from phosphorus ylids and azides. *J. Org. Chem.* 31: 1587–1590.

19 Holzer, W. and Ruso, K. (1992). Functionalisation of 1,2,3-triazole via lithiation of 1-[2-(trimethylsilyl)ethoxy]methyl-1*H*-1,2,3-triazole. *J. Heterocyclic Chem.* 29: 1203–1207.

20 Buckle, B.D.R. and Rockell, C.J.M. (1982). The 4-methoxybenzyl group, a versatile N-protecting group for the synthesis of N-unsubstituted v-triazoles. *J. Chem. Soc. Perkin Trans.* 1: 627–630.

21 Tanaka, Y., Velen, S.R., and Miller, S.I. (1973). Syntheses and properties of *H*-1,2,3-triazoles. *Tetrahedron* 29: 3271–3283.

22 Jankovič, D., Virant, M., and Gazvoda, M. (2022). Copper-catalyzed azide-alkyne cycloaddition of hydrazoic acid formed in situ from sodium azide affords 4-monosubstituted-1,2,3-triazoles. *J. Org. Chem.* 87: 4018–4028.

23 Loren, J.C., Krasiński, A., Fokin, V.V., and Sharpless, K.B. (2005). *NH*-1,2,3-triazoles from azidomethyl pivalate and carbamates: base-labile N-protecting groups. *Synlett* 18: 2847–2850.

24 Banert, K. (1989). Synthesis of 1,2,3-triazoles from propargyl azides by rearrangement of the azido group. Indication of short-lived allenyl azides and triazafulvenes. *Chem. Ber.* 122: 911–918.

25 Loren, J.C. and Sharpless, K.B. (2005). The Banert cascade: a synthetic sequence to polyfunctional *NH*-1,2,3-triazoles. *Synthesis* 9: 1514–1520.

26 Alexander, J.R., Kevorkian, P.V., and Topczewski, J.J. (2021). Intercepting the Banert cascade with nucleophilic fluorine: direct access to α-fluorinated *NH*-1,2,3-triazoles. *Chem. Commun.* 57: 5024–5027.

27 Thomas, J., Jana, S., Liekens, S., and Dehaen, W. (2016). A single-step acid catalyzed reaction for rapid assembly of *NH*-1,2,3-triazoles. *Chem. Commun.* 52: 9236–9239.

28 Medvedeva, A.S., Demina, M.M., Vu, T.D. et al. (2016). β-Cyclodextrin-catalyzed three-component synthesis of 4,5-disubstituted 1,2,3-(*NH*)-triazoles from propynals, trimethylsilyl azide and malononitrile in water. *Mendeleev Commun.* 26: 326–328.

29 Medvedeva, A.S., Demina, M.M., Kon'kova, T.V. et al. (2017). Microwave assisted solvent- and catalyst-free three-component synthesis of *NH*-1,2,3-triazoloimines. *Tetrahedron* 73: 3979–3985.

30 Hui, R., Zhao, M., Chen, M. et al. (2017). One-pot synthesis of 4-aryl-*NH*-1,2,3-triazoles through three-component reaction of aldehydes, nitroalkanes and NaN3. *Chin. J. Chem.* 35: 1808–1812.

31 Quan, X.-J., Ren, Z.-H., Wang, Y.-Y., and Guan, Z.-H. (2014). p-Toluenesulfonic acid mediated 1,3-dipolar cycloaddition of nitroolefins with NaN3 for synthesis of 4-Aryl-*NH*-1,2,3-triazoles. *Org. Lett.* 16: 5728–5731.

32 Efimov, I.V. (2019). Recent methods for the synthesis of *NH*-1,2,3-triazoles. *Chem. Heterocycl. Compd.* 55: 28–30.

33 Dutta, B., Hazarika, P.K., Saika, P. et al. (2023). [DDQM][TFSI]: a room temperature ionic liquid as an active electrode material for supercapacitor devices and a catalyst for rapid synthesis of 4-aryl-*NH*-1,2,3-triazoles under microwave irradiation. *New J. Chem.* https://doi.org/10.1039/D3NJ00862B.

34 Vergara-Arenas, B.I., Lomas-Romero, L., Ángeles-Beltrán, D. et al. (2017). Multicomponent synthesis of 4-aryl-*NH*-1,2,3-triazoles in the presence of Al-MCM-41 and sulfated zirconia. *Tetrahedron Lett.* 58: 2690–2694.

35 Shu, W.-M., Zhang, X.-F., Zhang, X.-X. et al. (2019). Metal-free cascade [4 + 1] cyclization access to 4-aryl-*NH*-1,2,3-triazoles from *N*-tosylhydrazones and sodium azide. *J. Org. Chem.* 84: 14919–14925.

36 Ren, M.-T., Li, M., Wang, A.-J. et al. (2020). Iodine-mediated condensation–cyclization of α-azido ketones with p-toluenesulfonyl hydrazide for synthesis of 4-aryl-*NH*-1,2,3-triazoles. *Eur. J. Org. Chem.* 2020: 2233–2236.

37 Huang, C., Geng, X., Zhao, P. et al. (2021). Direct synthesis of 4-aryl-1,2,3-triazoles via I2-promoted cyclization under metal- and azide-free conditions. *J. Org. Chem.* 86: 13664–13672.

38 Ishihara, K., Ishihara, K., Kato, Y. et al. (2022). A solvent-free synthesis of 4-aryl-*NH*-1,2,3-triazoles from ketones utilizing diphenyl phosphorazidate. *Synth. Commun.* 52: 1326–1335.

39 Yang, L., Wu, Y., Yang, Y. et al. (2018). Catalyst-free synthesis of 4-acyl-*NH*-1,2,3-triazoles by water-mediated cycloaddition reactions of enaminones and tosyl azide. *Beilstein J. Org. Chem.* 14: 2348–2353.

40 Opsomer, T. and Dehaen, W. (2021). Metal-free syntheses of N-functionalized and *NH*-1,2,3-triazoles: an update on recent developments. *Chem. Commun.* 57: 1568–1590.

41 Gribanov, P.S., Atoian, E.M., Philippova, A.N. et al. (2021). One-pot synthesis of 5-amino-1,2,3-triazole derivatives via dipolar azide−nitrile cycloaddition and Dimroth rearrangement under solvent-free conditions. *Eur. J. Org. Chem.* 2021: 1378–1384.

42 Beryozkina, T.V., Efimov, I.V., Fabian, W.M.F. et al. (2015). Reactivity of 1,2,3-triazoles towards sulfonyl chlorides. A novel approach to 1- and 2-sulfonyl-4-azolyl-1,2,3-triazoles. *Tetrahedron* 71: 6189–6195.

43 Rostovtsev, V.V., Green, L.G., Fokin, V.V., and Sharpless, K.B. (2002). A stepwise Huisgen cycloaddition process: copper(I)-catalyzed regioselective "ligation" of azides and terminal alkynes. *Angew. Chem. Int. Ed.* 41: 2596–2599.

44 Tornøe, C.W., Christensen, C., and Meldal, M. (2002). Peptidotriazoles on solid phase: [1,2,3]-triazoles by regiospecific copper(I)-catalyzed 1,3-dipolar cycloadditions of terminal alkynes to azides. *J. Org. Chem.* 67: 3057–3064.

45 De Nino, A., Maiuolo, L., Costanzo, P. et al. (2021). Recent progress in catalytic synthesis of 1,2,3-triazoles. *Catalysts* 11: 1120–1167.

46 Meldal, M. and Tornøe, C.W. (2008). Cu-Catalyzed azide-alkyne cycloaddition. *Chem. Rev.* 108: 2952–3015.

47 Hein, J.E. and Fokin, V.V. (2010). Copper-catalyzed azide-alkyne cycloaddition (CuAAC) and beyond: new reactivity of copper(I) acetylides. *Chem. Soc. Rev.* 39: 1302–1315.

48 Yoo, E.J., Ahlquist, M., Kim, S.H. et al. (2007). Copper-catalyzed synthesis of *N*-sulfonyl-1,2,3-triazoles: controlling selectivity. *Angew. Chem. Int. Ed.* 46: 1730–1733.

49 Yoo, E.J., Ahlquist, M., Bae, I. et al. (2008). Mechanistic studies on the Cu-catalyzed three-component reactions of sulfonyl azides, 1-alkynes and amines, alcohols, or water: dichotomy via a common pathway. *J. Org. Chem.* 73: 5520–5528.

50 Wang, F., Fu, H., Jiang, Y., and Zhao, Y. (2008). Copper-catalyzed cycloaddition of sulfonyl azides with alkynes to synthesize *N*-sulfonyltriazoles "on water" at room temperature. *Adv. Synth. Catal.* 350: 1830–1834.

51 Cano, I., Nicasio, M.C., and Pérez, P.J. (2010). Copper(I) complexes as catalysts for the synthesis of *N*-sulfonyl-1,2,3-triazoles from *N*-sulfonylazides and alkynes. *Org. Biomol. Chem.* 8: 536–538.

52 Raushel, J. and Fokin, V.V. (2010). Efficient synthesis of 1-sulfonyl-1,2,3-triazoles. *Org. Lett.* 12: 4952–4955.

53 Liu, Y., Wang, X., Xu, J. et al. (2011). Highly controlling selectivity of copper(I)-catalyzed azide/alkyne cycloaddition (CuAAC) between sulfonyl azids and normal alkynes or propynoates. *Tetrahedron* 67: 6294–6299.

54 Lazreg, F. and Cazin, C.S.J. (2018). Copper(I)-NHC complexes as efficient catalysts for the synthesis of 1,4-disubstituted 1,2,3-sulfonyltriazoles in air. *Organometallics* 37: 679–683.

55 Meza-Avina, M.E., Patel, M.K., Lee, C.B. et al. (2011). Selective formation of 1,5-substituted sulfonyl triazoles using acetylides and sulfonyl azides. *Org. Lett.* 13: 2984–2987.

56 Ramachary, D.B., Ramakumar, K., and Narayana, V.V. (2008). Amino acid-catalyzed cascade [3 + 2]-cycloaddition/hydrolysis reactions based on the push-pull dienamine platform: Synthesis of highly functionalized *NH*-1,2,3-triazoles. *Chem. Eur. J.* 14: 9143–9147.

57 Rajasekar, S. and Anbarasan, P. (2019). A general proline-catalyzed synthesis of 4,5-disubstituted *N*-sulfonyl-1,2,3-triazoles from 1,3-dicarbonyl compounds and sulfonyl azide. *Chem. Asian J.* 14: 4563–4567.

58 Zhang, W. and Xu, W. (2016). Synthesis of 1-arylsulfonyl-1,2,3-triazoles from (*Z*)-arylvinyl bromides by sequential elimination–cycloaddition reaction. *Chem. Heterocycl. Compd.* 52: 192–195.

59 Zheng, X., Wan, Y., Ling, F., and Ma, C. (2017). Copper-catalyzed tandem reaction of terminal alkynes and sulfonyl azides for the assembly of substituted aminotriazoles. *Org. Lett.* 19: 3859–3862.

60 Kwon, Y.-J., Lee, S.-G., and Kim, W.-S. (2023). Continuous flow synthesis of *N*-sulfonyl-1,2,3-triazoles for tandem relay Cu/Rh dual catalysis. *J. Org. Chem.* 88: 1200–1214.

61 Coe, P.L. and Holton, A.G. (1977). Some reactions of diazomethane with polyfluoroazaolefins. *J. Fluor. Chem.* 10: 553–564.

62 Zhang, R.Z., Zhang, R.X., Wang, S. et al. (2022). An *N*-trifluoromethylation/cyclization strategy for accessing diverse *N*-trifluoromethyl azoles from nitriles and 1,3-dipoles. *Angew. Chem. Int. Ed.* 61: e202110749.

63 Blastik, Z.E., Voltrová, S., Matoušek, V. et al. (2017). Azidoperfluoroalkanes: synthesis and application in copper(I)-catalyzed azide–alkyne cycloaddition. *Angew. Chem. Int. Ed.* 56: 346–349.

64 Wu, Y.-M., Deng, J., Li, Y., and Chen, Q.-Y. (2005). Regiospecific synthesis of 1,4,5-trisubstituted-1,2,3-triazole via one-pot reaction promoted by copper(I) salt. *Synthesis* 8: 1314–1318.

65 Li, L., Zhang, G., Zhu, A., and Zhang, L. (2008). A convenient preparation of 5-iodo-1,4-disubstituted-1,2,3-triazole: multicomponent one-pot reaction of azide and alkyne mediated by CuI-NBS. *J. Org. Chem.* 73: 3630–3633.

66 Li, L., Li, Y., Li, R. et al. (2011). A new synthetic protocol for one-pot preparations of 5-halo-1,4-disubstituted-1,2,3-triazoles. *Aust. J. Chem.* 64: 1383–1389.

67 Bogdan, A.R. and James, K. (2011). Synthesis of 5-iodo-1,2,3-triazole-containing macrocycles using copper flow reactor technology. *Org. Lett.* 13: 4060–4063.

68 Li, L., Ding, S., Yang, Y. et al. (2017). Multicomponent aqueous synthesis of iodo-1,2,3-triazoles: single-step models for dual modification of free peptide and radioactive iodo labeling. *Chem. Eur. J.* 23: 1166–1172.

69 Li, L., Xing, X., Zhang, C. et al. (2018). Novel synthesis of 5-iodo-1,2,3-triazoles using an aqueous iodination system under air. *Tetrahedron Lett.* 59: 3563–3566.

70 Hein, J.E., Tripp, J.C., Krasnova, L.B. et al. (2009). Copper(I)-catalyzed cycloaddition of organic azides and 1-iodoalkynes. *Angew. Chem. Int. Ed.* 48: 8018–8021.

71 Brotherton, W.S., Clark, R.J., and Zhu, L. (2012). Synthesis of 5-iodo-1,4-disubstituted-1,2,3-triazoles mediated by in situ generated copper(I) catalyst and electrophilic triiodide ion. *J. Org. Chem.* 77: 6443–6455.

72 Barsoum, D.N., Brassard, C.J., Deeb, J.H.A. et al. (2013). Synthesis of 5-iodo-1,2,3-triazoles from organic azides and terminal alkynes: ligand acceleration effect, substrate scope, and mechanistic insights. *Synthesis* 45: 2372–2386.

73 Barsoum, D.N., Okashah, N., Zhang, X., and Zhu, L. (2015). Mechanism of copper(I)-catalyzed 5-iodo-1,2,3-triazole formation from azide and terminal alkyne. *J. Org. Chem.* 80: 9542–9551.

74 Wang, D., Chen, S., and Chen, B. (2014). "Green" synthesis of 1,4-disubstituted 5-iodo-1,2,3-triazoles under neat conditions, and an efficient approach of construction of 1,4,5-trisubstituted 1,2,3-triazoles in one pot. *Tetrahedron Lett.* 55: 7026–7028.

75 Lal, S., Rzepa, H.S., and Díez-González, S. (2014). Catalytic and computational studies of N-heterocyclic carbene or phosphine-containing copper(I) complexes for the synthesis of 5-iodo-1,2,3-triazoles. *ACS Catal.* 4: 2274–2287.

76 Zhang, J., Chen, W., Wang, B. et al. (2015). One-pot three-component synthesis of 1,4,5-trisubstituted 5-iodo-1,2,3-triazoles from 1-copper(I) alkyne, azide and molecular iodine. *RSC Adv.* 5: 14561–14566.

77 Gribanov, P.S., Topchiy, M.A., Karsakova, I.V. et al. (2017). General method for the synthesis of 1,4-disubstituted 5-halo-1,2,3-triazoles. *Eur. J. Org. Chem.* 2017: 5225–5230.

78 Govdi, A.I., Danilkina, N.A., Ponomarev, A.V., and Balova, I.A. (2019). 1-Iodobuta-1,3-diynes in copper-catalyzed azide-alkyne cycloaddition: a one-step route to 4-ethynyl-5-iodo-1,2,3-triazoles. *J. Org. Chem.* 84: 1925–1940.

79 Gharpure, S.J., Naveen, S., Chavan, R.S., and Padmaja (2020). Regioselective synthesis of halotriazoles and their utility in metal catalyzed coupling reactions. *Eur. J. Org. Chem.* 2020: 6870–6886.

80 Arenas, J.L. and Crousse, B. (2021). An overview of 4- and 5-halo-1,2,3-triazoles from cycloaddition reactions. *Eur. J. Org. Chem.* 2021: 2665–2679.

81 Jones, G. (2002). The chemistry of the triazolopyridines: an update. *Adv. Heterocycl. Chem.* 83: 1–70.

82 Jones, G. and Abarca, B. (2010). The chemistry of the [1,2,3]triazolo[1,5-*a*] pyridines: an update. *Adv. Heterocycl. Chem.* 100: 195–252.

83 Deady, L.W. and Devine, S.M. (2006). Novel annulated products from aminonaphthyridinones. *Tetrahedron* 62: 2313–2320.

84 Xing, Y., Sheng, G., Wang, J. et al. (2014). Preparation of triazoloindoles via tandem copper catalysis and their utility as α-imino rhodium carbene precursors. *Org. Lett.* 16: 1244–1247.

85 Sheng, G., Huang, K., Chi, Z. et al. (2014). Preparation of 3-diazoindolin-2-imines via cascade reaction between indoles and sulfonylazides and their extensions to 2,3-diaminoindoles and imidazo[4,5-*b*]indoles. *Org. Lett.* 16: 5096–5099.

86 Katritzky, A.R., Lan, X., Yang, J.Z., and Denisko, O.V. (1998). Properties and synthetic utility of N-substituted benzotriazoles. *Chem. Rev.* 98: 409–548.

87 Hurd, C.D. and Mori, R.I. (1955). On acylhydrazones and 1,2,3-thiadiazoles. *J. Am. Chem. Soc.* 77: 5359–5364.

88 Kurandina, D. and Gevorgyan, V. (2016). Rhodium thiavinyl carbenes from 1,2,3-thiadiazoles enable modular synthesis of multisubstituted thiophenes. *Org. Lett.* 18: 1804–1807.

89 Duncia, J.V., Pierce, M.E., and Santella, J.B. III, (1991). Three synthetic routes to a sterically hindered tetrazole. A new one-step mild conversion of an amide into a tetrazole. *J. Org. Chem.* 56: 2395–2400.

90 Huff, B.E. and Staszak, M.A. (1993). A new method for the preparation of tetrazoles from nitriles using trimethylsilylazide/trimethylaluminum. *Tetrahedron Lett.* 34: 8011–8014.

91 Wittenberger, S.J. and Donner, B.G. (1993). Dialkyltin oxide mediated addition of trimethylsilyl azide to nitriles. A novel preparation of 5-substituted tetrazoles. *J. Org. Chem.* 58: 4139–4141.

92 Koguro, K., Oga, T., Mitsui, S., and Orita, R. (1998). Novel synthesis of 5-substituted tetrazoles from nitriles. *Synthesis* 1998: 910–914.

93 Curran, D.P., Hadida, S., and Kim, S.-Y. (1999). Tris(2-perfluorohexylethyl)tin azide: A new reagent for preparation of 5-substituted tetrazoles from nitriles with purification by fluorous/organic liquid-liquid extraction. *Tetrahedron* 55: 8997–9006.

94 Demko, Z.P. and Sharpless, K.B. (2001). Preparation of 5-substituted 1*H*-tetrazoles from nitriles in water. *J. Org. Chem.* 66: 7945–7950.

95 Himo, F., Demko, Z.P., Noodleman, L., and Sharpless, K.B. (2002). Mechanisms of tetrazole formation by addition of azide to nitriles. *J. Am. Chem. Soc.* 124: 12210–12216.

96 Demko, Z.P. and Sharpless, K.B. (2002). An expedient route to the tetrazole analogues of α-amino acids. *Org. Lett.* 4: 2525–2527.

97 Himo, F., Demko, Z.P., Noodleman, L., and Sharpless, K.B. (2003). Why is tetrazole formation by addition of azide to organic nitriles catalyzed by zinc(II) salts? *J. Am. Chem. Soc.* 125: 9983–9987.

98 Amantini, D., Beleggia, R., Fringuelli, F. et al. (2004). TBAF-Catalyzed synthesis of 5-substituted 1*H*-tetrazoles under solventless conditions. *J. Org. Chem.* 69: 2896–2898.

99 Ek, F., Manner, S., Wistrand, L.-G., and Frejd, T. (2004). Synthesis of fused tetrazole derivatives via a tandem cycloaddition and N-allylation reaction and parallel synthesis of fused tetrazole amines. *J. Org. Chem.* 69: 1346–1352.

100 Kantam, M.L., Kumar, K.B.S., and Sridhar, C. (2005). Nanocrystalline ZnO as an efficient heterogeneous catalyst for the synthesis of 5-substituted 1*H*-tetrazoles. *Adv. Synth. Catal.* 347: 1212–1214.

101 Kantam, M.L., Balasubrahmanyam, V., and Kumar, K.B.S. (2006). Zinc hydroxyapatite-catalyzed efficient synthesis of 5-substituted 1*H*-tetrazoles. *Synth. Commun.* 36: 1809–1814.

102 Jin, T., Kitahara, F., Kamijo, S., and Yamamoto, Y. (2008). Copper-catalyzed synthesis of 5-substituted 1*H*-tetrazoles via the [3 + 2] cycloaddition of nitriles and trimethylsilyl azide. *Tetrahedron Lett.* 49: 2824–2827.

103 Aali, E., Gholizadeh, M., and Noroozi-Shad, N. (2022). 1-Disulfo-[2,2-bipyridine]-1,1-diium chloride ionic liquid as an efficient catalyst for the green synthesis of 5-substituted 1*H*-tetrazoles. *J. Mol. Struct.* 1247: 131289–131295.

104 Abdollahi-Alibeik, M. and Moaddeli, A. (2015). Multi-component one-pot reaction of aldehyde, hydroxylamine and sodium azide catalyzed by Cu-MCM-41 nanoparticles: a novel method for the synthesis of 5-substituted 1*H*-tetrazole derivatives. *New J. Chem.* 39: 2116–2122.

105 Abadi, S.S.A.D.M. and Zarchi, M.A.K. (2017). A novel route for the synthesis of 5-substituted 1-*H* tetrazoles in the presence of polymer-supported palladium nanoparticles. *New J. Chem.* 41: 10397–10406.

106 Halder, M., Islam, M.M., Singh, P. et al. (2018). Sustainable generation of $Ni(OH)_2$ nanoparticles for the green synthesis of 5-substituted 1*H*-tetrazoles: a competent turn on fluorescence sensing of H_2O_2. *ACS Omega* 3: 8169–8180.

107 Swami, S., Sahu, S.N., and Shrivastava, R. (2021). Nanomaterial catalyzed green synthesis of tetrazoles and its derivatives: a review on recent advancements. *RSC Adv.* 11: 39058–39086.

108 Ghorbani-Choghamarani, A., Shiri, L., and Azadi, G. (2016). The first report on the eco-friendly synthesis of 5-substituted 1*H*-tetrazoles in PEG catalyzed by Cu(II) immobilized on Fe^3O^4@SiO^2@l-arginine as a novel, recyclable and non-corrosive catalyst. *RSC Adv.* 6: 32653–32660.

109 Movaheditabar, P., Javaherian, M., and Nobakht, V. (2022). Igneous rock powder as a heterogeneous multi-oxide nano-catalyst for the synthesis of 5-substituted-1*H*-tetrazoles in polyethylene glycol. *J. Iran. Chem. Soc.* 19: 1805–1816.

110 Mashhoori, M.-S. and Sandaroos, R. (2022). New ecofriendly heterogeneous nano-catalyst for the synthesis of 1-substituted and 5-substituted 1*H*-tetrazole derivatives. *Sci. Rep.* 12: 15364–15379.

111 Sajjadi, M., Nasrollahzadeh, M., Ghafuri, H. et al. (2022). Modified chitosan-zeolite supported Pd nanoparticles: a reusable catalyst for the synthesis of 5-substituted-1*H*-tetrazoles from aryl halides. *Int. J. Biol. Macromol.* 209: 1573–1585.

112 Bharathi, M., Indira, S., Vinoth, G. et al. (2022). A green one-pot three-component syntheses of tetrazole under microwave irradiation using Ni Schiff base complex of mixed ligands on MCM-41 as an efficient heterogeneous catalyst. *J. Porous. Mater.* 30: 691–703.

113 Reddy, K.S., Iyengar, D.S., and Bhalerao, U.T. (1983). The reaction of pyridine N-oxide and its benzoanalogues with arenesulfonyl azides: novel synthesis of tetrazoloazines. *Chem. Lett.* 1983: 1745–1748.

114 Boyer, J.H. and Schoen, W. (1956). 2,3-Ψ-Dinitrosopyridines. *J. Am. Chem. Soc.* 78: 423–425.

115 Keith, J.M. (2006). One-step conversion of pyridine N-oxides to tetrazolo [1,5-*a*]pyridines. *J. Org. Chem.* 71: 9540–9543.

116 Tamba, S., Nitani, M., Seo, T. et al. (2022). Tetrazolo[1,5-*a*]pyridine-containing π-conjugated systems: synthesis, properties, and semiconducting characteristics. *Org. Lett.* 24: 3792–3796.

117 Liu, S., Lentz, D., and Tzschucke, C.C. (2014). Conversion of pyridine N-oxides to tetrazolopyridines. *J. Org. Chem.* 79: 3249–3254.

118 Paronikyan, E.G., Sirakanyan, S.N., and Noravyan, A.S. (1997). Synthesis of derivatives of tetrazolo[1,5-*a*]- and oxazolo[4,5-*b*]pyrano(thiopyrano) [3,4-c]pyridines. *Chem. Heterocycl. Compd.* 33: 1101–1103.

119 Laha, J.K. and Cuny, G.D. (2008). Synthesis of tetrazolo[1,5-*a*]pyridines utilizing trimethylsilyl azide and tetrabutylammonium fluoride hydrate. *Synthesis* 2008: 4002–4006.

120 Roy, S., Khatua, H., Das, S.K., and Chattopadhyay, B. (2019). Iron(II)-based metalloradical activation: switch from traditional click chemistry to denitrogenative annulation. *Angew. Chem. Int. Ed.* 58: 11439–11443.

121 Khatua, H., Das, S.K., Roy, S., and Chattopadhyay, B. (2021). Dual reactivity of 1,2,3,4-tetrazole: manganese-catalyzed click reaction and denitrogenative annulation. *Angew. Chem. Int. Ed.* 60: 304–312.

122 Abdollahi-Basir, M.H., Mirhosseini-Eshkevari, B., Zamani, F., and Ghasemzadeh, M.A. (2021). Synthesis of tetrazolo[1,5-*a*]pyrimidine-6-carbonitriles using HMTA-BAIL@MIL-101(Cr) as a superior heterogeneous catalyst. *Sci. Rep.* 11: 5109–5120.

123 Porter, T.C., Smalley, R.K., Teguiche, M., and Purwono, B. (1997). Tetrazolo[1,5-*a*]quinolines and 1,2,3-triazolo[1,5-*a*]quinazolines by the action of cyanocarbanions on 2-azidoarylcarbonyl compounds. *Synthesis* 1997: 773–777.

124 Bekhit, A.A., El-Sayed, O.A., Aboulmagd, E., and Park, J.Y. (2004). Tetrazolo[1,5-*a*]quinoline as a potential promising new scaffold for the synthesis of novel anti-inflammatory and antibacterial agents. *Eur. J. Med. Chem.* 39: 249–255.

125 Meth-Cohn, O., Narine, B., and Tarnowski, B. (1981). A versatile new synthesis of quinolines and related fused pyridines. Part 5. The synthesis of 2-chloroquinoline-3-carbaldehydes. *J. C. S. Perkin I. Perkin* I: 1520–1530.

126 Kategaonkar, A.H., Labade, V.B., Shinde, P.V. et al. (2010). Synthesis and antimicrobial activity of tetrazolo[1,5-*a*]-quinoline-4-carbonitrile derivatives. *Monatsh. Chem.* 141: 787–791.

127 Chakroborty, S., Ramírez-Lopez, S.C., and Unnamatla, M.V.B. (2022). Synthesis of alkoxy-alkyl- tetrazolo[1,5-*a*] quinoline & tetrazolo[1,5-*a*] quinoline-4-carbaldehyde derivatives under green conditions. *Mater. Today Proc.* 57: 2423–2427.

128 Khaikate, O., Soorukram, D., Leowanawat, P. et al. (2019). Azide-triggered bicyclization of o-alkynylisocyanobenzenes: synthesis of tetrazolo[1,5-*a*]quinolines. *Eur. J. Org. Chem.* 2019: 7050–7057.

129 Liu, T., Ji, Y.-G., and Wu, L. (2019). tert-Butyl nitrite-mediated radical cyclization of tetrazole amines and alkynes toward tetrazolo[1,5-*a*]quinolines. *Org. Biomol. Chem.* 17: 2619–2623.

130 Cui, J.J., Araldi, G.L., Reiner, J.E. et al. (2002). Non-covalent thrombin inhibitors featuring P3-heterocycles with P1-bicyclic arginine surrogates. *Bioorg. Med. Chem. Lett.* 12: 2925–2930.

131 Halley, F. and Sava, X. (1997). Synthesis of 5-cyanoindazole and 1-methyl and 1-aryl-5-cyanoindazoles. *Synth. Commun.* 27: 1199–1207.

132 Orsini, P., Menichincheri, M., Vanotti, E., and Panzeri, A. (2009). Highly efficient synthesis of 5-benzyl-3-aminoindazoles. *Tetrahedron Lett.* 50: 3098–3100.

133 Xu, L., Peng, Y., Pan, Q. et al. (2013). Assembly of substituted 3-aminoindazoles from 2-bromobenzonitrile via a CuBr-catalyzed coupling/condensation cascade process. *J. Org. Chem.* 78: 3400–3404.

134 Devkate, C.G., Patil, A.M., Warad, K.D. et al. (2017). Ceric(IV) ammonium nitrate catalyzed highly efficient synthesis of 3-aminoindazole and their antibacterial screening. *Int. J. Res. Pharm. Chem.* 7: 513–517.

135 Zhang, C., Zhao, H., Li, Z. et al. (2020). Rapid access to 3-aminoindazoles from nitriles with hydrazines: a strategy to overcome the basicity barrier imparted by hydrazines. *Chem. Commun.* 56: 9521–9524.

136 Lefebvre, V., Cailly, T., Fabis, F., and Rault, S. (2010). Two-step synthesis of substituted 3-aminoindazoles from 2-bromobenzonitriles. *J. Org. Chem.* 75: 2730–2732.

137 Lohou, E., Collot, V., Stiebing, S., and Rault, S. (2011). Direct access to 3-aminoindazoles by Buchwald–Hartwig C—N coupling reaction. *Synthesis* 16: 2651–2663.

138 Cyr, P., Régnier, S., Bechara, W.S., and Charette, A.B. (2015). Rapid access to 3-aminoindazoles from tertiary amides. *Org. Lett.* 17: 3386–3389.

139 Sajadi, M.S., Darehkordi, A., and Hosseini, S.M.S. (2021). Synthesis of *N*-aryl-3*H*-indazol-3-imine and *N*-aryl-1*H*-indazol-3-amine via Na_2WO_4/H_2O_2 mediated by intramolecular N—N coupling. *Tetrahedron* 84: 132023–132028.

140 Zhu, W.F., Krämer, A., Knapp, S. et al. (2022). Cascade synthesis of kinase-privileged 3-aminoindazoles via intramolecular N—N bond formation. *J. Org. Chem.* 87: 3856–3862.

141 Wang, G.-L., Chen, X., Chang, Y.-N. et al. (2015). Synthesis of 1,2,3-benzotriazin-4-one derivatives containing spirocyclic indoline-2-one moieties and their nematicidal evaluation. *Chin. Chem. Lett.* 26: 1502–1506.

142 Fiorino, F., Caliendo, G., Perissutti, E. et al. (2005). Synthesis by microwave irradiation and antidiarrhoeal activity of benzotriazinone and saccharine derivatives. *Arch. Pharm. Chem. Life Sci.* 338: 548–555.

143 Chen, X., Jia, H., Li, Z., and Xu, X. (2019). Synthesis and nematicidal evaluation of 1,2,3-benzotriazin-4-one derivatives containing piperazine as linker against Meloidogyne incognita. *Chin. Chem. Lett.* 30: 1207–1213.

144 Chang, Y., Zhang, J., Chen, X. et al. (2017). Synthesis and nematicidal activities of 1,2,3-benzotriazin-4-one derivatives containing thiourea and acylthiourea against Meloidogyne incognita. *Bioorg. Med. Chem. Lett.* 27: 2641–2644.

145 Zhang, R., Guo, W., Wang, G. et al. (2020). Synthesis and nematicidal activities of 1,2,3-benzotriazin-4-one derivatives containing benzo[d][1,2,3]thiadiazole against Meloidogyne incognita. *Bioorg. Med. Chem. Lett.* 30: 127369–127372.

146 Caliendo, G., Fiorino, F., Grieco, P. et al. (1999). Preparation and local anaesthetic activity of benzotriazinone and benzoyltriazole derivatives. *Eur. J. Med. Chem.* 34: 1043–1051.

147 Moghimi, S., Goli-Garmroodi, F., Pilali, H. et al. (2016). Synthesis and anti-acetylcholinesterase activity of benzotriazinone-triazole systems. *J. Chem. Sci.* 128: 1445–1449.

148 Liu, C., Li, X., Zhang, J. et al. (2023). DNA-compatible benzotriazinone formation through aryl diazonium intermediates. *J. Org. Chem.* 88: 6565–6572.

149 Rocha-Alonzo, F., Chávez, D., Ochoa-Terán, A. et al. (2017). A novel synthesis of 1,2,3-benzotriazinones from 2-(o-aminophenyl)oxazolines. *J. Chemother.* 2017: https://doi.org/10.1155/2017/2384780.

150 El Rayes, S.M., Ali, I.A.I., Fathalla, W., and Mahmoud, M.A.A. (2020). Synthesis and biological activities of some new benzotriazinone derivatives based on molecular docking; promising HepG2 liver carcinoma inhibitors. *ACS Omega* 5: 6781–6791.

151 Miura, T., Yamauchi, M., Kosaka, A., and Murakami, M. (2010). Nickel-catalyzed regio- and enantioselective annulation reactions of 1,2,3,4-benzothiatriazine-1,1(2*H*)-dioxides with allenes. *Angew. Chem. Int. Ed.* 49: 4955–4957.

2

Transannulation of *N*-Sulfonyl-1,2,3-triazoles with Polar Multiple Bonds

Chuan-Ying Li and Ze-Feng Xu

School of Chemistry and Chemical Engineering, Zhejiang Sci-Tech University, Xiasha West Higher Education District, Hangzhou, China

2.1 Introduction

In 2008, Gevorgyan and Fokin reported that readily available *N*-sulfonyl-1,2,3-triazoles **2.1a** could generate α-diazo imines **2.1b** via a Dimroth-type equilibrium, which, upon treatment with rhodium(II) catalysts formed α-imino rhodium carbene complexes **2.1c**. These complexes underwent [3 + 2] cycloaddition with nitriles to give imidazoles **2.1d** (Scheme 2.1) [1].

This pioneering work by Gevorgyan and Fokin inaugurated *N*-sulfonyl-1,2,3-triazoles as new precursors for metal carbene. Structurally, this new type of carbene features an electrophilic carbene carbon and nucleophilic imine nitrogen, thus serving as 1,3-dipole synthons for various cyclization reactions with unsaturated bonds. Consequently, their application in organic synthesis has attracted tremendous interest from organic synthetic chemists. After over a decade of thriving development by many researchers, α-imino rhodium carbene has become one of the most important synthetic building blocks for nitrogen-containing compounds, especially nitrogen-heterocycles [2]. This chapter focuses on recent advances in reactions of *N*-sulfonyl-1,2,3-triazoles with polar multiple bonds such as carbonyls, nitriles, etc.

2.2 Reaction of *N*-Sulfonyl-1,2,3-triazoles with Carbonyls

The reaction of *N*-sulfonyl-1,2,3-triazoles with carbonyls is initiated by the nucleophilic addition of the carbonyl oxygen to the electrophilic carbon of α-imino rhodium carbene, resulting in the formation of a carbonyl oxygen ylide **2.2c**. This ylide is a zwitterion that possesses multiple reactive sites. The outcome of the reaction is determined by the structure of the ylide and the reagents used, often leading to the formation of heterocyclic compounds containing multiple heteroatoms (Scheme 2.2).

Denitrogenative Transformation of Nitrogen Heterocycles: Synthesis, Reactions and Applications,
First Edition. Edited by Pazhamalai Anbarasan.

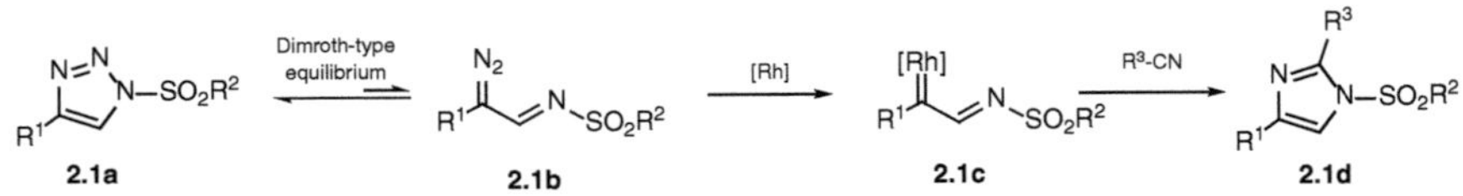

Scheme 2.1 Discovery of *N*-sulfonyl-1,2,3-triazole as carbene precursor.

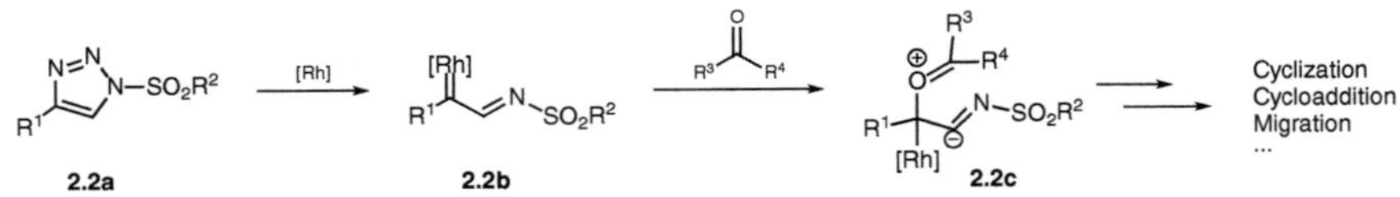

Scheme 2.2 General reaction mode of *N*-sulfonyl-1,2,3-triazoles with carbonyls.

In 2013, the Fokin group reported the asymmetric cyclization reaction of *N*-sulfonyl-1,2,3-triazoles **2.3a** with aldehydes **2.3b** at room temperature, achieving high enantioselectivity in the synthesis of multifunctionalized oxazoline derivatives **2.3c** (Scheme 2.3) [3]. Through the screening of various chiral rhodium catalysts, the optimal catalyst Rh_2(*S*-NTTL)$_4$ was identified for this reaction. However, under certain reaction conditions, the C—O bond in *N,O*-aminals was found to be unstable, leading to ring-opening reactions of the oxazoline ring and affecting the overall efficiency of the reaction. Additionally, 4-alkyl-substituted triazoles did not yield the desired products due to undesired side reactions. Furthermore, by altering the reaction conditions, this reaction could be extended to the cyclization of triazoles with aldimines, enabling the rapid and efficient synthesis of multifunctionalized imidazoles.

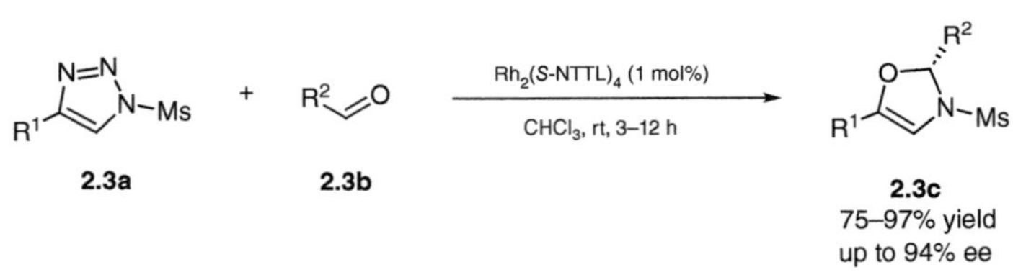

Scheme 2.3 Asymmetric synthesis of oxazolines.

In 2020, Li and Xu reported the rhodium-catalyzed transannulation-aromatization tandem cyclization reaction of *N*-sulfonyl triazoles **2.4a** with aryl aldehydes **2.4b**, leading to the synthesis of 2,5-diaryl-substituted oxazole derivatives **2.4c** (Scheme 2.4) [4]. By utilizing triazoles and aromatic aldehydes with different

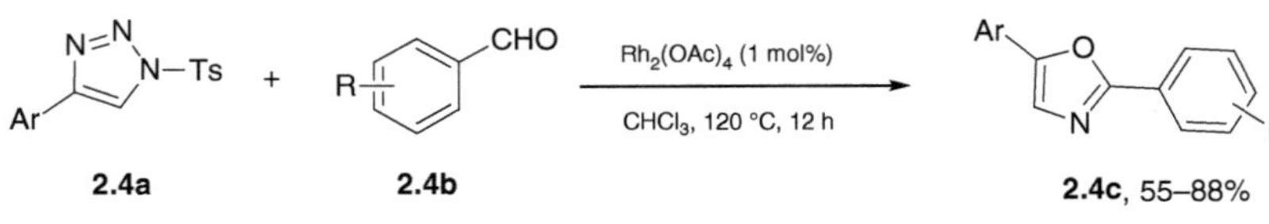

Scheme 2.4 Synthesis of oxazoles from *N*-sulfonyl-1,2,3-triazoles and aldehydes.

substituents, a wide range of 2,5-diaryl-substituted oxazoles can be readily accessed. This reaction requires elevated temperatures to facilitate the desulfonylative aromatization. While nitriles can also react with rhodium carbenes, aldehydes exhibit higher reactivity and excellent chemoselectivity in this transformation. Furthermore, this reaction can be applied efficiently for the synthesis of several bioactive molecules containing 2,5-diaryl-substituted oxazole motifs. Similar to the previous reactions, this process is incompatible with alkyl triazoles.

Although the ring-opening reaction of the *N,O*-aminals motif in oxazoline limits the application of the above reaction to some extent, it can be strategically utilized to synthesize other nitrogen-containing heterocyclic compounds. For instance, Miura and Murakami disclosed a rhodium-catalyzed cyclization reaction of α,β-unsaturated aldehydes and *N*-sulfonyl-1,2,3-triazoles. By taking advantage of the instability of the oxazoline intermediate, they achieved highly stereoselective synthesis of multifunctionalized *trans*-2,3-dihydropyrroles **2.5c** through the ring-opening-cyclization of the oxazoline intermediate **2.5d** (Scheme 2.5) [5]. The configurations of aldehydes did not influence on the reaction, affording the desired products in high yields. Based on control experiments, the authors proposed the depicted mechanism. Initially, the aldehyde reacts with the α-imino rhodium carbene to form the *oxa*-ylide intermediate, which then undergoes intramolecular cyclization to generate the oxazoline intermediate **2.5d**. Due to the instability of the C—O bond, the oxazoline ring opens to produce a zwitterionic intermediate **2.5e**, which further undergoes stereoselective intramolecular *aza*-Michael addition to provide the dihydropyrrole products **2.5c**. Additionally, the authors developed a one-pot procedure for this reaction by employing terminal alkynes, sulfonyl azide, α,β-unsaturated aldehydes, CuTC, and a rhodium catalyst. This one-pot multistep sequence enables the efficient synthesis of dihydropyrroles. It is worth noting that this transformation is limited to α,β-unsaturated aldehydes.

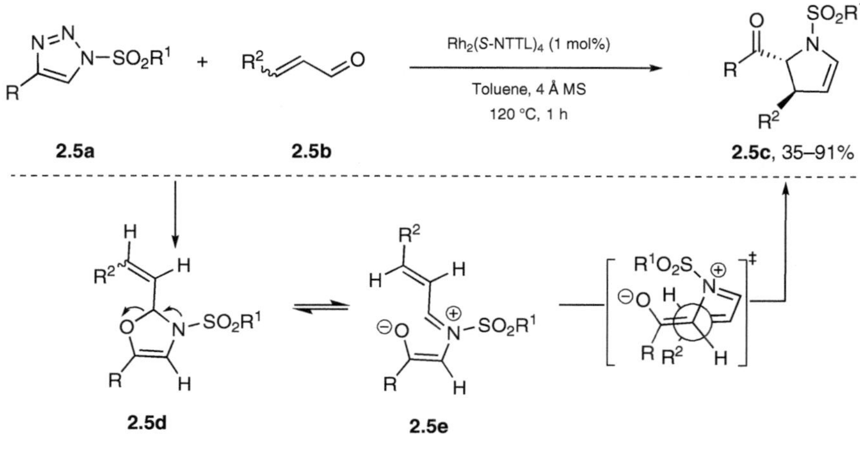

Scheme 2.5 Synthesis of dihydropyrroles.

Wang and co-workers achieved the transannulation reaction of *N*-sulfonyltriazoles with α,β-unsaturated ketones to synthesize multifunctionalized dihydropyrroles **2.6a** (Scheme 2.6) [6]. Through the screening of various rhodium catalysts, the authors identified $Rh_2(S\text{-}PTV)_4$ as the optimal catalyst for this transformation. When cyclic α,β-unsaturated ketones, such as benzofuro[3,2-*b*]pyran-3(2H)-one derivatives, were used as substrates, polycyclic fused heterocyclic compounds, specifically 2,3-dihydro-1*H*-benzofuro[3,2-*b*]pyrroles **2.6b**, could be rapidly constructed. The mechanism of this reaction is similar to the aforementioned reaction of *N*-sulfonyltriazole with unsaturated aldehydes. Mechanistically, this reaction relies on the nucleophilicity of the carbonyl group, and electron-donating substituents on the α,β-unsaturated ketones facilitate the process. Experimental evidence supports this observation, indirectly confirming the plausibility of the proposed mechanism.

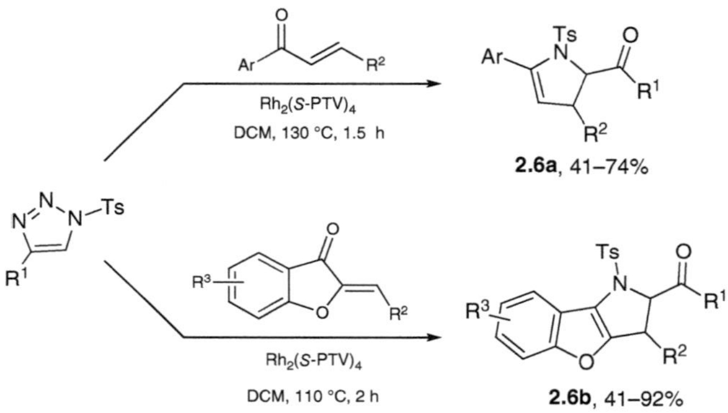

Scheme 2.6 Synthesis of dihydropyrrole derivatives.

Similarly, Fu and colleagues utilized other electron-rich arenes **2.7a** to capture the carbonyl ylide and construct piperidin-3-ones (Scheme 2.7) [7]. This method allows for highly selective access to *trans*-disubstituted 6-arylpiperidin-3-ones **2.7b** in good yields. One possible mechanism begins with the addition of the carbonyl oxygen to the α-imino rhodium carbene, generating a ylide intermediate **2.7c**. The electron-rich arene **2.7a** then undergoes nucleophilic addition to the ylide intermediate, forming a zwitterionic intermediate **2.7d**. Intermediate **2.7d** releases a proton, resulting in the formation of a rhodium-coordinated anionic species **2.7e**. Subsequently, Lewis acid coordination occurs on the oxygen, leading to the cleavage of the C—O bond facilitated by the electron-rich aryl substituent, and generating a new zwitterionic intermediate **2.7f**. Nucleophilic addition of nitrogen to the benzyl carbon takes place, accompanied by the departure of the rhodium catalyst and hydrolysis, ultimately providing the desired product **2.7b**.

Based on the previous works, Bi and Fu have also reported the reaction of alkyl aldehyde-tethered triazoles **2.8a** with alcohols to synthesize 6-alkoxy piperidin-3-ones **2.8b** (Scheme 2.8) [8]. The proposed mechanism involves the initial capture of the carbonyl by the α-imino rhodium carbene, forming a ylide intermediate **2.8c**. Nucleophilic attack by the alcohol on this species generates

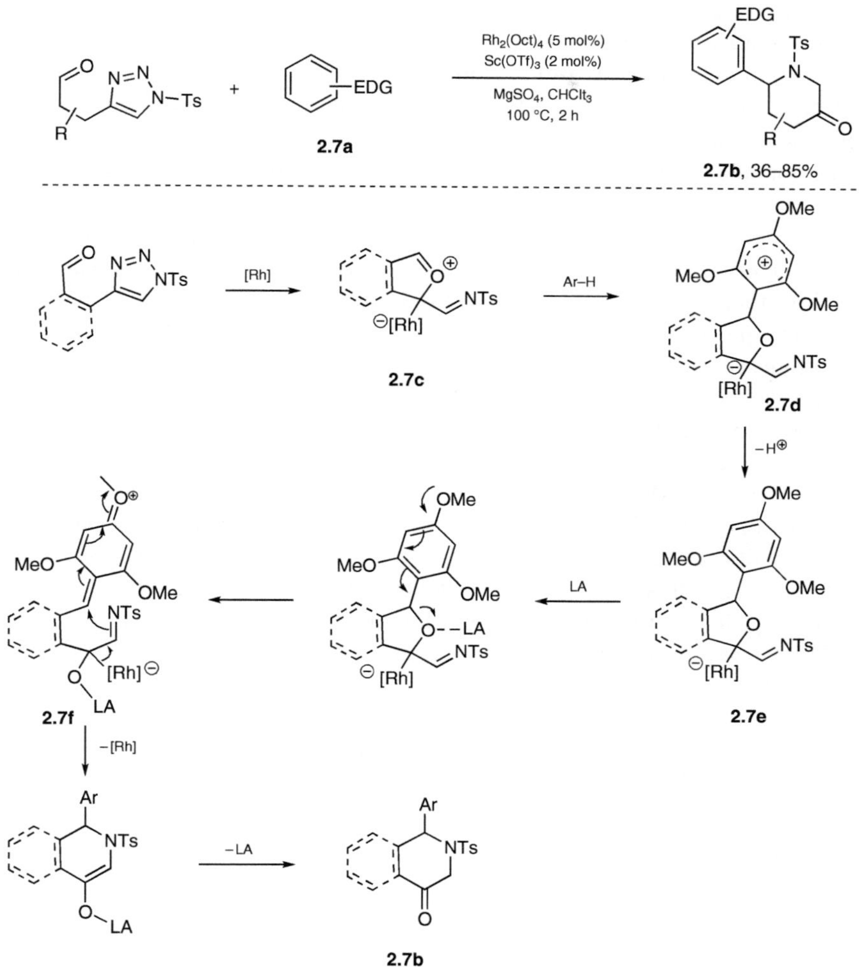

Scheme 2.7 Synthesis of 6-substituted piperidin-3-ones.

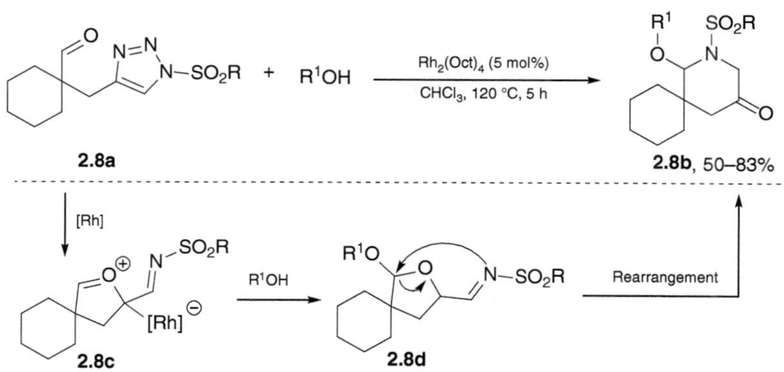

Scheme 2.8 Synthesis of 6-alkoxy piperidin-3-one.

a tetrahydrofuran derivative **2.8d**, which undergoes a skeletal rearrangement to yield the 6-alkoxy piperidin-3-one product **2.8b**. It is noteworthy that the use of chiral alcohols, such as (+)-pinanol, can provide enantioenriched 6-alkoxy piperidin-3-ones with an 82% yield and 97% ee (enantiomeric excess). This method offers a new strategy for accessing chiral 6-substituted piperidin-3-ones, demonstrating the potential of the protocol in drug synthesis without the need for protecting groups or complex synthetic operations.

Furthermore, Gong and Yang developed a stereoselective [3 + 2] cycloaddition for polycycle synthesis based on a similar ylide intermediate (Scheme 2.9) [9]. They utilized this protocol to accomplish the synthesis of various oxabicyclo[2.2.1]heptene motifs. The procedure involves the intramolecular nucleophilic attack of the carbonyl group on the α-imino rhodium carbene, forming a five-membered carbonyl *oxa*-ylide key intermediate. This is followed by intramolecular or intermolecular [3 + 2] cycloaddition reactions. Through systematic screening of rhodium catalysts, the authors discovered that this transformation proceeded with up to 97% yield when catalyzed by $Rh_2(OAc)_4$. Furthermore, using the chiral catalyst Rh_2(*S*-DOSP)$_4$, they achieved up to 83% ee, demonstrating good stereoselectivity. This work showcases the strategy of efficiently constructing polycyclic scaffolds through tandem rhodium-catalyzed reactions. It also highlights the selective formation of different products by controlling the reaction conditions, providing an effective approach for the stereoselective synthesis of complex polycyclic compounds.

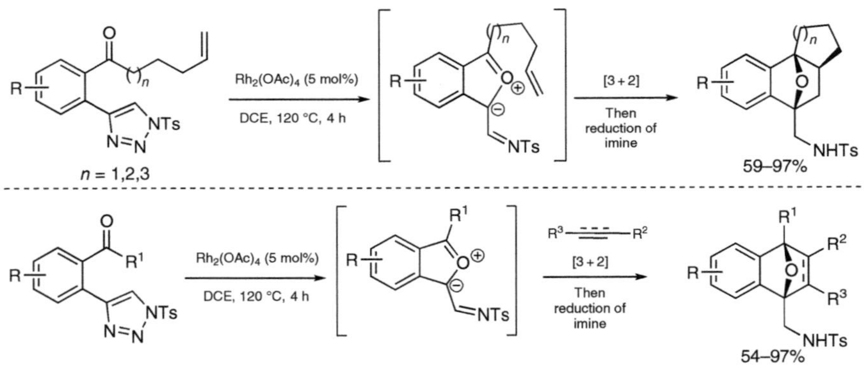

Scheme 2.9 Diastereoselective synthesis of oxabicyclo[2.2.1]heptene core.

By varying the tether chain between the carbonyl and triazole moieties, structurally complex fused heterocyclic compounds can be flexibly constructed. Tang and Shi developed an intramolecular cyclization reaction of triazoles to synthesize *N,O*-heterobridged cyclic compounds **2.10b** (Scheme 2.10) [10]. This reaction exhibits a broad substrate scope and tolerates many functional groups with diverse electronic and steric properties. The reaction proceeds by first generating a carbonyl ylide intermediate **2.10c**, which then undergoes ring-opening to produce a zwitterion intermediate **2.10d**. The two sequential cyclizations of the zwitterion

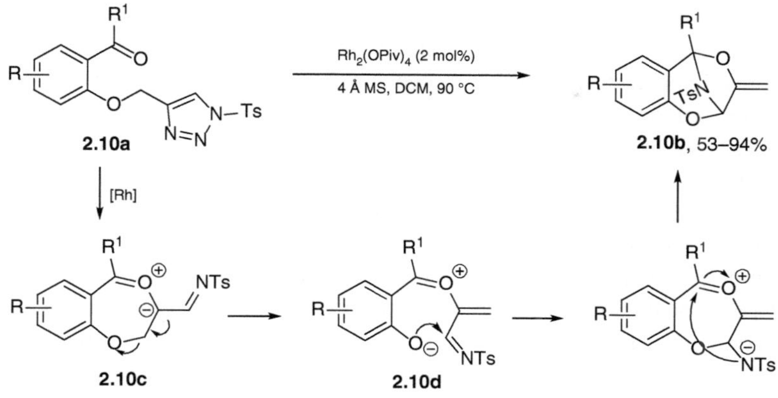

Scheme 2.10 Synthesis of heterobridged benzodioxepine derivatives.

intermediate **2.10d** provide the desired heterobridged cyclic product **2.10b**. This reaction offers an effective and valuable approach to constructing *N,O*-bridged ring systems, providing alternative insights into the design of drug molecules containing fused heterocyclic systems.

In 2015, Li and colleagues reported a Rh-catalyzed tandem reaction to prepare 2,5-epoxy-1,4-benzoxazepine derivatives **2.11c** from 1,2,3-triazoles **2.11a** and salicylaldehydes **2.11b** (Scheme 2.11) [11a]. Various triazoles and salicylaldehydes provided the desired products in moderate to good yields. The author proposed two mechanistic pathways. The first hypothesis involves O—H insertion and subsequent stepwise [3 + 2] cyclization (path *a*). Alternatively, oxazoline **2.11d** is obtained via formal [3 + 2] cycloaddition, which undergoes intramolecular

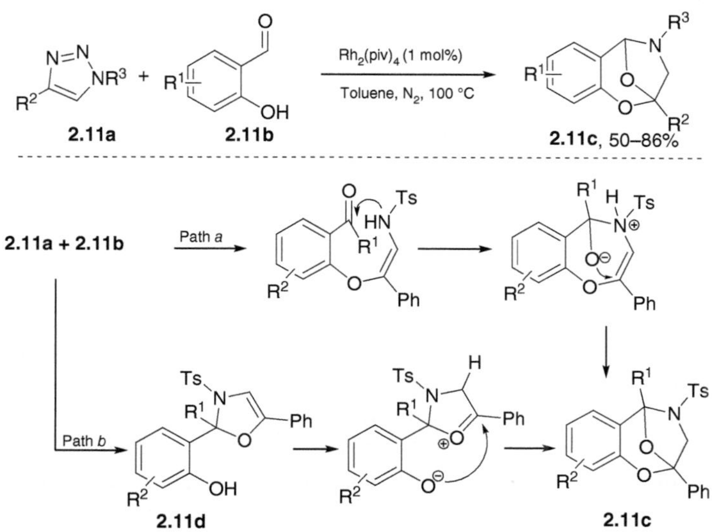

Scheme 2.11 Synthesis of 2,5-epoxy-1,4-benzoxazepines.

hydroalkoxylation to afford final heterocycles **2.11c** (path *b*). Based on the control experiments, path *b* is more convincing. Notably, *ortho*-hydroxy acetophenone was also incorporated successfully in this protocol. In 2016, the Yu group reported a similar transformation [11b].

In 2023, Xu and Li reported a straightforward process for the synthesis of 3-aminoquinolines **2.12a**, which starts from easily available triazoles **2.12b** and 2-aminobenzaldehydes **2.12c** (Scheme 2.12) [12]. The catalyst loading is 0.5 mol%, and the transformation shows good functional group compatibility. N—H insertion of the rhodium carbene into 2-aminobenzaldehyde **2.12c** yields an enamine **2.12d**, subsequent Lewis acid (LA)-catalyzed intramolecular annulation, proton transfer, and dehydration furnishes the target 3-aminoquinoline **2.12a**. With a halide on the 2-aryl, 3-aminoquinoline can be easily modified, which enables the convenient synthesis of quindoline and cryptolepine. This demonstrates the potential of this method in bioactive molecule and drug synthesis.

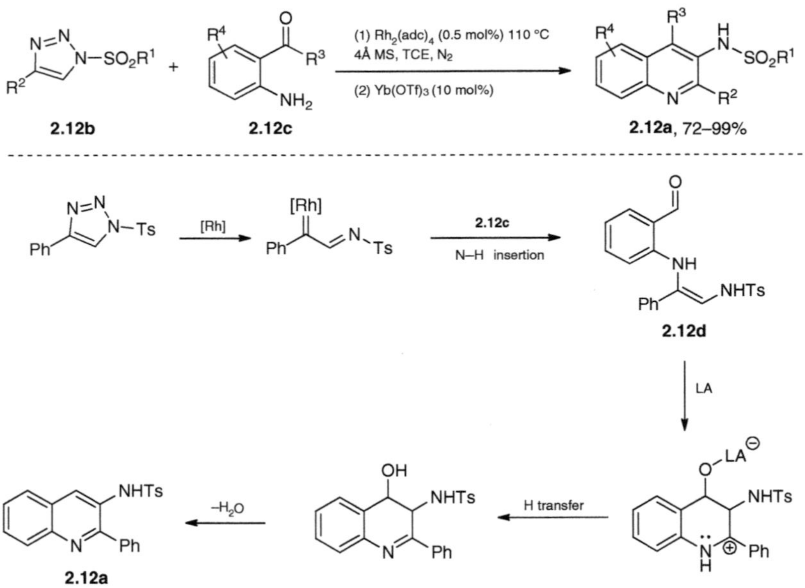

Scheme 2.12 Synthesis of 3-aminoquinolines.

Recently, Dawande and co-workers achieved an Rh-catalyzed tandem reaction for the synthesis of azepino-fused diindoles **2.13a** (Scheme 2.13) [13]. In this reaction, isatin-tethered 1,2,3-triazoles **2.13b** undergo an intramolecular cyclization reaction to generate a carbonyl ylide **2.13c**, which then undergoes a [3 + 2] cycloaddition reaction with indole **2.13d** to afford the oxygen-bridged adduct **2.13e**. Subsequently, the oxygen-bridged intermediate undergoes a sequential semipinacol rearrangement, ring expansion, and oxidation reaction to ultimately yield various azepino-fused diindoles **2.13a** in good to high yields.

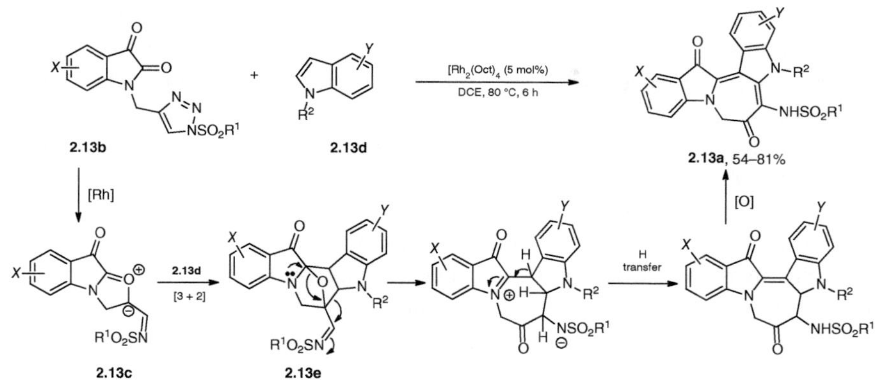

Scheme 2.13 Synthesis of azepino-fused diindoles.

Diketones could also be employed to capture the reactive α-imino rhodium carbene. Xu and Li have developed a rhodium-catalyzed reaction between *N*-sulfonyl-1,2,3-triazoles **2.14a** and 1,3-diones **2.14b** to synthesize diversely functionalized 2-carbonyl pyrroles **2.14c** (Scheme 2.14) [14]. This method enables convenient and efficient access to a range of multiply substituted 2-carbonyl pyrroles in yields up to 94% under mild conditions using readily accessible starting materials. When benzoyl acetone derivatives are used as trapping agents, the corresponding pyrrole products are obtained with high regioselectivity. The proposed mechanism

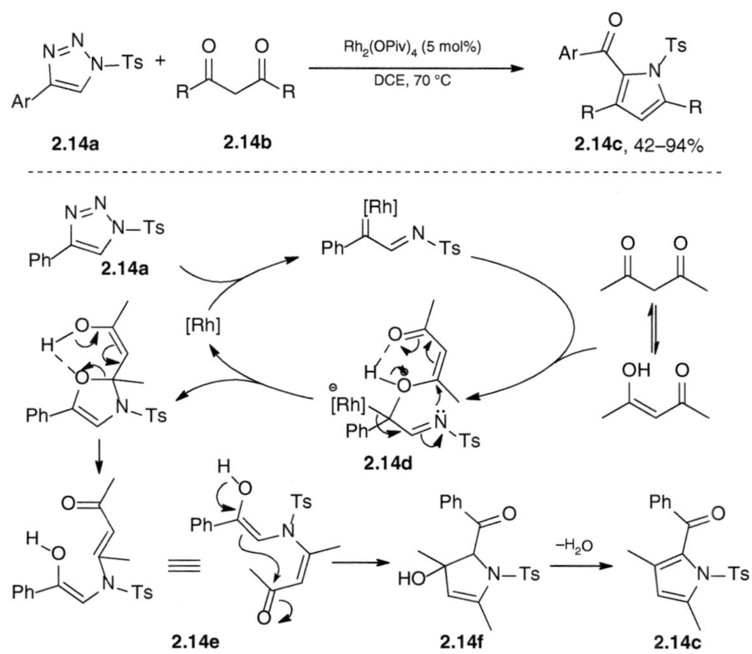

Scheme 2.14 Reaction of *N*-sulfonyl-1,2,3-triazoles with 1,3-diketones.

begins with the enol tautomer of the 1,3-dione **2.14b** undergoing nucleophilic attack on the rhodium carbene to generate a zwitterion intermediate **2.14d**. Subsequent addition-elimination reactions provide a ketone intermediate **2.14e**, which then undergoes an intramolecular aldol reaction to form a dihydropyrrole intermediate **2.14f**. Finally, aromatic ring construction furnishes the target product **2.14c**. Overall, this strategy offers a new and efficient route to polysubstituted pyrroles without harsh conditions or limitations on functional group tolerance.

Zhang and Bi have also reported reactions of *N*-sulfonyl-1,2,3-triazoles with 1,3-diones to form pyrroles and fused pyrroles determined by substrate control (Scheme 2.15) [15]. When reacted with noncyclic symmetric 1,3-diones, multiply substituted pyrroles were obtained through an O—H insertion-condensation reaction. Notably, excellent regioselectivity for asymmetric diketones was achieved. Cyclic ketones underwent C—H insertion-condensation to generate fused pyrroles. However, reactions with 2-substituted-1,3-diones selectively produced α-aminoketones instead of pyrroles. Thus, by modifying the structure of the diketone, the reaction outcome could be controlled to selectively form either pyrroles or α-aminoketones. This method enables efficient access to diverse pyrrole-based compounds, highlighting the versatility of triazoles in synthesis.

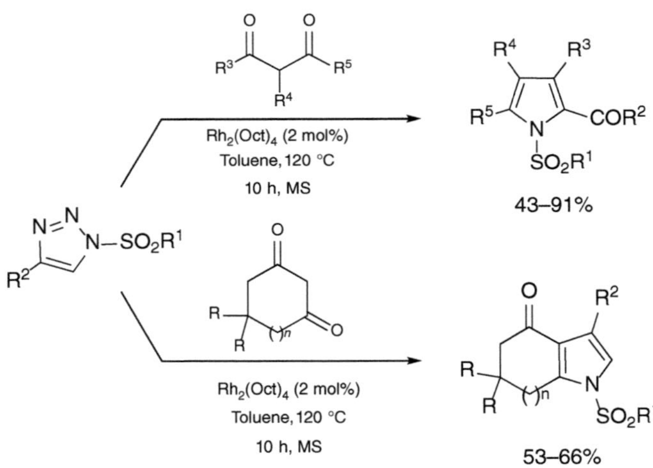

Scheme 2.15 Substrate-controlled synthesis of (fused)pyrroles.

Besides aldehydes and ketones, amides could also participate in the reactions with triazoles (Scheme 2.16) [16]. In 2014, the groups of Lee and Murakami independently reported the Rh-catalyzed reaction of *N*-sulfonyl-1,2,3-triazole with formamide to stereoselectively synthesize diaminoenone **2.16a** in good to excellent yields [16a]. During the reaction, the carbonyl oxygen in formamide undergoes nucleophilic attack on the rhodium carbene, followed by cyclization to generate an oxazoline derivative **2.16b**. After two rounds of ring opening and rearrangement via intermediate **2.16c**, the target product is obtained. Meanwhile, Yoo and colleagues used a similar strategy to achieve the stereoselective synthesis of a fully substituted

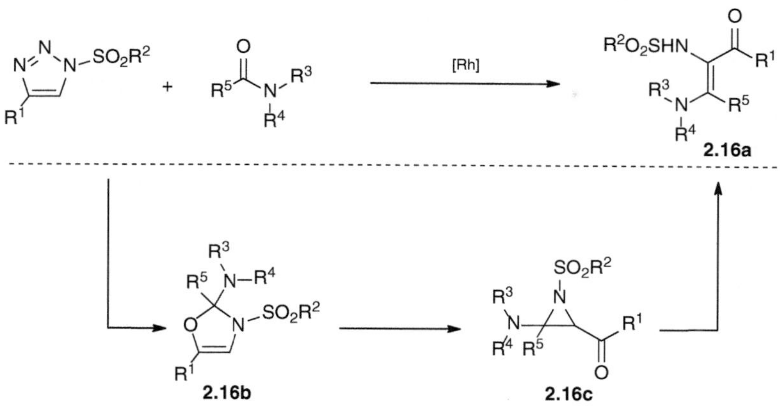

Scheme 2.16 Rh-catalyzed stereoselective synthesis of diaminoenones.

enaminone in good to excellent yields through the $Rh_2(esp)_2$-catalyzed reaction of *N*-sulfonyl-1,2,3-triazole with benzamide [16b].

Li et al. reported a Rh(II)/Co(III) dual-catalytic system to promote a cascade reaction of benzamide-tethered triazoles for stereoselective synthesis of biologically active spiroindolones (Scheme 2.17) [17]. The reaction proceeds via a Pictet–Spengler-type cyclization of the ylide initially generated from carbonyl nucleophilic addition to the rhodium carbene, followed by an intramolecular cyclization catalyzed by the cobalt(III) catalyst. The dual-catalytic cascade reaction provides a new and convenient strategy for the stereoselective construction of spiroindolones in good yields with high stereoselectivity, as well as a broad substrate scope. The design of the cyclization precursor controls the stereochemical outcome. By regulating the reactivity of triazoles, this approach expands the routes to complex drug-like molecules.

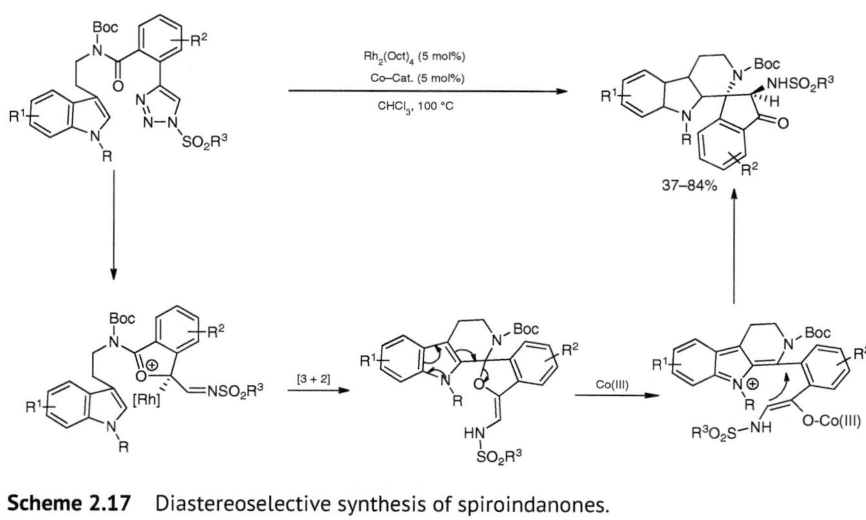

Scheme 2.17 Diastereoselective synthesis of spiroindanones.

Li and Zhai reported a Rh-catalyzed [3 + 2] dearomatization reaction of indoles **2.18a** containing *N*-sulfonyl-1,2,3-triazole (Scheme 2.18) [18]. This protocol enabled the construction of a series of biologically active indole alkaloid skeletons **2.18b–2.18d** with moderate to good yields. To avoid hydrolysis of imine **2.18b** to aldehyde **2.18c**, $NaBH_4$ was used for one-pot reduction of the imine to give stable secondary sulfonamides **2.18d** as a single isomer. The protocol was then employed in the concise synthesis of the vindoline skeleton.

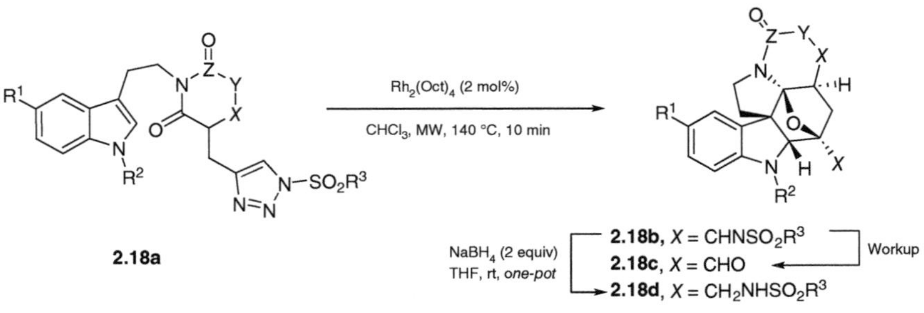

Scheme 2.18 Synthesis of indole alkaloid scaffolds.

2.3 Reaction of *N*-Sulfonyl-1,2,3-triazoles with Nitriles

Although the first reported rhodium-catalyzed transannulation of *N*-sulfonyl-1,2,3-triazole involved its reaction with nitriles, there are fewer examples involving nitriles compared to carbonyls, as discussed in Section 2.2. The basic reaction process is almost the same: the reaction is initiated by the nucleophilic addition of the nitrogen of the nitrile to the electrophilic carbon of α-imino rhodium carbene, resulting in the formation of a nitrogen ylide, which is a zwitterion with multiple reaction sites. However, only direct cyclizations of such zwitterions have been reported, leading to the formation of imidazole cores as the sole products.

In 2008, Gevorgyan and Fokin demonstrated that *N*-sulfonyl-1,2,3-triazoles can undergo rhodium-catalyzed ring-closing reactions with nitriles to form variously substituted imidazoles **2.19a** in good to excellent yields (Scheme 2.19) [1]. Optimization studies revealed that electron-rich rhodium catalysts were more effective than electron-poor rhodium. Additionally, a convenient one-pot, two-step imidazole synthesis was achieved starting from terminal alkynes, *p*-tosylazide, and nitriles. Two possible pathways were proposed to rationalize the formation of imidazole: in path *A*, a [3 + 2] cycloaddition of α-imino rhodium carbene with the nitrile formed a rhodium-tethered zwitterionic intermediate **2.19b**, which aromatized to imidazole **2.19a** upon rhodium departure. Alternatively, in path *B*, a ylide intermediate was generated through the addition of the nitrile to the carbene carbon **2.19c**, followed by a 1,3-Rh shift to generate a new rhodium carbene **2.19d**, which then cyclized and eliminated rhodium to yield the product.

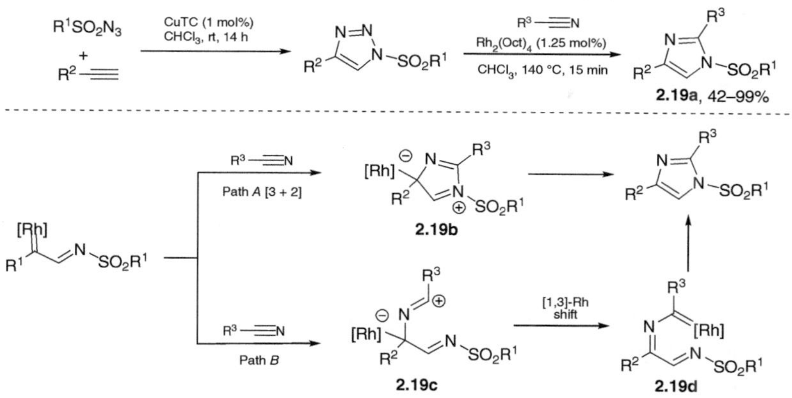

Scheme 2.19 Rh(II)-catalyzed transannulation of *N*-sulfonyl-1,2,3-triazoles with nitriles.

In 2015, Lee et al. reported a convenient method for synthesizing 2-bromo-1-sulfonylimidazoles **2.20a** through rhodium-catalyzed cyclization of *N*-sulfonyl-1,2,3-triazoles with the bromocyanide **2.20b** (Scheme 2.20) [19]. Various substituted triazoles underwent the transformation to yield the desired products, irrespective of electronic and steric effects. The broad substrate scope and mild reaction conditions make this approach useful for synthesizing diverse imidazole derivatives. Importantly, this rhodium-catalyzed cyclization provides convenient access to versatile 2-bromo-1-sulfonylimidazole intermediates, which can be further functionalized through Suzuki coupling to give 2-arylated imidazoles **2.20c** and Sonogashira coupling to give 2-alkynylated imidazoles **2.20d**.

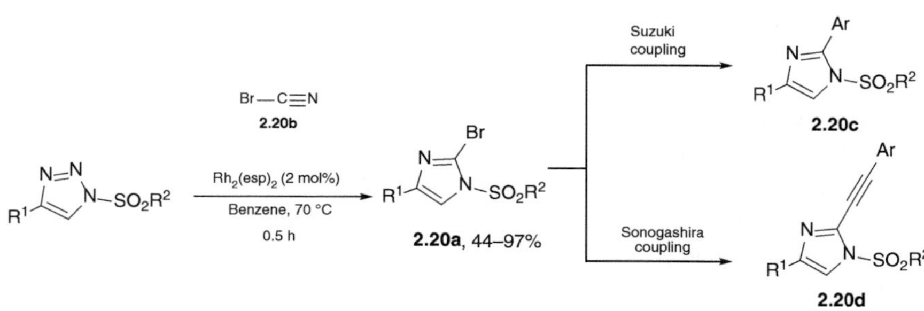

Scheme 2.20 Synthesis of 2-bromo-1-sulfonyl imidazole and their utilization.

Similarly, the Lee group synthesized *N*-imidazolyl sulfoximines **2.21a** from triazole and *N*-cyanosulfoximines **2.21b** (Scheme 2.21) [20]. A one-pot procedure was also achieved starting from terminal alkynes, sulfonylazide, and *N*-cyanosulfoximines **2.21b**. This involved a Cu-catalyzed [3 + 2] cycloaddition followed by Rh-catalyzed cyclization of the triazole with the *N*-cyanosulfoximines. The proposed mechanism suggests that the *N*-cyanosulfoximine **2.21b** attacks the α-imino rhodium carbene to form a zwitterionic intermediate which is very similar

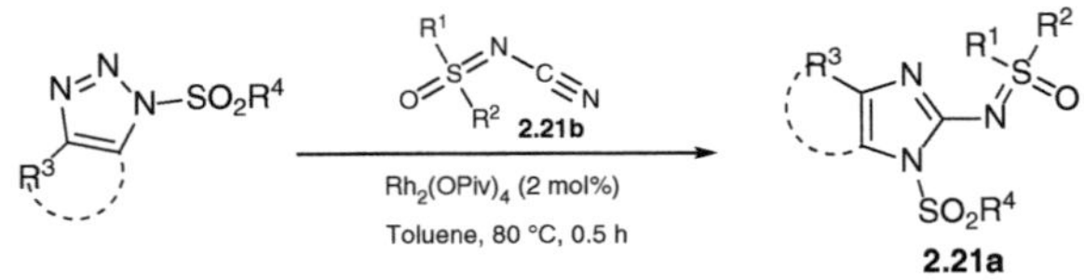

Scheme 2.21 Annulation of triazoles and *N*-cyanosulfoximines.

to **2.19c**. The subsequent ring closure occurs via the attack of the imino nitrogen on the cyano carbon, resulting in the formation of the desired imidazole product. This protocol provides a convenient one-pot procedure for synthesizing *N*-imidazolyl sulfoximines using readily available starting materials.

In 2018, Xu and Li reported a rhodium-free method for synthesizing imidazoles from *N*-sulfonyl-1,2,3-triazoles and nitriles using $BF_3 \cdot OEt_2$ as a promoter (Scheme 2.22) [21]. Under optimized conditions, various substituted triazoles were converted to the corresponding imidazoles in yields of up to 99%. Interestingly, aromatic nitriles exhibited lower reactivity compared to alkyl nitriles in this transformation. This method provides a simple and efficient rhodium-free alternative for synthesizing imidazoles from *N*-sulfonyltriazoles and nitriles. The broad substrate scope and excellent yields make it a valuable approach, particularly considering the avoidance of expensive rhodium catalysts.

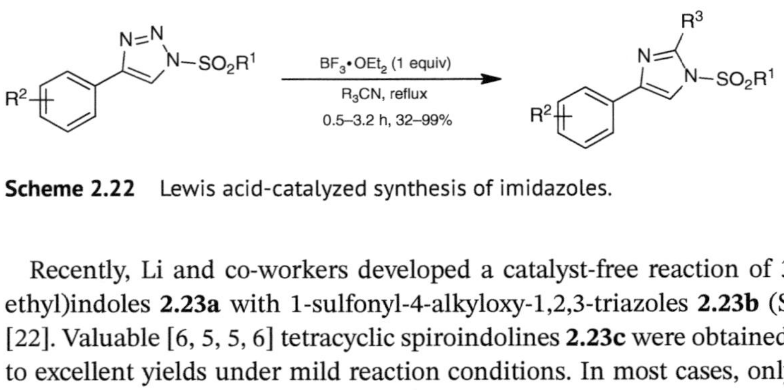

Scheme 2.22 Lewis acid-catalyzed synthesis of imidazoles.

Recently, Li and co-workers developed a catalyst-free reaction of 3-(2-isocyano ethyl)indoles **2.23a** with 1-sulfonyl-4-alkyloxy-1,2,3-triazoles **2.23b** (Scheme 2.23) [22]. Valuable [6, 5, 5, 6] tetracyclic spiroindolines **2.23c** were obtained in moderate to excellent yields under mild reaction conditions. In most cases, only one isomer was isolated. Denitrogenation of 4-alkyloxytriazole affords a zwitterion **2.23d**; the following isocyano attack and subsequent intramolecular nucleophilic cyclization lead to the formation of the final polycycles **2.23c**.

2.4 Reaction of *N*-Sulfonyl-1,2,3-triazoles with Polar Multiple Bonds Beyond Carbonyls and Nitriles

In 2013, Fokin, Chuprakov, and Kwok reported a transannulation reaction of *N*-mesyl-1,2,3-triazoles with isocyanates **2.24a** and isothiocyanates **2.24b** to efficiently synthesize imidazolones **2.24c** and thiazoles **2.24d** at room temperature (Scheme 2.24) [23]. Screening of Rh(II) catalysts identified $Rh_2(S\text{-NTTL})_4$ as the

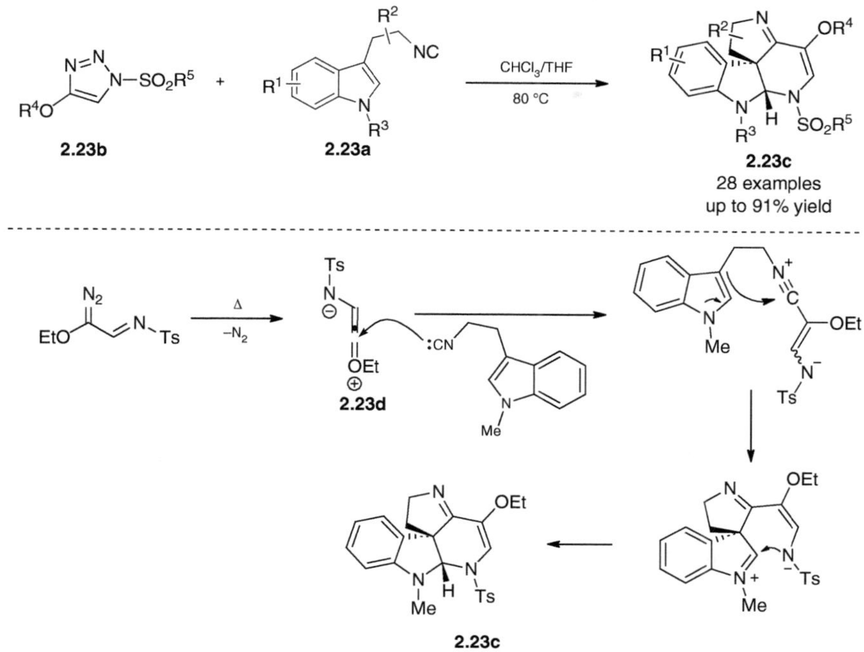

Scheme 2.23 Metal-free synthesis of [6,5,5,6]-tetracyclic spiroindolines.

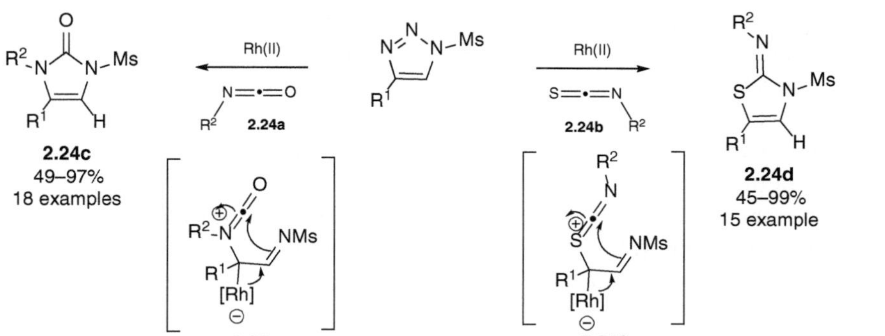

Scheme 2.24 Reactions of *N*-sulfonyl-1,2,3-triazoles with isocyanates and isothiocyanates.

optimal catalyst, enabling the synthesis of various imidazolones and thiazoles in good to excellent yields. Due to the difference in nucleophilicity, the electrophilic carbene carbon is attacked by the nitrogen atom of the isocyanates and subsequently cyclizes to form imidazolones via nitrogen ylide **2.24e**. On the other hand, isothiocyanates participate in the reaction with C=S, leading to the formation of thiazoles via sulfur ylide **2.24f**.

The Li group reported the rhodium-catalyzed reaction of *N*-sulfonyl-1,2,3-triazole with nitrosoarenes **2.25a** to synthesize *N*-acylamidine **2.25b** in high yields (Scheme 2.25) [24]. Two possible pathways were proposed. In path *a*, the α-imino

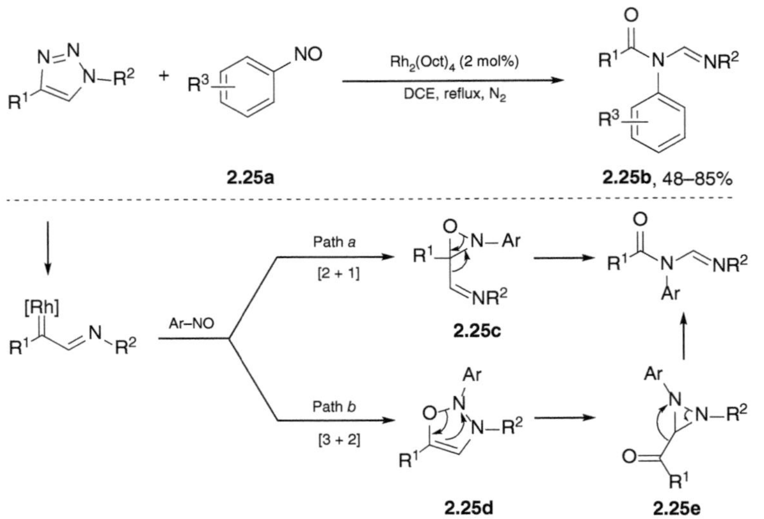

Scheme 2.25 Synthesis of *N*-acylamidines.

rhodium carbene undergoes a [2 + 1] cycloaddition with the nitrosoarene, forming a 1,2-oxaziridine intermediate **2.25c**. Subsequent cleavage of the N—O bond, followed by migration of the imino group, leads to the desired product **2.25b**. In path *b*, a [3 + 2] cycloaddition initially generates a 2,3-dihydro-1,2,3-oxadiazole **2.25d**, which then rearranges to a diaziridine intermediate **2.25e**. Ring-opening and migration of the acyl group then give the *N*-acylamidine product **2.25b**.

Miura and Murakami reported a Rh(II)-catalyzed transannulation of *N*-sulfonyl-1, 2,3-triazoles with thionoesters **2.26a**, which produced 2,5-disubstituted thiazoles **2.26b** in yields of up to 99% (Scheme 2.26) [25]. Initially, a [3 + 2] cyclization reaction of the α-imino rhodium carbene with C=S gives thiazolines **2.26c**, which are then aromatized to 2,5-disubstituted thiazoles **2.26b** with the assistance of acidic silica gel. Subsequently, this protocol was applied to the late-stage transformation of bioactive molecules and the highly selective synthesis of oligomeric arylene compounds.

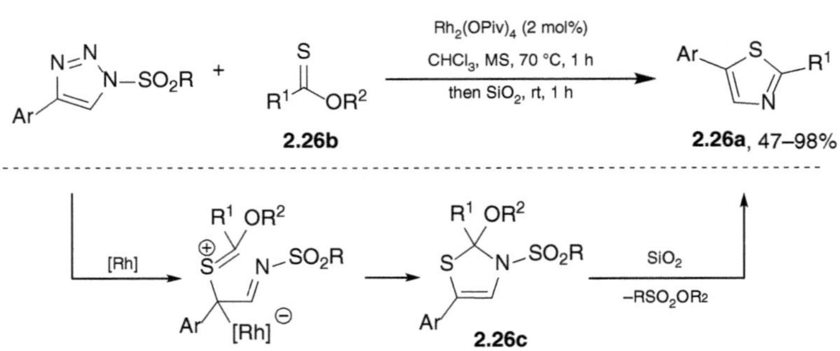

Scheme 2.26 Synthesis of thiazoles from *N*-sulfonyl-1,2,3-triazoles and thionoesters.

Novikov et al. developed a Rh(II)-catalyzed reaction of 1,2,3-triazoles with isoxazoles **2.27a** to synthesize pyrroles **2.27b** and pyrazines **2.27c** (Scheme 2.27) [26]. Initially, using $Rh_2(OPiv)_4$ in α,α,α-trifluorotoluene produced a mixture of both pyrroles and pyrazines. Further screening revealed that dilute conditions favored dihydropyrazine formation, indicating that pyrrole formation involves an intermolecular process. The optimal conditions were found to be refluxing toluene for selective pyrazine formation, while chloroform gave preferable pyrrole formation.

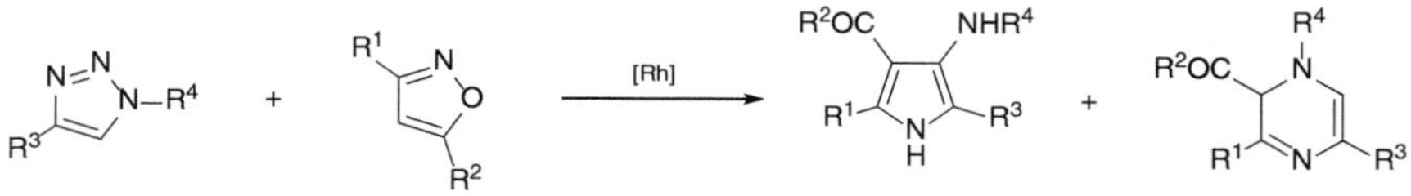

Scheme 2.27 Synthesis of pyrroles and pyrazines.

2.5 Conclusion

In summary, the rhodium-catalyzed reaction of *N*-sulfonyl-1,2,3-triazoles with polar multiple bonds offers a simple, mild, and modular route to synthesize diverse (hetero)cycles, poly(hetero)cycles, and even more complex heterobridges. This provides new strategies for cycle construction and new opportunities for drug design and drug discovery. However, there are still some issues that need to be addressed by future researchers in this field: (i) ketones exhibit limited reactivity in the rhodium-catalyzed transannulation of *N*-sulfonyl-1,2,3-triazoles; (ii) asymmetric transformations are still limited in scope; (iii) the catalyst is largely limited to rhodium, and the development of more nonmetal catalysts is desired; (iv) the exploration of additional useful reaction modes involving nitrogen ylides generated from nitriles and α-imino rhodium carbenes is desired.

References

1 Horneff, T., Chuprakov, S., Chernyak, N. et al. (2008). Rhodium-catalyzed transannulation of 1,2,3-triazoles with nitriles. *J. Am. Chem. Soc.* 130: 14972–14974.

2 (a) Davies, H.M.L. and Alford, J.S. (2014). Reactions of metallocarbenes derived from *N*-sulfonyl-1,2,3-triazoles. *Chem. Soc. Rev.* 43: 5151–5162. (b) Jiang, Y., Sun, R., Tang, X.-Y., and Shi, M. (2016). Recent advances in the synthesis of heterocycles and related substances based on α-imino rhodium carbene complexes derived from *N*-sulfonyl-1,2,3-triazoles. *Chem. Eur. J.* 22: 17910–17924. (c) Li, Y., Yang, H., and Zhai, H. (2018). The expanding utility of rhodium-iminocarbenes: recent advances in the synthesis of natural products and related scaffolds. *Chem. Eur. J.* 22: 17910–17924. (d) Li, W. and Zhang, J. (2020). Synthesis of heterocycles through denitrogenative cyclization of triazoles and benzotriazoles. *Chem. Eur. J.* 26: 11931–11945. (e) Pal, K. and Volla, C.M.R. (2021). Catalytic insertion reactions of α-imino carbenoids. *Chem. Rec.* 21: 4032–4058. (f) Akter, M., Rupa, K., and Anbarasan, P. (2022). 1,2,3-Triazole and its analogues: new surrogates for diazo compounds. *Chem. Rev.* 122: 13108–13205.

3 Zibinsky, M. and Fokin, V.V. (2013). Sulfonyl-1,2,3-triazoles: convenient synthones for heterocyclic compounds. *Angew. Chem. Int. Ed.* 52: 1507–1510.

4 Li, J., Zhu, S.-R., Xu, Y. et al. (2020). Synthesis of 2,5-diaryloxazoles through rhodium-catalyzed annulation of triazoles and aldehydes. *RSC Adv.* 10: 24795–24799.

5 Miura, T., Tanaka, T., Hiraga, K. et al. (2013). Stereoselective synthesis of 2,3-dihydropyrroles from terminal alkynes, azides, and α,β-unsaturated aldehydes via *N*-sulfonyl-1,2,3-triazoles. *J. Am. Chem. Soc.* 135: 13652–13655.

6 Ma, X., Liu, L., Wang, J. et al. (2018). Rhodium-catalyzed annulation of α-imino carbenes with α,β-unsaturated ketones: construction of multisubstituted 2,3-dihydropyrrole/pyrrole rings. *J. Chem. Phys.* 83: 14518–14526.

7 Li, Y., Zhang, R., Ali, A. et al. (2017). Synthesis of 6-substituted piperidin-3-ones via Rh(II)-catalyzed transannulation of *N*-sulfonyl-1,2,3-triazoles with electron-rich aromatic nucleophiles. *Org. Lett.* 19: 3087–3090.

8 Zhang, R., Li, Y., Ali, A. et al. (2017). Rh(II)-catalyzed tandem cyclization of aliphatic *N*-sulfonyl-1,2,3-triazoles and alcohols for the synthesis of 6-alkoxy piperidin-3-ones. *Adv. Synth. Catal.* 359: 4057–4061.

9 Yuan, H., Gong, J., and Yang, Z. (2016). Stereoselective synthesis of oxabicyclo[2.2.1]heptenes via a tandem dirhodium(II)-catalyzed triazole denitrogenation and [3 + 2] cycloaddition. *Org. Lett.* 18: 5500–5503.

10 Zhang, Y.-S., Tang, X.-Y., and Shi, M. (2014). Unprecedented synthesis of *aza*-bridged benzodioxepine derivatives using a tandem Rh(II)-catalyzed 1,3-rearrangement/[3 + 2] cycloaddition of carbonyltriazoles. *Chem. Commun.* 50: 15971–15974.

11 (a) Shi, Y., Yu, X., and Li, C.-Y. (2015). Rhodium-catalyzed synthesis of 2,5-epoxybenzo[f][1,4]oxazepines by tandem reaction of 1-sulfonyl-1,2,3-triazoles and salicylaldehydes. *Eur. J. Org. Chem.* 6429–6433. (b) Meng, J., Ding, X., Yu, X., and Deng, W.-P. (2016). Synthesis of 2,5-epoxy-1,4-benzoxazepines via rhodium(II)-catalyzed reaction of 1-tosyl-1,2,3-triazoles and salicylaldehydes. *Tetrahedron* 72: 176–183.

12 Li, J., Feng, J., Chen, T. et al. (2023). Synthesis of 3-aminoquinolines from α-imino rhodium carbenes and 2-aminobenzaldehydes. *Org. Biomol. Chem.* 21: 5935–5938.

13 Kahar, N., Jadhav, P., Reddy, R.V.R., and Dawande, S.A. (2020). Rhodium(II) catalysed domino synthesis of azepino fused diindoles from isatin tethered: *N*-sulfonyl-1,2,3-triazoles and indoles. *Chem. Commun.* 56: 1207–1210.

14 Cheng, W., Tang, Y., Xu, Z.-F., and Li, C.-Y. (2016). Synthesis of multi-functionalized 2-carbonylpyrrole by rhodium-catalyzed transannulation of 1-sulfonyl-1,2,3-triazole with β-diketone. *Org. Lett.* 18: 6168–6171.

15 Mi, P., Yuan, H., Wang, H. et al. (2017). Divergent reactions between α-imino rhodium carbenoids and 1,3-diketones: substrate-controlled O—H versus C—H insertion. *Eur. J. Org. Chem.* 2017: 1289–1293.

16 (a) Jung, D.J., Jeon, H.J., Kim, J.H. et al. (2014). DMF as a source of oxygen and aminomethine: stereoselective 1,2-insertion of rhodium(II) azavinyl carbenes into the C=O bond of formamides for the synthesis of *cis*-diamino enones. *Org. Lett.* 16: 2208–2211. (b) Miura, T., Funakoshi, Y., Tanaka, T., and Murakami, M. (2014). Direct production of enaminones from terminal alkynes via rhodium-catalyzed reaction of formamides with *N*-sulfonyl-1,2,3-triazoles. *Org. Lett.* 16: 2760–2763. (c) Lee, D.J., Shin, J., and Yoo, E.J.A. (2014). Rh-catalyzed regio- and stereoselective route to enamides: benzamides as assembling reagents. *Chem. Commun.* 50: 6620–6622.

17 Liu, Z., Du, Q., Zhai, H., and Li, Y. (2018). Relay catalysis of Rh(II) and cobaloxime: stereoselective synthesis of spiroindanones from *N*-sulfonyl-1,2,3-triazoles. *Org. Lett.* 20: 7514–7517.

18 Li, Y., Zhang, Q., Du, Q., and Zhai, H. (2016). Rh-catalyzed [3 + 2] cycloaddition of 1-sulfonyl-1,2,3-triazoles: access to the framework of aspidosperma and kopsia indole alkaloids. *Org. Lett.* 18: 4076–4079.

19 Lee, E., Ryu, T., Shin, E. et al. (2015). Synthesis of 2-bromoimidazoles from alkynes, *N*-sulfonylazides, and bromocyanides. *Org. Lett.* 17: 2470–2473.

20 Kim, S., Kim, J.E., Lee, J., and Lee, P.H. (2015). *N*-Imidazolylation of sulfoximines from *N*-cyano sulfoximines, 1-alkynes, and *N*-sulfonyl azides. *Adv. Synth. Catal.* 357: 3707–3717.

21 Yang, D., Shan, L., Xu, Z.-F., and Li, C.-Y. (2018). Metal-free synthesis of imidazole by $BF_3{\cdot}Et_2O$ promoted denitrogenative transannulation of *N*-sulfonyl-1,2,3-triazole. *Org. Biomol. Chem.* 16: 1461–1464.

22 Chen, C., Chen, J., Wang, H. et al. (2023). Catalyst-free synthesis of polycyclic spiroindolines by cascade reaction of 3-(2-isocyanoethyl)indoles with 1-sulfonyl-1,2,3-triazoles. *J. Org. Chem.* 88: 9543–9553.

23 Chuprakov, S., Kwok, S.W., and Fokin, V.V. (2013). Transannulation of 1-sulfonyl-1,2,3-triazoles with heterocumulenes. *J. Am. Chem. Soc.* 135: 4652–4655.

24 Ran, R.-Q., Xiu, S.-D., and Li, C.-Y. (2014). Synthesis of *N*-acylamidines via rhodium-catalyzed reaction of nitrosobenzene derivatives with *N*-sulfonyl-1,2,3-triazoles. *Org. Lett.* 16: 6394–6396.

25 Miura, T., Funakoshi, Y., Fujimoto, Y. et al. (2015). Facile synthesis of 2,5-disubstituted thiazoles from terminal alkynes, sulfonyl azides, and thionoesters. *Org. Lett.* 17: 2454–2457.

26 Rostovskii, N.V., Ruvinskaya, J.O., Novikov, M.S. et al. (2017). Switchable synthesis of pyrroles and pyrazines via Rh(II)-catalyzed reaction of 1,2,3-triazoles with isoxazoles: experimental and DFT evidence for the 1,4-diazahexatriene intermediate. *J. Org. Chem.* 82: 256–268.

3

Transannulation of *N*-Sulfonyl-1,2,3-triazoles with Nonpolar Multiple Bonds such as Alkenes, Alkynes, and Allenes

Prasiddha Nagarajan, Akhil Rampally, and Sudam G. Dawande

Department of Chemistry, Indian Institute of Technology Madras, Chennai, Tamil Nadu, India

3.1 Introduction

The carbocycles and heterocycles are ubiquitous molecular frameworks embedded in a large variety of natural products, biologically active molecules, pharmaceutical agents, and fine chemicals. For example, cyclopropanes, [1–3] pyrroles, pyridines, and their saturated derivatives [4–6] azepines [7] are present in various bioactive molecules, pharmaceutical agents, and serve as pivotal precursors in numerous intended organic syntheses. Owing to their unique activities, the synthesis of such motifs has always been an area of interest for synthetic chemists. In a broader sense, the intermolecular and intramolecular cycloaddition of a variety of synthons culminates into the carbocycles and heterocycles. Among various precursors, in 2009, Fokin and co-workers discovered that the bench stable *N*-sulfonyl-1,2,3-triazoles **3.1a** could undergo ring chain tautomerization to generate α-imino diazo **3.1b** compound, which in the presence of metal catalyst results in α-imino metal carbene intermediate **3.1c** via denitrogenation (Scheme 3.1). The metal carbene intermediate further leads to *in situ* transformations with a variety of competent reacting partners. These mainly include transannulation, C—H insertion, *X*—H (*X* = N, O, Si, S) insertion, ylide formation, subsequent rearrangements, etc.[8–13] In this, the metal-catalyzed transannulations of triazoles have gained immense attention as these transformations deliver a wide variety of carbocyclic and heterocyclic molecules based on the structure and reactivity of the partner. As the α-imino metal carbene intermediates provide an electrophilic carbene center and nucleophilic nitrogen atom, it undergoes cycloadditions with species having polar as well as nonpolar π bonds.

The current chapter describes the metal-catalyzed transannulations of the *N*-sulfonyl-1,2,3-triazoles with substrates having nonpolar multiple bonds. Here, we have mainly discussed the metal-catalyzed synthetic transformations of triazoles with alkenes, 1,3-dienes, alkynes, and allenes useful toward the synthesis of the cyclopropanes, (2,3-dihydro)pyrroles, azepines, pipyridines, quinolones, etc.

Denitrogenative Transformation of Nitrogen Heterocycles: Synthesis, Reactions and Applications,
First Edition. Edited by Pazhamalai Anbarasan.

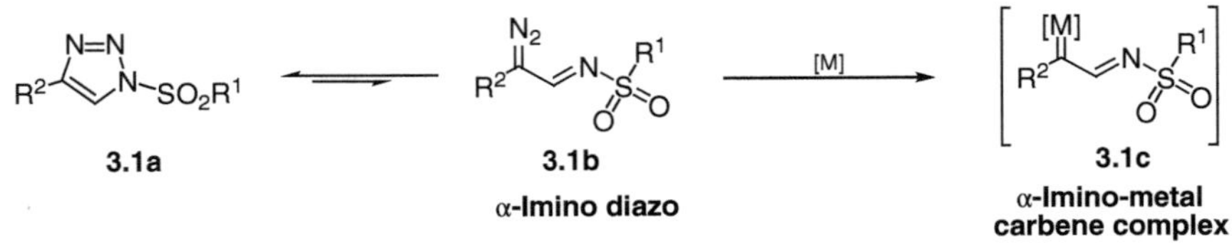

Scheme 3.1 Generation of metal-catalyzed α-imino metal carbene intermediate.

3.2 Transannulation of *N*-Sulfonyl-1,2,3-triazoles with Alkenes

3.2.1 Intermolecular Transannulation with Alkenes

Alkenes are among the substrates of choice for many cycloaddition reactions which include Diels–Alder reactions, radical cyclization reactions, and dipolar cycloaddition reactions due to the presence of a C—C double bond which offers a weak π bond. Specifically, for the synthesis of cyclopropanes, the [2 + 1] cycloaddition reaction of diazo-derived metal carbene intermediate with alkenes has emerged as a powerful strategy.

In 2009, Fokin et al. reported the first cyclopropanation reaction of *N*-sulfonyl-1,2,3-triazoles with alkenes with chiral dirhodium catalysts, which resulted in high enantioselectivity. Their study described that the aryl group present on the α-imino carbene, which is obtained from the triazole, acts as a donor substituent while the *N*-sulfonyl imino group as the acceptor substituent and as a combined effect, the controlled reactivity of the metal carbene intermediate gives higher selectivity for cyclopropanation reactions (Scheme 3.2) [14]. The use of 0.5 mol% tetracarboxylate based chiral $Rh_2[(S)\text{-NTTL}]_4$ catalyst delivers the best results in terms of yield, diastereo-, and enantioselectivity after basic hydrolysis of iminocyclopropane. Further, Fokin's group extended their work to *NH* triazoles, which involves *in-situ* transformation to *N*-triflyl triazoles using triflic anhydride in the presence of 2,6-ditertiarybutyl-4-methyl-pyridine (DTBMP) (**3.2a**) followed by sequential treatment with alkenes in the presence of $Rh_2[(S)\text{-NTTL}]_4$ as catalyst [15]. Electronically

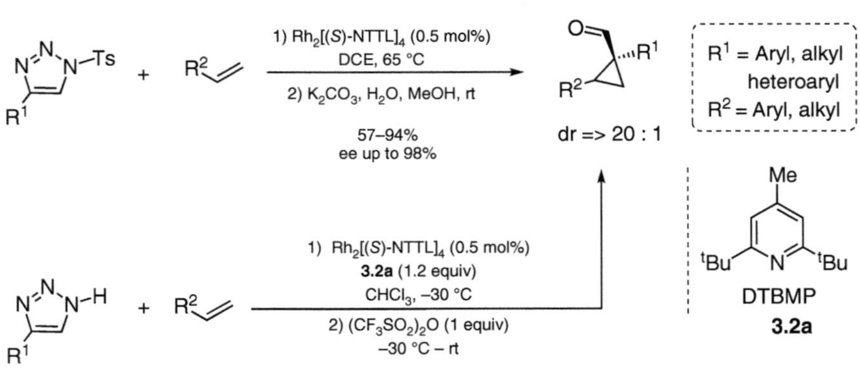

Scheme 3.2 Asymmetric cyclopropanation reactions of triazoles with alkenes.

and sterically divergent monosubstituted aryl and alkyl olefins as well as triazoles, react smoothly. Furthermore, heteroaryl-(thiophenyl) and alkenyl triazoles are also potential substrates for this transformation.

The reactions of 1,2-disubstituted alkenes such as (*E*)-1-phenyl-1-propenes with triazole furnished products with good yield and excellent enantioselectivity. On the contrary, (*Z*)-1-phenyl-1-propenes, results in an almost racemic mixture. In addition, authors noted that, in independent reactions, triazoles bearing electron-withdrawing group at C4 position as well as 4-methoxy styrene result in the formation of 2,3-dihydropyrroles with lower enantioselectivity (Scheme 3.3).

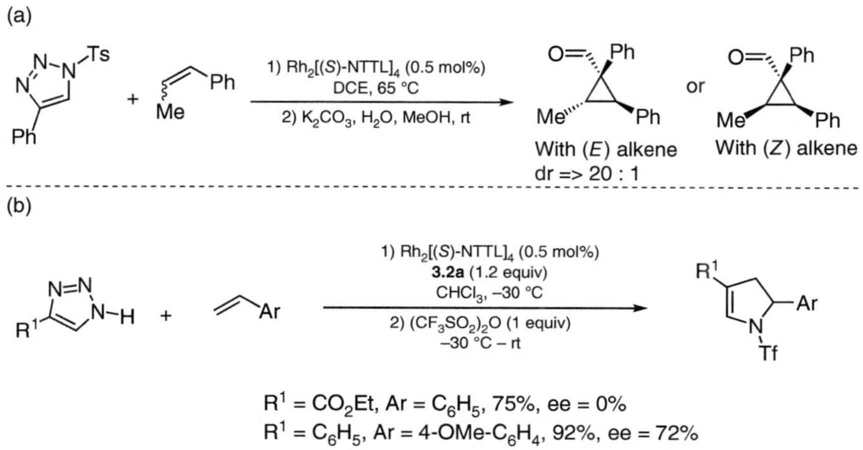

Scheme 3.3 Synthesis of dihydropyrroles from *NH* triazoles. (a) Cyclopropanation with 1,2-disubstituted olefin. (b) Effect of electronic nature of triazole and styrene.

The *N*-sulfonyl-imine cyclopropanes are unstable and require further *in situ* transformation into corresponding cyclopropyl aldehyde. To resolve this limitation, Fokin's group further demonstrated a chiral rhodium-catalyzed asymmetric cyclopropanation reaction of styrene with *N*-sulfamoyl triazole obtained from sulfonyl transfer agents such as 2,3-dimethyl-1*H*-imidazolium triflate and alkyne in the presence of CuTC catalyst (Scheme 3.4) [16].

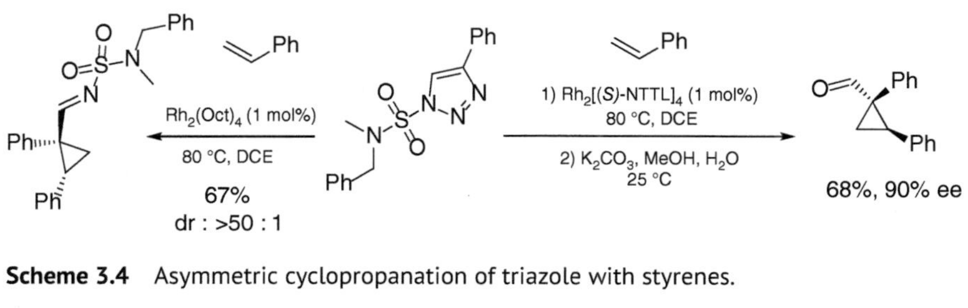

Scheme 3.4 Asymmetric cyclopropanation of triazole with styrenes.

Similarly, Davies et al. recently studied an asymmetric cyclopropanation through transannulation of 4-aryloxy-1-sulfonyl-1,2,3-triazoles with alkenes using $Rh_2[(S)\text{-NTTL}]_4$ and $Rh_2[(S)\text{-DPCP}]_4$ as catalysts with satisfactory yield

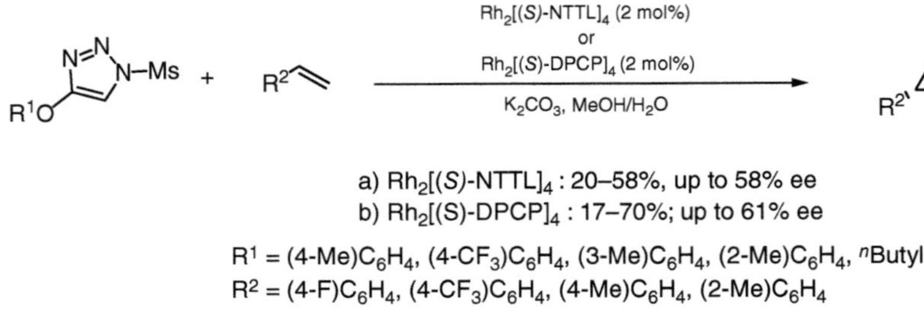

Scheme 3.5 Asymmetric cyclopropanation of aryloxy triazoles with styrenes.

and enantioselectivities (Scheme 3.5). This study describes the compatibility of the heteroatom donor groups such as oxygen present on the metal carbene intermediate toward enantioselective cyclopropanation [17]. The induced enantioselectivity indicates that the cyclopropanation reaction proceeds through chiral rhodium-bound carbene intermediate rather than free carbene. The presence of electron donating para substituent on the triazole gives lesser yield, diastereo- and enantioselectivity in comparison with electron-withdrawing para substituents. Electron-deficient styrenes and alkyl olefins react efficiently to furnish products with good diastereoselectivity; however, latter gives lower yields.

The reaction of triazoles with pinacol allyl boronates in the presence of chiral rhodium catalysts also results in corresponding cyclopropane derivatives (Scheme 3.6). The reactions proceed efficiently in the presence of 2.5 mol% of $Rh_2[(S)\text{-NTTL}]_4$ to furnish *trans*-cyclopropyl imine derivatives, which on reduction using DIBAL-H affords the cyclopropylamino species with good yield and secured stereochemistry [18]. The electron-deficient C4 substituent results in a lower yield compared to the electron-rich substituents.

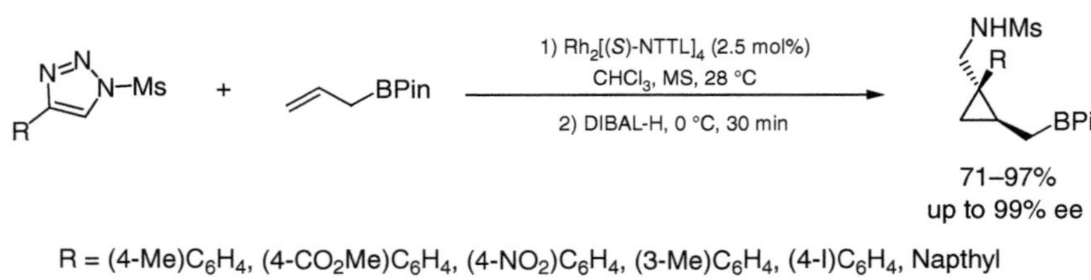

Scheme 3.6 Asymmetric cyclopropanation of triazoles with pinacol allyl boronates.

Murakami and co-workers also explored the $Rh_2[(S)\text{-NTTL}]_4$-catalyzed cyclopropanation reactions of specially designed 1-methylsulfonyl-4-(4-vinylphenyl)-1,2,3-triazoles for the synthesis of C3-symmetric triangular macrocycles with high enantioselectivities (Scheme 3.7) [19]. The 0.5 mol% loading of $Rh_2[(S)\text{-NTTL}]_4$ efficiently catalyzes the transformation to give the cyclic tri-imine, which, on basic hydrolysis using Al_2O_3, gives the desired macrocycle with a small amount of tetramer.

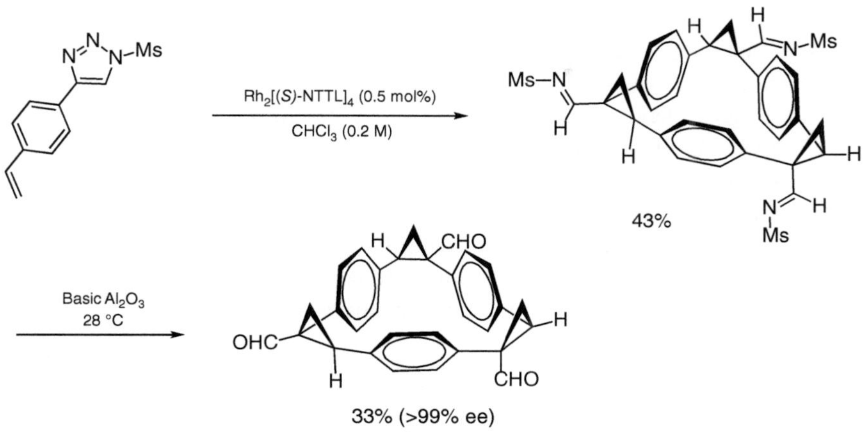

Scheme 3.7 Enantioselective synthesis of C3-symmetric triangular macrocycles.

The catalyst-free thermal decomposition of 4-phthalimido *N*-sulfonyl triazoles at 55 °C leads to denitrogenative transannulation with alkenes, which on hydrolysis using wet silica delivers cyclopropane aldehydes with excellent diastereoselectivity (Scheme 3.8) [20].

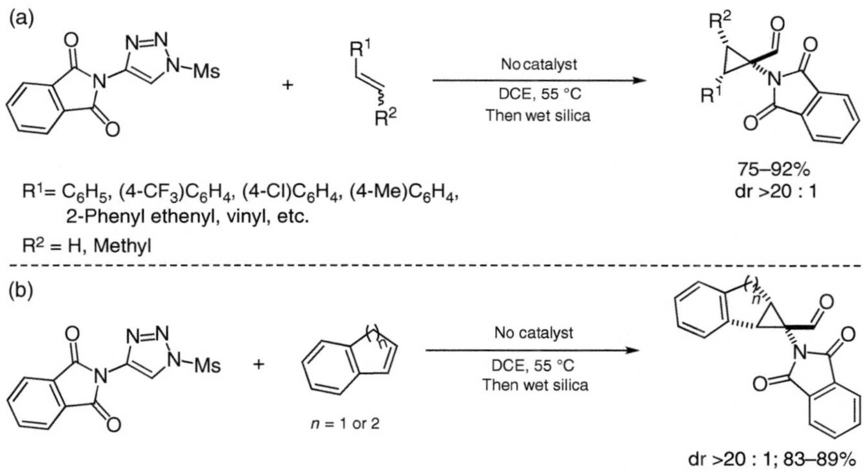

Scheme 3.8 Catalyst-free diastereoselective cyclopropanation of 4-phthalimido-*N*-sulfonyl triazoles with (a) acyclic olefins, (b) cyclic olefins.

The substituted acyclic and cyclic olefins react smoothly to afford cyclopropanes with good to excellent yields and diastereoselectivities. The authors proposed two possible reaction pathways. In the first pathway, a pyrazoline intermediate is formed by the 1,3-dipolar cycloaddition, followed by the formation of cyclopropane by thermal expulsion of the dinitrogen moiety. In the second pathway, a thermal decomposition of a diazo compound to give free carbene is proposed, which can then undergo cyclopropanation directly.

Fokin and co-workers examined the reaction between electron-rich styrenes such as *p*-methoxy styrene with *in situ* generated 4-aryl-*N*-triflated triazoles in the presence of $Rh_2[(S)\text{-NTTL}]_4$ catalysts to obtain 2,3-dihydropyrroles with moderate to good yield, however with moderate enantioselectivity (Scheme 3.9) [21]. However, the use of other chiral tetracarboxylate dirhodium catalysts such as $Rh_2[(S)\text{-4-Br-NTTL}]_4$, $Rh_2[(S)\text{-TFPTTL}]_4$ results in lower yields and enantioselectivities. The triazoles having an electron-deficient aryl group at the C4 position gave better yields and enantioselectivities. In addition, the presence of CO_2Et as a substituent at the C4 position of triazole resulted in poor yield of the 2,3-dihydropyrrole without any induction of enantioselectivity.

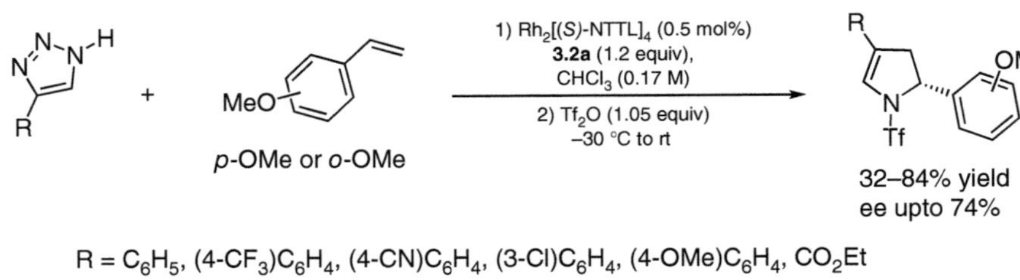

Scheme 3.9 Rhodium(II)-catalyzed dihydropyrrole synthesis from methoxystyrenes.

There are two pathways through which this reaction is predicted to proceed (Scheme 3.10). In **Pathway *A***, the olefin, which is electron-rich, reacts first with the carbene to yield cyclopropylaldimine **3.10a**. The cyclopropane ring is then cleaved, which is triggered by electron push–pull due to the presence of a triflate group, which is strongly electron-withdrawing and an electron-donating methoxy group, to generate a highly stabilized zwitterionic intermediate **3.10b**. This finally collapses to give the 2,3-dihydropyrrole. Initially, in this pathway, the stereo control is exerted by the chiral rhodium catalyst by influencing the enantioselectivity of the cyclopropane intermediate and in turn the final product. However, further rearrangement results in a certain loss of enantiomeric excess, which is determined by the comparison between the relative rate of rotation about the C_α—C_β bond and the rate at which the ring of **3.10b** closes to offer the cyclic motif.

In **Pathway *B***, the olefin, which is electron-rich, reacts first with rhodium carbene species and forms the zwitterionic species, which is still bound to rhodium **3.10c**. This is followed by the cyclization and further release of the rhodium catalyst from **3.10d** to yield the required product. Thus, according to this mechanism, the final stereoselectivity is controlled by the chiral rhodium catalyst.

Similarly, the triazoles react with *N*-vinyl indoles in the presence of rhodium(II) catalyst to furnish corresponding *N*-substituted 2,3-dihydropyrrolyl indole derivatives through [3 + 2] transannulation reaction (Scheme 3.11) [22]. On addition of 0.5 equivalent of 4-aryl-triazoles with respect to *N*-vinyl indole, the transannulation results in the formation of 2,3-dihydropyrrole **3.11c** with 2 mol% of $Rh_2(esp)_2$ at 80 °C. While using the 2.5 equivalent of 4-aryl-triazole, the 2,3-dihydropyrrolyl

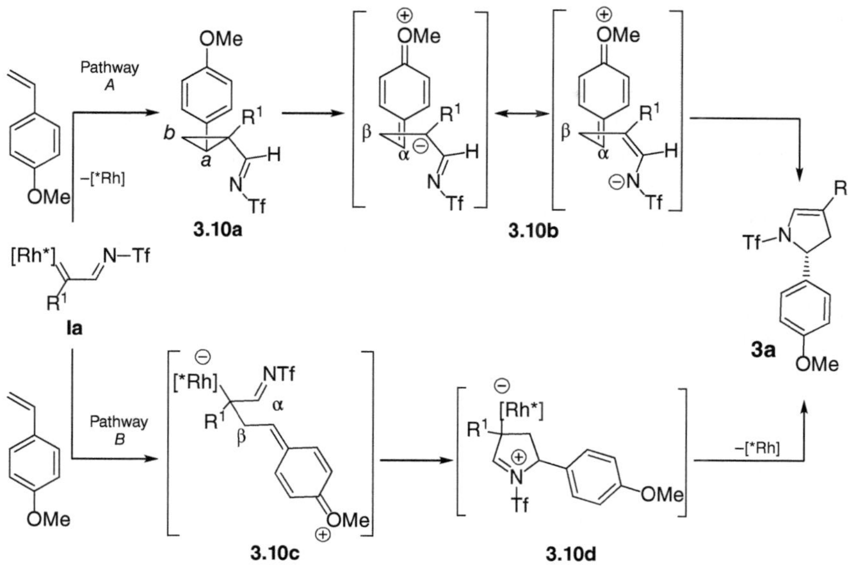

Scheme 3.10 Mechanism of Rh(II)-catalyzed dihydropyrrole synthesis from electron-rich styrene and triazoles.

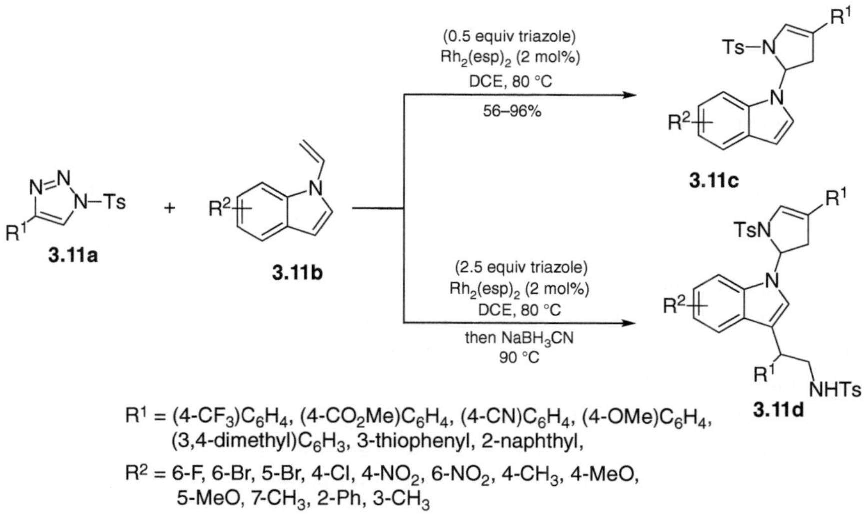

Scheme 3.11 Rh(II)-catalyzed transannulation of triazoles with *N*-vinyl indoles.

indole further undergoes C3 carbene insertion, which on reduction with $NaBH_3CN$ gives product **3.11d**.

The triazoles with electronically and sterically versatile C4 aryl substituents are tolerated to provide good yields. The electron-deficient indoles, possessing groups like nitro afford the product in good yield, while the electron-rich indoles with donating groups like 4-methoxy give a lower yield.

The rhodium-catalyzed transannulation of triazoles with cyclic vinyl ethers with rhodium(II) catalyst for highly diastereoselective synthesis of fused tetrahydrofurano–dihydropyrroles (THF-DPs) and tetrahydropyrano–dihydropyrroles (THP-DPs) which contain a syn *N′*,O-acetal linkage (Scheme 3.12) [23]. The resulting THF-DPs and THP-DPs formed after transannulation undergoes acid-catalyzed ring opening to afford corresponding C3 hydroxy alkyl pyrroles on treatment with 10 mol% of TfOH. Due to *syn* geometry at *N,O*-acetal linkage, the products obtained from cyclic enol ethers are more stable than those obtained from acyclic enol ethers.

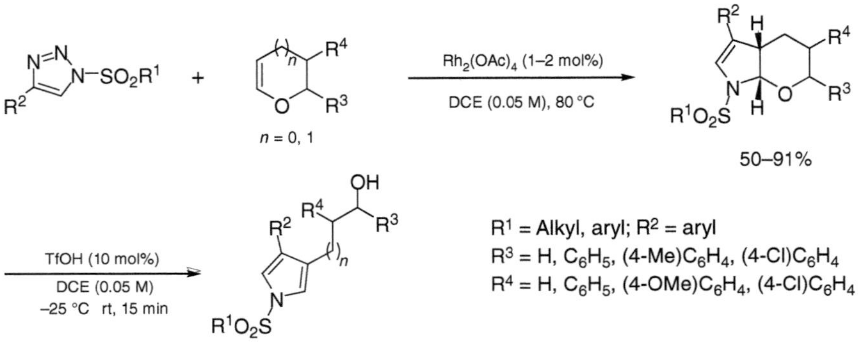

Scheme 3.12 Rh(II)-catalyzed transannulation of triazoles with cyclic vinyl ethers.

Electronically diverse 4-aryl triazoles react efficiently with substituted dihydrofurans and tetrahydropyrans in the presence of rhodium catalyst to give respective THF-DPs and THP-DPs in 52–90% yield and exclusive diastereoselectivity. In addition, the functionalities present in transannulation products are tolerated in the acid-catalyzed ring opening reaction.

Reversible tautomerization of the triazole and α-imino rhodium(II) carbenoid (**3.13a**) formation in the presence of rhodium followed by a nucleophilic attack of **3.13b** to the carbene center of **3.13a** which is electrophilic affords the zwitterionic intermediate **3.13c**. Then, electron migration from the anionic rhodium to the imine motif, increases the nitrogen nucleophilicity to facilitate the reaction with the cationic oxacarbenium center, which affords the 2,3-dihydropyrrole (**3.13f**) with the regeneration of rhodium catalyst. Another pathway was also proposed by authors where α-imino rhodium(II) carbenoid (**3.13a**) reacts with the alkene to give intermediate **3.13d**. Then ring-opening yields **3.13e** followed by the formation of the dihydropyrrole **3.13f**. However, the second pathway could not be confirmed by NMR studies (Scheme 3.13).

The rhodium(II)-catalyzed transannulation of triazoles with acyclic vinyl ethers for the synthesis of substituted pyrroles was reported by Anbarasan in 2014 (Scheme 3.14a) [24]. The 4-aryl-*N*-sulfonyl triazoles with distinct electronic nature react satisfactorily with vinyl ethers having a variety of aryl and alkyl substituents

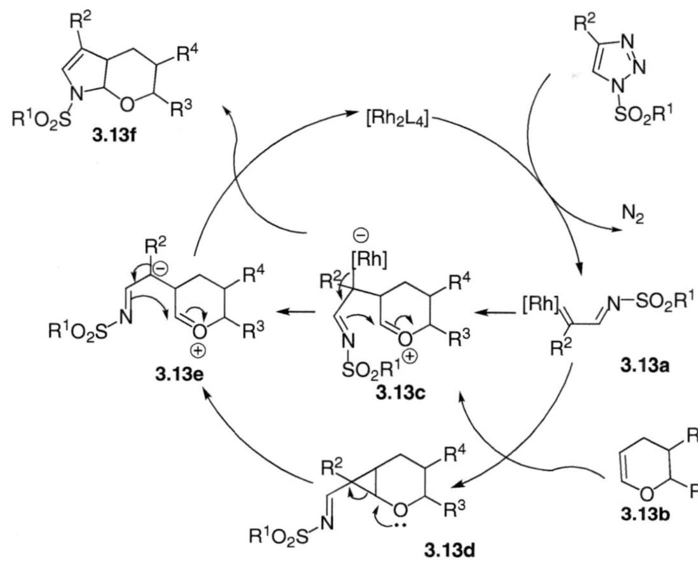

Scheme 3.13 Mechanism for Rh(II)-catalyzed synthesis of THF-DPs and THP-DPs.

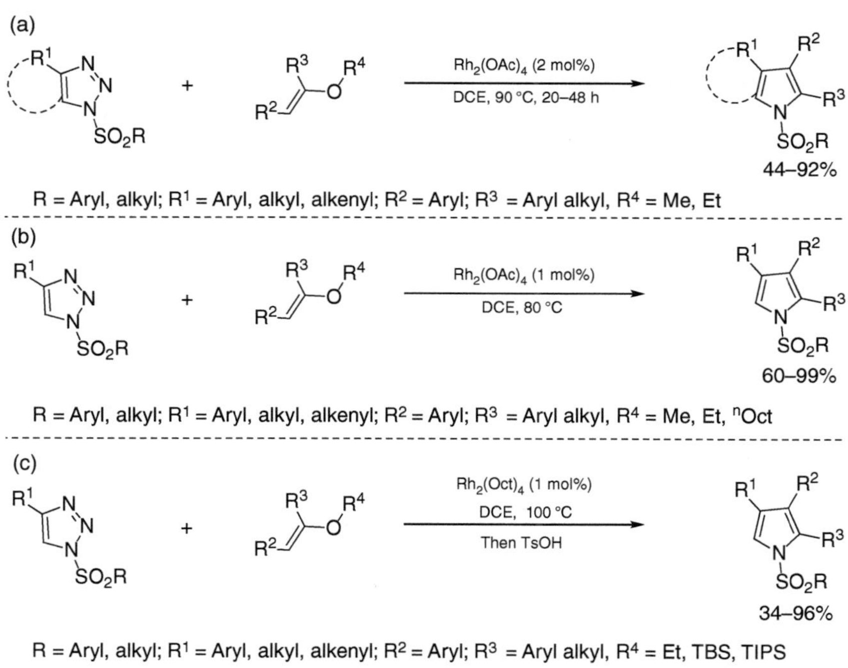

Scheme 3.14 Rh(II)-catalyzed transannulation of triazoles with acyclic vinyl ethers. Sources: (a) Adapted from Anbarasan [24]. (b) Lee [25]/American Chemical Society. (c) Adapted from Tang [26].

in the presence of 2 mol% of $Rh_2(OAc)_4$ catalyst to give pyrroles. In addition, C4 and C5 substituted triazoles react under reaction conditions smoothly to give tetrasubstituted pyrroles with moderate yields. Also, Lee's group reported similar reaction independently [25]. Notably, the enol ethers having electron deficient substituents at α and β position gave dihydropyrroles as products, which, on treatment with TMSOTf, gave the corresponding pyrrole. Similarly, cyclic enol ethers reacted efficiently to afford the tetra-substituted pyrroles in good yields (Scheme 3.14b). Further, Tang's group described the synthesis of substituted pyrroles through rhodium(II)-catalyzed transannulation of 4-aryl-*N*-tosyl-1,2,3-triazoles with alkyl or silyl enol ethers followed by treatment with TsOH at room temperature (Scheme 3.14c) [26]. The reactions of electronically and sterically variable 4-aryl triazoles with both alkyl and silyl enol ethers proceed efficiently to give substituted pyrroles; however, *p*-anisyl substituted enol ethers result in poor yield. Under similar reaction conditions, transannulation of 4-aryltriazoles were also studied with cyclic enol ethers and cycloalkenyl ethers, which resulted in corresponding 3-alkyl pyrroles and fused pyrroles as products, respectively.

The mechanism of reaction for the synthesis of pyrroles involves the formation of an α-diazoimine (**3.15a**) from triazole followed by denitrogenative metal incorporation to generate reactive rhodium carbenoid **3.15b**. Then nucleophilic addition of **3.15c** generates a zwitterionic species **3.15d** which undergoes further cyclization to form the intermediate **3.15e** along with the regeneration of reactive rhodium species. An alternate pathway for the formation of **3.15e** could also be achieved by the formation of an imino cyclopropane **3.15f** from imino carbenoid **3.15b** and **3.15c** followed by Cloke rearrangement. Finally, the aromatization of **3.15e** occurs by eliminating a molecule of alcohol to give pyrrole **3.15g** (Scheme 3.15).

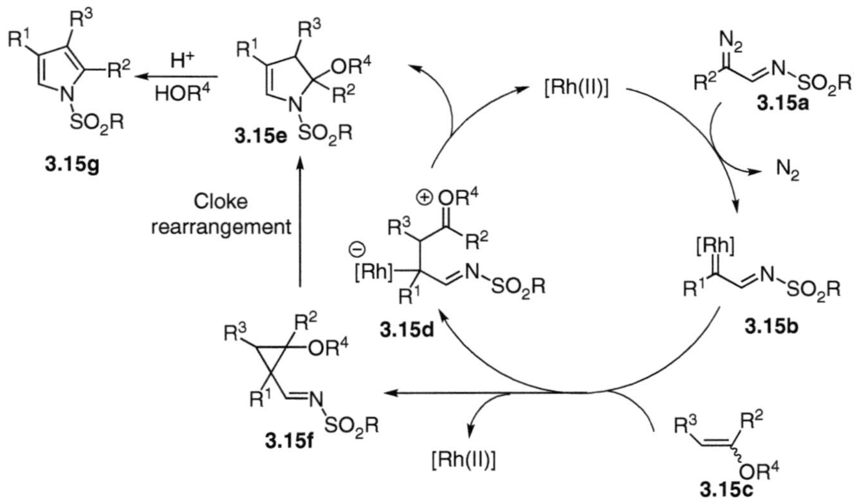

Scheme 3.15 Mechanism of Rh(II)-catalyzed transannulation for pyrrole synthesis.

The reactivity of rhodium(II) imino carbene intermediates formed from *N*-methanesulfonyl-1,2,3-triazoles has been explored with methylenecyclopropanes obtained from *cis*-cyclooctene for the synthesis of piperidine-fused *trans*-cycloalkenes. The reaction involves one-pot sequential enantioselective cyclopropanation and thermal skeletal rearrangement (Scheme 3.16) [27]. The one-pot two-step process involves highly enantioselective cyclopropanation of 4-aryl or 4-alkyl-*N*-methanesulfonyl-1,2,3-triazoles with 4-*cis*-cyclooctene in presence of $Rh_2[(S)\text{-NTTL}]_4$ as chiral dirhodium catalyst followed by microwave irradiation (MW) at 120 °C to give piperidine-fused *trans*-cycloalkenes via [3,3] sigmatropic rearrangement.

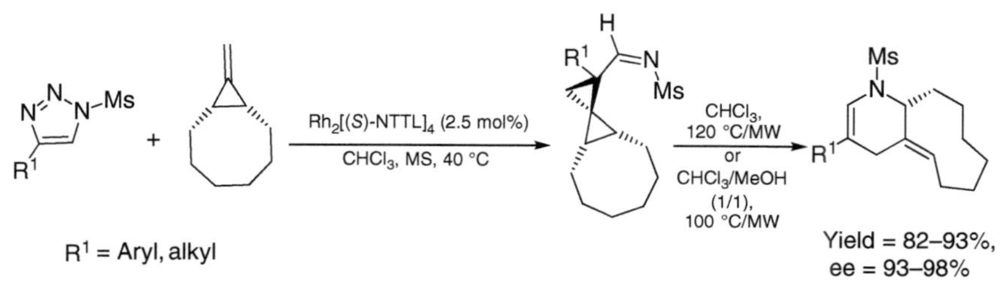

Scheme 3.16 Rh(II)-catalyzed piperidine-fused *trans*-cycloalkenes synthesis from triazoles and MCPs.

The authors proposed a transition state model based on the retro-Claisen rearrangement, in which the interconversion of allyl vinyl ethers and γ,δ-unsaturated aldehydes occurs via [3,3] sigmatropic rearrangement involving chair-like six-membered transition state model **3.17a** (Scheme 3.17). Similarly, α-imino spiropentane intermediate resulting in rhodium catalyzed reaction of triazole with methylenecyclopropanes is structurally analogous with the γ,δ-unsaturated aldehydes. Therefore, the transition state model **3.17c** is proposed for the thermal rearrangement of the α-imino spiropentane to piperidine-fused *trans* cycloalkenes. Model **3.17c** also supports the obtained stereochemistry of the product.

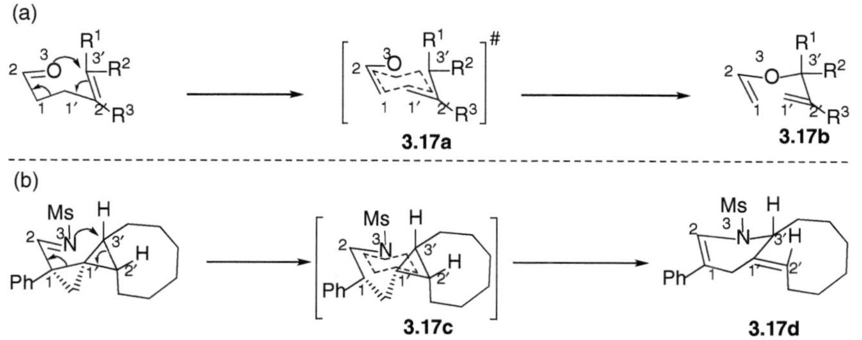

Scheme 3.17 Mechanism of Rh(II)-catalyzed synthesis of piperidine-fused *trans*-cycloalkene. (a) Retro Claisen rearrangement. (b) Sigmatropic rearrangement.

In 2014, Min Shi's research group reported rhodium(II)-catalyzed tandem cycloisomerization of *ortho*-substituted methylenecyclopropyl-4-aryl-*N*-sulfonyl-1,2,3-triazoles to synthesize highly substituted polycyclic nitrogen heterocycles through an α-imino rhodium carbenoid intermediate [28]. Triazoles possessing electronically distinct substituents react efficiently in the presence of 2 mol% of $Rh_2(OPiv)_4$ as the catalyst to deliver the polycyclic nitrogen heterocycles as the major product in good to excellent yields (Scheme 3.18).

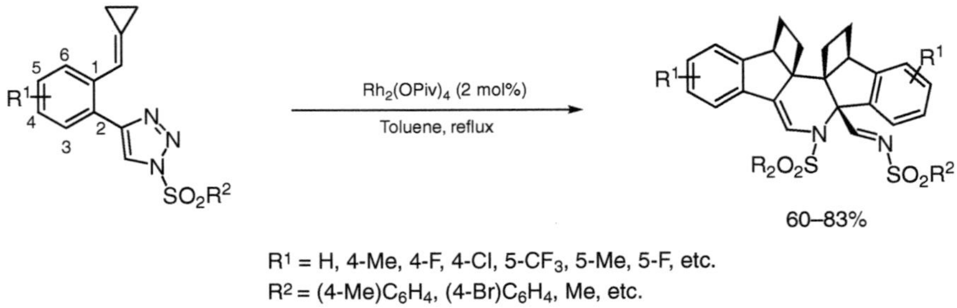

Scheme 3.18 Rh(II)-catalyzed tandem cycloisomerization of methylenecyclopropyl triazoles.

However, other structural variations in triazoles result in different products depending on the structure of the corresponding triazoles (Scheme 3.19). In this, the reactions of triazoles obtained from thiophene- or cycloalkane-tethered MCPs result in corresponding fused indolines as products, possibly due to the higher rate of thermally induced rearrangements compared to the generation of respective rhodium(II) imino carbenoids. In addition, triazoles tethered with MCP's bearing phenyl substituents produce [3 + 3] cyclization products. The reaction of triazoles

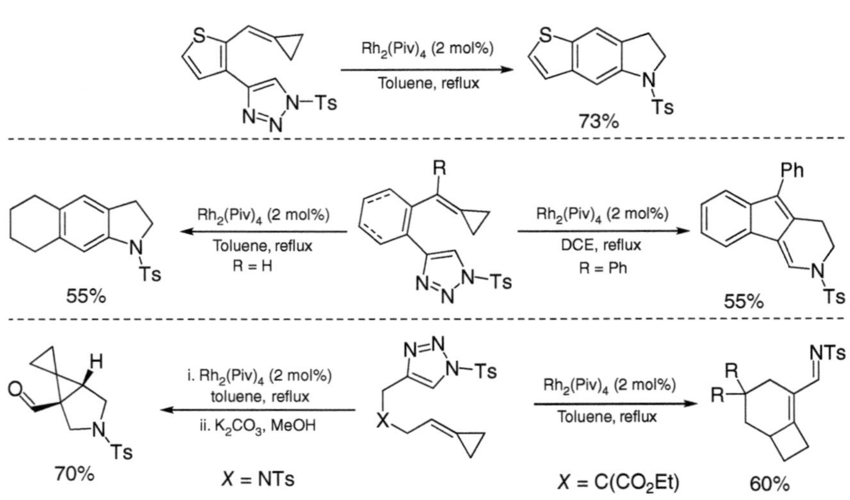

Scheme 3.19 Rh(II)-catalyzed transannulation of triazoles tethered with MCP.

tethered to MCP through nitrogen atom followed by base-mediated hydrolysis furnishes tricyclic aldehyde via sequential cyclopropanation and detosylation. In the reaction of triazoles tethered to MCP through the alkyl chain, the cyclization product is formed in good yield.

The controlled experiments and crossover experiments indicate the desired product forms via intermolecular Diels–Alder reaction (Scheme 3.20). The MCP tethered *N*-sulfonyl triazole undergoes ring chain tautomerization to diazoimine **3.20a**. This with rhodium results in rhodium(II) azavinyl carbene **3.20b**, which is in equilibrium with an electrophilic intermediate. Further intramolecular nucleophilic attack of MCP on metal carbene generates intermediate **3.20c**. Then ring expansion and removal of metal results in intermediate **3.20d**, which is highly reactive and instantly undergoes a [4 + 2] cycloaddition with another intermediate **3.20d** to give *N*-heterocycles **3.20e′** and **3.20e**. In addition, it has been observed that at reflux temperature, the product **3.20e′** transforms into thermodynamically more stable **3.20e** as the sole product via retro Diels–Alder reaction.

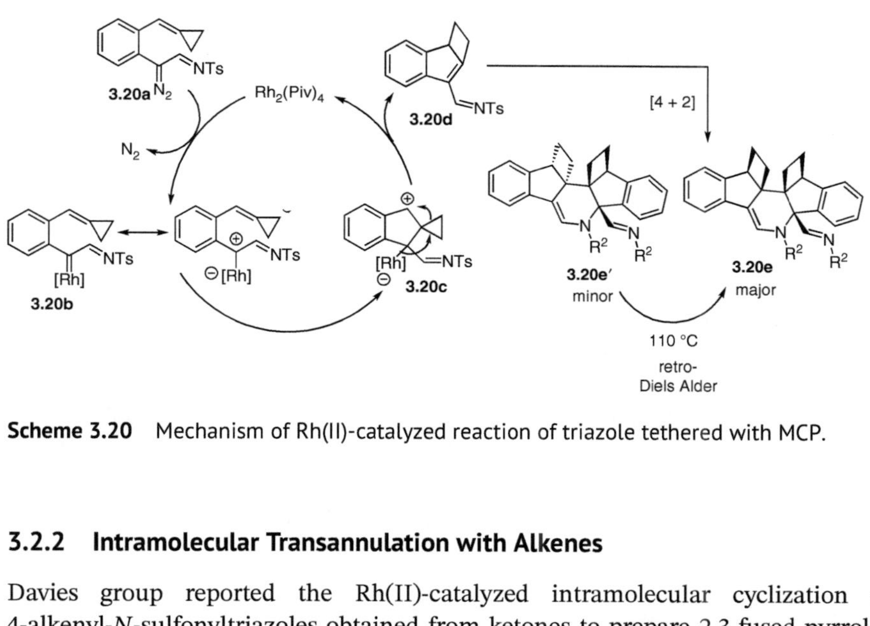

Scheme 3.20 Mechanism of Rh(II)-catalyzed reaction of triazole tethered with MCP.

3.2.2 Intramolecular Transannulation with Alkenes

Davies group reported the Rh(II)-catalyzed intramolecular cyclization of 4-alkenyl-*N*-sulfonyltriazoles obtained from ketones to prepare 2,3-fused pyrroles as well as indoles through one pot DDQ oxidation of the resulting 2,3-fused pyrroles furnished from corresponding cyclohexenyl derivatives (Scheme 3.21) [29]. Among various Rh(II) salts, $Rh_2(esp)_2$ gives optimal yield; however, $Rh_2(TFA)_4$ fails to deliver any product even after prolonged reaction time. The triazoles obtained from cyclic ketone bearing a variety of substituents react efficiently to give the products with excellent yields.

The mechanistic rationale is displayed in Scheme 3.22. The reaction is initiated by heating triazole **3.22a** in the presence of rhodium(II) catalyst, which

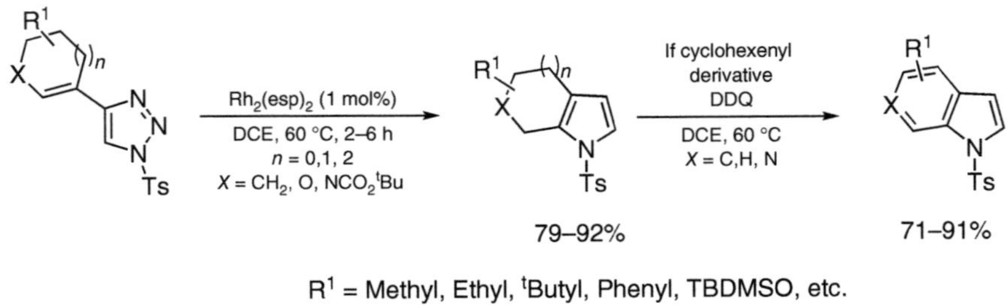

Scheme 3.21 Rh(II)-catalyzed intramolecular cyclization of 4-alkenyl-*N*-sulfonyltriazoles.

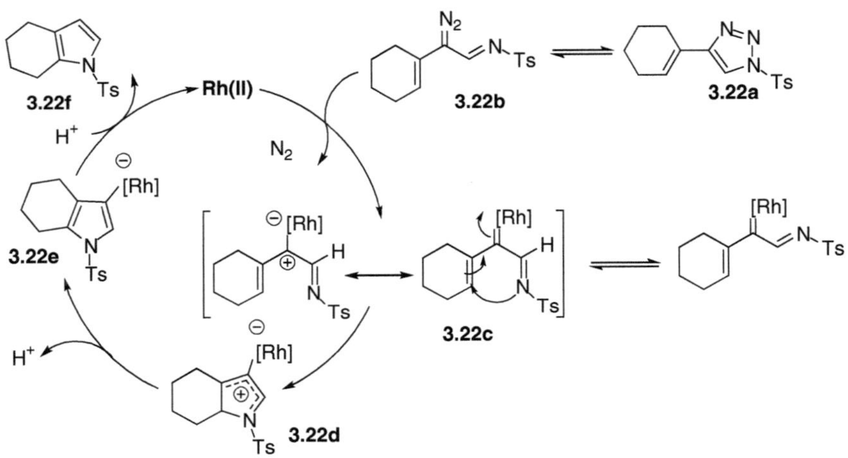

Scheme 3.22 Mechanism of Rh(II)-catalyzed intramolecular cyclization of 4-alkenyl-*N*-sulfonyltriazoles.

on ring chain isomerization followed by denitrogenation gives intermediate **3.22c**. This intermediate undergoes C—C bond rotation to gain *s*-trans geometry to form pyrrolylium cation **3.22d** via 4π electrocyclization. Next, deprotonation gives vinylrhodium intermediate **3.22e**. Finally, protonation of this intermediate results in the formation of product **3.22f** with regeneration of the catalyst.

3.3 Transannulation of *N*-Sulfonyl-1,2,3-triazoles with 1,3-Dienes

1,3-Dienes are the excellent substrates for the cycloaddition reactions, which usually result in [4 + 2], [4 + 3], [4 + 1], etc., cycloaddition reactions with suitable counterparts to deliver the corresponding carbocycles and heterocycles under metal-free conditions as well as in the presence of metal catalysts. In addition, it has been reported that the *N*-sulfonyl-1,2,3-triazoles also react with conjugated dienes in the presence of rhodium catalysts in intermolecular and intramolecular fashion to give corresponding transannulated products.

3.3.1 Intermolecular Transannulation with 1,3-Dienes

First rhodium(II)-catalyzed intermolecular transannulation of 4-Alkenyl-*N*-sulfonyl-1,2,3-triazoles with dienes for the asymmetric synthesis of fused polycyclic cycloheptadienes through a formal [4+3] cycloaddition is reported by Davies in 2013 (Scheme 3.23) [30].

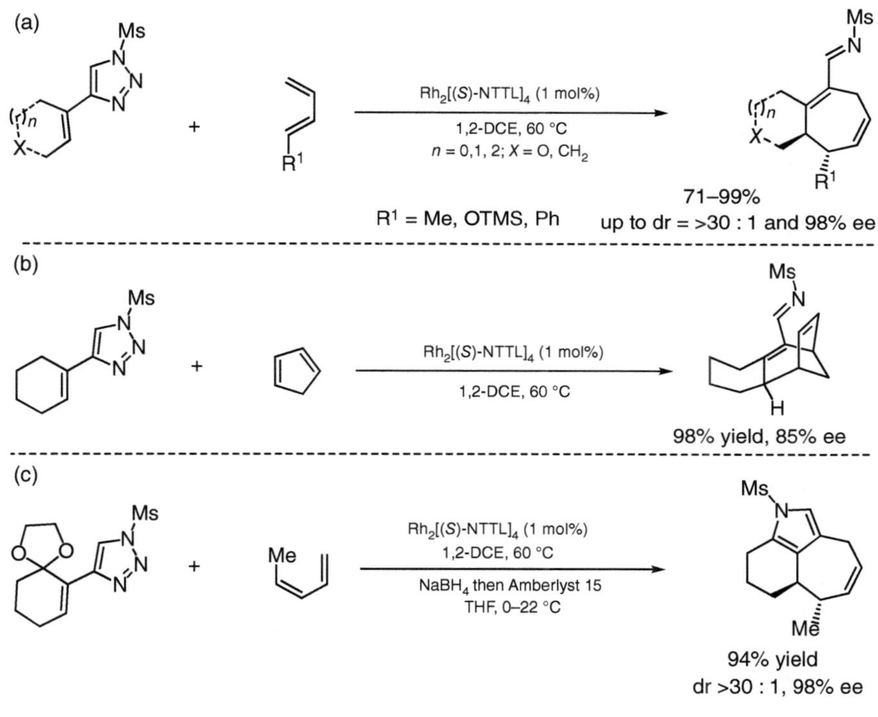

Scheme 3.23 Rh(II)-catalyzed transannulation of 4-alkenyl-triazoles with (a) acyclic-1,3-dienes, (b) cyclic-1,3-dienes, (c) substituted-1,3-dienes.

The use of 1 mol% $Rh_2[(S)\text{-NTTL}]_4$ provides the [4+3] cycloaddition product with the best yield, diastereo-and enantioselectivity at 60 °C in DCE as solvent. The substituted cycloalkenyl triazoles react efficiently with both (*Z*) and (*E*) alkenes to deliver the corresponding fused polycyclic cycloheptadienes in high yield and good to excellent enantioselectivity (Scheme 3.23a). In addition, the cyclic diene reacts smoothly with the triazole to deliver a bridged tricyclic structure with high yield and good enantioselectivity (Scheme 3.23b). The methodology is further extended to synthesize tetrahydroindole starting from the triazole with (*Z*)-penta-1,3-diene via rhodium-catalyzed transannulation followed by $NaBH_4$ reduction and then cyclization using Amberlyst 15 (Scheme 3.23c).

Similarly, in 2014 Tang's research group reported the first *aza* [4 + 3] cycloaddition of 1,3-dienes with triazoles to synthesize highly substituted 2,5-dihydroazepines (Scheme 3.24) [31].

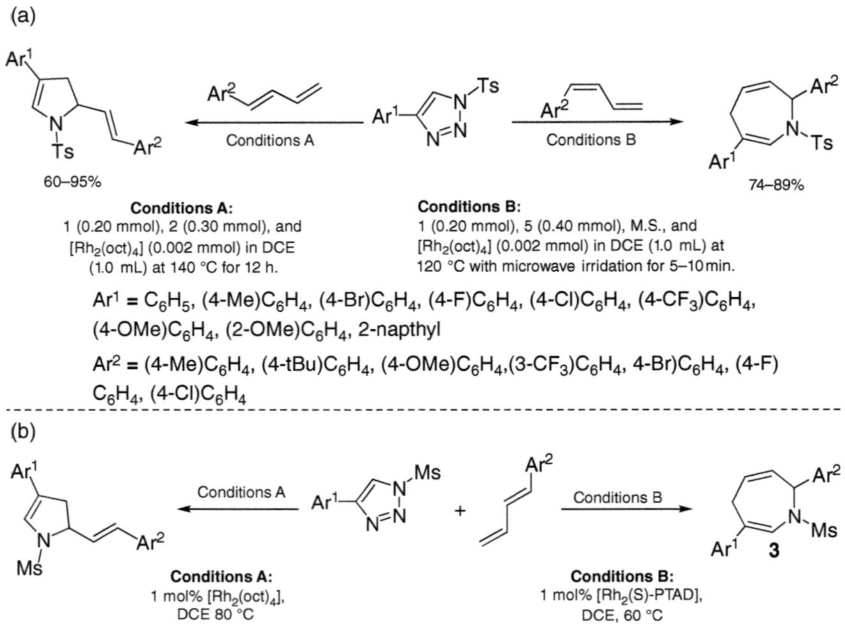

Scheme 3.24 Rh(II)-catalyzed transannulation of 4-aryl-*N*-sulfonyl-triazoles with 1,3-dienes. (a) Tang [31]. (b) Lee [32].

In addition, with *E*-alkene under modified reaction conditions, the 2,3-dihydropyrroles could be obtained as the major products through [3 + 2] cycloaddition reactions. The $Rh_2(oct)_4$ is found to be the catalyst of choice for the synthesis of both products. It is observed that the azepine formation required only 5–10 minutes under microwave irradiation, while dihydropyrrole formation required thermal conditions (140 °C) for 12 hours. It is identified that continuing the reaction for a longer time results exclusively in the formation of the dihydropyrrole product. Most interestingly, the geometry of the alkene influenced the reaction markedly. In the use of (*Z*)-1-aryl-1,3-dienes, and triazoles, the 2,5-dihydroazepine is obtained as the major product with the exception of 2-methoxyphenyl-1,3-diene. The reactions of 2-aryl-1,3-dienes with triazoles almost exclusively resulted in [4 + 3] cycloadditions except in some cases with little effect of the electronic nature of both reactants. In addition, the 2,5-dihydroazepines formed from 2-aryl-dienes are more stable and do not result in further transformation into corresponding 2,3-dihydropyrroles. Further, the transannulation of triazoles with 1,3-dienes having substituents such as 1,1-diphenyl-,1-phenyl-2-methyl-, and 1-TBSO gives [3 + 2] cycloaddition product as the major outcome.

Lee's group also noted that in the presence of Rh(II) catalyst, triazoles prepared from alkyl sulfonyl azides and aryl acetylenes react with 4-aryl-1,3-butadienes to give *aza*-[4 + 3] and *aza*-[3 + 2] cycloaddition reactions under mild conditions (Scheme 3.25) [32]. The reaction outcome is dependent on the rhodium catalyst. It is observed that the use of $Rh_2(Oct)_4$ gives *aza*-[3 + 2] cycloaddition product, while, on using $Rh_2[(S)\text{-PTAD}]_4$, the *aza*-[4 + 3] cycloaddition product is obtained. However, in this case, triazoles derived from aryl sulfonyl azides are unreactive. The mechanistic insights were gained by performing controlled experiments. The isolation of cyclopropylaldimine intermediates and further treatment at 140 °C in DCE as solvent results in the [4 + 3] and [3 + 2] cycloaddition reactions to give corresponding products. Based on this observation, the authors proposed a mechanism of reaction that shows that the rhodium(II) bound iminocarbene **3.25a** reacts with diene **3.25b** to give cyclopropylaldimine **3.25c** via [2 + 1] cycloaddition. This intermediate based on the nature of substituent on diene results in either *aza* Cope rearrangement to give 2,5-dihydroazepines (path *a*) (**3.25d**) or cyclopropylimine rearrangement to give 2,3-dihydropyrroles (path *b*) (**3.25e**).

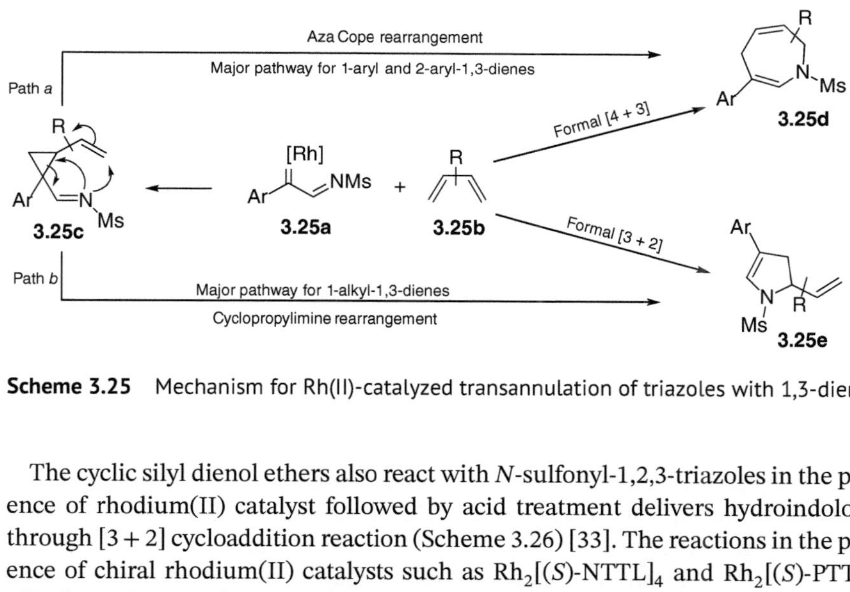

Scheme 3.25 Mechanism for Rh(II)-catalyzed transannulation of triazoles with 1,3-dienes.

The cyclic silyl dienol ethers also react with *N*-sulfonyl-1,2,3-triazoles in the presence of rhodium(II) catalyst followed by acid treatment delivers hydroindolones through [3 + 2] cycloaddition reaction (Scheme 3.26) [33]. The reactions in the presence of chiral rhodium(II) catalysts such as $Rh_2[(S)\text{-NTTL}]_4$ and $Rh_2[(S)\text{-PTTL}]_4$ affords products with good yields with lower enantioselectivities.

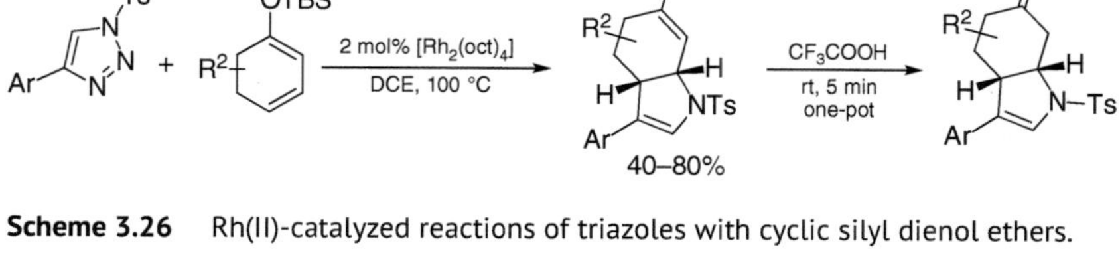

Scheme 3.26 Rh(II)-catalyzed reactions of triazoles with cyclic silyl dienol ethers.

3.3.2 Intramolecular Transannulation with 1,3-Dienes

Sarpong's research group developed a rhodium(II)-catalyzed intramolecular sequential cyclopropanation/cope rearrangement reaction of dienyl tethered triazoles with having trans geometry of double bonds to prepare fused dihydroazepine derivative (Scheme 3.27) [34]. The use of 1 mol% of $Rh_2(Adc)_4$ gives the best results in chloroform as a solvent at 60 °C. Irrespective of the aryl substitution, the reaction is facile, including phenyl, electron-rich, and deficient substrates.

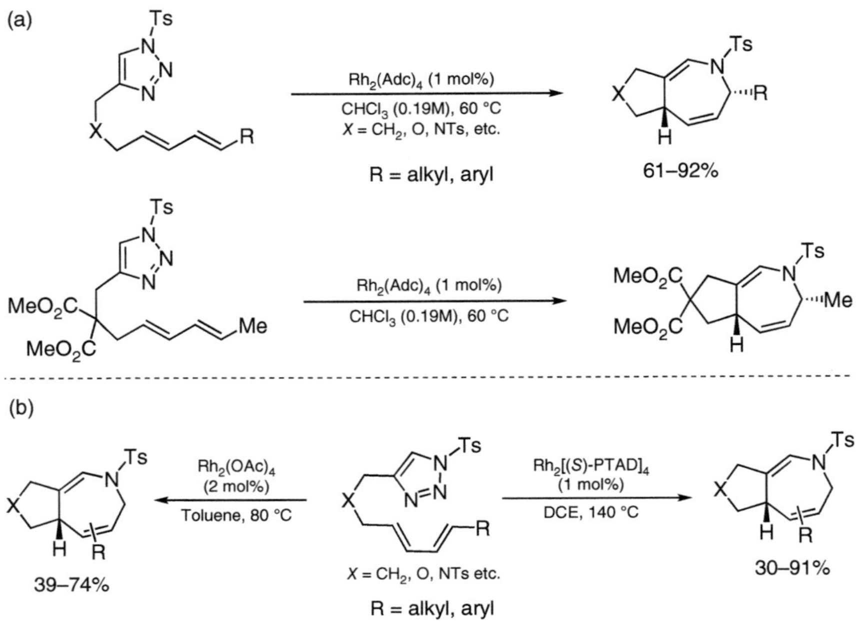

Scheme 3.27 Rh(II)-catalyzed intramolecular transannulation of triazoles tethered with 1,3-dienes. Sources: (a) Sarpong [34]/John Wiley & Sons. (b) Adapted from Shen and Tang [35, 36].

Internally substituted dienes also react under reaction conditions to afford the products, however, with lower yields. The diester-substituted triazole derivative results in lower yields, which could be attributed to side reactions due to intramolecular reaction of the proximal ester group on the carbene in the intermediate stage. In this case, an intramolecular cyclopropanation occurs, giving a 1-imino-2-vinylcyclopropane intermediate, which is transient and undergoes a 1-*aza*-cope rearrangement to afford the corresponding fused dihydroazepine moiety. Based on the stereochemistry of the product obtained, the proposed mechanism involves thermal ring chain isomerization of the triazole **3.28a** to give the α-imino diazo compound **3.28b**, which with rhodium(II) catalyst results in rhodium-iminocarbenoid intermediate **3.28c** (Scheme 3.28). Further, intramolecular [2 + 1] cycloaddition results in the iminovinylcyclopropane (IVC) **3.28d**. Then [3,3] sigmatropic rearrangement via transition state **3.28e** directly results in a dihydroazepine derivative.

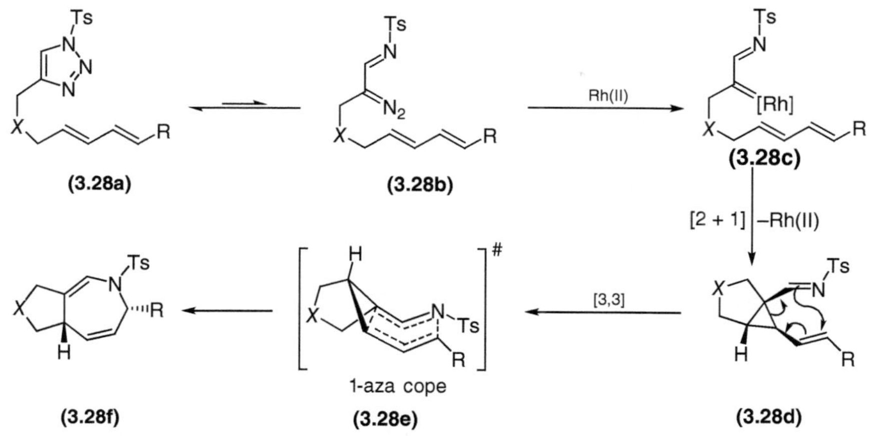

Scheme 3.28 Mechanism of Rh(II)-catalyzed intramolecular *aza*-Cope reaction of triazoles tethered with 1,3-dienes.

The intramolecular [4+3] cycloaddition reaction of dienyl-tethered triazoles with variable substitution patterns via iminovinylcyclopropane intermediate has also been explored for the other Rh(II) catalyst [35, 36]. It is found that the use of chiral catalysts like $Rh_2[(S)\text{-PTAD}]_4$ and $Rh_2[(S)\text{-DOSP}]_4$ gives products with good yields, but fails to achieve enantioselective transformation.

3.4 Transannulation of *N*-Sulfonyl-1,2,3-triazoles with Alkynes

In transition metal catalysis, alkynes display versatile reactivity as a π component in cycloaddition reactions to synthesize various heterocyclic molecules with competent reacting partners. Metal-catalyzed transannulation of *N*-sulfonyl-1,2,3-triazoles with alkynes as a π component gives a useful method for synthesizing pyrroles and other heterocyclic motifs.

3.4.1 Intermolecular Transannulation with Alkynes

In the metal-catalyzed transannulation of *N*-sulfonyl-1,2,3-triazoles with alkynes, reasonable progress has been made. The metal-catalyzed generation of the α-imino metal carbenoid intermediate from *N*-sulfonyl-1,2,3-triazole followed by *in situ* [3+2] cycloaddition with alkynes results in the formation of highly substituted pyrroles. Murakami and co-workers in 2009, reported the first example of nickel-catalyzed denitrogenative transannulation of internal alkynes with *N*-sulfonyl-1,2,3-triazoles for synthesizing trisubstituted pyrroles (Scheme 3.29) [37]. In this, a catalytic combination of $[Ni(cod)_2]/P(^{n}Bu)Ad_2$ along with $AlPh_3$ as Lewis acid in catalytic amount efficiently catalyzes the transannulation between triazoles and internal alkynes. 4-aryl-*N*-sulfonyl-1,2,3-triazoles reacts efficiently

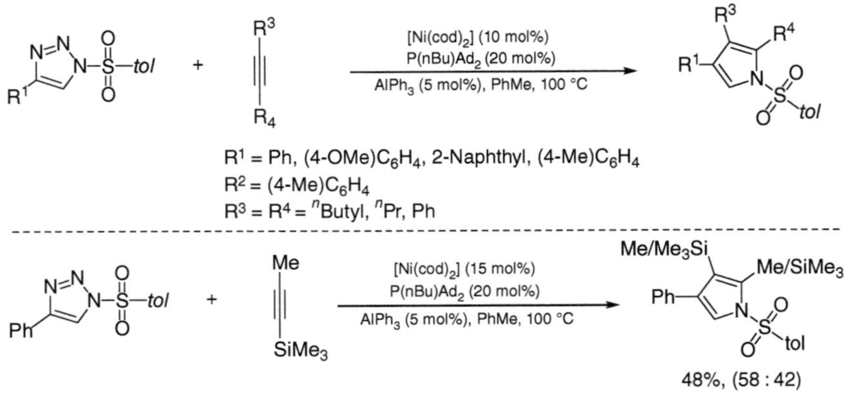

Scheme 3.29 Nickel-catalyzed transannulation of N-sulfonyl-1,2,3-triazoles with alkynes.

with symmetrical aliphatic alkynes; however, the triazoles obtained from aliphatic terminal alkyne failed to deliver the desired product. The unsymmetrical alkynes on transannulation with triazoles result in a regioisomeric mixture of the tetrasubstituted pyrroles. Meanwhile, the terminal alkynes are unsuccessful in affording the transannulation product with triazoles in the presence of $Ni(0)/PR_3$ as a catalytic system. Mechanistically, the ring chain tautomerization of the triazole **3.30a** leads to the generation of α-imino diazo compound **3.30b**, which undergoes denitrogenation in the presence of nickel catalyst to generate Ni-carbenoid **3.30c**, then intermediate **3.30c** undergoes cyclization to produce azanickelacycle **3.30d**, followed by alkyne **3.30e** insertion gives six-membered nickelacycle **3.30f**, which, upon subsequent reductive elimination, regenerates the Ni(0) species along with the desired product **3.30g** (Scheme 3.30). The Lewis acid is assumed to either promote the ring chain tautomerization or facilitate the reductive elimination.

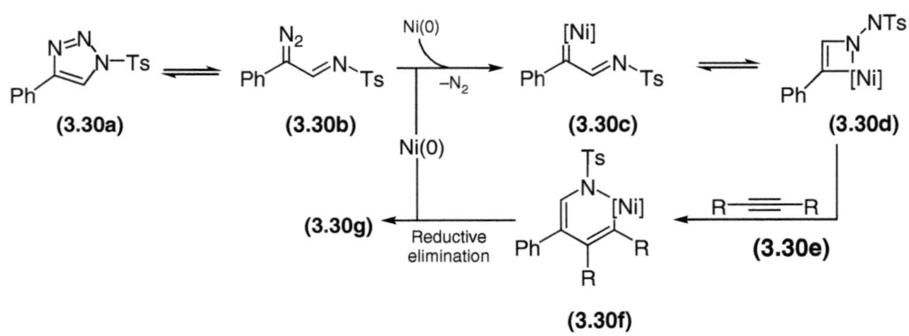

Scheme 3.30 Mechanism of Ni-catalyzed transannulation of N-sulfonyl-1,2,3-triazoles with internal alkynes.

The deprived reactivity of terminal alkynes toward transannulation with the triazole could be improved by the combination of $[Rh_2(oct)_4]/AgOCOCF_3$ as a catalytic system to produce 1,2,4-trisubstitutedpyrroles (Scheme 3.31) [38]. The triazoles with

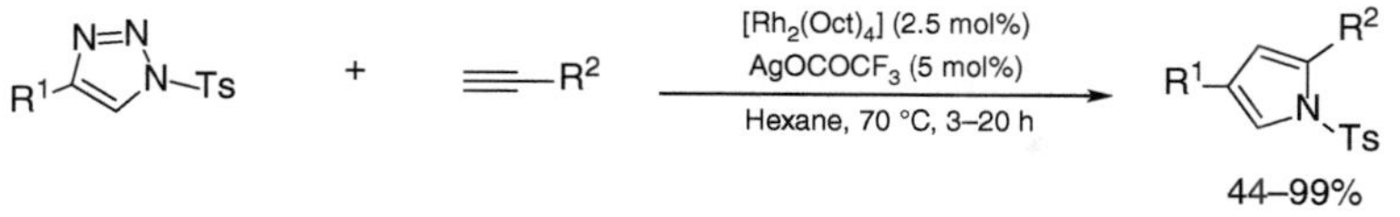

R^1 = C_6H_5, (4-CO_2Et)C_6H_4, (4-Me)C_6H_4, (4-Br)C_6H_4, *n*-Butyl, CO_2Et, H

R^2 = (2-Me)C_6H_4, (2-OMe)C_6H_4, (2-Me, 4-OMe)C_6H_3, (2-OPh)C_6H_4, (2, 3-Me)C_6H_3 etc.

Scheme 3.31 Rh(II)/Ag(I)-catalyzed transannulation of triazoles with alkynes.

4-aryl, alkyl, CO_2Et substituents with variable electronic nature react efficiently with electron-rich alkynes to afford corresponding trisubstituted pyrroles in moderate to excellent yields. Though the exact role of the silver catalyst is not clear, apparently, it acts as a Lewis acid to improve the electrophilicity of the Rh(II)-azavinyl carbene.

The denitrogenative insertion of rhodium(II) results in the formation of rhodium iminocarbene **3.32c** (Scheme 3.32). Then, the nucleophilic attack of alkyne produces ylide **3.32e** (path *a*) followed by cyclization leads to the formation of a cyclic zwitterionic species **3.32f**. Finally, the elimination of rhodium results in the product **3.32g**. On the contrary, an attack of silver acetylide on intermediate **3.32c** generates rhodium-bound propargylimine species **3.32h** (path *b*); this intermediate

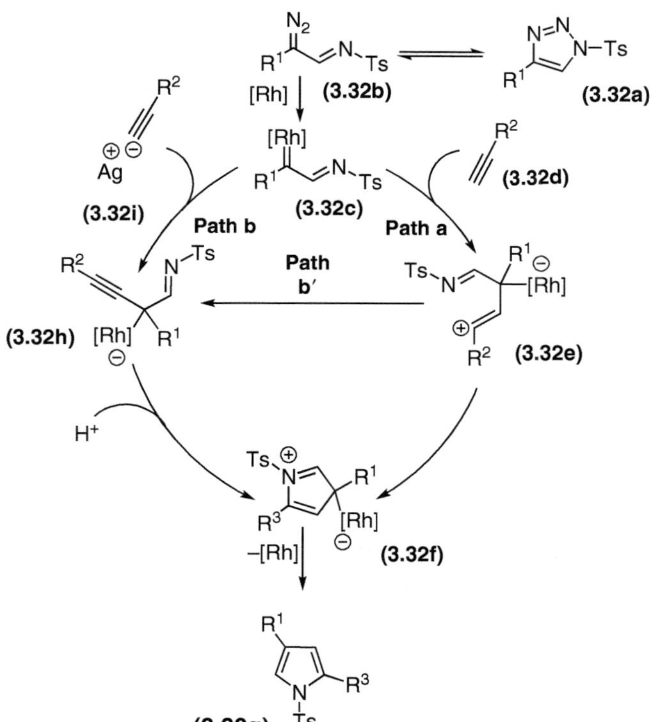

Scheme 3.32 Mechanism of Rh(II)/Ag(I)-catalyzed transannulation of triazoles with alkynes.

would undergo proton-assisted 5-endo-dig cyclization to deliver **3.32f** followed by elimination of rhodium to give product **3.32g**. However, based on deuterium labelling experiments the possibility of path ***b*** and path ***b′*** is eliminated by authors.

Li's research group also reported Rhodium(III)-catalyzed dual C(sp^2)—H bond activation of 4-aryl-*N*-sulfonyl-1,2,3-triazoles with internal alkynes leading to [3+2]/[5+2] cycloaddition reactions giving indeno-[1,7-cd]azepin-1-ols and 1-methyleneindeno[1,7-cd]azepines respectively (Scheme 3.33) [39]. It has been observed that the use of $AgSbF_6$ as an additive to activate the Rh(III) species and $Cu(OAc)_2$ as an oxidant plays a pivotal role in the efficient catalytic process. Aryl triazoles bearing electronically divergent substituents react smoothly with various diaryl alkynes to afford desired products with satisfactory yields. Under optimal reaction conditions, triazoles also react with alkyl-substituted internal alkynes to give 1-methyleneindeno[1,7-cd]azepines as products. The reactions are also successful with alcohols such as ethanol and benzyl alcohol as nucleophiles instead of water, resulting in corresponding products.

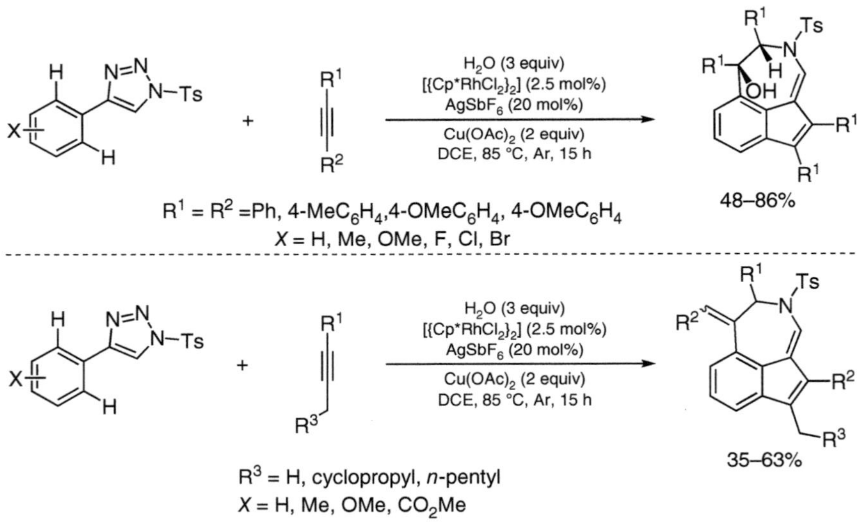

Scheme 3.33 Rh(III)-catalyzed dual C(sp^2)—H bond activation of 4-aryl-*N*-sulfonyl-1,2,3-triazoles with internal alkynes.

The catalytic cycle initiates with the generation of reactive Cp*RhX_2 species, which readily reacts with triazole to form corresponding rhodium carbenoid **3.34b** (Scheme 3.34). Further, alkyne **3.34c** addition to carbenoid generates an intermediate **3.34d**. This intermediate undergoes electrophilic cyclization with the triazole aryl ring. Then, further annulation of intermediate **3.34h** with the second molecule of alkyne gives Rh-coordinated intermediate **3.34f**. Trans addition of alkyne to intermediate **3.34f** results in species **3.34g**. At this stage, the outcome of the product is decided by the alkyne used. When 1,2-diphenyl acetylene is used, C—Rh bond cleaves through hydration by backside attack of water on C—Rh bond

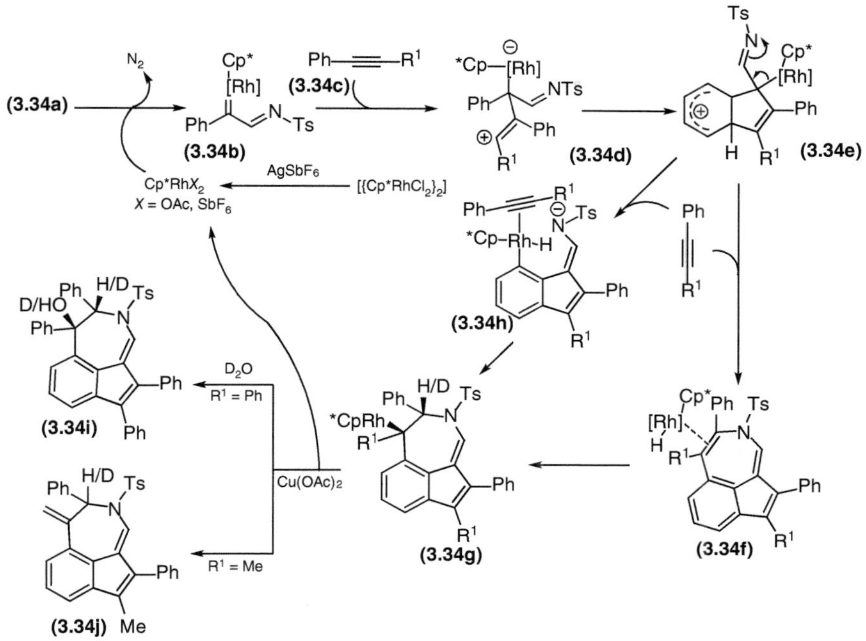

Scheme 3.34 Mechanism of Rh(III)-catalyzed dual C(sp^2)—H bond activation of triazoles with internal alkynes.

with the aid of $Cu(OAc)_2$ to give product **3.34i** and concurrent regeneration of active Cp*RhX_2 species. On the other hand, using phenylpropyne as alkyne results in selective β—H elimination with $Cu(OAc)_2$ and cleavage of the C—Rh bond occurs to give product **3.34j**. The formation of intermediate **3.34g** also supports the second annulation via C—H activation.

3.4.2 Intramolecular Transannulation with Alkynes

The *N*-sulfonyl-1,2,3-triazole units tethered with appropriately spaced alkynes with rhodium(II) as the catalyst undergo intramolecular transannulation to deliver corresponding 5,5-fused pyrroles (Scheme 3.35) [40]. The reaction proceeds through the formation of an α-imino-carbenoid intermediate, which further results in carbene-alkyne metathesis to generate another carbene intermediate. Then intramolecular nucleophilic attack of the nitrogen atom of imine on the carbene and further tautomerization results in fused pyrrole as a product. Substrates bearing substituted aryl group, alkyne, alkene, SPh, TMS, Br, at the alkyne terminal react efficiently to deliver corresponding 5,5-fused pyrroles. The substrates possessing a C3 tether such as ketones, nitriles, and protected alcohols, as well as when, X = NTs also react smoothly to give the respective products in good to excellent yields.

Li's research group disclosed that 4-(2-ethynylaryl)-*N*-sulfonyl-1,2,3-triazoles in the presence of silver catalyst react with *in situ* generated or external nucleophiles to give functionalized isoquinolines (Scheme 3.36) [41]. However, copper catalysts

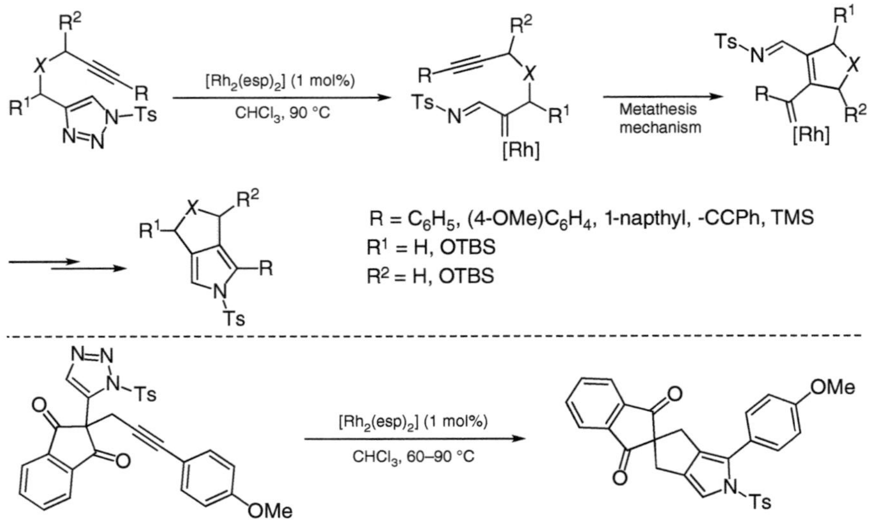

Scheme 3.35 Rh(II)-catalyzed intramolecular transannulation of triazoles tethered with alkynes.

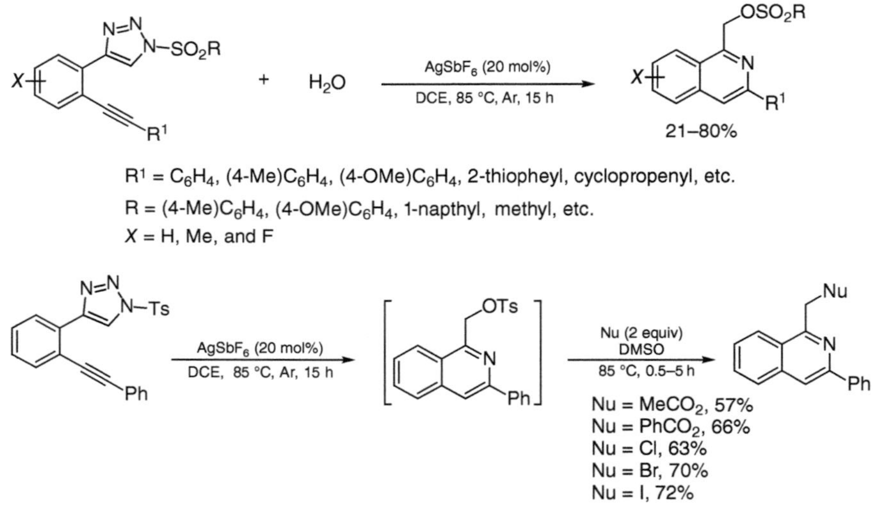

Scheme 3.36 Silver-catalyzed intramolecular cyclization of 4-(2-ethynylaryl)-*N*-sulfonyl-1,2,3-triazoles.

give products with lower yields, and rhodium(II) catalysts are ineffective in giving the desired products.

The reaction involves the opening of triazole in the presence of the silver catalyst under heating conditions, followed by nitrogen expulsion to form silver carbenoid **3.37b or 3.37c** (Scheme 3.37). Silver also coordinates with the alkyne to increase its electrophilicity. Then intermediate **3.37c** cyclizes with the addition of water to furnish **3.37d** and TsOH, which in turn could act as a nucleophile, followed by another cyclization with nucleophile to form three-membered cyclic intermediate

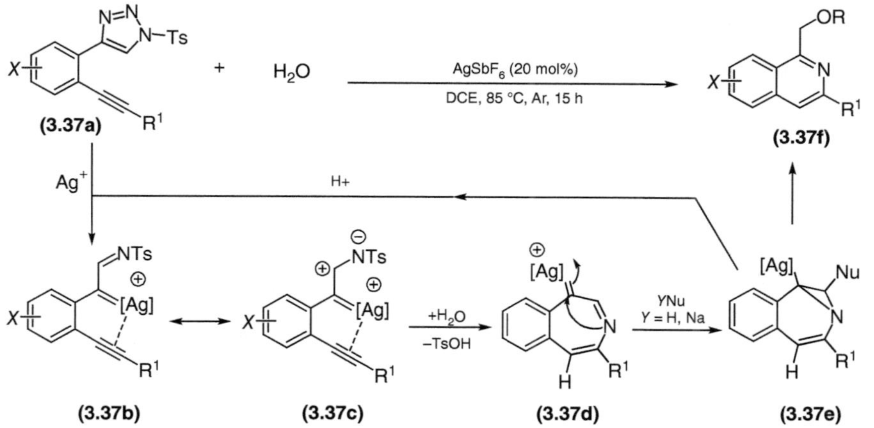

Scheme 3.37 Mechanism of silver-catalyzed intramolecular cyclization of 4-(2-ethynylaryl)-*N*-sulfonyl-1,2,3-triazoles.

3.37e. Then ring-opening and protonation furnishes isoquinoline **3.37f** with regeneration of active Ag(I) catalyst. Aryl triazoles substituted at the *ortho* position with aryl/alkyl alkynes react under the reaction condition to give isoquinolines. Nucleophile such as carboxylic acid, NaCl, NaBr, and NaI reacts efficiently under the reaction condition in DMSO as solvent.

3.5 Transannulation of *N*-Sulfonyl-1,2,3-triazoles with Allenes

Allenes display a wide range of reactions including cycloadditions. Allenes also react with *N*-sulfonyl-1,2,3-triazoles in the presence of Rh(II) and Ni(0) catalysts to give pyrroles and related molecular frameworks.

3.5.1 Intermolecular Transannulation with Allenes

Murakami and co-workers reported for the first time that the *N*-sulfonyl-1,2,3-triazoles readily react with allenes in the presence of Ni(0) catalyst to deliver the isopyrroles which undergo transposition of C—C double bond to form corresponding polysubstituted pyrroles using *p*-toluene sulfonic acid in catalytic amount (Scheme 3.38a) [42]. The optimal reaction conditions include the reaction of the triazole with allene in the presence of $Ni(cod)_2$ and rac-2,4-bis(diphenylphosphino)pentane (bdpp), which *in situ* generates Ni(0) as an active catalyst in 1,4-dioxane as solvent at 80 °C. Both 4-aryl and 4-alkyl substituted triazoles having aryl sulfonyl group with electron-donating and withdrawing substituents react with monosubstituted allenes to give pyrroles with good to excellent yields. However, sterically hindered allenes react sluggishly and phenyl-substituted allenes result in moderate yield. The 1,3-disubstituted allenes are unreactive under reaction conditions except cyclonona-1,2-diene, which gives corresponding fused pyrrole upon subsequent treatment with TsOH

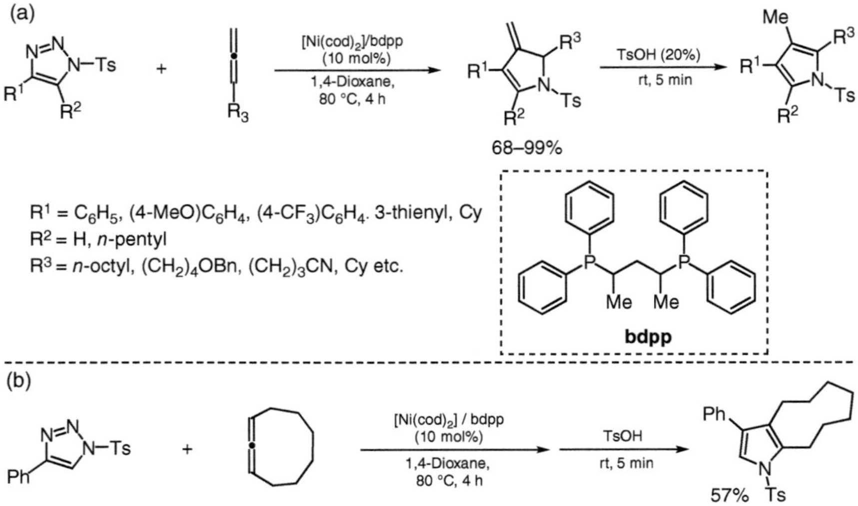

Scheme 3.38 Nickel-catalyzed transannulation of triazoles with (a) acyclic allenes, (b) cyclic allenes.

(Scheme 3.38b). Similarly, 5-substituted and 4,5-substituted *N*-sulfonyl triazoles are as effective as 4-susbtituted triazoles in giving respective poly-substituted pyrroles. The reactions of triazoles with 1,3-disubstituted allenes are unsuccessful, resulting in the deterioration of nickel catalyst.

The isopyrroles formed from nickel-catalyzed reaction when treated with suitable reagents that are electrophilic such as Eschenmoser's salt, diethylazodicarboxylate, and diethyl ketomalonate, gain aromaticity via Alder-ene reactions to give the pyrroles (Scheme 3.39).

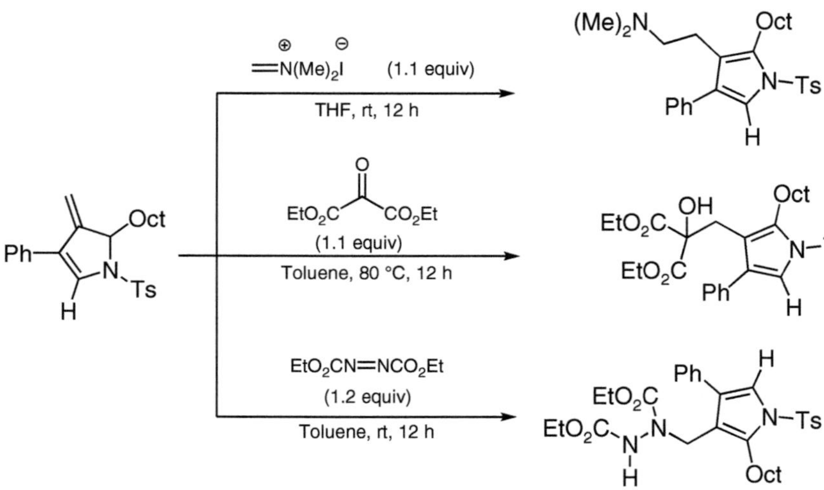

Scheme 3.39 Treatment of isopyrroles with electrophilic agents.

The reaction proceeds via the nucleophilic addition of allene to the electrophilic α-imino nickel carbene to generate the zwitterionic intermediate **3.40a** with an allylic cation fragment and anionic nickel. Then, nickel releases the pair of electrons, which makes the nitrogen more nucleophilic; this nucleophilic species adds on to more substituted carbon of allylic fragment to give isopyrrole (Scheme 3.40).

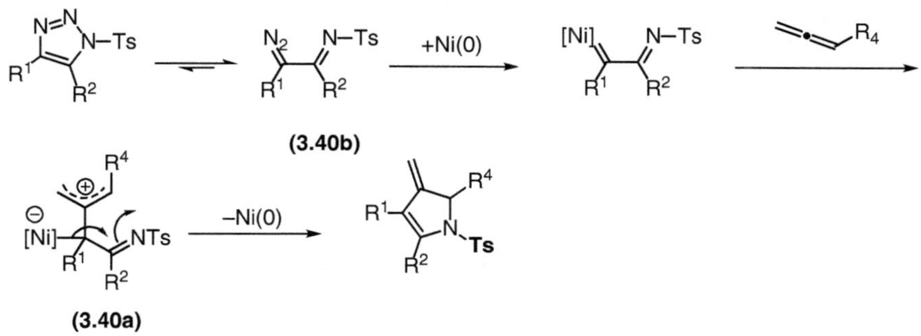

Scheme 3.40 Mechanism of Ni-catalyzed transannulation of triazoles with allenes.

In addition, in the presence of rhodium(II) complexes as catalysts tert-butyl substituted allenes react with *N*-sulfonyl-1,2,3-triazoles to give corresponding alkylidenecyclopropyl methanimine with high enantioselectivity (Scheme 3.41) [43]. This cyclopropyl derivative on heating to 70 °C, results in sequential cyclopropyl ring opening/shifting of a C—C double bond and nucleophilic attack of imine nitrogen selectively on exocyclic double bond in a concerted manner to give 3-methylene-2,3-dihydropyrrole as the major product with declined enantioselectivity.

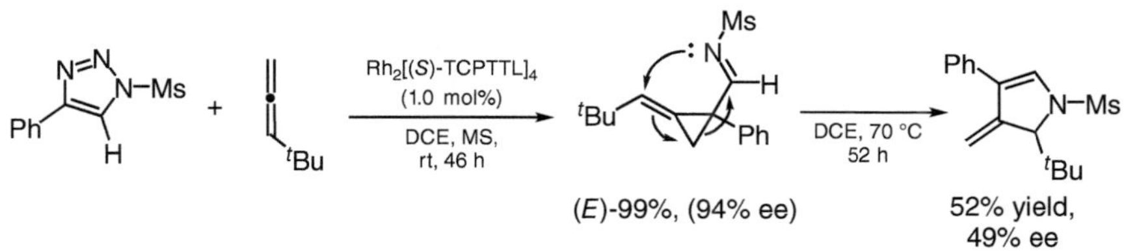

Scheme 3.41 Rh(II)-catalyzed transannulation of triazoles with allenes.

In 2017, Aclaide et al. reported rhodium(II)-catalyzed reactions of *N*-sulfonyl-1,2,3-triazoles with functionalized 3,3-disubstituted allenes bearing nucleophilic centers such as allenols display regiodivergent reactivity to give 2-pyrrolines (Scheme 3.42) [44].

The nucleophilic addition of the internal double bond of allenol **3.43c** onto the electrophilic rhodium carbenoid **3.43b** generates the zwitterion **3.43d** (Scheme 3.43). This further undergoes intramolecular cyclization with allylic carbocation through imine nitrogen to give pyrroline **3.43e** with regeneration of rhodium(II) species.

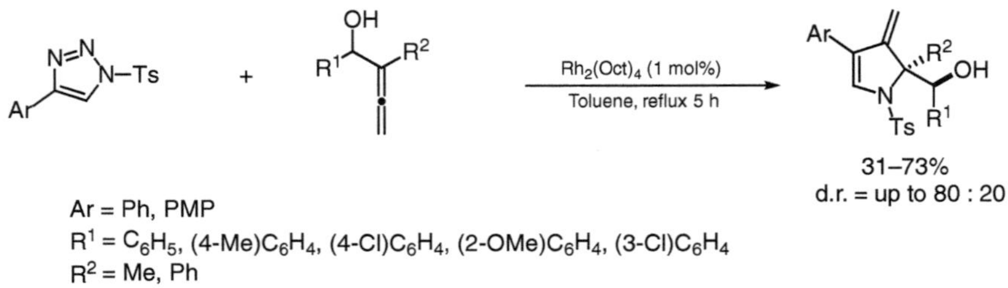

Scheme 3.42 Rh(II)-catalyzed reactions of triazoles with allenols.

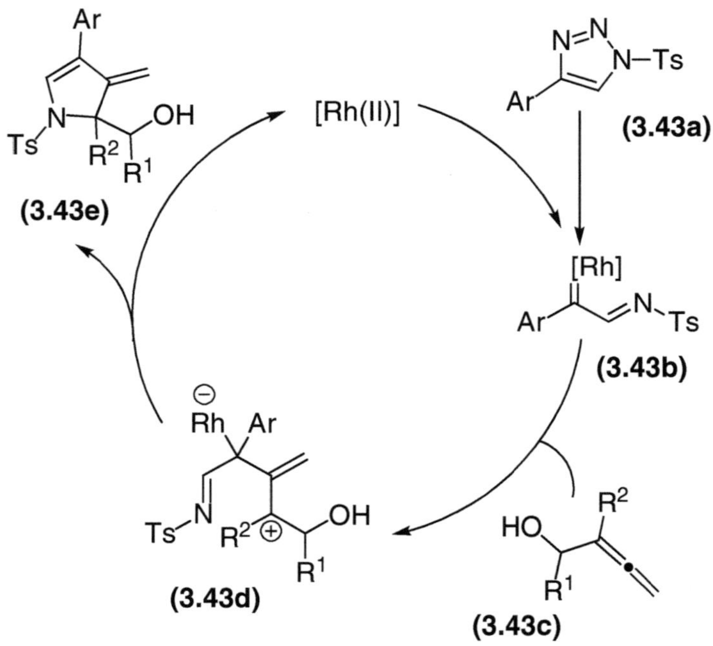

Scheme 3.43 Mechanism of Rh(II)-catalyzed reactions of allenols with triazoles.

3.5.2 Intramolecular Transannulation with Allenes

Sarpong reported that the interaction of allenes with rhodium(II) carbenoids generated by denitrogenation of *N*-sulfonyl-1,2,3-triazoles results in the Rh-bound trimethylenemethane **3.44a** variant which further results in intramolecular cyclization to afford functionalized 3,4-fused pyrroles (Scheme 3.44) [45].

N-sulfonyl-1,2,3-triazoles tethered with allenes bearing both electron-deficient and electron-rich aryl as well as alkyl substituents react in an intramolecular fashion with $Rh_2(Oct)_4$ catalyst efficiently under standard reaction conditions to deliver corresponding 3,4-fused pyrroles (Scheme 3.45).

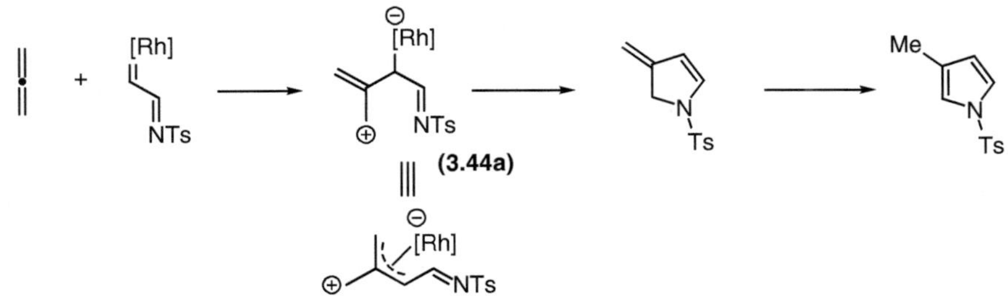

Scheme 3.44 Denitrogenation of *N*-sulfonyl-1,2,3-triazoles results in the Rh-bound trimethylenemethane.

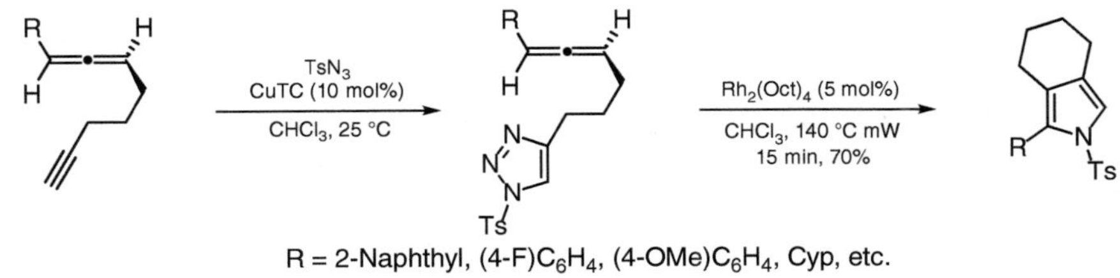

Scheme 3.45 Intramolecular transannulation of triazoles tethered with allenes.

3.6 Summary and Outlook

The *N*-sulfonyl-1,2,3-triazoles have been extensively explored as bench-stable surrogates of diazo compounds, leading to the generation of α-imino metal carbene intermediate through ring chain tautomerization and denitrogenation in the presence of metal salts. *In situ*, treatment of these intermediates displays versatile reactivity confined to the structural features of competent reactants. Predominantly, the α-imino metal carbene intermediates undergo transannulation, C—H and *X*—H insertion, carbonyl ylide formation, and subsequent rearrangements to provide access to desired molecular structures. Amidst the various possible reactions, transannulation of these intermediates is found to be highly attractive as these can readily react with various polar and nonpolar multiple bonds to produce carbocycles and heterocycles. The current chapter specifically emphasizes the accomplishments in the metal-catalyzed transannulation of *N*-sulfonyl-1,2,3-triazoles with substrates bearing nonpolar multiple bonds such as alkenes, dienes, alkynes, and allenes. The reactivity of α-imino metal carbene intermediates with nonpolar molecules has been exploited for the synthesis of various carbocycles and nitrogen heterocycles such as cyclopropanes, cycloheptadienes, pyrroles, dihydropyrroles, dihydropiperidines, azepines, and many other complex molecular motifs. Both intermolecular and intramolecular reactions of triazoles have been studied to furnish the synthesis of desired molecules with good yield and selectivity. However, the applications of *N*-sulfonyl triazoles are limited due to reasons such as the requirement of a rhodium catalyst for successful transformations, lesser reactivities of the intermediates with nonpolar multiple bonds compared to polar bonds, thus causing difficulties in using these molecules for late-stage modification. Therefore, further amendments in the triazole substituents and catalysts are required to achieve better utilities for these molecules.

References

1 Ma, S., Mandalapu, D., Wang, S., and Zhang, Q. (2022). Biosynthesis of cyclopropane in natural products. *Nat. Prod. Rep.* 39: 926–945.

2 Talele, T.T. (2016). The "cyclopropyl fragment" is a versatile player that frequently appears in reclinical/clinical drug molecules. *J. Med. Chem.* 59: 8712–8756.

3 Wu, W., Li, Z., and Jiang, H. (2018). Recent advances in the synthesis of cyclopropanes. *Org. Biomol. Chem.* 16: 7315–7329.

4 Fan, H.N., Peng, J., Hanmann, M.T., and Hu, J.F. (2008). Lamellarins and related pyrrole-derived alkaloids from marine organisms. *Chem. Rev.* 108: 264–287.

5 Bhardwaj, V., Gumber, D., Abbot, V. et al. (2015). Pyrrole: a resourceful small molecule in key medicinal hetero-aromatics. *RSC Adv.* 5: 15233–15266.

6 De, S., Ashok Kumar, S.K., Shah, S.K. et al. (2022). Pyridine: the scaffolds with significant clinical diversity. *RSC Adv.* 12: 15385–15406.

7 Pilli, R.A., Rosso, G.B., and de Oliveira Mda, C. (2010). The chemistry of stemona alkaloids: an update. *Nat. Prod. Rep.* 27: 1908–1937.

8 Chattopadhyay, B. and Gevorgyan, V. (2012). Transition-metal-catalyzed denitrogenative transannulation: converting triazoles into other heterocyclic systems. *Angew. Chem. Int. Ed.* 51: 862–872.

9 Gulevich, A.V. and Gevorgyan, V. (2013). Versatile reactivity of rhodium–iminocarbenes derived from *N*-sulfonyl triazoles. *Angew. Chem. Int. Ed.* 52: 1371–1373.

10 Davies, H.M.L. and Alford, J.S. (2014). Reactions of metallocarbenes derived from *N*-sulfonyl-1,2,3-triazoles. *Chem. Soc. Rev.* 43: 5151–5162.

11 Anbarasan, P., Yadagiri, D., and Rajasekar, S. (2014). Recent advances in transition-metal-catalyzed denitrogenative transformations of 1,2,3-triazoles and related compounds. *Synthesis* 46: 3004–3023.

12 Jiang, Y., Sun, R., Tang, X.-Y., and Shi, M. (2016). Recent advances in the synthesis of heterocycles and related substances based on α-imino rhodium carbene complexes derived from *N*-sulfonyl-1,2,3-triazoles. *Chem. Eur. J.* 22: 17910–17924.

13 Akter, M., Rupa, K., and Anbarasan, P. (2022). 1,2,3-Triazole and its analogues: new surrogates for diazo compounds. *Chem. Rev.* 122: 13108–13205.

14 Chuprakov, S., Kwok, S.W., Zhang, L. et al. (2009). Rhodium-catalyzed enantioselective cyclopropanation of olefins with *N*-sulfonyl 1,2,3-triazoles. *J. Am. Chem. Soc.* 131: 18034–18035.

15 Grimster, N., Zhang, L., and Fokin, V.V. (2010). Synthesis and reactivity of rhodium(II) *N*-triflyl azavinyl carbenes. *J. Am. Chem. Soc.* 132: 2510–2511.

16 Culhane, J.C. and Fokin, V.V. (2011). Synthesis and reactivity of sulfamoyl azides and 1-sulfamoyl-1,2,3-triazoles. *Org. Lett.* 13: 4578–4580.

17 Kubiak, R.W., Tracy, W.F., Alford, J.S., and Davies, H.M.L. (2022). Asymmetric cyclopropanation with 4-aryloxy-1-sulfonyl-1,2,3-triazoles: expanding the range of rhodium-stabilized donor/acceptor carbenes to systems with an oxygen donor group. *J. Org. Chem.* 87: 13517–13528.

18 Miura, T., Nakamuro, T., Nikishima, H., and Murakami, M. (2016). Asymmetric synthesis of cyclopropylmethanamines by rhodium-catalyzed cyclopropanation of pinacol allylboronate with *N*-sulfonyl-1,2,3-triazoles. *Chem. Lett.* 45: 1003–1005.

19 Miura, T., Nakamuro, T., Stewart, S.G. et al. (2017). Synthesis of enantiopure C3-symmetric triangular molecules. *Angew. Chem. Int. Ed.* 56: 3334–3338.

20 Alford, J.S. and Davies, H.M.L. (2012). Expanding the scope of donor/acceptor carbenes to *N*-phthalimido donor groups: diastereoselective synthesis of 1-cyclopropane α-amino acids. *Org. Lett.* 14: 6020–6023.

21 Kwok, S.W., Zhang, L., Grimster, N., and Fokin, V.V. (2014). Catalytic asymmetric transannulation of *NH*-1,2,3-triazoles with olefins. *Angew. Chem. Int. Ed.* 53: 3452–3456.

22 Jiang, B. and Shi, M. (2017). Rhodium(II)-catalyzed intermolecular [3 + 2] annulation of *N*-vinyl indoles with *N*-tosyl-1,2,3-triazoles via an *aza*-vinyl Rh carbene. *Org. Chem. Front.* 4: 2459–2464.

23 Kim, C.-E., Park, Y., Park, S., and Lee, P.H. (2015). Diastereoselective synthesis of tetrahydrofurano- and tetrahydropyrano-dihydropyrroles containing *N*,*O*-acetal moieties via rhodium-catalyzed transannulation of *N*-sulfonyl-1,2,3-triazoles with oxacycloalkenes. *Adv. Synth. Catal.* 357: 210–220.

24 Rajasekar, S. and Anbarasan, P. (2014). Rhodium-catalyzed transannulation of 1,2,3-triazoles to polysubstituted pyrroles. *J. Org. Chem.* 79: 8428–8434.

25 Kim, C.-E., Park, S., Eom, D. et al. (2014). Synthesis of pyrroles from terminal alkynes, *N*-sulfonyl azides, and alkenyl alkyl ethers through 1-sulfonyl-1,2,3-triazoles. *Org. Lett.* 16: 1900–1903.

26 Feng, J., Wang, Y., Li, Q. et al. (2014). Facile synthesis of pyrroles via Rh(II)-catalyzed transannulation of 1-tosyl-1,2,3-triazoles with silyl or alkyl enol ethers. *Tetrahedron Lett.* 55: 6455–6458.

27 Miura, T., Nakamura, T., Liang, C.-J., and Murakami, M. (2014). Synthesis of *trans*-cycloalkenes via enantioselective cyclopropanation and skeletal rearrangement. *J. Am. Chem. Soc.* 136: 15905–15908.

28 Chen, K., Zhu, Z.-Z., Zhang, Y.-S. et al. (2014). Rhodium(II)-catalyzed intramolecular cycloisomerizations of methylenecyclopropanes with n-sulfonyl 1,2,3-triazoles. *Angew. Chem. Int. Ed.* 53: 6645–6649.

29 Alford, J.S., Spangler, J.E., and Davies, H.M.L. (2013). Conversion of cyclic ketones to 2,3-fused pyrroles and substituted indoles. *J. Am. Chem. Soc.* 135: 11712–11715.

30 Parr, B.T. and Davies, H.M.L. (2013). Rhodium-catalyzed tandem cyclopropanation/cope rearrangement of 4-alkenyl-1-sulfonyl-1,2,3-triazoles with dienes. *Angew. Chem. Int. Ed.* 52: 10044–10047.

31 Shang, H., Wang, Y., Tian, Y. et al. (2014). The divergent synthesis of nitrogen heterocycles by rhodium(II)-catalyzed cycloadditions of 1-sulfonyl 1,2,3-triazoles with 1,3-dienes. *Angew. Chem. Int. Ed.* 53: 5662–5666.

32 Kim, S., Mo, J., Kim, J. et al. (2014). Aza-[4+3] and *aza*-[3+2] annulations for synthesis of dihydroazepines and dihydropyrroles from alkynes, sulfonyl azides, and 1,3-dienes. *Asian J. Org. Chem.* 3: 926–931.

33 Yang, H., Hou, S., Tao, C. et al. (2017). Rhodium-catalyzed denitrogenative [3+2] cycloaddition: access to functionalized hydroindolones and the framework of montanine-type amaryllidaceae alkaloids. *Chem. Eur. J.* 23: 12930–12936.

34 Schultz, E.E., Lindsay, V.N.G., and Sarpong, R. (2014). Expedient synthesis of fused azepine derivatives using a sequential rhodium(II)-catalyzed cyclopropanation/1-*aza*-cope rearrangement of dienyltriazoles. *Angew. Chem. Int. Ed.* 53: 9904–9908.

35 Xu, H.-D., Xu, K., Zhou, H. et al. (2015). Ring-strain-driven catalytic carbene formation–cyclopropanation–*aza*-cope rearrangement cascade: a facile entry to fused dihydroazepines from 1,3-dienyltriazoles. *Synthesis* 47: 641–646.

36 Tian, Y., Wang, Y., Shang, H. et al. (2015). Rhodium(ii)-catalyzed intramolecular formal [4+3] cycloadditions of dienyltriazoles: rapid access to fused 2,5-dihydroazepines. *Org. Biomol. Chem.* 13: 612–619.

37 Miura, T., Yamauchi, M., and Murakami, M. (2009). Nickel-catalysed denitrogenative alkyne insertion reactions of *N*-sulfonyl-1,2,3-triazoles. *Chem. Commun.* 12: 1470–1471.

38 Chattopadhyay, B. and Gevorgyan, V. (2011). Rh-catalyzed transannulation of n-tosyl-1,2,3-triazoles with terminal alkynes. *Org. Lett.* 13: 3746.

39 Yang, Y., Zhou, M.-B., Ouyang, X.-H. et al. (2015). Rhodium(III)-catalyzed [3+2]/[5+2] annulation of 4-aryl 1,2,3-triazoles with internal alkynes through dual C(sp^2)—H functionalization. *Angew. Chem. Int. Ed.* 54: 6595–6599.

40 Shi, Y. and Gevorgyan, V. (2013). Intramolecular transannulation of alkynyl triazoles via alkyne–carbene metathesis step: access to fused pyrroles. *Org. Lett.* 15: 5394–5396.

41 Yang, Y., Yu, J.-X., Ouyang, X.-H., and Li, J.-H. (2017). Complex annulations through silver carbenoid intermediate: an alternative entry to transformations of 1,2,3-triazoles. *Org. Lett.* 19: 3982–3985.

42 Miura, T., Hiraga, K., Biyajima, T. et al. (2013). Regiocontrolled synthesis of polysubstituted pyrroles starting from terminal alkynes, sulfonyl azides, and allenes. *Org. Lett.* 15: 3298.

43 Miura, T., Nakamura, T., Biyajima, T., and Murakami, M. (2015). Synthesis of α,β,γ,δ-unsaturated imines from *N*-sulfonyl-1,2,3-triazoles and allenes through rhodium-catalyzed cyclopropanation and thermal rearrangement. *Chem. Lett.* 44: 700–702.

44 Aclaide, B., Almendros, P., Cembellin, S. et al. (2017). Allenols versus allenones: rhodium-catalyzed regiodivergent and tunable allene reactivity with triazoles. *Chem. Eur. J.* 23: 13754–13759.

45 Schultz, E.R. and Sarpong, R. (2013). Application of in situ-generated Rh-bound trimethylenemethane variants to the synthesis of 3,4-fused pyrroles. *J. Am. Chem. Soc.* 135: 4696–4699.

4

Transannulations of *N*-Sulfonyl-1,2,3-triazoles with Carbo/Heterocycles

Mikhail S. Novikov and Nikolai V. Rostovskii

Institute of Chemistry, Saint Petersburg State University, St. Petersburg, Russia

4.1 Introduction

The transannulation reactions of *N*-sulfonyl-1,2,3-triazoles underlie an important strategy for the synthesis of a structurally diverse nitrogen heterocycles, including monocyclic molecules, *ortho*- and *ortho-peri*-fused systems, as well as bridged structures. Most of these denitrogenative reactions of 1-sulfonyltriazoles are catalyzed by rhodium(II) carboxylates at elevated temperature and proceed through the formation of unstable rhodium 2-azavinylcarbenes (Rh-AVCs) (Scheme 4.1, reaction 1). The electron-withdrawing R-sulfonyl group at the N1 atom of the triazole, which promotes the ring-opening to the valence α-iminodiazo isomer, is a prerequisite for the effective formation of the key intermediate of these reactions, the Rh-AVC complex. With the advent of the pioneering works of Gevorgyan [1] and Fokin [2], the chemistry of these nitrogen-containing rhodium(II) carbene complexes has become an active research area. A series of excellent review articles devoted to various aspects of their synthetic application was published in the past decade. In these reviews, the unique capabilities of Rh-AVCs to transform various carbo- and heterocyclic structures were reflected to one extent or another [3–8]. This chapter presents a general analysis of the transannulations of 1-sulfonyltriazoles with cyclic substrates including new advances in recent years in this field.

In general, the transannulation reactions of 1-sulfonyltriazoles with cyclic substrates can occur either with the retention of the ring of the substrate or its opening. Reactions of the first type most often represent a formal cycloaddition of Rh-AVCs, generated from the 1-sulfonyltriazoles, to an unsaturated cyclic molecule. In particular, Rh-AVCs can undergo cycloaddition in a (3 + 2) fashion to the endocyclic double bond of the substrate (Scheme 4.1, reaction 2). The cycloaddition can also involve the exocyclic atoms of substituents (Scheme 4.1, reaction 3). In addition, annulation can be carried out with the inclusion of only one of the Rh-AVC carbon atoms in the forming ring. This is achieved by using 1-sulfonyltriazoles with unsaturated

Denitrogenative Transformation of Nitrogen Heterocycles: Synthesis, Reactions and Applications, First Edition. Edited by Pazhamalai Anbarasan.

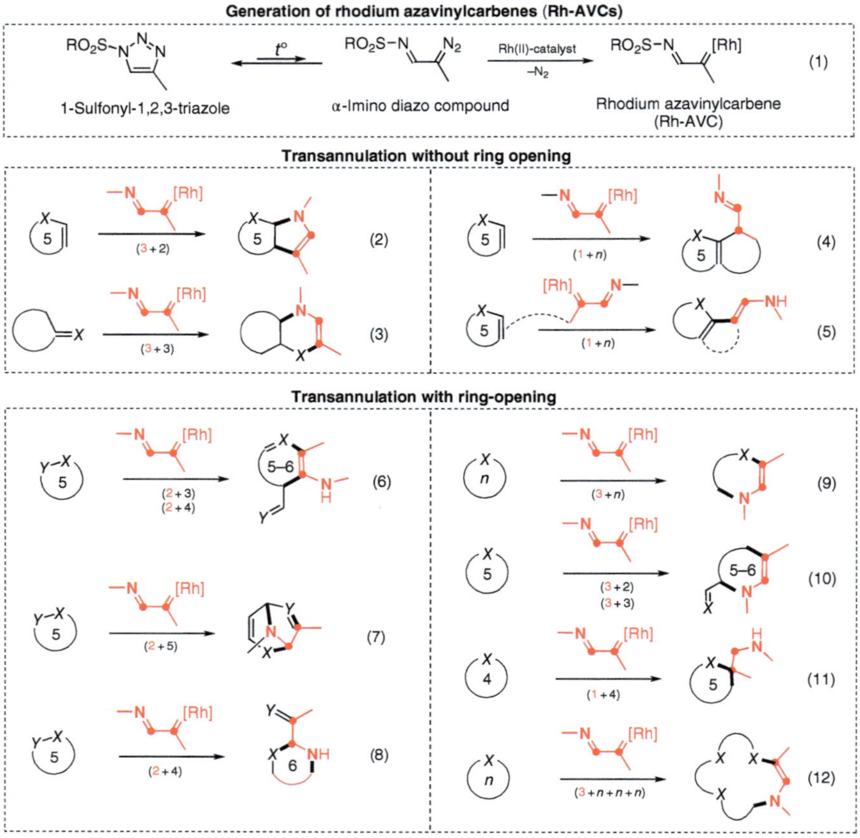

Scheme 4.1 Generation and transannulation reactions of Rh-AVCs.

substituents (Scheme 4.1, reaction 4) or by implementing the intramolecular version of the annulation (Scheme 4.1, reaction 5).

Cyclic molecules that are capable of undergoing the ring-opening in transannulation reactions of 1-sulfonyltriazoles are limited to heterocyclic compounds. These include N-, O-, S-heterocycles, as well as various polyheteroatomic cyclic molecules. Transannulation reactions of 1-sulfonyltriazoles with three- to six-membered heterocycles are known, among which five-membered heterocycles of both nonaromatic and aromatic series are most often used. Rh-AVC, forming a new heterocyclic structure with these heterocyclic substrates, most often acts as a two-atom C2 or CN synthon (Scheme 4.1, reactions 6–8), or a three-atom CCN synthon (Scheme 4.1, reactions 9, 10). And only in rare cases, it transfers only one carbon atom to the forming heterocycle (Scheme 4.1, reaction 11). The heterocyclic partner, in turn, can transfer part of the atoms of the cycle or all atoms. Moreover, there are examples in which several molecules of a heterocyclic substrate take part in the cycle formation (Scheme 4.1, reaction 12).

4.2 Transannulations with Ring Retention

4.2.1 Aromatic and Heteroaromatic Rings

All known transannulation reactions of 1-sulfonyltriazoles with arenes proceed with preservation of the six-membered carbocycle. Back in 2014, it was found that rhodium-catalyzed intramolecular dearomatizing (3 + 2) cycloaddition of 1-sulfonyltriazoles to arenes connected by a three-atom tether affords tricyclic molecules **4.2a** containing *cis*-fused 3a,7a-dihydroindole fragment (Scheme 4.2) [9]. The tether can be all-carbon or can include a heteroatom. Substrates possessing two- or four-carbon tether failed to undergo such a reaction. The annulation with *meta*-monosubstituted arenes occurred regioselectively at the less-hindered side. The reaction also proceeds with 3,5-dimethylphenyl substrate that allows to install a methyl group at the bridgehead position. The reaction can also be performed as a one-pot synthesis starting from 5-arylpent-1-ynes and sulfonylazides.

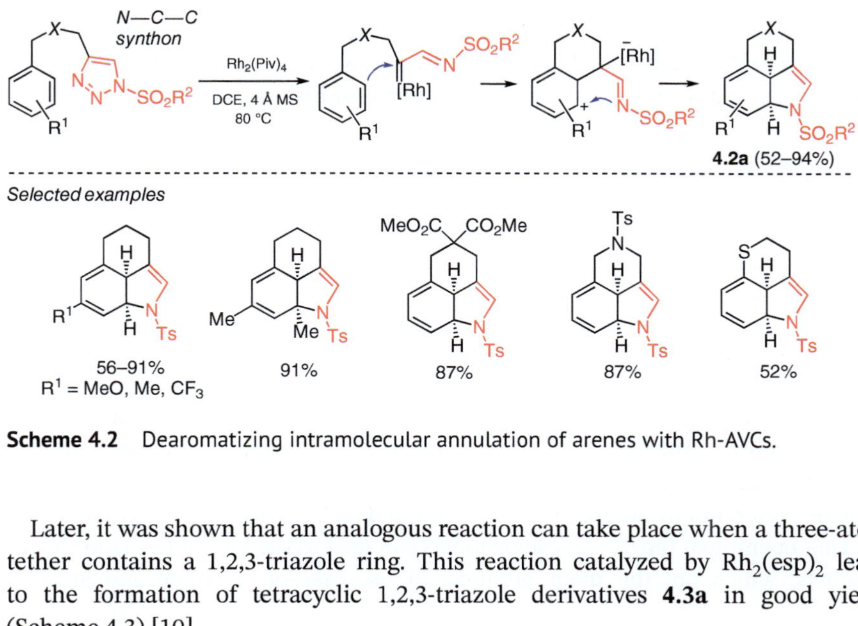

Scheme 4.2 Dearomatizing intramolecular annulation of arenes with Rh-AVCs.

Later, it was shown that an analogous reaction can take place when a three-atom tether contains a 1,2,3-triazole ring. This reaction catalyzed by $Rh_2(esp)_2$ leads to the formation of tetracyclic 1,2,3-triazole derivatives **4.3a** in good yields (Scheme 4.3) [10].

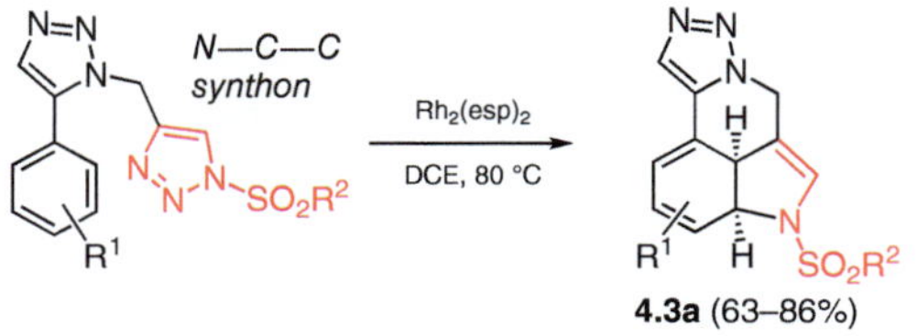

Scheme 4.3 Intramolecular annulation of arene-triazole-triazole hybrids.

The tether may contain a double bond of the benzene ring. In particular, it has been shown that diphenyl ether derivatives *ortho*-substituted by a 1,2,3-triazole fragment can be effectively transformed under Rh(II) catalysis into pyrrolo-fused xanthene derivatives **4.4a** in good yields (Scheme 4.4) [11].

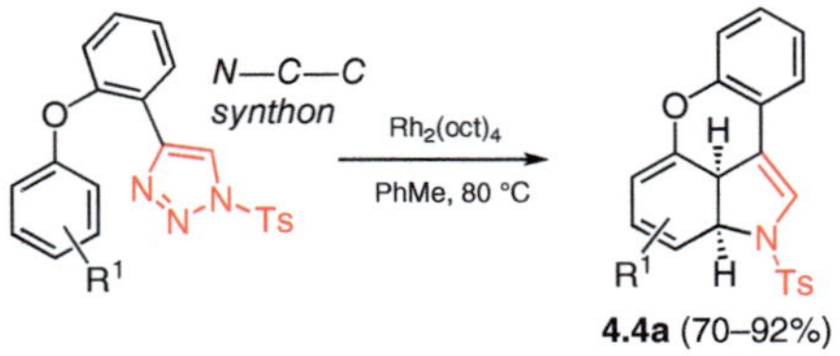

Scheme 4.4 Intramolecular annulation of diaryl ether-triazole hybrids.

Intermolecular (3 + 2) cycloaddition reaction of 1-sulfonyltriazoles with arenes is much more baffling. It was achieved under $Rh_2(esp)_2$ catalysis and microwave irradiation for alkylarenes when they were used in large excess, and even under these conditions, the yields of *cis*-fused 3a,7a-dihydroindoles **4.5a** were moderate (Scheme 4.5) [12]. Mono-alkylarenes underwent fusion mainly at the C3—C4 bond, *ortho*-xylene – at the C2—C3 bond, and tetralin – at the C4—C5 bond.

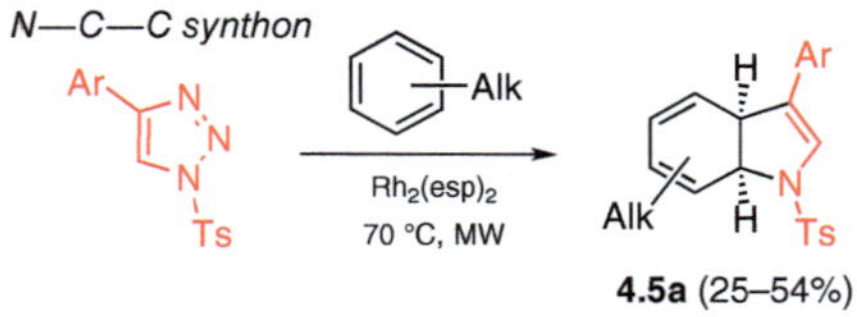

Scheme 4.5 Intermolecular annulation of arenes with 1-sulfonyl-1,2,3-triazoles.

Analogously, 4-phthalimido-1-sulfonyl-1,2,3-triazoles react upon heating with *p*-dialkylarenes resulting in (3 + 2)-cycloadducts, 3a,7a-dihydroindoles **4.6a**, as major products (Scheme 4.6) [13]. Cycloaddition occurs at the C1—C2 bond; carbene preferentially attacks methyl-substituted carbon. It was demonstrated that the ring-opening of the (3 + 2)-cycloadducts can lead to the formal C—H-insertion products, which are the main products in the case of *m*-xylene and anisole.

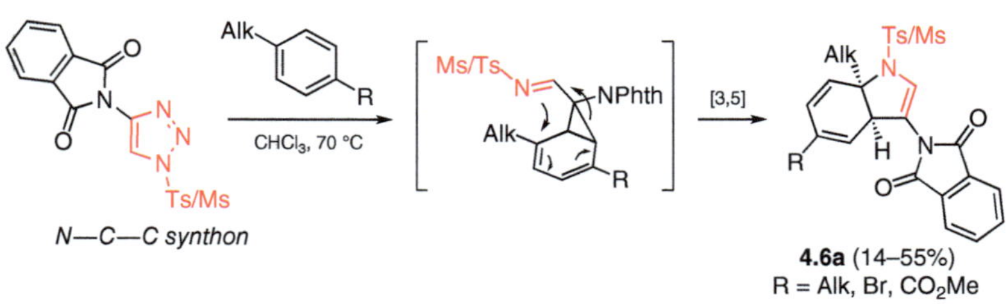

Scheme 4.6 Intermolecular annulation of arenes with 4-phthalimido-1-sulfonyl-1,2,3-triazoles.

Computational studies suggest that the cycloaddition takes places through a tandem arene cyclopropanation/3,5-sigmatropic rearrangement reaction.

Depending on the nature of the substitution of positions C2 and C3, indoles can give in the Rh(II)-catalyzed reactions with 1-sulfonyltriazoles either C—H-insertion products of the carbene or transannulation products. The reactions of C2-substituted indoles usually proceed via the first pathway [14, 15], while C3-substituted indoles preferably afford cycloadducts. Enantioselective transannulation of 4-aryl-1-sulfonyl-1,2,3-triazoles with C3-substituted indoles has been reported by Davies's group [16].

The rhodium-catalyzed intramolecular dearomatizing transannulation of 1-sulfonyltriazoles **4.7a** connected to the indole C3 position by a four-atom tether has been reported by Tang and Shi (Scheme 4.7) [17]. The products of this synthesis are polynitrogen tetracyclic derivatives **4.7b**. Despite using chiral catalyst, the author did not study enantioselectivity of the reaction. Reaction of indolyltriazoles **4.7c** with unsubstituted indole nitrogen atom gave another products – azepino[4,5-*b*]indoles **4.7d**, the additive of $Cu(OAc)_2$ fully shifts the reaction in this direction. It is proposed that the initial step in the formation of tetracyclic derivatives **4.7b** is cyclopropanation of the indole C2—C3 bond followed by ring opening and cyclization. Different reaction route in the presence of Cu(II) salt can be explained by coordination of the metal between indole NH and nitrogen of tosylimine. Moreover, $Cu(OAc)_2$ serves as an oxidant producing additional C=C bond in the product. A little later, similar results on the synthesis of tetracyclic spiroindolines **4.7b** using $Rh_2(Oct)_4$ as a catalyst were reported by Kang and co-authors [18]. In contrast to the previous work, a substrate with a free indole NH group also gave the spiroindoline derivative. The authors also tested enantioselective version of the reaction using $Rh_2(S\text{-BNP})_4$ as a catalyst.

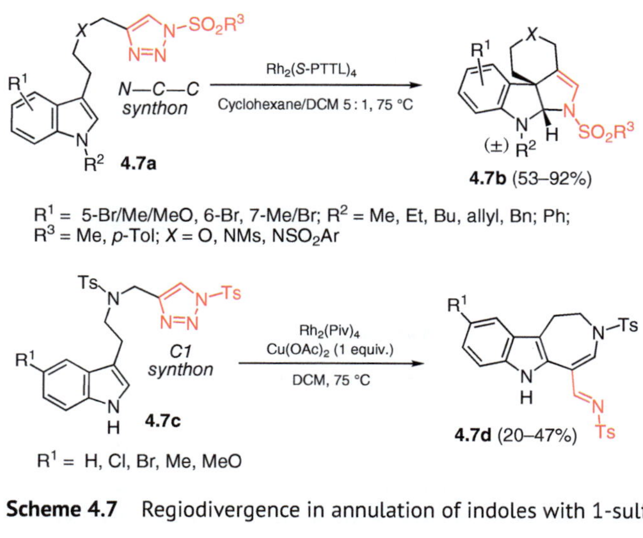

Scheme 4.7 Regiodivergence in annulation of indoles with 1-sulfonyl-1,2,3-triazoles.

Davies and Fu further investigated the scope of the reactions of indolyl- and pyrrolyl-tethered 1-sulfonyltriazoles [19]. They expanded the range of tetracyclic spiroindolines **4.7b** by synthesizing the cyclohexano- and azepano-fused analogs using Rh_2(*S*-PTTL)$_4$ as a catalyst (Scheme 4.8). The reaction with an ester-tethered substrate did not follow the transannulation pathway but provided the product with 1-azabuta-1,3-diene substituent formed from the 1,2,3-triazole moiety.

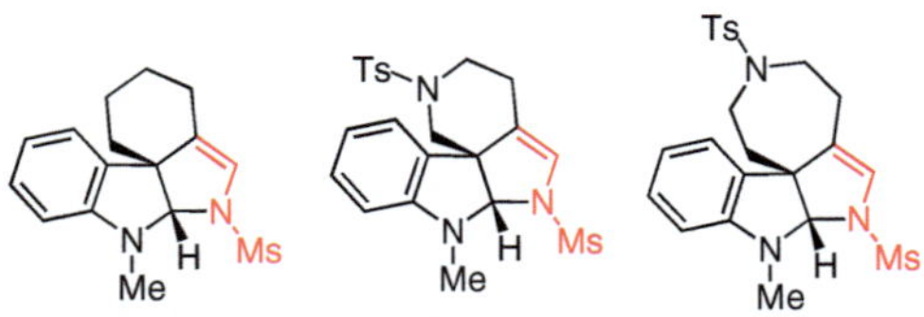

Scheme 4.8 Tetracyclic spiroindolines prepared by annulation of indoles with Rh-AVCs.

A substrate with the three-atom N-containing tether cannot adopt a proper conformation to achieve (3 + 2) cycloaddition. In this case, the formal intramolecular carbene insertion into the C—H bond becomes more favorable to give γ-carbolines **4.9a** (Scheme 4.9). This type of reactivity is also predominant for NH-indoles with the triazole moiety attached by a linker to the C3 indole atom. When the tether is attached to the indole/pyrrole C2 atom, the authors observed similar reactivity that leads to β-carbolines **4.9b** [19].

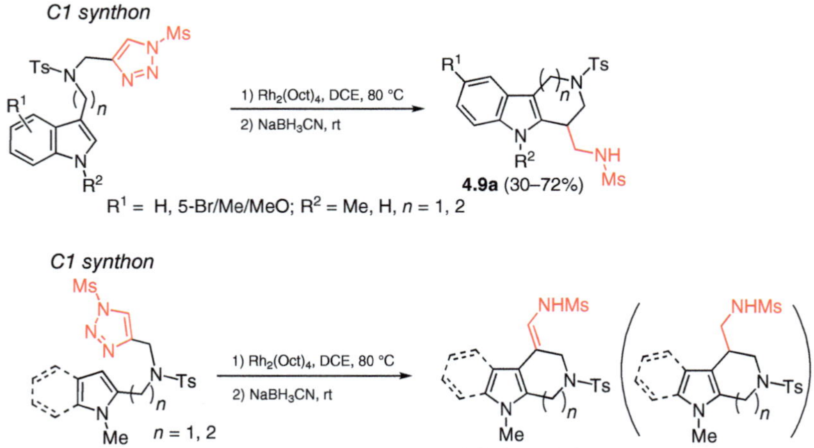

Scheme 4.9 Intramolecular Rh-AVCs insertion into the C—H bond of indoles.

The reactivity of electron-rich and electron-poor furans toward Rh-AVCs derived from 1-sulfonyltriazoles is rather different. The change in the reaction course Davies, Williams, and co-authors associated with a change in the reactivity of bicyclic intermediate **4.10c** (Scheme 4.10) [20]. In particular, the reaction of

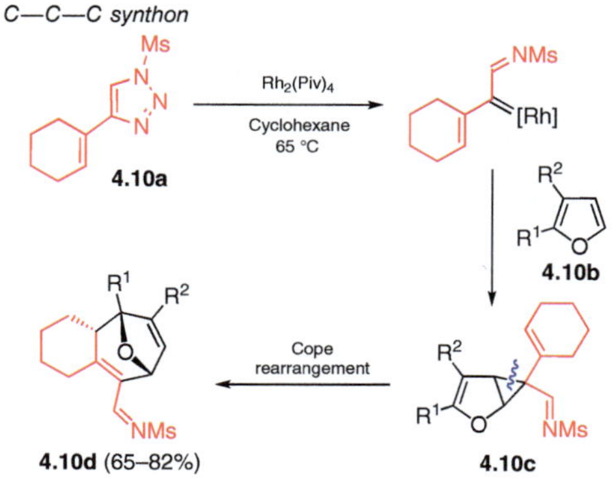

R^1 = H, R^2 = CO_2Et; R^1 = H, R^2 = Ac; R^1 = CO_2Me, R^2 = H

Scheme 4.10 Rh-AVCs in the "cyclopropanation/Cope rearrangement" reaction sequence.

1-mesyl-4-cyclohexenyl-1,2,3-triazole **4.10a** with acceptor-substituted furans **4.10b** affords in high yields adducts **4.10d** resulted from the Cope rearrangement of cyclopropafuran **4.10c**.

The same work presents an example of the transannulation of triazole **4.11a** by electron-deficient furans **4.11b**, which also occurs without the opening of the furan ring [20] (Scheme 4.11). The reaction proceeds in an unusual mode via the formation of the betaine intermediate **4.11c** followed by the cyclization at the cyclohexenyl moiety to give tetracyclic product **4.11d**. As a result, only one atom from the triazole moiety is incorporated into forming cyclopentene ring.

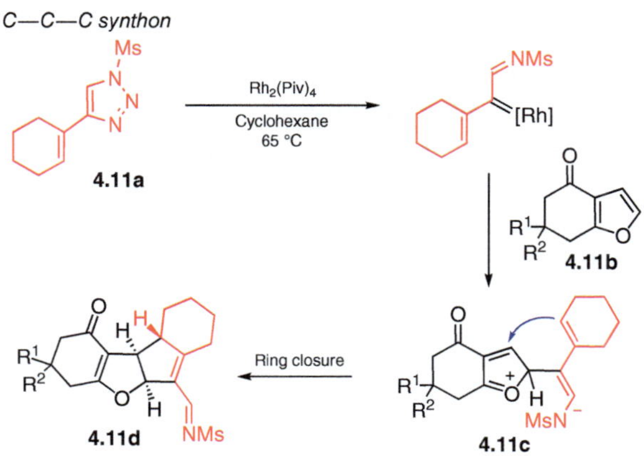

R^1 = R^2 = H (53%); R^1 = R^2 = Me (62%)

Scheme 4.11 Annulation of furans with cyclohexenyl-substituted Rh-AVCs.

4.2.2 Nonaromatic Rings

The transannulation of 1-sulfonyltriazoles with 8-aryl-8-azaheptafulvenes **4.12a** under $Rh_2(Oct)_4$ catalysis produces cyclohepta[*b*]pyrazines **4.12b** mostly in high yields (Scheme 4.12) [21]. It is supposed that the initially formed Rh-AVC attacks

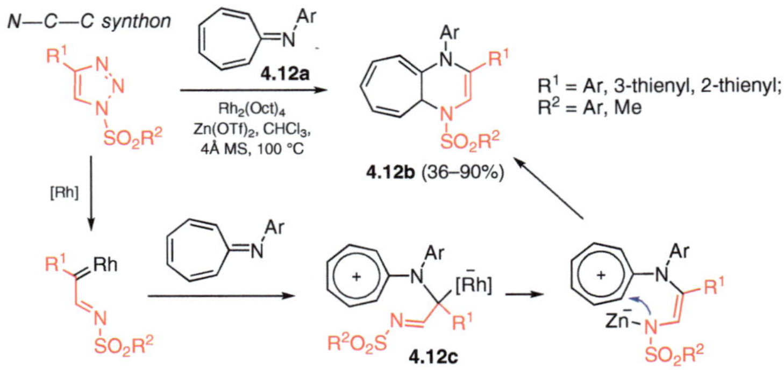

Scheme 4.12 Transannulation of 8-azaheptafulvenes with Rh-AVCs.

the nitrogen atom of the azaheptafulvene to provide a rhodium-bound zwitterionic intermediate **4.12c** possessing tropylium cation moiety. Subsequent ring closure under the assistance of a Lewis acid occurs to yield the cycloheptapyrazine **4.12b**. It should be mentioned that Rh-AVCs generated from 1-sulfonyltriazoles were also successively used for the synthesis of tricyclic imidazol-2-one derivatives from benzo[*d*][1,3]oxazin-2(4*H*)-ones [22].

4.3 Transannulations with Ring Opening

4.3.1 Heteroaromatic Rings

The reactions of furans with Rh-AVCs derived from 1-sulfonyltriazoles can proceed both with retention of the furan ring (see Schemes 4.10 and 4.11) and through its opening. The outcome of the reaction is highly dependent on the substitution pattern of the furan system. An electron-donating substituent in position 2 and/or 5 favors the furan ring-opening-cyclization reactions with the formation of a new heterocyclic system. It is assumed that in this case, the reaction can proceed through one of three intermediates, **4.13a–4.13c**, in which the cleavage of the C—O bond of the furan ring occurs (Scheme 4.13) [20].

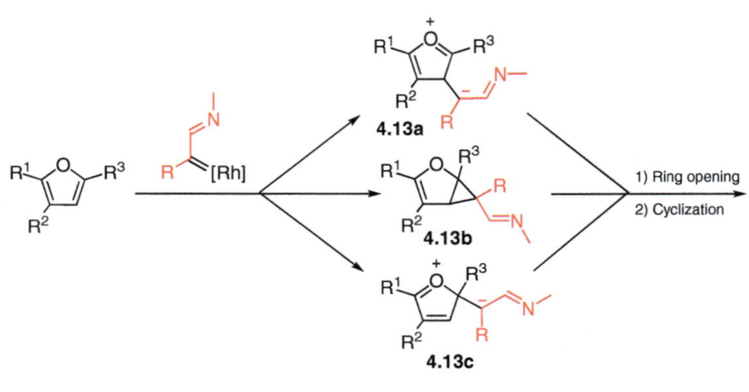

Scheme 4.13 Possible routes for the furan ring opening under action of Rh-AVCs.

The transformation of donor-substituted furans **4.14a** into highly functionalized pyrroles **4.14b** and **4.14c** using Rh-AVCs was reported by Davies and co-authors in 2013 (Scheme 4.14) [23]. This was achieved by the reaction of 1-sulfonyltriazoles with 2-mono and 2,5-disubstituted furans under Rh(II) catalysis. This reaction may be referred to (3 + 2) annulation of 1,2,3-triazoles with furans to give pyrroles where the triazole acts as a N—C—C synthon. Of the ten rhodium carboxylates examined in the reaction of 4-phenyl-1-tosyl-1,2,3-triazole with 2,5-dimethylfuran, the highest product yield (77%) was observed for $Rh_2(S\text{-DOSP})_4$. $Rh_2(OAc)_4$-catalyzed reaction gave the product in only 41% yield, while the catalysts with fluorinated carboxylate ligands proved to be inactive at all. The symmetrical furans in this reaction yield a single product in excellent yield. The reactions of asymmetrically substituted 2,5-disubstituted furans in many cases provide mixtures of regioisomers in ratios from 2 : 1 to >10 : 1. It is noteworthy that the smaller substituent from furan ring is transferred to the pyrrole ring of the major regioisomer.

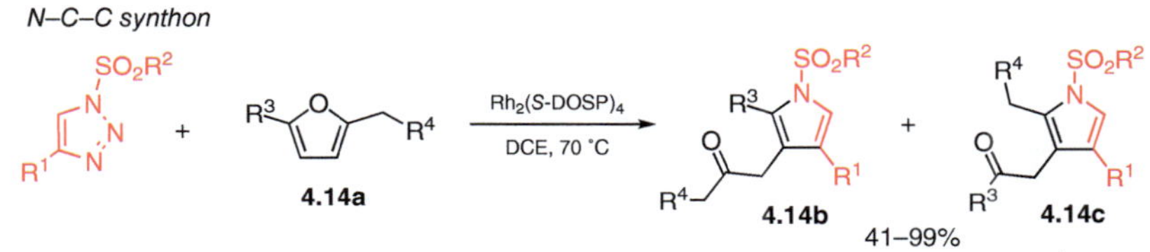

R^1 = Ar, cyclohexenyl; R^2 = Me, Et, *p*-Tol; R^3 = H, Me, Et; R^4 = H, Alk, alkenyl, $^i Pr_3SiO$

Scheme 4.14 Synthesis of pyrroles by transannulation of furans with 1-sulfonyl-1,2,3-triazoles.

The authors suggested that the reaction involves an attack of the Rh-AVC at the β-position of the furan ring to give intermediate **4.15a**, which further cyclizes to furopyrrole **4.15b** (Scheme 4.15). The opening of the dihydrofuran ring, which presumably occurs under mildly acidic conditions, leads to the formation of the final pyrrole **4.15c**. The requirement for Rh-AVC to attack the β-position of the furan ring explains why 2,5-disubstituted substrates are most suitable for this synthesis. The unsuitability of furan itself for this synthesis and the modest product yield in

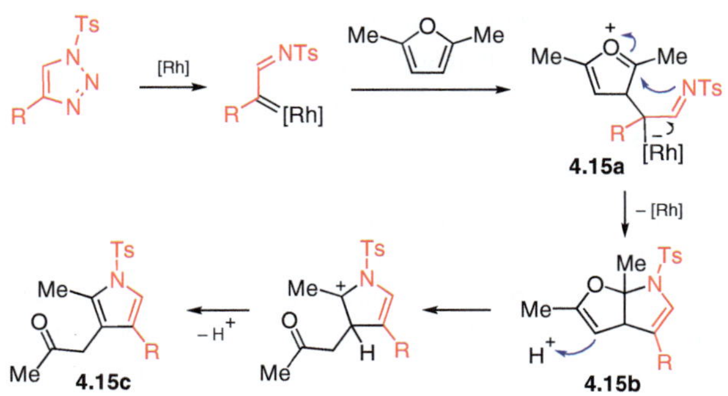

Scheme 4.15 Mechanism for the transannulation of furans to pyrroles.

the reaction of 2-methylfuran were rationalized in terms of competitive attack of the Rh-AVC at the α-position of the furan ring leading to other products.

Miura and co-authors described the synthesis of 1,2-dihydropyridines via the transannulation of 1-sulfonyltriazoles with 2-(siloxy)furans **4.16a** (Scheme 4.16) [24]. It is assumed that the reaction proceeds through stereoselective cycloaddition of the Rh-AVC to less sterically hindered C=C bond of the furan to give cyclopropafuran intermediate **4.16b**, which undergoes the ring-opening to (1*E*,3*E*,5*Z*)-1-azahexa-1,3,5-triene **4.16c**. If it contains a C4-alkyl substituent, it easily cyclizes to 1,2-dihydropyridine **4.16d** under the reaction conditions. Azahexatriene **4.16c** unsubstituted at the C4 position, on the contrary, is a stable compound that can be isolated in good yield, especially if $Rh_2(S\text{-NTTL})_4$ is used as a catalyst. These stable azahexatrienes can cyclize to 1,2-dihydropyridine **4.16e** in one-pot mode by heating with pyridine in chloroform at 70 °C.

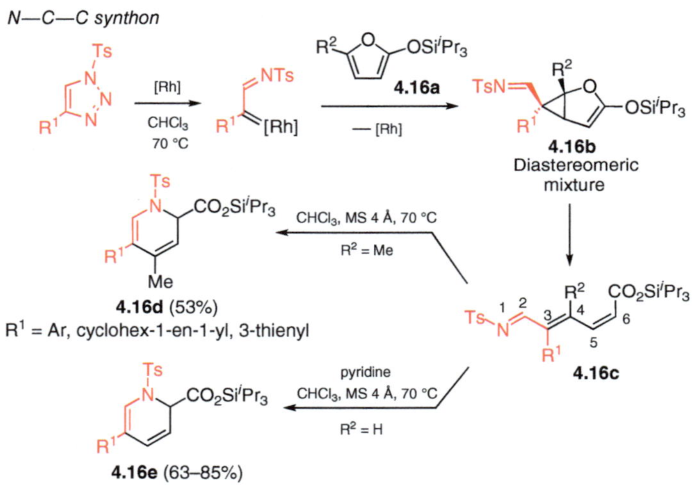

Scheme 4.16 Synthesis of 1,2-dihydropyridines from furans and 1-sulfonyl-1,2,3-triazoles.

An example of intramolecular version of rhodium-catalyzed reaction of triazoles **4.17a** with a tethered furan moiety has been reported by Hashmi and co-authors (Scheme 4.17) [25]. The mechanism of the reaction includes the formation of 1-azahexa-1,3,5-triene intermediate **4.17b**. The reaction provides 2-formyl- and 2-acetylpyridines **4.17c–4.17e** fused with the substituted pyrrolidine, piperidine, and tetrahydrofuran rings. An important advantage of the method is the possibility of synthesizing these compounds in one-pot starting from easily accessible (furan-2-ylmethyl)propargyl amines without isolation of the triazoles.

The first rhodium(II)-catalyzed reaction of 1-sulfonyl-1,2,3-triazoles with isoxazoles **4.18a** was reported by He, Tang, and co-authors in 2015 (Scheme 4.18) [26]. It was found that the triazoles smoothly react with isoxazoles bearing alkyl substituents at the C3 and C5 positions in the presence of $Rh_2(esp)_2$ in

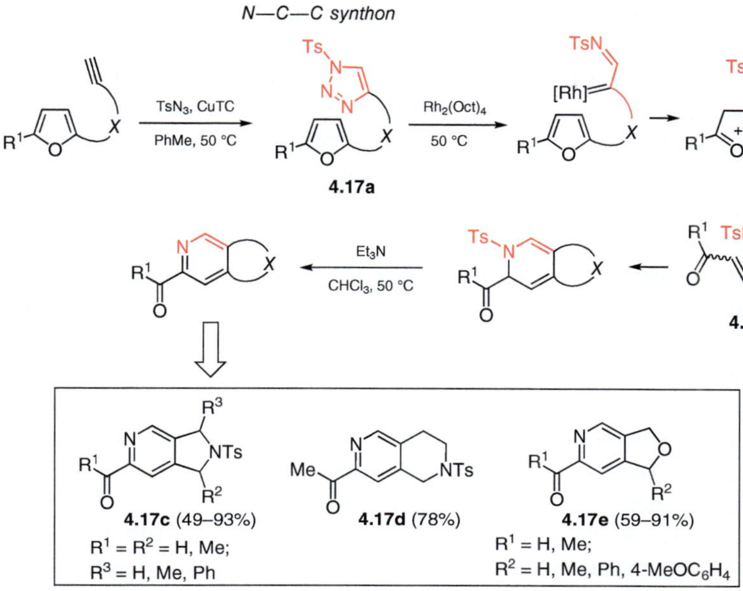

Scheme 4.17 Synthesis of fused pyridines by Rh-AVC-mediated transannulation of furans.

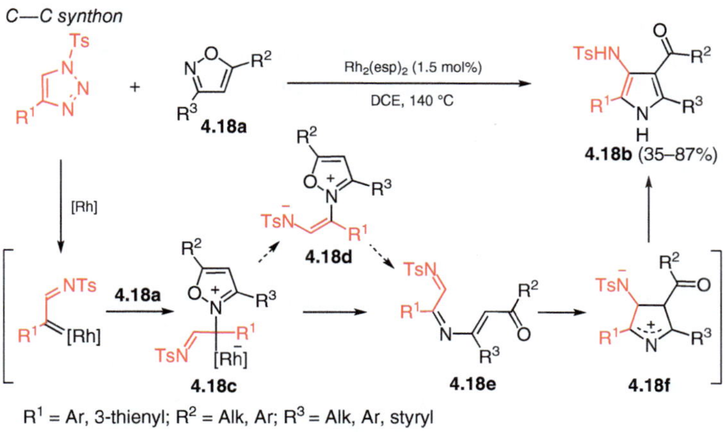

Scheme 4.18 Mechanism for the transannulation of isoxazoles to pyrroles.

(2 + 3) annulation fashion to give 4-acyl-3-sulfonamidopyrroles **4.18b**. A one-pot synthesis of the pyrroles from terminal alkynes, tosyl azide, and isoxazoles was also successfully accomplished (54–82%), greatly simplifying this isoxazole-pyrrole transformation. The choice of rhodium catalyst in the synthesis is essential. In particular, it was shown that $Rh_2(oct)_4$, $Rh_2(S\text{-DOSP})_4$, and $Rh_2(S\text{-PTAD})_4$ displayed poor activity in this reaction. The triazoles bearing electron-donating and electron-withdrawing substituents were well tolerated with this protocol, although the former generally give slightly higher product yields than the latter.

The authors suggested the mechanism for the formation of pyrroles **4.18b** involving the formation of intermediate isoxazolium ylide **4.18d**, its isomerization into 1,4-diazahexa-1,3,5-triene **4.18e**, and 5-*exo-trig* cyclization, thus affording zwitterionic intermediate **4.18f** (Scheme 4.18) [26]. The subsequent proton transfer and tautomerization lead to pyrrole **4.18b**. Later, using DFT calculations, it was shown that rhodium-free isoxazolium ylides of the type **4.18d** are absent on the PES of reactions of isoxazoles with rhodium carbenes, and the opening of the isoxazole ring occurs in rhodium-bound carbene complexes of the type **4.18c** [27]. As further studies of this reaction have shown, the azapolyene **4.18f** cyclization proceeds, as a rule, not so unambiguously, and not so high yields of pyrroles **4.18f** that often observed can be explained by the occurrence of a competing cyclization of the diazatriene intermediate into a dihydropyrazine derivative.

This problem of competition between two reactions was studied in detail in the work [27] (Scheme 4.19). In particular, it was found that the reaction of 1-sulfonyl-1,2,3-triazoles with 5-alkoxyisoxazoles **4.19a**, carried out in the presence of a rhodium(II) catalyst generally leads to two isomeric products: pyrazine-2-carboxylate **4.19b** and 4-aminopyrrole-3-carboxylate **4.19c**. It is noteworthy that a similar reaction of 3,5-diphenylisoxazole also gives rise to a mixture of these products in the ratio of ca. 7 : 1. The reaction carried out in chloroform at 100 °C under $Rh_2(OAc)_4$ catalysis provides mainly 4-aminopyrrole-3-carboxylates **4.19c**, which can be isolated in good yield. The use of $Rh_2(Piv)_4$ in refluxing toluene, however, results in the formation of 1,2-dihydropyrazine-2-carboxylates **4.19b** as the main products. The latter can be converted by a one-pot procedure to the pyrazine-2-carboxylates **4.19d** by heating with catalytic amounts of TsOH.

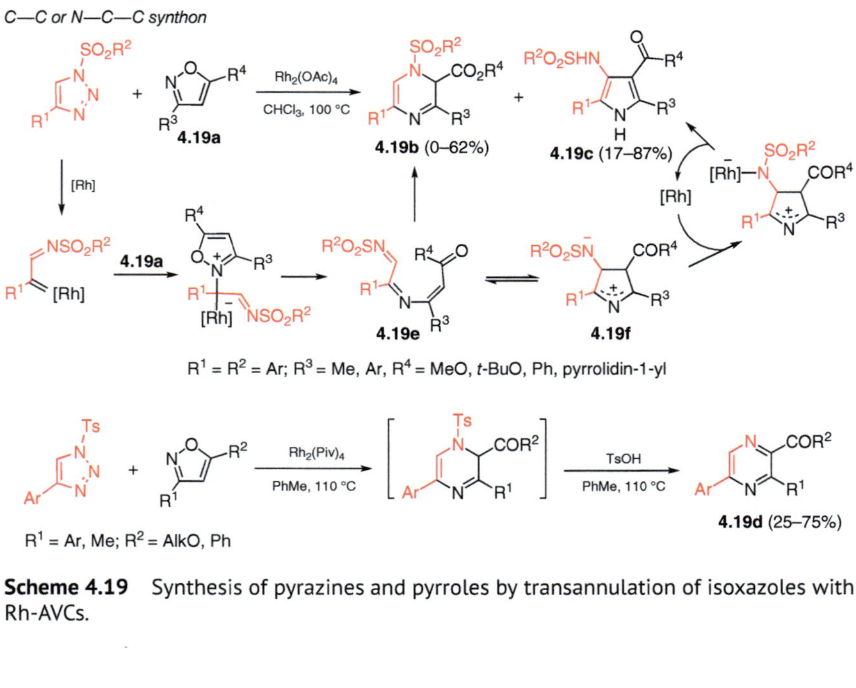

Scheme 4.19 Synthesis of pyrazines and pyrroles by transannulation of isoxazoles with Rh-AVCs.

It has been established that compounds **4.19b** and **4.19c** are formed from a common precursor, (5*Z*)-1,4-diazahexa-1,3,5-triene **4.19e**, which was detected by ^{1}H NMR spectroscopy. The influence of the nature of the catalyst on the product distribution is rationalized in terms of the complexation by the rhodium catalyst of intermediate zwitterion **4.19f**, the primary product of 1,4-diazahexatriene 1,5-cyclization. A decrease in the size of a ligand of the rhodium catalyst leads to an increase in the efficiency of its complexation with betain **4.19f** and, accordingly, to a shift in the equilibrium of diazatriene **4.19e** – betain **4.19f** toward the latter, which leads, in turn, to an increase of **4.19c**/**4.19b** ratio.

Rh-catalyzed reactions of 1-sulfonyl-1,2,3-triazoles with 3,4,5-trisubstituted isoxazoles generate 1,4-diazahexa-1,3,5-trienes substituted at the positions C5 and C6. The appearance of a substituent at C6 of an 1,4-diazahexa-1,3,5-triene can change dramatically its reactivity. As a result, instead of the 1,2-dihydropyrazine and pyrrole derivatives, other products are formed. In 2021, the reaction of 1-tosyl-1,2,3-triazoles with fully substituted isoxazoles under rhodium(II) catalysis was reported [28]. It was found that 1-sulfonyl-1,2,3-triazoles react with isoxazoles **4.20a** in the presence of $Rh_2(Piv)_4$ in a (2+4) annulation fashion to give pyrimidines **4.20b** (Scheme 4.20). It is noteworthy that a prerequisite to the success of this synthesis is the methyl substituent at C5 of isoxazole. The triazole acts as a two-atom C—N synthon in this transformation. The isoxazole provides four of its atoms into the final pyrimidine ring, while the oxygen of its ring is transferred to the side chain of the product. The reaction is catalyzed by various rhodium carboxylates, but $Rh_2(Piv)_4$ proved to be the catalyst of choice. Of all *N*-sulfonyl-1,2,3-triazoles tested, the maximum yields of pyrimidines were achieved with the *N*-tosyl derivative. The reaction conditions are tolerant of various 4-aryl substituents in the triazole ring, but, in general, some decrease in yield was observed when a strong electron-withdrawing substituent was introduced in this ring. According to DFT calculations, this reaction proceeds via the 1,6-cyclization of 6-acyl-1,4-diazahexatriene **4.20c** to 1,3-oxazine **4.20d** followed by a cascade of intramolecular aza-Diels-Alder and retro-aza-Diels-Alder reactions. The rate-determining step was found to be an aza-Diels-Alder reaction. This mechanism was also supported by the detection of the oxazine and oxazahexatriene intermediates by the ^{1}H NMR spectroscopy.

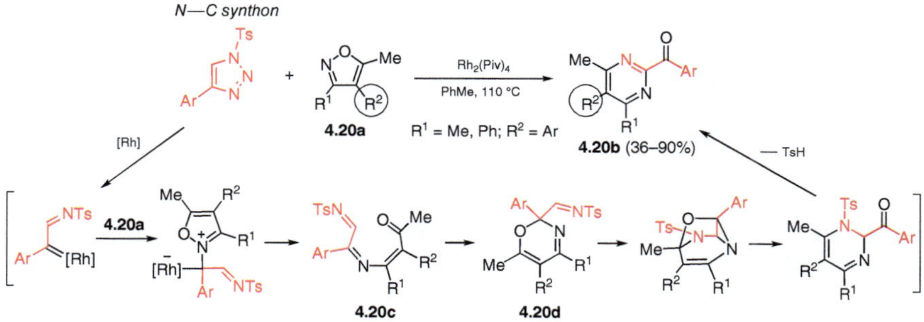

Scheme 4.20 Synthesis of pyrimidines by transannulation of isoxazoles with Rh-AVCs.

Surprisingly, replacing the 4-Me substituent for ethyl or phenyl group in the isoxazole (isoxazole **4.21a**) redirects the reaction toward dihydropyrazine **4.21b** (Scheme 4.21). At the same time, the closure of the substituents at C4 and C5 into the carbocyclic system (isoxazole **4.21c**) does not interfere with the formation of the pyrimidine product (compound **4.21d**). Probably, the role of substituents at C4 and C5 is associated with their influence on the conformational equilibrium of the intermediate 1,4-diazatriene and, first of all, the acyl fragment at C6.

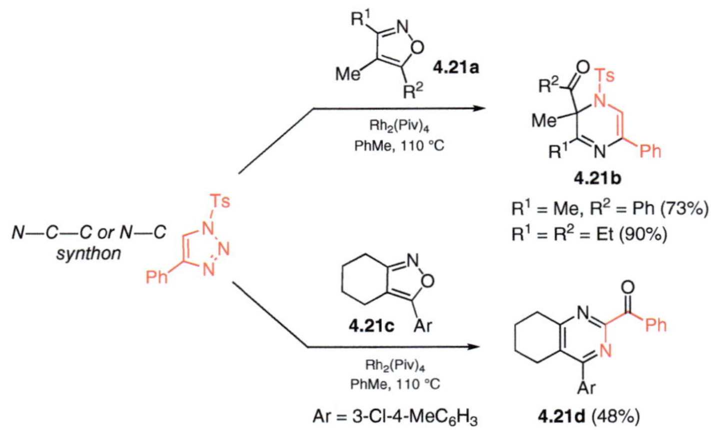

Scheme 4.21 Regiodivergence in annulation of isoxazoles with 1-sulfonyl-1,2,3-triazoles.

Benzo[*c*]isoxazoles **4.22a** reacts with 1-sulfonyl-1,2,3-triazoles similarly to their hydrogenated analogs mentioned above, tetrahydrobenzisoxazoles **4.21c**. This reaction reported five years earlier [29] proved to be a convenient method for the synthesis of quinazoline derivatives **4.22b** (Scheme 4.22). The limitation of the method is sensitivity to the nature of a C3-substituent of benzo[*c*]isoxazoles. In particular, both C3-unsubstituted and 3-phenyl-substituted substrates failed to give the corresponding quinazolines.

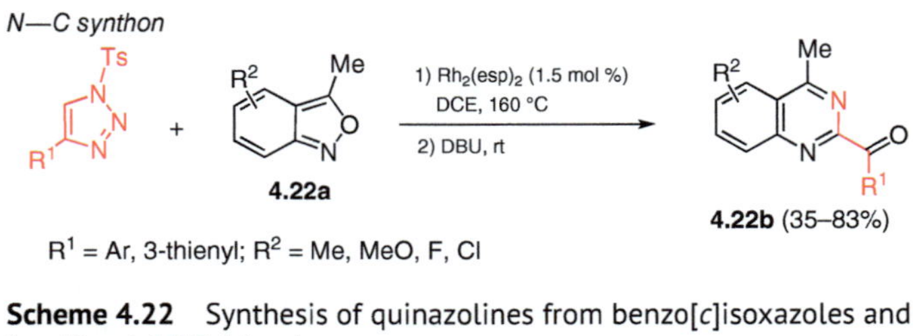

Scheme 4.22 Synthesis of quinazolines from benzo[*c*]isoxazoles and 1-sulfonyl-1,2,3-triazoles.

The authors suggested the mechanism of the reaction involving an aza-Diels–Alder reaction of the benzisoxazole to the C=N bond of the Rh-AVC generated from

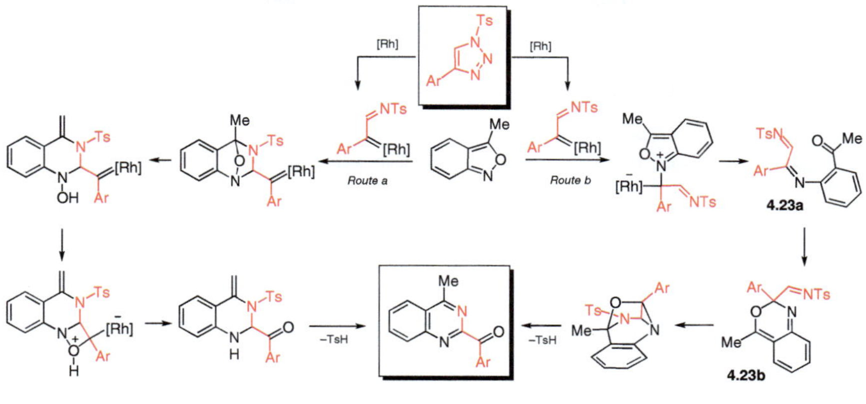

Scheme 4.23 Mechanism for the formation of quinazolines from benzo[*c*]isoxazoles.

the triazole (Scheme 4.23, route a). However, it is more likely that this reaction proceeds through the same path as the reactions of nonfused isoxazoles **4.21c**, that is, through the formation of diazatriene **4.23a**, cyclization to 1,3-oxazine **4.23b** and subsequent aza-Diels–Alder – retro-aza-Diels–Alder cascade (route b).

The isomers of benzo[*c*]isoxazoles, benzo[*d*]isoxazoles **4.24a**, also react with 1-sulfonyl-1,2,3-triazoles under the same conditions, but the final products of the reaction are imidazoles **4.24b** (Scheme 4.24) [29]. The triazole acts in this transformation as a C—C—N synthon, while only two atoms from the benzo[*d*]isoxazole are transferred to the imidazole ring. It is believed that the final product is formed via the opening of a weak N—O bond in the tricyclic intermediate **4.24d**, which resulted from the cyclization of the rhodium-bound benzisoxazolium ylide **4.24c**.

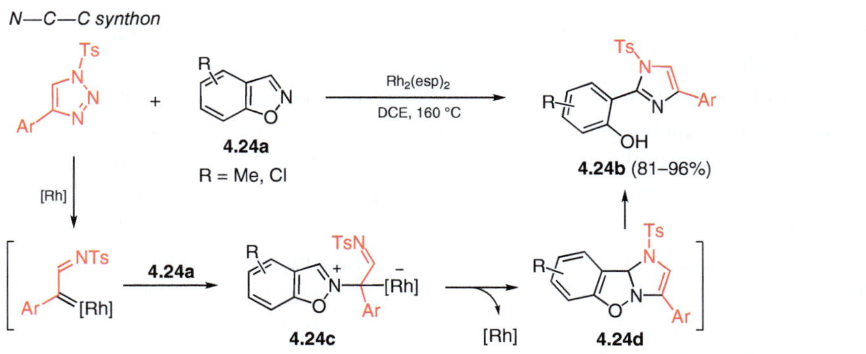

Scheme 4.24 Synthesis of imidazoles by transannulation of benzo[*d*]isoxazoles with Rh-AVCs.

The reaction of 1-sulfonyl-1,2,3-triazoles with 5-alkoxy- and 5-dialkylamino-1,2,4-oxadiazoles **4.25a**, in contrast to the similar reaction of isoxazoles, proceeds unambiguously regardless of the structure of the oxadiazole and leads to the formation of

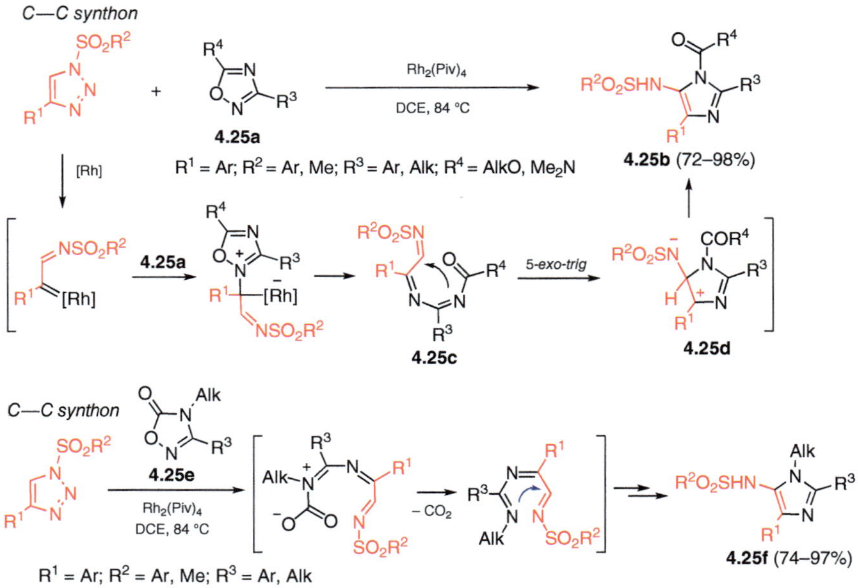

Scheme 4.25 Synthesis of imidazoles from 1,2,4-oxadiazole derivatives and 1-sulfonyl-1,2,3-triazoles.

imidazoles **4.25b** as the sole products (Scheme 4.25) [30]. A key stage in this process is the regioselective 5-*exo-trig* cyclization of azapolyene **4.25c** to imidazolium betaine **4.25d**, which undergoes rapid prototropic shift to form the final product. This transannulation reaction of 1-sulfonyl-1,2,3-triazoles is characterized by excellent yields of imidazoles **4.25b** and low sensitivity of the yields to the nature of the substituents in both the triazole and oxadiazole. The Rh(II)-catalyzed reactions of 1-sulfonyl-1,2,3-triazoles with 1,2,4-oxadiazol-5-ones **4.25e** proceed similarly to give 5-sulfonamidoimidazoles **4.25f** with an alkyl substituent at N1.

In addition to N—O azoles, N—N azoles can also be involved in Rh-catalyzed transannulation reactions of 1-sulfonyl-1,2,3-triazoles. These reactions proceed through the cleavage of the N—N bond under the action of Rh-AVCs generated from 1-sulfonyl-1,2,3-triazoles. The key azapolyene intermediate formed in this case, depending on the structure, can cyclize into one or another azaheterocycle. In 2021, an effective method for the preparation of 2,6,8-triazabicyclo[3.2.1]octa-3,6-dienes **4.26b** by the $Rh_2(Piv)_4$-catalyzed reaction of 1-sulfonyl-1,2,3-triazoles with 1,4-disubstituted pyrazoles **4.26a** has been developed (Scheme 4.26) [31]. It turned out that compounds **4.26b** can be obtained, and in most cases in almost quantitative yield, only from *N*-aryl- or *N*-heteroarylpyrazoles. The exception is *N*-alkylindazoles **4.26c**, which react with the triazoles to give benzo-fused analogs **4.26d**. *N*-Alkyl-substituted pyrazoles **4.26e** also smoothly react with 1-sulfonyl-1,2,3-triazoles under the same conditions but give rise to another product, imidazole derivatives **4.26f**. 2,6,8-Triazabicyclo[3.2.1]octa-3,6-dienes are intermediates on the way to imidazoles **4.26f**, and are converted into them in high

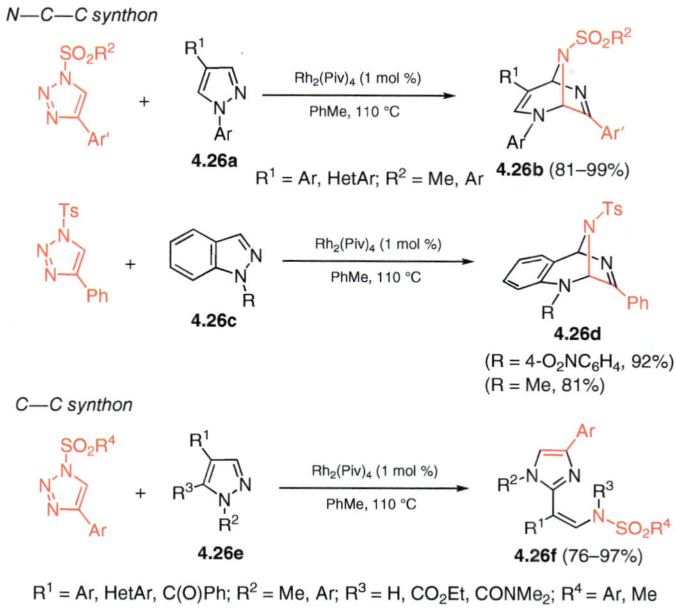

Scheme 4.26 Synthesis of 2,6,8-triazabicyclo[3.2.1]octa-3,6-dienes and (2-aminovinyl)imidazoles.

yields under the reaction conditions (R^2 = Alk) or under prolonged heating or under acid catalysis conditions (R^2 = Ar).

DFT calculations are in good agreement with the reaction mechanism involving the formation of 1,4,8-triazaocta-1,3,5,7-tetraene **4.27a** followed by intramolecular stepwise aza-Diels–Alder cycloaddition (Scheme 4.27). The extraordinary behavior of the conjugated azapolyene, namely its ability to almost spontaneously self-organize into cyclic and bicyclic structures through a sequence of valence isomerizations, is reminiscent of the famous "Rubik's snake" puzzle, from which intricate 3D objects are folded. The transformation of bicyclic compound **4.27b** to imidazole **4.27c** proceeds via its acid-catalyzed degenerative isomerization followed by the ring-opening and 1,4-prototropic shift.

The transannulation of 1-sulfonyl-1,2,3-triazoles into 3-sulfonamidopyrroles can be accomplished not only with isoxazoles, as it was described above, but also with 1,2,3-triazoles itself. In 2021, the $Rh_2(Piv)_4$-catalyzed reaction of 1-sulfonyl-1,2,3-triazoles with 1-alkyl-5-aryl-1,2,3-triazoles **4.28a** leading to *N*-alkylated 3-sulfonamidopyrroles **4.28b** was reported (Scheme 4.28) [32]. In this reaction, accompanied by the extrusion of two dinitrogen molecules, the 1-sulfonyltriazole is used as a C—C synthon, and 1-alkyltriazole introduces a triatomic C—C—N fragment into the resulting pyrrole ring. The scope of the reaction is restricted to triazoles **4.28a** having an alkyl group at N1 and an aryl group at C4. 1,4-Diaryltriazoles proved to be inactive in the reaction, while 4-alkyl-substituted triazoles reacted with the retention of the triazole ring to give

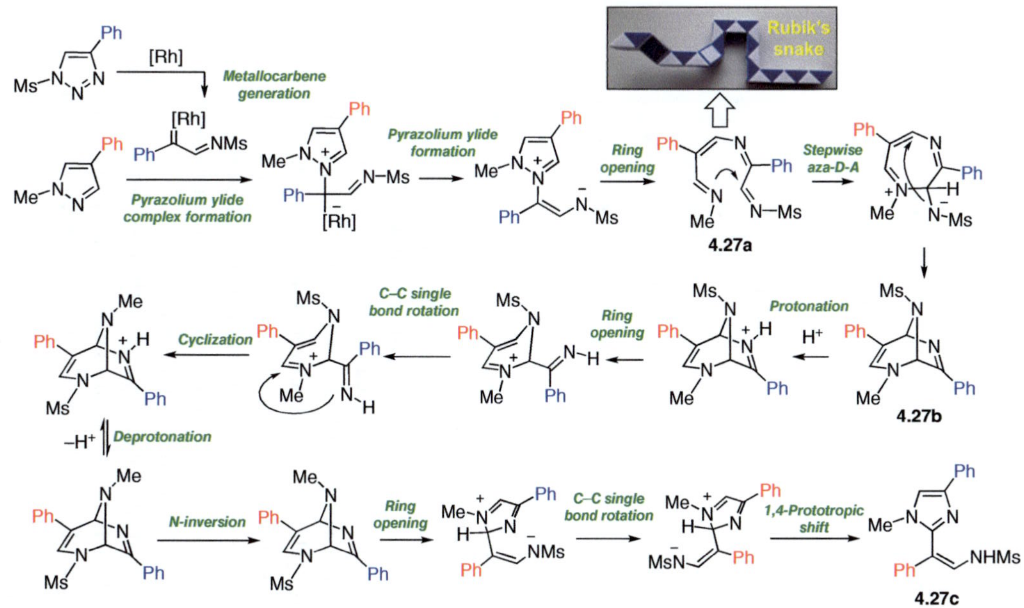

Scheme 4.27 Mechanism of the catalytic reaction of pyrazoles with 1-sulfonyl-1,2,3-triazoles.

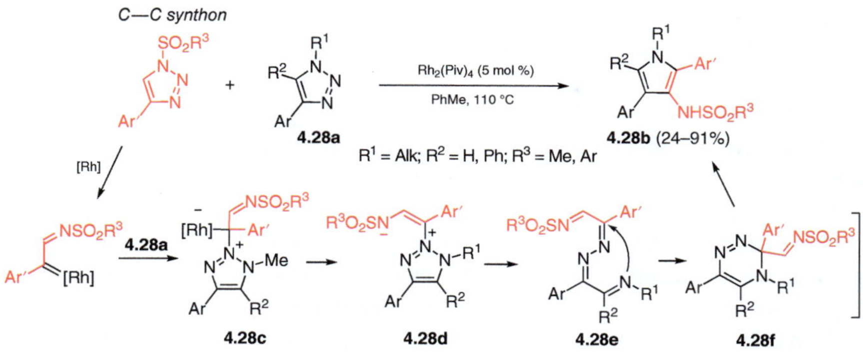

Scheme 4.28 Synthesis of pyrroles by transannulation of 1,2,3-triazoles with Rh-AVCs.

stable 1,2,3-triazol-3-ium ylides. According to DFT calculations, the reaction proceeds via the attack of the Rh-AVC at the N2 atom of the 1-alkyl-4-aryl-1,2,3-triazole and the successive formation of the rhodium-bound 1,2,3-triazol-2-ium ylide **4.28c**, metal-free 1,2,3-triazol-2-ium ylide **4.28d**, 1,4,5,8-tetraazaocta-1,3,5,7-tetraene **4.28e**, and 3-(2-azavinyl)-3,4-dihydro-1,2,4-triazine **4.28f**. The concerted denitrogenative ring contraction of the latter, followed by 1,2-prototropic shift, gives the final pyrrole. This protocol enables the preparation of 3-sulfonamidopyrroles in 24–91% yield.

4.3.2 Nonaromatic Rings

Rh-AVCs, generated from 1-sulfonyl-1,2,3-triazoles, are capable of reacting with saturated oxa- and polyoxaheterocycles of two types: small rings and rings containing an acetal moiety.

Rhodium(II)-catalyzed transannulation of 1-tosyl-1,2,3-triazoles with aryloxiranes **4.29a** to give 3,4-dihydro-2*H*-1,4-oxazines **4.29b** in moderate yields was described (Scheme 4.29) [33]. $Rh_2(S\text{-NTTL})_4$ and $Rh_2(S\text{-PTV})_4$ were found to

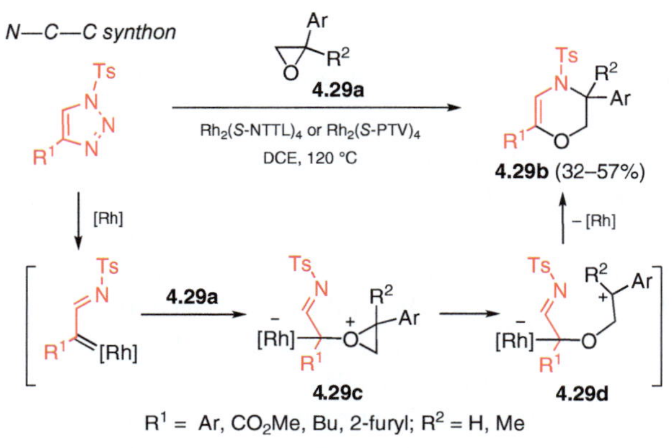

Scheme 4.29 Rh(II)-catalyzed reaction of 1-tosyl-1,2,3-triazoles with aryloxiranes.

be the most efficient catalysts. The reaction probably involves the formation of oxonium ylide **4.29c**, ring-opening to give secondary carbocation **4.29d** followed by cyclization.

Analogous reaction of glycidols **4.30a**, catalyzed by $Rh_2(Piv)_4$, proceeds differently depending upon substitution of an oxirane ring (Scheme 4.30). When glycidol contains an additional substituent at C2, the products of the reaction are 1,4-oxazepines **4.30b** which were obtained in 41–69% yields [34]. First, O—H-insertion of the Rh-AVC takes place, subsequent Lewis acid-catalyzed regioselective nucleophilic oxirane ring-opening across the O—C3 bond leads to the final product. The reaction of unsubstituted glycidol **4.30c** gave two regioisomeric (hydroxymethyl)-3,4-dihydro-2*H*-1,4-oxazines **4.30d** and **4.30e** formed formally via ring-opening of the oxirane across both the O—C2 and O—C3 bond with preference for the O—C2 cleavage. The isolated yields of 3-(hydroxymethyl)-3,4-dihydro-2*H*-1,4-oxazines **4.30e** are 35–76%. The authors proposed that the major isomer **4.30e** is formed via the Rh-AVC O—H-insertion reaction, while the minor isomer **4.30d** – via the formation of oxonium ylide **4.30f** [34].

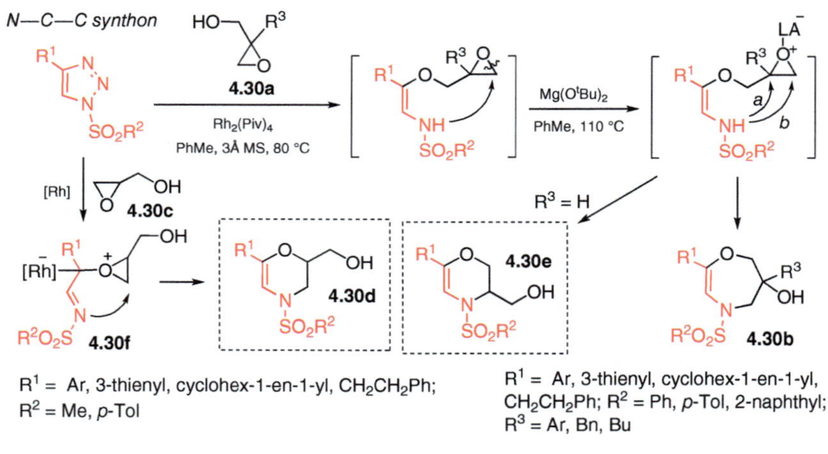

Scheme 4.30 Rh(II)-catalyzed reaction of 1-sulfonyl-1,2,3-triazoles with glycidols.

Several types of Rh(II)-catalyzed transannulations of 1-sulfonyl-1,2,3-triazoles with oxetanes **4.31a** have been reported by Lacour's group (Scheme 4.31) [35]. The best catalyst for these reactions was found to be Rh_2(*S*-TCPTTL)$_4$. Under "dilute" conditions, 2-iminotetrahydrofurans are formed by a (1 + 4) cycloaddition, which were reduced to aminomethyl derivatives **4.31b**. Under high concentrations of oxetanes, 13- and 15-membered oxazamacrocycles **4.31c** and **4.31d** as *Z*-isomers can be obtained via formal (5 + 4 + 4) and (3 + 4 + 4 + 4) cycloadditions of Rh-AVC with two/three oxetane molecules, respectively. Thermally activated phthalimido-triazoles gave a *Z*/*E* mixture of 15-membered macrocycles. It was proposed that the reactions proceed via the formation of oxonium ylide, which undergoes a 1,2-shift

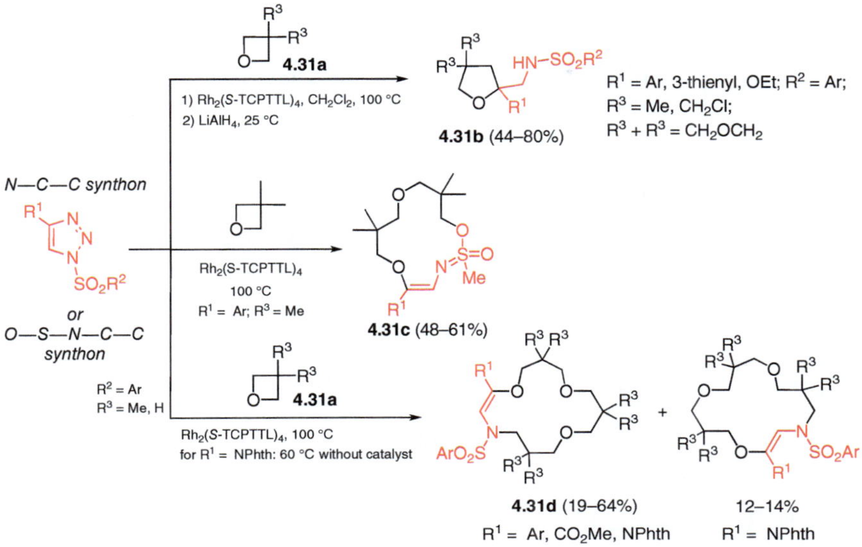

Scheme 4.31 Rh(II)-catalyzed reaction of 1-sulfonyl-1,2,3-triazoles with oxetanes.

to form tetrahydrofuran adduct. At high concentration, the strained oxonium ylide can be cleaved again and again by oxetane to give the final macrocycle.

$Rh_2(OAc)_4$-catalyzed transannulations of 1-sulfonyl-1,2,3-triazoles with 1,3-dioxolane and 1,3-dioxane have also been reported by Lacour and co-authors (Scheme 4.32) [36]. The products of these reactions are 2,3,7,8-tetrahydro-1,6,3-dioxazocines **4.32a** and 3,7,8,9-tetrahydro-2*H*-1,6,3-dioxazonines **4.32b**, respectively. Analogously, the reaction of 1-sulfonyl-1,2,3-triazoles with 1,3,5-trioxane catalyzed by $Rh_2(Piv)_4$ led to the formation of 6,7-dihydro-1,3,5,7-trioxazonines **4.32c** [37]. According to the mechanism reported [36], the reaction proceeds through the formation of rhodium-bound oxonium ylide **4.32d**. The subsequent ring-opening across the C—O bond results in oxo-carbenium ion **4.32e**, which is then trapped by the nitrogen atom of the sulfonylimino group to form the final medium-size heterocycle.

Among small heterocycles, in addition to oxiranes, 2*H*-azirines are distinguished by their unusual reactivity toward Rh-AVCs. The rhodium-catalyzed reactions of 1-sulfonyl-1,2,3-triazoles with azirines turned out to be very synthetically fruitful, and have been studied in detail by several research groups. Depending on the reaction conditions and substitution pattern of the substrates, these transannulation reactions lead to 3-aminopyrrole or pyrazine derivatives (Scheme 4.33).

In particular, Tang reported the selective formation of both these products (Scheme 4.33a) [38]. The reaction selectivity for some 2,3-disubstituted azirines can be switched from 1,2-dihydropyrazine to 3-aminopyrrole by the addition of catalytic amounts of chloroacetic acid. 1,2,3-Triazoles bearing an anisyl substituent in the sulfonyl group react with azirine-2-carboxylic esters with the predominant

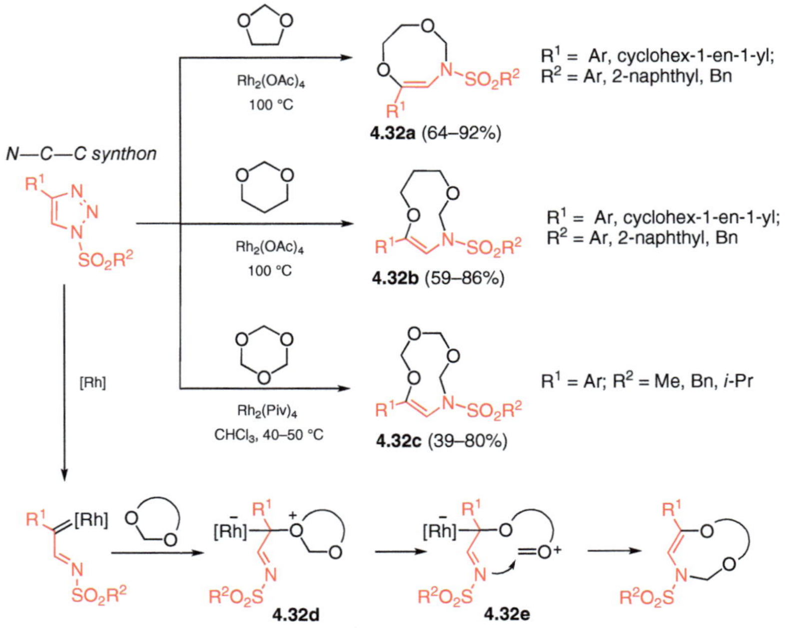

Scheme 4.32 Rh(II)-catalyzed reaction of 1-sulfonyl-1,2,3-triazoles with cyclic polyethers.

formation of pyrazine derivatives (Scheme 4.33b) [39]. Tang and Shi reported the exclusive formation of 3-aminopyrroles in most cases in similar reactions (Scheme 4.33c) [40]. In the work of Zhang and co-authors, on the contrary, the formation of only 1,2-dihydropyrazines was observed (Scheme 4.33d) [41].

Obviously, the reaction can be affected by different factors and has a complex mechanism which is summarized in Scheme 4.34. This mechanism in general is similar to that for the reaction of 1-sulfonyl-1,2,3-triazoles with isoxazoles (Scheme 4.19). The key intermediates are azirinium ylide **4.34a** and 1-sulfonyl-1,4-diazahexa-1,3,5-triene **4.34b**, which can undergo 6π-electrocyclization leading to dihydropyrazine **4.34c** or 5-*exo-trig* cyclization leading to 3-aminopyrrole **4.34d**.

The $Rh_2(OAc)_4$-catalyzed transannulation of 1-sulfonyl-1,2,3-triazoles with trisubstituted spiro-2*H*-azirines (Scheme 4.35) proceeds through the same path to give initially spiro-1,2-dihydropyrazines **4.35a** which under prolonged heating transform to 4-amino-spiro-3*H*-pyrroles **4.35b** via the reversible generation of diazahexatriene intermediates [42].

The reversibility of the 1,2-dihydropyrazine formation was also demonstrated in the Rh(II)-catalyzed reaction of 1-sulfonyl-1,2,3-triazoles with 2,2-diaryl-2*H*-azirines **4.36a** (Scheme 4.36). The primary transannulation products, 1,2-dihydropyrazines **4.36b**, can be isolated in good yields and are stable compounds at room temperature. However, upon heating, they undergo the ring-opening to 1,4-diazatrienes **4.36c** cyclizing to (*Z*)-*N*-(2-aminovinyl)indoles **4.36e**. The synthesis of indoles **4.36e** was successfully carried out in one-pot from the azirines and

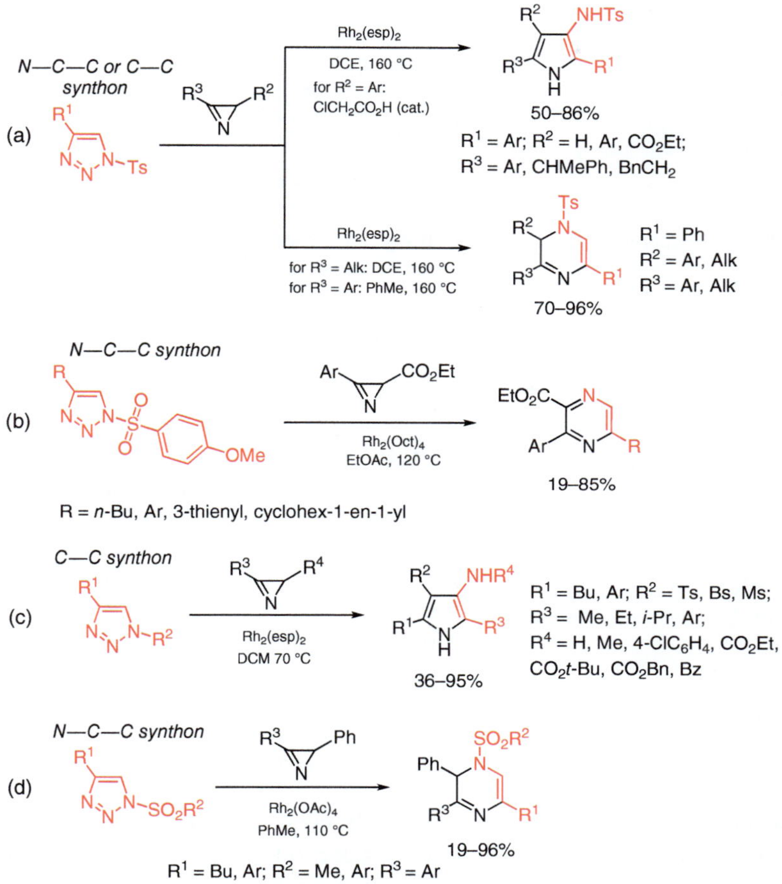

Scheme 4.33 Rh(II)-catalyzed reactions of 1-sulfonyl-1,2,3-triazoles with 2*H*-azirines (a–d).

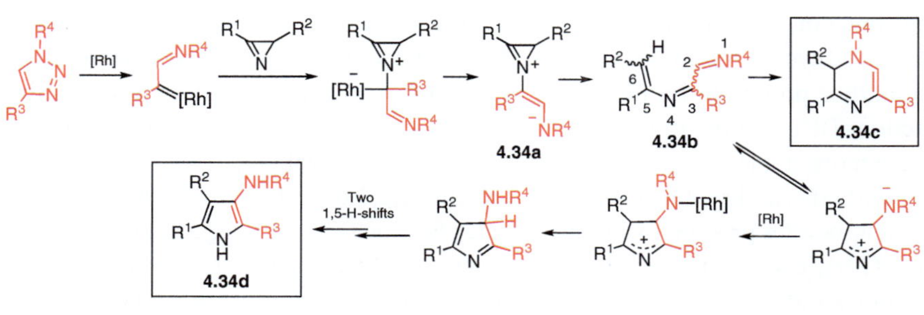

Scheme 4.34 Mechanism of the catalytic reaction of 1-sulfonyl-1,2,3-triazoles with 2*H*-azirines.

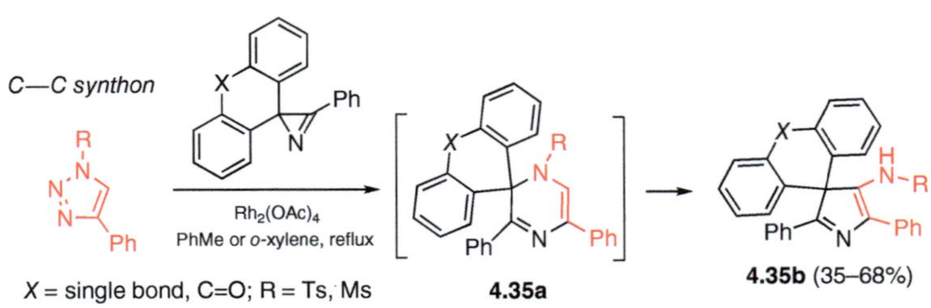

Scheme 4.35 Rh(II)-catalyzed reaction of 1-sulfonyl-1,2,3-triazoles with spiro-2*H*-azirines.

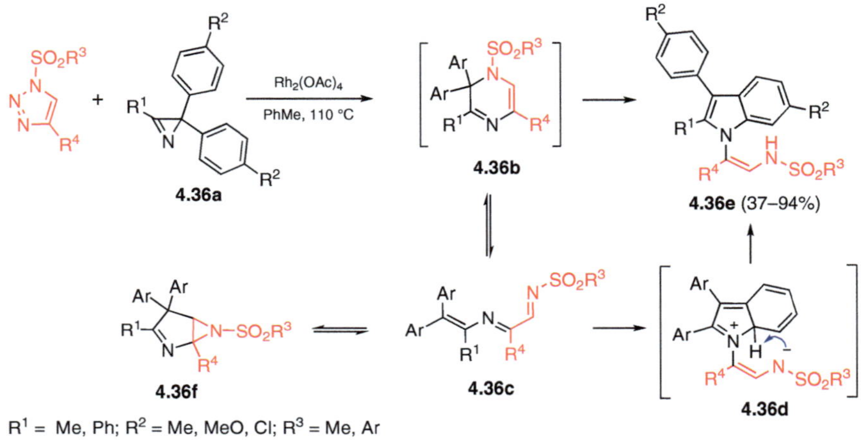

Scheme 4.36 Synthesis of indoles from 2*H*-azirines and 1-sulfonyl-1,2,3-triazoles.

triazoles without isolation of intermediate dihydropyrazines **4.36b**. According to DFT calculations, the cyclizations of diazatriene **4.36c** to 1,2-dihydropyrazine **4.36b** and diazabicyclohexene **4.36f** are low-barrier and reversible processes. At the same time, its cyclization to indole **4.36e**, via extremely unstable 7a*H*-indolium ylide **4.36d**, has higher barrier but proceeds irreversibly [43].

The $Rh_2(esp)_2$-catalyzed transannulation of 1-sulfonyl-1,2,3-triazoles with bicyclic aziridines **4.37a** unexpectedly leads to three-atom expansion of a carbamate ring, but not an aziridine ring, to give in good yields aziridines fused to nine-membered cycle – 6,7,8,9-tetrahydro-1,3,6-oxadiazonin-2-ones **4.37b** (Scheme 4.37). The author proposed the formation of metal-free aziridinium ylides **4.37c** in the reaction. DFT computations showed that the carbamate ring expansion proceeds from a more stable aziridinium ylide under kinetic control [44].

The Rh(II)-catalyzed reactions of 1-sulfonyl-1,2,3-triazoles with saturated nitrogen heterocycles proceed in a more complex fashion than with oxygen heterocycles.

$Rh_2(Piv)_4$-catalyzed reactions of 1-sulfonyl-1,2,3-triazoles with *N,N′*-diarylimidazolidines **4.38a** has been studied by Poblador-Bahamonde, Lacour, and co-authors [45]. Hexahydropyrazinoindoles **4.38b** were prepared in high yields when the aryl substituent did not bear an electron-withdrawing group (Scheme 4.38). In contrast, the exclusive formation of triazocine derivative **4.38c** was observed when one of the

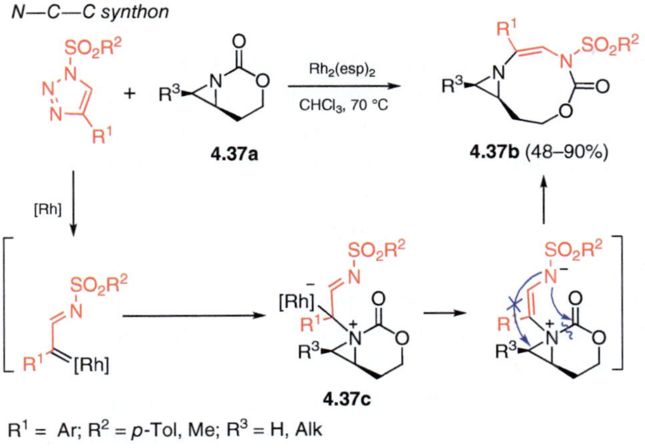

Scheme 4.37 Rh(II)-catalyzed reaction of 1-sulfonyl-1,2,3-triazoles with bicyclic aziridines.

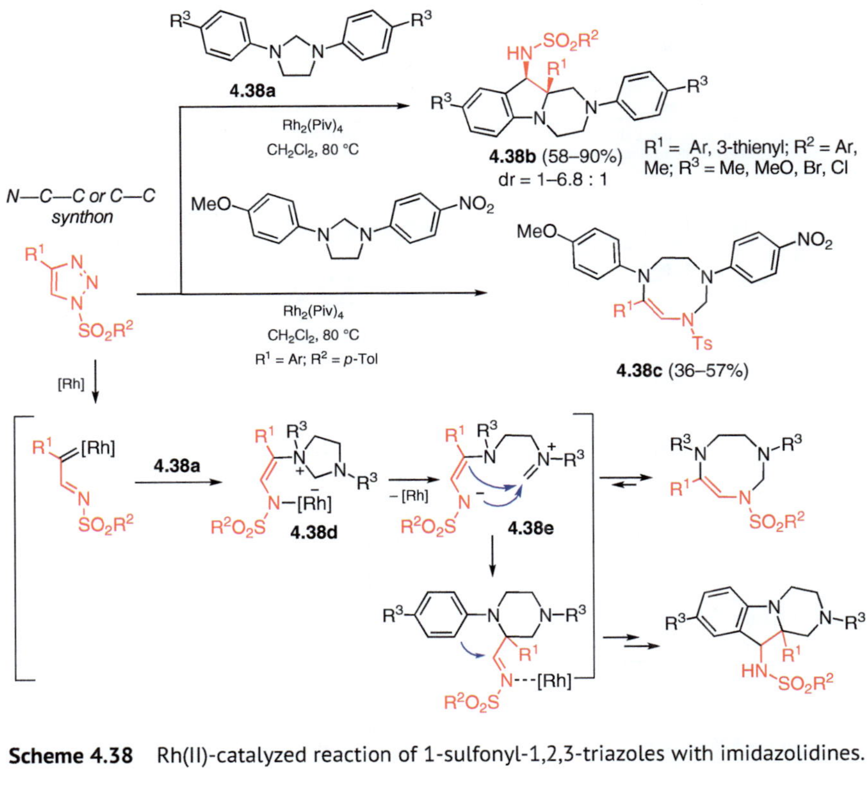

Scheme 4.38 Rh(II)-catalyzed reaction of 1-sulfonyl-1,2,3-triazoles with imidazolidines.

aromatic rings contained a nitro group. Based on DFT calculations, the authors proposed explanations for this regiodivergence. Initially, the Rh-AVC reacts with the imidazolidine to form ammonium ylide **4.38d** that undergoes the ring-opening and cycloaminomethylation of the sulfonamide nitrogen of betain **4.38e** affording triazocine derivative as a product of kinetic control. These compounds, in the absence of an electron-withdrawing aryl substituent, are unstable and under the reaction conditions undergo reverse ring opening to betaine **4.38e**. Cycloaminomethylation of carbon atom followed by Friedel-Crafts cyclization gives pyrazinoindoles.

The $Rh_2(Piv)_4$-catalyzed transannulation reaction of 1-sulfonyl-1,2,3-triazoles with Tröger bases **4.39a** occurs in an unusual fashion affording polycyclic heterocycles **4.39b**, containing indoline, benzodiazepine and hexahydropyrimidine cores [46] (Scheme 4.39). These compounds are formed in high diastereoselectivity (d.r. > 49 : 1) and mostly in good yields. The reaction with unsymmetrical nitro-containing Tröger bases **4.39c** proceeds regioselectively to give a single regioisomer. The reaction mechanism presumably involves the generation of ammonium ylide complex **4.39d** and subsequent cascade of [1,2]-Stevens rearrangement, Friedel-Crafts reaction, Grob fragmentation, and cyclization. Further expansion of the cycloaminal moiety in obtained compounds **4.39b** under the action of Rh-AVCs is also possible but featuring moderate product yields. In this case, a

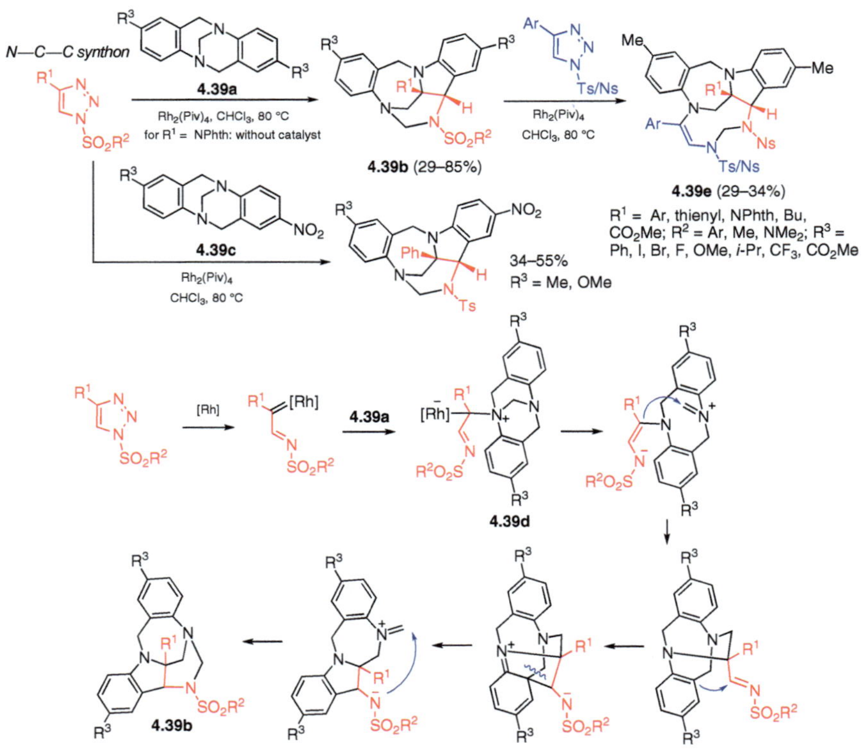

Scheme 4.39 Rh(II)-catalyzed reaction of 1-sulfonyl-1,2,3-triazoles with Tröger bases.

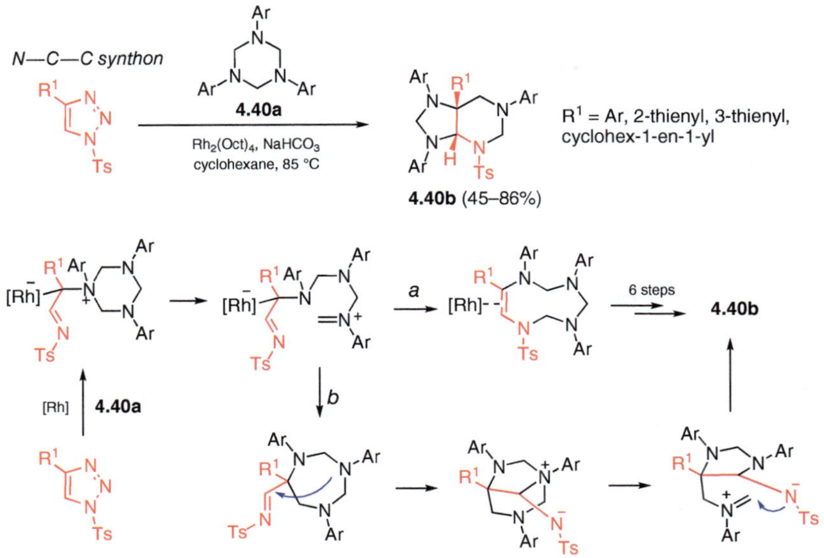

Scheme 4.40 Rh(II)-catalyzed reaction of 1-sulfonyl-1,2,3-triazoles with 1,3,5-triazinanes.

selective three-atom ring expansion of hexahydropyrimidine core occurs to form nine-membered ring – triazonanes **4.39e**, via initial attack of Rh-AVC onto the bridgehead nitrogen of **4.39b**.

The $Rh_2(Oct)_4$-catalyzed intermolecular transannulation of 1-sulfonyl-1,2,3-triazoles with *N*,*N*′,*N*″-triaryl-1,3,5-triazinanes **4.40a** leading to the octahydropurine derivatives **4.40b** in good yields has been described by Bao's group (Scheme 4.40) [47]. The highest product yields were achieved when 2 equiv. of $NaHCO_3$ were added; however, the authors did not present any explanation of this phenomenon. The plausible reaction mechanism, supported by DFT calculations data, involves a formal (6 + 3) cycloaddition reaction of 1,3,5-triazinane **4.40a** with the Rh-AVC followed by a cascade of six non-trivial steps (route *a*). In our opinion, the mechanism including formal [1,2]-Stevens rearrangement, nucleophilic attack of nitrogen at the activated C=N bond, aminal ring-opening, and recyclization (route *b*) is more feasible.

Aryltiiranes **4.41a** react with 1-sulfonyl-1,2,3-triazoles under Rh(II) catalysis analogously to aryloxiranes, 3,4-dihydro-2*H*-1,4-thiazines **4.41b** were obtained in 31–93% yields in this case (Scheme 4.41) [48]. The feature of the reaction is that an acetic acid additive accelerates the process and increases the product yields, probably due to enhancing the Rh-AVC electrophilicity. In contrast, alkyltiiranes under these conditions gave a complex mixture of products.

A formal (6 + 3) transannulation of 1-sulfonyl-1,2,3-triazoles with thiochromones **4.42a** has been reported by Xu and Li [49]. This reaction provides a facile and efficient method for the construction of nine-membered N,S-heterocycles – benzo[*h*][1,4]thiazonine derivatives **4.42b** (Scheme 4.42). The authors proposed the following scheme for the reaction course: the Rh-AVC attacks the sulfur atom of the substrate

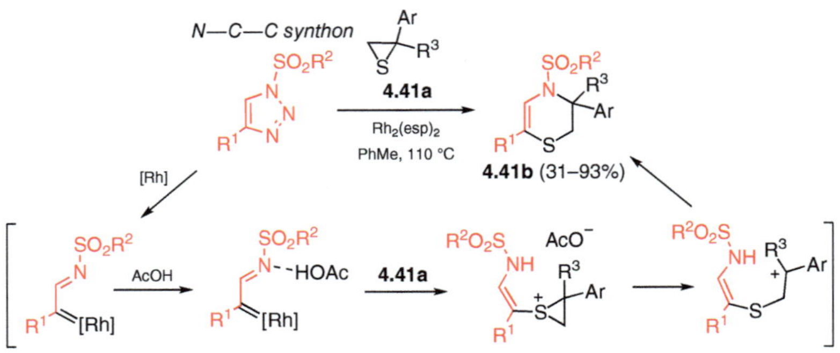

R^1 = Ar, 2-thienyl, 2-bromoethyl, 2-pyridyl; R^2 = Me, Ar; R^3 = H, Me, Et

Scheme 4.41 Rh(II)-catalyzed reaction of 1-sulfonyl-1,2,3-triazoles with tiiranes.

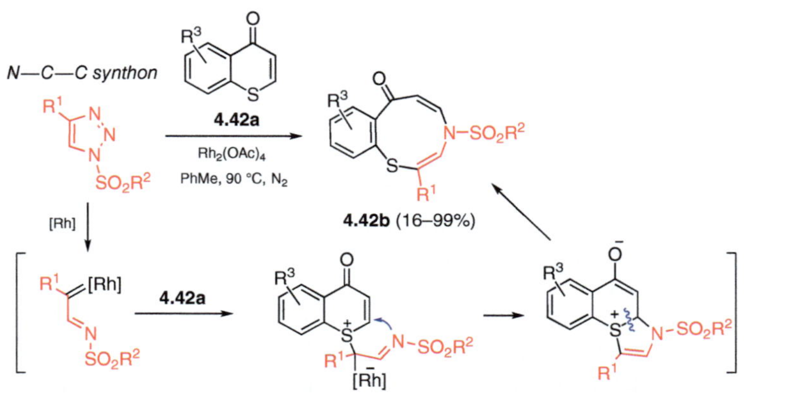

R^1 = Ar, 3-thienyl, 1-naphthyl; R^2 = Ar, Me, 2-naphthyl; R^3 = H, 6-Me/MeO/F/Cl/Br, 8-Me

Scheme 4.42 Rh(II)-catalyzed reaction of 1-sulfonyl-1,2,3-triazoles with thiochromones.

to form the sulfonium ylide complex, followed by intramolecular Michael addition and ring cleavage across the C—S bond. The presence of a sulfur atom is essential to the reaction, since an oxygen analog, chromone, failed to react with triazole under these conditions.

4.4 Summary and Outlook

This chapter provides a comprehensive account of methods for the synthesis of heterocycles via transannulation of 1-sulfonyl-1,2,3-triazoles with hetero- and carbocyclic compounds. This methodology provides a truly enormous structural diversity of heterocyclic products, which is due to the unique reactivity of Rh-AVCs, which are formed during the thermocatalytic decomposition of sulfonyltriazoles. These carbene complexes can participate in the formation of cyclic structures through selective interaction with both π-systems and nonbonding electrons of heteroatoms, and also give products of formal insertion into C—H bonds. As a result, there is a choice of how to construct the desired mono- or polycyclic system, either through formal cycloaddition reaction, or cyclization, or one of the ring-opening–cyclization sequences. The last of the approaches, realized through the opening of the ring of an annulating reagent, has been very actively developed in the last few years, and it is here that, most likely, we should expect the most striking achievements in the coming years. Some of the considerable challenges to be addressed in this area include (i) catalytic transannulation reactions of 1-sulfonyl-1,2,3-triazoles with azoles having an N—S bond, (ii) transannulation reactions with nonaromatic azole derivatives having weak N—O, N—N, and N—S bonds, (iii) new transannulation reactions via C—H insertion of Rh-AVCs, and (iv) the use of Rh-AVCs in the synthesis of stable azolium ylides as building blocks for the preparation of complex heterocycles. Finally, the problem of the transition from rhodium azavinylcarbene complexes to similar complexes of cheaper metals, such as copper, is also awaiting a solution.

References

1 Chuprakov, S., Hwang, F.W., and Gevorgyan, V. (2007). Rh-catalyzed transannulation of pyridotriazoles with alkynes and nitriles. *Angew. Chem. Int. Ed.* 46: 4757–4759.

2 Horneff, T., Chuprakov, S., Chernyak, N. et al. (2008). Rhodium-catalyzed transannulation of 1,2,3-triazoles with nitriles. *J. Am. Chem. Soc.* 130: 14972–14974.

3 Das, S.K., Roy, S., and Chattopadhyay, B. (2023). Transition-metal-catalyzed denitrogenative annulationto access high-valued N-heterocycles. *Angew. Chem. Int. Ed.* 62: e2022109.

4 Akter, M., Rupa, K., and Anbarasan, P. (2022). 1,2,3-Triazole and its analogues: new surrogates for diazo compounds. *Chem. Rev.* 122: 13108–13205.

5 Li, W. and Zhang, J. (2020). Synthesis of heterocycles through denitrogenative cyclization of triazoles and benzotriazoles. *Chem. Eur. J.* 26: 11931–11945.

6 Jiang, Y., Sun, R., Tang, X.-Y., and Shi, M. (2016). Recent advances in the synthesis of heterocycles and related substances based on a-imino rhodium carbene complexes derived from *N*-sulfonyl-1,2,3-triazoles. *Chem. Eur. J.* 22: 17910–17924.

7 Wang, Y., Lei, X., and Tang, Y. (2015). Rhodium-azavinylcarbene: a versatile synthon-enabling divergent synthesis of nitrogen heterocycles. *Synlett* 26: 2051–2059.

8 Davies, H.M.L. and Alford, J.S. (2014). Reactions of metallocarbenes derived from *N*-sulfonyl-1,2,3-triazoles. *Chem. Soc. Rev.* 43: 5151–5162.

9 Miura, T., Funakoshi, Y., and Murakami, M. (2014). Intramolecular dearomatizing [3 + 2] annulation of α-imino carbenoids with aryl rings furnishing 3,4-fused indole skeletons. *J. Am. Chem. Soc.* 136: 2272–2275.

10 Jana, S., Vroemans, R., and Dehaen, W. (2017). Synthesis of polycyclic dihydroindoles by selective decomposition of bis(1,2,3-triazoles) mediated by rhodium-catalysis. *Adv. Synth. Catal.* 359: 3085–3089.

11 Qiao, H., Bai, J., Zhu, M. et al. (2020). Rhodium (II)-catalyzed synthesis of tetracyclic 3,4-fused indoles and dihydroindoles. *Catalysts* 10: 920.

12 Kim, H., Kim, S.E., Jeon, W.H. et al. (2017). Synthesis of 3a,7a-dihydroindoles via rhodium-catalyzed intermolecular formal [3+2] transannulation reactions of triazoles with alkyl-substituted benzene derivatives. *Bull. Korean Chem. Soc.* 38: 1299–1305.

13 Wilkerson-Hill, S.M., Haines, B.E., Musaev, D.G., and Davies, H.M.L. (2018). Synthesis of [3a,7a]-dihydroindoles by a tandem arene cyclopropanation/3,5-sigmatropic rearrangement reaction. *J. Org. Chem.* 83: 7939–7949.

14 Miura, T., Zhao, Q., Funakoshi, Y., and Murakami, M. (2015). Site-selective introduction of an enamido group at the C(3)-position of Indoles. *Heterocycles* 91: 1579–1584.

15 Rajasekar, S. and Anbarasan, P. (2019). One-pot transannulation of *N*-sulfonyl-1,2,3-triazoles to dihydro-β-carbolines and dihydroisoquinolines via rhodium-catalyzed C—H insertioncum-base-mediated aza-Michael reaction. *J. Org. Chem.* 84: 7747–7761.

16 Spangler, J.E. and Davies, H.M.L. (2013). Catalytic asymmetric synthesis of pyrroloindolines via a rhodium(II)-catalyzed annulation of indoles. *J. Am. Chem. Soc.* 135: 6802–6805.

17 Zhang, Y.-S., Tang, X.-Y., and Shi, M. (2015). Divergent synthesis of indolefused polycycles via Rh(II)-catalyzed intramolecular [3+2] cycloaddition and C—H functionalization of indolyltriazoles. *Org. Chem. Front.* 2: 1516–1520.

18 Bora, P.P., Luo, Z.-L., Chen, L., and Kang, Q. (2016). Rh(II)-catalyzed intramolecular dearomatizing annulation of *N*-sulfonyl-1,2,3-triazoles: synthesis of polycyclic spiroindolines. *Tetrahedron* 72: 1467–1471.

19 Fu, L. and Davies, H.M.L. (2017). Scope of the reactions of indolyl- and pyrrolyl-tethered *N*-sulfonyl-1,2,3-triazoles: rhodium(II)-catalyzed synthesis of indole- and pyrrole-fused polycyclic compounds. *Org. Lett.* 19: 1504–1507.

20 Bettencourt, C.J., Krainz, T., Chow, S. et al. (2022). Unearthing the subtleties of rhodium(II)-catalyzed carbenoid cycloadditions to furans with an *N*-sulfonyl-1,2,3-triazole probe. *Org. Lett.* 24: 9290–9295.

21 Chen, W., Bai, Y.-L., Luo, Y.-C., and Xu, P.-F. (2017). Rh(II) catalyzed high order cycloadditions of 8-azaheptafulvenes with *N*-sulfonyl 1,2,3-triazloes or α-oxo diazocompounds. *Org. Lett.* 19: 364–367.

22 Pal, K., Sontakke, G.S., and Volla, C.M.R. (2019). Rh(II)-catalyzed highly diastereoselective cascade transannulation of *N*-sulfonyl-1,2,3-triazoles and vinyl benzoxazinanones. *Org. Lett.* 21: 3716–3720.

23 Parr, B.T., Green, S.A., and Davies, H.M.L. (2013). Rhodium-catalyzed conversion of furans to highly functionalized pyrroles. *J. Am. Chem. Soc.* 135: 4716–4718.

24 Funakoshi, Y., Miura, T., and Murakami, M. (2016). Synthesis of penta-2,4-dien-1-imines and 1,2-dihydropyridines by rhodium-catalyzed reaction of *N*-sulfonyl-1,2,3-triazoles with 2-(siloxy)furans. *Org. Lett.* 18: 6284–6287.

25 Makarov, A.S., Uchuskin, M.G., and Hashmi, A.S.K. (2019). Intramolecular azavinyl carbene-triggered rearrangement of furans. *Chem. Sci.* 10: 8583–8588.

26 Lei, X., Li, L., He, Y.-P., and Tang, Y. (2015). Rhodium(II)-catalyzed formal [3 + 2] cycloaddition of *N*-sulfonyl-1,2,3-triazoles with isoxazoles: entry to polysubstituted 3-aminopyrroles. *Org. Lett.* 17: 5224–5227.

27 Rostovskii, N.V., Ruvinskaya, J.O., Novikov, M.S. et al. (2017). Switchable synthesis of pyrroles and pyrazines via Rh(II)-catalyzed reaction of 1,2,3-triazoles with isoxazoles: experimental and DFT evidence for the 1,4-diazahexatriene intermediate. *J. Org. Chem.* 82: 256–268.

28 Tiuftiakov, N.Y., Strelnikova, J.O., Filippov, I.P. et al. (2021). Rhodium-catalyzed synthesis of 2-aroylpyrimidines via cascade heteropolyene rearrangement. *Org. Lett.* 23: 6998–7002.

29 Lei, X., Gao, M., and Tang, Y. (2016). Rh(II)-catalyzed transannulation of *N*-sulfonyl-1,2,3-triazoles with 2,1-benzisoxazoles or 1,2-benzisoxazoles. *Org. Lett.* 18: 4990–4993.

30 Strelnikova, J.O., Rostovskii, N.V., Starova, G.L. et al. (2018). Rh(II)-catalyzed transannulation of 1,2,4-oxadiazole derivatives with 1-sulfonyl-1,2,3-triazoles: regioselective synthesis of 5-sulfonamidoimidazoles. *J. Org. Chem.* 83: 11232–11244.

31 Strelnikova, J.O., Koronatov, A.N., Rostovskii, N.V. et al. (2021). Rhodium-catalyzed denitrogenative diazole–triazole coupling toward *aza*-bridged structures and imidazole-based chelating ligands. *Org. Lett.* 23: 4173–4178.

32 Koronatov, A.N., Afanaseva, K.K., Sakharov, P.A. et al. (2021). Rh(II)-catalyzed denitrogenative 1-sulfonyl-1,2,3-triazole-1-alkyl-1,2,3-triazole cross-coupling as a route to 3-sulfonamido-1*H*-pyrroles and 1,2,3-triazol-3-ium ylides. *Org. Chem. Front.* 8: 1474–1481.

33 Ma, X., Pan, S., Wang, H., and Chen, W. (2014). Rhodium-catalyzed transannulation of *N*-sulfonyl-1,2,3-triazoles and epoxides: regioselective synthesis of substituted 3,4-dihydro-2*H*-1,4-oxazines. *Org. Lett.* 16: 4554–4557.

34 Ko, Y.O., Jeon, H.J., Jung, D.J. et al. (2016). Rh(II)/Mg(OtBu)$_2$-catalyzed tandem one-pot synthesis of 1,4-oxazepines and 1,4-oxazines from *N*-sulfonyl-1,2,3-triazoles and glycidols. *Org. Lett.* 18: 6432–6435.

35 Guarnieri-Ibáñez, A., Medina, F., Besnard, C. et al. (2017). Diversity-oriented synthesis of heterocycles and macrocycles by controlled reactions of oxetanes with α-iminocarbenes. *Chem. Sci.* 8: 5713–5720.

36 Medina, F., Besnard, C., and Lacour, J. (2014). One-step synthesis of nitrogen-containing medium-sized rings via α-imino diazo intermediates. *Org. Lett.* 16: 3232–3235.

37 Pospech, J., Ferraccioli, R., Neumann, H., and Beller, M. (2015). Rhodium(II)-catalyzed annulation of azavinyl carbenes through ring-expansion of 1,3,5-trioxane: rapid access to nine-membered 1,3,5,7-trioxazonines. *Chem. Asian J.* 10: 2624–2630.

38 Wang, Y., Lei, X., and Tang, Y. (2015). Rh(II)-catalyzed cycloadditions of 1-tosyl 1,2,3-triazoles with 2*H*-azirines: switchable reactivity of rhazavinylcarbene as [2C]- or aza-[3C]-synthon. *Chem. Commun.* 51: 4507–4510.

39 Ryu, T., Baek, Y., and Lee, P.H. (2015). Synthesis of pyrazines from rhodium-catalyzed reaction of 2*H*-azirines with *N*-sulfonyl 1,2,3-triazoles. *J. Org. Chem.* 80: 2376–2383.

40 Zhao, Y.-Z., Yang, H.-B., Tang, X.-Y., and Shi, M. (2015). Rh^{II}-catalyzed [3 + 2] cycloaddition of 2*H*-azirines with *N*-sulfonyl-1,2,3-triazoles. *Chem. Eur. J.* 21: 3562–3566.

41 Ding, H., Hong, S., and Zhang, N. (2015). Rhodium(II)-catalyzed transannulation of 1-sulfonyl-1,2,3-triazoles with 2*H*-azirines: a new method to dihydropyrazines. *Tetrahedron Lett.* 56: 507–510.

42 Khaidarov, A.R., Rostovskii, N.V., Starova, G.L. et al. (2018). Synthesis of spirocyclic 3*H*-pyrrol-4-amines from 2*H*-azirines and 1-sulfonyl-1,2,3-triazoles. *Chem. Heterocycl. Compd.* 54: 946–950.

43 Khaidarov, A.R., Rostovskii, N.V., Zolotarev, A.A. et al. (2019). Synthesis of 1-(2-aminovinyl)-indoles and 1,3′-biindoles by reaction of 2,2-diaryl-substituted 2*H*-azirines with α-imino Rh(II) carbenoids. *J. Org. Chem.* 84: 3743–3753.

44 Dequina, H.J., Eshon, J., Schmid, S.C. et al. (2022). Re-evaluation of product outcomes in the Rh-catalyzed ring expansion of aziridines with *N*-sulfonyl-1,2,3-triazoles. *J. Org. Chem.* 87: 10902–10907.

45 Guarnieri-Ibáñez, A., de Aguirre, A., Besnard, C. et al. (2021). Regiodivergent synthesis of pyrazinoindolines vs. triazocines via α-imino carbenes addition to imidazolidines. *Chem. Sci.* 12: 1479–1485.

46 Bosmani, A., Guarnieri-Ibáñez, A., Goudedranche, S. et al. (2018). Polycyclic indoline-benzodiazepines by electrophilic additions of α-imino carbenes to Tröger bases. *Angew. Chem. Int. Ed.* 57: 7151–7155.

47 Ge, J., Wu, X., and Bao, X. (2019). Rhodium(II)-catalyzed annulation of *N*-sulfonyl-1,2,3-triazoles with 1,3,5-triazinanes to produce octahydro-1*H*-purine derivatives: a combined experimental and computational study. *Chem. Commun.* 55: 6090–6093.

48 Lu, X.-L., Liu, Y.-T., Wang, Q.-X. et al. (2016). Straightforward regioselective construction of 3,4-dihydro-2*H*-1,4-thiazine by rhodium catalysed [3 + 3] cycloaddition of thiirane with 1-sulfonyl-1,2,3-triazole: a pronounced acid additive effect. *Org. Chem. Front.* 3: 725–729.

49 Jablasone, S.T., Ye, Z., Duan, S. et al. (2021). Synthesis of benzothiazonine by rhodium-catalyzed denitrogenative transannulation of 1-sulfonyl-1,2,3-triazole and thiochromone. *Org. Biomol. Chem.* 19: 5758–5761.

5

Insertion of Azavinyl Carbenes to C–H/*X*–H Bond

Yury N. Kotovshchikov and Gennadij V. Latyshev

Chemistry Department, M. V. Lomonosov Moscow State University, Moscow, Russian Federation

5.1 Abbreviations

AVC	Azavinyl carbene
bdpp	*rac*-2,4-Bis(diphenylphosphino)pentane
CuAAC	Cu-catalyzed azide-alkyne cycloaddition
CuTC	Copper(I) thiophene-2-carboxylate
EDG	Electron-donating group
EWG	Electron-withdrawing group
MBH	Morita–Baylis–Hillman
PPTS	Pyridinium *p*-toluenesulfonate
ST	1-Sulfonyl-1,2,3-triazole

Common rhodium carboxylates

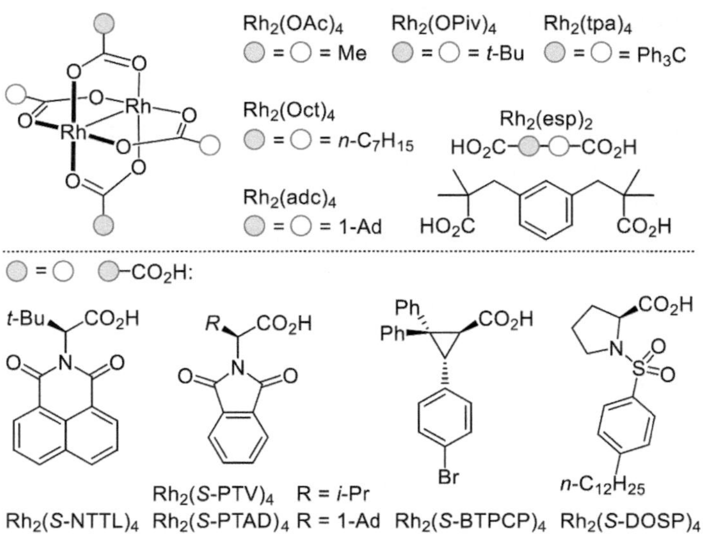

Denitrogenative Transformation of Nitrogen Heterocycles: Synthesis, Reactions and Applications, First Edition. Edited by Pazhamalai Anbarasan.

5.2 Introduction

The denitrogenative decomposition of diazo compounds upon action of transition metals is a common route to metal carbenes, which are highly reactive valuable intermediates. Among diverse transformations of these species, insertion to C—H and heteroatom—H bonds is well-studied and recognized as one of the most synthetically useful [1–11]. The emerged metal-stabilized AVCs, produced from 1,2,3-triazoles capable of undergoing electrocyclic ring-opening to isomeric α-diazoimines, significantly enriched the classic carbene chemistry. Apart from standard 1,1-insertion, these bifunctional reagents can afford the products of formal 1,3-insertion (Scheme 5.1). Moreover, the functionalization of two adjacent carbons was also achieved in a few rare examples. This chapter covers various C/*X*—H insertion reactions of AVCs, generated from STs, and further cascades triggered by these processes. The insertion reactions involving 5-iodotriazoles as diazo precursors are summarized in the last section.

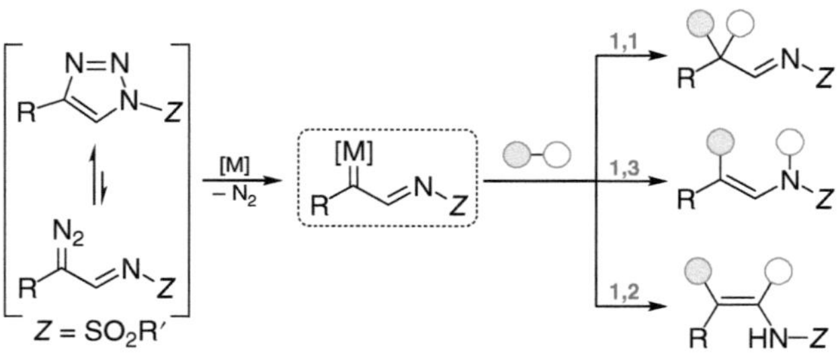

Scheme 5.1 Types of C/*X*—H insertion reactions of AVCs.

5.3 Insertion Reactions of *N*1-Sulfonyl-1,2,3-triazoles

5.3.1 C–H Insertion

The first example of intermolecular insertion of STs to C—H bonds in alkanes was demonstrated by Fokin's group in 2011 (Scheme 5.2) [12]. The decomposition of 4-aryl-STs in the presence of chiral Rh(II) catalyst (Rh_2(*S*-NTTL)$_4$ or Rh_2(*S*-PTAD)$_4$) and *ca*. 20-fold excess of cycloalkane with subsequent reduction of *N*-sulfonylimine

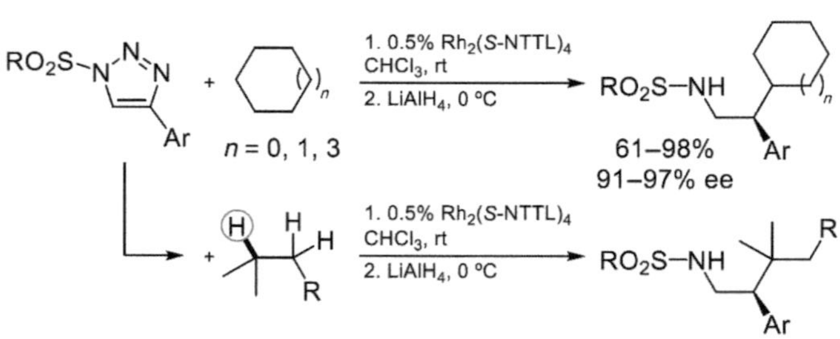

Scheme 5.2 Insertion of Rh-AVCs to nonactivated aliphatic C—H bonds.

affords moderate to high yield of β-aryl-β-cycloalkylamines with excellent enantioselectivity (up to 97% ee). The unfavorable triazole-diazoimine equilibrium in STs allows to maintain low concentration of Rh-AVCs in the reaction mixture, avoiding concurrent carbene dimerization. The reaction tolerates both electron-withdrawing and electron-donating substituents, though the yields are slightly lower with latter due to increased rate of decomposition of parent STs. Notably, the approach can be extended to alkanes with various types of C—H bonds, where Rh-AVCs demonstrate remarkably high selectivity (up to 5 : 1) of insertion into 3° C—H bonds over 2° while maintaining high enantioselectivity.

Intramolecular transannular proximity-promoted insertion into 2° C—H of cyclooctyne-derived STs was reported both in racemic and asymmetric variants (Scheme 5.3) [13]. $Rh_2(OAc)_4$-catalyzed reaction readily takes place in toluene at 90 °C, affording [3.3.0]-bicyclic scaffold with complete diastereocontrol in mediocre to good yields. The solvent plays a crucial role in achieving good results in the enantioselective process, both in terms of yield and enantioselectivity. The optimal conditions comprise heating ST at a slightly lower temperature (50 °C) in C_6F_6 in the presence of 5% $Rh_2(S\text{-NTTL})_4$.

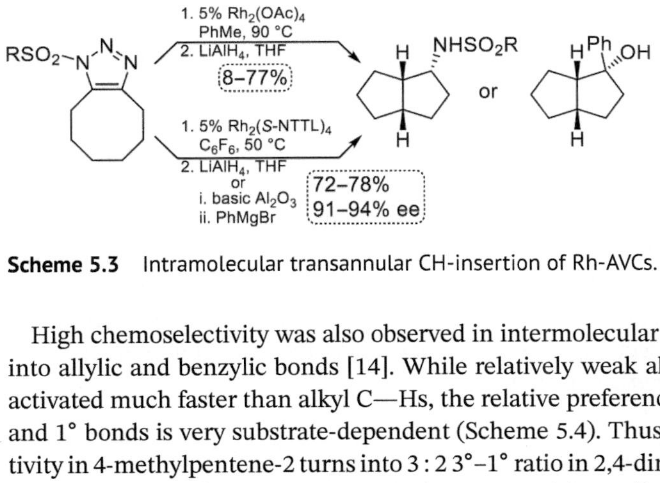

Scheme 5.3 Intramolecular transannular CH-insertion of Rh-AVCs.

High chemoselectivity was also observed in intermolecular insertion of Rh-AVCs into allylic and benzylic bonds [14]. While relatively weak allylic C—H bonds are activated much faster than alkyl C—Hs, the relative preference of 3° versus both 2° and 1° bonds is very substrate-dependent (Scheme 5.4). Thus, high 3° chemoselectivity in 4-methylpentene-2 turns into 3 : 2 3°–1° ratio in 2,4-dimethylpentene-2, and, finally, into high 1° preference in α-terpinene. Notwithstanding, the enantioselectivity of CH-insertion remains very high (94–96% ee) even for activation of 1° C—H.

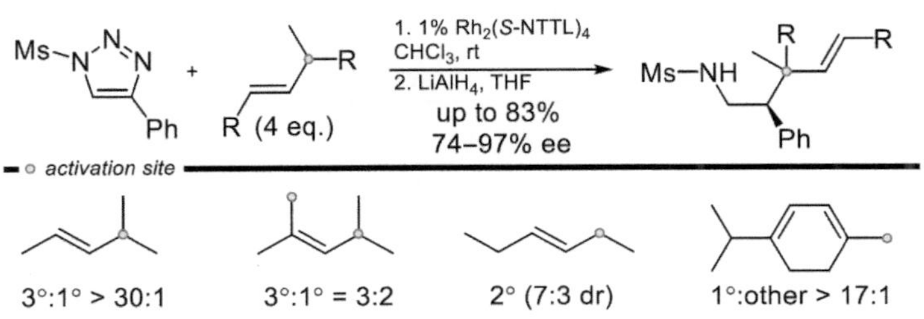

Scheme 5.4 Insertion of Rh-AVCs to allylic C–H bonds.

The same catalyst can activate benzylic C—H bonds, although with somewhat inferior enantioselectivity and much lower yields (Scheme 5.5). The preference for 3° over 1° benzylic position is moderate (3°: 1° = 4 : 1); thus, the activation of 4-methylanisole is possible. The absence of *p*-substituent in phenyl ring triggers a concurrent electrophilic attack of Rh-carbenoid in this position, affording formal (3+2)-cycloaddition of AVC to benzene.

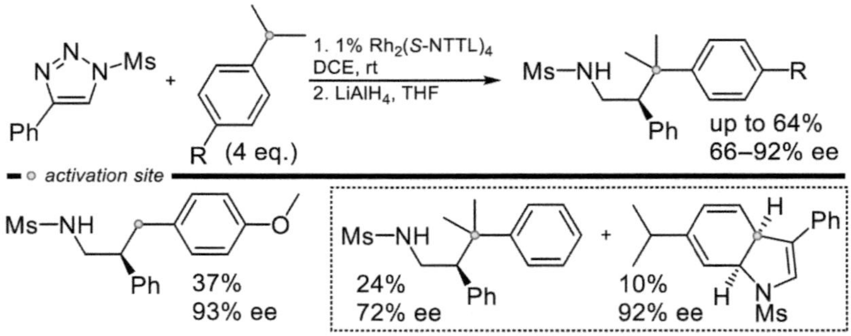

Scheme 5.5 Insertion of Rh-AVCs to benzylic C—H bonds.

The successful activation of 3° C—H in 4-methylpentene-2 allows for a short enantioselective route to complex β-arylpyrrolidines (Scheme 5.6, R = Me) [15]. The 4-step procedure, including CH-insertion, imine reduction, ozonolysis, and reductive amination, does not require chromatographic purification of the intermediates. The overall yield with various 4-aryl-STs is fair, but the enantioselectivity is pretty good (74–96% ee). The intermediate hemiaminal can be trapped with other nucleophiles such as allyltrimethylsilane and *N*-methylindole with varying levels of diastereoselectivity. The preferential insertion into 2° C—H over 1° in (*E*)-pentene-2 was also explored (Scheme 5.6, R = H), though the whole approach was complicated by modest diastereoselectivity of insertion, leading to the formation of inseparable mixtures of stereoisomeric pyrrolidines.

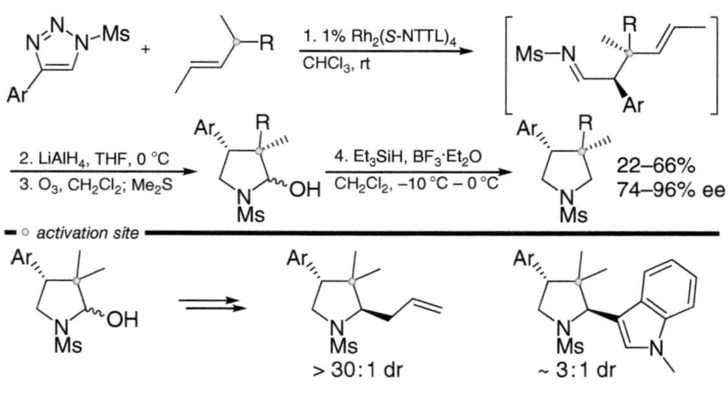

Scheme 5.6 Approach to β-arylpyrrolidines starting from insertion of Rh-AVCs to allylic C—H bonds.

The presence of TBSO-group in an allylic position allows for unprecedented regio- and stereocontrol in Rh-catalyzed CH-activation. Silyloxy group in allyloxysilyl ethers directs Rh catalyst to the "distal" 2° or 3° allylic position (Scheme 5.7) [16]. The diastereoselectivity for 2° allyl ethers (R′ = H) strongly depends on the configuration of allyl ether, with (*E*)-alkenes affording *ca.* 4 : 1 diastereomeric mixtures, while (*Z*)-alkene leading exclusively to single diastereomer. Similarly, "distal" C—H activation was observed for benzylic TBS-ethers.

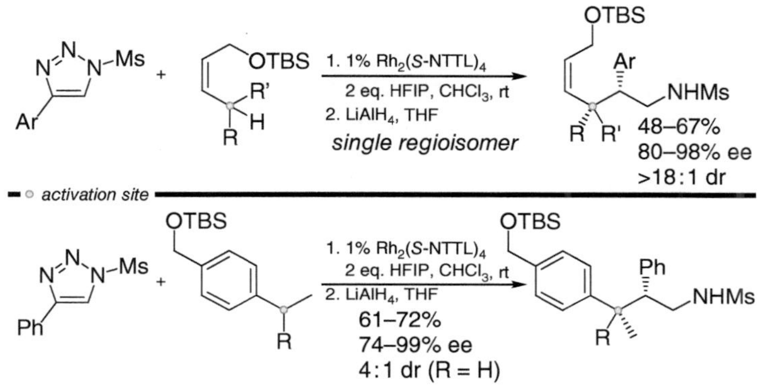

Scheme 5.7 Insertion of Rh-AVCs to distal allylic/benzylic C—H bonds.

The balance of steric and electronic factors is known to deliver carbene species to β-carbon of organosilanes [17–19]. Thus, silacyclobutane derivatives deliver β-C—H insertion products when treated with 4-aryl-STs in the presence of Rh_2(*S*-NTTL)$_4$ with moderate to good yields and enantioselectivities (Scheme 5.8) [20]. The extension of the procedure to both larger silacycloalkanes and aliphatic organosilanes is possible, though the formation of diastereomers in a typical 3–5 : 1 ratio complicates the course of the reaction. Nevertheless, even the selective modification of 1° C—H of methyl group in both Et_4Si and *t*-$BuOSiEt_3$ is possible, with yields ranging from low to moderate.

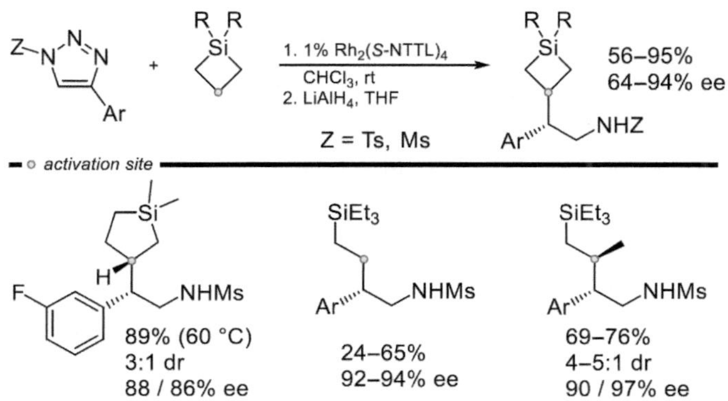

Scheme 5.8 Insertion of Rh-AVCs to β-C—H bonds in silanes.

While the intermolecular insertion of Rh-AVC into α-C—H bonds of ethers is unusual [21–23], the intramolecular variant provides a convenient route to various furan derivatives. Thus, benzofuran imines were obtained in the insertion into 2° α-C—H bond of aryl(alkyl)ethers with concomitant *in situ* Cu-catalyzed oxidation by O_2 (Scheme 5.9) [24]. The catalytic system comprising $Rh_2(Oct)_4$ and CuTC displayed the optimal performance, affording moderate to good yields of the products. Similar protocol employing the increased amount of $Rh_2(Oct)_4$ and separate oxidation step was reported later [25]. The intermediate 3-alkylidenebenzofurans can be isolated in moderate to high yields (61–91%) in the reaction catalyzed by Rh_2(*S*-PTV)$_4$, which was claimed to perform better, than $Rh_2(Oct)_4$ [26]. The products can be transformed either to 3-aminoalkylbenzofurans by the action of H_2/Pd/C, or dehydrogenated by the same catalyst (Pd/C) to benzofuran imines.

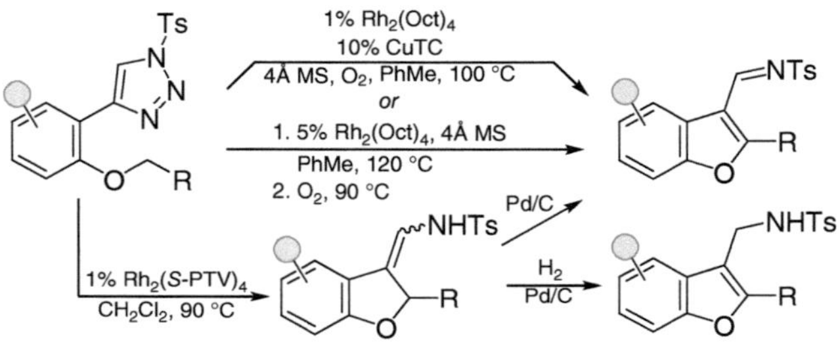

Scheme 5.9 Intramolecular insertion of Rh-AVCs to α-C–H bonds of ethers.

The similar chemistry was reported with *N*-benzylanilines, which undergo intramolecular insertion to benzylic C—H bond affording indole derivatives (Scheme 5.10). The reaction is quite sensitive to the substitution pattern on aniline nitrogen, with $Rh_2(OAc)_4$ providing good yields for *N*,*N*-dibenzylanilines [27] while $Rh_2(Oct)_4$ is more appropriate for *N*-acetylanilines [24, 25]. These conditions can be used even for activation of 1° C—H in methyl group. The addition of molecular sieves allows for a significant increase in the yield of indole due to the inhibition of concurrent carbenoid hydration.

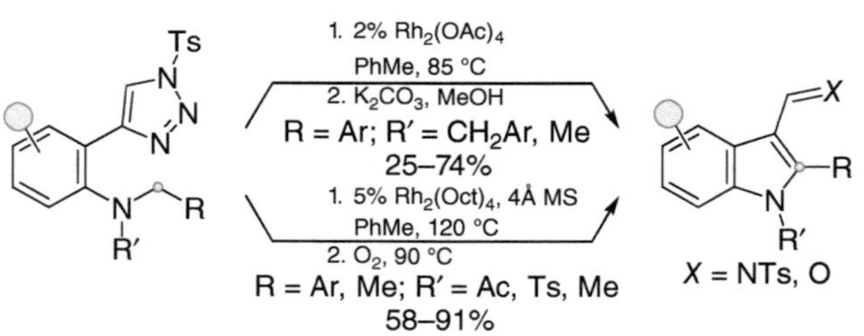

Scheme 5.10 Intramolecular insertion of Rh-AVCs to C–H bonds of *N*-benzylanilines.

The activation of a distal C—H bond in 4-alkyl substituted STs was also explored to provide access to 2,3-disubstituted pyrrolidines (Scheme 5.11) [28]. Moderate to

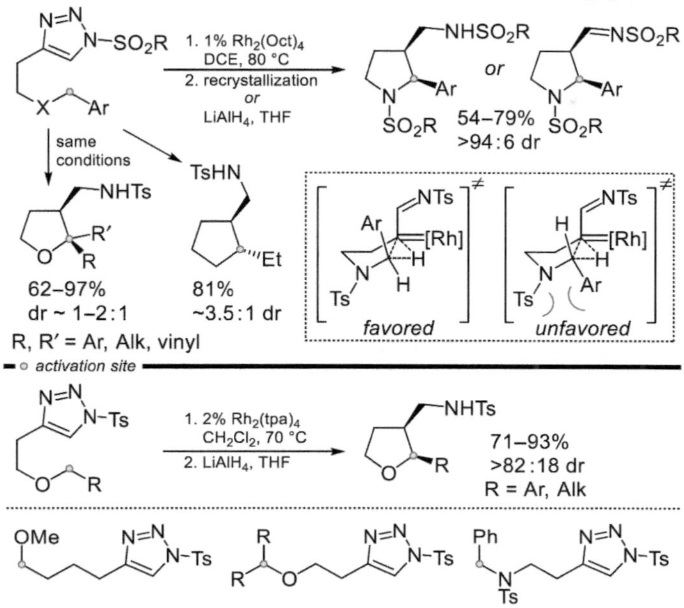

Scheme 5.11 Synthesis of saturated five-membered (hetero)cycles *via* intramolecular CH-insertion of Rh-AVCs.

good yields and high diastereoselectivities were observed for *N*-benzyl sulfonamides in $Rh_2(Oct)_4$-catalyzed reaction. The predominant formation of *cis*-pyrrolidine was explained in terms of the preferential axial orientation of aryl group in a chair-like transition state. While high yields were observed for the structurally related ethers, the diastereoselectivity of *cis*-tetrahydrofuran formation drops nearly to zero. Fortunately, the use of $Rh_2(tpa)_4$ [29] allows to restore high levels of diastereoselection. While the selective formation of *trans*-2,3-disubstituted tetrahydrofuran was not specifically targeted, the promising initial results (93%, *trans*:*cis* = 78 : 22) was obtained when the combination of $Rh_2(OAc)_4$ and 2-picoline have been used to generate Rh-AVC. The predominant formation of *trans*-isomer was also observed in $Rh_2(Oct)_4$-catalyzed insertion into C—H bond of 4-hexyl-ST [28].

The $Rh_2(OAc)_4$-catalyzed decomposition of 4-aryl-STs in the presence of an excess of *N*,*N*-dialkylanilines (4 eq.) affords 4-aminovinyl-anilines with yields ranging from low to good (Scheme 5.12) [30]. The reaction is believed to follow S_EAr-like mechanism. The substituents *ortho*- to amine break the conjugation of nitrogen

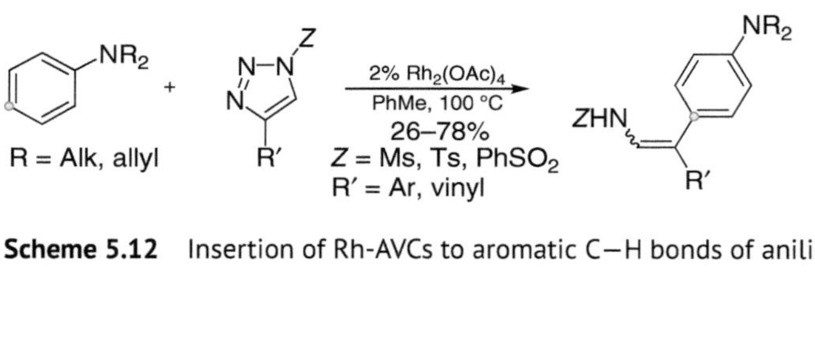

Scheme 5.12 Insertion of Rh-AVCs to aromatic C—H bonds of anilines.

lone pair with aromatic ring and, therefore, are not tolerated. The presence of the allyl group doesn't interfere with the C—H insertion.

The reaction of Rh-AVCs with azulenes proceeds as insertion in most nucleophilic first position affording 1-(aminovinyl) derivatives as a mixture of *E*:*Z* isomers with good to excellent yields (Scheme 5.13) [31]. The reaction tolerates various aryl and vinyl-STs and alkyl- and carboxy-substituted [32, 33] azulene derivatives. The decrease in the yield was observed for 4-alkyl-STs. The Rh-catalyzed aminovinylation procedure can even be applied to the aminovinyl-bearing azulene with moderate yield. The most probable mechanism includes S_EAr-type substitution in azulene by electrophilic Rh-AVC with concomitant intramolecular proton abstraction by imine nitrogen/Rh(II) dissociation, since neither cyclopropane, nor sulfonylimine intermediates were not observed by NMR.

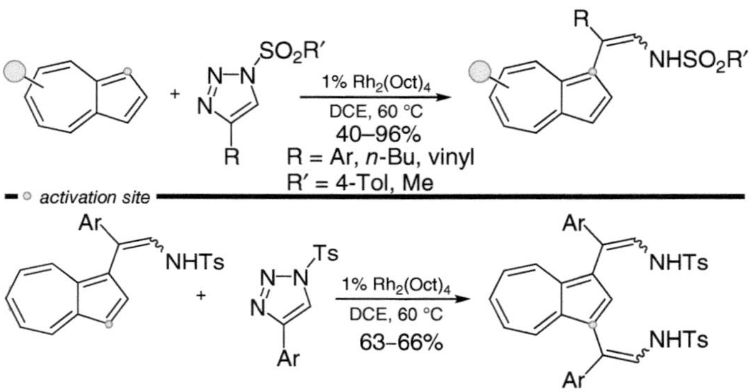

Scheme 5.13 Insertion of Rh-AVCs to C–H bonds of azulenes.

Less nucleophilic anisole derivatives, which are only barely active in $Rh_2(OAc)_4$-catalyzed process [30], undergo CH-insertion in the presence of Rh-AVC, generated from $Rh_2(esp)_2$ (Scheme 5.14) [34]. Good yields are achieved when reaction is conducted in arene acting both as a nucleophile and a solvent. The activation of *p*-position was also demonstrated in intramolecular variant.

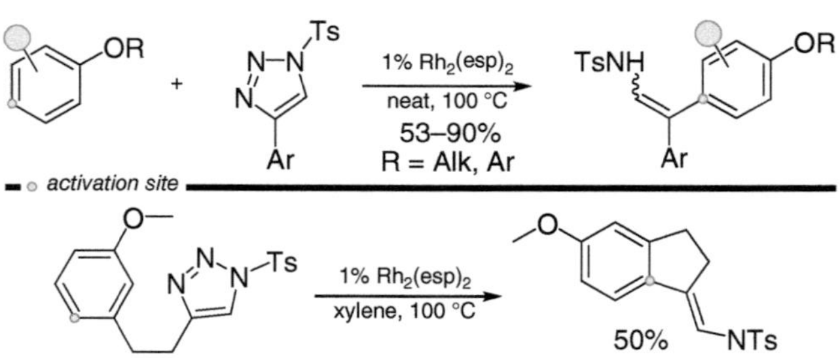

Scheme 5.14 Insertion of Rh-AVCs to aromatic C–H bonds of aryl ethers.

4-Acyl-STs, which generate much more electrophilic Rh-AVCs, are capable of inserting into C—H aromatic bonds of a more wide range of substrates

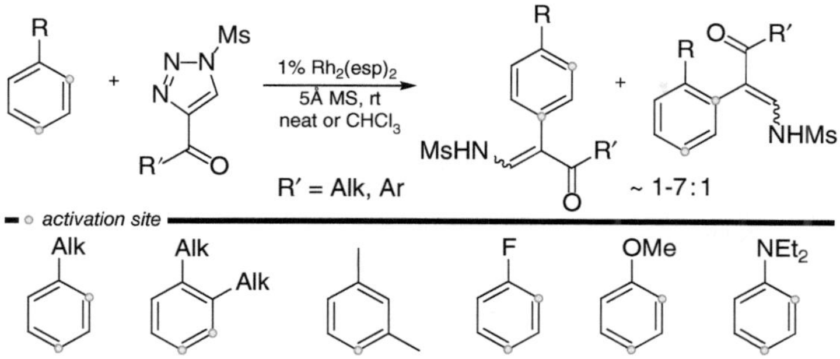

Scheme 5.15 Insertion of electrophilic Rh-AVCs to C—H bonds of arenes.

(Scheme 5.15) [35], including fluorobenzene and various alkylbenzenes. However, the regioselectivity of insertion is only moderate and the formation of *p*- and *o*-product mixtures in up to 7 : 1 ratio is usually observed. The ρ value of −3.59 and small inverse secondary KIE of leaving H ($k_H/k_D = 0.90$) are consistent with S_EAr-type mechanism.

Acceptor/acceptor type Rh-AVCs, generated from 4-acyl-STs, prefer to insert into aromatic C—H bond in the presence of benzylic C—H. Thus, no intermolecular activation of benzylic C—H in cumene was observed [35]. This preference holds true for intramolecular processes involving 4-acyl-STs [36]. The CH-insertion in STs containing equidistant aromatic and benzylic C—H bonds is directed toward the activation of aromatic bonds for acyl-ST and toward benzylic insertion for alkyl-ST (Scheme 5.16). However, this preference can be overridden by the size of the forming cycle, affording clean benzylic activation for acyl-ST, bearing 4-carbon linker between phenyl and triazole moieties.

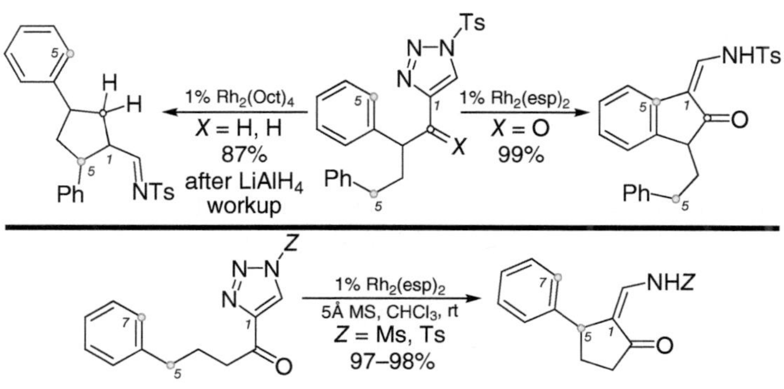

Scheme 5.16 Chemoselective intramolecular insertion of Rh-AVCs to aromatic or benzylic C—H bonds.

Overall, 4-acyl-STs are capable of generating very electrophilic Rh-AVCs, which undergo intramolecular insertion into C—H bonds of electron-donating, electron-neutral, and even slightly electron-withdrawing aromatics with the formation of indanones and tetralones in good to excellent yields (Scheme 5.17)

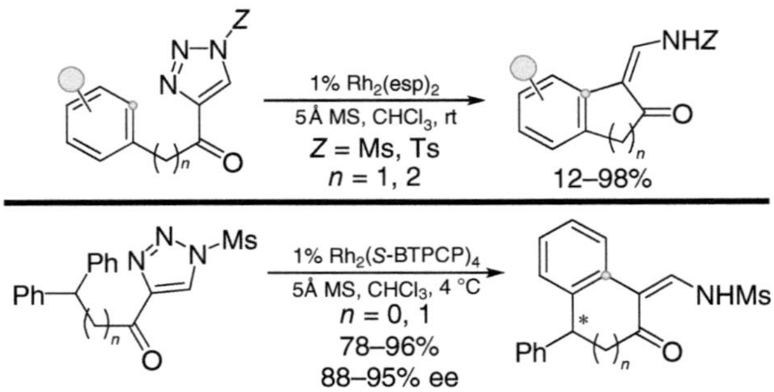

Scheme 5.17 Synthesis of indanones and tetralones *via* intramolecular CH-insertion of Rh-AVCs.

[36]. The enantiodiscrimination of bis(phenyl) derivatives is also possible. The reaction is not very regioselective (*ca.* 2 : 1 *p*:*o* ratio) for *m*-substituted aromatics and affords low yields of indanones upon reaction with arenes, bearing strong electron acceptors (CF_3, NO_2).

The intramolecular CH-insertion of Rh-AVC in *o*-CH of anilines allows access to 3-substituted indoles (Scheme 5.18) [37]. The yields are generally good, provided that the aniline nitrogen is protected with an alkyl group. The oxidation of intermediately formed dihydroindole moiety is attributed to β-hydride elimination in anionic Rh-complex.

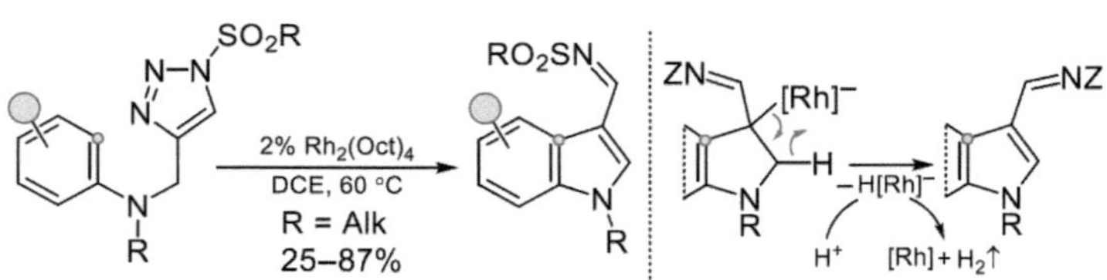

Scheme 5.18 Approach to indoles *via* intramolecular insertion of Rh-AVCs to aniline C–H bonds.

The detailed investigation of the process (Scheme 5.19) [38] revealed rather low selectivity of *o*-/*p*-attack on 3-substituted benzene moiety. More intriguing, the generated Rh-AVC prefers to activate more electron-deficient aromatic rings of *N*,*N*-diarylamines. The protecting benzyl group on *N* is tolerated, while the presence of allyl moiety affords the cyclopropanation products predominantly, unless very strong EWG is attached to a double bond.

While the presence of electron-deficient protecting groups on nitrogen usually aborts CH-insertion process, Shi et al. have managed to control hydride shift, leading to the decomposition of Rh-AVC (Scheme 5.20) [39]. The addition of LiCl (5 eq.) to $Rh_2(OPiv)_4$–catalyzed reaction allows to obtain dihydroindole derivatives in moderate yields.

Amazingly, the change of heteroatom from N to O in linker, tying aromatic ring and ST moiety, leads to the unexpected 2-aminobenzofuran derivatives

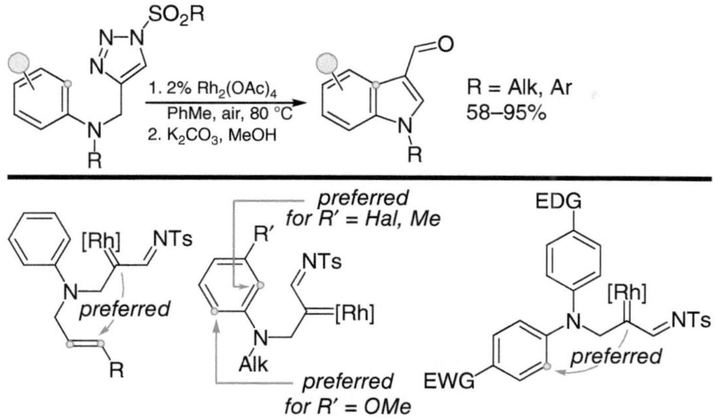

Scheme 5.19 Chemoselectivity of indole preparation *via* intramolecular insertion of Rh-AVCs.

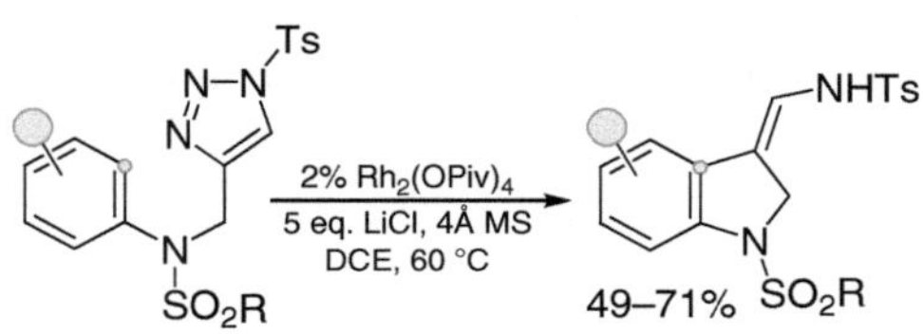

Scheme 5.20 Synthesis of dihydroindoles *via* intramolecular insertion of Rh-AVCs.

(Scheme 5.21) [39]. The fragmentation of the alkylrhodium complex with concomitant 1,2-addition of phenol to imine is believed to afford the rearranged product. The heating of methylenedihydrofuran at high temperatures yields more stable benzofuran.

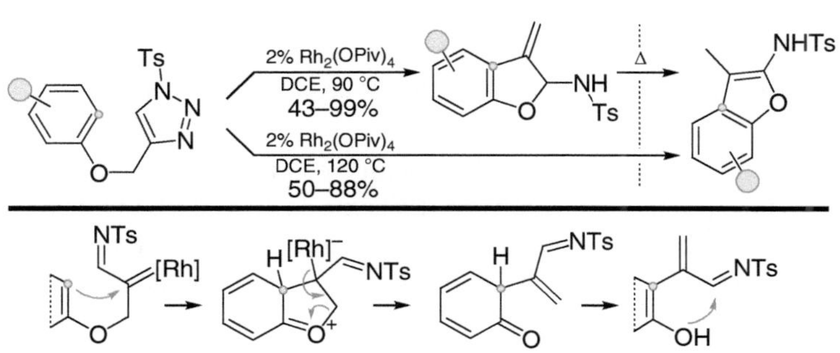

Scheme 5.21 Approach to 2-aminobenzofurans *via* cascade triggered by intramolecular attack of Rh-AVC on aryl ether.

The intramolecular CH-insertion in diarylamine-derived STs was employed in the development of the approach to acridones (Scheme 5.22) [40]. The activation of C—H bond proceeds in an inert atmosphere, while the base-mediated oxidation of enamine is conducted in air. A significant decrease in the yield

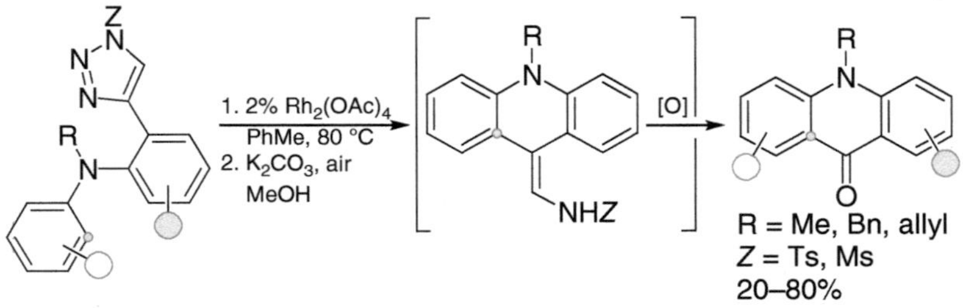

Scheme 5.22 Approach to acridones *via* intramolecular CH-insertion of Rh-AVCs.

is observed for allyl-substituted substrates, as well as in the presence of strongly-electron-withdrawing substituents in reacting aniline ring.

An efficient synthetic approach to fluorenes and related derivatives was developed utilizing intramolecular CH-insertion in biaryl-tethered STs (Scheme 5.23) [41]. The annulation can be also applied for the preparation of fused furan and thiophene derivatives. The observed regioselectivity for *m*-substituted arenes is unusually low, though it can be substantially increased by using $Rh_2(S\text{-PTAD})_4$ as a catalyst.

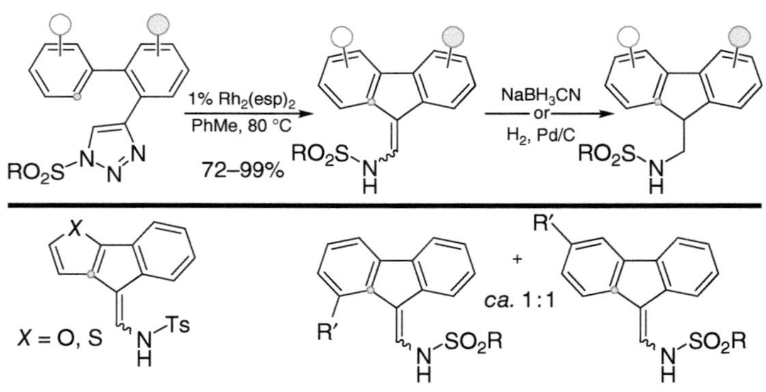

Scheme 5.23 Approach to fluorenes *via* intramolecular CH-insertion of Rh-AVCs.

A similar strategy for the formation of five-membered ring was evaluated on styryl-substituted STs to produce benzofulvenes (Scheme 5.24) [42]. The success of the transformation is critically dependent on the structure of the substrate, since neither disubstituted alkenes, nor ST bridged to phenyl by 2-carbon saturated chain, does not afford cyclic products.

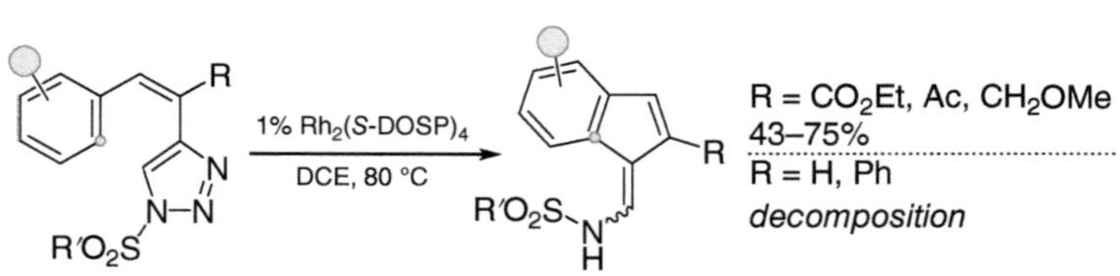

Scheme 5.24 Approach to benzofulvenes *via* intramolecular CH-insertion of Rh-AVCs.

The non-catalyzed insertion of AVC, generated from thermally-labile 4-alkoxy-STs, into C—H bonds of indole and pyrrole was demonstrated by Davies and Alford (Scheme 5.25) [43]. The intermediate vinyl ethers were transformed to α-(sulfonylamino)ketones by mild acidic hydrolysis. The transformation was carried out as one-pot procedure starting from alkoxyalkynes and sulfonyl azide. The mechanism is believed to include the intermediate formation of cyclopropanated heterocycle with concomitant cleavage of donor-acceptor cyclopropane. The formal CH-insertion proceeded regioselectively and furnished good to high yields for 3-unsubstituted indoles, and both 2- and 3-unsubstituted pyrroles, while the reaction with 3-methylindole derivatives afforded (3+2)-annulation product.

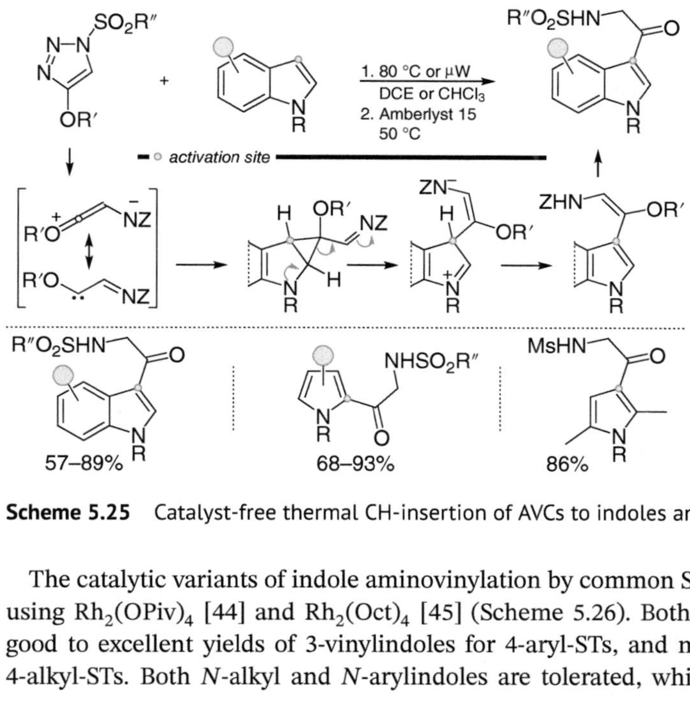

Scheme 5.25 Catalyst-free thermal CH-insertion of AVCs to indoles and pyrroles.

The catalytic variants of indole aminovinylation by common STs were developed using $Rh_2(OPiv)_4$ [44] and $Rh_2(Oct)_4$ [45] (Scheme 5.26). Both procedures afford good to excellent yields of 3-vinylindoles for 4-aryl-STs, and moderate yields for 4-alkyl-STs. Both *N*-alkyl and *N*-arylindoles are tolerated, while the presence of

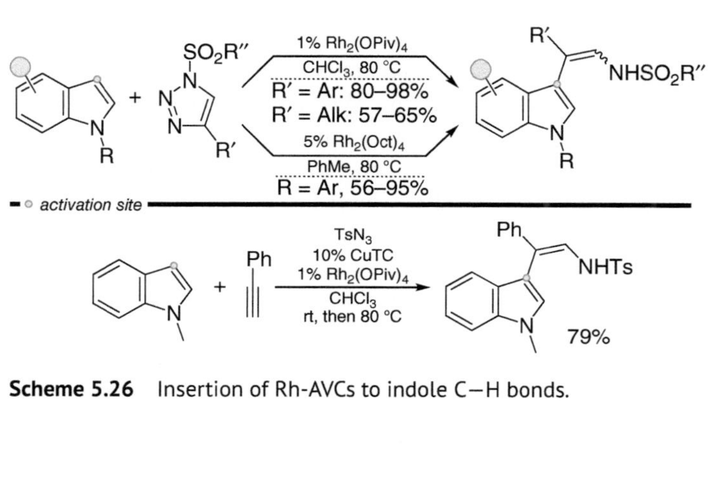

Scheme 5.26 Insertion of Rh-AVCs to indole C—H bonds.

electron-withdrawing groups on indole nitrogen reduces the yields by 20–30%. A similar reaction was also achieved for 2-methoxythiophene, which afforded the product of formal CH-insertion into the fifth position in good yield (74%) [45]. The aminovinylation is rather robust and can be carried out in one-pot manner starting from terminal alkyne, indole, and tosyl azide.

The potential of the Rh-catalyzed aminovinylation of indoles as a route to tryptamine derivatives was explored in detail by Anbarasan's group [46]. The efficient CH-insertion of AVCs was achieved using simple $Rh_2(OAc)_4$ as a catalyst (Scheme 5.27). The enamine reduction was performed in one-pot manner by the addition of $NaBH_3CN$ to the crude vinylindole and continuation of the heating. The method provides moderate to good yields of various β-aminoethylated indoles, including 7-azaindole. Interestingly, 3-methylindole is also reactive and gives the product of C2—H insertion in 47% yield. The approach is applicable to the functionalization of some pyrroles affording C2-substituted products selectively for *N*-benzylpyrroles and mixtures of C2/C3-regioisomers for *N*-aryl and *N*-Boc analogs.

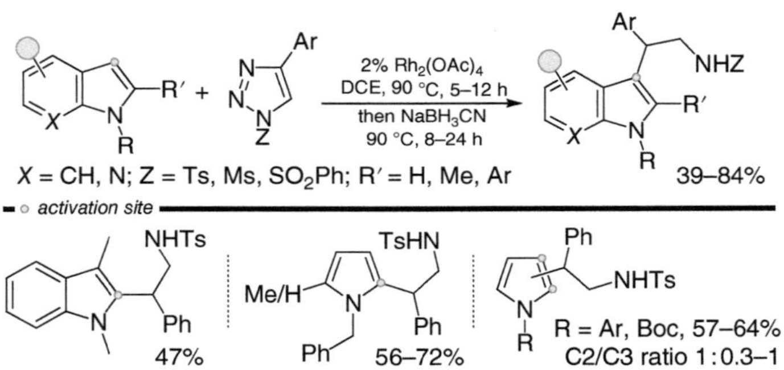

Scheme 5.27 Approach to tryptamine analogs by the reaction of Rh-AVCs with indoles and pyrroles.

The regioselectivity of the AVC CH-insertion can be switched by the use of directing group-assisted Rh(III) catalysis [47]. Thus, the attachment of 2-pyrimidyl or 2-pyridyl substituent to the indole nitrogen enabled the efficient functionalization of C2—H bond using 4-acyl-STs in the presence of $[Cp^*RhCl_2]_2$, $AgSbF_6$, and pivalic acid (Scheme 5.28). In contrast, the common catalytic system comprising $Rh_2(OAc)_4$ provided the standard attack of the AVC to the more nucleophilic C3 position. The same strategy was applied to *N*-pyridylisoquinolones affording the corresponding 3-vinylated heterocycles in moderate to high yields. Moreover, this approach was extended to the *o*-directed CH-insertion of the Rh-AVCs to ethyl benzimidate. The addition of $Zn(OAc)_2$ promoted the tandem heterocyclization leading to isoquinolines.

The C2-functionalization of *N*-(2-pyrimidyl)indoles by the reaction with fused STs using chiral spirocycle-based Rh(III) catalyst allowed to achieve atroposelective transformation with high ee values up to 97% (Scheme 5.29) [48]. The excess of AgOAc was employed for the oxidation of the insertion product to the naphthalene.

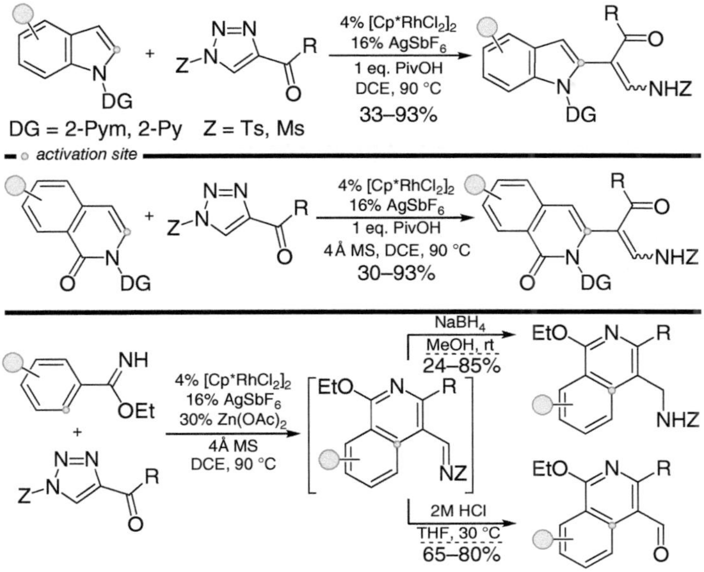

Scheme 5.28 Directing group-assisted Rh(III) catalysis for the regioselective CH-insertion of AVCs.

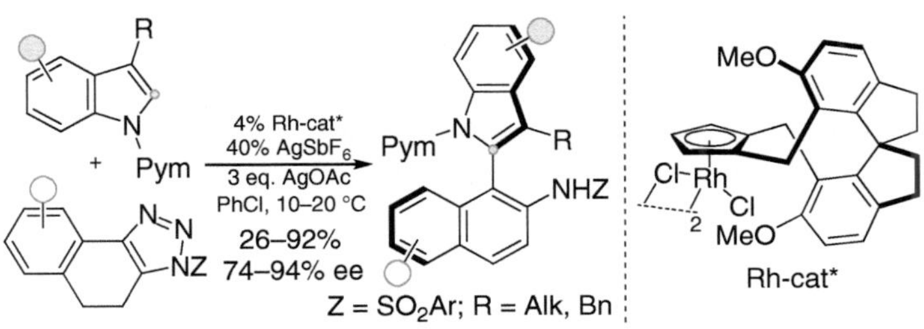

Scheme 5.29 Atroposelective Rh(III)-catalyzed CH-insertion of AVCs to indoles.

Apart from indoles and pyrroles, other electron-rich azaheterocycles can also undergo the analogous CH-insertion in the reaction with the Rh-AVCs (Scheme 5.30). Thus, indolizines can be readily transformed into 3-vinylated derivatives in generally good yields [49]. *N*-ethyl and *N*-phenylcarbazoles react smoothly with 4-aryl-STs affording 3-substituted derivatives in moderate to good yields [50]. The presence of electron-withdrawing groups as well as unprotected NH on carbazole is not tolerated. Electron-rich 2-aminopyridines and 2-aminoisoquinolines afford aminovinylated products when treated with $Rh_2(OAc)_4$ and 4-aryl-STs [51]. The reaction doesn't proceed with 3- and 4-aminopyridines and 2-aminopyridines with 2° NH.

The developed protocols for the attachment of β-aminovinyl fragment to the indole core were utilized for the annulation of a new ring *via* divergent reaction pathways. Thus, performing this transformation with β-indol-2-ylacrylates enabled further intramolecular *aza*-Michael addition which was initiated by heating the

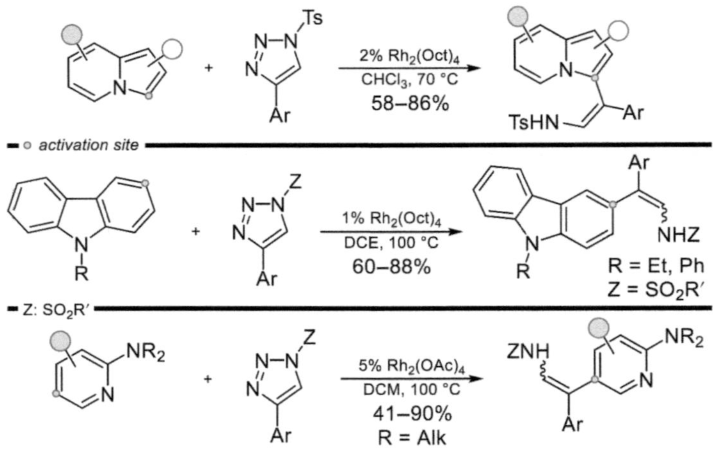

Scheme 5.30 Insertion of Rh-AVCs to C—H bonds of electron-rich azaheterocycles.

crude insertion product with an amine base (Scheme 5.31) [52]. Among the Michael acceptors vinylketones were also found to be appropriate substrates, whereas highly electron-deficient olefins bearing nitro and cyano groups didn't undergo the CH-insertion. The reaction with α-substituted ketones proceeded in a diastereoselective fashion, albeit giving the target heterocycles in lower yields (44–51%). The same annulation strategy combining CH-insertion of the Rh-AVC with *aza*-Michael addition was applied to the reaction with 3-vinylanilines producing 1,2-dihydroisoquinolines in 42–66% yields (Scheme 5.31).

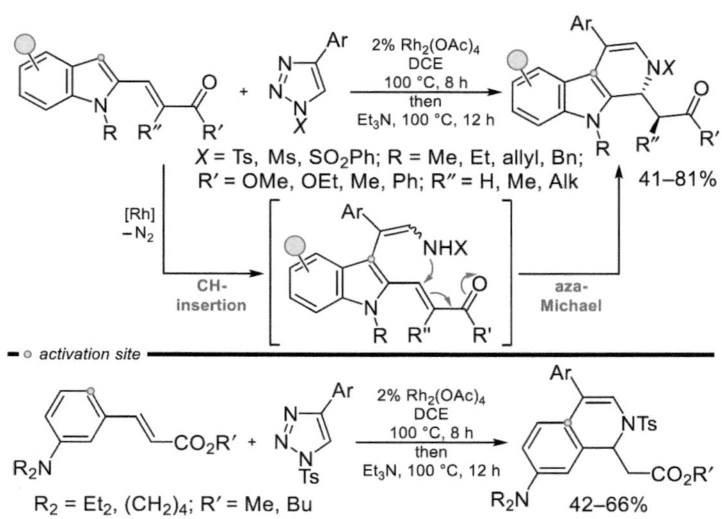

Scheme 5.31 Heteroannulation *via* Rh-AVC CH-insertion followed by intramolecular *aza*-Michael addition.

The asymmetric synthesis of 1,2-dihydro-β-carbolines was developed by performing *aza*-Michael addition in the presence of the chiral squaramide derived from cinchonidine (Scheme 5.32) [53]. The activation of both enone and enamide through hydrogen bonding with the bifunctional organocatalyst is believed to provide the achieved high level of enantiocontrol. The annulation was carried out efficiently in one-pot manner with the initial Rh-catalyzed β-aminovinylation of the indole followed by the organocatalyzed cyclization.

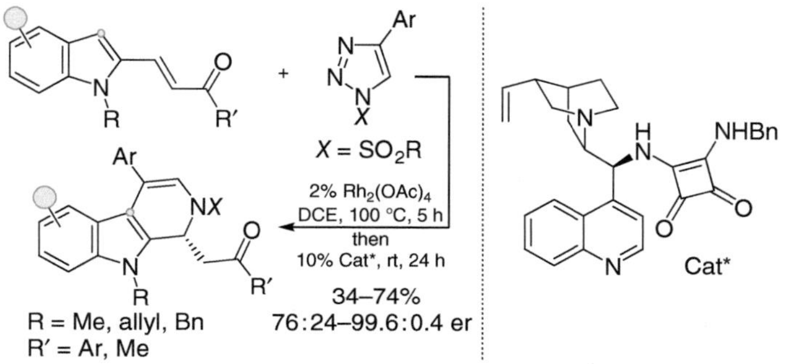

Scheme 5.32 Asymmetric approach to 1,2-dihydro-β-carbolines based on organocatalytic cyclization of Rh-AVC CH-insertion products.

2,3-Divinylindoles prepared by the Rh-AVC CH-insertion were involved in another type of annulation giving access to 3-aminocarbazoles (Scheme 5.33) [54]. Although heating the starting materials did not evoke the intended 6π-electrocyclization, treating them with DDQ allowed to achieve this transformation accompanied by oxidative aromatization. The procedure can be performed either in one-pot without isolation of the enamide or as two separate steps affording target carbazoles in comparable yields *ca.* 40–60%. The method is applicable to 2-vinylindoles bearing either electron-withdrawing (ketone, ester) or electron-neutral (Me, Ph) groups attached to the alkene.

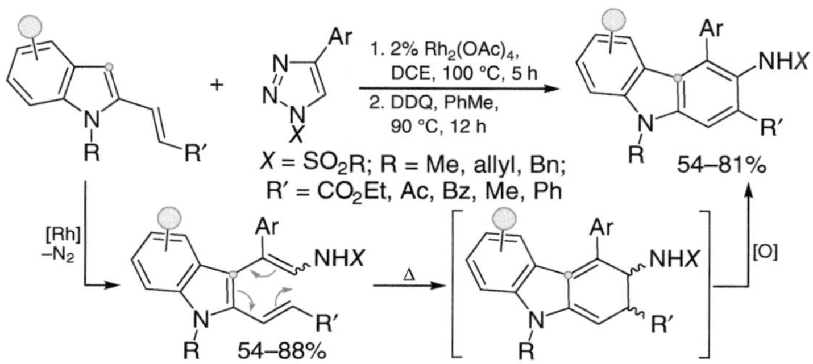

Scheme 5.33 Approach to 3-aminocarbazoles *via* 6π-electrocyclization of Rh-AVC CH-insertion products.

Li and co-workers utilized the regioselective CH-insertion of the Rh-AVCs to the position 3 of indoles for the assembly of azepino[5,4,3-*cd*]indole skeleton (Scheme 5.34) [55]. The one-pot construction of seven-membered ring was achieved *via* the intramolecular N-alkylation of the obtained enamine by a pre-installed allyl alcohol. The delayed addition of PPTS as an acidic additive promoted the cyclization process. The method delivers good to excellent yields for NH- as well as *N*-alkyl indoles, while EWGs at *N*1 suppress the reaction completely. The proposed strategy was applied as a key step in the total synthesis of alkaloid (±)-aurantioclavine (Scheme 5.34).

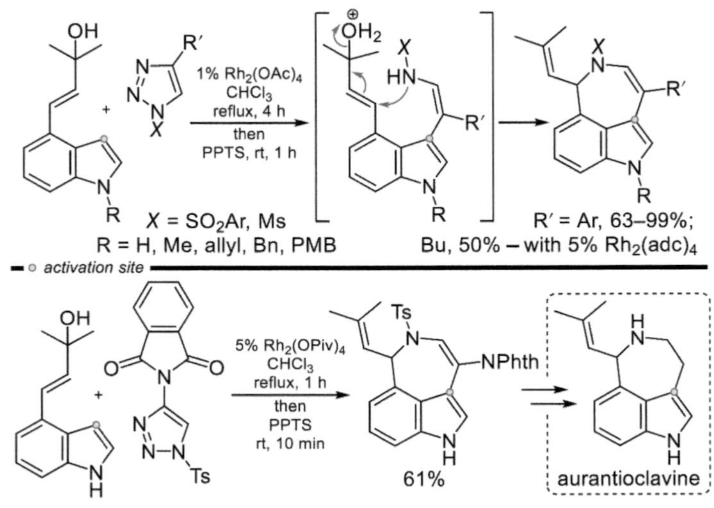

Scheme 5.34 One-pot conversion of Rh-AVC CH-insertion products to azepino[5,4,3-*cd*]indoles.

The intramolecular insertion of the Rh-AVCs to a C—H bond of indole or pyrrole ring was extensively investigated by several research groups in order to elaborate efficient routes to various fused heterocyclic systems, including alkaloid-like scaffolds. In 2014, Tang and Shi et al. studied the reactivity of the STs tethered to the N atom of indoles and pyrroles [56]. The denitrogenative annulation of the seven-membered ring *via* selective C2—H insertion proceeded smoothly in the presence of almost any Rh(II) carboxylates in DCE with $Rh_2(Oct)_4$ being slightly more active than the others (Scheme 5.35). The mild reduction of the obtained enamines afforded target fused diazepanes in good overall yield.

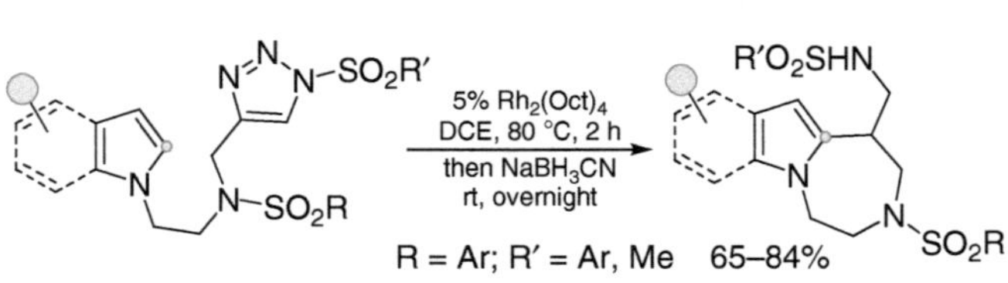

Scheme 5.35 Diazepane annulation *via* intramolecular Rh-AVC CH-insertion.

The same strategy was applied to the STs tethered to the indole core at the position 2 by means of either sulfonamide [57] or ether [58] linker providing access to tetrahydro-β-carbolines or their *oxa*-analogs, respectively (Scheme 5.36). In this case, the annulation of the six-membered cycle occurred through the formal insertion of the Rh-AVC to the C3—H bond. The synthetic utility of the approach was illustrated by the formal total synthesis of oxopropaline G [57].

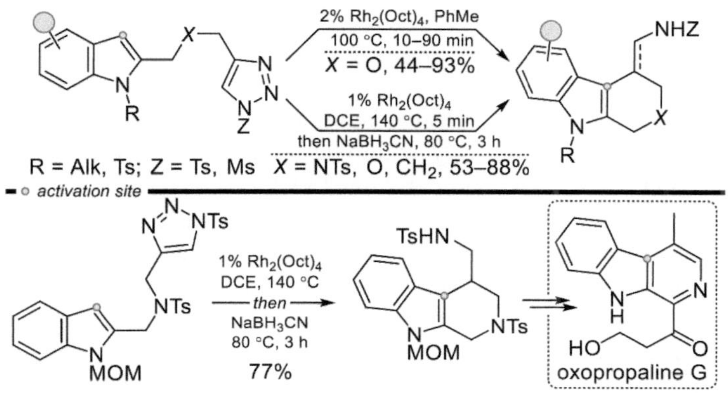

Scheme 5.36 Intramolecular Rh-AVC insertion to C3—H bond of indole.

Rh-catalyzed denitrogenation of STs tethered to C3 position of the indole evokes cyclization *via* competing pathways [59, 60]. Depending on the linker length and the nature of the substituent on the indole nitrogen, either (3+2)-cycloaddition or CH-insertion could occur. The insertion path is favored for NH-indoles undergoing annulations with the formation of six- or seven-membered rings (Scheme 5.37). The addition of 1 eq. $Cu(OAc)_2$ could increase the selectivity toward CH-insertion products albeit providing only rather low yields of the fused azepanes [59].

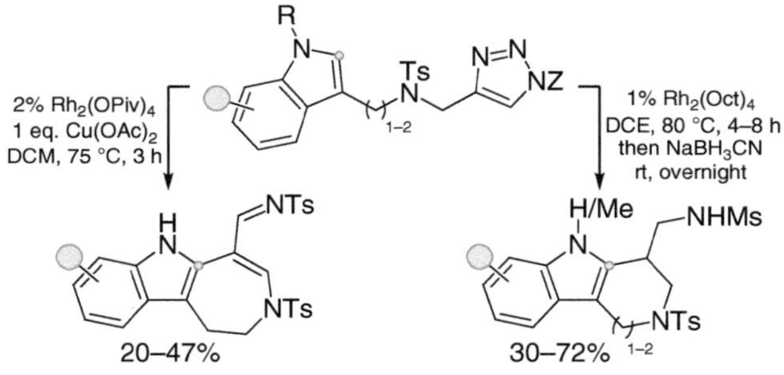

Scheme 5.37 Intramolecular Rh-AVC insertion to C2—H bond of indole.

5.3.2 O—H Insertion

The insertion of Rh-AVCs to an O—H bond was pioneered in 2012 by Murakami and co-workers using the reaction with water affording α-(sulfonylamino)ketones (Scheme 5.38) [61]. The observed transformation was rationalized by the initial

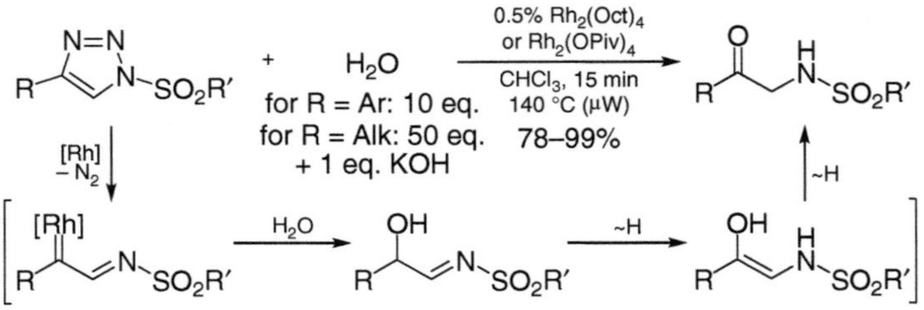

Scheme 5.38 Rh-catalyzed denitrogenative hydration of STs.

formation of *α*-hydroxyimine followed by imine-enamine and keto-enol tautomerizations. The hydration proceeded quickly within 15 minutes being conducted under microwave irradiation at a high temperature (140 °C). The use of KOH and the increased amounts of water (50 instead of 10 eq.) allowed to achieve high yields with alkyl-substituted STs by suppressing the competing β-hydride migration side process. In addition, the transformation was carried out efficiently using water as the only solvent in the presence of the nonionic amphiphile PTS, although a higher loading of the Rh-catalyst was required (2.5 instead of 0.5%).

The Rh-catalyzed denitrogenative hydration of STs was utilized by several research groups as a part of further domino processes [62, 63]. The obtained *α*-aminoketones were involved in the intramolecular Michael addition furnishing 2-aminoindanones in moderate to good yields (Scheme 5.39). Notably, the chemoselective cyclization with a C—C bond formation was observed, despite the presence of O and N nucleophilic centers. The *trans*-isomer was obtained as a major product, although the chromatographic purification often a bit lowered the diastereoselectivity due to an epimerization of the α-aminoketone.

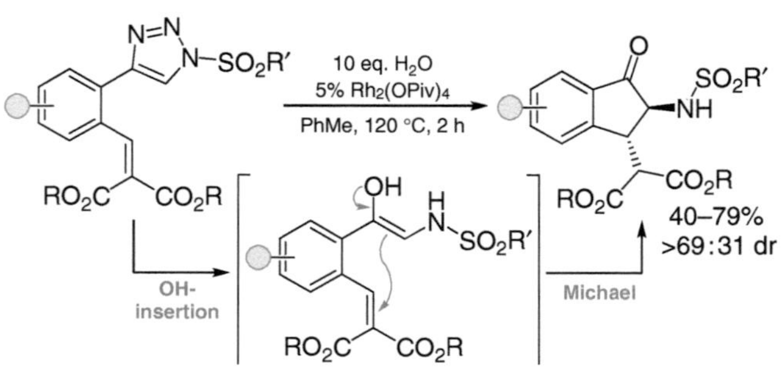

Scheme 5.39 Approach to indanones *via* Rh-AVC hydration followed by intramolecular Michael addition.

The reaction between STs, paraformaldehyde, and 2-substituted indoles afforded the coupling products in reasonable yields (Scheme 5.40) [63]. According to the putative mechanism, the α-aminoenol generated *via* hydration of the Rh-AVC is trapping by a vinyliminium cation intermediate produced by an electrophilic methylenation of the indole. The use of some other carbonyl compounds as well as an imine instead of the formaldehyde appeared inefficient.

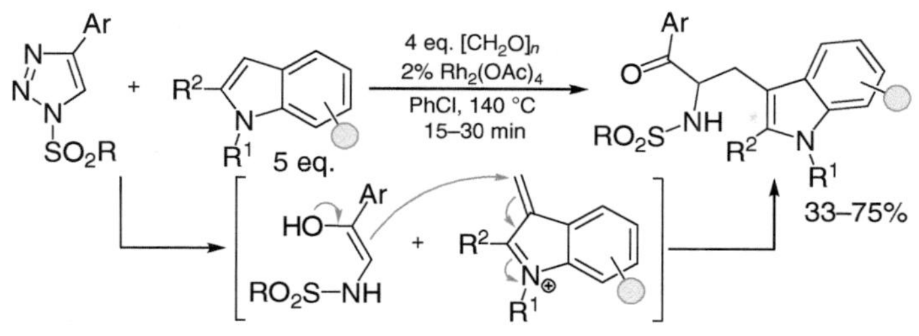

Scheme 5.40 Rh-catalyzed assembly of tryptamines from STs, paraformaldehyde, and indoles.

In 2014, Fokin et al. significantly extended the substrate scope of Rh-AVC OH-insertion by performing the reaction with carboxylic acids, alcohols, phenols, and silanols (Scheme 5.41) [64]. The mechanism involving ylide formation followed by intramolecular proton transfer was proposed to explain *Z*-selective formal 1,3-insertion furnishing various β-oxygenated enamines. All types of substrates reacted well under rather mild conditions upon heating at 75 °C. Whereas aromatic and some bulky carboxylic acids afforded typical β-acyloxy enamines, the other analogous products underwent a facile 1,4-acyl migration converting to *N*-acyl α-aminoketones. The insertion of AVCs to the O—H bond of propargyl alcohols induced a facile [3,3]-sigmatropic rearrangement and furnished *α*-allenylketones. Interestingly, the substrates bearing iodo- and chloroethynyl moieties also underwent this cascade transformation smoothly.

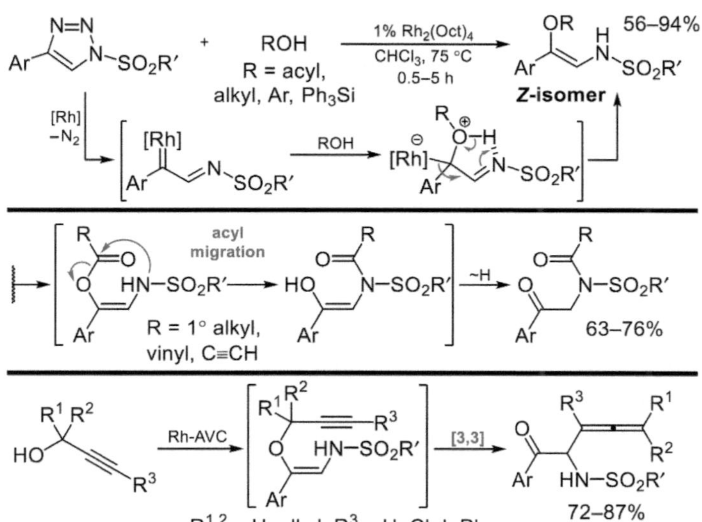

Scheme 5.41 Formal 1,3-insertion of Rh-AVCs to various O—H bonds.

(*Z*)-Silyloxy enamines obtained from the OH-insertion of the Rh-AVCs to silanols were efficiently involved in the $TiCl_4$-mediated Mukaiyama aldol reaction with aldehydes (Scheme 5.42) [65]. The condensation was conducted under mild conditions

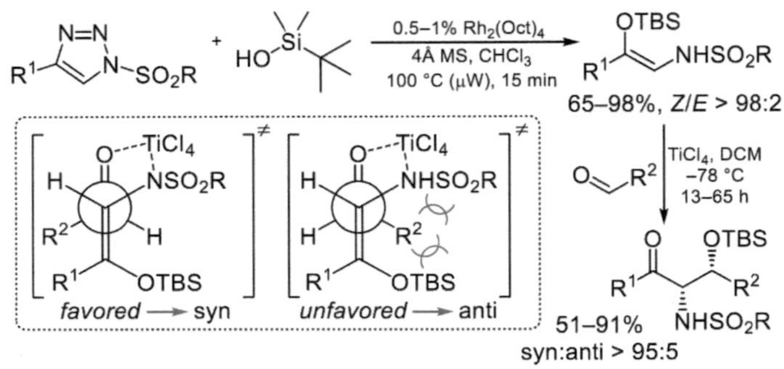

Scheme 5.42 Rh-AVC OH-insertion to silanols followed by Mukaiyama aldol reaction.

at −78 °C, delivering protected α-amino-β-oxyketones with excellent *syn*-selectivity. The stereochemical outcome was explained through the open transition state with Ti(IV) chelated by the carbonyl and sulfonamide groups. The gauche arrangement of the bulky substituents is believed to disfavor the path leading to the *anti*-adduct.

The approach to naphtho[2,1-*b*]furans based on the oxidative annulation of the furan ring in the reaction between STs and β-naphthols under a tandem Rh(II)/Yb(III) catalysis was established (Scheme 5.43) [66]. The cascade starts with the OH-insertion of the Rh-AVC to the naphthol affording β-aryloxyenamine. Then, the action of $Yb(OTf)_3$ as a Lewis acid initiates an imine-enamine tautomerization followed by an intramolecular Friedel-Crafts-type aminoalkylation. The oxidative dehydrogenation in the oxygen atmosphere furnishes the final fused aromatic heterocycle. An array of β-naphthols afforded the products in low to high yields up to 93%, whereas phenols appeared less reactive, giving the corresponding benzofurans in yields 29–38%.

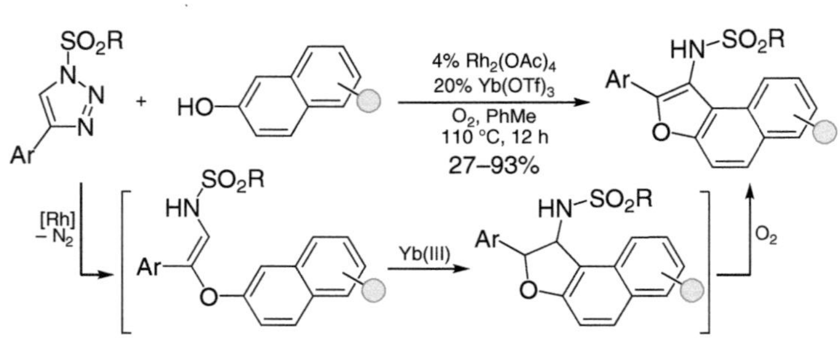

Scheme 5.43 Oxidative annulation of AVCs to β-naphthols under tandem Rh(II)/Yb(III) catalysis.

The 1,3-OH-insertion of Rh-AVCs to allyl alcohols was investigated [67, 68]. The produced allyl β-aminovinyl ethers readily underwent the Claisen rearrangement affording α-allyl-α-aminoketones (Scheme 5.44). The one-pot combination of the CuAAC and the Rh-catalyzed denitrogenative cascade also appeared viable. The developed approach was applied not only to primary allyl alcohols, including β- and γ-substituted ones, but also to secondary and tertiary substrates. The analogous reaction between STs and 2,3-alkadienols afforded α-aminoketones bearing a conjugated 1,3-diene system. The transformation demonstrated an appreciable level of stereocontrol provided by a chair-like transition state. Moreover, the enantiomerically enriched *sec*-allyl alcohol afforded (*E*)-product exclusively with only a moderate decrease in the enantiopurity. This strategy was successfully applied to the formal total synthesis of alkaloid (−)-α-conhydrine (Scheme 5.44) [68]. The asymmetric versions of the cascade transformation involving OH-insertion of Rh-AVCs to allyl alcohols and subsequent Claisen rearrangement catalyzed by chiral Lewis acids are discussed in Chapter 9 [69, 70].

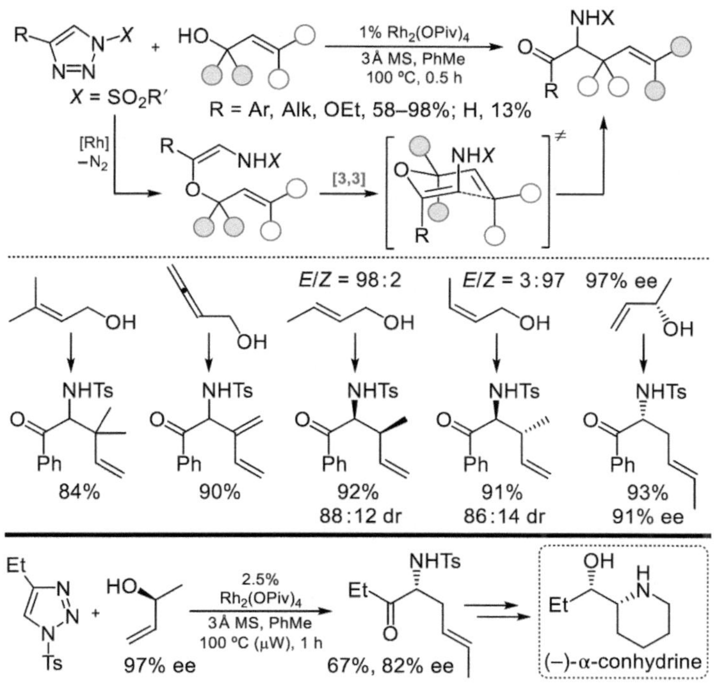

Scheme 5.44 OH-insertion of Rh-AVCs to allyl alcohols followed by Claisen rearrangement.

The reaction of Rh-AVCs with allenols was further studied by Alcaide and Almendros et al. [71]. The *sec*-benzyl alcohols bearing a cumulated double bond system in the allylic position afforded 1,2,3-trisubstituted conjugated (*Z*)-1,3-dienes in moderate to good yields (Scheme 5.45). The nature of the substituents in the position 4 of the triazole ring was found to affect the transformation chemoselectivity. Thus, replacing an acyl moiety with an aryl group led to another reactivity pattern and a formal (3+2)-cycloaddition of the AVC to the alkene [72].

The same strategy was applied to the functionalized allyl alcohols [73, 74]. In particular, the Rh-catalyzed denitrogenation of STs in the presence of (*E*)-γ-hydroxyacrylates enabled the preparation of (*E*)-α-vinylated γ-oxo-β-aminoesters

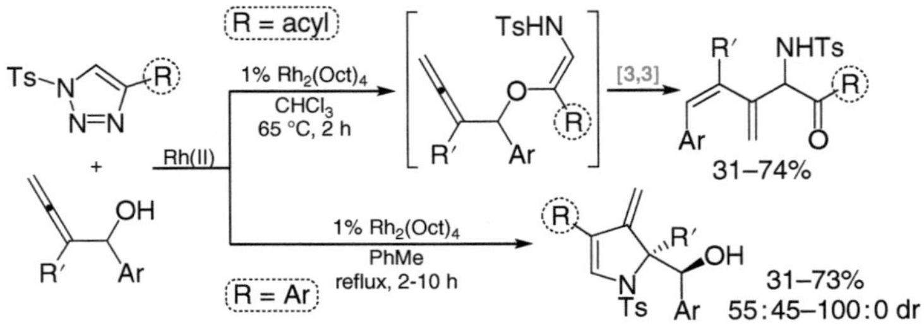

Scheme 5.45 Divergent substrate-dependent reactions of Rh-AVCs with allenols.

in generally high yields with good diastereoselectivity toward *syn*-product (Scheme 5.46). Likewise, the reaction with MBH adducts afforded α-alkylidene-δ-oxo-γ-aminoesters in very high yields. As opposed to the stereochemical outcome for alkyl-substituted *sec*-allyl alcohols which produced (*E*)-products exclusively, the analogous transformation of the MBH adducts appeared less stereoselective and delivered mixtures containing (*E*)-isomers predominantly (*E*/*Z* ratio from 56 : 44 to 88 : 12). The observed moderate selectivity was confirmed by the DFT calculations indicating a small energy difference (*ca.* 1.5–2 kcal/mol) between the chair-like transition states for the [3,3]-sigmatropic rearrangement leading to (*E*)- and (*Z*)-isomers, respectively. The favorability of the axial alignment for substituent R^2 providing the preferential formation of (*E*)-product may be attributed to the increased steric repulsion from the neighboring ester group.

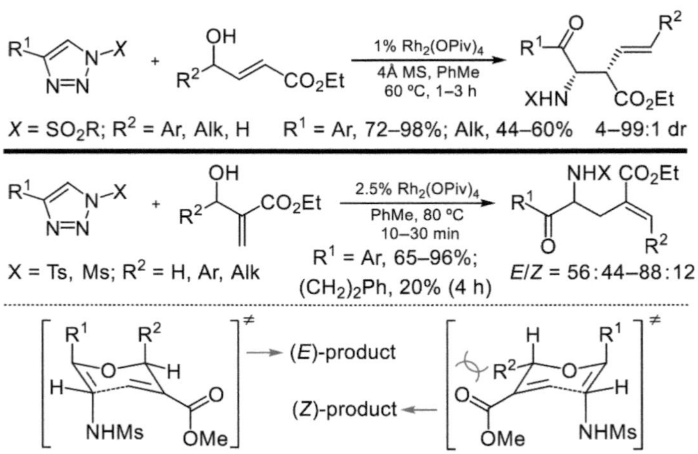

Scheme 5.46 OH-insertion of Rh-AVCs to γ-hydroxyacrylates and MBH adducts.

The similar Rh-catalyzed reaction between STs and the MBH adducts, prepared from aldehydes and cyclic enones, was performed [75]. The obtained mixture of diastereomers was treated with DBU upon heating at 75 °C in toluene to assemble 3,4-fused pyrroles in good yields (Scheme 5.47). The pyrrole ring formation is believed to occur *via* a stepwise mechanism including intramolecular *aza*-Michael

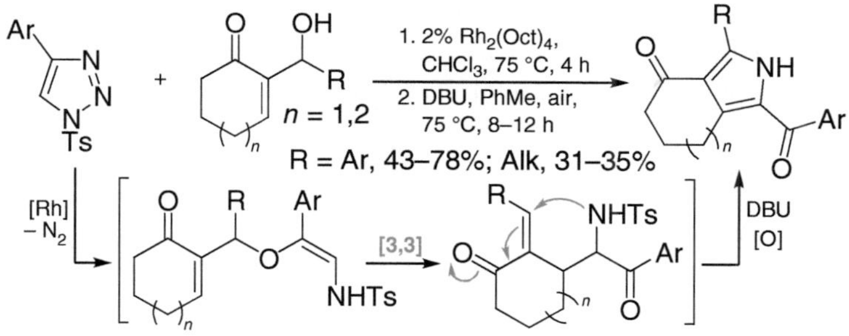

Scheme 5.47 Oxidative cyclization of the coupling products obtained from Rh-AVCs and MBH adducts.

addition, base-promoted elimination of *p*-toluenesulfinic acid, tautomerization, and oxidative aromatization in the presence of air.

In 2020, Xu and Li et al. investigated the Rh-catalyzed denitrogenation of STs in the presence of benzoquinone-derived allyl alcohols [76]. The previously developed cascade combining the Rh-AVC 1,3-OH-insertion and the Claisen rearrangement was followed by a keto-enol tautomerization delivering *o*-substituted phenols (Scheme 5.48).

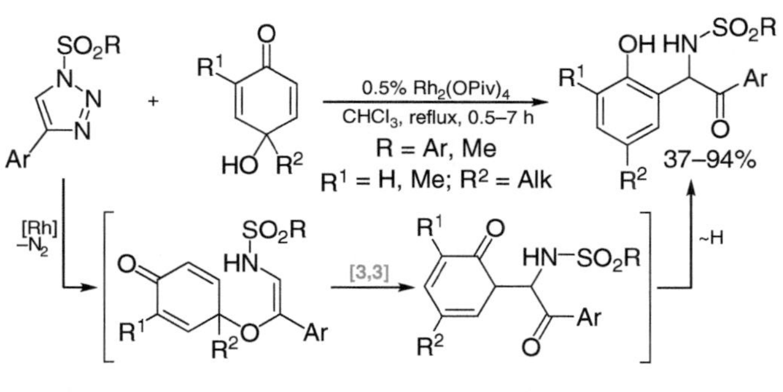

Scheme 5.48 OH-insertion of Rh-AVCs to benzoquinone derived allyl alcohols.

The reaction between STs and 2-(hydroxymethyl)allyl carbonates under the dual Rh(II)/Pd(0) relay catalysis was reported [77]. The well-established sequence of steps, involving Rh-AVC OH-insertion and [3,3]-sigmatropic rearrangement, was followed by the Pd-catalyzed heterocyclization (Scheme 5.49). The oxidative addition of Pd(0) to the allyl carbonate leads to the decarboxylation and the generation of zwitterionic 1,5-dipole comprising π-allyl Pd(II) complex. The last step represents an intramolecular N-allylation affording final 2-acyl-4-alkylidenepyrrolidines. The nucleophilic attack on the nonsymmetrical π-allyl system proceeds regioselectively at the electronically preferred more crowded position. The method provides an agile tool for a stereoselective formation of the pyrrolidine ring. Thus, the reaction with *sec*-allyl alcohols affords *cis*-2,5-disubstituted heterocycles predominantly. Moreover, the choice of *Z*/*E*-configuration of the starting alkene allows controlling the relative stereochemistry of the contiguous pyrrolidine carbons 2 and 3. A straightforward approach for the conversion of the obtained *N*-tosylpyrrolidines to NH pyrroles upon heating with DBU was established (Scheme 5.49) [78]. The aromatization was achieved through the base-mediated elimination of *p*-toluenesulfinic acid and subsequent proton transfer.

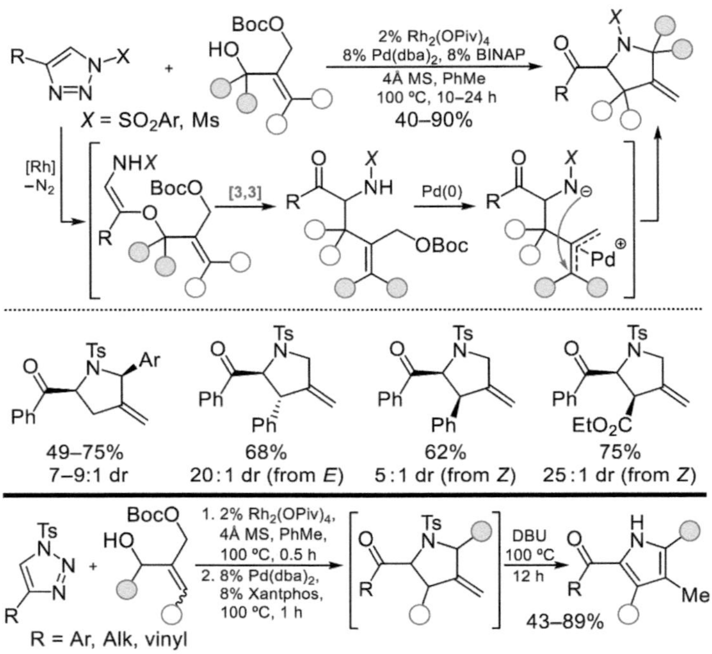

Scheme 5.49 Approach to pyrrolidines from STs and 2-(hydroxymethyl)allyl carbonates under tandem Rh(II)/Pd(0) catalysis.

An efficient route to 2-acyl-2,5-dihydropyrroles was developed using the OH-insertion of Rh-AVCs to propargyl alcohols (Scheme 5.50) [79]. [3,3]-Sigmatropic rearrangement of the *in situ* generated β-propargyloxyenamines smoothly afforded α-allenyl-α-aminoketones. The delayed addition of Au(I) catalyst induced 5-*endo-trig* hydroamination furnishing target compounds under mild conditions in good to excellent yields. The reaction with a chiral *sec*-propargyl alcohol gave a 3 : 1 mixture of *trans*- and *cis*-2,5-disubstituted pyrrolines, respectively, with complete retention of enantiomeric purity. The change of the Au(I) catalyst and, especially, the counter anion can switch the chemoselectivity of the heterocyclization (Scheme 5.50) [80]. Thus, the use of $(Ph_3P)AuNTf_2$ for the cycloisomerization of α-allenyl-α-aminoketones evoked nucleophilic attack of the carbonyl oxygen instead of the sulfonamide to the double bond and furnished 3-aminofurans in good yields.

The reactivity of Rh-AVCs toward 2-furfuryl alcohols was investigated [81]. The initial OH-insertion products smoothly underwent the [3,3]-sigmatropic rearrangement affording α-aminoketones, bearing 3-furyl moiety, after rearomatization

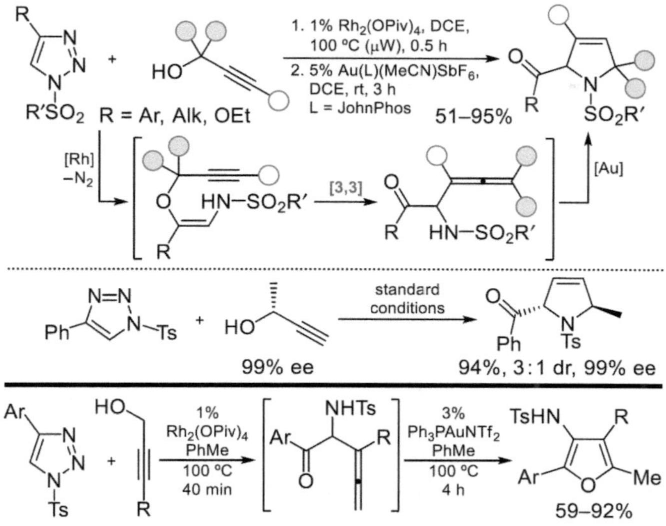

Scheme 5.50 OH-insertion of Rh-AVCs to propargyl alcohols.

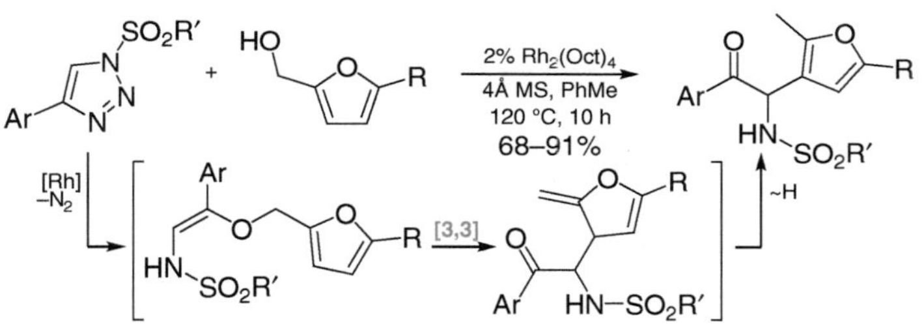

Scheme 5.51 OH-insertion of Rh-AVCs to 2-furfuryl alcohols.

(Scheme 5.51). The method is applicable mainly to primary alcohols allowing the preparation of a variety of 2-methylfurans in good yields. Whereas diverse functional groups in the fifth position of the furan ring were well-tolerated, 3- and 4-substituted substrates gave complex mixtures.

In 2016, Bi and co-workers disclosed the Rh-catalyzed denitrogenative reaction of STs with benzyl alcohols [82]. The OH-insertion of the AVCs leading to β-benzyloxyenamines was followed by a formal [1,3]-alkyl shift furnishing α-benzyl-α-aminoketones (Scheme 5.52). The rearrangement was found to occur only in the presence of the Rh catalyst. A wide scope of the transformation, including various 1°, 2°, and 3° benzyl alcohols, was demonstrated. The asymmetric reaction between STs and benzyl alcohols under Rh(II)/In(III) relay catalysis is discussed in Chapter 9 [83].

The feasibility of the intramolecular 1,3-OH-insertion of the Rh-AVC to a pendant alcohol was studied [84]. Whereas the use of an aliphatic linker between the ST and the hydroxyl moiety gave poor results, a phenylene tether appeared more suitable and the insertion to benzyl alcohols afforded phthalan derivatives in high

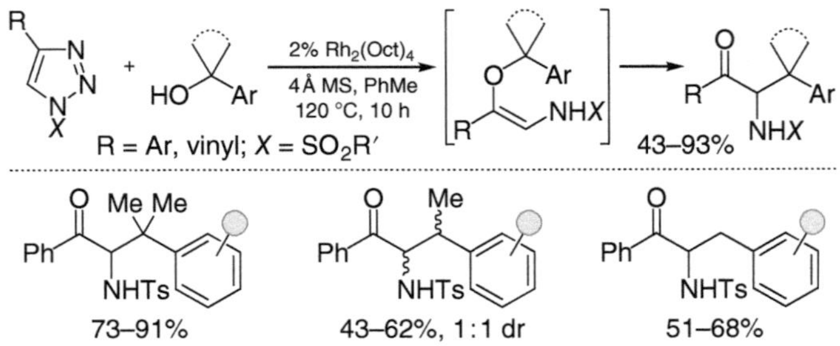

Scheme 5.52 OH-insertion of Rh-AVCs to benzyl alcohols.

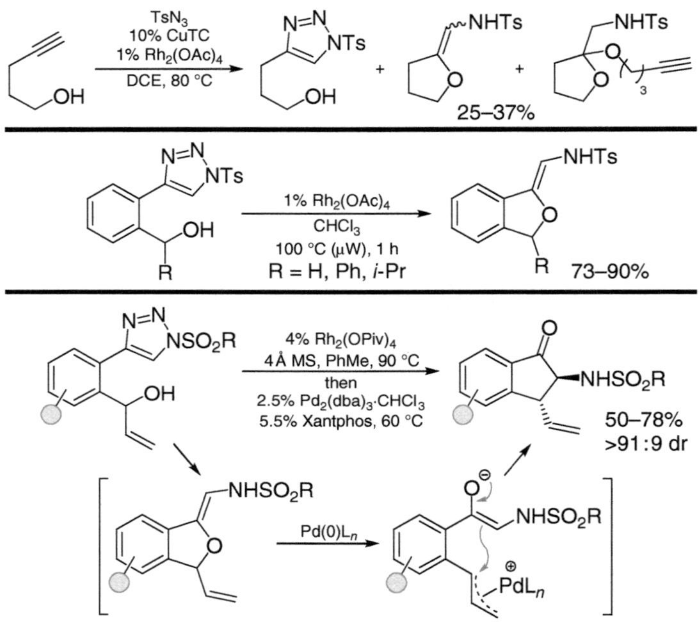

Scheme 5.53 Intramolecular OH-insertion of Rh-AVCs.

yields (Scheme 5.53). The analogous transformation with *sec*-allyl alcohols was also elaborated [85]. In contrast to the aforementioned β-allyloxy enamines, the structurally related phthalans were found stable and did not undergo the Claisen rearrangement (Scheme 5.53). In turn, the addition of a Pd(0) complex caused the cleavage of the allylic C—O bond followed by a cycloisomerization leading to 2-amino-3-vinylindanones. The two-step procedure utilizing Rh and Pd catalysts was carried out in one pot and afforded target compounds in moderate to good yields with high *trans*-diastereoselectivity.

Shortening the distance between the hydroxyl group and the carbene center allowed to disclose the novel domino transformations [86, 87]. Thus, the Rh-catalyzed denitrogenation of the STs bearing a 2-hydroxyethyl fragment furnished 2-aminocyclobutanones in good to almost quantitative yields (Scheme 5.54).

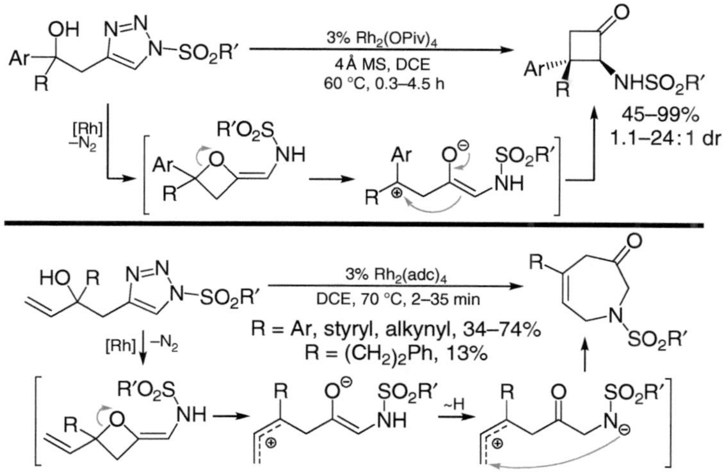

Scheme 5.54 Domino reactions involving intramolecular OH-insertion of Rh-AVCs and oxetane rearrangements.

The method is applicable to a wide range of benzyl and benzhydryl alcohols giving the products with variable but generally rather low diastereoselectivity. According to the proposed mechanism, the Rh-AVC 1,3-insertion to the OH in the β-position produces a strained oxetane intermediate. The breaking of a C—O bond gives a zwitter-ion containing enolate and stabilized benzyl cation moieties, which further undergoes a C—C coupling. The described transformation can be regarded as an intramolecular variant of the aforementioned formal 1,3-shift in β-benzyloxy enamines [82].

The introduction of the double bond to the substrate structure and, thus, using an allyl alcohol moved the reaction to another pathway, leading to 5,6-dehydroazepan-3-ones (Scheme 5.54) [87]. The transformation proceeded very quickly reaching the completion within several minutes and afforded the target heterocycles in moderate yields. The plausible mechanism involves the Rh-AVC intramolecular OH-insertion followed by an ionization of the oxetane by a heterolytic cleavage of a C—O bond. Then, the proton transfer is believed to convert the enolate to a sulfonamide anion. The formation of a seven-membered ring occurs through the nucleophilic nitrogen attack to a terminal carbon of the delocalized allyl cation.

The reaction of STs with glycidols was studied under a relay Rh(II)/Mg(II) catalysis [88]. The chemoselectivity was determined by substituent R attached to the epoxide ring (Scheme 5.55). Thus, 2-substituted glycidols (R ≠ H) were converted to 1,4-oxazepines in moderate yields by means of a two-step one-pot procedure. The OH-insertion of the Rh-AVC was followed by a ring-opening of the epoxide by a sulfonamide nitrogen promoted by the addition of Mg *tert*-butoxide acting as a Lewis acid. The nucleophilic attack occurred exclusively at the less crowded terminal carbon leading to the seven-membered ring formation. In contrast, the use of unsubstituted glycidols (R = H) switched the chemoselectivity, affording

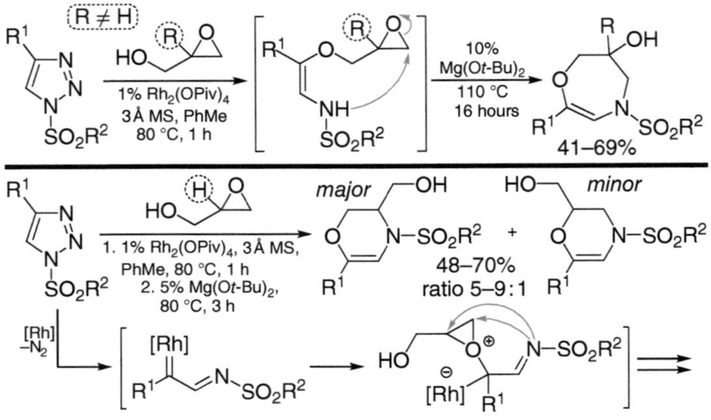

Scheme 5.55 OH-insertion of Rh-AVCs to glycidols.

mixtures of regioisomeric oxazine derivatives. An alternative reaction path leading to the minor isomer implied the formation of an ylide intermediate through the initial interaction of the Rh-AVC with the epoxide oxygen.

A straightforward route to assemble 3,4-dihydro-2*H*-1,4-oxazines through the Rh-catalyzed reaction of STs with halohydrins was developed [89]. At first, the OH-insertion of the Rh-AVC affords the standard (*Z*)-β-alkoxyenamine. Then, heating the crude product with K_2CO_3 evokes the intramolecular S_N2-alkylation of the sulfonamide nitrogen, leading to the heterocycle formation (Scheme 5.56). Notably, performing the reaction as a domino-process by a simultaneous addition of the Rh-catalyst and the base appeared viable, moreover, sometimes improving yields. The method was applied in the total synthesis of morpholine alkaloid (±)-chelonin C using a diastereoselective reduction of the enamine followed by N-detosylation. In 2023, the same strategy enabled access to a related natural product (−)-chelonin A starting from the enantioenriched bromohydrin [90].

The initial 1,3-OH-insertion of the Rh-AVCs can be accompanied by a rearrangement of β-oxyenamines to α-aminoketones (Scheme 5.57). This isomerization proceeds through a cyclization to an oxazoline intermediate followed by a ring-opening with a C—O bond cleavage. The intramolecular nucleophilic addition of sulfonamide nitrogen to a C(sp^2) is favorable for electron-deficient moieties, such as carbonyls, Michael acceptors, imines, *etc.* The versatility of this general strategy was illustrated by the reactions with diverse substrates providing elegant efficient routes to valuable structural motifs. One of the first examples was disclosed in 2016 by Lee's group in the Rh-catalyzed reaction of STs with tropolones (Scheme 5.58) [91]. A series of 2-aminotropones were prepared in good to excellent yields. The rearrangement of the isolated β-oxyenamine was shown to proceed upon heating in the absence of the Rh catalyst.

1,3-Diketones existing predominantly in the enol form can also participate in the OH-insertion upon the interaction with the Rh-AVCs [92, 93]. Thus, the reaction with 2-methylcyclohexane-1,3-dione affords β-aminoenone *via* the

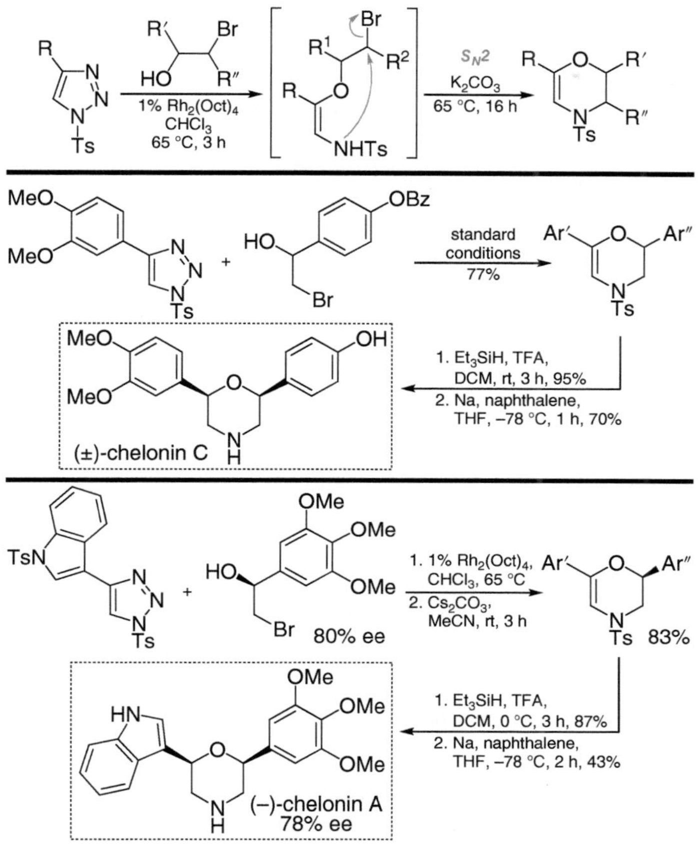

Scheme 5.56 OH-insertion of Rh-AVCs to halohydrins.

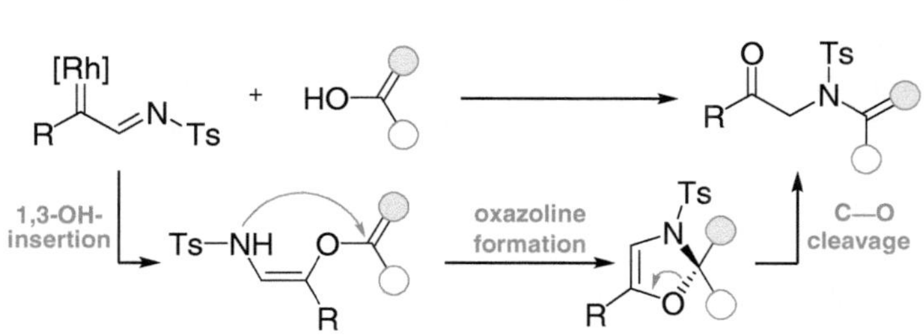

Scheme 5.57 Rearrangement of β-oxyenamines to α-aminoketones.

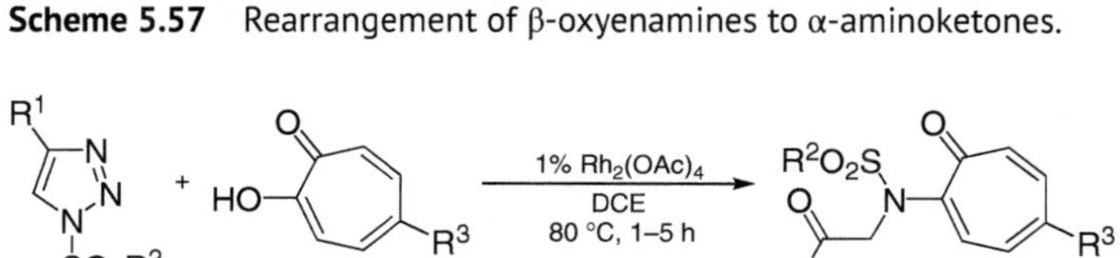

R^1 = Ar, Alk, vinyl; R_2 = Ar, Alk; R_3 = H, Br, Ar 48–98%

Scheme 5.58 Rh-AVC insertion to tropolones.

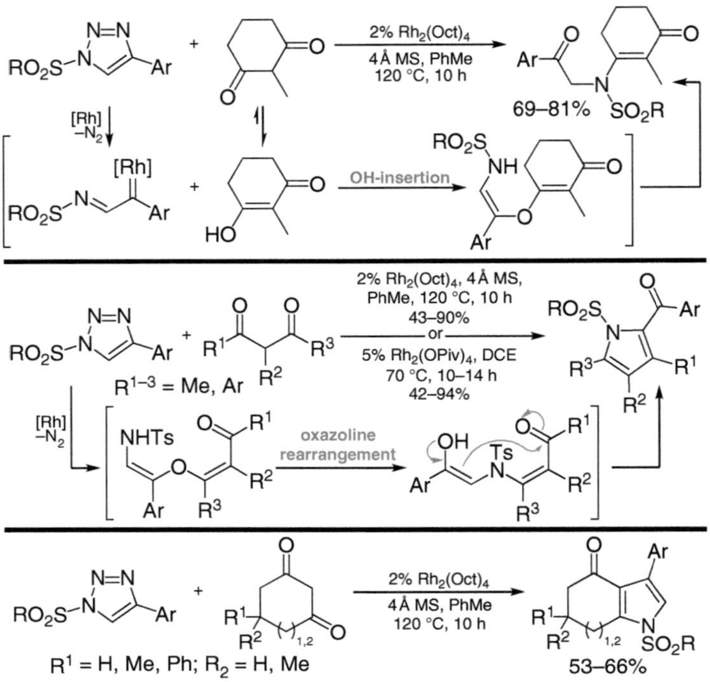

Scheme 5.59 Rh-AVC insertion to enol form of 1,3-diketones.

OH-insertion/rearrangement sequence (Scheme 5.59). Under the same conditions acyclic diketones are converting to 2-acylpyrroles in moderate to excellent yields. The heterocyclization is believed to occur through the intramolecular condensation of the corresponding β-aminoenone with an embedded ketone fragment. Non-symmetrical diketones bearing a phenacyl group react regioselectively affording 3-arylpyrrols. The reaction with 2-unsubstituted cyclic 1,3-diketones produces 2,3-fused pyrroles in reasonable yields. Although the cyclization of β-aminoenone could be assumed as a possible pathway also in this case, the authors consider an alternative mechanism involving a CH-insertion of the Rh-AVC to a methylene group [93].

The Rh-catalyzed denitrogenation of the STs in the presence of *N*-acylhydrazones affords labile amidine-containing compounds [94], presumably, through the OH-insertion of the Rh-AVC to an imidic acid tautomer of the hydrazide and the subsequent rearrangement (Scheme 5.60). The obtained compounds were shown to be convenient precursors for the divergent preparation of 1,2,4-triazines *via* cyclization under either acidic or basic conditions.

The combination of the formal OH-insertion of the Rh-AVC followed by O to N migration through the oxazoline intermediate was extensively studied by Volla et al. [95, 96]. This approach was shown to be a useful tool for the derivatization of a plethora of diverse azaheterocycles capable of converting to lactim tautomer (Scheme 5.61). The list of applicable substrates includes isatoic anhydrides,

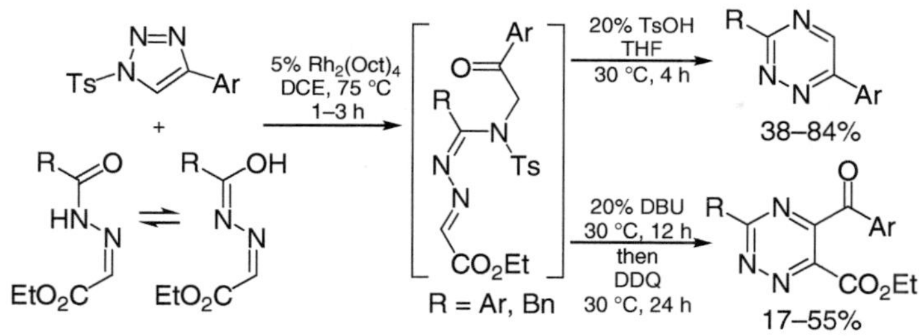

Scheme 5.60 Rh-AVC insertion to *N*-acylhydrazones.

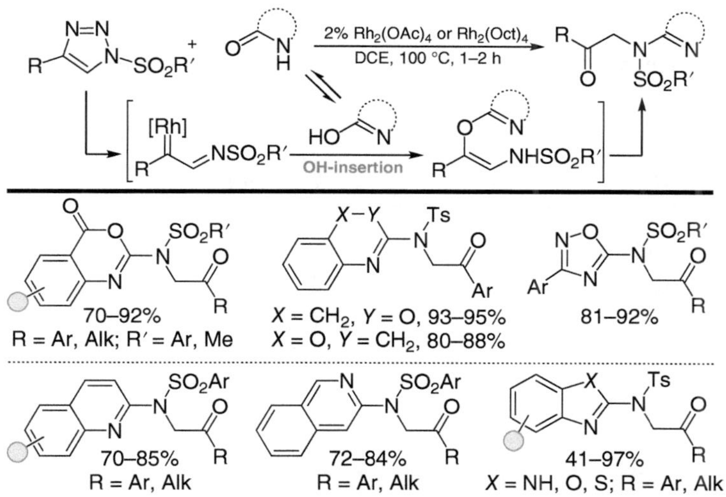

Scheme 5.61 Rh-AVC insertion to lactim form of various azaheterocycles.

oxadiazolones [95], quinolones, and isoquinolones [96]. This synthetic strategy was extended using 2-hydroxysubstituted benzannulated azoles [97].

The analogous reaction with 4-vinyl benzoxazinanones also involves the OH-insertion/rearrangement sequence [98]. The intermediate aminoketone undergoes Rh-catalyzed cleavage of the allylic C—O bond, which evokes cyclic hemiaminal formation followed by the second ring closure *via* an intramolecular allylic etherification (Scheme 5.62). The method enables an efficient assembly of tricyclic-fused imidazolidinones in a diastereoselective fashion.

In 2017, Deng's and Volla's groups independently reported the unusual Rh-catalyzed reaction of the STs with isatins, leading to formal 1,2-difunctionalization of the AVCs [99, 100]. The method furnishes interesting indigo-like α-amino enaminones in good yields with complete (*E*)-selectivity (Scheme 5.63). Although the mechanism remains ambiguous, one of the putative pathways involves the OH-insertion/rearrangement cascade. The further N-to-C 1,2-shift may occur through the formation of a spiro-aziridine intermediate followed by its ring opening. Deng and co-workers also achieved the analogous transformation using

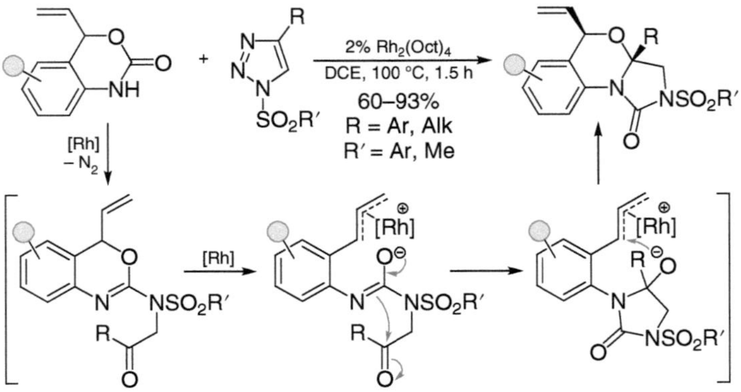

Scheme 5.62 Approach to fused imidazolidinones *via* domino reaction of Rh-AVCs with 4-vinyl benzoxazinanones.

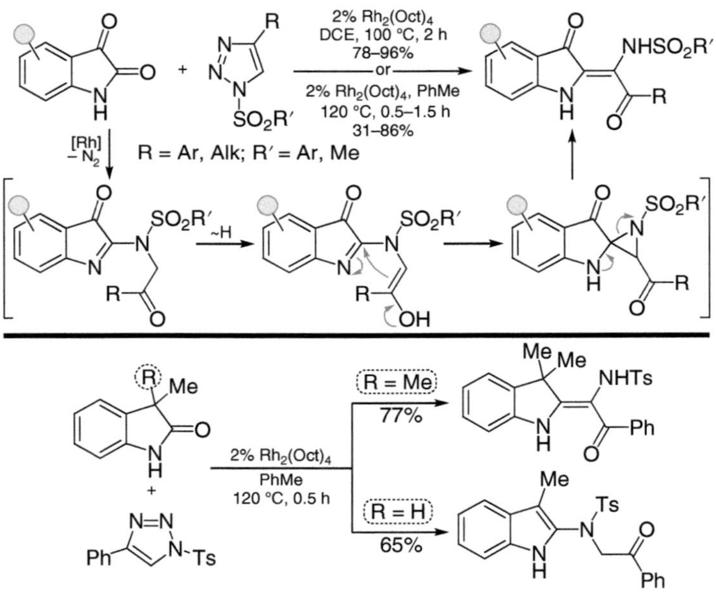

Scheme 5.63 Rh-catalyzed reaction of STs with isatins.

3,3-dimethyloxindole, whereas the reaction with enolizable 3-methyloxindole stopped after typical O to N migration, affording the corresponding 2-aminoindole.

The reactivity of the Rh-AVCs toward 3-benzylideneoxindoles was studied [101]. The initial formal OH-insertion affording β-oxyenamine led to the subsequent annulation *via* two competing reaction pathways (Scheme 5.64). The formation of pyrroloindole arises from the O-to-N migration through the spiro-oxazoline intermediate which is followed by a Michael addition. Alternatively, β-oxyenamine does not undergo rearrangement and cyclizes to 1,4-oxazepinoindole. The pyrroloindole was obtained as a major or even a single product in most experiments, although the

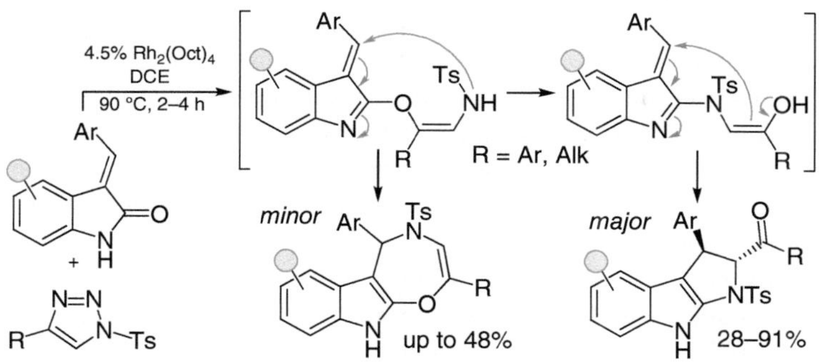

Scheme 5.64 Rh-catalyzed reaction of STs with 3-arylideneoxindoles.

chemoselectivity was found highly sensitive to the electronic effects of substituents in the starting materials.

5.3.3 N–H Insertion

The examples of N—H insertion of AVCs are not so numerous, in comparison with the analogous processes involving C—H and O—H bonds. The first contribution was made in 2014 by Fokin's group which performed the Rh-catalyzed formal 1,3-hydroamination, giving access to alkene-based 1,2-diamine derivatives [64]. The exclusive formation of (*Z*)-isomer was attributed to ylide-mediated mechanism and intramolecular proton transfer (Scheme 5.65). This strategy was successfully applied to a wide variety of weak NH nucleophiles, including primary and secondary

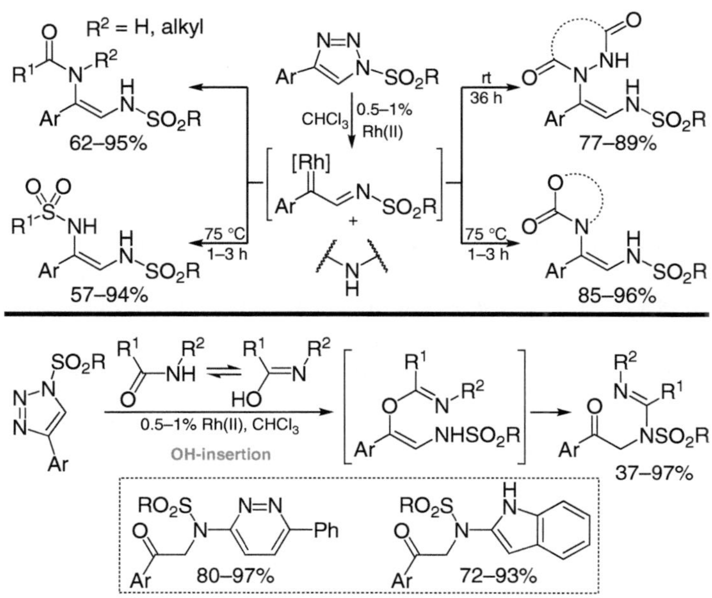

Scheme 5.65 Formal 1,3-insertion of Rh-AVCs to various N−H bonds.

amides, carbamates, sulfonamides, and hydrazides. In the case of secondary amides, the formation of isomeric ketones was identified as a competing process, which accounted for the OH-insertion to imidic acid tautomer followed by rearrangement. This pathway was more pronounced for the substrates with the increased steric congestion of the N—H bond. The exclusive selectivity toward ketone formation was also observed in the reactions with oxindole and pyridazinone due to the favorable generation of heteroaromatic ring systems. Further computational studies of Bao and co-workers revealed that the initially proposed mechanism, based on insertion to the imidic acid O—H bond, is unlikely [102]. Instead, a pathway involving the nucleophilic addition of the carbonyl oxygen to Rh-AVC and formal (3+2)-cycloaddition, followed by ring-opening, was suggested.

In 2015, Lee and Yoo accomplished the Rh-catalyzed denitrogenative insertion of STs to the N—H bond of 9*H*-carbazoles (Scheme 5.66) [103]. The obtained enamines were involved in the chemoselective intramolecular oxidative CH-amination under Pd catalysis, affording valuable C—N coupled heterobiaryls comprising carbazole and indole moieties. The required *cis*-orientation of amino and aryl groups is believed to be provided *via* Pd-catalyzed double bond isomerization of (*Z*)-enamine, originally produced by the Rh-AVC 1,3-insertion.

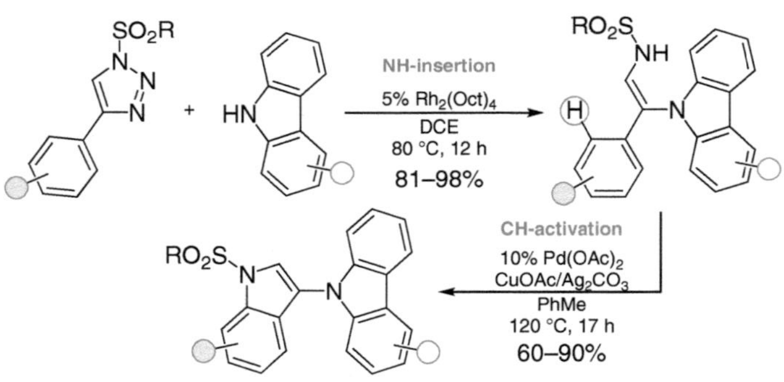

Scheme 5.66 NH-insertion of Rh-AVCs to carbazoles.

The scope of nitrogen nucleophiles capable of undergoing NH-insertion was extended to diarylamines and *N*-aryl glycine esters [104]. The reaction proceeded well using simple $Rh_2(OAc)_4$, albeit rather high catalyst loadings (4%) were employed (Scheme 5.67). A similar protocol was utilized to perform the analogous reaction with α-(arylamino)ketones [105]. The obtained enamines were

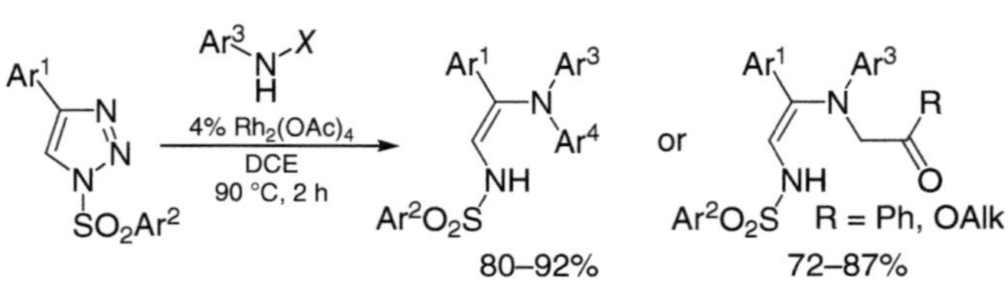

Scheme 5.67 NH-insertion of Rh-AVCs to anilines.

subjected to BF_3-promoted cyclodehydration furnishing 3-(sulfonylamino)pyrroles (Scheme 5.68). The three-step approach, starting from the CuAAC, was efficiently telescoped into a one-pot process, which not only improved the overall yield of the target pyrroles, but also allowed a decrease in the catalyst amounts (both Cu and Rh).

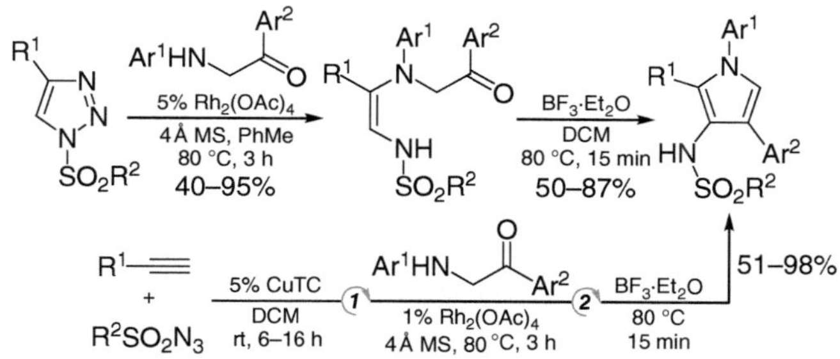

Scheme 5.68 Assembly of 3-aminopyrroles based on NH-insertion of Rh-AVCs.

The Rh-catalyzed NH-insertion of AVCs to β-enamino esters initiates a domino-process leading to 3-imidazolines (Scheme 5.69) [106]. The C—N bond formation is followed by enamine-imine tautomerization and intramolecular *aza*-Michael addition. Notably, the reaction with tetrasubstituted β-enamino esters exhibited the exceptional diastereoselectivity. Interestingly, the related insertion of Rh-AVCs to β-enaminones evoked heterocyclization *via* divergent reaction pathways affording either imidazoles or pyrroles [107]. The chemoselectivity was controlled by the nature of substituent R in β-enaminone (Scheme 5.70). In the case of a methyl group, the cyclization proceeded through *aza*-Michael addition. However, the intermediate imidazoline underwent a facile aromatization *via* a retro-Mannich C—C bond cleavage with acetophenone elimination. In contrast, more bulky phenyl group suppressed conjugate addition and switched the reactivity toward the pyrrole ring formation through condensation and subsequent elimination of tosylamide.

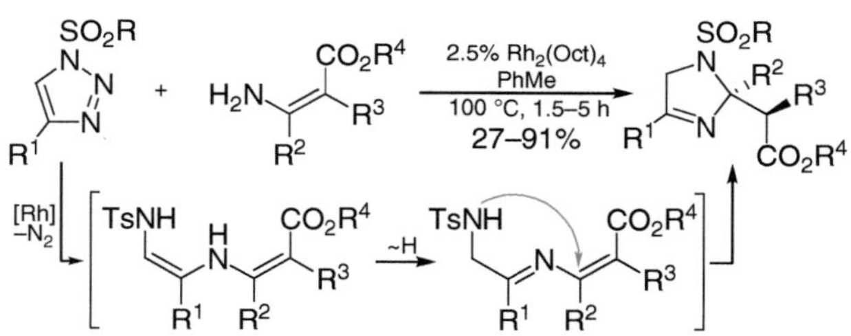

Scheme 5.69 NH-insertion of Rh-AVCs to β-enamino esters.

The Rh-AVC formal 1,3-insertion to the N—H bond of carbamates was combined with intramolecular S_N2-alkylation of the enamine nitrogen (Scheme 5.71) [108].

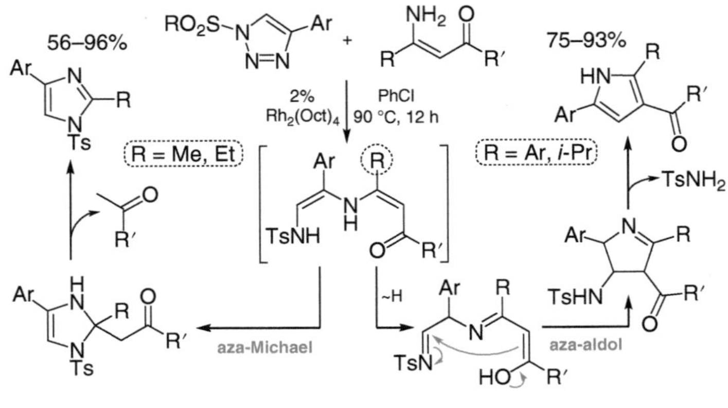

Scheme 5.70 NH-insertion of Rh-AVCs to β-enaminones.

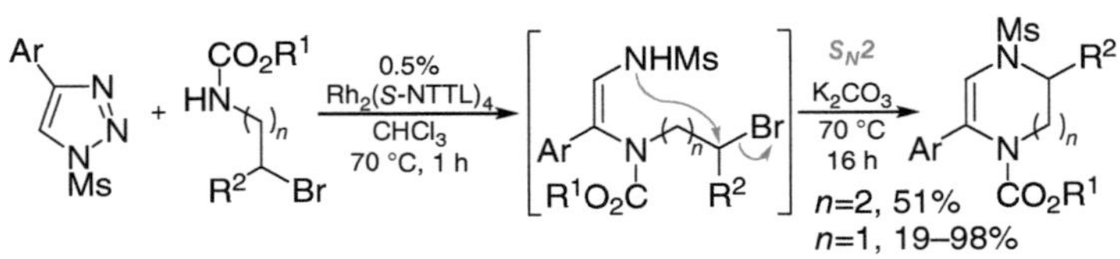

Scheme 5.71 Approach to 2,3-dehydropiperazines involving Rh-AVC NH-insertion.

The cyclization step was promoted by the addition of K_2CO_3 affording a series of 2,3-dehydropiperazines. The method was also applied to the preparation of a tetrahydrodiazepine in a moderate yield.

Domino-reactions initiated by the Rh-catalyzed C—N coupling of AVCs with anilines represent a fruitful strategy for a modular assembly of various azaheterocyclic scaffolds. The *in situ* generated *N*-aryl enamines can be efficiently trapped by some pendant functional groups in the *o*-position of the benzene ring. The first implementation of this approach was described in 2016 by Anbarasan et al., who developed a route to indolines from STs and 2-vinylanilines [109]. The annulation step is believed to occur *via* either *aza*-ene- or metallo-ene-type reaction (Scheme 5.72). The formation of a C—C bond between two contiguous quaternary carbon atoms proceeded *via* a chair-like transition state with the exclusive diastereoselectivity. The prepared indoline-imines were subjected to the reduction followed by the hypervalent iodine-mediated oxidative C—N coupling, leading to the annulation of the imidazolidine ring also in a stereoselective fashion.

The same group reported an alternative variant of heterocyclization initiated by the NH-insertion of Rh-AVCs to anilines [110]. The introduction of the benzyl alcohol group to the *o*-position afforded dihydro-3,1-benzoxazines predominantly in the form of *trans*-2,4-diaryl derivatives with substrate-dependent diastereomeric ratio (Scheme 5.73). According to the putative mechanism, the enamine tautomerization evokes nucleophilic addition of the alcohol to the iminium cation giving the cyclic hemiaminal ether.

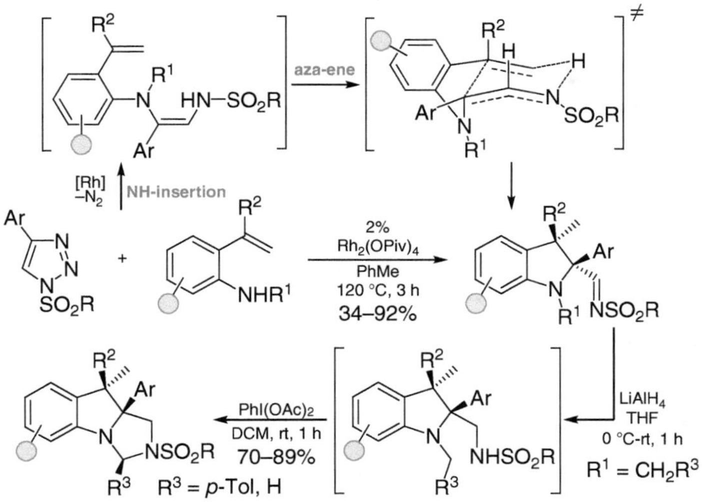

Scheme 5.72 NH-insertion of Rh-AVCs to 2-vinylanilines.

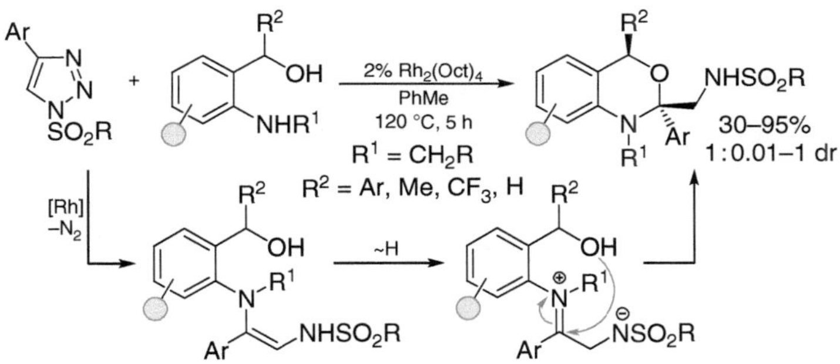

Scheme 5.73 NH-insertion of Rh-AVCs to 2-(hydroxymethyl)anilines.

A route to 3-aminoquinolines from STs and 2-acylanilines was reported by two groups independently [111, 112]. The Rh-catalyzed NH-insertion of AVC is followed by the intramolecular trapping of the enamine with a carbonyl group, leading to the annulation of a pyridine ring (Scheme 5.74). Anbarasan et al. established that the use of *N*-PMB-protected anilines is more productive in comparison with the substrates bearing a free amino-group [111]. The aromatization of the heterocyclic system is achieved in this case by a dealkylation step under the reaction conditions. It was shown that the NH-insertion proceeded much faster than the subsequent pyridine ring formation, which required a prolonged heating at 110 °C [112]. The delayed addition of 10% $Yb(OTf)_3$ acting as a Lewis acid greatly facilitated the condensation and enabled the one-pot process to reach completion within an hour. Interestingly, 4-cyanoquinoline was isolated in the reaction with the corresponding benzoyl cyanide, which more often demonstrates reactivity as an acylating reagent.

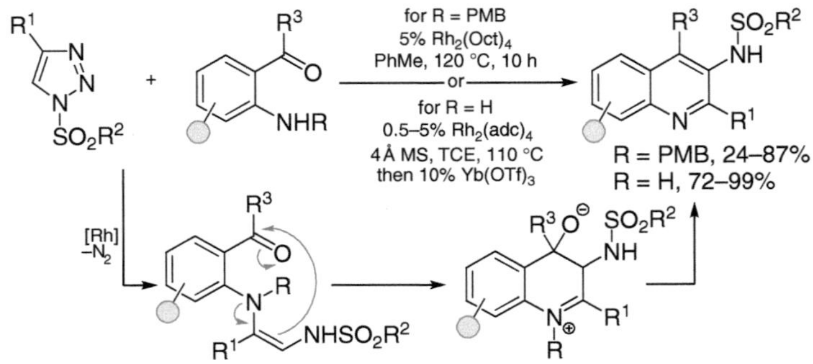

Scheme 5.74 NH-insertion of Rh-AVCs to 2-acylanilines.

5.3.4 B–H Insertion

The denitrogenative reaction of STs with hydroborating reagents was first described in 2016 by Miura and Murakami [113]. The interaction of Rh-AVC with 9-BBN led to 1,3-insertion to a B—H bond, producing boron *aza*-enolate with complete (*E*)-stereoselectivity (Scheme 5.75). The proposed mechanism involves *in situ* formation of ate complex between imine nitrogen and electrophilic borane followed by hydride shift from the boron atom to the carbene. The obtained boron *aza*-enolates appeared very prone to hydrolysis upon chromatographic purification, and, therefore, were mixed in the crude form with diverse aldehydes, furnishing the products of *aza*-aldol reaction. The observed *syn* selectivity was rationalized using a boat-like transition state. To simplify the isolation process, the imino group was reduced by DIBAL-H, affording a series of stereodefined 1,3-aminoalcohols.

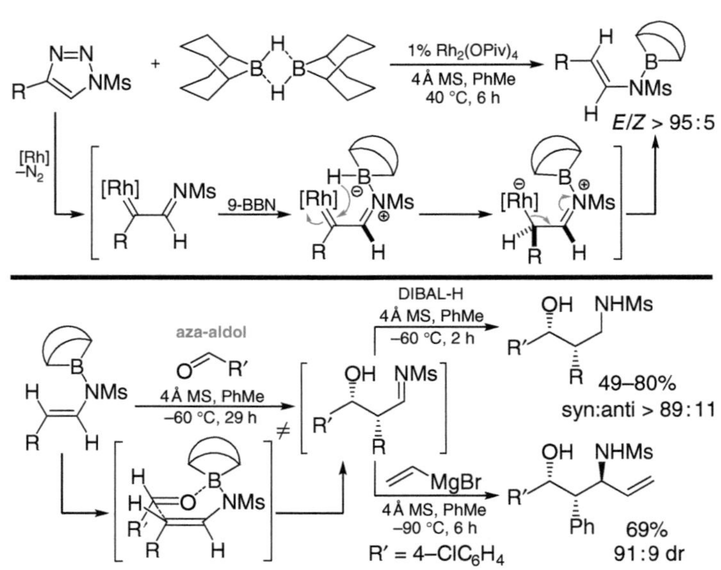

Scheme 5.75 Conversion of STs to boron *aza*-enolates and subsequent *aza*-aldol reaction.

Furthermore, the diastereoselective addition of Grignard reagent to imine allowed assembling of structures bearing three contiguous stereocenters.

This approach was extended to the reaction of STs with less reactive hydroborating reagent HBpin [114]. Previously applied Rh catalyst failed to perform the analogous transformation; however, the catalytic system comprising Ni(cod)$_2$ and bdpp in THF at 80 °C allowed to prepare the corresponding boron *aza*-enolate with high (*E*)-selectivity (Scheme 5.76). Likewise, further *aza*-aldol reactions delivered *syn* adducts, which were reduced to 1,3-aminoalcohols. 2,3-Diarylsubstituted *aza*-aldols underwent facile dehydration and were converted to (*E*)-allylamines after reduction.

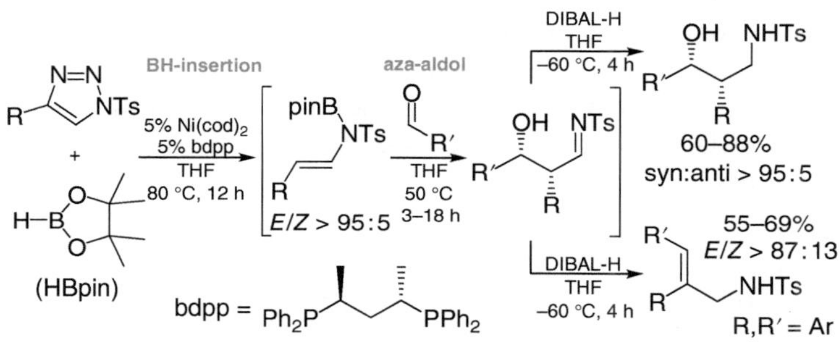

Scheme 5.76 Ni-catalyzed denitrogenative hydroboration of STs with HBpin.

5.3.5 Si–H and P–H Insertion

The insertion of AVCs to a Si—H bond was only reported once [115]. The Rh-catalyzed hydrosilylation was achieved using 1% Rh$_2$(Oct)$_4$ in chloroform at 60 °C (Scheme 5.77). The choice of the silylating reagent appeared crucial for providing sufficient reaction efficiency. Thus, Ph$_3$SiH was found to be the appropriate reaction partner, whereas no product was obtained in the case of some other silanes, such as Et$_3$SiH, Me$_2$PhSiH, and Ph$_2$SiH$_2$. Due to the hydrolytic instability of

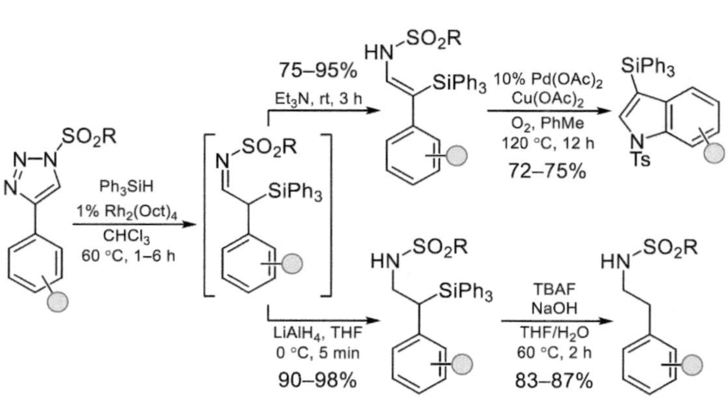

Scheme 5.77 Insertion of Rh-AVCs to Si–H bond.

initially forming silylated imines, these intermediates were smoothly transformed to isomeric enamines under Et_3N-mediated conditions or reduced with $LiAlH_4$ in a one-pot manner. The obtained enamines were involved in Pd-catalyzed oxidative cyclization in the presence of $Cu(OAc)_2$ and oxygen, affording 3-silylated indoles. The possibility to remove the Ph_3Si group in phenethylamines was demonstrated by treatment with TBAF and NaOH.

A unique $Ni(cod)_2$-catalyzed denitrogenative phosphorylation of STs was reported in 2019 by Jiang and co-workers [116]. The process represents a very rare example of formal 1,2-insertion of AVC to the P—H bond of *H*-phosphine oxides (Scheme 5.78). According to the proposed mechanism, the rearrangement of Ni-AVC to Ni-bound ketenimine intermediate is followed by the addition of P nucleophile and subsequent proton transfer. Reductive elimination recovers Ni(0) catalyst, affording the final α-phosphorylated enamine with excellent (*E*)-stereoselectivity. The protocol demonstrates good results for diarylphosphine oxides, while being less efficient for phosphinate and arylalkylphosphine oxide (34–38% yields), and inapplicable to dialkylphosphine oxides.

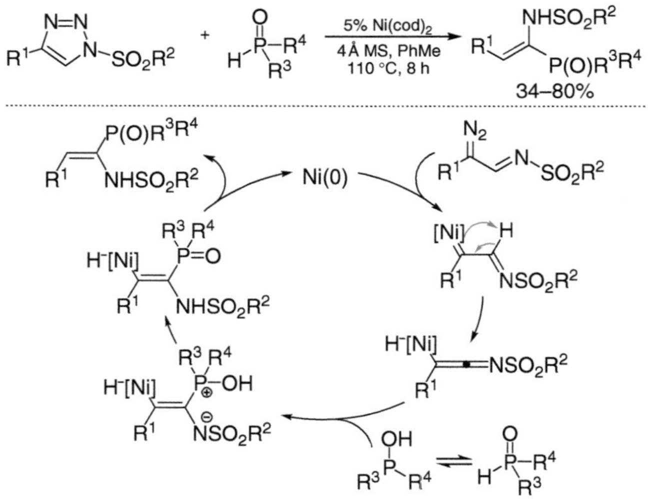

Scheme 5.78 Ni-catalyzed denitrogenative reaction of STs with *H*-phosphine oxides.

5.4 Insertion Reactions of 5-Iodo-1,2,3-triazoles

In spite of the remarkable achievements of the synthetic methodology exploiting triazole-diazoimine equilibrium, this chemistry is primarily associated with a limited range of substrates. In particular, STs and triazolo[1,5-*a*]pyridines were extensively investigated in this field. Therefore, further progress is largely dependent on the search for some other triazole derivatives capable of releasing the diazoimine species.

In 2018, Kotovshchikov and Latyshev et al. disclosed an alternative approach for the *in situ* generation of diazoimines utilizing 5-iodo-1,2,3-triazoles [117].

These compounds are readily available by the Cu-catalyzed cycloaddition between azides and 1-iodoalkynes [118] and, in general, demonstrate rather low reactivity typical for heteroaryl halides. However, tethering 5-iodotriazoles to an appropriate nucleophile gives an opportunity to assemble precursors of various triazole-fused systems [119–121]. In turn, the annulation of a new ring *via* the intramolecular iodine substitution can boost the triazole strain and induce the electrocyclic opening to diazoimine (Scheme 5.79). For instance, this scenario was realized for 5-iodotriazoles bearing a pendant phenol moiety. Thus, heating these substrates in the presence of Et_3N in dioxane at 100 °C evoked the base-mediated cyclization followed by a cleavage of an N—N bond and generation of 2-(1-diazoalkyl)benzoxazoles. In comparison with the widely employed STs, benzoxazole-fused triazoles are much less stable and exist predominantly in the form of diazo tautomer.

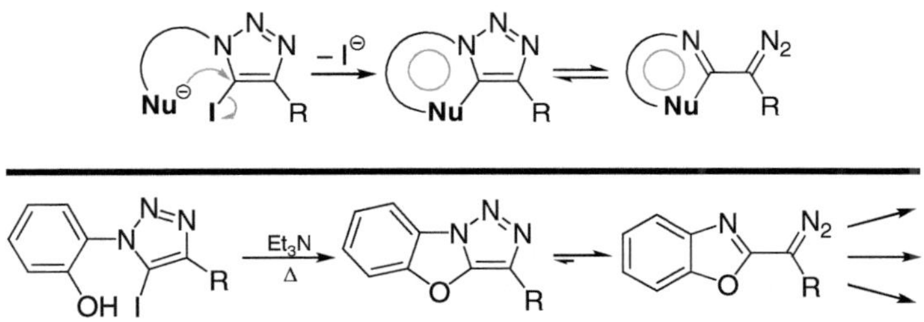

Scheme 5.79 Annulation-triggered ring-opening of 5-iodo-1,2,3-triazoles.

Trapping the diazo group by the Cu catalyst allowed achieving several efficient transformations leading to the attachment of nitrogen and sulfur nucleophiles by formal NH- and SH-insertions of the corresponding carbene. Thus, the subjection to aliphatic amines in the presence of 10% CuI afforded a series of α-aminoalkyl benzoxazoles in moderate to good yields [117]. Notably, highly abundant Cu catalysts appeared more efficient than very expensive Rh(II) carboxylates, which are almost indispensable in the related denitrogenative reactions of the STs. Mechanistic studies revealed alkyl iodide to be an intermediate in the cascade amination process (Scheme 5.80). This fact was rationalized by the migratory insertion of the iodide to the copper carbenoid which proceeds with a low activation barrier *ca.* 2 kcal/mol, according to DFT calculations.

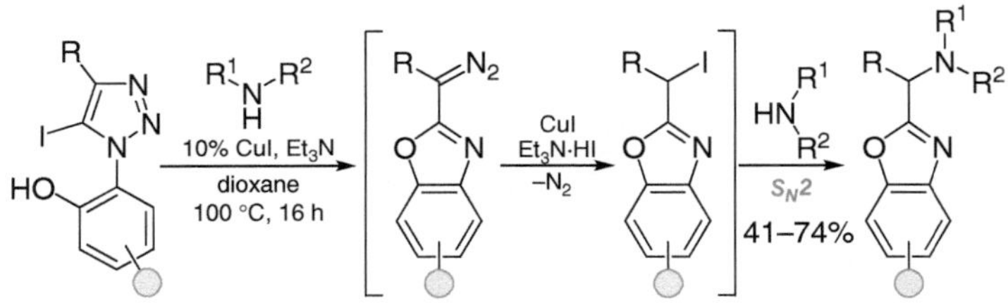

Scheme 5.80 Cu-catalyzed capture of 2-(1-diazoalkyl)benzoxazoles with amines.

Almost the same conditions were found efficient for performing reactions with sulfur nucleophiles [122, 123]. The annulation-triggered release of the diazo group by the base-mediated cyclization of 2-(5-iodotriazolyl)phenols was followed by the denitrogenative formal SH-insertion to heterocyclic thiols in the presence of 10% $(Ph_3P)_3CuBr$ (Scheme 5.81). While various azole- and azine-derived thiols, which are known to predominantly exist in tautomeric thioamide form, furnished the coupling products in moderate to high yields, the reaction with thiophenol led only to a reductive cleavage of the C—I bond in iodotriazole, thus quenching the cascade. To overcome this limitation, the introduction of arylthio group was achieved by either the delayed thiol addition or the use of pre-formed copper(I) thiophenolate with the latter variant being more productive. The addition of carbon disulfide to the Cu-catalyzed reaction with amines completely suppressed the formal NH-insertion pathway and afforded dithiocarbamates instead [123]. The protocol gave good yields for secondary amines, especially diverse saturated azaheterocycles, whereas poor results were obtained for barely nucleophilic anilines and primary amines due to isothiocyanate formation.

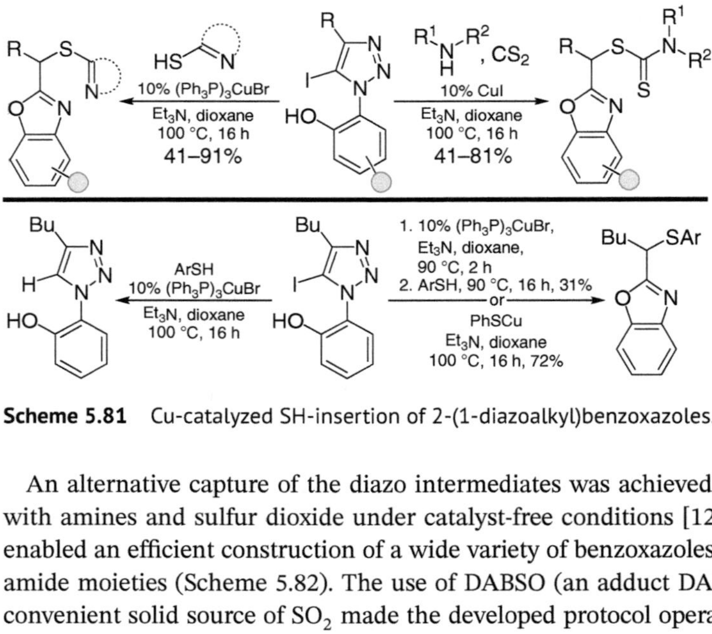

Scheme 5.81 Cu-catalyzed SH-insertion of 2-(1-diazoalkyl)benzoxazoles.

An alternative capture of the diazo intermediates was achieved by the reaction with amines and sulfur dioxide under catalyst-free conditions [124]. The method enabled an efficient construction of a wide variety of benzoxazoles bearing sulfonamide moieties (Scheme 5.82). The use of DABSO (an adduct DABCO·$2SO_2$) as a convenient solid source of SO_2 made the developed protocol operationally simple.

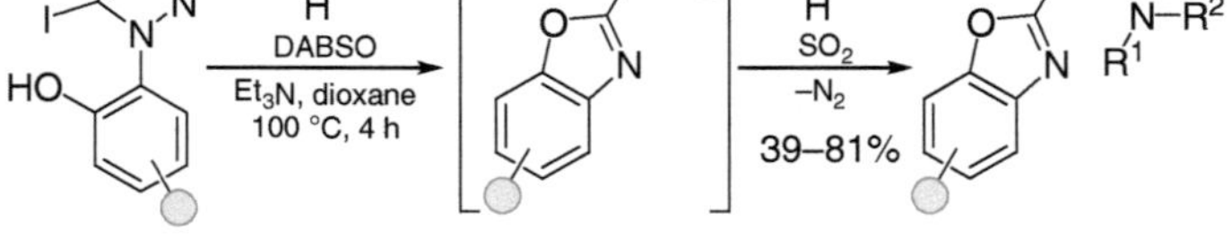

Scheme 5.82 Noncatalytic sulfonylative trapping of 2-(1-diazoalkyl)benzoxazoles.

In spite of a sluggish reaction with aqueous ammonia, the preparation of a primary sulfonamide was accomplished with $(NH_4)_2CO_3$ in good yield.

The synthetic strategy implying the annulation-triggered diazo release from 5-iodotriazoles was extended by changing the phenol to a carboxyl group [125]. The cyclization of the corresponding benzoic acids occurs smoothly under similar reaction conditions upon heating at 100 °C in the presence of Et_3N (Scheme 5.83). In comparison with benzoxazole-fused triazoles, the related lactones are less prone to spontaneous electrocyclic isomerization to diazoimines. Nevertheless, the Cu catalysis made it possible to intercept the diazo form with hetaryl thiols delivering sulfur-containing benzoxazinones. Another approach to the release of the diazo group involves a facile cleavage of the lactones by amines producing 5-hydroxytriazoles, which undergo tautomerization to α-diazoamides. Thus, the Cu-catalyzed reaction with amines afforded conformationally rigid pseudopeptides based on anthranilamides. Interestingly, the addition of CS_2 to the reaction led to a chemoselective incorporation of this molecule giving diamides bearing a dithiocarbamate moiety [123]. In addition, a two-step one-pot procedure was elaborated by the initial Cu-catalyzed thiolative trapping of the diazoimine followed by a nucleophilic ring-opening of the obtained benzoxazinones by amines [125].

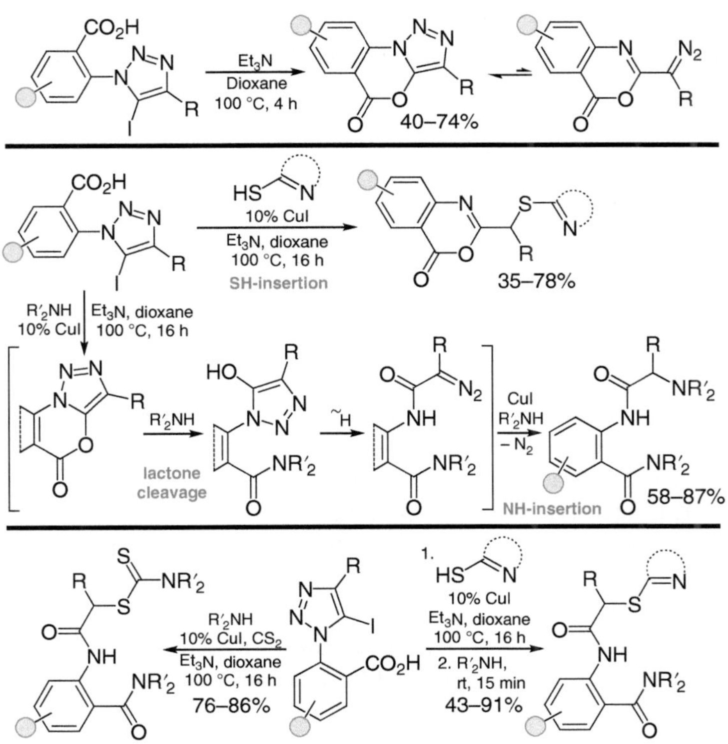

Scheme 5.83 NH- and SH-insertions triggered by cyclization of 2-(5-iodotriazolyl)benzoic acids.

5.5 Summary and Outlook

The review clearly highlights the versatility and reliability of AVC insertion reactions to C/*X*—H bonds for the efficient preparation of various compounds, including valuable synthetic building blocks and drug-like azaheterocyclic scaffolds. The initial AVC insertion to O—H and N—H bonds often evokes subsequent rearrangements, greatly expanding the product diversity. The Rh-catalyzed denitrogenative insertion of the STs can be combined with further tandem or one-pot transformations triggered by transition metals (Pd, Au) or Lewis acids. Notably, most reactions proceed in regio- and diastereoselective fashion, and a high level of enantiocontrol is achievable with chiral catalysts. Although the developed approaches were applied in a few syntheses of natural products, this promising research area seems to be in its infancy. Furthermore, whereas CH, OH, and NH insertions are well-studied, the reactivity of AVCs toward other XH bonds is definitely underexplored. The search for earth-abundant metal catalysts as an alternative to expensive Rh carboxylates could also boost the application of this methodology.

Acknowledgment

This work was supported by the Russian Science Foundation (grant number 23-73-10043).

References

1 Gillingham, D., and Fei, N. (2013). Catalytic X—H insertion reactions based on carbenoids. *Chem. Soc. Rev.* 42: 4918–4931.

2 Davies, H.M.L., and Beckwith, R.E.J. (2003). Catalytic enantioselective C—H activation by means of metal–carbenoid-induced C–H insertion. *Chem. Rev.* 103: 2861–2904.

3 Doyle, M.P., Duffy, R., Ratnikov, M., and Zhou, L. (2010). Catalytic carbene insertion into C—H bonds. *Chem. Rev.* 110: 704–724.

4 Santiago, J.V., and Machado, A.H.L. (2016). Enantioselective carbenoid insertion into C(sp^3)—H bonds. *Beilstein J. Org. Chem.* 12: 882–902.

5 Keipour, H., Carreras, V., and Ollevier, T. (2017). Recent progress in the catalytic carbene insertion reactions into the silicon–hydrogen bond. *Org. Biomol. Chem.* 15: 5441–5456.

6 Buckley, A.M., Brouder, T.A., Ford, A., and Maguire, A.R. (2019). Stereoselective C—C bond-forming reactions through C(sp^3)—H bond insertion of metal carbenoids. In: *C-H Activation for Asymmetric Synthesis*, 1e, 1–49. Wiley.

7 Bergstrom, B.D., Nickerson, L.A., Shaw, J.T., and Souza, L.W. (2021). Transition metal catalyzed insertion reactions with donor/donor carbenes. *Angew. Chem. Int. Ed.* 60: 6864–6878.

8 Huang, M.-Y., and Zhu, S.-F. (2021). Uncommon carbene insertion reactions. *Chem. Sci.* 12: 15790–15801.

9 Pal, K., and Volla, C.M.R. (2021). Catalytic insertion reactions of α-imino carbenoids. *Chem. Rec.* 21: 4032–4058.

10 Zhao, X., Wang, G., and Hashmi, A.S.K. (2021). Carbene B—H insertion reactions for C−B bond formation. *ChemCatChem* 13: 4299–4312.

11 He, Y., Huang, Z., Wu, K. et al. (2022). Recent advances in transition-metal-catalyzed carbene insertion to C—H bonds. *Chem. Soc. Rev.* 51: 2759–2852.

12 Chuprakov, S., Malik, J.A., Zibinsky, M., and Fokin, V.V. (2011). Catalytic asymmetric C—H insertions of rhodium(II) azavinyl carbenes. *J. Am. Chem. Soc.* 133: 10352–10355.

13 Williams, M.B., Wells, R.J., and Boyer, A. (2022). Synthesis and reactivity of 1-sulfonylcyclooctatriazoles. *Chem. Commun.* 58: 12495–12498.

14 Kubiak, R.W., Mighion, J.D., Wilkerson-Hill, S.M. et al. (2016). Enantioselective intermolecular C—H functionalization of allylic and benzylic sp^3 C—H bonds using *N*-sulfonyl-1,2,3-triazoles. *Org. Lett.* 18: 3118–3121.

15 Kubiak, R.W., and Davies, H.M.L. (2018). Rhodium-catalyzed intermolecular C—H functionalization as a key step in the synthesis of complex stereodefined β-arylpyrrolidines. *Org. Lett.* 20: 3771–3775.

16 Vaitla, J., Boni, Y.T., and Davies, H.M.L. (2020). Distal allylic/benzylic C—H functionalization of silyl ethers using donor/acceptor rhodium(II) carbenes. *Angew. Chem. Int. Ed.* 59: 7397–7402.

17 Seyferth, D., Washburne, S.S., Attridge, C.J., and Yamamoto, K. (1970). Halomethylmetal compounds. XXXIV. Insertion of phenyl(bromodichloro-methyl)mercury-derived dichlorocarbene into carbon-hydrogen bonds. Tetraalkylsilicon and tetraalkyltin compounds. *J. Am. Chem. Soc.* 92: 4405–4417.

18 Ando, W., Konishi, K., and Migita, T. (1974). Reactions of carboalkoxycarbenes with alkylsilanes. *J. Org. Chem.* 67: C7–C9.

19 Hatanaka, Y., Watanabe, M., Onozawa, S. et al. (1998). Rhodium-catalyzed insertion of carbenoids into β C—H bonds of silacycloalkanes: a facile and general approach to functionalized silacycloalkanes. *J. Org. Chem.* 63: 422–423.

20 Garlets, Z.J., and Davies, H.M.L. (2018). Harnessing the β-silicon effect for regioselective and stereoselective rhodium(II)-catalyzed C—H functionalization by donor/acceptor carbenes derived from 1-sulfonyl-1,2,3-triazoles. *Org. Lett.* 20: 2168–2171.

21 Horneff, T., Chuprakov, S., Chernyak, N. et al. (2008). Rhodium-catalyzed transannulation of 1,2,3-triazoles with nitriles. *J. Am. Chem. Soc.* 130: 14972–14974.

22 Medina, F., Besnard, C., and Lacour, J. (2014). One-step synthesis of nitrogen-containing medium-sized rings via α-imino diazo intermediates. *Org. Lett.* 16: 3232–3235.

23 Pospech, J., Ferraccioli, R., Neumann, H., and Beller, M. (2015). Rhodium(II)-catalyzed annulation of azavinyl carbenes through ring-expansion of 1,3,5-trioxane: rapid access to nine-membered 1,3,5,7-trioxazonines. *Chem. Asian J.* 10: 2624–2630.

24 Li, L., Xia, X.-H., Wang, Y. et al. (2015). Synthesis of benzofurans *via* tandem rhodium-catalyzed C(sp^3)—H insertion and copper-catalyzed dehydrogenation. *Adv. Synth. Catal.* 357: 2089–2097.

25 Shen, H., Fu, J., Yuan, H. et al. (2016). Synthesis of 2,3-disubstituted indoles and benzofurans by the tandem reaction of rhodium(II)-catalyzed intramolecular C—H insertion and oxygen-mediated oxidation. *J. Org. Chem.* 81: 10180–10192.

26 Ma, X., Wu, F., Yi, X. et al. (2015). One-pot synthesis of 2,3-disubstituted dihydrobenzofurans and benzofurans via rhodium-catalyzed intramolecular C—H insertion reaction. *Chem. Commun.* 51: 6862–6865.

27 Shen, M.-H., Pan, Y.-P., Jia, Z.-H. et al. (2015). An efficient approach to 1,2,3-trisubstituted indole via rhodium catalyzed carbene C_{sp3}—H bond insertion. *Org. Biomol. Chem.* 13: 4851–4854.

28 Senoo, M., Furukawa, A., Hata, T., and Urabe, H. (2016). Rhodium-catalyzed intramolecular C—H bond activation with triazoles: preparation of stereodefined pyrrolidines and other related cyclic compounds. *Chem. Eur. J.* 22: 890–895.

29 Lindsay, V.N.G., Viart, H.M.-F., and Sarpong, R. (2015). Stereodivergent intramolecular C(sp^3)—H functionalization of azavinyl carbenes: synthesis of saturated heterocycles and fused N-heterotricycles. *J. Am. Chem. Soc.* 137: 8368–8371.

30 Yadagiri, D., and Anbarasan, P. (2014). Rhodium catalyzed direct arylation of α-diazoimines. *Org. Lett.* 16: 2510–2513.

31 Park, S., Yong, W.-S., Kim, S., and Lee, P.H. (2014). Diastereoselective *N*-sulfonylaminoalkenylation of azulenes from terminal alkynes and azides via *N*-sulfonyl-1,2,3-triazoles. *Org. Lett.* 16: 4468–4471.

32 Maeng, C., Son, J.-Y., Lee, S.C. et al. (2020). Expansion of azulenes as non-benzenoid aromatic compounds for C—H activation: rhodium- and iridium-catalyzed oxidative cyclization of azulene carboxylic acids with alkynes for the synthesis of azulenolactones and benzoazulenes. *J. Org. Chem.* 85: 3824–3837.

33 Maeng, C., Yun, Y., Son, J., and Lee, P.H. (2022). Palladium-catalyzed oxidative cyclization of azulene-2-carboxylic acids with 1,3-dienes for the synthesis of alkenyl azulenolactones. *Adv. Synth. Catal.* 364: 2859–2864.

34 Shin, S., Park, Y., Kim, C.-E. et al. (2015). Synthesis of 2-alkoxyaryl-2-aryl enamines via tandem copper-catalyzed cycloaddition and rhodium-catalyzed alkoxyarylation from alkynes, *N*-sulfonyl azides, and aryl ethers. *J. Org. Chem.* 80: 5859–5869.

35 Miura, T., Zhao, Q., and Murakami, M. (2017). Selective functionalization of aromatic C(sp^2)—H bonds in the presence of benzylic C(sp^3)—H bonds by electron-deficient carbenoids generated from 4-acyl-1-sulfonyl-1,2,3-triazoles. *Angew. Chem. Int. Ed.* 56: 16645–16649.

36 Zhao, Q., Miura, T., and Murakami, M. (2019). Cyclization reaction of 4-acyl-1-sulfonyl-1,2,3-triazoles possessing phenyl rings through generation of electron-deficient carbenoids. *Chem. Lett.* 48: 510–512.

37 Rajagopal, B., Chou, C.-H., Chung, C.-C., and Lin, P.-C. (2014). Synthesis of substituted 3-indolylimines and indole-3-carboxaldehydes by rhodium(II)-catalyzed annulation. *Org. Lett.* 16: 3752–3755.

38 Xu, H.-D., Xu, K., Jia, Z.-H. et al. (2014). Substituent enabled divergent synthesis of N-heterocycles: a metal carbene approach involving intramolecular carbene interception. *Asian J. Org. Chem.* 3: 1154–1158.

39 Tang, X.-Y., Zhang, Y.-S., He, L. et al. (2015). Intramolecular annulation of aromatic rings with *N*-sulfonyl 1,2,3-triazoles: divergent synthesis of 3-methylene-2,3-dihydrobenzofurans and 3-methylene-2,3-dihydroindoles. *Chem. Commun.* 51: 133–136.

40 Xu, H.-D., Pan, Y.-P., Ren, X.-T. et al. (2015). A one-pot construction of acridones by rhodium catalyzed reaction of *N*-phenyl-2-(1-sulfonyl-1*H*-1,2,3-triazol-4-yl)aniline. *Tetrahedron Lett.* 56: 6734–6737.

41 Seo, B., Jeon, W.H., Kim, J. et al. (2015). Synthesis of fluorenes via tandem copper-catalyzed [3 + 2] cycloaddition and rhodium-catalyzed denitrogenative cyclization in a 5-*exo* mode from 2-ethynylbiaryls and *N*-sulfonyl azides in one pot. *J. Org. Chem.* 80: 722–732.

42 Shin, S., Son, J.-Y., Choi, C. et al. (2016). Synthesis of benzofulvenes through rhodium-catalyzed transannulation of enynyl triazoles. *J. Org. Chem.* 81: 11706–11715.

43 Alford, J.S., and Davies, H.M.L. (2014). Mild aminoacylation of indoles and pyrroles through a three-component reaction with Ynol ethers and sulfonyl azides. *J. Am. Chem. Soc.* 136: 10266–10269.

44 Miura, T., Murakami, M., Zhao, Q., and Funakoshi, Y. (2015). Site-selective introduction of an enamido group at the C(3)-position of indoles. *Heterocycles* 91: 1579.

45 Huang, J., Yang, Y., and Chen, Z. (2016). Rhodium(II)-catalyzed regioselective carbenoid insertion reaction of simple indoles with *N*-sulfonyltriazoles: a rapid access to tryptamine vinylogues. *Adv. Synth. Catal.* 358: 201–206.

46 Rajasekar, S., Yadagiri, D., and Anbarasan, P. (2015). One-pot aminoethylation of indoles/pyrroles with alkynes and sulfonyl azides. *Chem. Eur. J.* 21: 17079–17084.

47 Tian, M., Liu, B., Sun, J., and Li, X. (2018). Rh(III)-catalyzed C—C coupling of diverse arenes and 4-acyl-1-sulfonyltriazoles via C—H activation. *Org. Lett.* 20: 4946–4949.

48 Zou, Y., Wang, P., Kong, L., and Li, X. (2022). Rhodium-catalyzed atroposelective C—H arylation of (hetero)arenes using carbene precursors as arylating reagents. *Org. Lett.* 24: 3189–3193.

49 Kahar, N.M., Nabar, K.U., Jadhav, P.P., and Dawande, S.G. (2019). Rhodium(II)-catalyzed highly stereoselective C3 functionalization of indolizines with *N*-sulfonyl-1,2,3-triazoles. *Asian J. Org. Chem.* 8: 79–82.

50 Samala, S., Shin, J., Shim, J.Y., and Yoo, E.J. (2017). Regioselective C3 functionalizations of 9*H*-carbazoles via C(sp^2)—H insertions of Rh^{II} carbenoids. *Asian J. Org. Chem.* 6: 998–1002.

51 Luo, Z., Wang, C., Jiang, H., and Zeng, W. (2023). Amine-directed regioselective pyridyl Csp^2—H aminovinylation. *Org. Lett.* 25: 7020–7024.

52 Rajasekar, S., and Anbarasan, P. (2019). One-pot transannulation of *N*-sulfonyl-1,2,3-triazoles to dihydro-β-carbolines and dihydroisoquinolines via rhodium-catalyzed C—H insertion-cum-base-mediated *aza*-Michael reaction. *J. Org. Chem.* 84: 7747–7761.

53 Rajasekar, S., and Anbarasan, P. (2019). Tandem Rh(II) and chiral squaramide relay catalysis: enantioselective synthesis of dihydro-β-carbolines via insertion to C—H bond and *aza*-Michael reaction. *Org. Lett.* 21: 3067–3071.

54 Rajasekar, S., Ramachandran, K., and Anbarasan, P. (2022). Efficient synthesis of 3-aminocarbazoles from *N*-sulfonyl-1,2,3-triazoles and 2-alkenylindole. *J. Chem. Sci.* 134: 121.

55 Duan, S., Xue, B., Meng, H. et al. (2021). Rhodium(II)-catalyzed [4 + 3] cyclization of triazoles with indole derivatives and its application in the total synthesis of (±)-aurantioclavine. *Chin. J. Chem.* 39: 1145–1152.

56 Yang, J., Zhu, C., Tang, X., and Shi, M. (2014). Rhodium(II)-catalyzed intramolecular annulation of 1-sulfonyl-1,2,3-triazoles with pyrrole and indole rings: facile synthesis of N-bridgehead azepine skeletons. *Angew. Chem. Int. Ed.* 53: 5142–5146.

57 Shang, H., Tian, Y., Luo, J. et al. (2016). Rhodium(II)-catalyzed intramolecular annulation of 1-sulfonyl-1,2,3-triazoles with indoles: facile synthesis of functionalized tetrahydro-β-carbolines. *RSC Adv.* 6: 30835–30839.

58 Xie, H., Yang, J.-X., Bora, P.P., and Kang, Q. (2016). Rh(II)-catalyzed intramolecular annulation of *N*-sulfonyl 1,2,3-triazoles with indole derivatives: a new method for synthesis pyranoindoles. *Tetrahedron* 72: 3014–3021.

59 Zhang, Y.-S., Tang, X.-Y., and Shi, M. (2015). Divergent synthesis of indole-fused polycycles via Rh(II)-catalyzed intramolecular [3 + 2] cycloaddition and C—H functionalization of indolyltriazoles. *Org. Chem. Front.* 2: 1516–1520.

60 Fu, L., and Davies, H.M.L. (2017). Scope of the reactions of indolyl- and pyrrolyl-tethered *N*-sulfonyl-1,2,3-triazoles: rhodium(II)-catalyzed synthesis of indole- and pyrrole-fused polycyclic compounds. *Org. Lett.* 19: 1504–1507.

61 Miura, T., Biyajima, T., Fujii, T., and Murakami, M. (2012). Synthesis of α-amino ketones from terminal alkynes via rhodium-catalyzed denitrogenative hydration of *N*-sulfonyl-1,2,3-triazoles. *J. Am. Chem. Soc.* 134: 194–196.

62 Xing, S., Xia, H., Wang, X. et al. (2019). Diastereoselective access to 2-aminoindanones *via* the rhodium(II)-catalyzed tandem reaction involving O—H insertion and Michael addition. *Org. Chem. Front.* 6: 3121–3126.

63 Liu, S., Yao, W., Liu, Y. et al. (2017). A Rh(II)-catalyzed multicomponent reaction by trapping an a-amino enol intermediate in a traditional two-component reaction pathway. *Sci. Adv.* 3: e1602467.

64 Chuprakov, S., Worrell, B.T., Selander, N. et al. (2014). Stereoselective 1,3-insertions of rhodium(II) azavinyl carbenes. *J. Am. Chem. Soc.* 136: 195–202.

65 Miura, T., Nakamuro, T., Hiraga, K., and Murakami, M. (2014). The stereoselective synthesis of α-amino aldols starting from terminal alkynes. *Chem. Commun.* 50: 10474–10477.

66 Kaladevi, S., Kamalraj, M., Altia, M. et al. (2019). Cascade Rh(II) and Yb(III) catalyzed synthesis of substituted naphthofurans *via* transannulation of *N*-sulfonyl-1,2,3-triazoles with β-naphthols. *Chem. Commun.* 55: 4507–4510.

67 Miura, T., Tanaka, T., Biyajima, T. et al. (2013). One-pot procedure for the introduction of three different bonds onto terminal alkynes through *N*-sulfonyl-1,2,3-triazole intermediates. *Angew. Chem. Int. Ed.* 52: 3883–3886.

68 Miura, T., Tanaka, T., Zhao, Q. et al. (2017). Synthesis of 2-substituted 2-amino ketones by rhodium-catalyzed reaction of *n*-sulfonyl-1,2,3-triazoles with 2-alkenols. *Helv. Chim. Acta* 100: e1600320.

69 Chen, Y., Dong, S., Xu, X. et al. (2018). Bimetallic rhodium(II)/indium(III) relay catalysis for tandem insertion/asymmetric Claisen rearrangement. *Angew. Chem. Int. Ed.* 57: 16554–16558.

70 Liu, Y., Chen, Y., Yihuo, A. et al. (2022). Diastereodivergent synthesis of chiral α-aminoketones via a catalytic O—H insertion/Barnes–Claisen rearrangement reaction. *ACS Catal.* 12: 1784–1790.

71 Alcaide, B., Almendros, P., Fernández, I. et al. (2019). Chemoselectivity switching in the rhodium-catalyzed reactions of 4-substituted-1-sulfonyl-1,2,3-triazoles with allenols: noticeable differences between 4-acyl- and 4-aryl-triazoles. *Adv. Synth. Catal.* 361: 1160–1165.

72 Alcaide, B., Almendros, P., Cembellín, S. et al. (2017). Allenols versus allenones: rhodium-catalyzed regiodivergent and tunable allene reactivity with triazoles. *Chem. Eur. J.* 23: 13754–13759.

73 Jung, D.J., Jeon, H.J., Lee, J.H., and Lee, S. (2015). Cu^{I}/Rh^{II}-catalyzed tandem convergent multicomponent reaction for the regio- and stereocontrolled synthesis of γ-oxo-β-amino esters. *Org. Lett.* 17: 3498–3501.

74 Jeon, H.J., Kwak, M.S., Jung, D.J. et al. (2016). Synthesis of α-methylene-δ-oxo-γ-amino esters via Rh(II)-catalyzed coupling of 1-sulfonyl-1,2,3-triazoles with Morita–Baylis–Hillman adducts. *Org. Biomol. Chem.* 14: 11238–11243.

75 Jia, R., Meng, J., Leng, J. et al. (2018). Rhodium(II)-catalyzed reaction of 1-tosyl-1,2,3-triazoles with Morita–Baylis–Hillman adducts: synthesis of 3,4-fused pyrroles. *Chem. Asian J.* 13: 2360–2364.

76 Xu, Z., Wang, W., Cen, M. et al. (2020). Synthesis of α-(2-hydroxyphenyl)-α-aminoketones by rhodium-catalyzed tandem reaction of 1-sulfonyl-1,2,3-triazoles and benzoquinone-derived alcohols. *Adv. Synth. Catal.* 362: 2888–2893.

77 Huang, L.-Z., Xuan, Z., Park, J.-U., and Kim, J.H. (2022). Dual Rh(II)/Pd(0) relay catalysis involving sigmatropic rearrangement using *N*-sulfonyl triazoles and 2-hydroxymethylallyl carbonates. *Org. Lett.* 24: 6951–6956.

78 Park, J.-U., Huang, L.-Z., Cho, H.-J. et al. (2023). One-pot synthesis of unprotected 2-acylpyrroles from 1,2,3-triazoles and 2-hydroxymethylallyl carbonates. *J. Org. Chem.* 88: 585–593.

79 Miura, T., Tanaka, T., Matsumoto, K., and Murakami, M. (2014). One-pot synthesis of 2,5-dihydropyrroles from terminal alkynes, azides, and propargylic alcohols by relay actions of copper, rhodium, and gold. *Chem. Eur. J.* 20: 16078–16082.

80 Cheng, X., Yu, Y., Mao, Z. et al. (2016). Facile synthesis of substituted 3-aminofurans through a tandem reaction of *N*-sulfonyl-1,2,3-triazoles with propargyl alcohols. *Org. Biomol. Chem.* 14: 3878–3882.

81 Mi, P., Wang, H., Zhao, R., and Song, J. (2018). Tandem O—H insertion/[3,3]-sigmatropic rearrangement of rhodium carbenoids with 2-furfuryl alcohols: a strategy to access polysubstituted furans. *Eur. J. Org. Chem.* 2018: 759–762.

82 Mi, P., Kiran Kumar, R., Liao, P., and Bi, X. (2016). Tandem O—H insertion/[1,3]-alkyl shift of rhodium azavinyl carbenoids with benzylic alcohols: a route to convert C—OH bonds into C—C bonds. *Org. Lett.* 18: 4998–5001.

83 Chen, Y., Liu, Y., Li, Z. et al. (2020). Tandem insertion-[1,3]-rearrangement: highly enantioselective construction of α-aminoketones. *Angew. Chem. Int. Ed.* 59: 8052–8056.

84 Bennett, J.M., Shapiro, J.D., Choinski, K.N. et al. (2017). Synthesis of phthalan and phenethylamine derivatives via addition of alcohols to rhodium(II)-azavinyl carbenoids. *Tetrahedron Lett.* 58: 1117–1122.

85 Lee, K.R., Jung, D.J., Ahn, S. et al. (2021). Catalyst-controlled divergent transformations of *N*-sulfonyl-1,2,3-triazoles into isoquinolin-3-ones and 2-aminoindanones. *Org. Biomol. Chem.* 19: 5093–5097.

86 Feng, Z., Wang, W., Yan, Z. et al. (2022). Synthesis of α-amino cyclobutanones via formal 1,3-hydroxy migration triggered by formation of α-imino rhodium carbene. *Org. Lett.* 24: 2950–2954.

87 Feng, Z., Jiao, H., Ye, Z. et al. (2022). Synthesis of azepane derivatives via formal 1,3-migration of hydroxy and acyloxy groups and selective annulation. *Org. Lett.* 24: 5254–5259.

88 Ko, Y.O., Jeon, H.J., Jung, D.J. et al. (2016). Rh(II)/$Mg(O^tBu)_2$-catalyzed tandem one-pot synthesis of 1,4-oxazepines and 1,4-oxazines from *N*-sulfonyl-1,2,3-triazoles and glycidols. *Org. Lett.* 18: 6432–6435.

89 Jones, K.D., Nutt, M.J., Comninos, E. et al. (2020). A one-pot reaction of α-imino rhodium carbenoids and halohydrins: access to 2,6-substituted dihydro-2*H*-1,4-oxazines. *Org. Lett.* 22: 3490–3494.

90 Gunawan, N., Nutt, M.J., Bissember, A.C. et al. (2023). *N*-Tosyl-1,2,3-triazoles as scaffolds for morpholines: the total synthesis of (−)-chelonin A. *Synlett* 34: 1524–1528.

91 Seo, B., Jeon, W.H., Kim, C.-E. et al. (2016). Sequential functionalization of the O—H and C(sp^2)—O bonds of tropolones by alkynes and *N*-sulfonyl azides. *Adv. Synth. Catal.* 358: 1078–1087.

92 Cheng, W., Tang, Y., Xu, Z.-F., and Li, C.-Y. (2016). Synthesis of multifunctionalized 2-carbonylpyrrole by rhodium-catalyzed transannulation of 1-sulfonyl-1,2,3-triazole with β-diketone. *Org. Lett.* 18: 6168–6171.

93 Mi, P., Yuan, H., Wang, H. et al. (2017). Divergent reactions between α-imino rhodium carbenoids and 1,3-diketones: substrate-controlled O—H versus C—H insertion. *Eur. J. Org. Chem.* 2017: 1289–1293.

94 Meng, J., Wen, M., Zhang, S. et al. (2017). Unexpected O—H insertion of rhodium-azavinylcarbenes with *N*-acylhydrazones: divergent synthesis of

3,6-disubstituted- and 3,5,6-trisubstituted-1,2,4-triazines. *J. Org. Chem.* 82: 1676–1687.

95 Pal, K., Hoque, A., and Volla, C.M.R. (2018). Rh-catalyzed denitrogenative reaction of *N*-sulfonyl-1,2,3-triazoles with isatoic anhydrides and oxadiazolones. *Chem. Eur. J.* 24: 2558–2564.

96 Pandey, S., Parveen, S., and Volla, C.M.R. (2023). Rh(II)-catalyzed denitrogenative reaction of *N*-sulfonyl-1,2,3-triazoles with quinolones and isoquinolones. *Chem. Asian J.* 18: e202300614.

97 Kook, G.Y., Kim, D., Chae, M.K., and Ko, H.M. (2022). Rhodium(II)-catalyzed highly selective 1,3-insertion reactions using *N*-sulfonyl-1,2,3-triazoles with heteroaryl ethers or heteroaryl alcohols. *J. Org. Chem.* 87: 7253–7263.

98 Pal, K., Sontakke, G.S., and Volla, C.M.R. (2019). Rh(II)-catalyzed highly diastereoselective cascade transannulation of *N*-sulfonyl-1,2,3-triazoles and vinyl benzoxazinanones. *Org. Lett.* 21: 3716–3720.

99 Meng, J., Jia, R., Leng, J. et al. (2017). Carbon–carbon bond formation by reaction of rhodium azavinylcarbenes with secondary amides: access to indigo analogues from isatins. *Org. Lett.* 19: 4520–4523.

100 Pal, K., Shukla, R.K., and Volla, C.M.R. (2017). Rh(II)-catalyzed denitrogenative reaction of *n*-sulfonyl-1,2,3-triazoles with isatins for the construction of indigoids. *Org. Lett.* 19: 5764–5767.

101 Ma, X., Xie, X., Liu, L. et al. (2018). Facile synthesis of pyrroloindoles *via* a rhodium(II)-catalyzed annulation of 3-benzylidene-indolin-2-ones and α-imino carbenes. *Chem. Commun.* 54: 1595–1598.

102 Ogunlana, A.A., Zou, J., and Bao, X. (2019). Insights into the mechanisms and chemoselectivities of carbamates and amides in reactions involving Rh(II)-azavinylcarbene: a computational study. *J. Org. Chem.* 84: 8151–8159.

103 Lee, D.J., and Yoo, E.J. (2015). Efficient synthesis of C—N-coupled heterobiaryls by sequential N—H functionalization reactions. *Org. Lett.* 17: 1830–1833.

104 He, X., Wu, Y., Zhou, T. et al. (2020). Rh-catalyzed C—N coupling of *N*-sulfonyl-1,2,3-trizales with secondary amines for regioselective synthesis of phenylvinyl-1,2-diamines. *Synth. Commun.* 50: 2685–2697.

105 Williams, M.B., and Boyer, A. (2022). Modular synthesis of highly substituted 3-azapyrroles by Rh(II)-catalyzed N—H bond insertion and cyclodehydration. *J. Org. Chem.* 87: 16139–16156.

106 Jeon, H.J., Jung, D.J., Kim, J.H. et al. (2014). From triazoles to imidazolines through the sequential N—H insertion of α-imino rhodium–carbenes into β-enamino esters/enamine–imine tautomerization/conjugate addition cascade. *J. Org. Chem.* 79: 9865–9871.

107 Wang, H., Zhou, T., Wu, M. et al. (2023). Substituent-controllable cascade regioselective annulation of β-enaminones with *N*-sulfonyl triazoles for modular access to imidazoles and pyrroles. *Molecules* 28: 4416.

108 Nutt, M.J., Annear, J.W., Jones, K.D. et al. (2023). Dirhodium-catalyzed transannulation of *N*-sulfonyl-1,2,3-triazoles to 2,3-dehydropiperazines. *J. Org. Chem.* 88: 11968–11979.

109 Yadagiri, D., Reddy, A.C.S., and Anbarasan, P. (2016). Rhodium catalyzed diastereoselective synthesis of 2,2,3,3-tetrasubstituted indolines from *N*-sulfonyl-1,2,3-triazoles and ortho-vinylanilines. *Chem. Sci.* 7: 5934–5938.

110 Rupa, K., Yadagiri, D., Bagavathi, R., and Anbarasan, P. (2023). Synthesis of dihydro-3,1-benzoxazine derivatives from 1,3-amino alcohols and *N*-sulfonyl-1,2,3-triazole. *Org. Lett.* 25: 3375–3379.

111 Rupa, K., Yadagiri, D., and Anbarasan, P. (2023). Rhodium-catalyzed [4 + 2]-annulation of *o*-acylanilines with *N*-sulfonyl-1,2,3-triazoles: synthesis of 3-aminoquinolines. *J. Org. Chem.* 88: 9077–9086.

112 Li, J., Feng, J., Chen, T. et al. (2023). Synthesis of 3-aminoquinolines from α-imino rhodium carbenes and 2-aminobenzaldehydes. *Org. Biomol. Chem.* 21: 5935–5938.

113 Miura, T., Nakamuro, T., Miyakawa, S., and Murakami, M. (2016). A *syn*-selective *aza*-aldol reaction of boron *aza*-enolates generated from *N*-sulfonyl-1,2,3-triazoles and 9-BBN-H. *Angew. Chem. Int. Ed.* 55: 8732–8735.

114 Miura, T., Miyakawa, S., Nakamuro, T., and Murakami, M. (2019). Generation of boron *aza*-enolates by a nickel-catalyzed reaction of triazoles with pinacolborane and their addition to aldehydes. *Chem. Lett.* 48: 965–967.

115 Wang, H., Qiao, H., Zhang, H. et al. (2015). Rhodium-catalyzed hydrosilylation reaction of *N*-sulfonyl-1,2,3-triazoles with triphenylsilane: access to diverse compounds. *Eur. J. Org. Chem.* 2015: 4471–4480.

116 Liu, Y., Xie, P., Li, J. et al. (2019). Nickel-catalyzed coupling of *N*-sulfonyl-1,2,3-triazole with *H*-phosphine oxides: stereoselective and site-selective synthesis of α-aminovinylphosphoryl derivatives. *Org. Lett.* 21: 4944–4949.

117 Kotovshchikov, Y.N., Latyshev, G.V., Navasardyan, M.A. et al. (2018). Annulation-induced cascade transformation of 5-iodo-1,2,3-triazoles to 2-(1-aminoalkyl)benzoxazoles. *Org. Lett.* 20: 4467–4470.

118 Hein, J.E., Tripp, J.C., Krasnova, L.B. et al. (2009). Copper(I)-catalyzed cycloaddition of organic azides and 1-iodoalkynes. *Angew. Chem. Int. Ed.* 48: 8018–8021.

119 Tatevosyan, S.S., Kotovshchikov, Y.N., Latyshev, G.V. et al. (2020). A route to triazole-fused sultams via metal-free base-mediated cyclization of sulfonamide-tethered 5-iodotriazoles. *J. Org. Chem.* 85: 7863–7876.

120 Tatevosyan, S.S., Kotovshchikov, Y.N., Latyshev, G.V. et al. (2022). Facile access to triazole-fused 3,1-benzoxazines enabled by metal-free base-promoted intramolecular C—O coupling. *Synthesis* 54: 369–377.

121 Kotovshchikov, Y.N., Tatevosyan, S.S., Latyshev, G.V. et al. (2023). Divergent cyclization of 2-(5-iodo-1,2,3-triazolyl)benzamides toward triazole-fused lactams and cyclic imidates. *New J. Chem.* 47: 12239–12247.

122 Kotovshchikov, Y.N., Latyshev, G.V., Kirillova, E.A. et al. (2020). Assembly of thiosubstituted benzoxazoles via copper-catalyzed coupling of thiols with 5-iodotriazoles serving as diazo surrogates. *J. Org. Chem.* 85: 9015–9028.

123 Kotovshchikov, Y.N., Sultanov, R.H., Latyshev, G.V. et al. (2022). Domino assembly of dithiocarbamates *via* Cu-catalyzed denitrogenative thiolation of iodotriazole-based diazo precursors. *Org. Biomol. Chem.* 20: 5764–5770.

124 Gevondian, A.G., Kotovshchikov, Y.N., Latyshev, G.V. et al. (2021). Domino construction of benzoxazole-derived sulfonamides *via* metal-free denitrogenation of 5-iodo-1,2,3-triazoles in the presence of SO_2 and amines. *J. Org. Chem.* 86: 5639–5650.

125 Voloshkin, V.A., Kotovshchikov, Y.N., Latyshev, G.V. et al. (2022). Annulation-triggered denitrogenative transformations of 2-(5-iodo-1,2,3-triazolyl)benzoic acids. *J. Org. Chem.* 87: 7064–7075.

6

Insertion of Azavinyl Carbenes to C—*X*/*X*—*X* Bond

Claire Empel and Rene M. Koenigs

Institute of Organic Chemistry, RWTH Aachen University, Aachen, Germany

6.1 Introduction

Carbenes are versatile reactive intermediates and can exhibit electrophilic, nucleophilic, or radical character depending on the nature of the substituents attached to the carbene center and spin state. Their reactivity typically involves the reaction at the carbene center, while substituents usually play a spectator role. As such, carbenes find a plethora of applications in organic chemistry, of which the reaction with heteroatom nucleophiles is of particular interest to give ylide intermediates. These ylide intermediates can give rise to a multitude of rearrangement or *X*—H insertion reactions depending on the substitution pattern of the heteroatom nucleophile [1–3].

The most commonly used precursors to carbenes are diazo compounds, or building blocks that mask the diazo functional group. In this context, the triazole heterocycle **6.1a** takes up a distinct position and can unveil an α-imino diazo compound **6.1b** upon thermally induced Dimroth rearrangement [3, 4]. As a direct consequence, triazole delivers an α-imino carbene intermediate **6.1c**, which itself can react both at the carbene center and at the imine nitrogen. This in turn, results in multifold reactivity patterns of α-imino carbenes with heteroatom nucleophiles, which will be discussed in this chapter in the context of insertion reactions into C—*X* or *X*—*X* bonds (Scheme 6.1). It should be noted that this reactivity pattern can be employed across a variety of further transformations, such as transannulation reactions. The reader is referred to the respective chapters for an overview on reactions that go beyond the insertion reaction into C—C or C—*X* bonds.

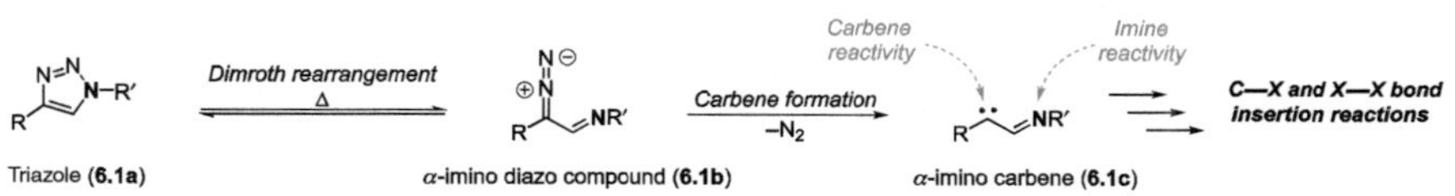

Scheme 6.1 Formation and reactivity of α-imino carbenes.

Denitrogenative Transformation of Nitrogen Heterocycles: Synthesis, Reactions and Applications, First Edition. Edited by Pazhamalai Anbarasan.

6.2 1,3-Difunctionalization

The thermal reaction of triazoles in the presence of suitable metal catalysts proceeds *via* an initial Dimroth rearrangement to expose the diazo functional group followed by nitrogen extrusion and formation of a metal-bound imino carbene. The latter typically shows electrophilic reactivity and can react – similar to other electrophilic metal carbene complexes – with heteroatom nucleophiles. In this context, a seminal contribution by Murakami describes the one-pot reaction of a terminal alkyne with sulfonyl azides to give a triazole heterocycle in the presence of a Cu(I) catalyst at room temperature. Addition of a Rh(II) catalyst and water as a reagent to the reaction mixture and heating to 120 °C furnishes the desorbed α-amino carbonyl compound, where the former carbene center is formally oxidized to a carbonyl group and the sulfonyl-protected imino functional group is formally reduced to a sulfonamide [5]. Although this transformation displays an example of O—H functionalization, this initial report already features the main topic of this chapter, which focuses on the insertion of the imino carbene fragment into C—*X* or *X*—*X* bonds. Specifically, two different types of reaction pathways should be distinguished: In a more classic sense of carbene reactivity, the carbene center is formally inserted into a C—*X* bond, which can be regarded as a 1,1-difunctionalization reaction. In contrast, in the aforementioned case of imino carbenes, the entire three-atom imino carbene fragment is formally inserted into a C—*X* bond, which can be seen as a 1,3-difunctionalization reaction of imino carbenes or as an enamine homologation reaction.

The concept of such 1,3-difunctionalization reactions was further explored by Murakami and co-workers in the reaction with allylic alcohol. Following the same line of initial reaction steps, a rhodium-bound imino carbene is formed, which can be intercepted by allylic alcohol to give the formal product of 1,3-difuntionalization. Here, however, an allyl vinyl ether is formed, which is prone toward Claisen rearrangement and results in translocation of the allyl substituent to the alpha position of a newly formed carbonyl group. In summary, this process describes the 1,2,3-trifunctionalization of the imino carbene, where an O—H bond and a C—O bond of the allylic alcohol are broken [6]. Miura and Murakami further exemplified this strategy in the reaction with silanols, which furnishes silyl enol ethers that themselves can be used in Mukaiyama aldol reactions [7].

Almost at the same time, Fokin and co-workers reported on related reactions using triazoles and nitrogen-based nucleophiles, such as amides or carbamates. In this case, a formal N—H functionalization reaction occurred, where the nitrogen atom is installed on the carbene center, and the hydrogen atom is transferred to the former imine functional group. However, when oxindole or other cyclic oxygenated N-heterocycles were employed, an intriguing insertion of the imino carbene fragment into the C=O double bond was observed. From a mechanistic perspective, **6.4b** adds to the carbene center of a rhodium imino carbene complex under the formation of *Z*-configured olefinic intermediate **6.4d**, where the sulfonamide group and incoming **6.4b** are on the same face of the newly formed double bond. This orientation then facilitates a nucleophilic substitution reaction and results in translocation

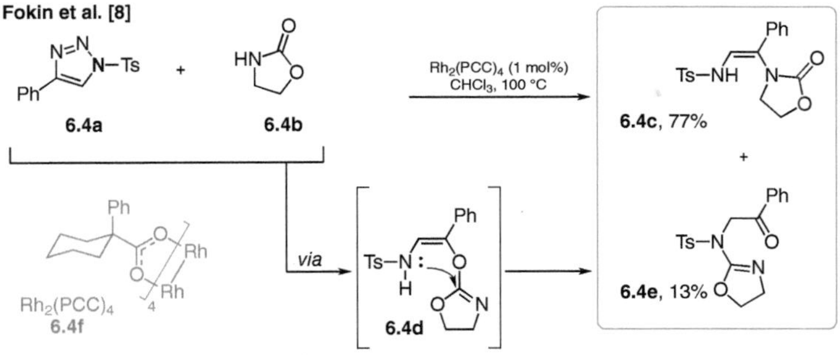

Scheme 6.2 Reaction of triazoles with cyclic oxygen N-heterocycles. Source: Adapted from Fokin et al. [8].

of the heterocycle to the sulfonamide nitrogen atom and formation of a new C=O double bond (Scheme 6.2). This type of reactivity was further explored with regards to the nucleophile, where the authors could demonstrate that carboxylic acids react in a similar reaction under the scission of a C—O bond [8].

This formal insertion of an enamine fragment into C—O bonds is exemplary of reactions that will be further discussed throughout this chapter. An important aspect of such transformation lies within the formation of a ylide intermediate, which can be typically accessed with various heteroatom nucleophiles. In this context, strained saturated oxygen heterocycles, such as epoxides and oxetanes, are of high interest and such 1,3-difunctionalization reactions can proceed with the release of ring strain and the introduction of the enamine fragment into a new heterocycle. Wang, Chen, and co-workers described such a rhodium-catalyzed process. In this case, the rhodium-catalyzed reaction of triazoles **6.3a** initially gives rhodium carbene complex **6.3d**. This complex undergoes ylide formation. The latter features a nucleophilic nitrogen atom in proximity to an electrophile. Intramolecular addition reactions result in ring-opening of the epoxide and formation of a dihydromorpholine ring system (Scheme 6.3a) [9]. This transformation represents an example of a transannulation reaction; however, it can also be regarded as a formal insertion reaction of the imino carbene into the C—O bond of an epoxide. In a related fashion, the groups Xu and Lee followed up on this synthesis strategy using glycidols or episulfides to furnish the corresponding dihydromorpholines [10] or dihydrothiazines (Scheme 6.3b) [11]. Such transformations can similarly be achieved using azirines. These strained N-heterocycles undergo a closely related C—N bond insertion reaction of the enamine fragment as demonstrate by Lee [12], Tang [13], and Shi [14]. These groups disclosed that the reaction of triazoles with carboxylate-substituted azirines yields trisubstituted pyrazines **6.3k** under loss of the sulfonyl-protecting group (Scheme 6.3c) [12]. Alkyl- or H-substituted azirines, however, react to give a pyrrole heterocycle [13, 14]. These differences in reactivity can be rationalized from the reactivity of the azirine-based ylide intermediate. Depending on the electronic properties of the azirine, a cascade of ring-opening and formation of a cationic acyclic intermediate occurs that cyclizes back to give a six-membered ring. In the case of monosubstituted azirines **36.3n** such stabilization

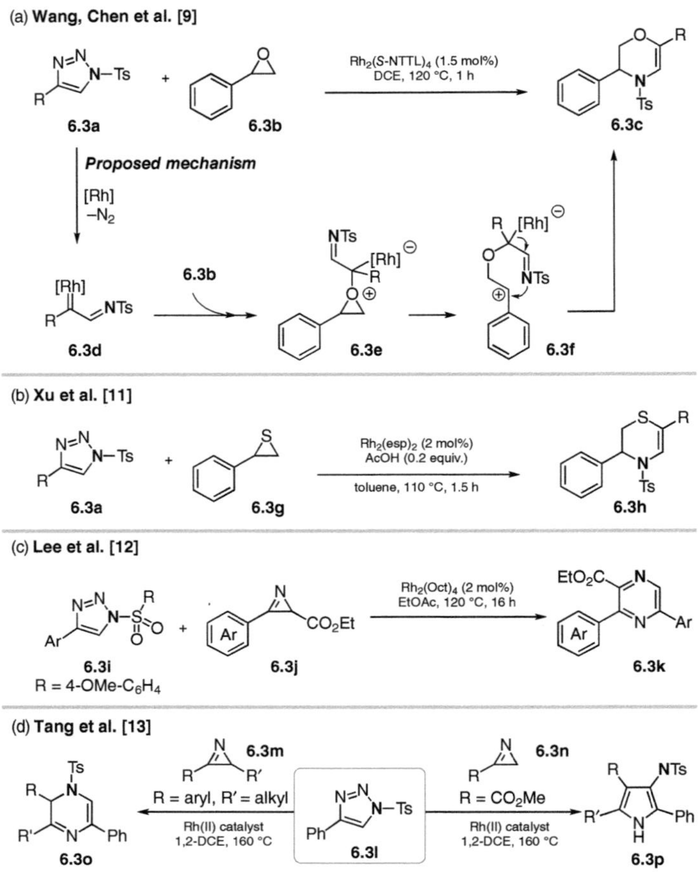

Scheme 6.3 (a) Synthesis of dihydromorpholine ring systems. Source: Adapted from Wang Chen et al. [9]. (b) Episulfides for the synthesis of dihydrothiazines. Source: Adapted from Xu et al. [11]. (c) Synthesis trisubstituted pyrazines from strained *N*-heterocycles. Source: Adapted from Lee et al. [12]. (d) Reactivity of mono- and disubstituted azirines. Source: Adapted from Tang et al. [13].

is not possible, which leads to direct cyclization to the neighboring tosylimine (Scheme 6.3d) [13].

Lacour and co-workers described the reaction of oxetanes **6.4b**, which can give rise to a multifold reactivity pattern, depending on the concentration of the oxetane. Following a similar initial ylide formation pathway, the oxetane-derived ylide intermediate can either undergo a [1,2]-sigmatropic rearrangement leading to tetrahydrofurans **6.4c** or formal insertion of the carbene fragment into the C—O bond of the oxetane. At high concentrations of the oxetane, a second molecule of oxetane can react as a nucleophile followed by cyclization with the nitrogen atom of the former imino carbene to give larger ring systems **6.4d** and **6.4g**. The size of these depends on the number of oxetane units incorporated in these reaction steps (Scheme 6.4) [15].

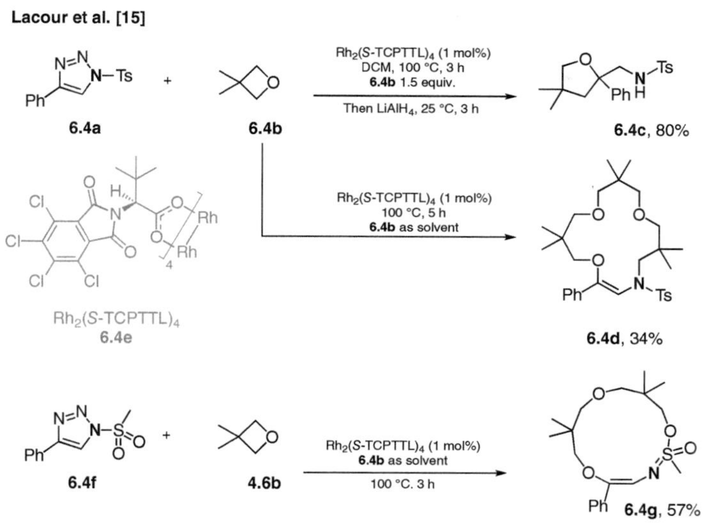

Scheme 6.4 Synthesis of tetrahydrofuranes and larger ring systems from oxetanes. Source: Adapted from Lacour et al. [15].

Such 1,3-difunctionalization reactions – or C—*X* bond insertion reactions – have also been studied in the context of acetals and related heterocycles as nucleophiles. Lacour and co-workers described the reaction of cyclic acetals or mixed acetals derived from formaldehyde. Here, following ylide formation, the enamine fragment undergoes formal insertion into the C—O bond to forge medium-sized ring systems that bear two oxygen atoms and one nitrogen atom (Scheme 6.5a, b) [16, 17]. Beller and co-workers reported on a related reaction using 1,3,5-trioxane **6.5k**, which ring expands under formal insertion of the enamine fragment into one of the six C—O bonds (Scheme 6.5c) [18]. Despite the close resemblance, the related 1,3,5-triazinane ring system **6.5n** experiences a different reactivity and a complex cascade reaction occurs, where an octahydro-purine derivative **6.5o** is ultimately formed (Scheme 6.5d) [19].

The above reactions described the formal C—*X* bond insertion reaction of an imino carbene into the heterocyclic ring system with concomitant ring expansion of the heterocyclic system by either a carbon atom or an enamine fragment.

The process of C—*X* bond insertion reactions was further studied with a range of different nucleophilic acyclic reactants. In analogy to previously described processes, the triazole initially forms a metal-bound imino carbene that reacts in ylide formation. The close proximity of the imine functional and an electrophilic site next to the ylide forming heteroatom can result in a nucleophilic substitution reaction. This concept resembles the process of the reaction of imino carbenes with water, where a proton is migrating. In the context of C—*X* bond insertion reactions, Miura and Murakami disclosed the rhodium-catalyzed reaction of triazoles with acyl sulfides. This reaction proceeds *via* the intermittent formation of a sulfonium ylide, where the cationic sulfur atom is substituted with an acyl group. The latter can be readily translocated to the neighboring nitrogen atom and formation of β-sulfanyl

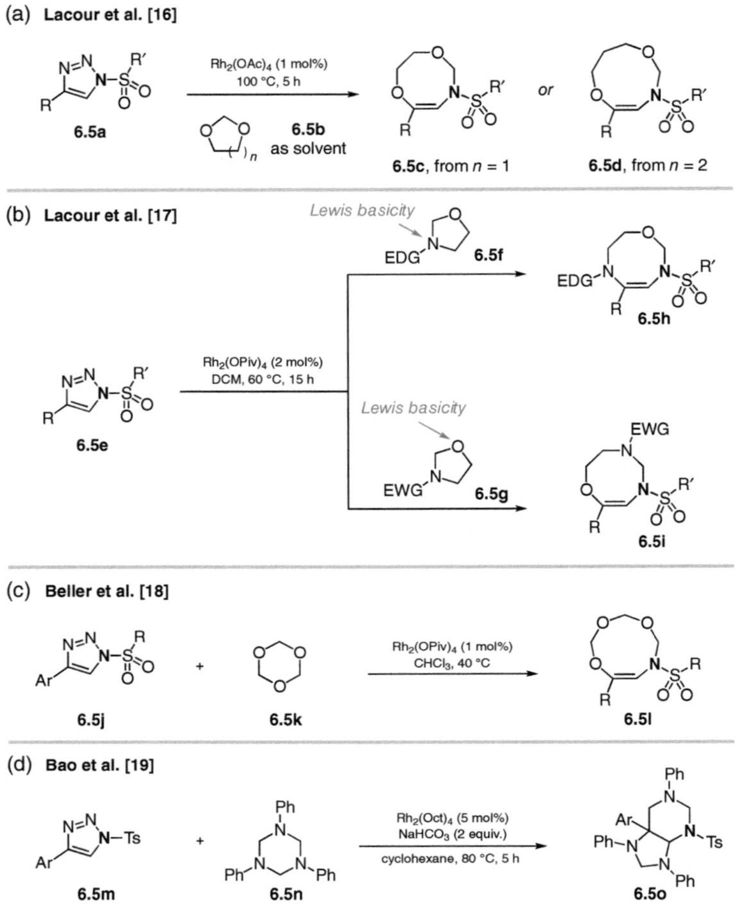

Scheme 6.5 (a) Reaction of cyclic acetals with triazoles. Source: Adapted from Lacour et al. [16]. (b) Reaction of mixed acetals with triazoles. Source: Adapted from Lacour et al. [17]. (c) Reaction of 1,3,5-trioxane with triazoles. Source: Adapted from Beller et al. [18]. (d) Reaction of 1,3,5-triazinane with triazoles. Source: Bao et al. [19].

enamide **6.6c** as a reaction product (Scheme 6.6a) [20]. Importantly, in the case of thionoester **6.6h**, a related reaction occurs, yet with a strikingly different reaction outcome. In this case, a non-classical ylide intermediate is formed, where the positive charge is delocalized across the thionoester, which then leads to the formation of a thiazole after chromatography on silica gel (Scheme 6.6b) [21].

Li and co-workers further described the reaction with vinylogous systems, where a related reaction occurs. Under Rh(II)-catalyzed reaction conditions β-(methylthio)-α,β-unsaturated ketone **6.7b**, reacts with triazole 6.7a to give β-amino-α,β-unsaturatedketone **6.7c** (Scheme 6.7a) [22]. Furthermore, Li and co-workers reported on the synthesis of multifunctionalized benzothiazonine **6.7h** from thiochromaone **6.7g** (Scheme 6.7b) [23].

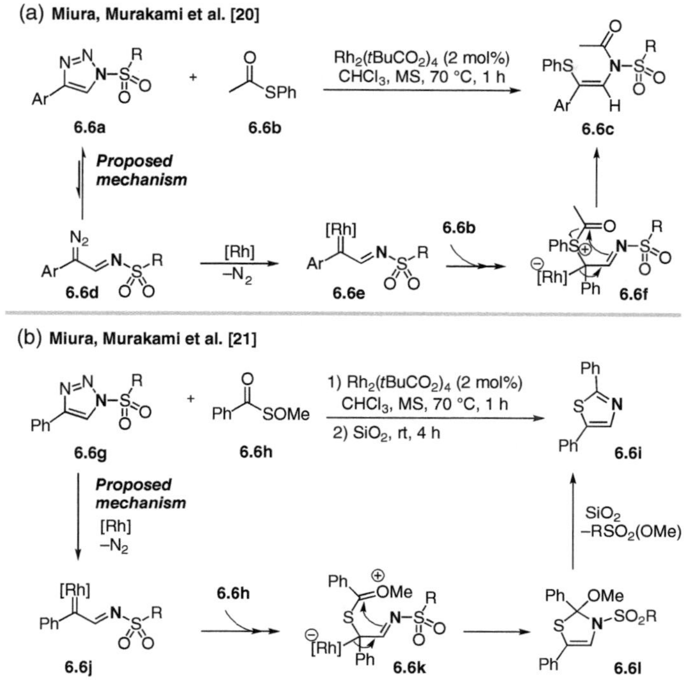

Scheme 6.6 (a) Reaction of triazoles with acyl sulfides. (b) Reaction of thioesters with triazoles *via* a nonclassic ylide intermediate. Source: Adapted from Murakami et al. [20, 21].

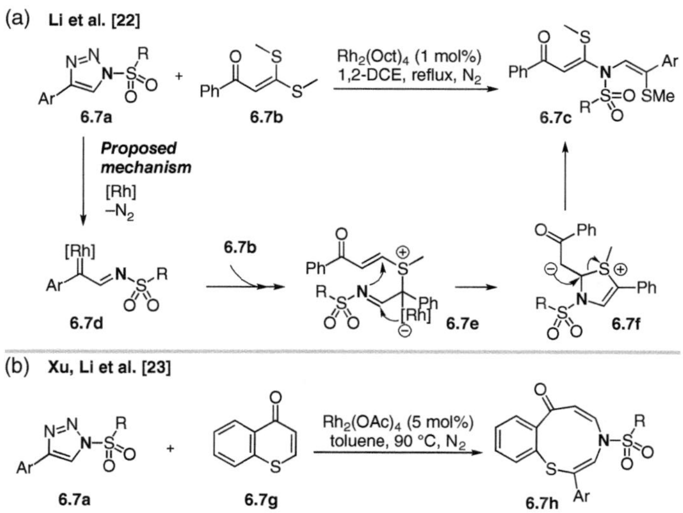

Scheme 6.7 (a) Reactions of **6.7b** with triazoles. Source: Adapted from Li et al. [22]. (b) Reactions of **6.7g** with triazoles. Source: Adapted from Li et al. [23].

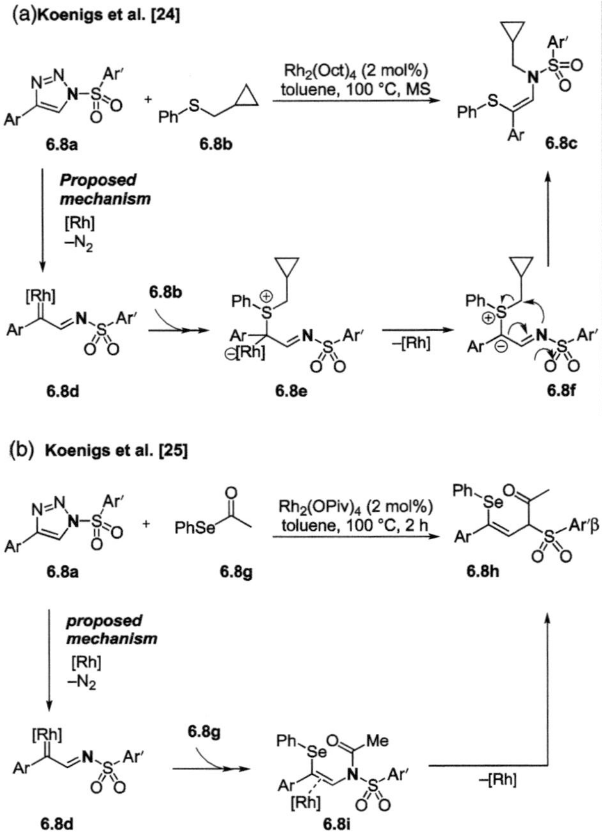

Scheme 6.8 (a) Reactions of **6.8b** with triazoles. (b) Reaction of acyl selenides with triazoles. Source: Adapted from Koenigs et al. [24, 25].

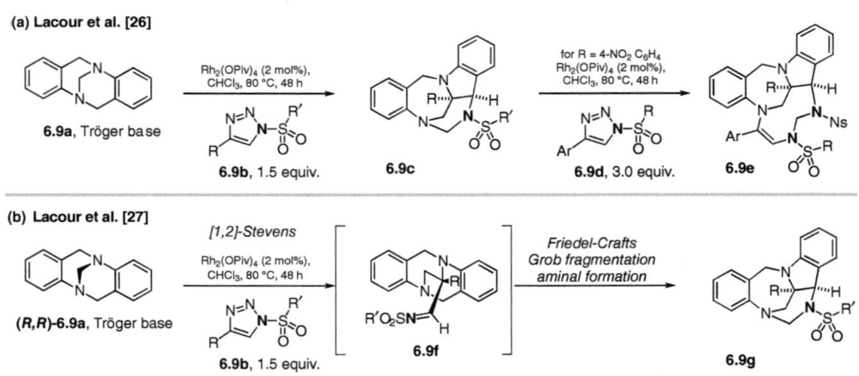

Scheme 6.9 (a) Synthesis of polycyclic heterocycle **6.9c** and benzodiazepine-indolinetriazonane **6.9e** from Tröger base. Source: Adapted from Lacour et al. [26]. (b) Reaction of enantiopure Tröger base with triazoles. Source: Adapted from Lacour et al. [27].

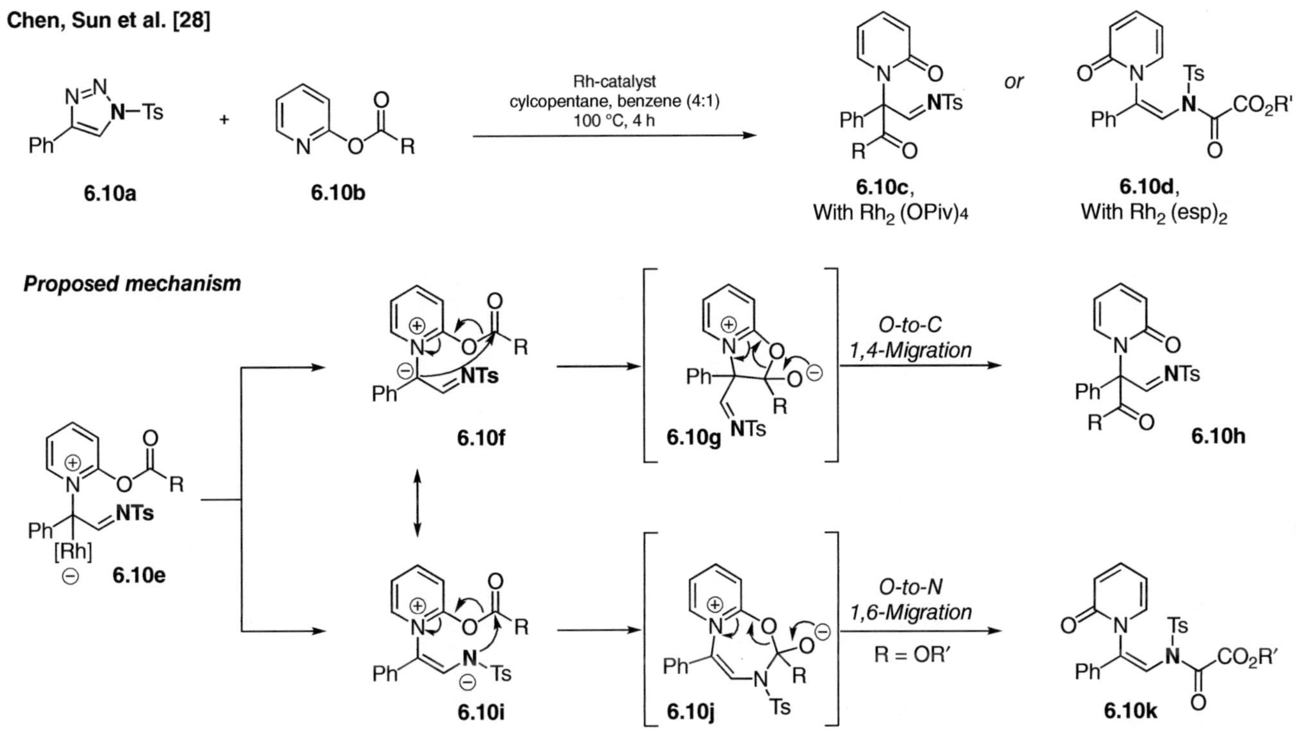

Scheme 6.10 Reaction of 2-oxopyridines with triazoles. Source: Adapted from Sun et al. [28].

The Koenigs group more recently explored this type of transformations using aliphatic sulfides, where the imino carbene undergoes a formal enamine insertion reaction into a C—S bond, where the carbon is an alkyl chain. The authors studied the migratory aptitude of different alkyl groups, and could observe that a cyclopropyl methyl group, an acetonitrile, or a cinnamyl group undergo such formal enamine insertion reactions (Scheme 6.8a) [24]. Koenigs and co-workers further studied the rhodium-catalyzed reaction of acyl selenides, where an intermittent selenium ylide is formed, which experiences a similar migration of the acyl group to give a β-selenyl enamide (Scheme 6.8b) [25].

The reaction of Tröger base **6.9a** with triazoles under Rh(II)-catalyzed reaction conditions was investigated by Lacour and co-workers. In the first reaction step, polycyclic heterocycle **6.9c** is formed as a single diastereomer followed by a cascade reaction involving [1,2]-Stevens reaction, Friedel-Crafts reaction, Grob fragmentation, and aminal formation. The obtained product **6.9c** can further react with an excess of triazole **6.9b** to form benzodiazepine-indolinetriazonane **6.9e** (Scheme 6.9a) [26]. When applying the enantiopure Tröger base, a full racemization is observed (Scheme 6.9b) [27].

Chen, Sun, and co-workers studied the reactivity of 2-oxopyridines **6.10b** with triazoles **6.10a** under Rh(II)-catalyzed reaction conditions. In the case of O-carboxylic ester pyridines, a selective O-to-C 1,4-migration occurs, while for 2-pyridyl carboxylates, a O-to-N 1,6-migration was observed (Scheme 6.10) [28].

6.3 [2,3]-Sigmatropic Rearrangement Reactions

Rearrangement reactions represent one of the most fundamental reactivity of carbene reagents. Specifically, upon metal-catalyzed decomposition of a diazo compound or related precursors, an electrophilic metal carbene is formed that forms a ylide intermediate, which itself can engage in rearrangement reactions [1]. In this context, the [2,3]-sigmatropic rearrangement reaction displays one of the key transformations, which has been explored with a variety of nucleophiles for diazo compounds and the reader is referred to reviews in this research area [2]. Herein, we describe recent advances on such [2,3]-sigmatropic rearrangement reactions with triazoles as surrogates of diazo compounds and discuss similarities and differences in the reactivity of imino carbenes and related carbenes from diazo compounds.

Xu, Shen, and co-workers reported in this context on the intramolecular reaction of triazoles with a neighboring allyl amine functional group. In the presence of a rhodium catalyst, the triazole undergoes the formation of a rhodium imino carbene intermediate and subsequent formation of cyclic ammonium ylide **6.11e**. This ammonium ylide reacts in a classic [2,3]-sigmatropic rearrangement reaction to form five-membered pyrrolidine ring **6.11f** bearing a fully substituted carbon atom next to the nitrogen atom. The addition of a copper catalyst to the reaction mixture leads to Friedel-Crafts cyclization to furnish a benzopyrrolizidine ring system **6.11c**. Without the addition of the copper catalyst, the imine functional

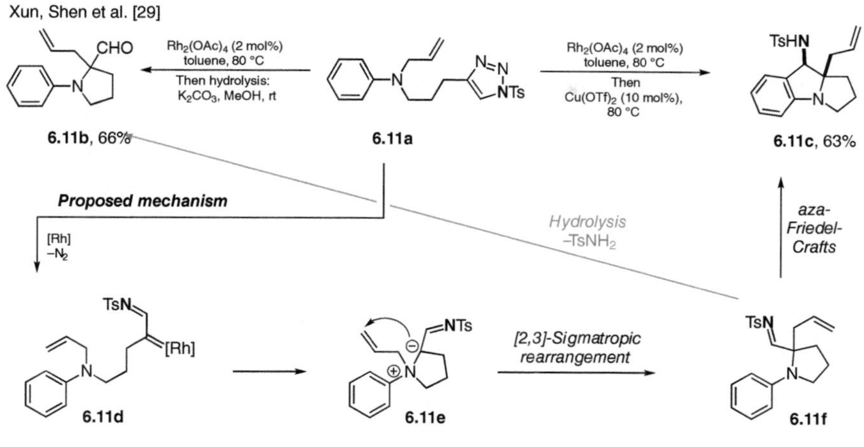

Scheme 6.11 [2,3]-sigmatropic rearrangement reaction of **6.11e** followed by hydrolysis reaction or *aza*-Friedel-Crafts reaction. Source: Xu, Shen et al. [29].

group experiences hydrolysis during workup to expose an aldehyde functional group (Scheme 6.11) [29]. The observed reaction closely resembles the reaction of ammonium ylides derived from donor/acceptor diazo compounds and carbene intermediates as recently disclosed by Koenigs and co-workers under catalyst free conditions [30].

The rearrangement reaction of allylic ethers with triazoles was studied in detail by Boyer during the same period. Boyer studied the intramolecular reaction of allylic ethers tethered to a triazole core. Depending on the substitution pattern of

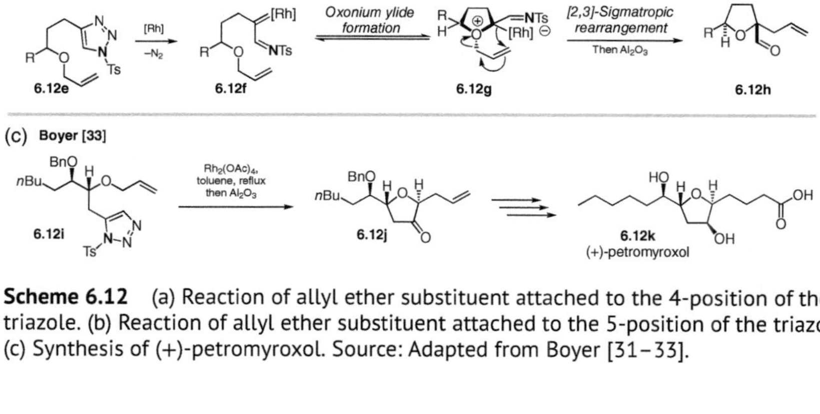

Scheme 6.12 (a) Reaction of allyl ether substituent attached to the 4-position of the triazole. (b) Reaction of allyl ether substituent attached to the 5-position of the triazole. (c) Synthesis of (+)-petromyroxol. Source: Adapted from Boyer [31–33].

the triazole heterocycle two related [2,3]-sigmatropic rearrangement reactions were observed. In the case the allyl ether substituent was attached to the fifth-position of the triazole core **6.12a**, the product of the rearrangement reaction is a furan heterocycle, where an aldehyde (from the former imino functional group) and further substituents are located on the furane ring (Scheme 6.12a) [31]. This reaction resembles the previously discussed reactivity of allylamines [29]. When the allyl ether group is attached to the 4-position of the triazole core **6.12e**, the reaction product diverges in the connectivity and a furan imine is formed (Scheme 6.12b) [32]. These differences in reactivity can be explained by two distinct ylide intermediates that are formed. Boyer further explored the application of this approach in the synthesis of the natural product (+)-Petromyroxol **6.12k** that contains a central 2,3,5-trisubstituted furan ring. Importantly, the substituents at the 2 and 5 positions are located on opposite faces of the ring system, which thus represent a very suitable target to assess the intramolecular [2,3]-sigmatropic rearrangement of oxonium ylides in total synthesis (Scheme 6.12c) [33].

Previously described [2,3]-sigmatropic rearrangement reactions thus represent a transformation, where the imino carbene fragment reacts with allylic amines or ethers under the scission of a C—*X* bond. Allylic sulfides react similarly with triazoles in the presence of a rhodium catalyst and Murakami and Anbarasan described their respective approaches almost at the same time, which gives access to a fully substituted carbon atom (Scheme 6.15a, b) [34, 35]. Xiao and co-workers further described this process in the reaction of 3,3-difluoroallyl sulfide **6.13h** (Scheme 6.15c) [36]. Organoselenium compounds initially display a similar reactivity and allylic selenides react with triazoles under the formation of selenonium ylide **6.13n** followed by [2,3]-sigmatropic rearrangement as described by Koenigs and co-workers. The reaction product represents an example of an *α*-seleno imine, which can easily undergo homolytic cleavage of the C—Se bond and formation of two radical species that can engage in a downstream radical cyclization chain reaction with a second molecule of the initial rearrangement product. The related process for C—S bond cleavage is less favored (Scheme 6.13d) [37].

Anbarasan and co-workers reported on a Rh(II)-catalyzed Sommelet–Hauser-type rearrangement reaction. *α*-Thioester **6.14b** reacted in the presence of triazole **6.14a** to furnish the desired enamide **6.14c**. Interestingly, α-thioester **6.14b** can be replaced by α-thioamides and *α*-thioketones. The reaction proceeds *via* the formation of ylide intermediate **6.14e** followed by a proton transfer to form intermediate **6.14f**. In the last step, the *aza*-[2,3]-sigmatropic rearrangement (Sommelet–Hauser-type rearrangement) takes place to deliver product **6.14c** (Scheme 6.14). Anbarasan and co-workers were also able to furnish compound **6.14c** in a one-pot synthesis starting from phenylacetylene and tosyl azide in the presence of a Cu(I)-catalyst [38].

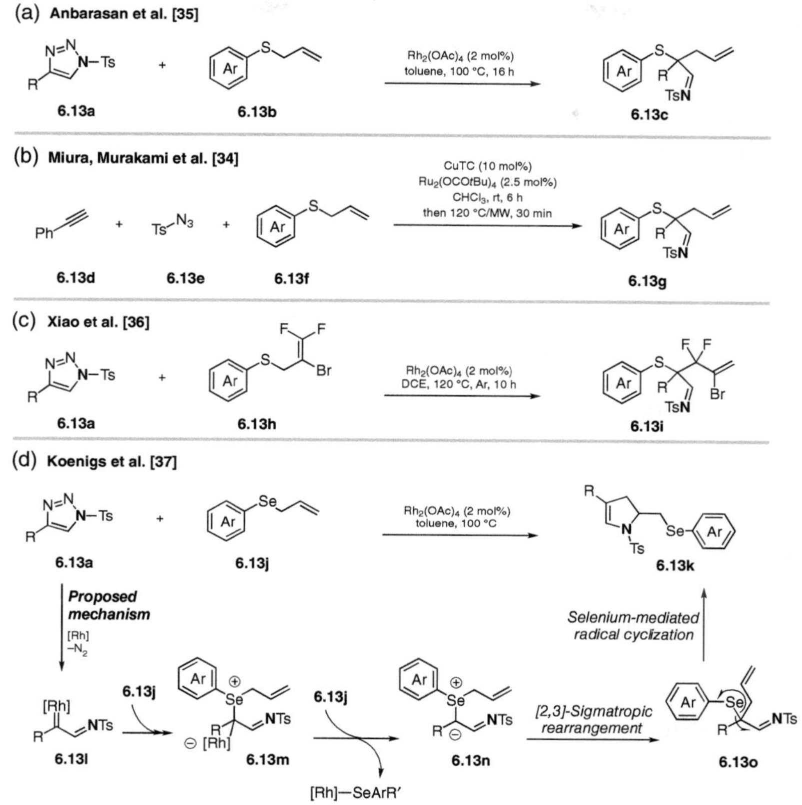

Scheme 6.13 (a) Allylic sulfides in the reaction with triazoles. Source: Adapted from Anbarasan et al. [35]. (b) One-pot reaction of alkynes, azides, and allylic sulfides. Source: Adapted from Miura et al. [34]. (c) Reaction of 3,3-difluoroallyl sulfide with triazoles. Source: Adapted from Xiao et al. [36]. (d) Allylic selenides in the reaction with triazoles. Adapted from Koenigs et al. [37].

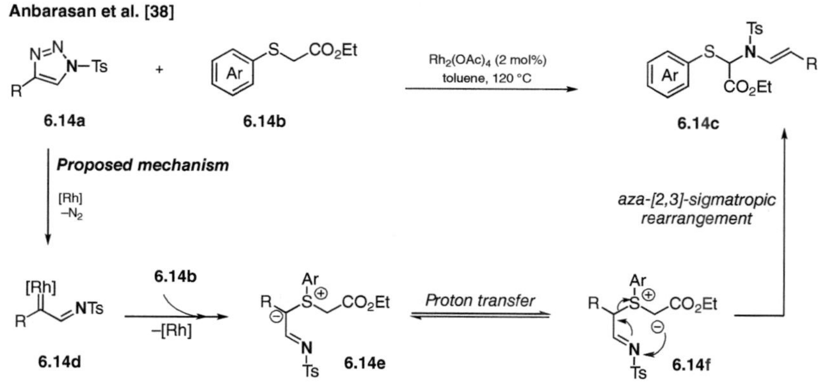

Scheme 6.14 Sommelet--Hauser-type rearrangement reaction of α-Thioesters with triazoles. Source: Adapted from Anbarasan et al. [38].

6.4 Insertion into *X*–*X* bonds

The thioacetilization of triazoles was described by Fu and co-workers. The reaction proceeds *via* the insertion of the carbene fragment into the S—S bond due to the low stability of product **6.15c**; subsequent hydrolysis was performed to form product **6.15d**. Under reductive reaction conditions, product **6.15e** was observed (Scheme 6.15) [39].

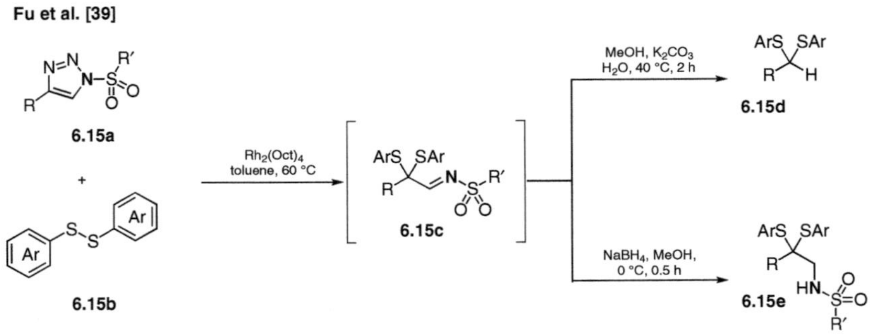

Scheme 6.15 Synthesis of thioacetals *via* S–S bond insertion reaction. Source: Adapted from Fu et al. [39].

6.5 Conclusion

In this chapter, we presented the various applications of triazoles as precursors for α-imino carbenes. Such α-imino carbenes have been utilized in various ring-expansion reactions of both small, strained ring systems such as epoxides and aziridines, as well as less strained scaffolds such as the Tröger base, 1,3,5-trioxanes, or 1,3,5-triazinanes. In addition, α-imino carbenes can also engage in rearrangement reactions to afford ring-open products. Future studies on α-imino carbenes could further shorten synthetic routes for the synthesis of exotic N-heterocycles and broaden the scope of such structures.

References

1 Jana, S., Guo, Y., and Koenigs, R.M. (2021). Recent perspectives on rearrangement reactions of ylides *via* carbene transfer reactions. *Chem. Eur. J.* 27: 1270–1281.

2 Zhang, Y. and Wang, J. (2010). Catalytic [2,3]-sigmatropic rearrangement of sulfur ylide derived from metal carbene. *Coord. Chem. Rev.* 254: 941–953.

3 Akter, M., Rupa, K., and Anbarasan, P. (2022). 1,2,3-Triazole and its analogues: new surrogates for diazo compounds. *Chem. Rev.* 122: 13108–13205.

4 Davies, H.M.L. and Alford, J.S. (2014). Reactions of metallocarbenes derived from *N*-sulfonyl-1,2,3-triazoles. *Chem. Soc. Rev.* 43: 5151–5162.

5 Miura, T., Biyajima, T., Fujii, T., and Murakami, M. (2012). Synthesis of α-amino ketones from terminal alkynes *via* rhodium-catalyzed denitrogenative hydration of *N*-sulfonyl-1,2,3-triazoles. *J. Am. Chem. Soc.* 134: 194–196.

6 Miura, T., Tanaka, T., Biyajima, T. et al. (2013). One-pot procedure for the introduction of three different bonds onto terminal alkynes through *N*-sulfonyl-1,2,3-triazole intermediates. *Angew. Chem. Int. Ed.* 52: 3883–3886.

7 Miura, T., Nakamuro, T., Hiraga, K., and Murakami, M. (2014). The stereoselective synthesis of α-amino aldols starting from terminal alkynes. *Chem. Commun.* 50: 10474–10477.

8 Chuprakov, S., Worrell, B.T., Selander, N. et al. (2014). Stereoselective 1,3-insertions of rhodium(II) azavinyl carbenes. *J. Am. Chem. Soc.* 136: 195–202.

9 Ma, X., Pan, S., Wang, H., and Chen, W. (2014). Rhodium-catalyzed transannulation of *N*-sulfonyl-1,2,3-triazoles and epoxides: regioselective synthesis of substituted 3,4-dihydro-2*H*-1,4-oxazines. *Org. Lett.* 16: 4554–4557.

10 Ko, Y.O., Jeon, H.J., Jung, D.J. et al. (2016). Rh(II)/Mg($O^tBu)_2$-catalyzed tandem one-pot synthesis of 1,4-oxazepines and 1,4-oxazines from *N*-sulfonyl-1,2,3-triazoles and glycidols. *Org. Lett.* 18: 6432–6435.

11 Lu, X.L., Liu, Y.T., Wang, Q.X. et al. (2016). Straightforward regioselective construction of 3,4-dihydro-2*H*-1,4-thiazine by rhodium catalysed [3 + 3] cycloaddition of thiirane with 1-sulfonyl-1,2,3-triazole: a pronounced acid additive effect. *Org. Chem. Front.* 3: 725–729.

12 Ryu, T., Baek, Y., and Lee, P.H. (2015). Synthesis of pyrazines from rhodium-catalyzed reaction of 2*H*-azirines with *N*-sulfonyl 1,2,3-triazoles. *J. Org. Chem.* 80: 2376–2383.

13 Wang, Y., Lei, X., and Tang, Y. (2015). Rh(II)-catalyzed cycloadditions of 1-tosyl 1,2,3-triazoles with 2*H*-azirines: switchable reactivity of Rh-azavinylcarbene as [2C]- or *aza*-[3C]-synthon. *Chem. Commun.* 51: 4507–4510.

14 Zhao, Y.Z., Yang, H.B., Tang, X.Y., and Shi, M. (2015). Rh(II)-catalyzed [3 + 2] cycloaddition of 2 *H*-azirines with *N*-sulfonyl-1,2,3-triazoles. *Chem. Eur. J.* 21: 3562–3566.

15 Guarnieri-Ibáñez, A., Medina, F., Besnard, C. et al. (2017). Diversity-oriented synthesis of heterocycles and macrocycles by controlled reactions of oxetanes with α-iminocarbenes. *Chem. Sci.* 8: 5713–5720.

16 Medina, F., Besnard, C., and Lacour, J. (2014). One-step synthesis of nitrogen-containing medium-sized rings *via* α-imino diazo intermediates. *Org. Lett.* 16: 3232–3235.

17 Viudes, O., Guarnieri-Ibáñez, A., Besnard, C., and Lacour, J. (2023). Regio-divergent synthesis of oxadiazocines *via* dirhodium-cata-lyzed reactivity of oxazolidines and α-imino carbenes. *Synlett* 34: 1472–1476.

18 Pospech, J., Ferraccioli, R., Neumann, H., and Beller, M. (2015). Rhodium(II)-catalyzed annulation of azavinyl carbenes through ring-expansion of 1,3,5-trioxane: rapid access to nine-membered 1,3,5,7-trioxazonines. *Chem. Asian J.* 10: 2624–2630.

19 Ge, J., Wu, X., and Bao, X. (2019). Rhodium(II)-catalyzed annulation of *N*-sulfonyl-1,2,3-triazoles with 1,3,5-triazinanes to produce octahydro-1-*H*-purine derivatives: a combined experimental and computational study. *Chem. Commun.* 55: 6090–6093.

20 Miura, T., Fujimoto, Y., Funakoshi, Y., and Murakami, M. (2015). A reaction of triazoles with thioesters to produce β-sulfanyl enamides by insertion of an enamine moiety into the sulfur–carbonyl bond. *Angew. Chem. Int. Ed.* 54: 9967–9970.

21 Miura, T., Funakoshi, Y., Fujimoto, Y. et al. (2015). Facile synthesis of 2,5-disubstituted thiazoles from terminal alkynes, sulfonyl azides, and thionoesters. *Org. Lett.* 17: 2454–2457.

22 He, J., Man, Z., Shi, Y., and Li, C.Y. (2015). Synthesis of β-amino-α,β-unsaturated ketone derivatives *via* sequential rhodium-catalyzed sulfur ylide formation/rearrangement. *J. Org. Chem.* 80: 4816–4823.

23 Jablasone, S.T., Ye, Z., Duan, S. et al. (2021). Synthesis of benzothiazonine by rhodium-catalyzed denitrogenative transannulation of 1-sulfonyl-1,2,3-triazole and thiochromone. *Org. Biomol. Chem.* 19: 5758–5761.

24 Li, F., Pei, C., and Koenigs, R.M. (2022). Rhodium-catalyzed enamine homologation of sulfides with triazoles as carbene precursor. *Org. Lett.* 22: 6816–6821.

25 Li, F., Pei, C., Quaranta, C., and Koenigs, R.M. (2021). 1,3-Difunctionalization of imino-carbenes via rhodium-catalyzed reactions of triazoles with acyl selenides. *Adv. Synth. Catal.* 363: 4365–4370.

26 Bosmani, A., Guarnieri-Ibáñez, A., Goudedranche, S. et al. (2018). Polycyclic indoline-benzodiazepines through electrophilic additions of α-imino carbenes to Tröger bases. *Angew. Chem. Int. Ed.* 57: 7151–7155.

27 Bosmani, A., Guarnieri-Ibáñez, A., and Lacour, J. (2019). Configurational lability of imino-substituted *Ethano Tröger* bases. Insight on the racemization mechanism. *Helv. Chim. Acta.* 102: e1900021.

28 Xu, G., Shao, Y., Tang, S. et al. (2020). Dearomative migratory rearrangement of 2-oxypyridines enabled by α-imino rhodium carbene. *Org. Lett.* 22: 9303–9307.

29 Xu, H.D., Jia, Z.H., Xu, K. et al. (2015). One-pot protocol to functionalized benzopyrrolizidine catalyzed successively by $Rh_2(OAc)_4$ and $Cu(OTf)_2$: a transition metal–Lewis acid catalysis relay. *Org. Lett.* 17: 66–69.

30 Li, F., He, F., and Koenigs, R.M. (2019). Catalyst-free [2,3]-sigmatropic rearrangement reactions of photochemically generated ammonium ylides. *Synthesis* 51: 4348–4358.

31 Boyer, A. (2014). Rhodium(II)-catalyzed stereocontrolled synthesis of 2-tetrasubstituted saturated heterocycles from 1-sulfonyl-1,2,3-triazoles. *Org. Lett.* 16: 5878–5881.

32 Boyer, A. (2014). Rhodium(II)-catalyzed stereocontrolled synthesis of dihydrofuran-3-imines from 1-tosyl-1,2,3-triazoles. *Org. Lett.* 16: 1660–1663.

33 Boyer, A. (2015). Enantioselective synthesis of (+)-petromyroxol, enabled by rhodium-catalyzed denitrogenation and rearrangement of a 1-sulfonyl-1,2,3-triazole. *J. Org. Chem.* 80: 4771–4775.

34 Miura, T., Tanaka, T., Yada, A., and Murakami, M. (2013). Doyle–Kirmse reaction using triazoles leading to one-pot multifunctionalization of terminal alkynes. *Chem. Lett.* 42: 1308–1310.

35 Yadagiri, D. and Anbarasan, P. (2013). Rhodium-catalyzed denitrogenative [2,3] sigmatropic rearrangement: an efficient entry to sulfur-containing quaternary centers. *Chem. Eur. J.* 19: 15115–15119.

36 Wang, J., Yu, J., Chen, J. et al. (2021). Doyle–Kirmse reaction using 3,3-difluoroallyl sulfide and *N*-sulfonyl-1,2,3-triazole: an efficient access to *gem*-difluoroallylated multifunctional quaternary carbon. *Org. Biomol. Chem.* 19: 6974–6978.

37 Li, F., Pei, C., and Koenigs, R.M. (2021). Rhodium-catalyzed cascade reactions of triazoles with organoselenium compounds – a combined experimental and mechanistic study. *Chem. Sci.* 12: 6362–6369.

38 Reddy, A.C.S., Ramachandran, K., Reddy, P.M., and Anbarasan, P. (2020). Rhodium-catalyzed Sommelet–Hauser type rearrangement of α-diazoimines: synthesis of functionalized enamides. *Chem. Commun.* 56: 5649–5652.

39 Zhang, H., Wang, H., Yang, H., and Fu, H. (2015). Rhodium-catalyzed denitrogenative thioacetalization of *N*-sulfonyl-1,2,3-triazoles with disulfides: an entry to diverse transformation of terminal alkynes. *Org. Biomol. Chem.* 13: 6149–6153.

7

Denitrogenative Rearrangements of *N*-Sulfonyl-1,2,3-triazoles

Om Prakash Dash, Shivam Pandey, and Chandra M. R. Volla

Department of Chemistry, Indian Institute of Technology Bombay, Powai, Mumbai, India

7.1 Introduction to Carbenes

In recent decades, carbene chemistry has become a topic of great interest within the chemical community due to its versatile reaction landscape. Carbenes are neutral species with a divalent carbon containing six valence electrons, where the carbon atom is covalently bonded to two other groups and the remaining two valence electrons are distributed between two nonbonding orbitals [1]. They are classified into two types: singlet carbenes and triplet carbenes, depending on the spin state of the nonbonding electrons [2]. In singlet carbenes, both the nonbonding electrons are located in a single sp^2-hybridized orbital with antiparallel spins, resulting in a bent geometry with a bond angle of 100°–110° (Figure 7.1). On the other hand, triplet carbenes have two unpaired electrons with the parallel spin and can be either sp^2-hybridized with a bent structure having a bond angle of 130°–150°, or sp-hybridized with a linear structure having an electron in each nonbonding orbital.

7.2 Metal Carbenoids

Over recent decades, metal carbenes have sparked tremendous interest in the area of synthetic chemistry as one of the key reactive intermediates [3]. Due to their versatile reactivity, they can be involved in a wide array of transformations, offering new avenues to access structurally complex molecules in an atom-economical manner. On the basis of electronic properties exerted by the substituents present on the carbenic carbon center, the metal carbenes are often distinguished into four groups: (i) Fischer carbenes, (ii) Shrock carbenes, (iii) NHC carbenes, and (iv) radical carbenes (Figure 7.2).

The Fischer-type carbenes are singlet carbenes named after Ernst Otto Fischer, a renowned German chemist. The σ-donation of the lone pair of electrons of carbene

Denitrogenative Transformation of Nitrogen Heterocycles: Synthesis, Reactions and Applications,
First Edition. Edited by Pazhamalai Anbarasan.

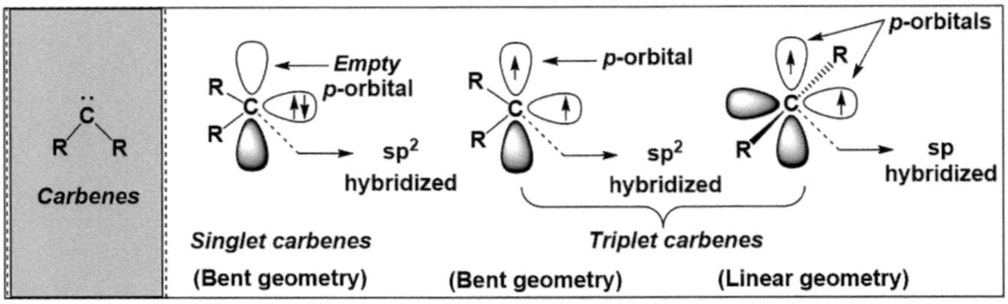

Figure 7.1 Structure and types of carbenes.

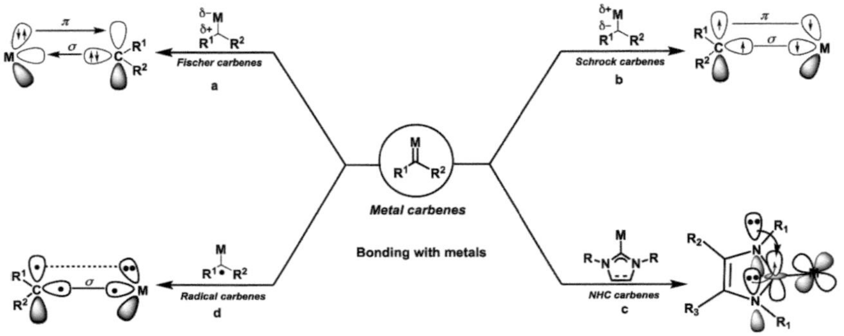

Figure 7.2 Bonding and types of metal carbenoids.

to the vacant *d*-orbital of the metal, followed by π-back donation from the metal *d*-orbital to the vacant *p*-orbital of carbenic carbon makes the carbenic carbon center electrophilic (Figure 7.2a) [4]. The presence of π-donating heteroatom such as *O*, *N*, or *S* on the carbenic carbon increases the stability and also augments their propensity to bind with late transition metals in their low oxidation state such as Fe(0), Mo(0), Cr(0), and W(0). On the other hand, Schrock-type carbenes named after Richard R. Schrock are generated by early transition metals in high oxidation states such as Ti(IV) and Ta(V), where two covalent bonds are formed with polarization toward the carbene carbon, making them nucleophilic in nature (Figure 7.2b) [5]. *N*-Heterocyclic carbenes (NHCs) were synthesized with sterically hindered substituents on the adjacent heteroatom to gain extra kinetic stability. The singlet state of NHCs is lower in energy than the triplet state due to the π-donation from adjacent heteroatoms into the *p*-orbital of carbene carbon, which makes them powerful σ-donors but weak π-acceptors (Figure 7.2c) [6]. Carbene radicals are a distinctive kind of organometallic species possessing an unpaired electron in carbene carbon and bear significant radical character (Figure 7.2d). Generally, carbene radicals were synthesized from Fischer-type carbenes *via* one-electron reduction using an external reducing agent or upon direct carbene formation from diazo compounds/diazo precursors with a planar, low-spin and open-shell paramagnetic transition metal complex [7].

7.3 Metal Carbenoids Derived from Diazo Compounds

Diazo compounds have an extensive account of applications in synthetic organic chemistry. Therefore, several methods such as diazotization, base-catalyzed decomposition of nitroso derivatives, hydrazone oxidation, Bamford-Stevens, and Regitz diazo transfer reactions have been devoted to the synthesis of diazo compounds from commercially available precursors (Figure 7.3a) [8]. The diazo compounds can be decomposed to their corresponding carbene intermediates in the reaction medium through various methods such as thermolysis, photolysis, or transition metal catalysis [9]. Previously, photolysis and thermolysis were the primary methods for generating carbene intermediates; however, these methods had limitations such as lower efficiency and uncontrollable reactions. Conversely, in recent years, several methods have been developed involving the decomposition of diazo compounds in the presence of transition metals such as Fe, Cu, Pd, Ni, Rh, Ru, Ag, and Au to synthesize stable carbenoids which provide novel and efficient methodologies to access important heterocyclic and biologically active motifs [10]. The carbenes get extra stability through a strong σ-bond formed *via* lone pair donation from carbenic carbon to metal followed by a weak back donation of electrons from metal *d*-orbital to *p*-orbital of carbenoid carbon. These carbenoids are highly electrophilic in nature and their reactivity depends on the functional groups attached to the carbenes as well as on the nature of the metal (Figure 7.3b).

Based on the functionality of these carbenoids, they can be categorized into three main classes: acceptor–acceptor, acceptor, and acceptor–donor [11]. Typical acceptor groups are keto, cyano, nitro, sulfonyl, and phosphonyl, while typical donor groups are aryl, alkynyl, vinyl, and heteroaryl. Acceptor groups in metal carbenoids make them more electrophilic and as a result, less selective, while the donor groups are affected in a reverse manner. In general, acceptor substituents make metal carbenes more electrophilic and reactive, whereas donor groups result in more stable and thus more selective.

7.4 α-Imino Metal Carbenoids and Their Reactivity

In academia, diazo compounds are commonly used as carbene precursors, but they have not gained significant attention in industry due to their instability, exothermic decomposition, and explosive behavior [12]. To address these issues, several other alternative compounds such as *N*-tosyl hydrazones and *N*-sulfonyl-triazoles have emerged as mild and stable diazo surrogates which enable the *in situ* generation of diazo compounds and can be effectively trapped by suitable transition metal catalysts for the formation of metal carbenoids even in large-scale reactions [13]. Notably, the ring-chain isomerism of *N*-sulfonyl-1,2,3-triazoles gives rise to a new class of diazo compounds known as α-diazoimines, which could be used to generate highly reactive carbene intermediates in the presence of transition metal catalysts. Over a century ago, the chemical community discovered the phenomenon of ring-chain tautomerism in 1,2,3-triazoles. It is well known that the suitably substituted 1,2,3-triazoles can rearrange to their constitutional isomeric form.

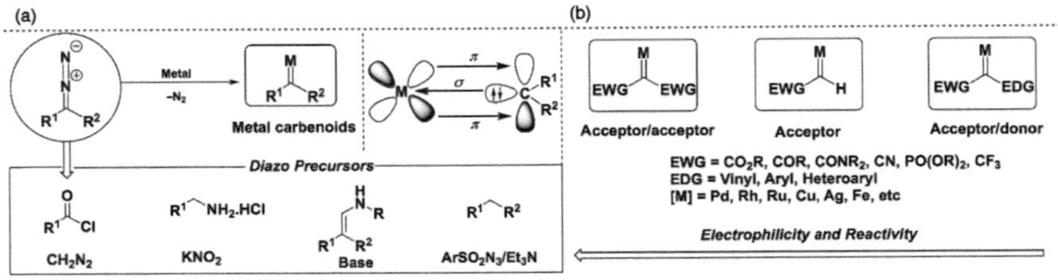

Figure 7.3 Metal carbenoids derived from diazo compounds and their classification. (a) Metal carbenoids from diazo compounds. (b) Classification of metal carbenoids.

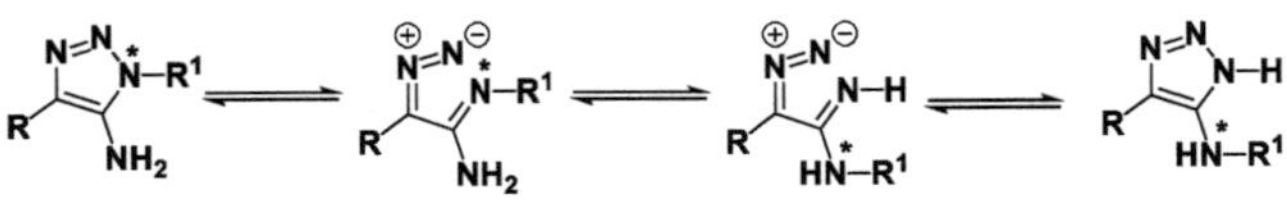

Figure 7.4 Dimroth rearrangement of 5-amino-1,2,3-triazoles.

Particularly, 1,2,3-triazoles bearing amine group at the C-5 position can undergo a ring-opening, ring-closure sequence, known as the Dimroth rearrangement, which proceeds *via* imino diazo intermediates (Figure 7.4) [14].

In the case of 1,2,3-triazoles having an electron-withdrawing *N*-sulfonyl group, facile denitrogenation occurs in the presence of transition metals to provide α-imino carbenoids (also known as azavinyl carbenoids). The *N*-sulfonyl-1,2,3-triazoles can be easily synthesized by Cu(I)-catalyzed (3 + 2)-cycloaddition reaction of azides and terminal alkynes [15]. In 2008, Gevorgyan and Fokin independently used the phenomenon of ring-chain tautomerization of *N*-sulfonyl-1,2,3-triazoles by trapping the open-chain isomer using transition metal catalyst to generate azavinyl metal carbenoid complexes [16]. Later, significant contributions came from the groups of Murakami, Davies, Sarpong, Shi, Anbarasan, and others showcasing the broad applicability of Rh–azavinyl carbenoid generated from *N*-sulfonyl-1,2,3-triazoles [17]. Metal catalysts that have been found to facilitate the decomposition of *N*-sulfonyl-1,2,3-triazoles are nickel(0)-ligand systems, dimer of rhodium(II) carboxylates, and silver(I) salts. Among all the catalysts, dirhodium(II) tetracarboxylates were proven to be the most versatile, enabling a wide variety of transformations with *N*-sulfonyl-1,2,3-triazoles. These carbenoid complexes have three different reactive sites: an electrophilic carbenoid carbon, an electrophilic imine carbon, and a nucleophilic nitrogen center. These carbene species can function as [C]-, [C,C,N]-, or [C,C]-synthons with suitable coupling partners and can further undergo diverse transformations such as cyclopropanation, insertion, cycloaddition, and rearrangement reactions to deliver a wide range of biologically important N-heterocyclic compounds (Figure 7.5) [18].

In the current book chapter, a comprehensive discussion of the various rearrangement reactions with α-imino carbenoids generated from *N*-sulfonyl-1,2,3-triazoles is presented and the sections are divided based on the intermediates involved in the rearrangement process.

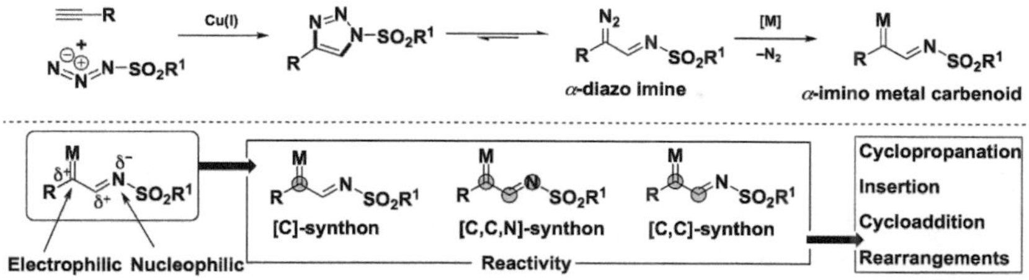

Figure 7.5 Synthesis and reactivity profile of *N*-sulfonyl 1,2,3-triazoles.

7.5 Denitrogenative Rearrangements of *N*-Sulfonyl-1,2,3-triazoles Through Azadiene Formation

In 2012, the groups of Murakami and Fokin independently reported facile and highly site-selective ring expansion reaction of α-imino rhodium(II) carbenoids to access enaminones (Scheme 7.1a) [19]. Along with ring-expansion, Fokin and co-workers also demonstrated rearrangement reactions of Rh(II)-carbenoids leading to 1-azadienes (Scheme 7.1b). Interestingly, it was found that the selectivity between ring expansion and rearrangement was completely influenced by the functional group present at the 4th position of triazole **7.1a**. In the case of cyclic and tertiary alcohol-substituted triazoles (Scheme 7.1a), rhodium carbenoid **7.1b** undergoes ring expansion to form intermediate **7.1c**, which further rearranges and finally delivers enaminones **7.1e** exclusively. On the other hand, rhodium carbenoid **7.1b** generated from triazoles having hydrogen, alkyl, acetoxy, or amine followed an intramolecular rearrangement to form intermediate **7.1f**, which finally led to the formation of 1-azadienes **7.1g** (Scheme 7.1b). A number of enaminones and 1-azadienes were synthesized in good to excellent yields from differently substituted triazoles. In addition, enaminones were efficiently transformed into other fascinating heterocycles and ketone derivatives.

In 2015, Shi group developed a Rh(II) catalyzed intramolecular 1,2-sulfur rearrangement with sulfur-tethered *N*-sulfonyl-1,2,3-triazoles **7.2a** to afford a variety of tetrahydropyridines **7.2d** [20a]. The reaction involves an intramolecular 1,2-sulfur migration initiated by the *aza*-vinyl Rh-carbenoid **7.2b** to generate thio-substituted 1-azadiene **7.2c** followed by intermolecular *aza*-Diels–Alder reaction. Independently, Anbarasan group also reported the similar sulfur migration and this methodology was successfully employed toward the synthesis of tetrahydropyridines **7.2d**, tetrahydropyridines **7.2f**, and cyclohexenes **7.2h** *via in situ* trapping of the intermediate, 1-azadiene, with various dienophiles such as alkene, alkyne, dienes, enol ether, enamine, and ketene *S*,*S*-acetal (Scheme 7.2) [20b].

In 2017, Li and Xu developed an intramolecular transannulation reaction of 4-(1-acetoxyallyl)-1-sulfonyl-1,2,3-triazoles **7.3a** for accessing 1,2-dihydropyridine derivatives **7.3e** in moderate yields (Scheme 7.3) [21]. The tandem reaction process involves the formation of Rh(II)-imino carbenoid **7.3b** followed by intramolecular nucleophilic attack of oxygen on Rh-carbenoid to form intermediate

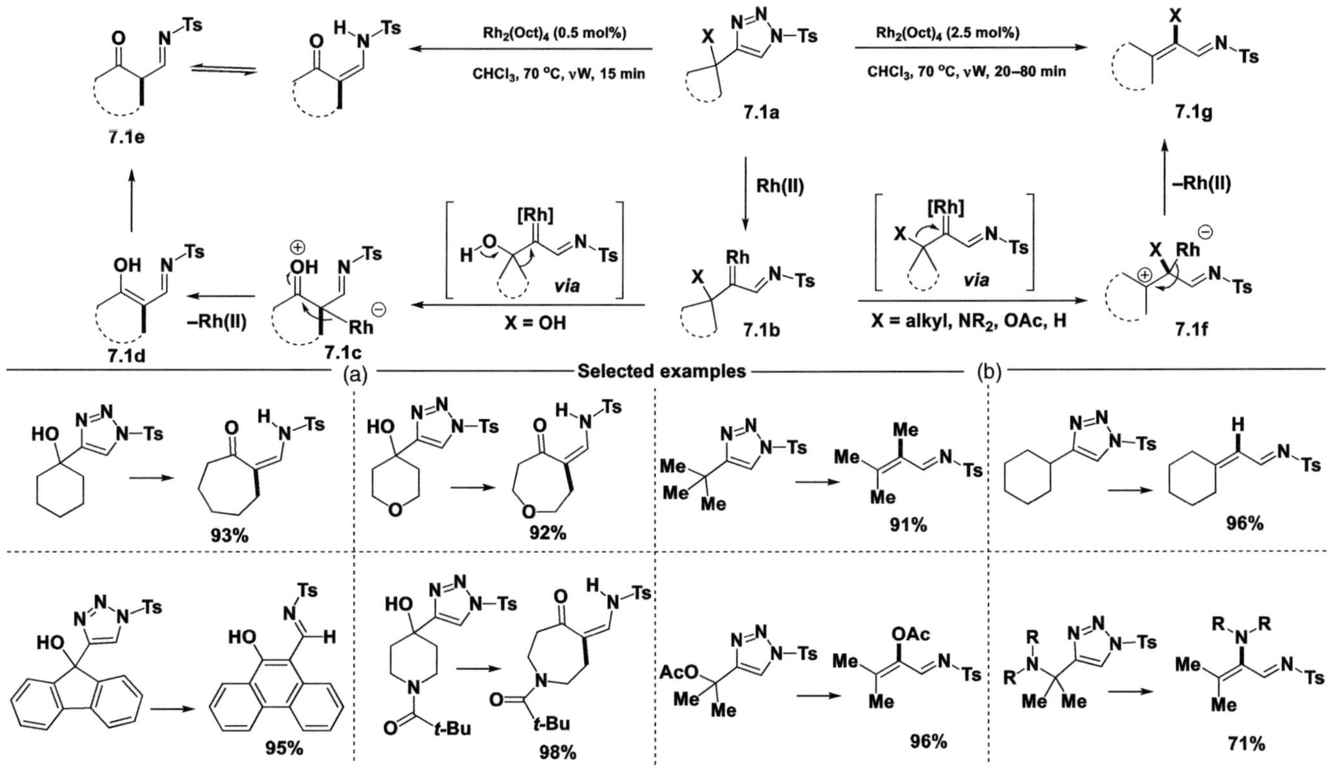

Scheme 7.1 Ring expansion and rearrangement reactions in Rh-AVCs. (a) Ring expansion. (b) Rearrangement.

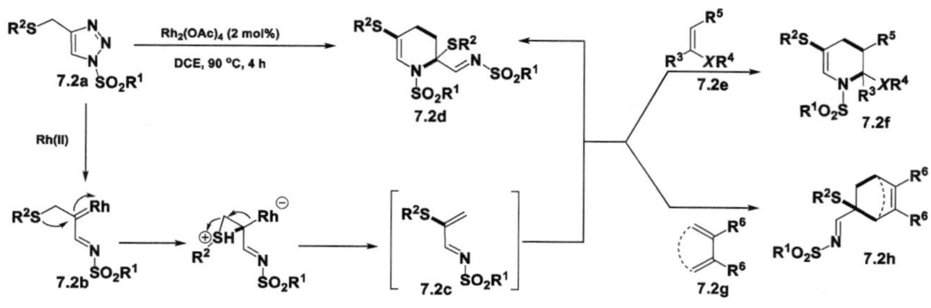

Scheme 7.2 1,2-Sulfur migration in Rh-AVCs.

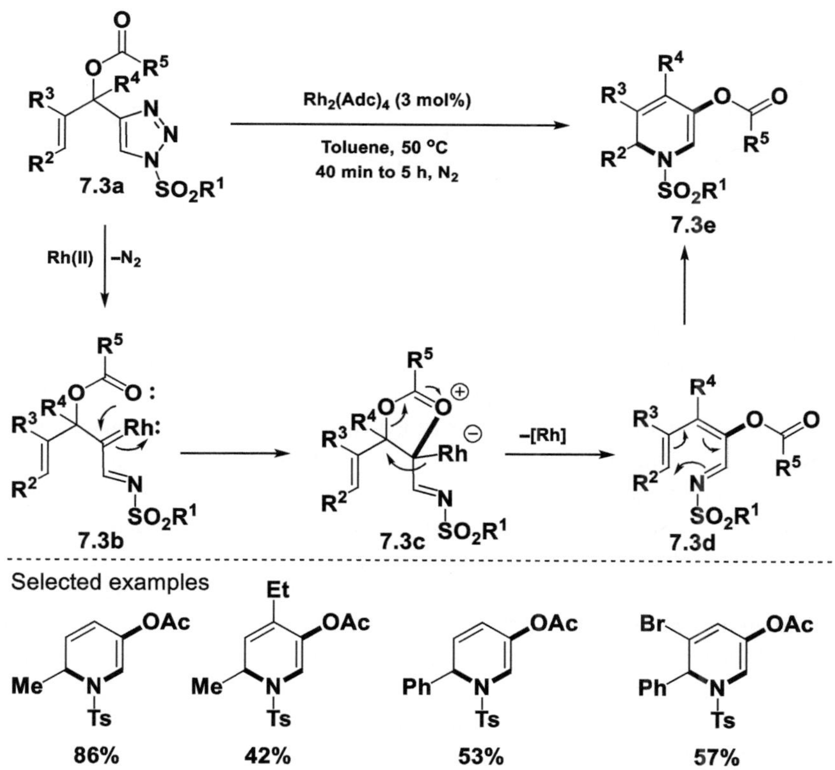

Scheme 7.3 Synthesis of 1,2-dihydropyridine derivatives.

7.3c. At this stage, a chemoselective 1,2-migration of an ester group generates 1-azadienes **7.3d** and finally, 6π-electrocyclic ring closure leads to the formation of dihydropyridines **7.3e**.

In 2022, Murakami and co-workers developed an elegant strategy for the synthesis of substituted 1-naphthols from 2-triazolyl-1-indanones **7.4a** *via* 1,2-acyl migration to the adjacent rhodium carbenoid (Scheme 7.4) [22]. α-Imino rhodium carbenoid **7.4b** from the triazolyl moiety undergoes 1,2-acyl migration, leading to zwitterionic

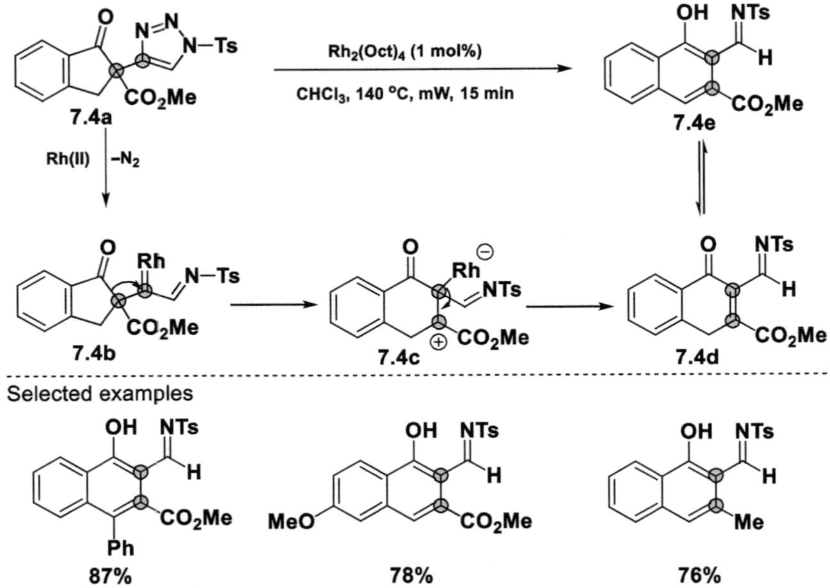

Scheme 7.4 1,2-acyl migration in Rh-AVCs.

intermediate **7.4c**. Release of Rh(II) catalyst forms intermediate **7.4d**, which tautomerizes to deliver substituted naphthol motifs **7.4e** in good to excellent yields.

In 2017, Volla and co-workers demonstrated a novel approach to furnish 4-aminooxazolidinones **7.5f** from 1,2,3-triazolyl-carbamates **7.5a** in excellent yields (Scheme 7.5) [23]. The mechanism involves the formation of diazo intermediate **7.5b** from the corresponding triazolyl moiety, which generates α-imino rhodium carbenoid in the presence of Rh(II) catalyst. Two possible pathways were proposed, which involve the attack of the oxygen atom (*via* **7.5c** or **7.5c′**) onto the Rh-carbenoid leading to the formation of either three-membered oxonium ylide **7.5d** (Path *A*) or five-membered oxonium ylide **7.5d′** (Path *B*). Later, Rh-catalyst is regenerated and C—O bond cleavage occurs to form azadiene intermediate **7.5e**. Intramolecular addition of carbamate nitrogen onto azadiene delivers the final product **7.5f**.

From 2018 to 2020, Xu and Li demonstrated a series of transformations by treating ester-tethered 1-sulfonyl-1,2,3-triazoles **7.6a** with different coupling partners such as enol ethers, substituted indoles, and strained azirine derivatives for accessing highly substituted heterocycles [24(a–d)]. The chemoselective tandem sequence proceeded through 1,2-migration of the ester group to yield 1-azadienes **7.6b** in the presence of Rh(II) catalyst followed by *aza*-Diels-Alder reaction with enol ethers **7.6c** and substituted indoles **7.6e** to furnish piperidines **7.6d** and pyrido[2,3-*b*]indole derivatives **7.6f** respectively in good to excellent yields (Scheme 7.6a, b). Along the same line, they have also tested indoles **7.6g** and strained 2*H*-azirines **7.6i** with *in situ* generated *aza*-1,3-dienes derived from ester-tethered *N*-sulfonyl-1,2,3-triazoles to access cyclopenta-[*b*] indoles **7.6h** and fused pyrimidine derivatives **7.6j** (Scheme 7.6c, d). Shortly thereafter, Volla and

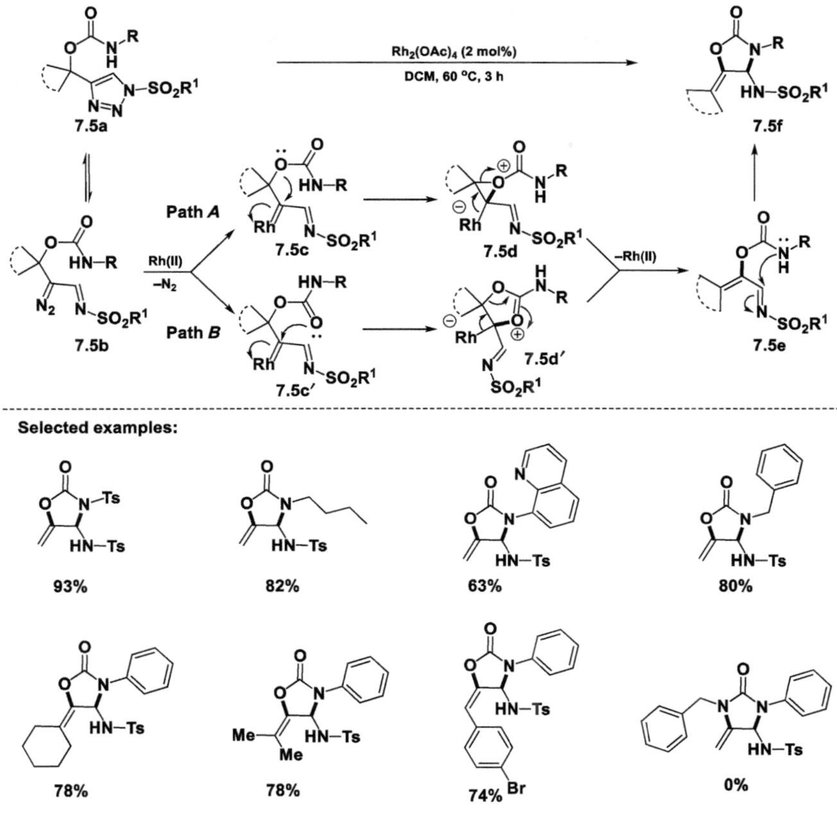

Scheme 7.5 Synthesis of 4-aminooxazolidinones.

co-workers reported a Rh(II)-catalyzed reaction of *N*-sulfonyl-1,2,3-triazolyl esters with either 1,3,5-trimethoxybenzene **7.6k** or *N*-methylindole **7.6m**, which proceeds *via* Rh(II)-catalyzed intramolecular rearrangement of triazoles into 1-azadienes followed by regioselective nucleophilic addition to afford structurally diverse allyl amines **7.6l** or homotryptamines **7.6n** respectively in good to moderate yields (Scheme 7.6e, f) [24e].

In 2021, Volla and co-workers described a Rh(II)-catalyzed denitrogenative cascade transannulation of 1,2,3-triazolyl arylpropiolates **7.7a** with indoles **7.6g** for accessing butenolide tethered homotryptamines in one-pot process (Scheme 7.7) [25]. The authors proposed that the α-iminocarbenoid **7.7b** generated in the presence of Rh(II) catalyst is attacked by the lone pair of carbonyl oxygen resulting in cyclic zwitterionic intermediate **7.7c**, which rearranges to give key 1-azadiene derivative **7.7d**. Nucleophilic addition of indoles **7.6g** in a Michael fashion produces enaminones **7.7e**. Intramolecular cyclization leads to butanolide intermediate **7.7f**, which on reduction results in **7.7g**. Interestingly, after prolonging the primary reaction time to 10 h, intermediate **7.7f** undergoes rearrangement to produce iminium cation **7.7h** and anion **7.7i**. The tautomeric form of **7.7i**, that is, **7.7j**,

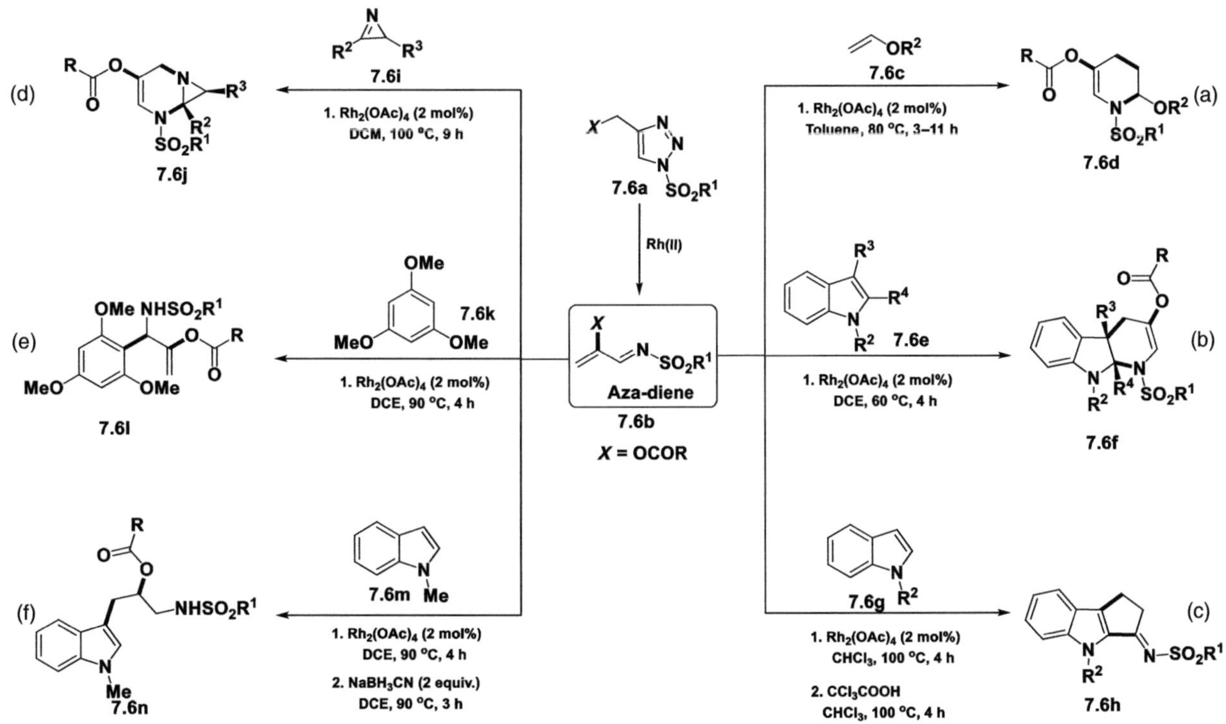

Scheme 7.6 Synthesis of various functionalized N-heterocycles. (a) Reaction with enol ether. (b) Reaction with 2,3-disubstituted indoles. (c) Reaction with unsubstituted indoles. (d) Reaction with aziridines. (e) Reaction with trimethoxybenzene. (f) Reaction with indole derivatives.

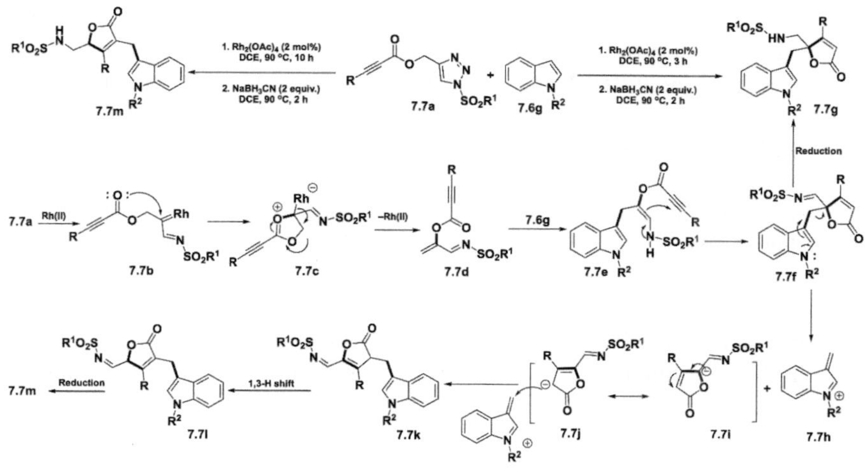

Scheme 7.7 Synthesis of butenolide tethered homotryptamine derivatives.

with a less sterically hindered nucleophilic site, recombines with **7.7h** to give the intermediate **7.7k**, which upon [1,3]-hydrogen shift followed by imine reduction provides highly functionalized butenolides **7.7m**.

In 2022, Volla and co-workers disclosed substrate-controlled denitrogenative rearrangements of 4-substituted *N*-sulfonyl-1,2,3-triazoles **7.8a** under Rh(II)-catalysis to access enaminone, pyrrol-3-one and azadiene derivatives (Scheme 7.8) [26]. The

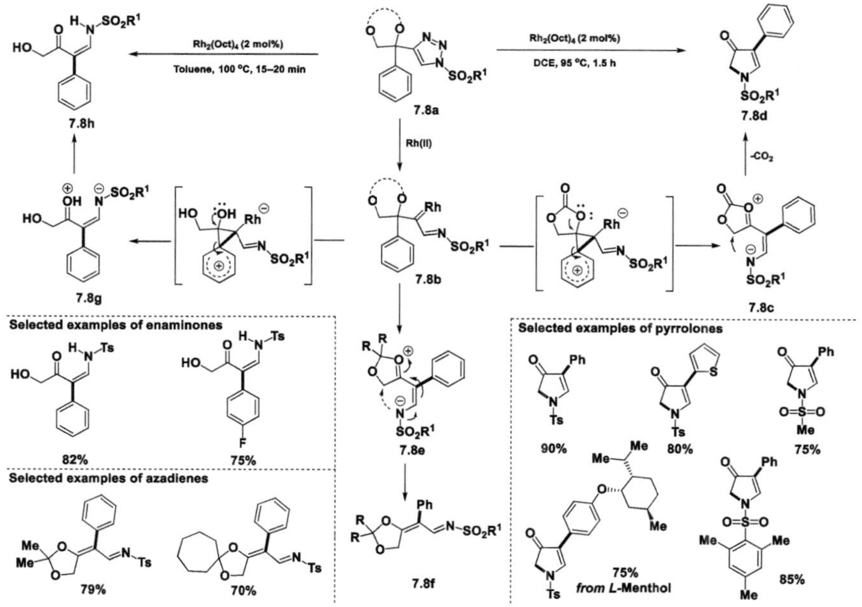

Scheme 7.8 1,2-Aryl migration in Rh-AVCs.

transformations are completely dependent on the substituent present at the C4 position of triazolyl moiety. The reaction pathways mainly proceed through the generation of α-imino Rh(II)-carbenoid **7.8b** followed by 1,2-aryl migration *via* phenonium ion intermediate to generate zwitterionic intermediate **7.8c** which undergo intramolecular nucleophilic cyclization to afford pyrrolone derivatives **7.8d**. In order to assess the importance of the leaving group, triazolyl cyclic dioxolane was employed in the developed protocol instead of carbonates. The corresponding intermediate **7.8e** undergoes rearrangement to azadiene **7.8f** instead of nucleophilic attack because of the weaker leaving group ability of acetals. Interestingly, engaging *N*-sulfonyl-1,2,3-triazolyl diols under slightly different conditions preferentially led to 1,2-aryl migration to form intermediate **7.8g**, which undergoes tautomerization to access (*Z*)-enaminones **7.8h** exclusively due to intramolecular hydrogen bonding between the NH and carbonyl group.

7.6 Denitrogenative Rearrangements of *N*-Sulfonyl-1,2,3-triazoles Through Zwitterionic Intermediate

The group of Li has disclosed 1,3-acyloxy and 1,3-hydroxy group migration with Rh-AVCs to access highly substituted cyclopropane and cyclobutanone derivatives respectively (Scheme 7.9) [27(a, b)]. α-Imino rhodium carbenoid **7.9b** generated from the designed triazoles **7.9a** underwent 1,3-group migration to generate corresponding zwitterionic intermediates (**7.9c** or **7.9e**), which further undergo intramolecular cyclization to deliver cyclopropane **7.9d** or cyclobutanone **7.9f** motifs in good yields. The developed protocol is operationally simple and chemoselective, ensuring the compatibility of many functional groups and a wide range of cyclopropanes and cyclobutanones were synthesized in excellent yields.

Later, the same group achieved similar 1,3-hydroxy and 1,3-acyloxy group migration with Rh-carbenoids employing slightly modified triazoles **7.10a** to construct

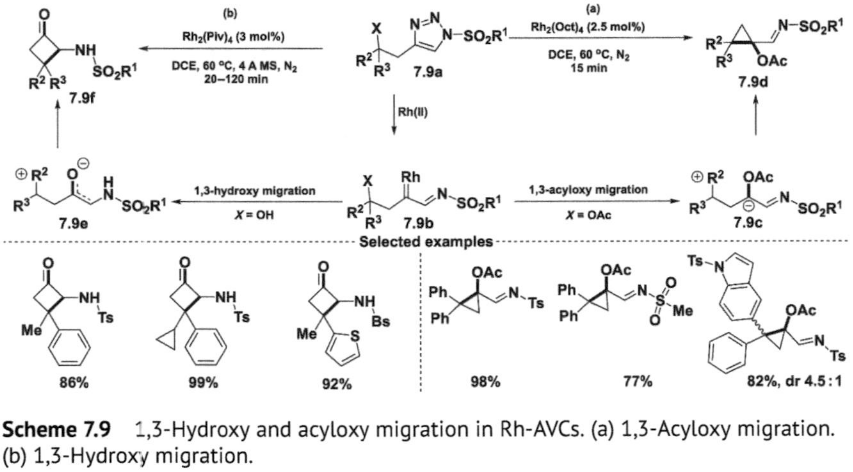

Scheme 7.9 1,3-Hydroxy and acyloxy migration in Rh-AVCs. (a) 1,3-Acyloxy migration. (b) 1,3-Hydroxy migration.

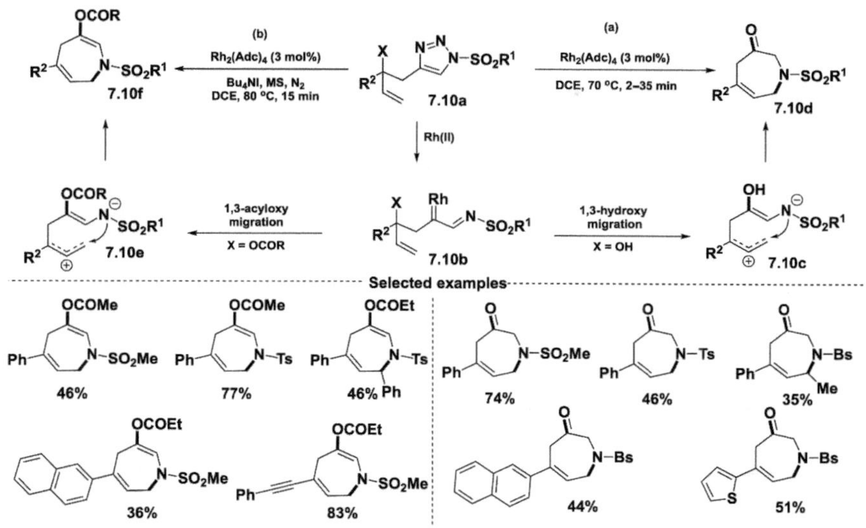

Scheme 7.10 Synthesis of azepane derivatives. (a) 1,3-Hydroxy migration. (b) 1,3-Acyloxy migration.

functionalized seven-membered *aza* heterocycles in good yields (Scheme 7.10) [28]. These transformations proceed through the formation of zwitterionic intermediates (**7.10c** or **7.10e**) *via* 1,3-group migration with *in situ* generated Rh-AVC **7.10b**, which further undergo annulation to forge azepane (**7.10d** or **7.10f**) derivatives. These derivatives were further transformed into other important heterocyclic compounds, which illustrate the excellent synthetic utility of the synthesized derivatives.

7.7 Denitrogenative Rearrangements of *N*-Sulfonyl-1,2,3-triazoles Through Ylide Formation

In 2014, Boyer and co-workers demonstrated the rearrangement of *N*-sulfonyl-1,2,3-triazoles **7.11b** involving oxonium ylide to synthesize a wide range of 2-tetrasubstituted heterocycles (Scheme 7.11) [29]. Alkyne substrate **7.11a** was initially converted to the corresponding *N*-sulfonyl-1,2,3-triazole **7.11b**, which undergoes denitrogenation to form the rhodium carbenoid **7.11c**. Intramolecular nucleophilic attack of oxygen onto the electrophilic carbenoid center results in the formation of oxonium ylide **7.11d**, which undergoes [2,3]-sigmatropic rearrangement followed by base-mediated hydrolysis to afford the desired 2-tetrasubstituted tetrahydrofuran derivatives **7.11e**.

In 2016, Li and co-workers illustrated the rearrangement of *N*-sulfonyl-1,2,3-triazoles involving bromonium ylide intermediate to furnish 4-bromo-1,2-dihydroisoquinolines (Scheme 7.12) [30]. Bromonium ylide **7.12b** is formed from *o*-benzyl bromide-containing *N*-sulfonyl-1,2,3-triazoles **7.12a** *via* the intramolecular nucleophilic attack of bromine at the strongly electrophilic Rh carbenoid center. Intermediate **7.12b** then undergoes cleavage of C—Br bond to form zwitterionic

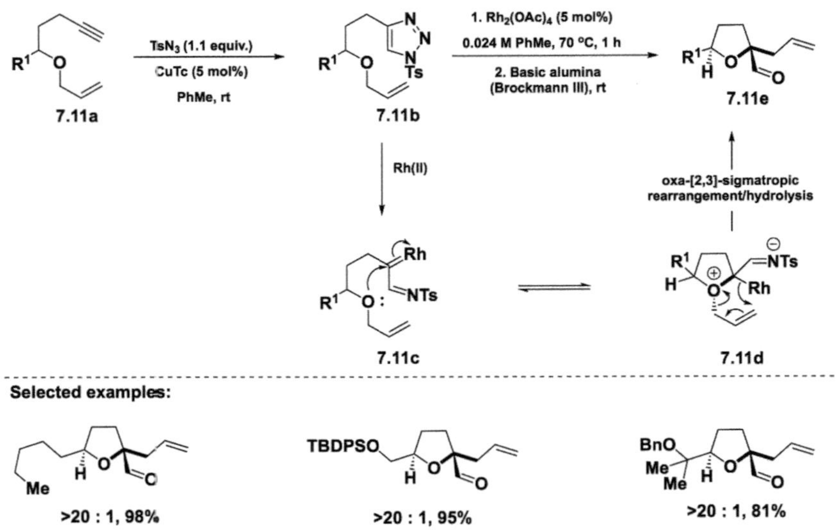

Scheme 7.11 Synthesis of 2-tetrasubstituted saturated heterocycles.

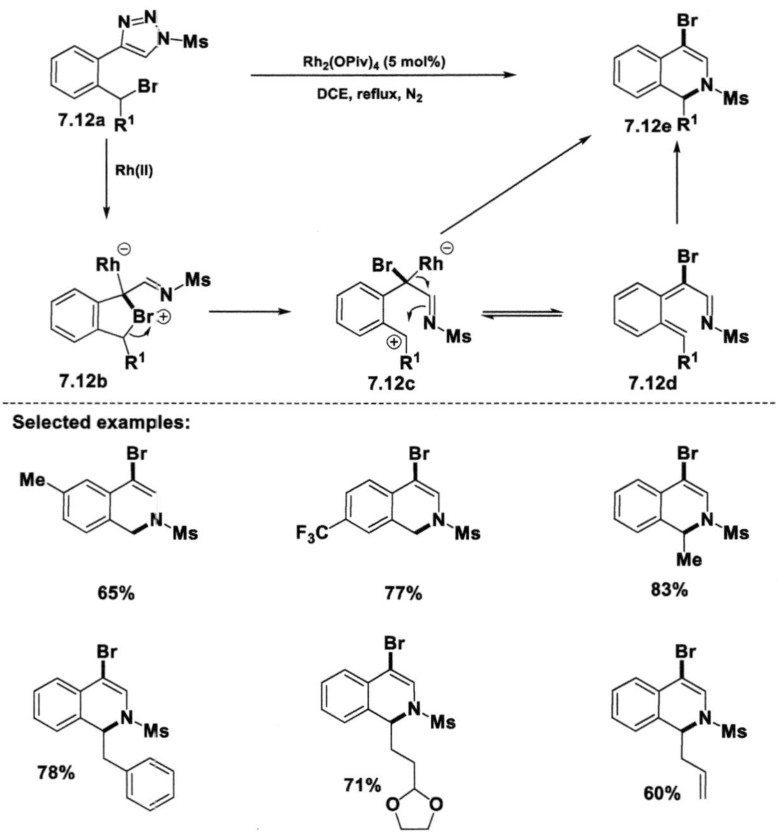

Scheme 7.12 Synthesis of 4-bromo-1,2-dihydroisoquinolines.

intermediate **7.12c**, which, on intramolecular nucleophilic addition, leads to **7.12e**. Alternatively, intermediate **7.12c** establishes an equilibrium with intermediate **7.12d** by displacement of the Rh catalyst. Finally, 6π electrocyclic ring closure produces the desired product **7.12e**.

Huang and co-workers in 2016 demonstrated the rearrangement of ether-tethered *N*-sulfonyl 1,2,3-triazoles **7.13a** to furnish dihydroisoquinoline **7.13d** and 2-aminoindanone **7.13h** derivatives *via* the formation of an oxonium ylide intermediate (Scheme 7.13) [31a]. Intramolecular nucleophilic attack of methoxy group at the electrophilic rhodium carbenoid **7.13b** generates the oxonium ylide intermediate **7.13c**. At this stage, ring closure with the nitrogen atom to the benzylic carbon center followed by the cleavage of adjacent C—O bond leads to the formation of 1,2-dihydroisoquinolines **7.13d**. Alternatively, 1,3-Rh migration may also take place to form intermediate **7.13e** which, then undergoes a skeletal rearrangement to provide carbocycle **7.13g** and its subsequent hydrolysis leads to the formation of

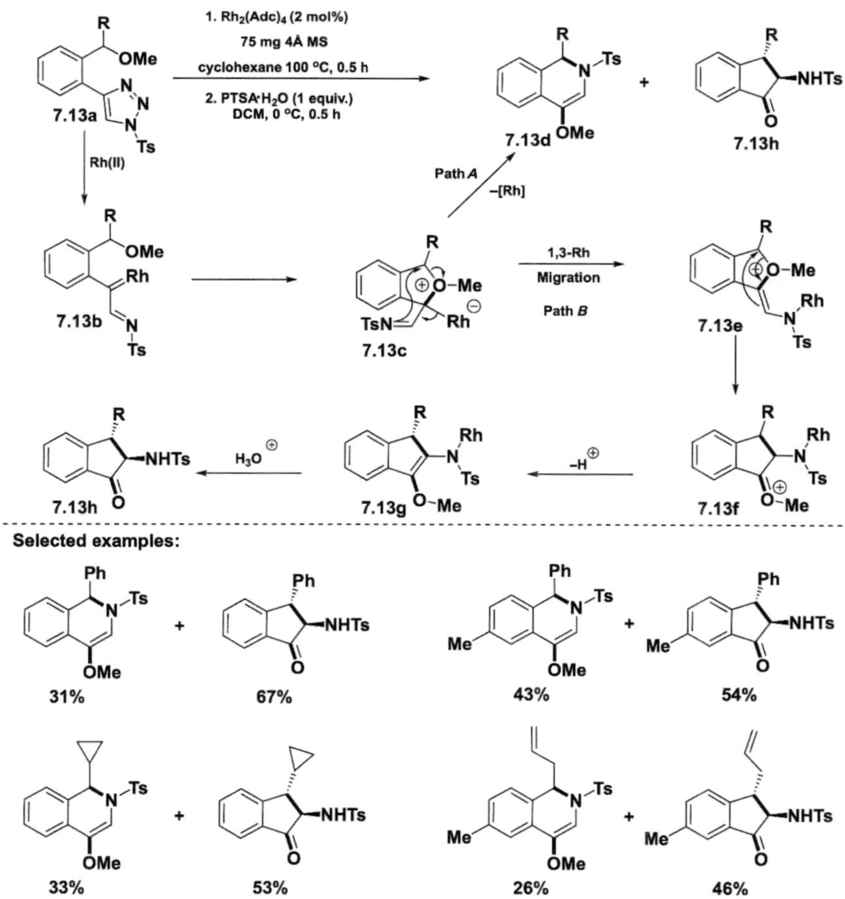

Scheme 7.13 Synthesis of dihydroisoquinoline and 2-aminoindanone derivatives.

2-aminoindanones **7.13h**. Similar rearrangement was also reported by Tang and Shi independently [31b].

Another elegant strategy involving oxonium ylides was reported by Gong and Yang in 2017 to synthesize indole-substituted indanones from carbonyl-tethered triazoles **7.14a** and indoles **7.6g** (Scheme 7.14) [32a]. The oxonium ylide **7.14b** formed in the first step undergoes nucleophilic addition of indole to form **7.14c**, which then releases rhodium catalyst to deliver **7.14d**. The presence of electron-rich indole moiety and the electron-withdrawing —Ts group in **7.14d** facilitates a skeletal rearrangement to generate intermediate **7.14e**, which then delivers the desired product **7.14f** *via* an intramolecular cyclization. The observed diastereoselectivity in the product is found to be consistent with the previous reports [32b].

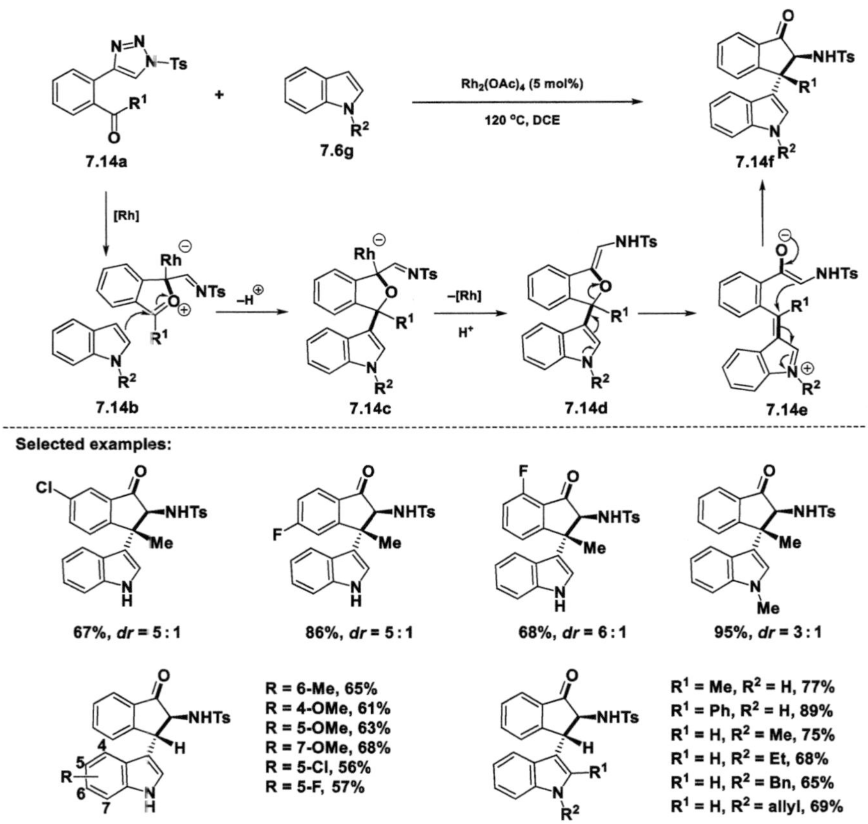

Scheme 7.14 Synthesis of indole-substituted indanones.

In 2020, Anbarasan and co-workers developed an efficient protocol to synthesize functionalized enamides *via* Sommelet–Hauser-type rearrangement of sulfur ylides (Scheme 7.15) [33]. Sulfur ylide **7.15d** is firstly generated by the nucleophilic addition of sulfur atom of α-thioester **7.15b** to the electrophilic rhodium carbenoid.

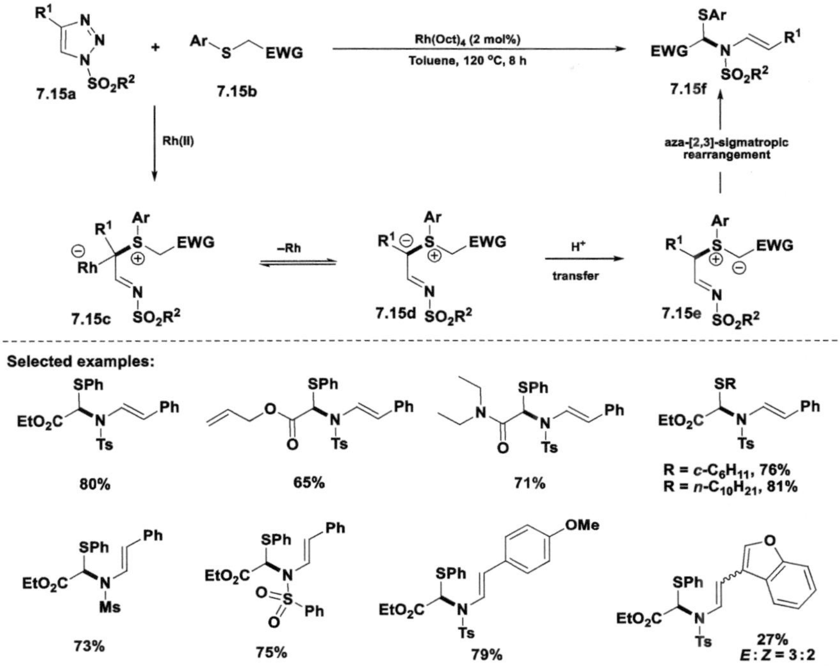

Scheme 7.15 Synthesis of functionalized enamides.

At this stage, rhodium catalyst is released to form **7.15d**, which undergoes proton transfer resulting in sulfur ylide **7.15e**. Subsequent *aza*-[2,3]-sigmatropic rearrangement produces enamide **7.15f**.

In 2020, Chen and Sun reported a novel dearomative migratory rearrangement reaction of 2-oxypyridines enabled by α-imino rhodium carbenoid (Scheme 7.16) [34]. The reaction of *o*-carboxylic ester substituted pyridines **7.16a** with *N*-sulfonyl-1,2,3-triazoles **7.15a** leads to ylide intermediate **7.16b**, which undergoes nucleophilic addition of carbanion to the acyl group followed by intramolecular rearrangement to affords *O*-to-*C* 1,4-migratory product **7.16d** through **7.16c** in presence of $Rh_2(OPiv)_4$. Alternatively, the addition of sulfonyl nitrogen atom to the acyl group in **7.16e** (resonant of **7.16b**) proceeded through a seven-membered intermediate **7.16f** and led to the formation of *O*-to-*N* 1,6-migratory product **7.16g** in the presence of $Rh_2(esp)_4$ in the same solvent mixture (cyclopentane-benzene, 4/1).

In 2021, Li and co-workers reported an efficient strategy to prepare multi-functionalized acrolein derivatives **7.17e** (Scheme 7.17) [35]. Rhodium carbenoid **7.17b** (formed from **7.17a**) undergoes nucleophilic attack of oxygen to α-imino rhodium carbenoid to form oxonium ylide **7.17c**. Later, Grob-type fragmentation of oxonium ylide **7.17c** followed by hydrolysis produces **7.17d** which on tautomerization affords acrolein **7.17e**.

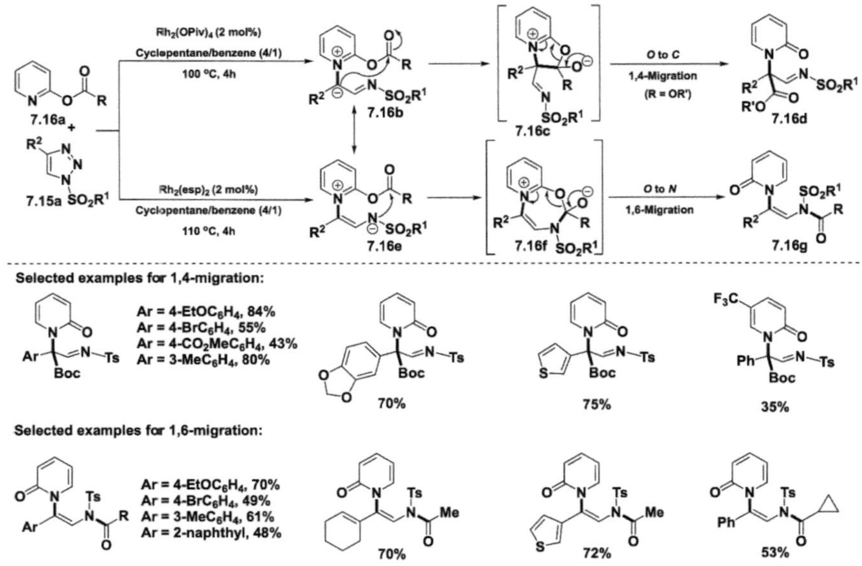

Scheme 7.16 Synthesis of *N*-substituted 2-pyridones.

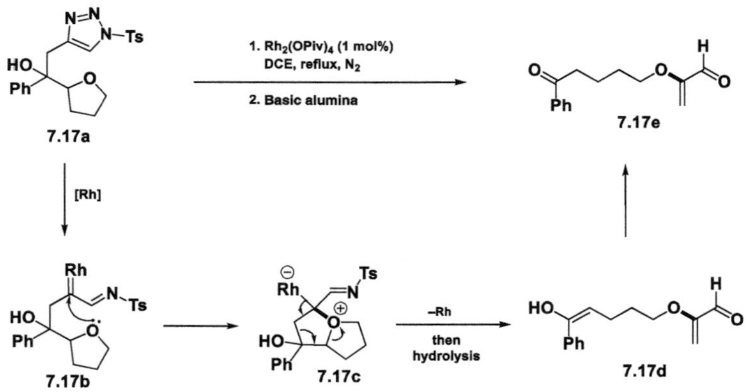

Scheme 7.17 Rh(II)-catalyzed Grob-type fragmentation of oxonium ylides.

7.8 Denitrogenative Rearrangements of *N*-Sulfonyl-1,2,3-triazoles Through Ketenimine Formation

In 2017, Li and co-workers reported a facile metal-free synthesis of 2-aminonaphthalenes from thermal cyclization of 1-sulfonyl-4-(2-alkenylphenyl)-1,2,3-triazoles **7.18a** *via* ketenimine intermediates (Scheme 7.18) [36]. Triazole moiety forms the corresponding diazo intermediate **7.18b**, which undergoes denitrogenation to form carbene intermediate **7.18c**. A 1,2-hydride shift in **7.18c**

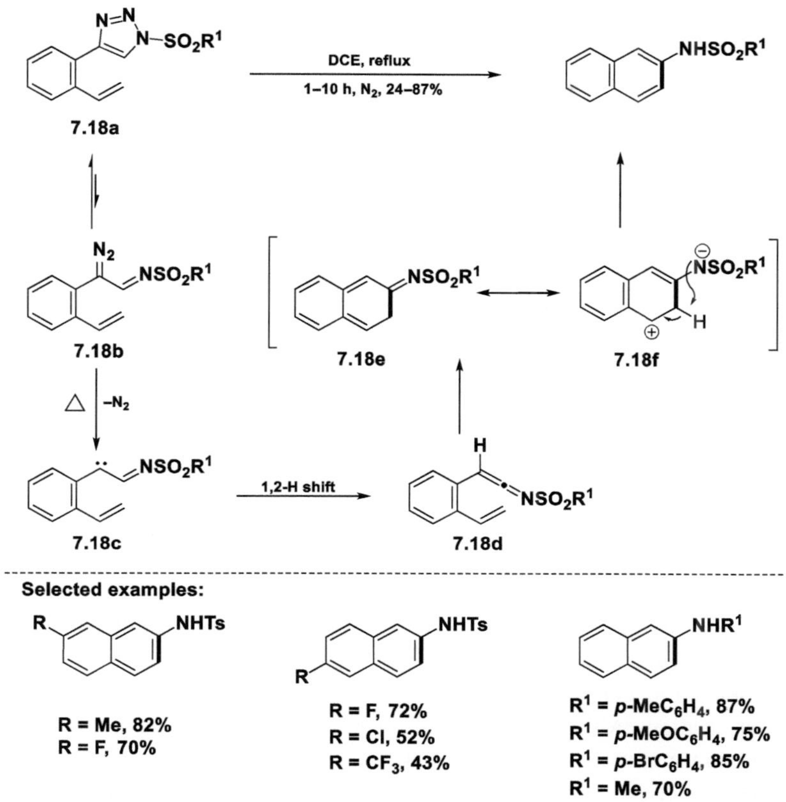

Scheme 7.18 Metal-free synthesis of 2-aminonaphthalenes.

leads to the ketenimine intermediate **7.18d**, which on subsequent 6π-electrocyclic ring closure produces **7.18e**. Finally, intermediate **7.18f** on aromatization furnishes the desired product **7.18g**. However, intermediate **7.18c** having electron-donating substituent at the vinylic carbon followed a different cyclization pathway (not shown here).

In 2020, Li and co-workers illustrated another metal-free thermal rearrangement of *N*-sulfonyl 1,2,3-triazoles **7.15a** to produce α-cyano sulfones **7.19c** (Scheme 7.19) [37]. The ketenimine intermediate **7.19b** was produced from denitrogenation of **7.15a** followed by a 1,2-hydride shift in **7.19a**. At this stage, homolytic cleavage of N—S bond delivered an intimate radical pair. Recombination of the two radicals results in 1,3-sulfonyl migration to furnish the desired product **7.19c**.

7.9 Conclusion

Over the past few decades, the use of *N*-sulfonyl-1,2,3-triazoles as precursors for the *in situ* generation of *α*-imino metal carbenoids in the presence of transition metals has opened novel routes to synthesize a wide variety of nitrogen containing

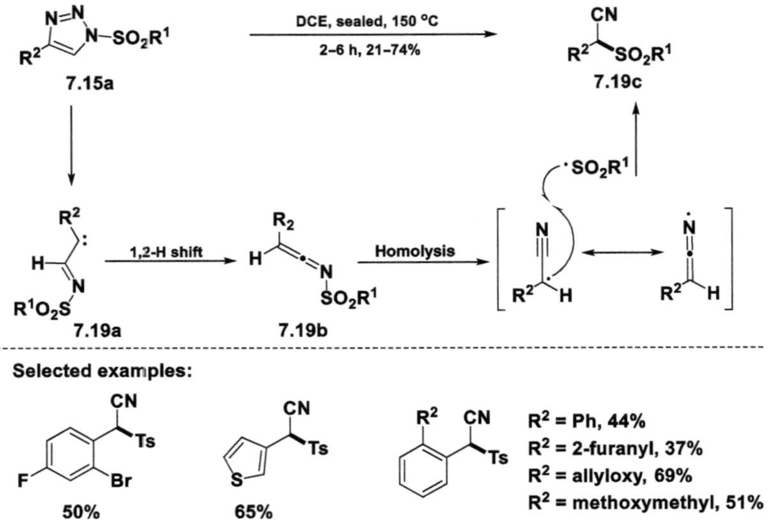

Scheme 7.19 Metal-free synthesis of α-cyano sulfones.

heterocycles that are not accessible from conventional α-oxo metallocarbenes. This chapter describes different types of rearrangement reactions of α-imino metal carbenoids and many of these developed protocols were found to proceed in a one-pot fashion. Although significant contributions have been made by the scientific community, a more diverse screening of cost-effective metal catalysts with newly designed templates could further expand the reaction scope of *N*-sulfonyl-1,2,3-triazoles and may lead to the discovery of new reactivity patterns.

Acknowledgments

The authors would like to thank the Science and Engineering Research Board (SERB, India, CRG/2023/004060) for supporting this work. O.P.D. and S.P. thank the Council for Scientific and Industrial Research (CSIR) and the University Grants Commission (UGC) for their fellowship.

References

1 (a) Moss, R.A. and Shieh, W.-C. (1978). *Tetrahedron Lett.* 19: 1935–1938. (b) Stang, P.J. (1978). *Chem. Rev.* 78: 383–405. (c) Sander, W., Bucher, G., and Wierlacher, S. (1993). *Chem. Rev.* 93: 1583–1621. (d) Mizuhata, Y., Sasamori, T., and Tokitoh, N. (2009). *Chem. Rev.* 109: 3479–3511. (e) Arduengo, A.J. and Bertrand, G. (2009). *Chem. Rev.* 109: 3209–3210.

2 (a) de Frémont, P., Marion, N., and Nolan, S.P. (2009). *Coord. Chem. Rev.* 253: 862–892. (b) Pauling, L. (1980). *J. Chem. Soc. Chem. Commun.* 688. (c) Grasse,

P.B., Brauer, B.E., Zupancic, J.J. et al. (1983). *J. Am. Chem. Soc.* 105: 6833–6845. (d) Tomioka, H. (1997). *Acc. Chem. Res.* 30: 315–321.

3 (a) Doyle, M.P. (1986). *Chem. Rev.* 86: 919–939. (b) Doyle, M.P. and Forbes, D.C. (1998). *Chem. Rev.* 98: 911–936.

4 (a) Fischer, E.O. and Maasböl, A. (1964). *Angew. Chem. Int. Ed. Eng.* 3: 580–581. (b) Dötz, K.H. and Stendel, J. Jr., (2009). *Chem. Rev.* 109: 3227–3274. (c) Raubenheimer, H.G. (2014). *Dalton Trans.* 43: 16959–16973.

5 (a) Schrock, R.R. (1975). *J. Am. Chem. Soc.* 97: 6577–6578. (b) Guggenberger, L.J. and Schrock, R.R. (1975). *J. Am. Chem. Soc.* 97: 6578–6579. (c) Tebbe, F.N., Parshall, G.W., and Reddy, G.S. (1978). *J. Am. Chem. Soc.* 100: 3611–3613. (d) Vyboishchikov, S.F. and Frenking, G. (1998). *Chemistry* 4: 1428–1438.

6 (a) Arduengo, A.J. III, Harlow, R.L., and Kline, M. (1991). *J. Am. Chem. Soc.* 113: 361–363. (b) Bourissou, D., Guerret, O., Gabbaï, F.P., and Bertrand, G. (2000). *Chem. Rev.* 100: 39–92. (c) Melaimi, M., Soleilhavoup, M., and Bertrand, G. (2010). *Angew. Chem. Int. Ed. Eng.* 49: 8810–8849.

7 (a) Epping, R.F.J., Vesseur, D., Zhou, M., and de Bruin, B. (2023). *ACS Catal.* 13: 5428–5448. (b) Dzik, W.I., Zhang, X.P., and de Bruin, B. (2011). *Inorg. Chem.* 50: 9896–9903.

8 (a) Ye, T. and McKervey, M.A. (1994). *Chem. Rev.* 94: 1091–1160. (b) Maas, G. (2009). *Angew. Chem. Int. Ed. Eng.* 48: 8186–8195. (b) Curtius, T. (1883). *Ber. Dtsch. Chem. Ges.* 16: 2230–2231. (c) Anrdt, F., Eistert, B., and Partale, W. (1927). *Ber. Dtsch. Chem. Ges.* 60B: 1364–1370. (d) Arndt, F., Eistert, B., and Amende, J. (1928). *Ber. Dtsch. Chem. Ges.* 61B: 1949–1953. (e) Bradley, W. and Robinson, R. (1928). *J. Chem. Soc.* 1310–1318. (f) Regitz, M. (1967). *Angew. Chem. Int. Ed. Eng.* 6: 733–749. (g) Cava, M.P., Litle, R.L., and Napier, D.R. (1958). *J. Am. Chem. Soc.* 80: 2257–2263.

9 (a) Ciszewski, Ł.W., Rybicka-Jasińska, K., and Gryko, D. (2019). *Org. Biomol. Chem.* 17: 432–448. (b) Xiang, Y., Wang, C., Ding, Q., and Peng, Y. (2019). *Adv. Synth. Catal.* 361: 919–944.

10 (a) Xia, Y., Zhang, Y., and Wang, J. (2013). *ACS Catal.* 3: 2586–2598. (b) Liu, L. and Zhang, J. (2016). *Chem. Soc. Rev.* 45: 506–516.

11 (a) Davies, H.M. and Morton, D. (2011). *Chem. Soc. Rev.* 40: 1857–1869. (b) Xiang, Y., Wang, C., Ding, Q., and Peng, Y. (2018). *Adv. Synth. Catal.* 361: 919–944.

12 (a) Green, S.P., Wheelhouse, K.M., Payne, A.D. et al. (2020). *Org. Process. Res. Dev.* 24: 67–84.

13 (a) Xiao, Q., Zhang, Y., and Wang, J. (2013). *Acc. Chem. Res.* 46: 236–247.

14 (a) Dimroth, O. and Liebigs, J. (1909). *Ann. Chem.* 364: 183–226. (b) Dimroth, O., Michaelis, W., and Liebigs, J. (1927). *Ann. Chem.* 459: 39–46.

15 (a) Rostovtsev, V.V., Green, L.G., Fokin, V.V., and Sharpless, K.B. (2002). *Angew. Chem. Int. Ed.* 41: 2596–2599. (b) Tornøe, C.W., Christensen, C., and Meldal, M. (2002). *J. Org. Chem.* 67: 3057–3064. (c) Meldal, M. and Tornøe, C.W. (2008). *Chem. Rev.* 108: 2952–3015.

16 Horneff, T., Chuprakov, S., Chernyak, N. et al. (2008). *J. Am. Chem. Soc.* 130: 14972–14974.

17 (a) Chattopadhyay, B. and Gevorgyan, V. (2011). *Angew. Chem. Int. Ed.* 51: 862–872. (b) Davies, H.M. and Alford, J.S. (2014). *Chem. Soc. Rev.* 43: 5151–5162. (c) Anbarasan, P., Yadagiri, D., and Rajasekar, S. (2014). *Synthesis* 46: 3004–3023. (d) Jiang, Y., Sun, R., Tang, X.-Y., and Shi, M. (2016). *Chem. Eur. J.* 22: 17910–17924. (e) Li, W. and Zhang, J. (2020). *Chem. Eur. J.* 26: 11931–11945.

18 Akter, M., Rupa, K., and Anbarasan, P. (2022). *Chem. Rev.* 122: 13108–13205.

19 (a) Miura, T., Funakoshi, Y., Morimoto, M. et al. (2012). *J. Am. Chem. Soc.* 134: 17440–17443. (b) Selander, N., Worrell, B.T., and Fokin, V.V. (2012). *Angew. Chem. Int. Ed.* 51: 13054–13057.

20 (a) Jiang, Y., Tang, X.-Y., and Shi, M. (2015). *Chem. Commun.* 51: 2122–2125. (b) Yadagiri, D. and Anbarasan, P. (2015). *Chem. Sci.* 6: 5847–5852.

21 Dai, H., Yu, S., Cheng, W. et al. (2017). *Chem. Commun.* 53: 6417–6420.

22 Miura, T., Moritani, S., Shiratori, Y., and Murakami, M. (2022). *Chem. Commun.* 58: 2710–2713.

23 Pal, K. and Volla, C.M.R. (2017). *Org. Chem. Front.* 4: 1380–1384.

24 (a) Yu, S., An, Y., Wang, W. et al. (2018). *Adv. Synth. Catal.* 360: 2125–2130. (b) Duan, S., An, Y., Xue, B. et al. (2020). *Adv. Synth. Catal.* 362: 1831–1835. (c) Duan, S., Zhang, W., Hu, Y. et al. (2020). *Adv. Synth. Catal.* 362: 3570–3575. (d) Xu, Z.-F., An, Y., Chen, Y., and Duan, S. (2019). *Tetrahedron Lett.* 60: 1849–1853. (e) Pal, K., Sontakke, G.S., and Volla, C.M.R. (2020). *Adv. Synth. Catal.* 362: 3627–3634.

25 Pal, K. and Volla, C.M.R. (2021). *Org. Lett.* 23: 4294–4299.

26 Sontakke, G.S., Pal, K., and Volla, C.M.R. (2022). *Org. Lett.* 24: 8796–8801.

27 (a) Xu, Z.-F., Liu, J., Chang, X. et al. (2020). *Org. Lett.* 22: 5163–5169. (b) Feng, Z., Wang, W., Yan, Z. et al. (2022). *Org. Lett.* 24: 2950–2954.

28 Feng, Z., Jiao, H., Ye, Z. et al. (2022). *Org. Lett.* 24: 5254–5259.

29 Boyer, A. (2014). *Org. Lett.* 16: 5878–5881.

30 He, J., Shi, Y., Cheng, W. et al. (2016). *Angew. Chem. Int. Ed.* 55: 4557–4561.

31 (a) Yu, Y., Zhu, L., Liao, Y. et al. (2016). *Adv. Synth. Catal.* 358: 1059–1064. (b) Sun, R., Jiang, Y., Tang, X.Y., and Shi, M. (2016). *Chem. Eur. J.* 22: 5727–5733.

32 (a) Yuan, H., Gong, J., and Yang, Z. (2017). *Chem. Commun.* 53: 9089–9092. (b) Zhang, J., Han, X., and Lu, X. (2016). *Org. Lett.* 18: 2898–2901.

33 Reddy, A.C.S., Ramachandran, K., Reddy, P.M., and Anbarasan, P. (2020). *Chem. Commun.* 56: 5649–5652.

34 Xu, G., Shao, Y., Tang, S. et al. (2020). *Org. Lett.* 22: 9303–9307.

35 Chen, T., Yan, Z., Duan, S. et al. (2021). *Org. Chem. Front.* 8: 6371–6376.

36 Xu, Z.F., Yu, X., Yang, D., and Li, C.Y. (2017). *Org. Biomol. Chem.* 15: 3161–3164.

37 Cen, M., Xiang, Q., Xu, Y. et al. (2020). *Org. Chem. Front.* 7: 596–601.

8

Denitrogenative Transformations of *NH*- and *N*-Fluoroalkyl-1,2,3-triazoles

Vladimir Motornov[1,2] *and Petr Beier*[1]

[1] *Institute of Organic Chemistry and Biochemistry, Czech Academy of Sciences, Prague 6, Czechia*
[2] *Georg-August-Universität Göttingen, Institut für Organische und Biomolekulare Chemie, Göttingen, Germany*

8.1 Introduction

1,2,3-Triazoles are five-membered nitrogen heterocycles with versatile reactivity [1–3] and high medicinal importance [4]. Since the discovery of azide–alkyne click reaction in 2002 [5, 6], triazole derivatives gained enormous attraction in organic, medicinal, biomolecular, and material sciences. Among them, 1,2,3-triazoles bearing an electron-withdrawing group at position N1 are of additional synthetic value because of their tendency for N1—N2 bond cleavage in denitrogenative triazole ring opening transformations [1]. *N*-sulfonyl-1,2,3-triazoles [7–11] and *N*-fluoroalkyl-1,2,3-triazoles [12, 13] are the most explored members of the family, which undergo ring cleavage under metal catalysis or using Lewis or Brønsted acids. This chapter is focused on denitrogenation strategies of *N*-fluoroalkyl-1,2,3-triazoles either thermally, or mediated by rhodium(II) catalysts, or Lewis or Brønsted acids and on denitrogenation of *NH*-1,2,3-triazoles enabled by in-situ temporal installation of an electron-withdrawing group on nitrogen of the triazole. These transformations proceed through vinyl cation or rhodium carbene intermediates and lead to the synthesis of various heterocyclic and *N*-alkenyl compounds (Scheme 8.1).

8.2 Denitrogenation of *NH*-1,2,3-Triazoles

Synthesis of *NH*-1,2,3-triazoles is covered in Chapter 1 of this book. Denitrogenation strategies of *NH*-1,2,3-triazoles are based on the attachment of an electron-withdrawing group on N1 of the triazole ring, followed by rhodium(II)-catalyzed or acid-mediated triazole ring opening and elimination of nitrogen molecule. Thus, *in situ* sulfonylation with triflic anhydride was used to generate reactive *N*-triflyl triazoles [14]. Their ring cleavage by a chiral Rh(II) catalyst in the presence of an excess of alkene yielded 2,3-dihydropyrroles with low to moderate stereoselectivity (Scheme 8.2).

Denitrogenative Transformation of Nitrogen Heterocycles: Synthesis, Reactions and Applications,
First Edition. Edited by Pazhamalai Anbarasan.

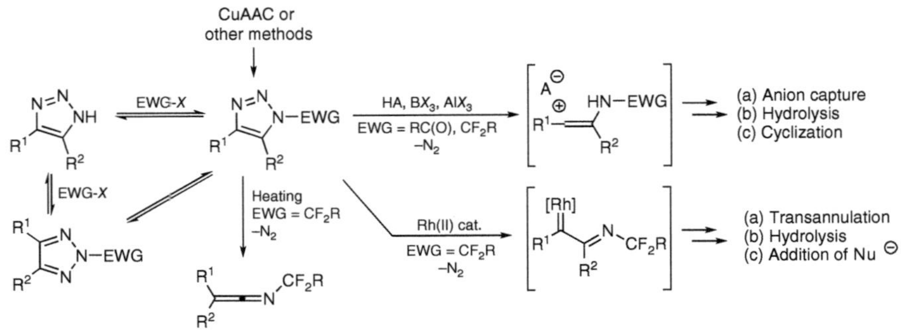

Scheme 8.1 Schematic representation of the utilization of *N*-EWG-substituted 1,2,3-triazoles in ring cleavage denitrogenative transformations.

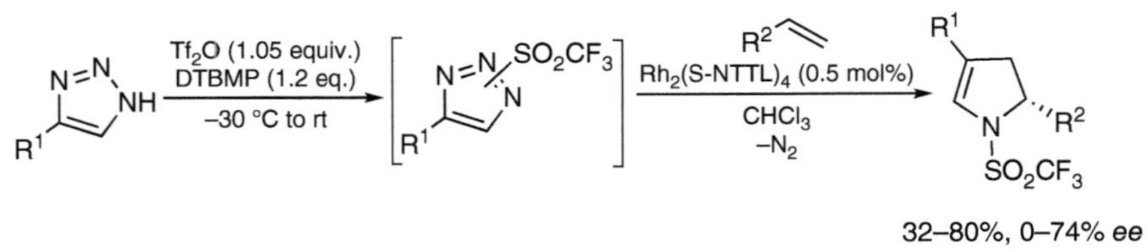

Scheme 8.2 Synthesis of 2,3-dihydropyrroles from *NH*-triazoles and triflic anhydride.

Another, more versatile strategy of temporal installation of an electron-withdrawing group on N1 of 1,2,3-triazoles is acylation [15]. Tandem acylation followed by the ring cleavage without the isolation of *N*-acyltriazoles was achieved using acyl halides or acid anhydrides. The mechanism of this transformation comprises the reversible formation of N1 and N2-acylated triazoles, followed by acid-mediated cleavage of N1-acylated isomer. Acid HX, a side-product of acylation, protonated the resulting acyltriazole to species **8.3a** and unstable **8.3b**. Species **8.3b** has the tendency for ring opening to diazonium intermediate **8.3c**, followed by elimination of a nitrogen molecule to vinyl cation **8.3d** in an irreversible step. The vinyl cation can undergo intramolecular cyclization to oxazoles **8.3e**, hydrolysis to acylaminoketones **8.3f**, or recombination to halogenated enamides **8.3g** as discussed further in this chapter (Scheme 8.3).

Cleavage of *NH*-1,2,3-triazoles with an excess of acyl chlorides or bromides under elevated temperature led to the formation of β-enamido halides **8.4a** in moderate to good yields (Scheme 8.4) [16]. β-Enamido halides are not easily accessed by other synthetic routes and are present in natural products, which highlights the utility of the synthetic method.

This process, in the presence of sodium triflate, was applied for the synthesis of β-enamido triflates (Scheme 8.5) [16]. An alternative method used for the synthesis of β-fluoroacylenamido triflates is based on the formation of N2-acyltriazoles *via* the *in-situ* acylation of *NH*-1,2,3-triazoles with fluorinated acid anhydrides followed by the treatment with triflic acid. The reaction proceeds through N2—N1 acyltriazole interconversion and the ring cleavage of the N1 isomer (Scheme 8.6) [15].

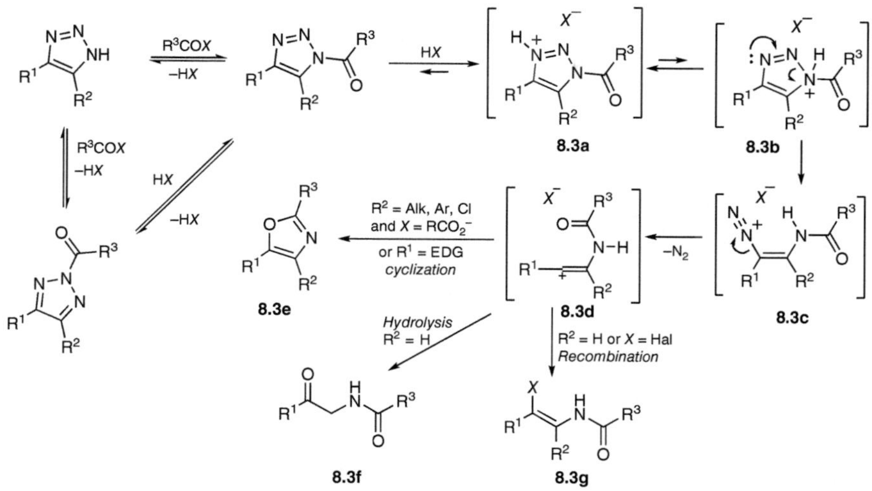

Scheme 8.3 Mechanism of *NH*-triazole cleavage after acylation.

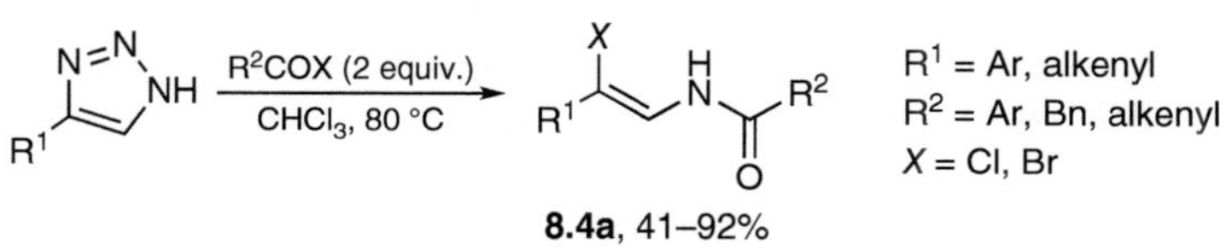

Scheme 8.4 Ring opening of *NH*-1,2,3-triazoles with acyl halides.

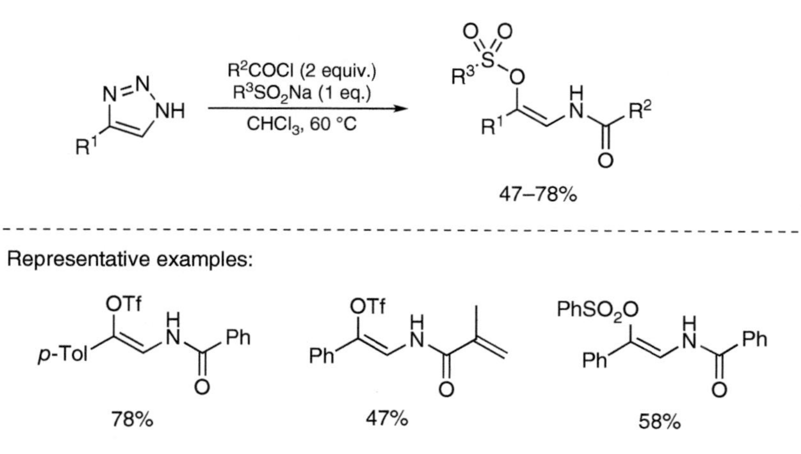

Scheme 8.5 Preparation of β-enamido triflates from sodium sulfonates, *NH*-1,2,3-triazoles, and acyl halides.

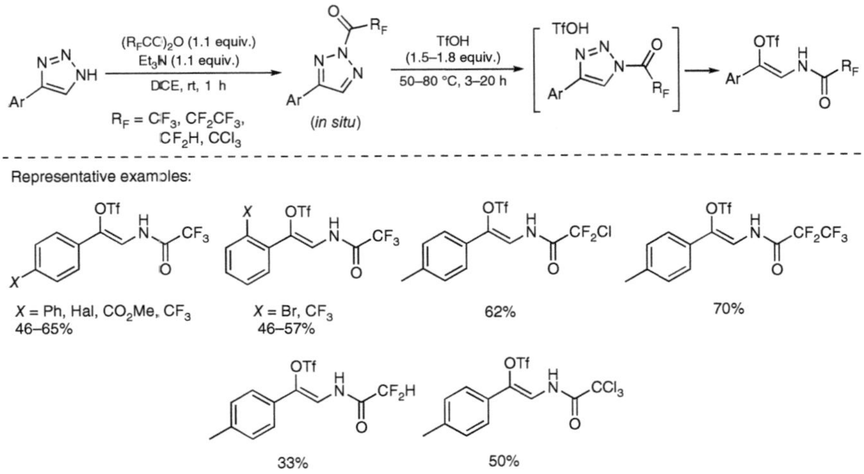

Scheme 8.6 One-pot two-step synthesis of β-fluoroacylenamido triflates.

β-Enamido triflates are useful building blocks that undergo Pd-catalyzed cross-coupling reactions of the triflate group to access highly functionalized enamide derivatives, which are useful drug candidates and synthetic intermediates [17, 18]. None of the above-mentioned methods was applicable for the less reactive alkyl-substituted acylating agents, such as acetic anhydride or acetyl chloride, since the formed acyltriazoles were stable to the ring cleavage even at elevated temperatures [15].

The cleavage of *NH*-1,2,3-triazoles with excess of fluoroalkylated acid anhydrides led to medicinally important 2-fluoroalkyl oxazoles **8.7a** (in the case of 4,5-disubstituted triazoles) or 2-acylaminoketones **8.7b** (for 4-substituted triazoles, R = H) (Scheme 8.7) [19]. In the first case, intramolecular cyclization took place, while in the second case, unstable β-acyloxyenamide **8.7c** formed, which underwent hydrolysis to 2-acylaminoketone **8.7b** after the treatment with aqueous base. The difference in reactivity is due to the increased vinyl cation stability of disubstituted

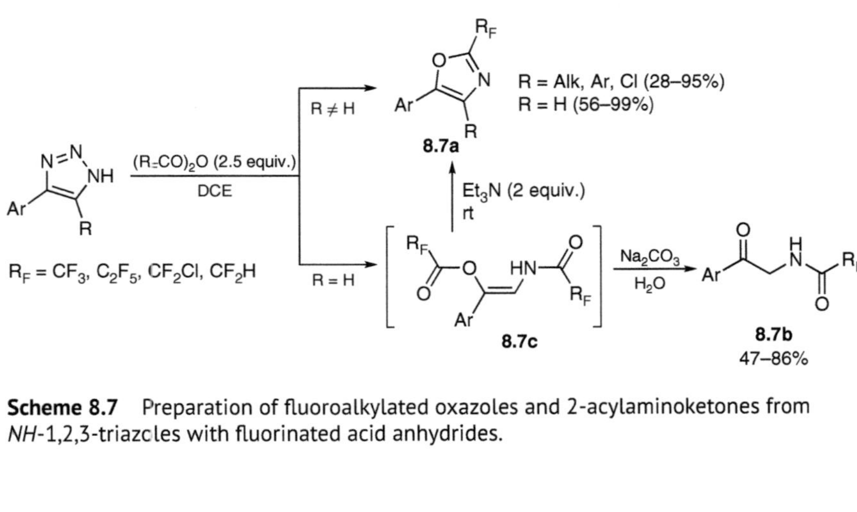

Scheme 8.7 Preparation of fluoroalkylated oxazoles and 2-acylaminoketones from *NH*-1,2,3-triazoles with fluorinated acid anhydrides.

examples, which makes them more susceptible to the intramolecular cyclization. Intermediate **8.7c** can be easily cyclized to oxazoles **8.7a** in mild conditions using a base (Et_3N) in a one-pot process (Scheme 8.7) [20].

A 4,5-disubstituted *NH*-1,2,3-triazole reacted with trichloroacetic anhydride to afford 2-unsubstituted oxazole **8.8a**. Because of the low stability of the trichloromethyl-substituted oxazole intermediate **8.8b** during silica gel column chromatography, hydrolysis and decarboxylation easily take place. In this reaction, trichloroacetic anhydride acted as a one-carbon building block (Scheme 8.8) [19].

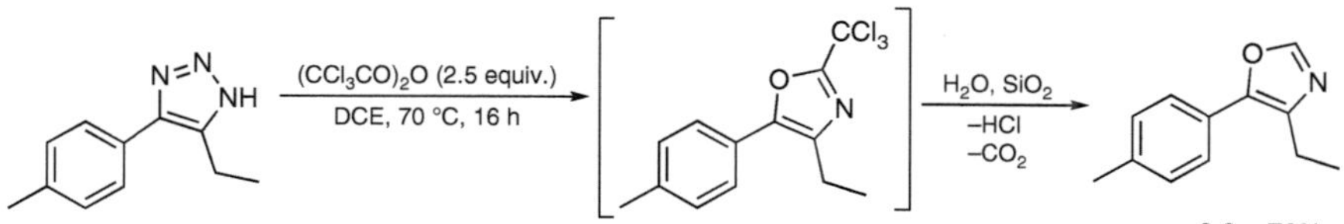

Scheme 8.8 Synthesis of 2-unsubstituted oxazole from *NH*-1,2,3-triazole and trichloroacetic anhydride.

Earlier report describes oxazole formation from unsubstituted 2-trimethylsilyl-1,2,3-triazole under harsh reaction conditions (Scheme 8.9) [21]. This transformation gained attention recently for medicinal chemistry studies, as multisubstituted oxazoles are important biologically active compounds, as reported in scattered examples in patent literature [22–26].

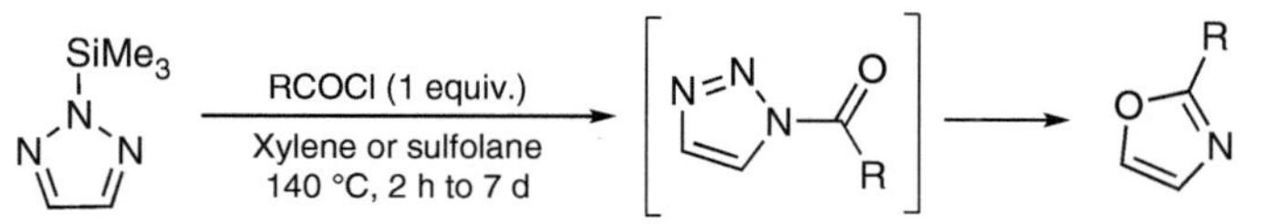

Scheme 8.9 Formation of oxazoles from 2-trimethylsilyl-1,2,3-triazoles and acyl chlorides.

The straightforward access to fluorinated 2-acylaminoketones **8.10b** was used in one-pot syntheses of fluoroalkylated heterocycles directly from *NH*-1,2,3-triazoles. 2-Fluoroalkyl imidazoles **8.10c** were prepared by the cleavage with trifluoroacetic or perfluoropropanoic anhydrides, followed by the treatment of the ketamide intermediate with primary amine (or ammonium acetate for R = H) under microwave conditions. The acid formed after hydrolysis of the enamide to 2-acylaminoketone promoted the Robinson–Gabriel cyclization. This transformation led to imidazoles **8.10c** in one-pot from triazoles (Scheme 8.10) [19]. Alternatively, condensation of 2-acylaminoketones **8.10b** with hydrazine hydrate provided fluoroalkylated 1,2,4-triazines **8.10d** (Scheme 8.10) [19].

The formation of vinyl cation intermediate in *N*-acyltriazole cleavage was supported by a solvent switch from chlorinated to nitrile solvents [27]. In a special case, the adducts **8.11a** (product of the Ritter's reaction) formed and hydrolyzed to bis(enamides) **8.11b** (Scheme 8.11). However, this reaction was not general and

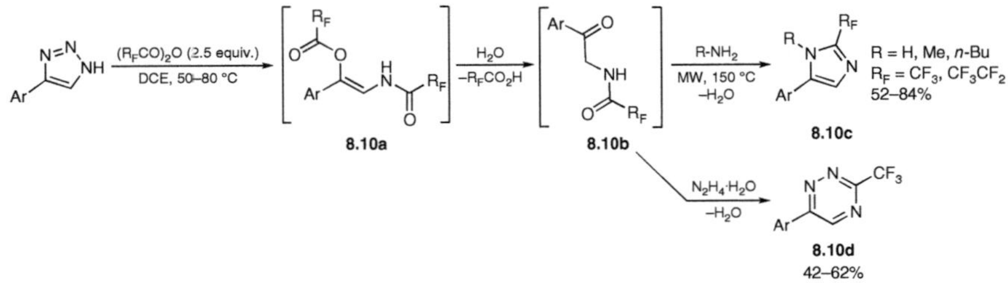

Scheme 8.10 One-pot synthesis of 2-fluoroalkyl-imidazoles and 3-fluoroalkyl-1,2,4-triazines from *NH*-1,2,3-triazoles.

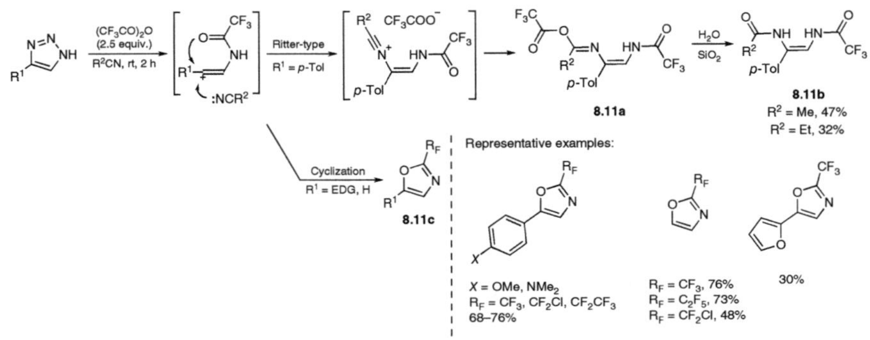

Scheme 8.11 Synthesis of bis(enamides) **8.11b** and oxazoles **8.11c** by denitrogenation of *NH*-1,2,3-triazoles with trifluoroacetic anhydride in nitrile solvent.

in the case of more electron-rich triazoles, the cyclization of vinyl cation to oxazoles **8.11c** took place. This route represents an alternative to the aforementioned (Scheme 8.7), applicable to electron-rich substrates. The formation of oxazoles **8.11c** rather than enamides **8.3g** in polar MeCN was rationalized by the decreased stability of the vinyl cation–trifluoroacetate anion contact ion pair, which prevented recombination and favored cyclization [27].

Several efficient 1,2,3-*NH*-triazole denitrogenation protocols were also developed for *NH*-benzotriazole, which was acylated on N1. The treatment of *N*-acylbenzotriazole with a Lewis acid facilitated benzotriazole ring cleavage leading to benzoxazoles under rather harsh conditions (Scheme 8.12) [28].

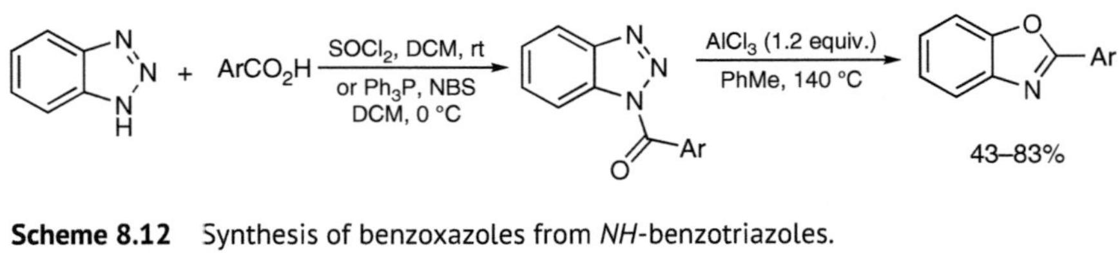

Scheme 8.12 Synthesis of benzoxazoles from *NH*-benzotriazoles.

Rare examples of *ortho*-iodoacetanilide formation in moderate yields from *NH*- and related *N*-acylbenzotriazole were reported when AlI_3/Ac_2O system or Al/I_2 were used [29]. These are the only known cases of *N*-acetyltriazole ring cleavage. The reaction of *N*-acetylbenzotriazole with $AlCl_3$ led only to the deacylation, and not to the desired denitrogenation product (Scheme 8.13) [28].

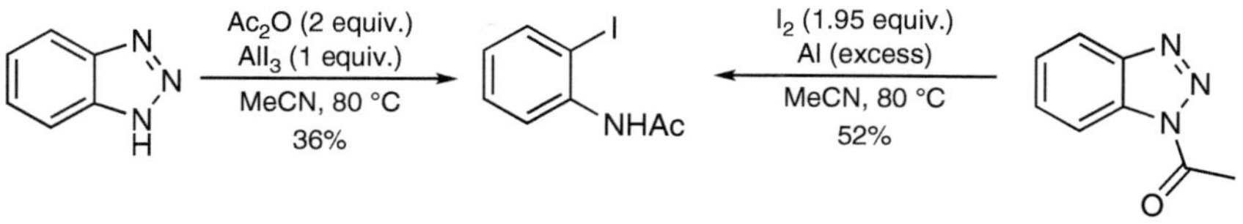

Scheme 8.13 Synthesis of *o*-iodoacetanilide by AlI_3-mediated cleavage of N1-acetylbenzotriazole.

Cleavage of electron-rich 4-arylated *NH*-1,2,3-triazoles with thiophosgene-formed vinyl isothiocyanates **8.14a** by elimination of hydrogen chloride from the vinyl chloride intermediate (Scheme 8.14) [30]. Vinyl isothiocyanate moiety is present in natural products with antifungal and antibacterial activity and is difficult to access by other methods. Changing from electron-rich aromatic *NH*-triazoles to unsubstituted *NH*-1,2,3-triazole gave product **8.14b** of addition of hydrogen chloride across the C=C double bond.

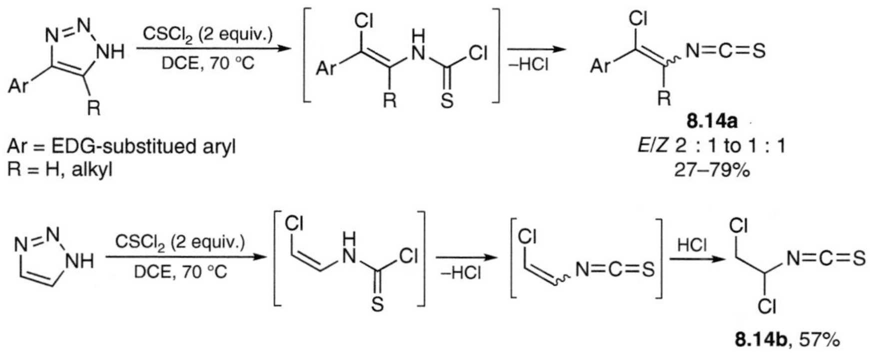

Scheme 8.14 Denitrogenation of *NH*-1,2,3-triazoles with thiophosgene.

An analogous reaction with triazoles containing electron-rich aryl or alkenyl substituent in position 4 took place with triphosgene [30]. The *in situ* formed carbamoyl chlorides **8.15a** were reacted with nucleophiles to provide multifunctional products **8.15b**, such as *N*-alkenyl carbamates, ureas, and thiocarbamates (Scheme 8.15).

Denitrogenative transformation of *NH*-1,2,3-triazoles was studied on more complex substrates such as triazoles **8.16a**. Cleavage with acyl chlorides in the presence of catalytic triflic acid afforded cyclic enaminones **8.16b** (Scheme 8.16) [20]. The reaction proceeded *via* denitrogenation of *N*-acyltriazole and semipinacol rearrangement cascade.

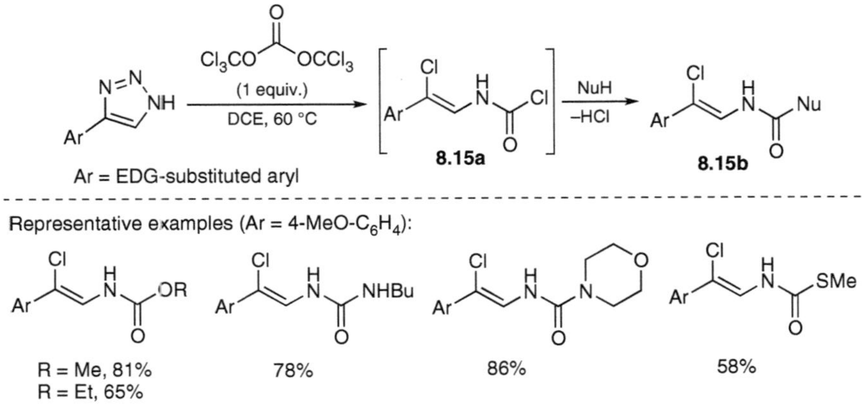

Scheme 8.15 Preparation of multifunctional *N*-alkenyl compounds by denitrogenation of *NH*-1,2,3-triazoles with triphosgene.

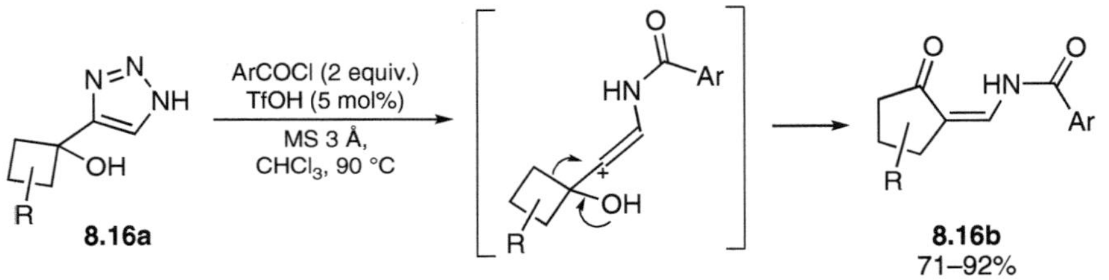

Scheme 8.16 Synthesis of cyclic enaminones by cleavage of 4-(1-hydroxycyclobutyl)-1,2,3-triazoles with acyl chlorides.

8.3 Denitrogenation of *N*-Fluoroalkyl-1,2,3-triazoles

N-Fluoroalkyl-1,2,3-triazoles are easily available by copper(I)-catalyzed cycloaddition of fluorinated azides with terminal alkynes or by organocatalyzed [3 + 2] cycloaddition of fluorinated azides with activated carbonyl compounds [31–38].

Thermal decomposition under microwave heating of *N*-fluoroalkyl-1,2,3-triazoles afforded unusually stable *N*-fluoroalkylated ketenimines **8.17a** (Scheme 8.17) [39]. The reaction proceeded by triazole ring opening, 1,2-shift of the R^2 group with concomitant dinitrogen elimination as confirmed by isotopic labeling study and theoretical calculations. The ketenimine products were isolated and found to undergo addition with oxygen-, sulfur-, or nitrogen-nucleophiles to *N*-fluoroalkyl substituted imidates, thioimidates and amidines, respectively. Alternatively, the ketenimines reacted with alkynes or alkenes to afford *N*-fluoroalkylated cyclobutane or cyclobutene imines, respectively (Scheme 8.17).

Similarly to *N*-sulfonyl-1,2,3-triazoles, *N*-fluoroalkyl-1,2,3-triazoles underwent various rhodium(II)-catalyzed reactions. Under microwave heating, the triazole ring opening took place with the elimination of dinitrogen and the formation of key rhodium(II) carbene reactive intermediates **8.18a** (Scheme 8.18) [34, 37, 40–42].

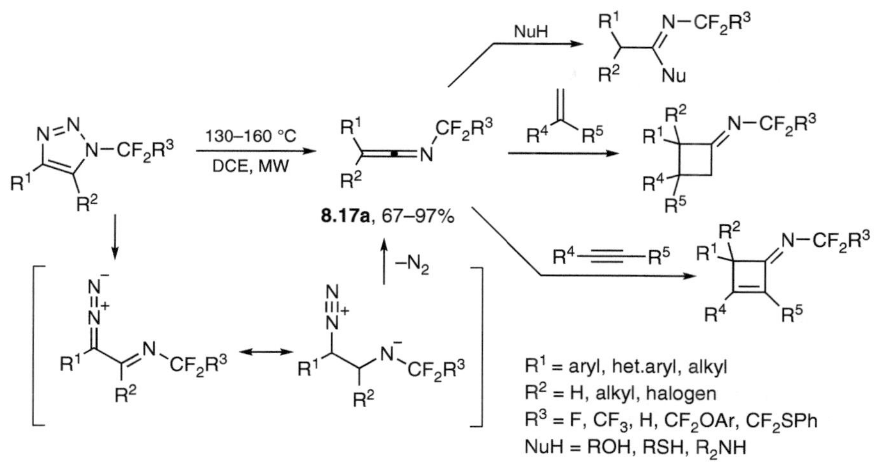

Scheme 8.17 Preparation and reactivity of *N*-fluoroalkyl ketenimines **8.17a**.

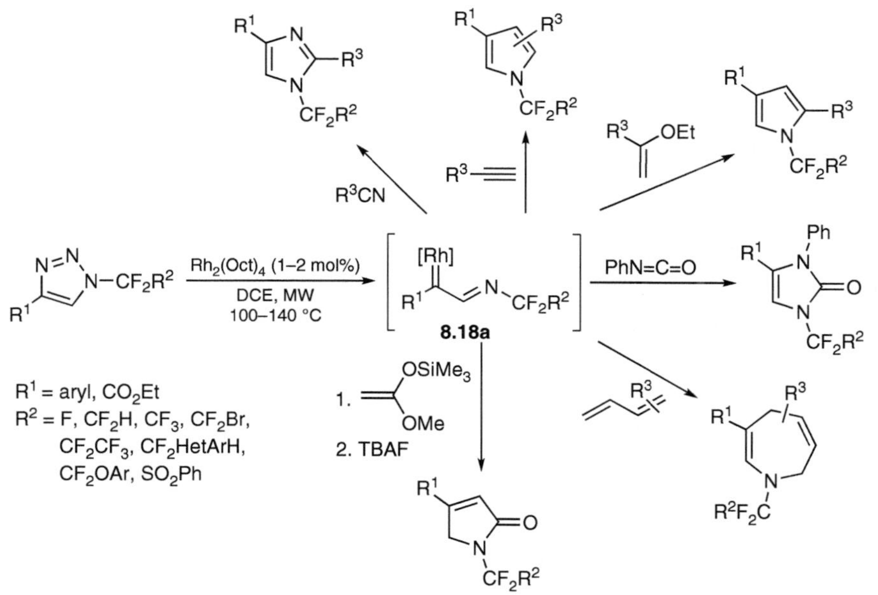

Scheme 8.18 Rhodium(II)-catalyzed reactions of *N*-fluoroalkyl-1,2,3-triazoles leading to the formation of new nitrogen heterocycles.

Their [3 + 2]-cycloaddition with nitriles afforded *N*-fluoroalkylated imidazoles, with alkynes [41] or vinyl ethers *N*-fluoroalkylated pyrroles were formed. In the case of butadienes as reaction partners *N*-fluoroalkylated dihydroazepines formed as a result of [3 + 4] cycloaddition [42]. All these nitrogen heterocycles feature a fluoroalkyl group on nitrogen. Recently, various N—CF_3 azoles were found to display favorable medicinal chemistry properties (such as increased lipophilicity) compared to *N*-methyl azoles [43].

Under rhodium(II) transannulation conditions, 4-cyclohexenyl-substituted *N*-fluoroalkyl-1,2,3-triazoles afforded *N*-fluoroalkylated fused pyrroles. Their oxidation resulted in the formation of *N*-fluoroalkylated indole derivatives. The two steps can be conveniently performed as a one-pot process (Scheme 8.19) [44].

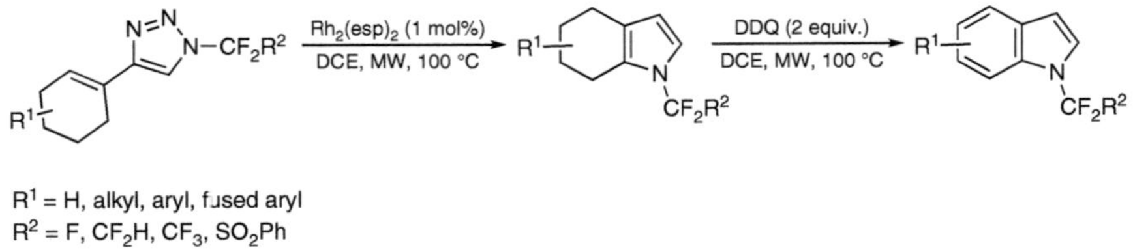

Scheme 8.19 Rhodium(II)-catalyzed reactions of 4-cyclohexenyl-substituted *N*-fluoroalkyl-1,2,3-triazoles leading to *N*-fluoroalkylated indoles.

The rhodium carbene intermediate was also reactive with heteroatom nucleophiles, such as water or amine nucleophile affording 2-fluoroalkyl oxazoles, thiazoles, or imidazoles (Scheme 8.20) [34, 45]. During the process, the initially nitrogen-bound CF_2 group was incorporated in the five-membered heterocycle ring after defluorination, which makes the whole process an interesting example of C—F activation. In all Rh(II)-catalyzed processes, the R group must be a strongly electron-withdrawing group, while *N*-difluoromethyl triazoles were unreactive.

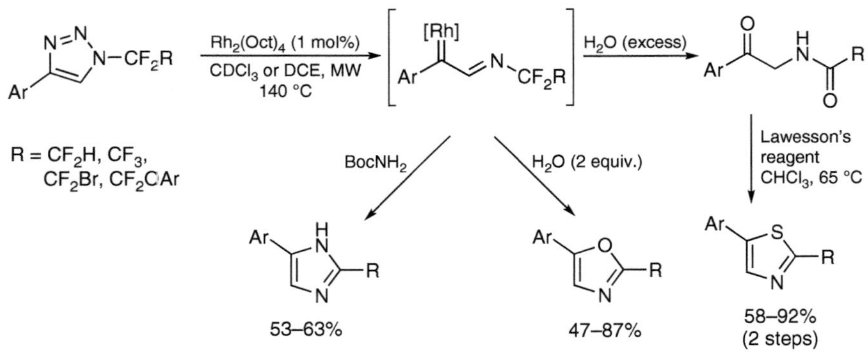

Scheme 8.20 Rh(II)-catalyzed ring opening and defluorinative annulation of *N*-fluoroalkyl-1,2,3-triazoles.

In the presence of strong Brønsted acid, *N*-fluoroalkyl-1,2,3-triazoles underwent denitrogenative ring cleavage. With triflic acid, the transformations led to β-enamido triflates and with fluorosulfonic acid, the products were β-enamido fluorosulfonates (Scheme 8.21) [46]. Mechanistic investigations and theoretical calculations suggested that the reaction proceeded *via* a vinyl cation. The triflate anion is bound to NH by hydrogen bonding and delivers the triflate group to the carbocation center from the same side of the C=C bond, providing the (*Z*)-stereoisomers of the products. The CF_2 group next to the nitrogen atom again undergoes

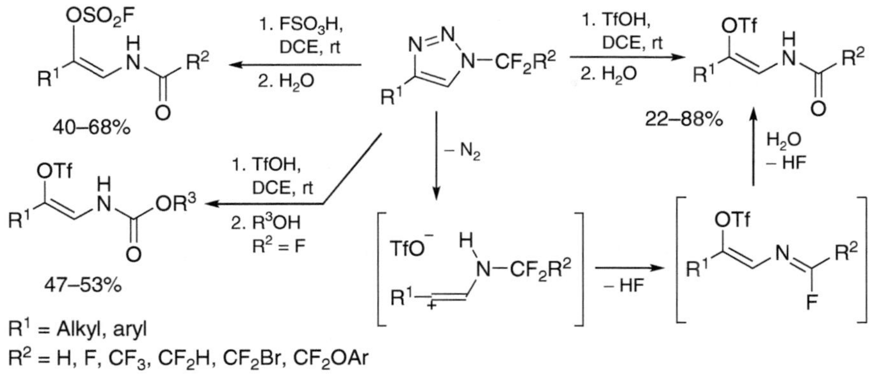

Scheme 8.21 Triflic and fluorosulfonic acid-mediated transformation of *N*-fluoroalkyl-1,2,3-triazoles into (*Z*)-β-enamido triflates or fluorosulfonates.

hydrolysis to the amide functionality. In the case of *N*—CF_3-triazole substrates, this hydrolysis can be used in further reaction with additional nucleophile, such as alcohol to give carbamate products [46]. β-Enamido triflates exhibited excellent reactivity in cross-coupling reactions to functionalized enamides with the control of stereochemistry on the C=C bond [18, 46] and in hetarylation reactions [17].

Lewis's acids, such as $BF_3{\cdot}OEt_2$ also promoted denitrogenation of *N*-fluoroalkylated triazoles. This time, β-enamido fluorides **8.22a** were formed (Scheme 8.22) [47]. Similarly, aluminum halides mediated the conversion of *N*-fluoroalkyl-1,2,3-triazoles into vinyl halides with an additional imidoyl halide moiety **8.23a** (Scheme 8.23) [48]. In both cases, the reaction proceeds via a vinyl cation with a nitrogen atom bound either to boron or aluminum. In the case of AlX_3, the *N*-difluoromethylene group did not hydrolyses to amide, but imidoyl halides were formed and isolated as stable products. Further evidence of the formation of vinyl cation intermediate was obtained from an experiment in the presence of an external electron-rich aromatic. In this case, intermolecular aromatic C-H insertion took place, leading to compound **8.24a**, in addition to vinyl halide product **8.24b** (Scheme 8.24) [49]. The highly functionalized imidoyl halides **8.23a** were found to undergo chemoselective halide substitution at the imidoyl group by nucleophiles, cyclization, or cross coupling on vinyl halide functionality as further transformations.

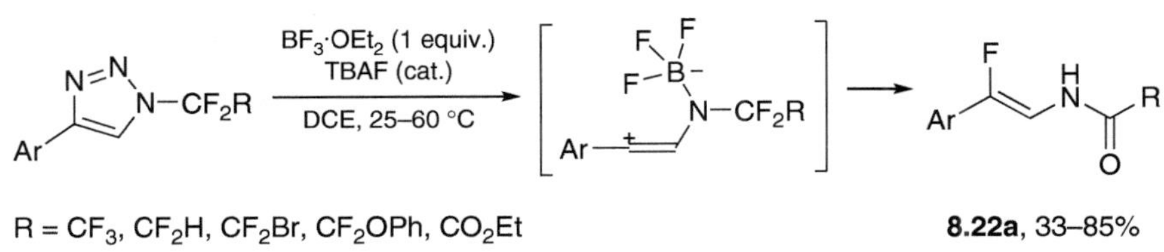

Scheme 8.22 Boron trifluoride-mediated conversion of *N*-fluoroalkyl-1,2,3-triazoles into (*Z*)-*β*-enamido fluorides.

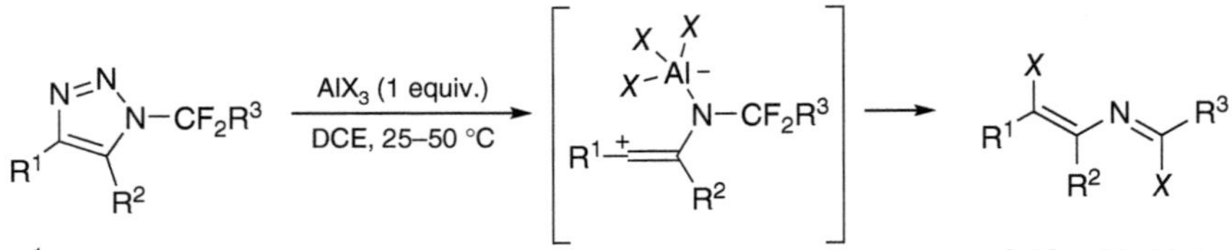

8.23a, 30–90%

R^1 = Alkyl, aryl
R^2 = Alkyl, aryl, alkynyl, Cl, Br, I
R^3 = CF_3, CF_2H, CF_2Br, CF_2OAr, CO_2Et
X = Cl, Br, I

Scheme 8.23 Aluminum trihalide-mediated conversion of *N*-fluoroalkyl-1,2,3-triazoles into (*Z*)-haloalkenyl imidoyl halides.

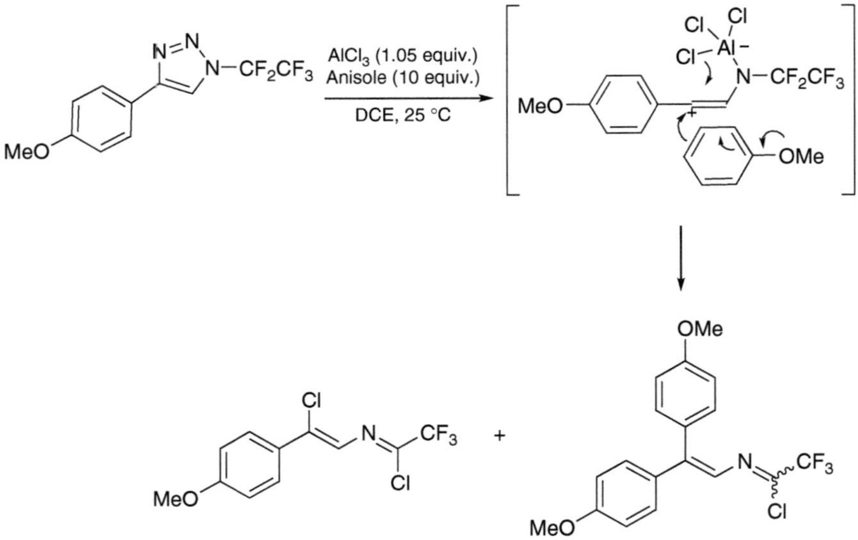

8.24b, 9% **8.24a**, 34%, *Z*/*E* 82:18

Scheme 8.24 Evidence of the vinyl cation intermediate in the formation of imidoyl halide **8.24a**.

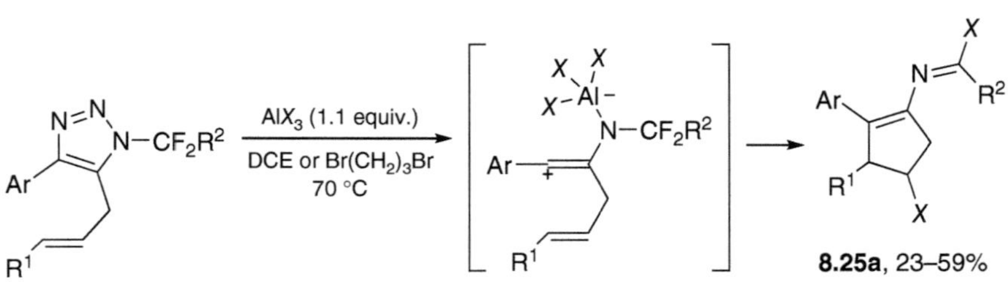

8.25a, 23–59%

R^1 = H, Me, Ph
R^2 = CF_3, CF_2H, CF_2Br, CF_2OPh
X = Cl, Br

Scheme 8.25 Aluminum trihalide-mediated transformation of *N*-fluoroalkyl-5-allyl-1,2,3-triazoles into halogenated cyclopentenes.

N-Fluoroalkyl-5-allyl-1,2,3-triazoles prepared by intercepted CuAAC from copper(I) acetylides, fluorinated azides, and allyl bromides or iodides were denitrogenated with aluminum chloride or bromide with concomitant cyclization of the vinyl cation to the C=C bond and halide capture to afford halogenated cyclopentenes **8.25a** (Scheme 8.25) [49].

8.4 Summary and Outlook

In summary, *NH*-1,2,3-triazoles, easily available basic chemicals, are amenable to denitrogenative ring cleavage transformations after temporal attachment of electron-withdrawing groups (e.g. acylation). This atom-efficient strategy was used in a variety of reactions leading to oxazoles, multifunctionalized *N*-alkenyl compounds, acylaminoketones, and heterocycles derived from them. *N*-Fluoroalkylated 1,2,3-triazoles, similarly to *N*-sulfonyl-1,2,3-triazoles, undergo rhodium(II)-catalyzed transannulation reactions to a variety of novel *N*-fluoroalkylated heterocycles. Additionally, their Lewis or Brønsted acid-mediated denitrogenative ring cleavage afforded key reactive intermediate vinyl cation, which was transformed to novel *N*-alkenyl compounds (enamides, imidoyl halides) or applied in cyclization. Thermal rearrangement of *N*-fluoroalkylated 1,2,3-triazoles provides *N*-fluoroalkyl ketenimines. Thus, denitrogenative strategies of *NH*-1,2,3-triazoles or *N*-fluoroalkyl-1,2,3-triazoles offer novel synthetic pathways to diverse classes of nitrogen heterocycles and *N*-alkenyl compounds which hold promise in the realms of medicinal chemistry and the advancement of industrial synthetic processes.

Acknowledgments

The work was financially supported by the Czech Academy of Sciences (Research Plan RVO: 61388963).

References

1 Dehaen, W. and Bakulev, V.A. (ed.) (2015). *Chemistry of 1,2,3-triazoles, in Topics in Heterocyclic Chemistry*, vol. 40. Springer.

2 Davies, H.M. and Alford, J.S. (2014). Reactions of metallocarbenes derived from *N*-sulfonyl-1,2,3-triazoles. *Chem. Soc. Rev.* 43: 5151.

3 Li, W. and Zhang, J. (2020). Synthesis of heterocycles through denitrogenative cyclization of triazoles and benzotriazoles. *Chem. Eur. J.* 26: 11931.

4 Kumar, S., Sharma, B., Mehra, V., and Kumar, V. (2021). Recent accomplishments on the synthetic/biological facets of pharmacologically active 1*H*-1,2,3-triazoles. *Eur. J. Med. Chem.* 212: 113069.

5 Tornøe, C.W., Christensen, C., and Meldal, M. (2002). Peptidotriazoles on solid phase: [1,2,3]-triazoles by regiospecific copper(I)-catalyzed 1,3-dipolar cycloadditions of terminal alkynes to azides. *J. Org. Chem.* 67: 3057.

6 Rostovtsev, V.V., Green, L.G., Fokin, V.V., and Sharpless, K.B. (2002). A stepwise Huisgen cycloaddition process: copper(I)-catalyzed regioselective "ligation" of azides and terminal alkynes. *Angew. Chem. Int. Ed.* 41: 2596.

7 Akter, M., Rupa, K., and Anbarasan, P. (2022). 1,2,3-Triazole and its analogues: new surrogates for diazo compounds. *Chem. Rev.* 122: 13108.

8 Chattopadhyay, B. and Gevorgyan, V. (2012). Transition-metal-catalyzed denitrogenative transannulation: converting triazoles into other heterocyclic systems. *Angew. Chem. Int. Ed.* 51: 862.

9 Anbarasan, P., Yadagiri, D., and Rajasekar, S. (2014). Recent advances in transition-metal-catalyzed denitrogenative transformations- of 1,2,3-triazoles and related compounds. *Synthesis* 46: 3004.

10 Xu, Z.-F., Dai, H.C., Shan, L.H., and Li, C.-Y. (2018). Metal-free synthesis of (*E*)-monofluoroenamine from 1-sulfonyl-1,2,3-triazole and $Et_2O{\cdot}BF_3$ via stereospecific fluorination of α-diazoimine. *Org. Lett.* 20: 1054.

11 Yang, D., Shan, L., Xu, Z.-F., and Li, C.-Y. (2018). Metal-free synthesis of imidazole by $BF_3{\cdot}Et_2O$ promoted denitrogenative transannulation of *N*-sulfonyl-1,2,3-triazole. *Org. Biomol. Chem.* 16: 1461.

12 Bakhanovich, O. and Beier, P. (2020). Synthesis, stability and reactivity of α-fluorinated azidoalkanes. *Chem. Eur. J.* 26: 773.

13 Markos, A., Matoušek, V., and Beier, P. (2022). Fluoroalkyl azides and triazoles: unlocking a novel chemical space. *Aldrichim. Acta* 55: 37.

14 Kwok, S.W., Zhang, L., Grimster, N.P., and Fokin, V.V. (2014). Catalytic asymmetric transannulation of *NH*-1,2,3-triazoles with olefins. *Angew. Chem. Int. Ed.* 53: 3452.

15 Motornov, V., Pohl, R., Klepetářová, B., and Beier, P. (2023). *N*-Acyl-1,2,3-triazoles – key intermediates in denitrogenative transformations. *Chem. Commun.* 59: 9364.

16 Wang, T., Tang, Z., Luo, H. et al. (2021). Access to (*Z*)-β-substituted enamides from N1-H-1,2,3-triazoles. *Org. Lett.* 23: 6293.

17 Lu, Q., Wang, T., Wu, Q. et al. (2022). C—H heteroarylation of aromatics via catalyst free S_N2' coupling cycloaromatization. *Green Chem.* 24: 4399.

18 Chvojka, T., Markos, A., Voltrová, S. et al. (2021). Ligand-dependent stereoselective Suzuki–Miyaura cross-coupling reactions of β-enamido triflates. *Beilstein J. Org. Chem.* 17: 2657.

19 Motornov, V. and Beier, P. (2022). Access to fluoroalkylated azoles and 2-acylaminoketones via fluorinated anhydride-mediated cleavage of *NH*-1,2,3-triazoles. *Org. Lett.* 24: 1958.

20 Xu, M., Liu, L., Wang, T. et al. (2022). Acid-catalyzed ring-expansion of 4-(1-hydroxycyclobutyl)-1,2,3-triazoles. *Org. Chem. Front.* 9: 1065.

21 Williams, E.L. (1992). A general, one-pot method for the synthesis of 2-substituted oxazoles. *Tetrahedron Lett.* 33: 1033.

22 Patent WO2019/43139, 2019, A1.

23 Patent WO2009/39135, 2009, A1.

24 Patent CN110698420, 2020, A.

25 Patent US2012/129843, 2012, A1.

26 Murugesan, N., Gu, Z., Stein, P.D. et al. (2000). 2. Discovery of 4′-oxazolyl biphenylsulfonamides as a new class of potent, highly selective ET_A antagonists. *J. Med. Chem.* 43: 3111.

27 Motornov, V. and Beier, P. (2022). One-pot synthesis of 4-substituted 2-fluoroalkyloxazoles from *NH*-1,2,3-triazoles and fluoroalkylated acid anhydrides. *New J. Chem.* 46: 14318.

28 Singh, A.S., Mishra, N., Kumar, D., and Tiwari, V.K. (2017). Lewis-acid-mediated benzotriazole ring cleavage (BtRC) strategy for the synthesis of 2-aryl benzoxazoles from *N*-acylbenzotriazoles. *ACS Omega* 2: 5044.

29 Dong, B., Liu, Y., Yang, P. et al. (2022). Denitrogenative cleavage of benzotriazoles and benzotriazinones, and selective *N*-desulfonylation of benzotriazoles by aluminum halides. *Tetrahedron Lett.* 103: 153965.

30 Motornov, V. and Beier, P. (2023). Synthesis of *N*-vinyl isothiocyanates and carbamates by the cleavage of *NH*-1,2,3-triazoles with one-carbon electrophiles. *Org. Biomol. Chem.* 21: 1147.

31 Blastik, Z.E., Voltrová, S., Matoušek, V. et al. (2017). Azidoperfluoroalkanes: synthesis and application in copper(I)-catalyzed azide–alkyne cycloaddition. *Angew. Chem. Int. Ed.* 56: 346.

32 Voltrová, S., Putovný, I., Matoušek, V. et al. (2018). Reintroducing azidodifluoromethane: synthesis, stability and [3 + 2] cycloadditions. *Eur. J. Org. Chem.* 2018: 5087.

33 Voltrová, S., Muselli, M., Filgas, J. et al. (2017). Synthesis of tetrafluoroethylene- and tetrafluoroethyl-containing azides and their 1,3-dipolar cycloaddition as synthetic application. *Org. Biomol. Chem.* 15: 4962.

34 Tichý, D., Koštál, V., Motornov, V. et al. (2020). Preparation of 1-azido-2-bromo-1,1,2,2-tetrafluoroethane and its use in the synthesis of *N*-fluoroalkylated nitrogen heterocycles. *J. Org. Chem.* 85: 11482.

35 Blastik, Z.E., Klepetářová, B., and Beier, P. (2018). Enamine-mediated azide-ketone [3 + 2] cycloaddition of azidoperfluoroalkanes. *ChemSelect* 3: 7045.

36 Voltrová, S., Filgas, J., Slavíček, P., and Beier, P. (2020). Azidofluoromethane: synthesis, stability and reactivity in [3 + 2] cycloadditions. *Org. Chem. Front.* 7: 10.

37 Ziabko, M., Klepetářová, B., and Beier, P. (2023). Synthesis of azidodifluoromethyl phenyl sulfone and its use as a synthetic equivalent of the azidodifluoromethyl anion. *J. Org. Chem.* 88: 6939.

38 Shaitanova, E., Matoušek, V., Herentin, T. et al. (2023). Synthesis and cycloadditions reactions of 1-azido-1,1,2,2-tetrafluoroethane. *J. Org. Chem.* 88: 14969.

39 Kubíčková, A., Markos, A., Voltrová, S. et al. (2023). Aza-Wolff rearrangement of *N*-fluoroalkyl triazoles to ketenimines. *Org. Chem. Front.* 10: 3201.

40 Motornov, V., Markos, A., and Beier, P. (2018). A rhodium-catalyzed transannulation of *N*-(per)fluoroalkyl-1,2,3-triazoles under microwave conditions – a general route to *N*-(per)fluoroalkyl-substituted five-membered heterocycles. *Chem. Commun.* 54: 3258.

41 Bakhanovich, O., Khutorianskyi, V., Motornov, V., and Beier, P. (2021). Synthesis of *N*-perfluoroalkyl-3,4-disubstituted pyrroles by rhodium-catalyzed

transannulation of *N*-fluoroalkyl-1,2,3-triazoles with terminal alkynes. *Beilstein J. Org. Chem.* 17: 504.

42 Motornov, V. and Beier, P. (2018). Chemoselective Aza-[4 + 3]-annulation of *N*-perfluoroalkyl-1,2,3-triazoles with 1,3-dienes: access to *N*-perfluoroalkyl-substituted azepines. *J. Org. Chem.* 83: 15195.

43 Schiesser, S., Chepliaka, H., Kollback, J. et al. (2020). *N*-trifluoromethyl amines and azoles: an underexplored functional group in the Medicinal Chemist's Toolbox. *J. Med. Chem.* 63: 13076.

44 Bakhanovich, O., Klepetářová, B., and Beier, P. (2023). Rhodium(II)-catalyzed transannulation approach to *N*-fluoroalkylated indoles. *Org. Biomol. Chem.* 21: 7924.

45 Motornov, V., Košťál, V., Markos, A. et al. (2019). General approach to 2-fluoroalkyl 1,3-azoles via the tandem ring opening and defluorinative annulation of *N*-fluoroalkyl-1,2,3-triazoles. *Org. Chem. Front.* 6: 3776.

46 Markos, A., Voltrová, S., Motornov, V. et al. (2019). Stereoselective synthesis of (*Z*)-β-enamido triflates and fluorosulfonates from *N*-fluoroalkylated triazoles. *Chem. Eur. J.* 25: 7640.

47 Markos, A., Janecký, L., Klepetářová, B. et al. (2021). Stereoselective synthesis of (*Z*)-β-enamido fluorides from *N*-fluoroalkyl- and *N*-sulfonyl-1,2,3-triazoles. *Org. Lett.* 23: 4224.

48 Markos, A., Janecký, L., Chvojka, T. et al. (2021). Haloalkenyl imidoyl halides as multifacial substrates in the stereoselective synthesis of *N*-alkenyl compounds. *Adv. Synth. Catal.* 363: 3258.

49 Janecký, L., Markos, A., Klepetářová, B., and Beier, P. (2023). Lewis-acid-mediated intramolecular cyclization of 4-aryl-5-allyl-1,2,3-triazoles to substituted cyclopentene derivatives. *J. Org. Chem.* 88: 1155.

9

Asymmetric Perspective on Denitrogenative Transformation of 1,2,3-Triazoles

Pokhriyal Yamini, Akanksha Babbar, and Dongari Yadagiri

Laboratory of Synthesis and Catalysis, Department of Chemistry, Indian Institute of Technology Roorkee, Roorkee 247667, Uttarakhand, India

9.1 Introduction

Enantiomerically pure compounds have broad applications in different fields, for example, biology, medicine, and material chemistry [1]. Due to that, the synthesis of these compounds using asymmetric catalysis is highly desirable. Many methods are available for the synthesis of chiral molecules from achiral starting material with chiral catalysts and achiral catalysts in combination with chiral ligands. Among them, one of the standout approaches is metallocarbene chemistry for the synthesis of enantioenrich molecules with the help of chiral catalysts [2]. Previously, the synthesis of chiral molecules was achieved via asymmetric carbene transfer reactions, which include insertion reactions, cyclopropanation reactions, and other transformations from carbenes, which are derived from diazo and related compounds (Scheme 9.1) [2, 3]. The difficulty associated with the carbenes derived from the diazo compounds is their explosive nature of the diazo group, low stability, and storability [3]. To avoid the limitation of these carbene precursors, such as diazo compounds, Fokin and Gevorgyan groups discovered the *N*-sulfonyl-1,2,3-triazoles as a surrogate of a new type of donor–acceptor carbenes [4].

The advantage of using the triazoles as carbene precursors are their stability, storability, and easy accessibility from azide-alkyne cycloaddition reactions. Suitably substituted-1,2,3-triazoles would undergo ring-chain isomerism, which is well known as Dimroth rearrangement (Scheme 9.1b) [5].

9.2 Rhodium-Catalyzed Cyclopropanation Reactions

Initially, *N*-sulfonyl 1,2,3-triazoles precursor for azavinyl carbenes was discovered by Fokin and Gevorgyan for the transannulation reaction with nitriles [4]. Later, many groups have explored the utilization of azavinyl carbenes for their functionalization for synthesizing synthetic intermediates and heterocyclic and carbocyclic compounds, but asymmetric catalysis needs to be developed [6]. For instance, the

Denitrogenative Transformation of Nitrogen Heterocycles: Synthesis, Reactions and Applications, First Edition. Edited by Pazhamalai Anbarasan.

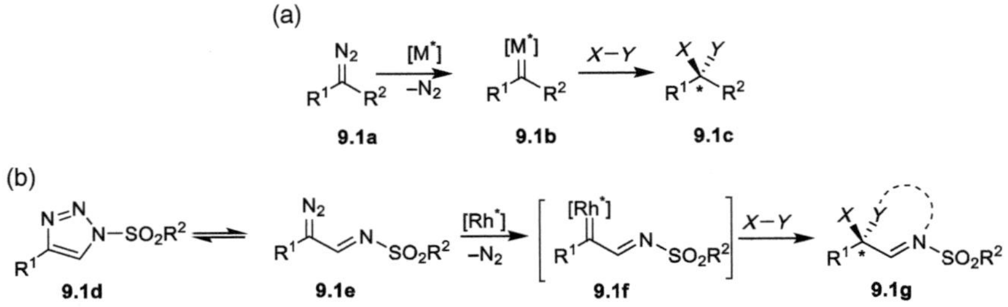

Scheme 9.1 Asymmetric functionalization of metallocarbenes. (a) Metallocarbenes derived from diazocompounds. (b) Rhodium azavinyl carbenes derived from *N*-sulfonyl-1,2,3-triazoles.

first asymmetric example of *N*-sulfonyl-1,2,3-triazoles for cyclopropanation with alkenes was shown by Fokin's group in 2009 [7] in the presence of chiral rhodium catalyst (Scheme 9.2). The authors showed asymmetric cyclopropanation between *N*-sulfonyl-4-phenyl-1,2,3-triazoles **9.2a** and alkenes **9.2b** using $Rh_2(S\text{-NTTL})_4$ as catalyst at 80 °C via in situ generated rhodium azavinyl carbenes followed by [2 + 1]-cycloaddition resulting in sulfonyl imines **9.2c**, which upon further hydrolysis in the presence of basic conditions lead to the formation of aldehydes

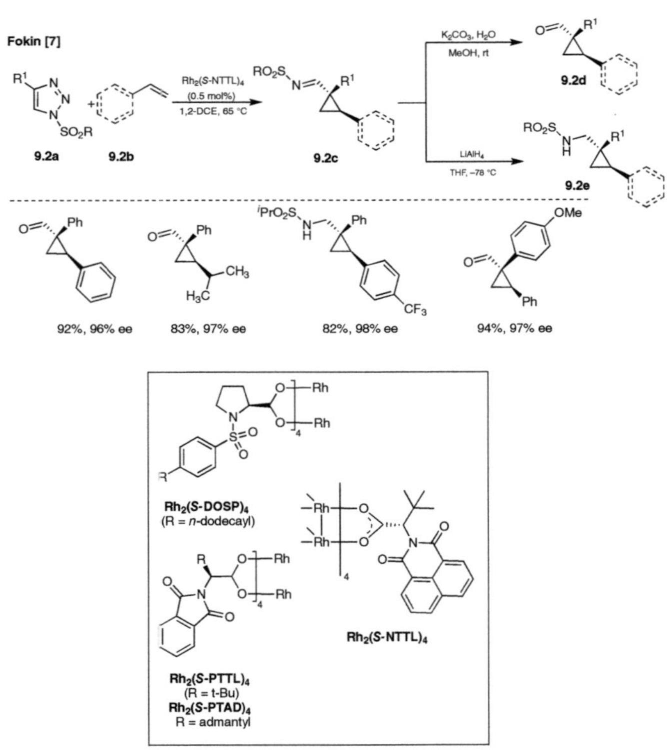

Scheme 9.2 Rh(II)-catalyzed asymmetric cyclopropanation. Source: Adapted from Fokin [7].

9.2d and then reduction with $LiAlH_4$ would deliver the corresponding amines **9.2e** in high enantioselective manner, respectively. Later, the same group reported asymmetric cyclopropanation reaction via in situ generated *N*-triflyl rhodium azavinyl carbenes [8] with *NH*-triazoles **9.2a** and triflic anhydride (Tf_2O) with the help of 2,6-di-*tert*-butyl-4-methyl pyridine (DTBP) **9.2e** as a base (Scheme 9.3). When authors used alkene as 4-methoxy styrene, they observed transannulation products such as 2,3-dihydropyrroles with excellent yield but moderated ee.

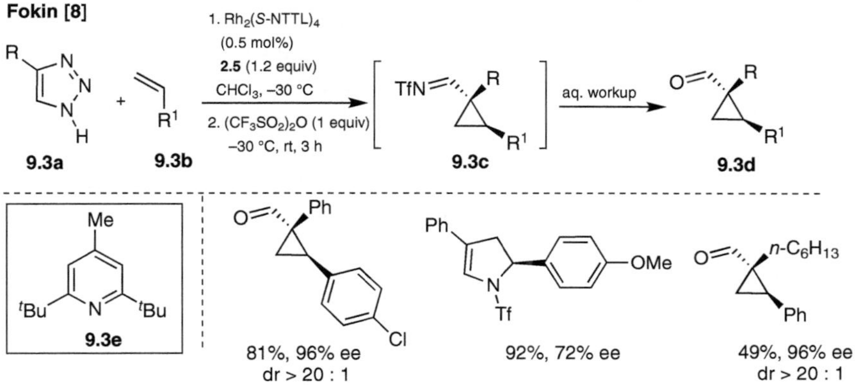

Scheme 9.3 Reactivity of *N*-triflyl azavinyl carbenes. Source: Adapted from Fokin [8].

In 2011, the Fokin group utilized the electron-deficient triazolyl substituted-1,2,3-triazole **9.4a** precursor for azavinyl carbene [9] instead of *N*-sulfony-1,2,3-triazoles **9.3a** for the asymmetric cyclopropanation reaction (Scheme 9.4). Utilized the electron-deficient heterocyclic group, such as the triazolyl group at the *N*-1 position of the triazole **9.4a**, behaved as an equivalent of the sulfonyl group of triazole

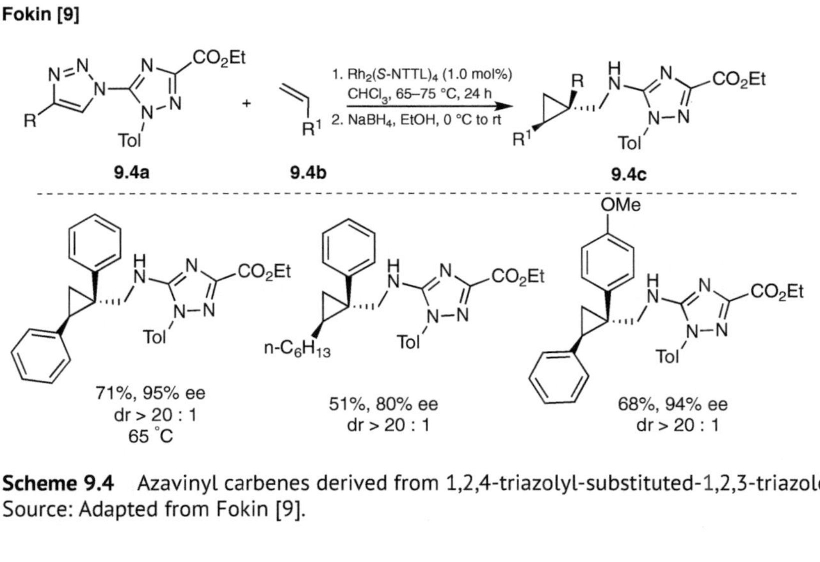

Scheme 9.4 Azavinyl carbenes derived from 1,2,4-triazolyl-substituted-1,2,3-triazoles. Source: Adapted from Fokin [9].

9.2a to generate azavinyl carbene. It shows similar reactivity for the asymmetric cyclopropanation reaction with alkenes **9.4b** followed by reduction with $NaBH_4$ to synthesize chiral secondary amines **9.4c**.

Later, the cyclopropanation strategy was utilized by Miura, Murakami group for the enantioselective synthesis of piperidine-fused *trans*-cycloalkenes in 2014 [10] from 1-mesyl-4-substituted-1,2,3-triazole **9.5a** and methylenecyclopropane **9.5b** in the presence of $Rh_2(S\text{-NTTL})_4$ (Scheme 9.5). In situ generated rhodium azavinyl carbene underwent [2 + 1]-cycloaddition reaction with methylenecyclopropane **9.5b** would form another cyclopropane ring **9.5c** onto the existing methylenecyclopropane ring of **9.5b**. Next, the thermal skeletal rearrangement upon microwave irradiation delivered enantioenrich piperidine-fused *trans*-cycloalkenes such as *trans*-cyclononene in a one-pot two-step protocol. They have also explored with eight-membered rings such as 8-methylenebicyclo[5.1.0]octane and then thermal skeletal rearrangement and observed the formation of *trans*-, *cis*-cyclooctene, and 1,3-diene. Interestingly, methylenecyclopropane nine-membered ring derived from *cis*-cyclononene underwent skeletal rearrangement at 40 °C and delivered the enantioenrich *trans*-cyclodecene with a high yield and enantioselectivity.

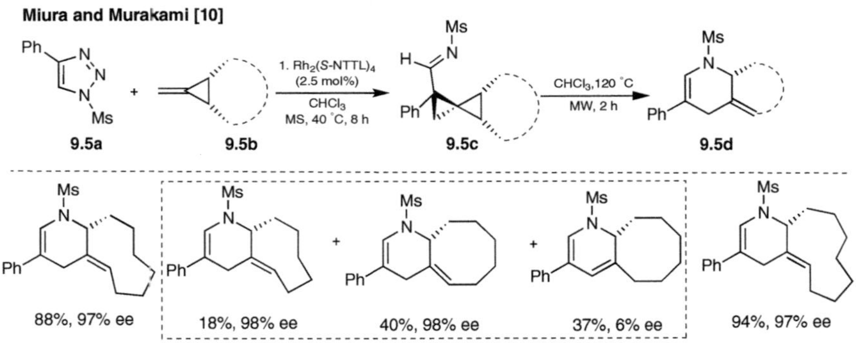

Scheme 9.5 Enantioselective synthesis of piperidine-fused *trans*-cycloalkanes. Source: Adapted from Miura and Murakami [11].

Furthermore, authors also showed a one-pot synthesis of enantioenrich *trans*-cyclononene from alkynes, mesyl azide, and methylenecyclopropane via the integration of CuTc, $Rh_2(S\text{-NTTL})_4$ followed by microwave-assisted skeletal rearrangement all steps in one-pot. The Miura, Murakami group showed the enantioselective synthesis of cyclopropylmethanamines **9.6d** motifs, which are pharmaceutically important molecules. Initial asymmetric cyclopropanation was observed with the reaction of *N*-sulfonyl-1,2,3-triazoles **9.6a** and pinacol allylboronate **9.6b** (Scheme 9.6) in the presence of a rhodium catalyst [11]. This one-pot synthesis allowed the preparation of cyclopropylmethanamines upon reduction with DIBAL-H.

In 2017, Miura, Murakami group expanded the asymmetric cyclopropanation reaction to enantioselective synthesis of macrocycle molecules such as C_3-symmetric triangular molecules (Scheme 9.7) from the specially designed

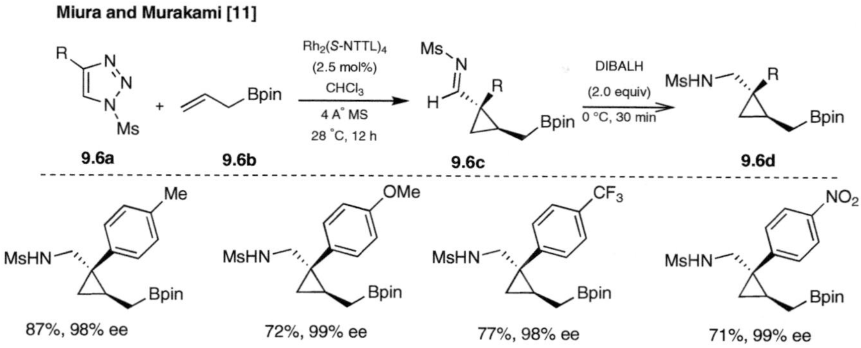

Scheme 9.6 Enantioselective synthesis of cyclopropylmethanamines. Source: Adapted from Miura and Murakami [12].

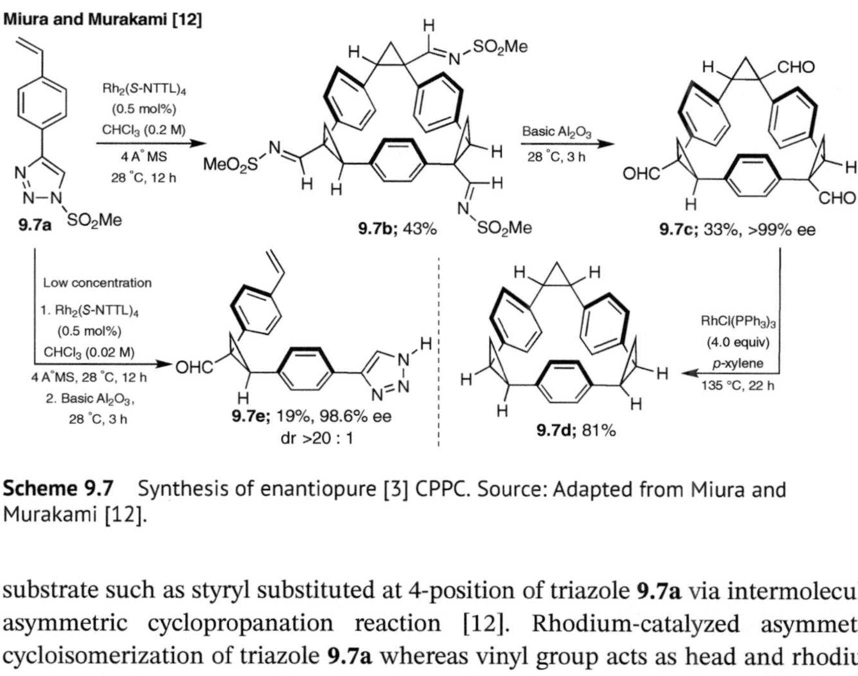

Scheme 9.7 Synthesis of enantiopure [3] CPPC. Source: Adapted from Miura and Murakami [12].

substrate such as styryl substituted at 4-position of triazole **9.7a** via intermolecular asymmetric cyclopropanation reaction [12]. Rhodium-catalyzed asymmetric cycloisomerization of triazole **9.7a** whereas vinyl group acts as head and rhodium azavinyl carbene act as tail or vice versa to form the cyclic triamine **9.7b** with 43% yield, but ee (enantiomeric excess) was not determined due to rapid hydrolysis of imine moiety under chiral HPLC conditions. In situ generated cyclic triimine **9.7b** hydrolyzed under basic Al_2O_3 to trialdehyde **9.7c** (Scheme 9.7) and determined the ee. Trialdehyde **9.7c** was recrystallized using acetone, forming a fine crystal. The authors have reported that the formation of trialdehyde **9.7c**, which required a high solvent concentration, while at a low concentration, dimeric form **9.7e** was obtained in only 19% yield (Scheme 9.7).

Furthermore, the authors explored the decarbonylation of the formyl group of trialdehyde **9.7c** mediated by $RhCl(PPh_3)_3$ to synthesize the tiara-like hydrocarbon

9.7d (Scheme 9.7). After decarbonylation, compound **9.7d** loses the chirality because it possesses the C_{3v}-symmetry; now, it is an achiral molecule.

The same group disclosed synthesizing chiral hydrocarbons termed crypto-optically active compounds. The reaction between triazole **9.8a** with alkene containing ^{13}C-carbon labeled **9.8b** gave the cyclopropane as chiral hydrocarbon **9.8c** [13], which further underwent decarbonylation would form the chiral cyclopropane **9.8d** (Scheme 9.8) The chirality of these hydrocarbons is present due to the asymmetrical distribution of ^{12}C/^{13}C isotopes in their structure. Authors have shown the hydrocarbon's chirality based upon the vibrational circular dichroism (VCD) in which the molecular vibrations between isotopomers are measured and assigned absolute configuration of the cyclopropane ring **9.8d**, which was further supported by DFT calculations. In addition, it also assigned the absolute configuration of C_3-symmetric chiral molecules such as [3]CPPC by VCD techniques. Furthermore, the authors have also shown the intramolecular asymmetric synthesis of chiral carbon isotope **9.9d** (Scheme 9.9).

Miura and Murakami [13]

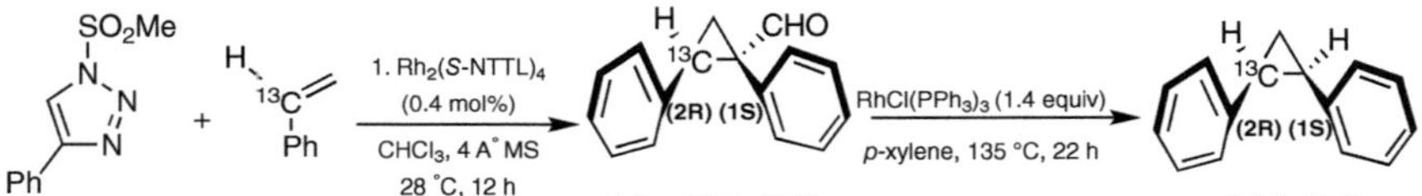

9.8a 9.8b 9.8c; 92%, 95% ee 9.8d; 66%

Scheme 9.8 Asymmetric synthesis of chiral carbon isotope hydrocarbons. Source: Adapted from Miura and Murakami [13].

Miura and Murakami [13]

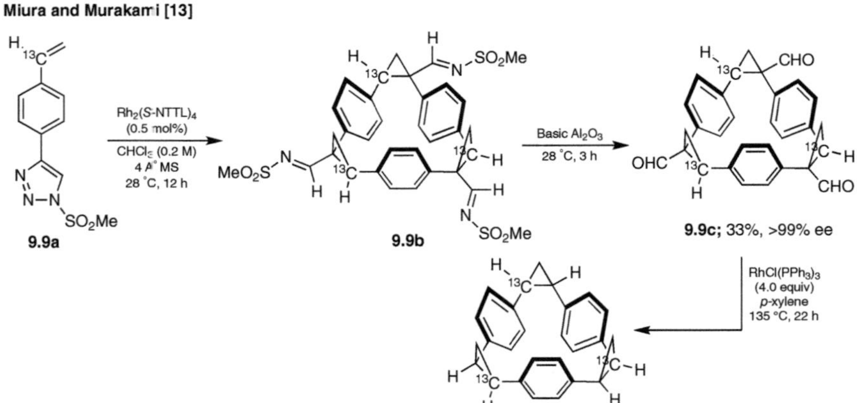

9.9a 9.9b 9.9c; 33%, >99% ee 9.9d

Scheme 9.9 Synthesis of crypto-optically active compounds. Source: Adapted from Miura and Murakami [13].

9.3 Rhodium-Catalyzed C–H Insertion Reactions

The Fokin group after exploration of asymmetric cyclopropanation reaction of alkenes with rhodium azavinyl carbene. In 2011, they reported the asymmetric C—H functionalization [14] of rhodium azavinyl carbene with unactivated sp^3 C—H bonds of alkanes (Scheme 9.10). Initially, authors showed the secondary C—H bond insertion of cyclohexane **9.10b** with 1-sulfonyl-1,2,3-triazoles **9.10a** in the presence of chiral rhodium catalyst gave the chiral sulfonyl imine **9.10c** which further on reduction produced the chiral *N*-sulfonylamines **9.10d** (Scheme 9.10). It was observed that aryl sulfonyl substituted triazoles underwent C—H insertion more efficiently in the presence of Rh_2(*S*-PTAD)$_4$ instead of Rh_2(*S*-NTTL)$_4$. Also, other alkanes **9.10e** were explored with mesyl-substituted triazoles for tertiary C—H bond insertion of alkanes (Scheme 9.10) to synthesize corresponding chiral amines **9.10f**. The scope of this reaction was explored with various functional groups containing triazoles and resulted in good to excellent yields with high enantioselectivity.

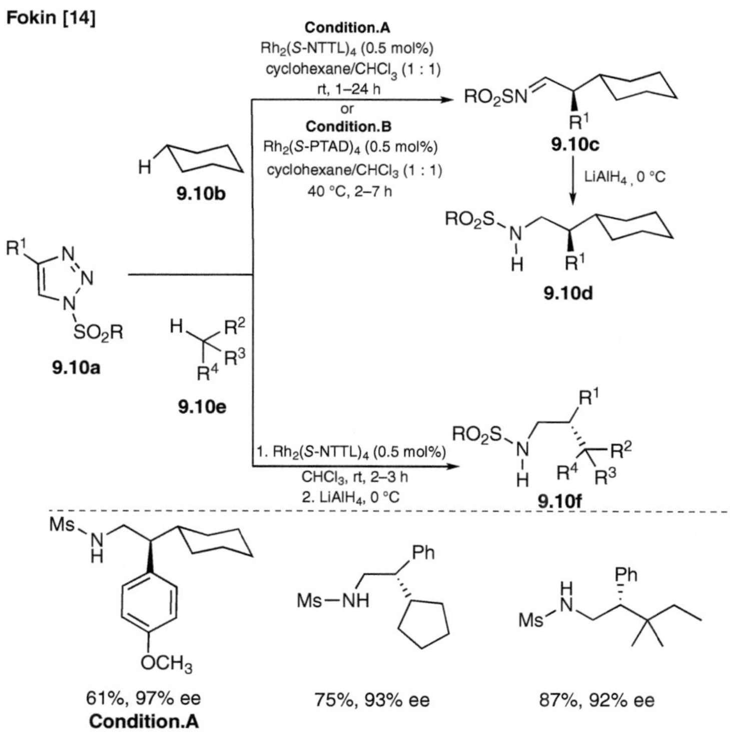

Scheme 9.10 Asymmetric C–H insertion of unactivated sp^3 C–H bonds. Source: Adapted from Fokin [14].

In 2016, the Davies group reported [15] the enantioselective C—H insertion of allylic and benzylic substrates (Scheme 9.11). Rhodium azavinyl carbene generated from 4-phenyl-*N*-sulfonyl-1,2,3-triazoles **9.11a** underwent a C—H insertion with

Davies [15]

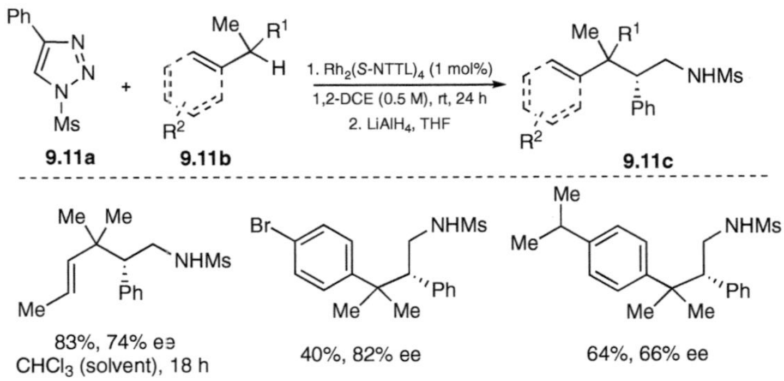

Scheme 9.11 Asymmetric C—H insertion of allylic and benzylic of sp^3 C—H bonds. Source: Adapted from Davies [15].

9.11b to furnish the product **9.11c** (Scheme 9.11). In the substrates containing primary and tertiary C—H bonds, the tertiary C—H insertion product was observed majorly with moderate enantioselectivity. Interestingly, no cyclopropanation product was observed with olefin-tethered C—H bonds of alkanes; only the C—H insertion product was observed.

In 2018, the Davies group utilized an enantioselective allylic C—H insertion [16] strategy to access the stereoselective β-arylpyrrolidines (Scheme 9.12). This reaction underwent a one-pot stepwise procedure to synthesize β-arylpyrrolidines **9.12c** and **9.12e**. At first, enantioselective allylic C—H insertion of rhodium azavinyl carbenes followed by imine reduction would form the amine followed by ozonolysis to give the cyclic hemiaminal, and then reductive amination to access the stereoselective β-arylpyrrolidines (Scheme 9.12). The chirality was always retained at every step, and chromatographic isolation was done at the last step. In the case of secondary C—H insertion of trans alkenes **9.12d**, two stereogenic centers were formed to give stereo-defined 3-aryl-4-alkylpyrrolidines **9.12e** (Scheme 9.12). Further studies were conducted to minimize enamine-induced epimerization using chloroform as a solvent and lowering the temperature for reduction, mediated by triethylsilane and boron trifluoride diethyl etherate. Various substituted alkenes were examined by varying the chain length and functional group to increase the diastereoselectivity and enantioselectivity of highly substituted arylpyrrolidines.

Davies group continued their interest in C—H insertion reactions. They have utilized the silicon-substituted alkane substrates for C—H insertion with azavinyl carbenes to synthesize organosilicon compounds [17], which are valuable compounds in various fields. Since the hydrogen at the β-position of silicon is quite liable, it is the most vulnerable position for C—H insertion in the case of both cyclic **9.13b** and aliphatic silicon substituted alkanes **9.13c**, which further helped to achieve high enantioselectivity (Scheme 9.13). However, using larger cyclic silanes (silacycles) gave the multiple chiral centers containing compounds leading to a high order of complexity. Regio and enantioselective C—H insertion product was observed due to the β-silyl effect.

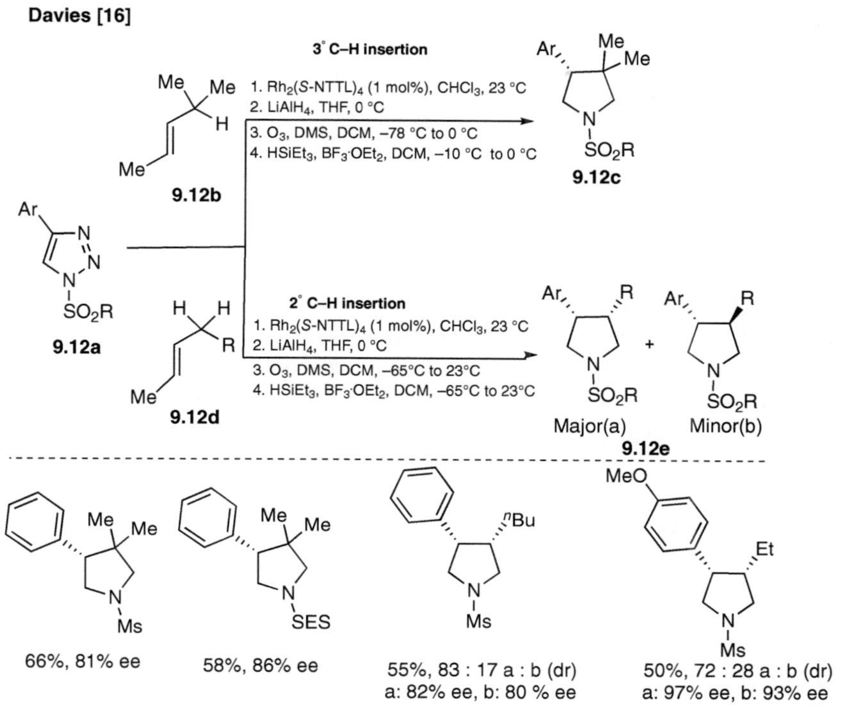

Scheme 9.12 Stepwise stereoselective synthesis of β-arylpyrrolidines. Source: Adapted from Davies [16].

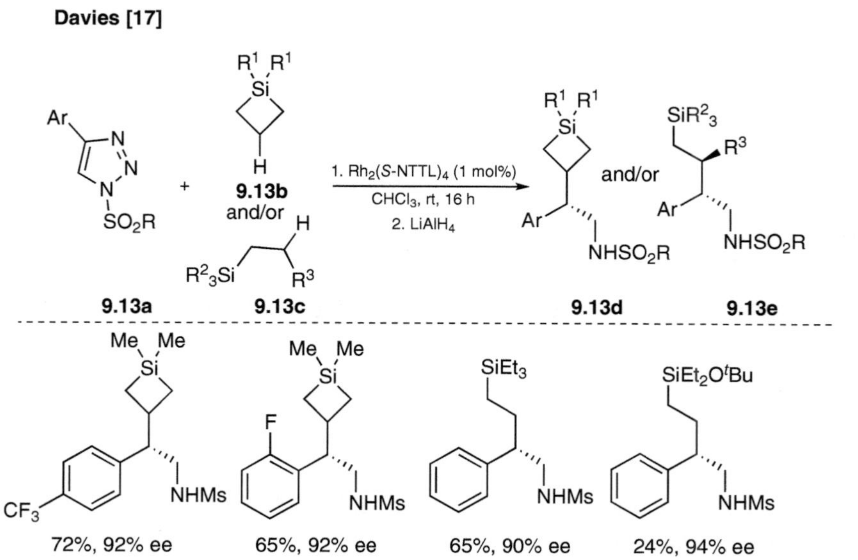

Scheme 9.13 Asymmetric C–H insertion of silicon-substituted alkanes. Source: Adapted from Davies [17].

In 2020, the Davies group explored the site-selective δ C—H insertion reaction at the distal position of silyl ethers **9.14b** in the presence of a more electronically preferred C—H bond α to oxygen [18]. These results offer a novel enantioselective approach toward distal δ C—H insertion of allylic alcohols instead of proximal allylic C—H functionalization, and no cyclopropanation product was observed.

Here, the classic rhodium azavinyl carbene, derived from mesyl-triazole **9.14a**, was utilized along with hexafluoroisopropanol (HFIP) as an additive (Scheme 9.14) to improve the yield and enantioselectivity. To avoid the ylide formation, the functional group at the ether should be sterically hindered or decrease the reactivity of oxygen of **9.14b** to enhance the C—H insertion product at the distal position. Similarly, the reaction was tried with various substituted diazo compounds, and a similar enantioselective distal δ C—H insertion product was obtained. Furthermore, the method was used in the synthesis of 3,4-disubstituted L-proline, an essential biological active molecule in phenyl kainic acid.

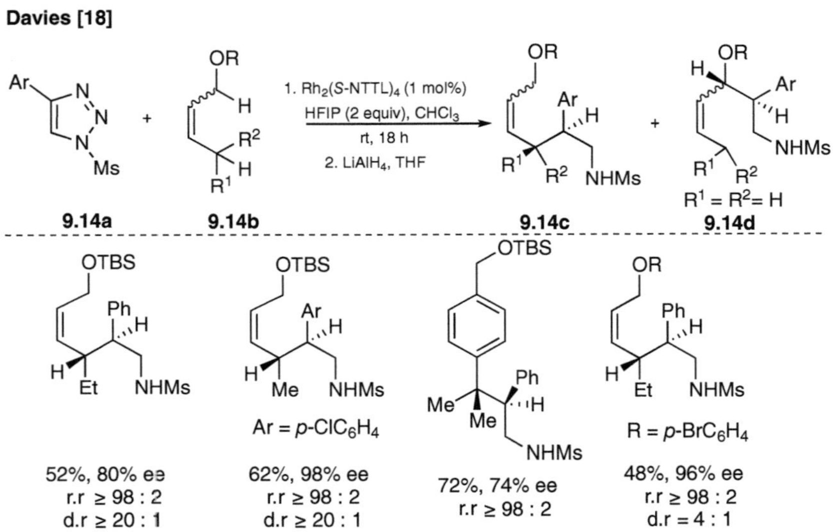

Scheme 9.14 Distal site-selective δ C—H insertion of allylic alcohols. Source: Adapted from Davies [18].

9.4 Rhodium-Catalyzed Transannulation Reactions

In 2013, the Fokin group reported the enantioselective synthesis of oxazolines [19] reaction between azavinyl carbene with aldehyde (Scheme 9.15). In situ generated rhodium-azavinyl carbene reacted with aldehyde **9.15b** to form the ylide **9.15d**, followed by intramolecular cyclization to deliver the oxazolines **9.15c** instead of the usual 1,3-dipolar cycloaddition reaction from ylide **9.15d** (Scheme 9.15). Various substituted chiral oxazolines **9.15c** were synthesized, containing various functional groups with excellent yield and moderate enantioselectivity.

Next, the Davies group reported the annulation of C3-substituted indole with azavinyl carbene for the enantioselective synthesis of pyrroloindolines

Fokin [19]

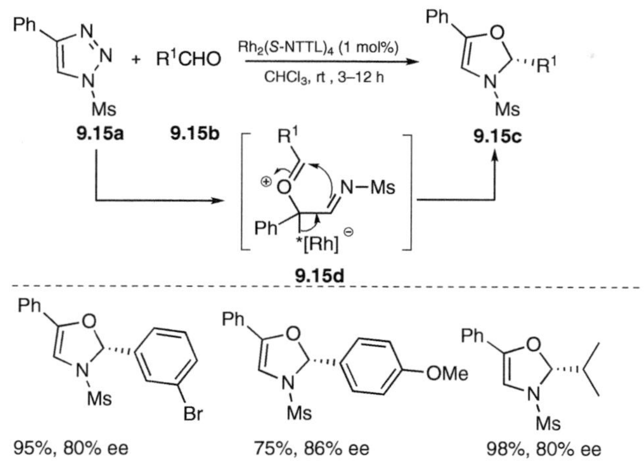

Scheme 9.15 Asymmetric synthesis of oxazolines. Source: Adapted from Davies [19].

(Scheme 9.16) via [3 + 2]-cycloaddition fashion similar to their previous report of formal [3 + 2]-cycloaddition [20] between substituted indole with vinylcarbenoids, which is derived from vinyl diazoacetates [21]. In this formal [3 + 2]-cycloaddition of C3-substituted indoles **9.16c** with *N*-sulfonyl-1,2,3-triazoles **9.16a** (Scheme 9.16), two pathways were proposed based on electronically substituted indolic nitrogen atom. In the case of path a, electron-rich *N*-alkyl indoles reacted with electrophilic rhodium carbenoids and delivered the observed product via zwitterionic intermediate **9.16d**, due to the considerable polarization of C2—C3 olefin. In the case of path *b*, a less established pathway, electron-deficient *N*-alkyl indoles underwent cyclopropanation at C2—C3 olefin to give cyclopropylindolines intermediate **9.16e**; further ring-opening and recombination gave the corresponding pyrroloindolines **9.16c** (Scheme 9.16).

The Fokin group, during their investigation enantioselective cyclopropanation of rhodium(II) *N*-triflyl azavinyl carbene with electron-rich styrenes, observed a transannulation product [8] (Scheme 9.3).

In this study, the same group explored various 4-substituted *NH*-triazoles in the presence of Tf_2O, DTMP, and Rh_2(*S*-NTTL)$_4$ for the in situ generation of rhodium(II)-*N*-triflyl azavinyl carbene with electron-rich styrenes **9.17b** to access the one-pot asymmetric synthesis of enantioselective 2,3-dihydropyrroles **9.17c** [22].

The authors observed that the enantioselectivity of the reaction was cumulatively dependent on the chiral rhodium catalyst's steric volume and electronic substitution properties at the C4 position of *NH*-triazole **9.17a** (Scheme 9.17). Two pathways were proposed for the enantioselective synthesis of 2,3-dihydropyrroles; in path a, chiral rhodium catalyst controls the stereoselectivity of product **9.17c** by determining the enantioselectivity of intermediate cyclopropylaldimine **9.18c** (Scheme 9.18). Further cleavage of cyclopropane ring due to the combined effect of electron-withdrawing group effect of triflyl and donating group of methoxy would

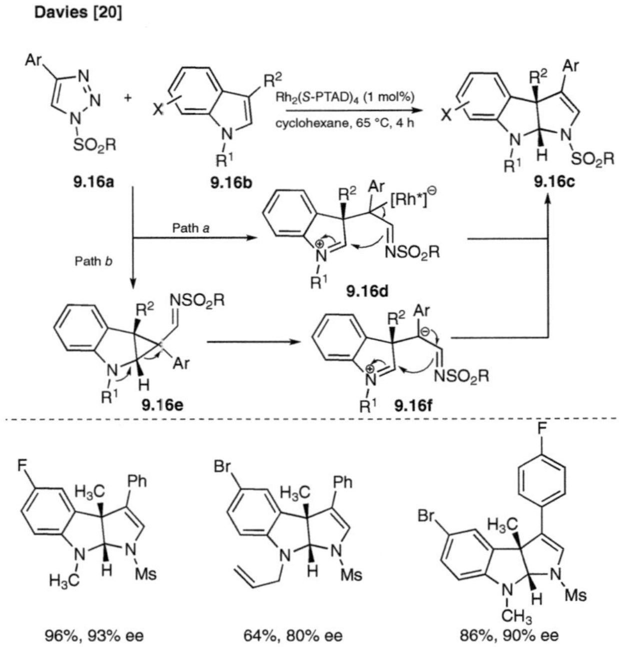

Scheme 9.16 Asymmetric synthesis of pyrroloindolines via formal [3 + 2]-cycloaddition. Source: Adapted from Davies [20].

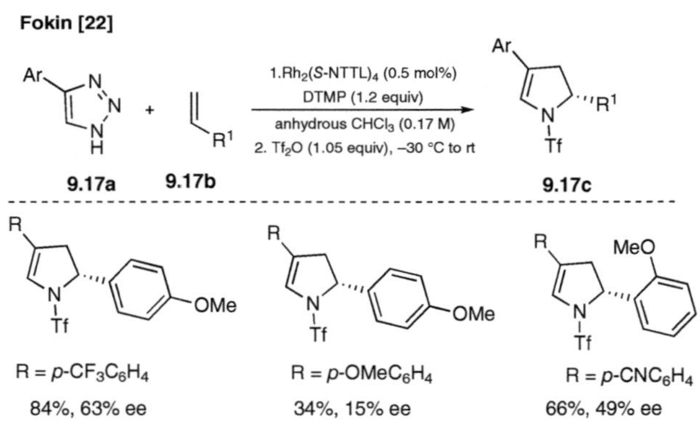

Scheme 9.17 One-pot enantioselective synthesis of 2,3-dihydropyrroles. Source: Adapted from Fokin [22].

generate zwitterionic intermediate **9.18d** and then cyclization to furnish the desired product **9.18e** (Scheme 9.18). In path b, the chiral rhodium catalyst directly controls the stereoselectivity of product **9.18e** by reacting electron-rich alkene **9.18b** with rhodium azavinyl carbene **9.18a** to form zwitterionic intermediate **9.18f** and further cyclization to give the intermediate **9.18g** followed by regeneration of Rh-catalyst to deliver the 2,3-dihydropyrroles **9.18e**. Based on various control experiments, they have hypothesized that introducing a chiral Bronsted acid with cyclopropylimines will give the enantioselective 2,3-dihydropyrroles.

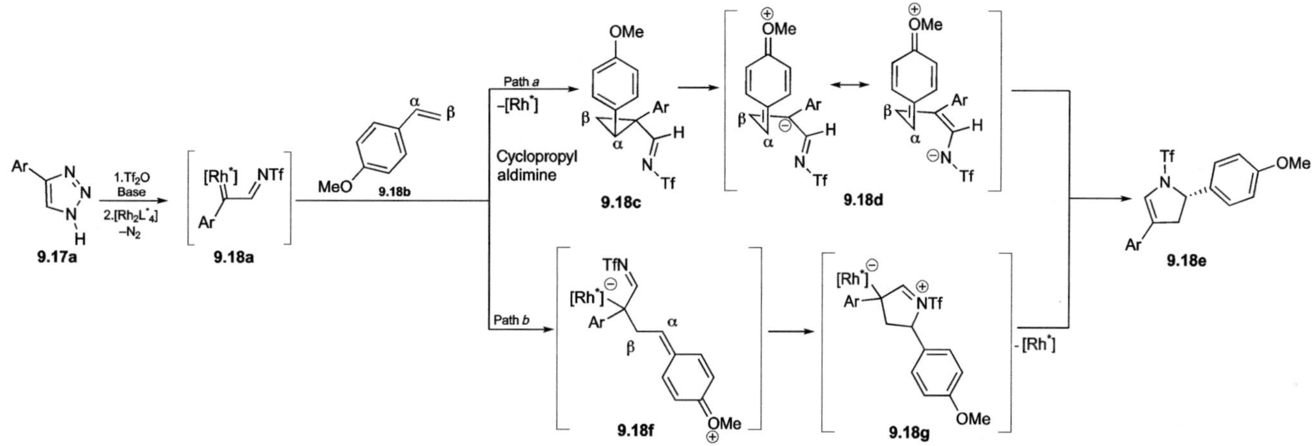

Scheme 9.18 Plausible mechanism for enantioselective synthesis of 2,3-dihydropyrroles.

9.5 Miscellaneous Reactions

9.5.1 Relay Catalysis

In 2018, the Feng group reported the bimetallic relay catalytic system for enantioselective synthesis of β/γ-amino acid derivatives by utilizing an achiral rhodium catalyst with a chiral ligand such as *N*, *N′*-dioxide–indium(III) complex [23]. The combination of a chiral In(III) Lewis-acid ligand and rhodium catalyst with the allyl alcohol esters **9.19b** helps to achieve high levels of enantioselectivity (Scheme 9.19) with *N*-sulfonyl-1,2,3-triazole **9.19a** via O—H insertion/asymmetric Claisen rearrangement. Another allylic alcohol ester derivative, Morita–Baylis–Hillman adduct **9.19d**, was also used for the bimetallic relay catalysis to furnish enantioriched δ-oxo-γ-amino esters **9.19e**.

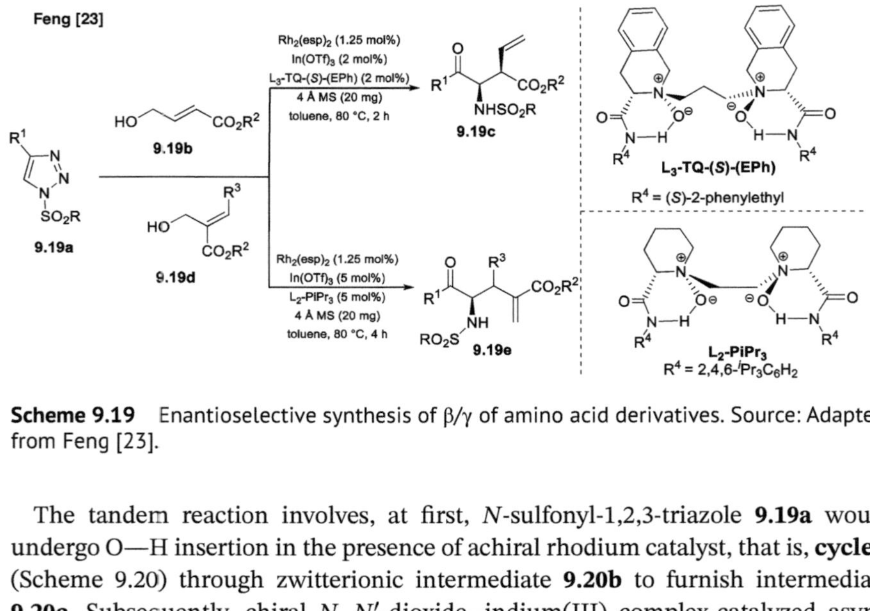

Scheme 9.19 Enantioselective synthesis of β/γ of amino acid derivatives. Source: Adapted from Feng [23].

The tandem reaction involves, at first, *N*-sulfonyl-1,2,3-triazole **9.19a** would undergo O—H insertion in the presence of achiral rhodium catalyst, that is, **cycle 1** (Scheme 9.20) through zwitterionic intermediate **9.20b** to furnish intermediate **9.20c**. Subsequently, chiral *N*, *N′*-dioxide–indium(III) complex-catalyzed asymmetric Claisen rearrangement, that is, **cycle 2** (Scheme 9.20). The asymmetric induction of Claisen type [3]-sigmatropic rearrangement was achieved through the bidentate binding of the allyl vinyl ether intermediate **9.20d** to In(III) with oxygen atoms of both the tosyl group and ester group via six-membered chair like transition state **9.20e**.

In 2020, the Dong and Feng group extended this work with benzylic alcohols [24] for the enantioselective synthesis of α-aminoketones **9.21c** (Scheme 9.21). The product formation is similar to allylic alcohol ester derivatives under bimetallic relay catalysis (Scheme 9.20) in first-step Rh-catalyzed reaction of *N*-sulfonyl-1,2,3-triazoles **9.21a** with benzylic alcohols **9.21b** gave the O—H insertion product **9.21f** followed by [1,3]-shift via the coordination of chiral *N*, *N′*-dioxide–indium(III) complex with oxygen atom of the enol ether through the

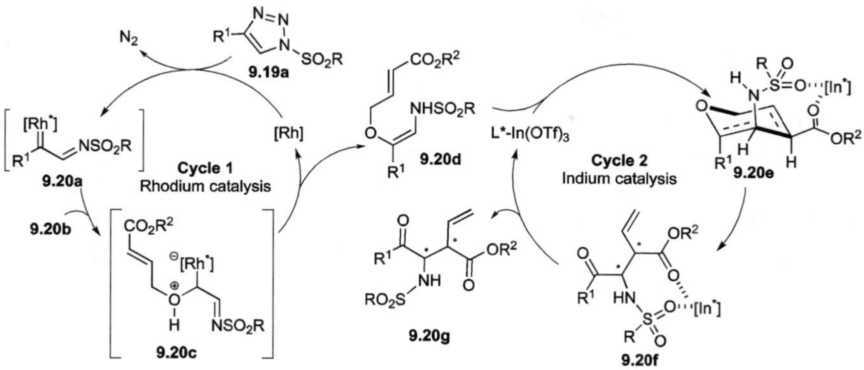

Scheme 9.20 Bimetallic relay catalysis mechanism.

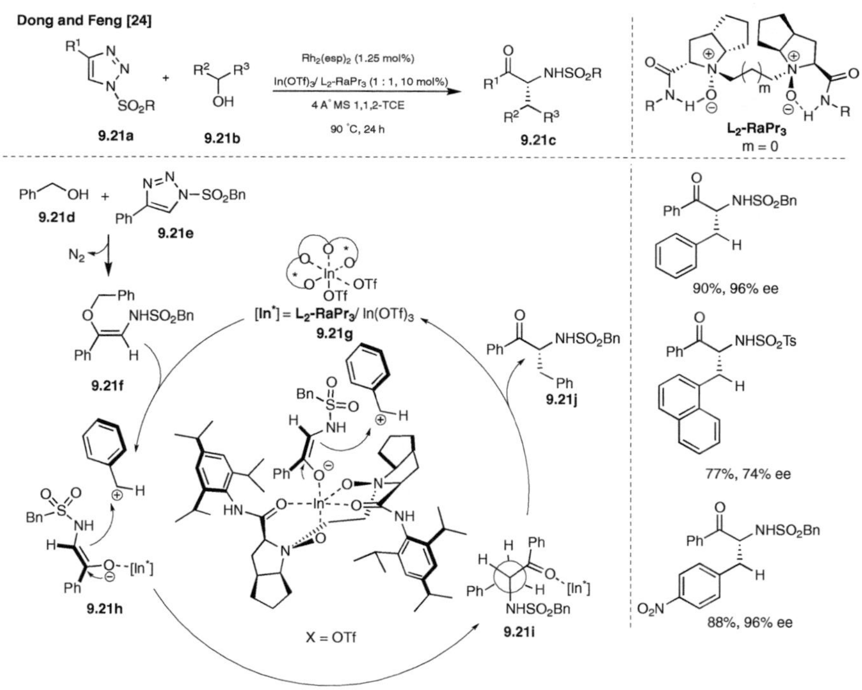

Scheme 9.21 Enantioselective synthesis of α-aminoketones. Source: Adapted from Dong and Feng [24].

weakness of O-Bn bond to facilitate the cleavage of the bond to form the ion-pair intermediate **9.21h**, which would lead to the formation of enantiomerically pure α-aminoketones **9.21j**. Based on the control studies, the dual catalyst is needed to form the enantioselective product **9.21c**. However, at elevated temperatures without using a chiral indium catalyst, they have observed the enantioselective α-aminoketones benzyl alcohol as a chiral starting material.

Anbarasan group reported the one-pot enantioselective synthesis of dihydro-β-carbolines derivatives via relay catalysis in the presence of catalytic amounts of rhodium and squaramide [25]. In situ generated rhodium azavinyl carbene first would undergo insertion of substituted indole **9.22b** at C3 position to form the enamide **9.22d** and then subsequent chiral squaramide catalyzed via re-face attack intermediate **9.22e** followed by intramolecular *aza*-Michael reaction gives the desired product **9.22c**. Substituted dihydro-β-carbolines **9.22c** were synthesized with a high enantiomeric ratio and moderate to good yields from the variety of functional groups containing corresponding starting materials (Scheme 9.22).

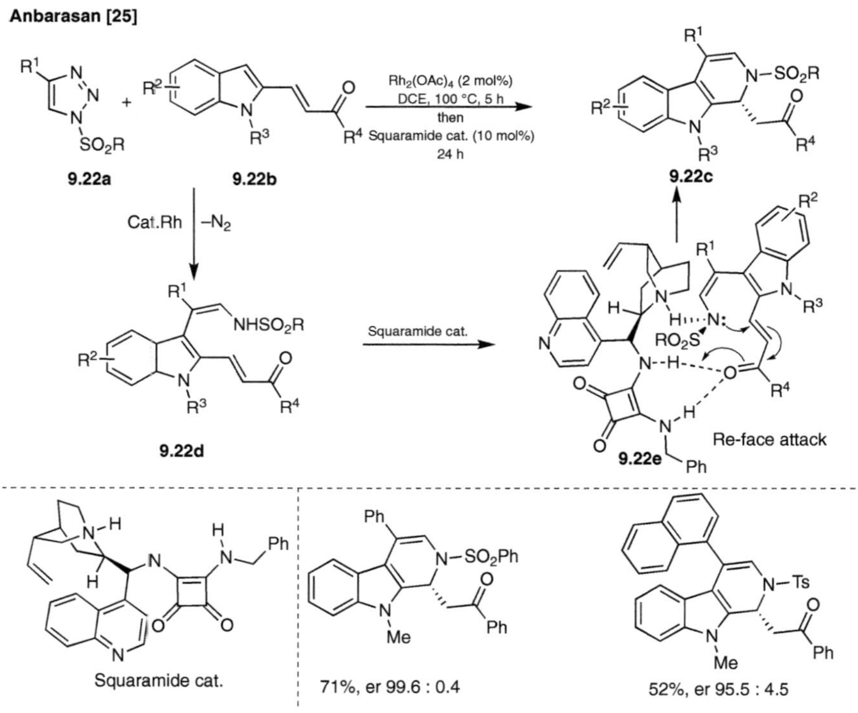

Scheme 9.22 Enantioselective synthesis of dihydro-β-carbolines. Source: Adapted from Anbarasan [25].

9.5.2 Atom Transfer Reactions

In 2012, the Fokin group reported catalytic asymmetric oxygen atom transfer reaction from the diazo sulfonylamidines to chiral sulfinylamidines [26]. Depending on the substitution, the starting triazole would undergo ring-chain isomerism to give the open form of the triazole. 5-Amino substituted triazole would experience the open state of diazo sulfonyl amidines or diazo amidine **9.23e**, a novel class of azavinyl carbene precursor (Scheme 9.23). It was synthesized from sulfonyl azide

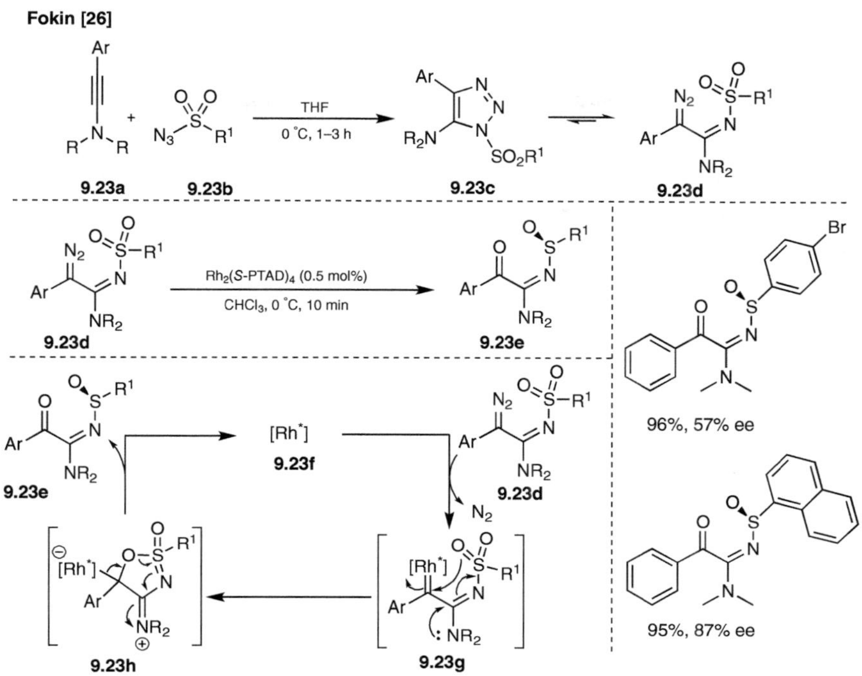

Scheme 9.23 Synthesis of chiral sulfinylamidines. Source: Adapted from Fokin [26].

9.23b and ynamines **9.23a** [27] via in situ formed triazoles **9.23c**, which is in equilibrium with the open-chain form of diazo sulfonylamidines **9.23d**. The diazo tautomer might be the predominating tautomer. Interestingly, the reaction of diazo sulfonylamidines **9.23d** with styrene did not observe the cyclopropanation product; instead, they observed the oxygen atom transfer product from, which resulted from the migration of sulfur to the carbon atom. Diazo sulfonylamidines **9.23d** in the presence of chiral rhodium-catalyst would form the rhodium azavinyl carbene **9.23g**, further intramolecular oxygen atom attack on electrophilic rhodium carbene with the aid of electron-rich amine source would give the intermediate **9.23h**. The enantioselectivity originated from the geometry of rhodium carbene intermediate **9.23g**. It is facilitated by one oxygen over the other, and subsequent asymmetric oxygen atom transfer would produce chiral sulfinylamidines **9.23e**.

9.6 Application in Natural Product and Total Synthesis

The chemistry of azavinyl carbenes has a broad application for synthesizing synthetic intermediates, small-membered heterocyclic compounds, and carbocyclic compounds in a single step to access the achiral and chiral molecules. In 2013, the Sarpong group reported the synthesis of diverse substituted 3,4-fused pyrroles via intramolecular transannulation of allene-tethered *N*-sulfonyl-1,2,3-triazoles

[28]. They have shown the synthesis of cycloprodigiosin natural products using the same strategy. Enantioenrich allenyl alkyne **9.24b** was prepared as a mixture of diastereomers from the corresponding chiral alkyne **9.24a**. Next, the allenyl alkyne **9.24b** was subjected to copper-catalyzed azide-alkyne cycloaddition followed by the addition of rhodium-catalyst. Intramolecular transannulation would give the 1 : 1 mixture 3,4-fused pyrrole **9.24c** and α, β-unsaturated imine **9.24d**. Further deprotection of the tosyl group of **9.24c** and then condensation of pyrrole gave the first enantioselective synthesis of the natural product (*R*)-cycloprodigiosin **9.24e** in 71% yield (Scheme 9.24), which shows the antitumor and immunosuppressor activities.

Sarpong [28]

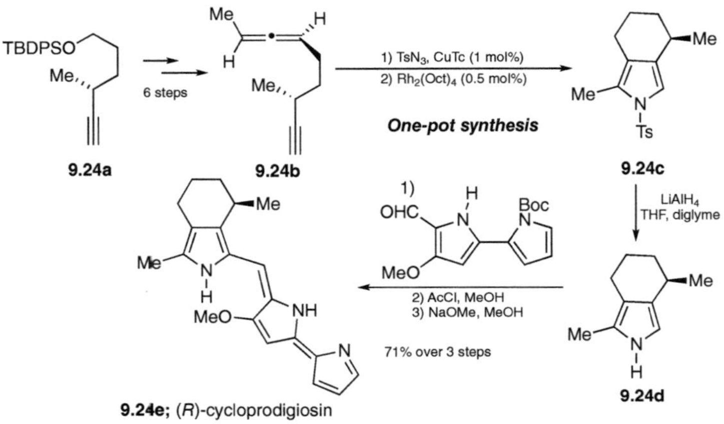

Scheme 9.24 Enantioselective synthesis of cycloprodigiosin. Source: Adapted from Sarpong [28].

In 2017, the Luo group reported the synthesis of ergot alkaloid (+)-lysergol and its analogs via [3 + 2] annulation strategy to access the pharmaceutical important molecules [29]. Initially, in the attempt to synthesize lysergol, the chiral homopropargylic alcohol **9.25a** was subjected to TBS protection, followed by silver-catalyzed cyclization to give **9.25b** (Scheme 9.25).

Next, the cyclopropanation followed by ring-opening of *gem*-dibromocyclopropane and then nucleophilic addition of compound **9.25c** provided the **9.25d.** Furthermore, chiral alkyne **9.25e** was synthesized in a couple of steps via the Suzuki cross-coupling and Seyferth–Gilbert homologation reaction. The chiral alkyne was converted to substituted triazole **9.25f** via copper-catalyzed [3 + 2]-cycloaddition with tosyl azide and then followed by rhodium-catalyzed intramolecular dearomative transannulation will give compound **9.25g** (Scheme 9.25). In the last step, deprotection of the tosyl groups will furnish (+)-Lysergol **9.25h**. Similarly, they have shown synthesis of its analogs **9.25i** and **9.25j**.

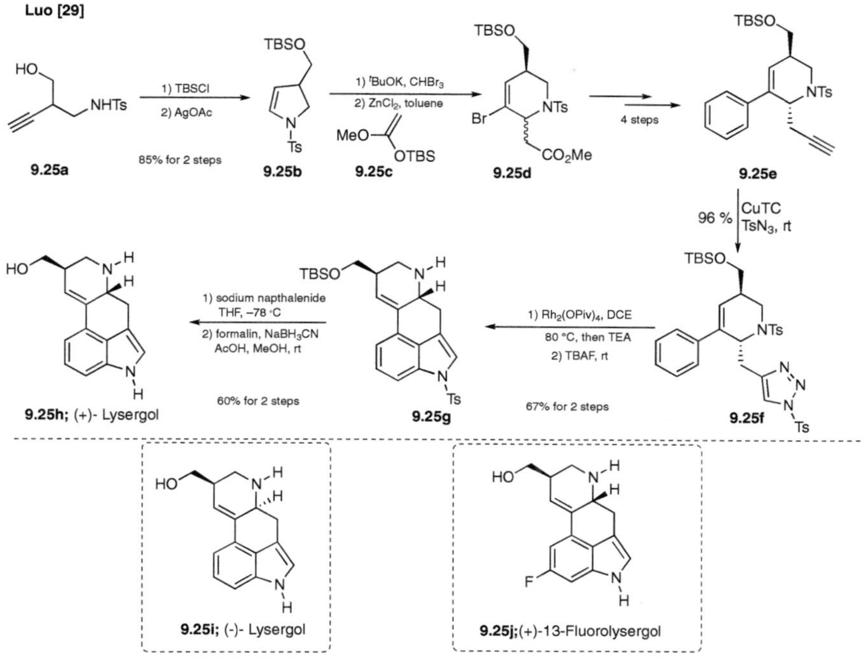

Scheme 9.25 Asymmetric synthesis of (+)-Lysergol. Source: Adapted from Luo [29].

In collaboration with the Houk group, the Li group reported intramolecular rhodium-catalyzed [3 + 2]-cycloaddition reaction for the synthesis of polycyclic systems, for example, biologically active molecules containing bridged medium-sized polycyclic compounds [30]. Most bridged medium-sized ring compound synthesis shows the achiral version via [3 + 2]-cycloaddition strategy. Authors extended their methodology for the asymmetric total synthesis of nakafuran-8 **9.26h**, an antifeedant against reef fishes, and also shows other biological activities [31]. The asymmetric synthesis of nakafuran-8 **9.26h** demonstrated through the intramolecular rhodium-catalyzed [3 + 2]-cycloaddition reaction of chiral triazoles **9.26e**, which is derived from chiral starting material **9.26a** in a couple of steps (Scheme 9.26). In the presence of $Rh_2(Oct)_4$, chiral *N*-sulfonyl triazoles **9.26e** underwent intramolecular [3 + 2]-cycloaddition with tethered alkene aldehyde moiety to yield the imine **9.26f**, which was further hydrolysis followed by reduction with $NaBH_4$ in a single step to give corresponding alcohol **9.26g** (Scheme 9.26) and then synthetic manipulation to complete the asymmetric synthesis of nakafuran-8 **9.26h**. Computational studies supported the cycloaddition product regioselectivity, chemoselectivity, and diastereoselectivity of the cycloaddition product.

Houk and Li [30]

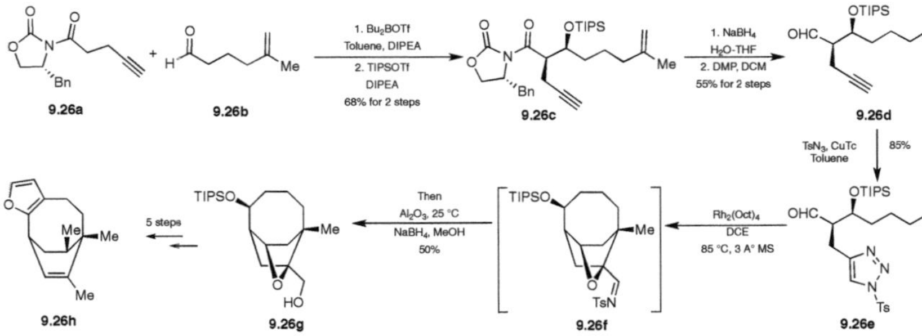

Scheme 9.26 Synthesis of nakafuran-8. Source: Adapted from Houk and Li [30].

9.7 Summary and Outlook

In summary, rhodium-catalyzed denitrogenative transformation of 1,2,3-triazoles paved the way to synthesize chiral compounds, heterocyclic compounds, natural products, and biologically active molecules in a single-step, one-pot manner, which would otherwise include multiple-step synthesis, without producing any byproduct except for nitrogen gas. However, there is ample room for further discoveries, such as: (i) designing of different chiral catalysts and ligands other than expensive rhodium-catalyst; (ii) Influencing the ring-chain isomerism of 1,2,3-triazoles is by changing the electronics of the substituents at *N*1-position of 1,2,3-triazoles, especially the highly withdrawing nature of sulfonyl group. These challenges will broaden the scope of denitrogenative transformations of 1,2,3-triazoles and offer plenty of opportunities for the asymmetric synthesis of complex molecules of diverse importance, bringing a new direction to this field.

Acknowledgments

We thank the DST-SERB-CRG (CRG/2022/000442), New Delhi, India, for financial support. P.Y. and A.B. sincerely thank the PMRF fellowship, respectively.

References

1 (a) Eliel, E.L., Wilen, S.H., and Mander, L.N. (1994). *Stereochemistry of Organic Compounds*. New York: Wiley. (b) Noyori, R. (1994). *Asymmetric Catalysis in Organic Synthesis*. New York: Wiley. (c) Federsel, H.-J. (2005). *Nat. Rev. Drug Discov.* 4: 685–697. (d) Blaser, H.U. and Schmidt, E. (ed.) (2004). *Asymmetric Catalysis on Industrial Scale: Challenges, Approaches and Solutions*. New York: Wiley. (e) Farina, V., Reeves, J.T., Senanayake, C.H., and Song, J.J. (2006). Asymmetric synthesis of active pharmaceutical ingredients. *Chem. Rev.* 106: 2734–2793.

(f) Yoon, T.P. and Jacobsen, E.N. (2003). Privileged chiral catalysts. *Science* 299: 1691–1693. (g) Doyle, M.P. and Forbes, D.C. (1998). Recent advances in asymmetric catalytic metal carbene transformations. *Chem. Rev.* 98: 911–935.

2 (a) El Ashry, E.S.H., Nadeem, S., Shah, M.R., and El Kilany, Y. (2010). Recent advances in the Dimroth rearrangement: a valuable tool for the synthesis of heterocycles. *Adv. Heterocycl. Chem.* 101: 161–228. (b) Kim, J.Y., Kang, B.C., and Ryu, D.H. (2017). Catalytic asymmetric roskamp reaction of silyl diazoalkane: synthesis of enantioenriched α-silyl ketone. *Org. Lett.* 19: 5936–5939. (c) Wang, Y., Wen, X., Cui, X. et al. (2017). Asymmetric radical cyclopropanation of alkenes with in situ generated donor-substituted diazo reagents via Co(II)-based metalloradical catalysis. *J. Am. Chem. Soc.* 139: 1049–1052. (d) Dian, L. and Marek, I. (2018). Asymmetric preparation of polysubstituted cyclopropanes based on direct functionalization of achiral three-membered carbocycles. *Chem. Rev.* 118: 8415–8434. (e) Wei, B., Sharland, J.C., Lin, P. et al. (2020). In situ kinetic studies of Rh(II)-catalyzed asymmetric cyclopropanation with low catalyst loadings. *ACS Catal.* 10: 1161–1170. (f) Kang, Z., Chang, W., Tian, X. et al. (2021). Ternary catalysis enabled three-component asymmetric allylic alkylation as a concise track to chiral α, α-disubstituted ketones. *J. Am. Chem. Soc.* 143: 20818–20827. (g) Wang, X., Ke, J., Zhu, Y. et al. (2021). Asymmetric radical process for general synthesis of chiral heteroaryl cyclopropanes. *J. Am. Chem. Soc.* 143: 11121–11129. (h) Gu, X., Mo, X., Bai, W.J. et al. (2023). Catalytic asymmetric P—H insertion reactions. *J. Am. Chem. Soc.* 145: 20031–20040.

3 (a) Ye, T. and McKervey, M.A. (1994). Organic synthesis with α-diazocarbonyl compounds. *Chem. Rev.* 94: 1091–1160. (b) Padwa, A. and Austin, D.J. (1994). Ligand effects on the chemoselectivity of transition metal catalyzed reactions of α-diazo carbonyl compounds. *Angew. Chem. Int. Ed.* 33: 1797–1815. (c) Padwa, A. and Weingarten, M.D. (1996). Cascade processes of metallo carbenoids. *Chem. Rev.* 96: 223–270. (d) Wang, T. and Hashmi, A.S.K. (2021). 1,2-Migrations onto gold carbene centers. *Chem. Rev.* 121: 8948–8978. (e) Liu, Z. and Wang, J. (2013). Cross-coupling reactions involving metal carbene: C=C/C—C bond formation to C—H bond functionalization. *J. Org. Chem.* 78: 10024–10030. (f) Xia, Y., Zhang, Y., and Wang, J. (2013). Catalytic cascade reactions involving metal carbene migratory insertion. *ACS Catal.* 3: 2586–2598. (g) Babbar, A., Yamini, P., Saleem, M., and Yadagiri, D. (2023). Transition metal-catalyzed reactivity of carbenes with boronic acid derivatives for arylation (alkylation) and beyond. *Org. Biomol. Chem.* 21: 7062–7078.

4 Horneff, T., Chuprakov, S., Chernyak, N. et al. (2008). Rhodium-catalyzed transannulation of 1,2,3-triazoles with nitriles. *J. Am. Chem. Soc.* 130: 14972–14974.

5 Dimroth, O. (1909). Ueber intramolekulare Umlagerungen. Umlagerungen in der Reihe des 1, 2, 3-Triazols. *Justus Liebigs Ann. Chem.* 364: 183–226.

6 (a) Chattopadhay, B. and Gevorgyan, V. (2012). Transition-metal-catalyzed denitrogenative transannulation: converting triazoles into other heterocyclic systems. *Angew. Chem. Int. Ed.* 51: 862–872. (b) Davies, H.M. and Alford, J.S. (2014). Reactions of metallocarbenes derived from N-sulfonyl-1,2,3-triazoles. *Chem. Soc.*

Rev. 43: 5151–5162. (c) Anbarasan, P., Yadagiri, D., and Rajasekar, S. (2014). Recent advances in transition-metal-catalyzed denitrogenative transformations of 1,2,3-traizoles and related compounds. *Synthesis* 46: 3004–3023. (d) Akter, M., Rupa, K., and Anbarasan, P. (2022). 1,2,3-Triazoleand its analogues: new surrogates for diazo compounds. *Chem. Rev.* 122: 13108–13205.

7 Chuprakov, S., Kwok, S.W., Zhang, L. et al. (2009). Rhodium-catalyzed enantioselective cyclopropanation of olefins with *N*-sulfonyl 1,2,3-triazoles. *J. Am. Chem. Soc.* 131: 18034–18035.

8 Grimster, N., Zhang, L., and Fokin, V.V. (2010). Synthesis and reactivity of rhodium(II) *N*-triflyl azavinyl carbenes. *J. Am. Chem. Soc.* 132: 2510–2511.

9 Zibinsky, M. and Fokin, V.V. (2011). Reactivity of *N*-(1,2,4-triazolyl)-substituted1, 2,3-triazoles. *Org. Lett.* 13: 4870–4872.

10 Miura, T., Nakamuro, T., Liang, C.J., and Murakami, M. (2014). Synthesis of *trans*-cycloalkenes via enantioselective cyclopropanation and skeletal rearrangement. *J. Am. Chem. Soc.* 136: 15905–15908.

11 Miura, T., Nakamuro, T., Nikishima, H., and Murakami, M. (2016). Asymmetric synthesis of cyclopropylmethanamines by rhodium-catalyzed cyclopropanation of pinacol allylboronate with *N*-sulfonyl-1,2,3-triazoles. *Chem. Lett.* 45: 1003–1005.

12 Miura, T., Nakamuro, T., Stewart, S.G. et al. (2017). Synthesis of enantiopure C3-symmetric triangular molecules. *Angew. Chem. Int. Ed.* 129: 3382–3386.

13 Miura, T., Nakamuro, T., Nagata, Y. et al. (2019). Synthesis of *trans*-cycloalkenes via enantioselective cyclopropanation and skeletal rearrangement. *J. Am. Chem. Soc.* 141: 13341–13345.

14 Chuprakov, S., Malik, J.A., Zibinsky, M., and Fokin, V.V. (2011). Catalytic asymmetric C—H insertions of rhodium(II) azavinyl carbenes. *J. Am. Chem. Soc.* 133: 10352–10355.

15 Kubiak, R.W., Mighion, J.D., Wilkerson-Hill, S.M. et al. (2016). Enantioselective intermolecular C—H functionalization of allylic and benzylic sp^3 C—H bonds using *N*-sulfonyl-1,2,3-triazoles. *Org. Lett.* 18: 3118–3121.

16 Kubiak, R.W. and Davies, H.M.L. (2018). Rhodium-catalyzed intermolecular C—H functionalization as a key step in the synthesis of complex stereodefined β-arylpyrrolidines. *Org. Lett.* 20: 3771–3775.

17 Garlets, Z.J. and Davies, H.M.L. (2018). Harnessing the β-silicon effect for regioselective and stereoselective rhodium(II)-catalyzed C—H functionalization by donor/acceptor carbenes derived from 1-sulfonyl-1,2,3-triazoles. *Org. Lett.* 20: 2168–2171.

18 Vaitla, J., Boni, Y.T., and Davies, H.M.L. (2020). Distal allylic/benzylic C–H functionalization of silyl ethers using donor/acceptor rhodium(II) carbenes. *Angew. Chem. Int. Ed.* 59: 7397–7402.

19 Zibinsky, M. and Fokin, V.V. (2013). Sulfonyl-1,2,3-triazoles: convenient synthones for heterocyclic compounds. *Angew. Chem. Int. Ed.* 125: 1547–1550.

20 Lian, Y. and Davies, H.M.L. (2010). Rhodium-catalyzed [3 + 2] annulation of Indoles. *J. Am. Chem. Soc.* 132: 440–441.

21 Spangler, J.E. and Davies, H.M.L. (2013). Catalytic asymmetric synthesis of pyrroloindolines via a rhodium(II)-catalyzed annulation of indoles. *J. Am. Chem. Soc.* 135: 6802–6805.

22 Kwok, S.W., Zhang, L., Grimster, N.P., and Fokin, V.V. (2014). Catalytic asymmetric transannulation of *NH*-1,2,3-triazoles with olefins. *Angew. Chem. Int. Ed.* 126: 3520–3524.

23 Chen, Y., Dong, S., Xu, X. et al. (2018). Bimetallic rhodium(II)/indium(III) relay catalysis for tandem insertion/asymmetric claisen rearrangement. *Angew. Chem. Int. Ed.* 57: 16554–16558.

24 Chen, Y., Liu, Y., Li, Z. et al. (2020). Tandem insertion–[1,3]-rearrangement: highly enantioselective construction of α-aminoketones. *Angew. Chem. Int. Ed.* 59: 8052–8056.

25 Rajasekar, S. and Anbarasan, P. (2019). Tandem Rh(II) and chiral squaramide relay catalysis: enantioselective synthesis of dihydro-β-carbolines via insertion to C—H bond and *aza*-Michael reaction. *Org. Lett.* 21: 3067–3071.

26 Selander, N. and Fokin, V.V. (2012). Rhodium(II)-catalyzed asymmetric sulfur(VI) reduction of diazo sulfonylamidines. *J. Am. Chem. Soc.* 134: 2477–2480.

27 (a) Harmon, R.E., Stanley, F. Jr., Gupta, S.K., and Johnson, J. (1970). *N*,*N*-dialkylamino-1,2,3-triazole-α-diazoamidine tautomers from substituted benzenesulfonyl azides and ynamines. *J. Org. Chem.* 35: 3444–3448. (b) Regitz, M. and Himbert, G. (1970). Über das 5-amino-triazol/diazoamidin-gleichgewicht. *Tetrahedron Lett.* 11: 2823–2826.

28 Schultz, E.E. and Sarpong, R. (2013). Application of in situ-generated Rh-bound Trimethylenemethane Variants to the Synthesis of 3,4-Fused Pyrroles. *J. Am. Chem. Soc.* 135: 4696–4699.

29 Yuan, H., Guo, Z., and Luo, T. (2017). Synthesis of (+)-lysergol and its analogues to assess serotonin receptor activity. *Org. Lett.* 19: 624–627.

30 Hou, B.L., Wong, J.J., Lv, N. et al. (2021). Facile generation of bridged medium-sized polycyclic systems by rhodium-catalysed intramolecular (3 + 2) dipolar cycloadditions. *Nat. Commun.* 12: 5239.

31 (a) Schulte, G., Scheuer, P.J., and McConnell, O.J.H. (1980). Two furanosesquiterpene marine metabolites with antifeedant properties. *Chim. Acta* 63: 2159–2167. (b) Huang, X.-C., Li, J., Li, Z.-Y. et al. (2008). Sesquiterpenes from the Hainan sponge Dysidea septosa. *J. Nat. Prod.* 71: 1399–1403.

10

Denitrogenative Transformations of 3-Diazoindolin-2-imines

Kavuri Rupa and Pazhamalai Anbarasan

Department of Chemistry, Indian Institute of Technology Madras, Chennai, Tamil Nadu, India

10.1 Introduction

Heterocyclic compounds are an indispensable class of organic molecules and play a pivotal role in the realm of drug and industrial chemistry. They are also present in a wide range of biologically active natural products [1]. Many of these heterocyclic compounds, natural products, and structurally complex molecules have been synthesized from metal carbenoid intermediates [2]. These metal carbenoids are traditionally derived from diazo compounds, which are toxic and explosive in nature. Hence, significant attention has been devoted in the recent past to identify the suitable surrogate for diazo compounds. In this context, *N*-sulfonyl-1,2,3-triazoles have emerged as a promising substitute for diazo compounds. They have the ability to generate α-imino metal carbenoids *in situ* on reaction with transition metals. Due to the presence of both nucleophilic imine nitrogen and electrophilic carbenoid carbon, they can undergo numerous transformations to yield a diverse array of heterocyclic compounds [3]. Inspired by this chemistry, Lu and Wang developed an efficient method for the synthesis and transformation of triazoloindole, which predominantly exists as 3-diazoindolin-2-imine (Figure 10.1). In the presence of transition metal catalysts, 3-diazoindolin-2-imines generate reactive α-imino metal carbenoids, which undergo various transformations in combination with suitable coupling partners to form indole-embedded complex molecules. In this chapter, all denitrogenative transformations of 3-diazoindolin-2-imines are compiled and arranged them according to the type of reaction.

10.2 Synthesis of 3-Diazoindolin-2-imines

Synthesis of *N*-sulfonyl-1,2,3-triazoles was challenging due to the poor stability of (*N*-sulfonyl-1,2,3-triazol-5-yl)copper intermediates, which readily decompose to *N*-sulfonylketenimines. After several modifications, Fokin and co-workers

Denitrogenative Transformation of Nitrogen Heterocycles: Synthesis, Reactions and Applications, First Edition. Edited by Pazhamalai Anbarasan.

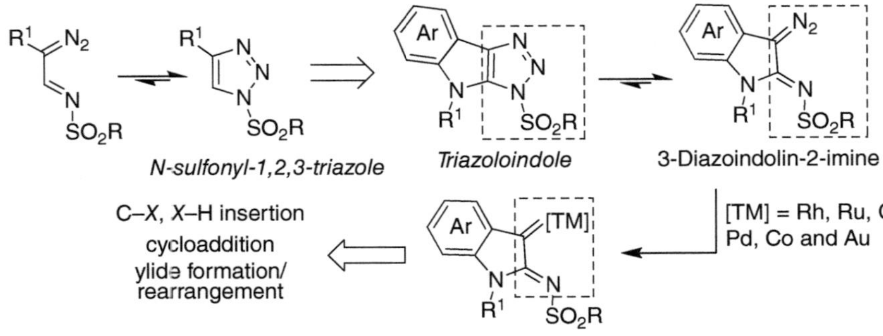

Figure 10.1 Indole embedded α-imino metal carbenoid.

accomplished *N*-sulfonyl-1,2,3-triazoles in excellent yields through intermolecular ligand stabilization of (*N*-sulfonyl-1,2,3-triazol-5-yl)copper species. Stimulated by this work, in 2014, Lu and Wang disclosed the synthesis of triazoloindoles **10.1d** from *o*-alkynyl anilines **10.1a** and sulfonyl azides **10.1b** in the presence of copper catalyst (Scheme 10.1).

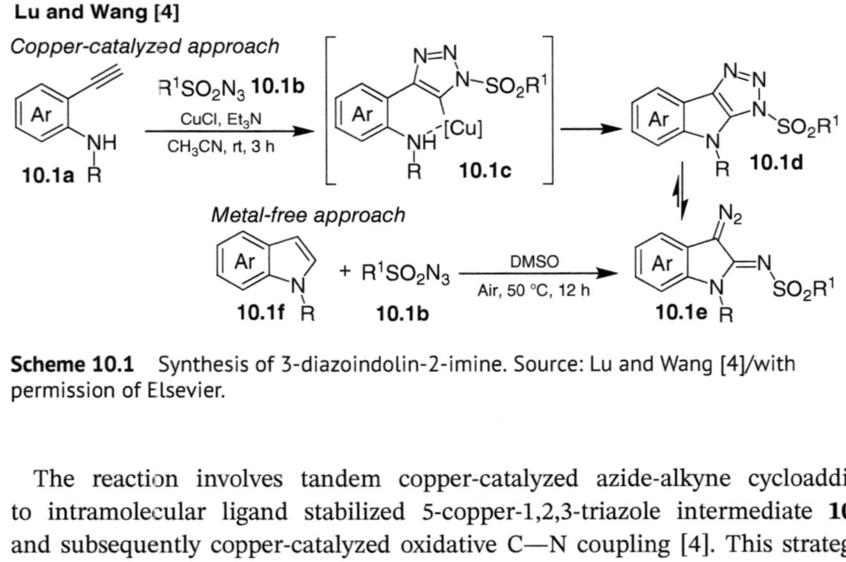

Scheme 10.1 Synthesis of 3-diazoindolin-2-imine. Source: Lu and Wang [4]/with permission of Elsevier.

The reaction involves tandem copper-catalyzed azide-alkyne cycloaddition to intramolecular ligand stabilized 5-copper-1,2,3-triazole intermediate **10.1c**, and subsequently copper-catalyzed oxidative C—N coupling [4]. This strategy is constrained by its limited substrate scope, and lack of scalability in the reaction as well as the use of a catalyst, which prompted the exploration and development of an alternative approach. Hence, the same group has developed an alternative catalyst-free approach for the synthesis of triazoloindoles **10.1d** through 1,3-dipolar cycloaddition of indoles **10.1f** with tosyl azides **10.1b** and subsequent dehydroaromatization followed by ring-opening [5]. Further single crystal X-ray analysis of synthesized triazoloindoles **10.1d** revealed that they predominantly exist as its open-chain form 3-diazoindolin-2-imines **10.1e**.

10.3 Insertion Reactions of 3-Diazoindolin-2-imines

10.3.1 Insertion into *X*—H Bonds

The generation of novel C—heteroatom bonds represents a fundamental and crucial aspect of organic chemistry. The introduction of *X*—H bonds (*X* = *O*, *N*, *S*, and *P*) into α-imino carbenes leads to the generation of new C—*X* bonds. Lu and Wang [5] reported 3-diazoindolin-2-imines **10.2a** insertion into N—H bonds of 2°-aromatic amines **10.2b** in the presence of a rhodium catalyst to form a variety of 2,3-diaminoindoles **10.2c** (Scheme 10.2). Similarly, the reaction between 3-diazoindolin-2-imines **10.2a** and β-enamino esters **10.2d** in the presence of rhodium catalyst resulted in 2,3-diaminoindoles **10.2e** [6]. Further, these compounds **10.2e** were converted to 5*H*-pyrazino [2,3-*b*]indoles **10.2f** through a sequential dehydrogenation, 6π-electrocyclic reaction, and aromatization.

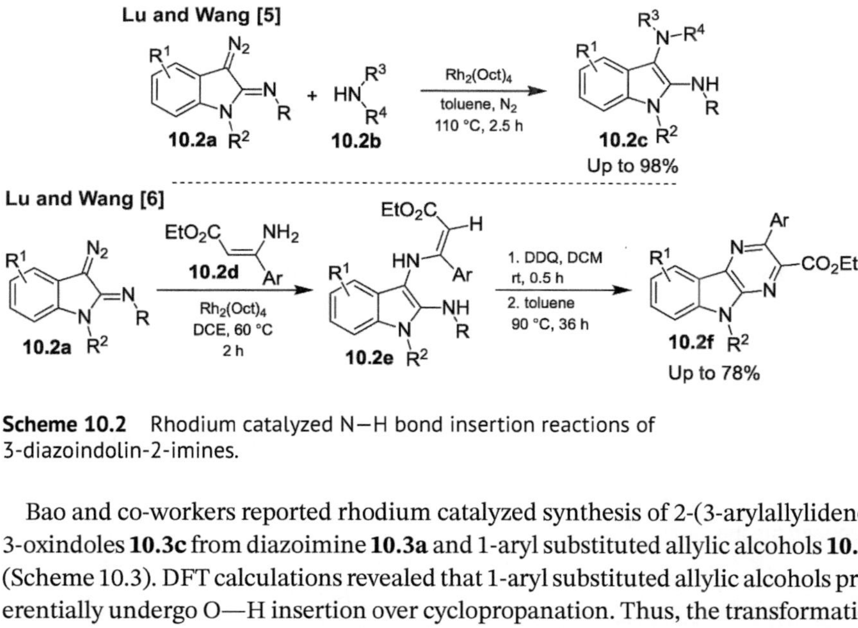

Scheme 10.2 Rhodium catalyzed N—H bond insertion reactions of 3-diazoindolin-2-imines.

Bao and co-workers reported rhodium catalyzed synthesis of 2-(3-arylallylidene)-3-oxindoles **10.3c** from diazoimine **10.3a** and 1-aryl substituted allylic alcohols **10.3b** (Scheme 10.3). DFT calculations revealed that 1-aryl substituted allylic alcohols preferentially undergo O—H insertion over cyclopropanation. Thus, the transformation involves a sequence of reactions such as O—H insertion, Claisen rearrangement followed by tosylamide elimination to afford 2-(3-arylallylidene)-3-oxindoles **10.3c**. Additionally, the photophysical properties of **10.3c** were studied at room temperature using UV/vis and photoluminescence spectroscopy [7].

Jiang and co-workers demonstrated a chemo and stereoselective P—H insertion of *H*-phosphine oxides **10.4b** into α-imino copper carbenes **10.4d** to yield P-stereogenic-3-phosphinoylindoles **10.4c** [8]. $Cu(CH_3CN)_4PF_6$ was chosen as the most efficient catalyst for this transformation under mild reaction conditions. The authors proposed two mechanistic pathways for this transformation. Path ***A*** involves the direct insertion of α-imino copper carbenes **10.4d** into P—H bonds of *H*-phosphine oxides **10.4b**, followed by tautomerization (Scheme 10.4). On

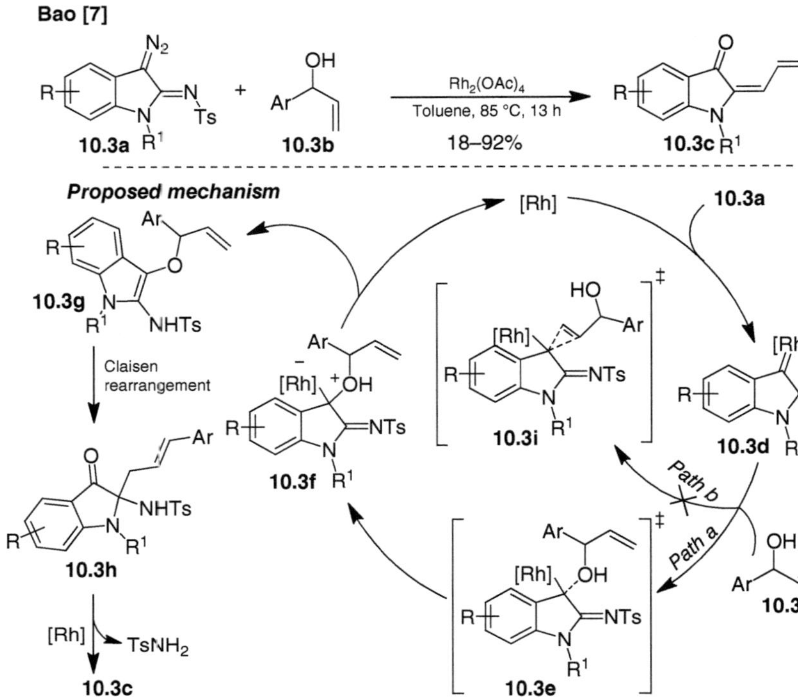

Scheme 10.3 Rhodium catalyzed synthesis of 2-(3-arylallylidene)-3-oxindoles. Source: Bao [7]/John Wiley & Sons.

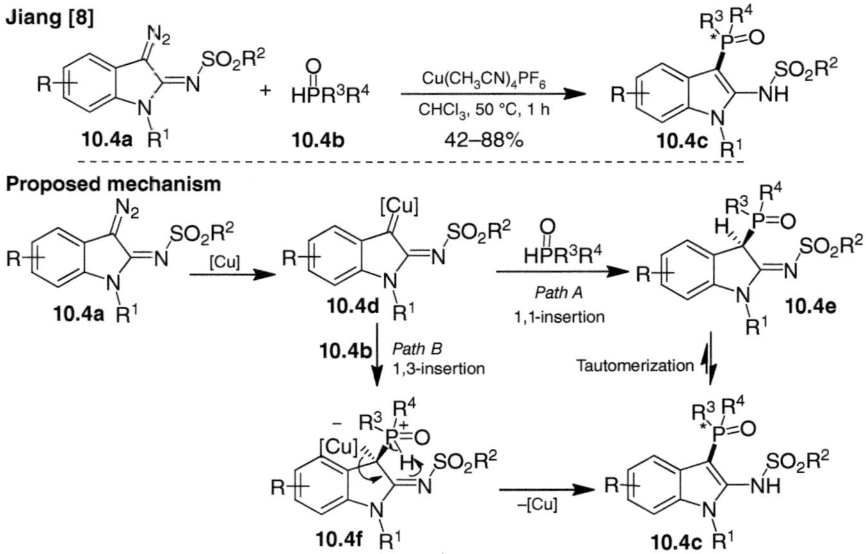

Scheme 10.4 Copper-catalyzed insertion of *H*-phosphine oxides into 3-diazoindolin-2-imine.

the other hand, Path ***B*** begins with the formation of a P-ylide intermediate **10.4f**, which subsequently undergoes copper dissociation and a proton transfer from the phosphorous to the imine moiety, ultimately yielding the 3-phosphinoylindoles **10.4c**. In 2021 Jiang and Xie provided evidence through DFT calculations and proposed path ***B*** as the most feasible route for this reaction [9].

S—H insertion reactions of diazo carbonyl compounds and thiols have been reported in various strategies including, metal catalysis, organocatalysis, heterogeneous catalysis, photolysis, and biocatalysis. In 2022, Muthusamy and co-workers demonstrated $BF_3{\cdot}OEt_2$ catalyzed chemoselective S—H insertion of thioamide **10.5c** and thiourea **10.5b** derivatives into 3-diazoindolin-2-imines **10.5a** to form new C—S bonds [10]. The reaction did not afford any C—O derived product when thiourea was replaced with *N,N'*-dimethyl urea under the optimized conditions (Scheme 10.5). Notably, this reaction is moisture and air-resistant, yielding the products effectively.

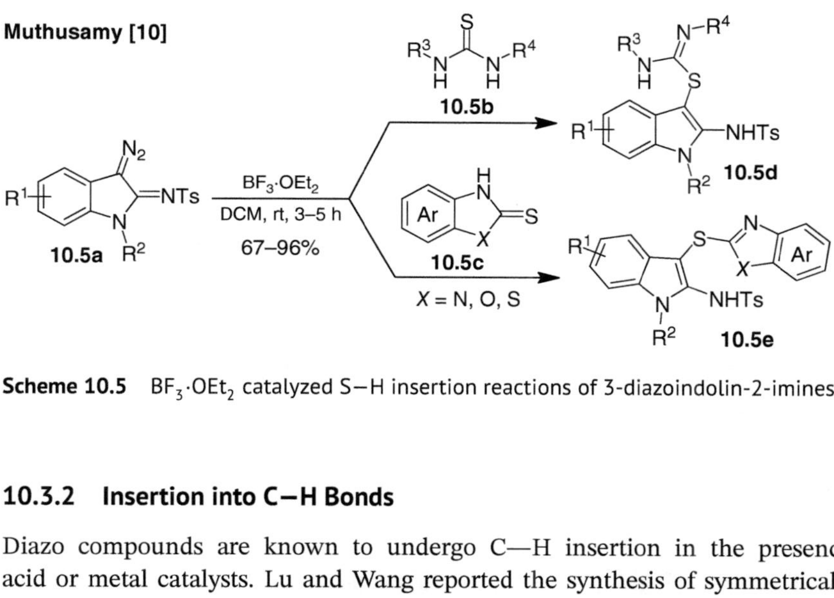

Scheme 10.5 $BF_3{\cdot}OEt_2$ catalyzed S—H insertion reactions of 3-diazoindolin-2-imines.

10.3.2 Insertion into C—H Bonds

Diazo compounds are known to undergo C—H insertion in the presence of acid or metal catalysts. Lu and Wang reported the synthesis of symmetrical and unsymmetrical 3,3-diaryl-2-imino-indoles **10.6c** through double arylation of **10.6a** in the presence of copper catalyst [11]. The molar ratio of the (hetero)arene used in the reaction controls the outcome of the product to either 3-aryl or 3,3-diaryl-2-imino-indoles **10.6c** (Scheme 10.6). Mechanistically, C—H insertion followed by Cu-catalyzed dehydrogenative cross coupling produced **10.6c**.

Lu and Wang reported the insertion of 3-diazoindolin-2-imine **10.7a** into metal-activated C—H bond of arenes **10.7b** to afford biaryls **10.7c** [12], which are privileged scaffolds in natural compounds, bioactive pharmaceuticals, and optoelectronic materials. Under the optimized conditions, various substituted 3-diazoindolin-2-imines **10.7a** were successfully inserted into C—H bond of 2-aryl-pyridines, 2-phenylpyrimidines, and 1-phenylpyrazoles to yield the corresponding 3-aryl-2-aminoindoles **10.7c**. Recently, they achieved the arylation

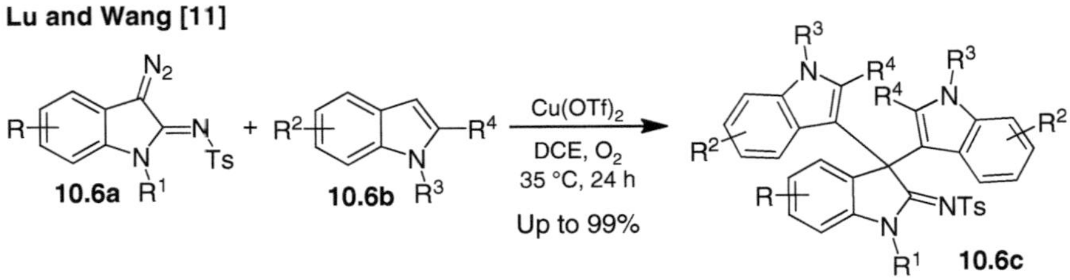

Scheme 10.6 Copper catalyzed synthesis of 3,3-diaryl-2-imino-indoles. Source: Lu and Wang [11]/American Chemical Society.

of **10.7a** with *N*-pyrimidinylindoles **10.7d** to afford biologically significant 2,3′-biindoles **10.7e** using Co(III) catalyst [13]. These reactions involved the directing group-assisted C—H bond activation followed by insertion (Scheme 10.7).

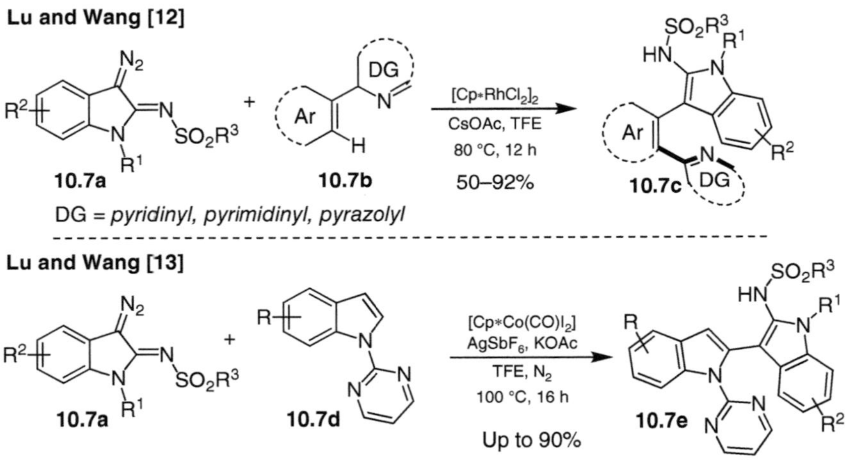

Scheme 10.7 Directing group-assisted C—H bond insertion reactions.

The same research group accomplished a metal-free C—H bond insertion of diazoimines **10.8a** into electron-rich arenes, yielding 3-aryl-2-aminoindoles **10.8b** and **10.8c** [14]. TfOH was selected as optimal catalyst for this reaction among various Brønsted acids examined including MsOH, TsOH, and $PhSO_3H$. Under optimized conditions, a broad spectrum of electron-rich arenes like thiophene, pyrrole, indole, substituted benzene, and naphthalene exhibited compatible reactivity, leading to the formation of a diverse array of 3-aryl-2-aminoindoles (Scheme 10.8).

Carbene insertion into C—H bonds of alkenes is challenging because alkenes are prone to undergo cyclopropanation when exposed to carbenes. In 2022, Yu and co-workers reported 3-diazoindolin-2-imines **10.9a** insertion into C—H bonds of α-oxo ketene dithioacetals **10.9b** through the cooperative use of HFIP and $B(C_6F_5)_3$ as the catalyst for the synthesis of 3-vinylated indoles **10.9d** [15]. According to the proposed mechanism, first, HFIP forms an adduct **10.9f** with $B(C_6F_5)_3$, which

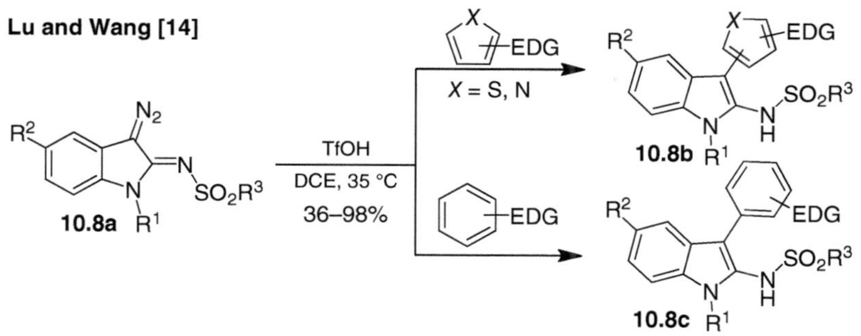

Scheme 10.8 TfOH-catalyzed reactions of 3-diazoindolin-2-imines with electron-rich arenes. Source: Lu and Wang [14]/American Chemical Society.

subsequently makes a hydrogen bond with the diazo compound **10.9a**. This interaction leads to the extrusion of dinitrogen, resulting in the generation of a carbene **10.9j**/carbenium species **10.9i**. Trapping of this species with **10.9b** affords 3-vinylated indoles **10.9d** (Scheme 10.9).

10.3.3 Insertion into C—*X* Bonds

C—S bond formation is important because many crucial pharmaceuticals, agrochemicals, materials, and natural products incorporate sulfur-containing groups. In 2018, Lu and Wang exploited thioesters **10.10b** and **10.10c** as nucleophiles to react with α-imino rhodium carbenes generated from 3-diazoindolin-2-imine **10.10a** [16]. This reaction led to the synthesis of 2-amino-3-(arylthio)indoles **10.10d** through the sequence of cascade reactions, such as C—S coupling followed by N—C coupling. Cyclic thioesters **10.10c** were subjected to these reaction conditions to afford 7,8-dihydrobenzo[7, 8][1, 4]thiazocino[3,2-*b*]indol-6(5*H*)-ones **10.10e** in good yields (Scheme 10.10).

Subsequently, an efficient approach for the synthesis of biologically active and pharmaceutically significant 3,3-disubstituted indoline derivatives has also been demonstrated. Copper-catalyzed reaction of allyl and propargyl aryl sulfides with 3-diazoindolin-2-imines **10.11a** led to the formation of 3-allyl- and 3-allenyl-3-arylthioindolin-2-imines (**10.11d** and **10.11e**), respectively [17]. Mechanistically, the *in situ* generated copper carbene from **10.11a** forms sulfonium ylides with allyl and propargyl aryl sulfides; subsequent [2, 3]-sigmatropic rearrangement resulted in the formation of 3,3-disubstituted indoline derivatives (Scheme 10.11).

Lu and Wang reported synthesis of various heterocyclic compounds by subjecting 3-diazoindolin-2-imines **10.12a** with various tertiary aromatic amines. The outcome of the product depends on the substituents present on the nitrogen atom. For instance, the reaction of *N*,*N*-dialkylanilines **10.12b** with **10.12a** results in the synthesis of 3-aryl-2-aminoindoles **10.12c** through a Friedel-Crafts-type arylation process [18]. In the case of *N*-allyl-*N*-alkylanilines, the formation of 3-allyl-3-amino-2-iminoindolines **10.12d** occurred through C—N coupling and subsequent 1,2- or 2,3-sigmatropic rearrangement. When *N*-propargyl-*N*-alkylanilines

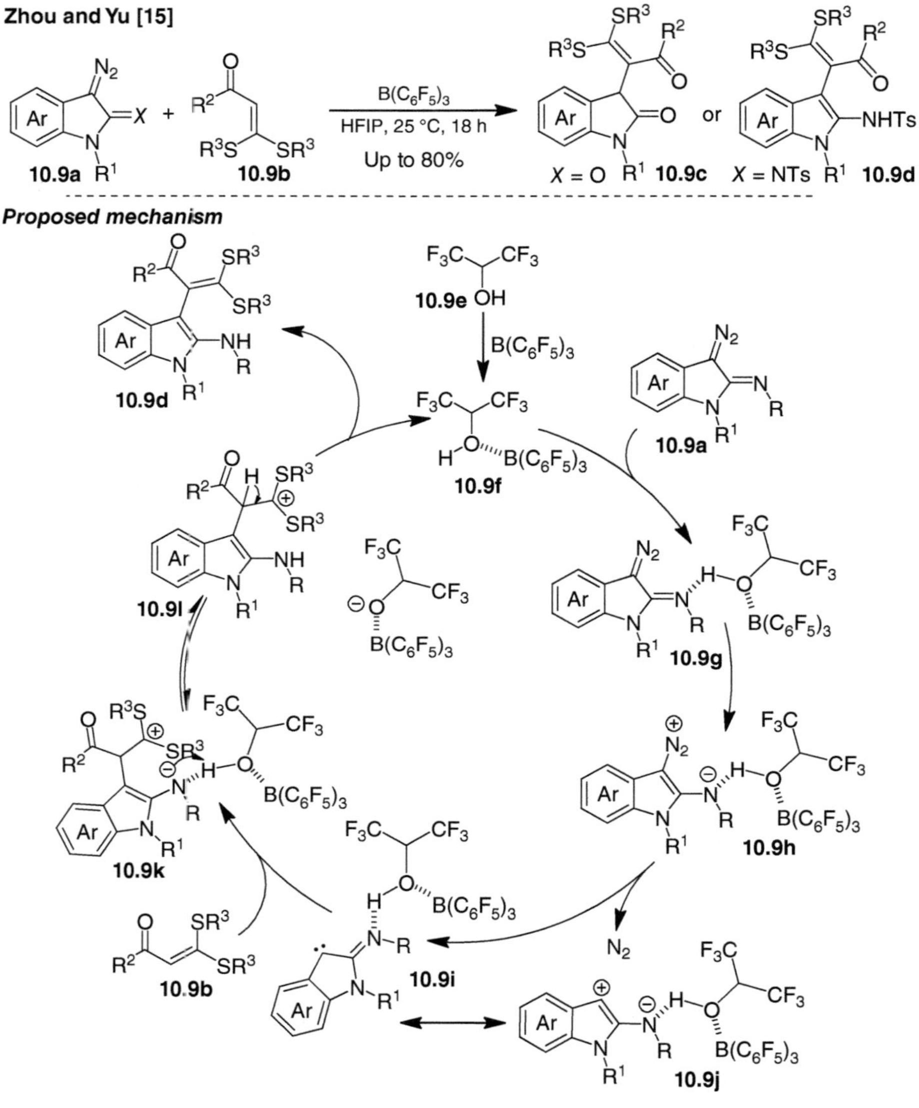

Scheme 10.9 Carbene insertion into internal alkenyl C—H bond of α-oxo ketene dithioacetals . Source: Adapted from Zhou and Yu [15].

10.12e were used as nucleophiles to trap α-imino rhodium carbenes, a cascade of reactions led to the production of tetrahydro[1,4]diazepino[2,3-*b*]indoles **10.12f** (Scheme 10.12).

Subsequently, they demonstrated C—*X* bonds (where *X* = Cl, Br, I) insertion into 3-diazoindolin-2-imines to form valuable 3-haloindol-2-amines. Insertion of allyl bromide into 3-diazoindolin-2-imine **10.13a** was accomplished using a $Pd_2(dba)_3$ in acetonitrile [19]. This reaction methodology proved to be versatile

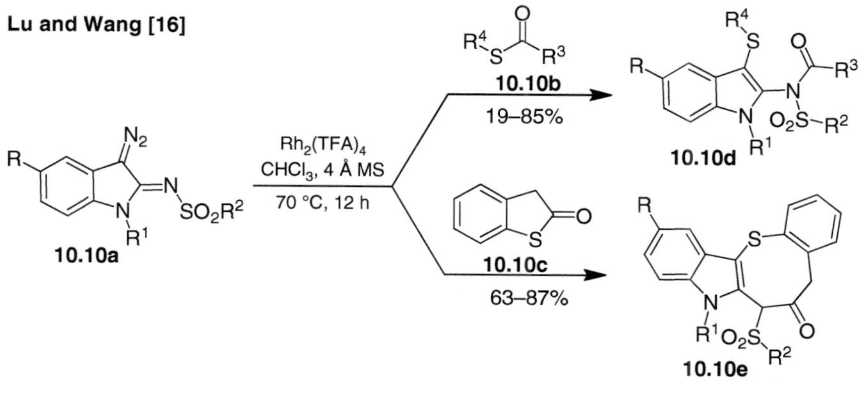

Scheme 10.10 Rh-catalyzed C—S bond insertion of 3-diazoindolin-2-imines with thioesters. Source: Adapted from Lu and Wang [16].

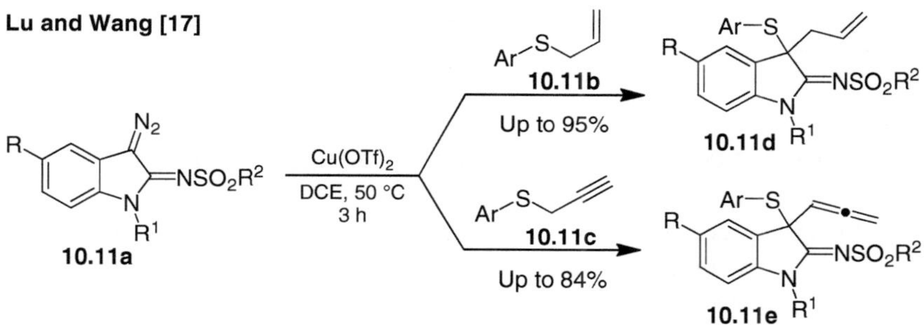

Scheme 10.11 Copper-catalyzed synthesis of 3,3-disubstituted indoline derivatives. Source: Adapted from Lu and Wang [17].

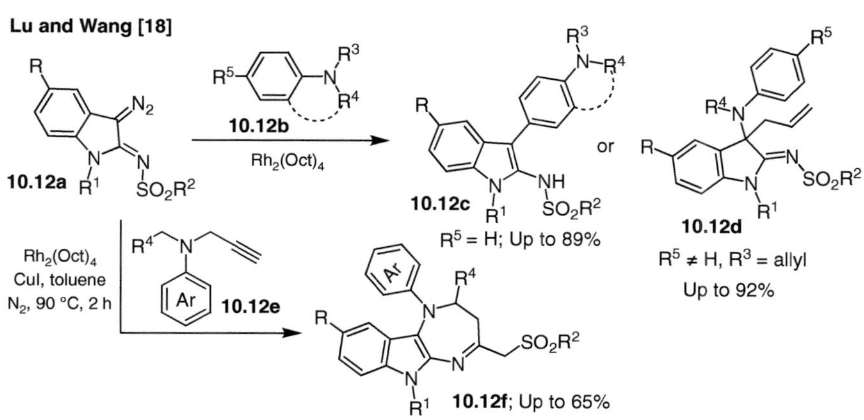

Scheme 10.12 Rh-catalyzed reactions of 3-diazoindolin-2-imines with aniline derivatives.

and yielded a diverse array of 3-bromoindol-2-amines **10.13b** from substrates such as allylic bromides, benzyl bromides, α-bromoketones, α-bromocarboxylic esters and α-bromonitriles. Under modified reaction conditions, the insertion of allylic chlorides and alkyl iodides into **10.13a** was also achieved (Scheme 10.13).

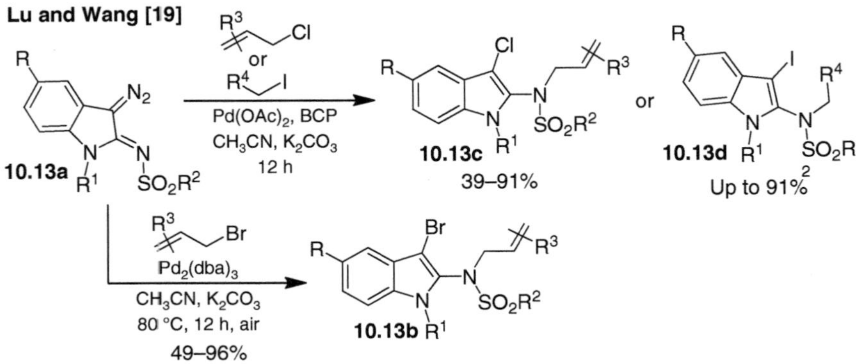

Scheme 10.13 Palladium-catalyzed synthesis of 3-haloindol-2-amines. Source: Adapted from Lu and Wang [19].

10.4 Annulation Reactions of 3-Diazoindolin-2-imines

Lu and Wang exposed 3-diazoindolin-2-imine **10.14a** to olefins and alkynes in the presence of rhodium catalyst to form spirocyclopropyl iminoindoles **10.14c** and pyrrolo[2,3-*b*]indoles **10.14b**, respectively [4]. On the other hand, substituted indoles on reaction with 3-diazoindolin-2-imine **10.14a** enabled the synthesis of a diverse range of valuable indole compounds such as 3,3′-biindoles **10.14d** and 2,3′-biindoles **10.14e** (Scheme 10.14).

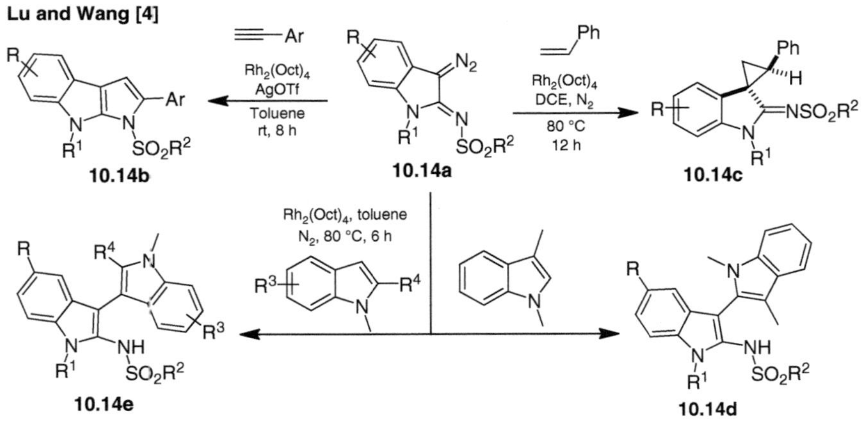

Scheme 10.14 Rhodium-catalyzed transformations of 3-diazoindolin-2-imines. Source: Adapted from Lu and Wang [4].

Later, in their extended study, synthesis of 9*H*-pyrido[2,3-*b*]indoles **10.15c** and tetrahydrofuro[3′,2′:4,5]pyrrolo[2,3-*b*]indoles **10.15e** was accomplished through nucleophilic addition of furans **10.15b** and dihydrofurans **10.15d** to α-imino rhodium carbenes, generated *in situ* from 3-diazoindolin-2-imines **10.15a** [20]. Unsymmetrical furans, including 2-ethyl, 2-methyl-5-benzyl, and 3-methylfuran, produced products with good regioselectivity (Scheme 10.15).

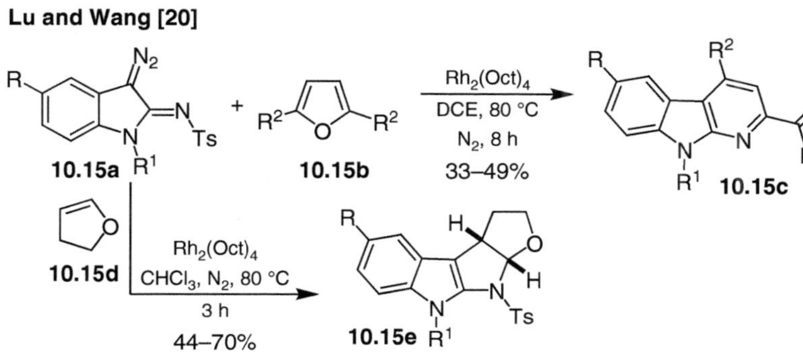

Scheme 10.15 Reactions of diazoimines with 2,5-disubstituted furan and dihydrofuran. Source: Lu and Wang [20]/American Chemical Society.

Isoquinoline-fused indoles exhibit notable biological activity and find versatile applications in organic optoelectronic devices. Li and co-workers developed an approach for the synthesis of indoloisoquinolines **10.16c** from readily accessible compounds such as imidates **10.16a** and 3-diazoindolin-2-imines **10.16b** under mild reaction conditions [21]. Thus, the reaction occurred through Rh(III)-catalyzed C—H activation followed by C—C/C—N coupling of imidates **10.16a** with **10.16b**. Notably, PivOH was a key component in this reaction to promote C—H activation and cyclization processes (Scheme 10.16).

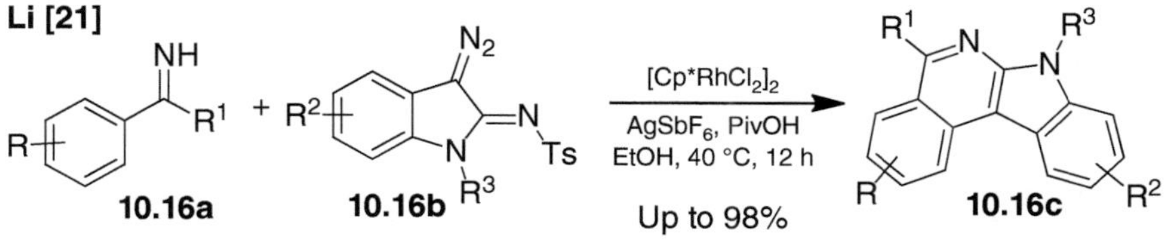

Scheme 10.16 Synthesis of isoquinoline-fused indoles. Source: Li [21]/American Chemical Society.

Meanwhile, Volla and co-workers constructed biologically important indolo[2,3-c] isoquinolin-5-ones **10.17c** by employing *N*-methoxybenzamides **10.17a** and 3-diazoindolin-2-imines **10.17b** through Rh(III)-catalyzed [4 + 2]-annulation [22]. The transformation proceeds *via* a sequence of cascade reactions such as Rh(III) catalyzed C—H bond activation, carbenoid formation, migratory insertion, and protonolysis followed by an intramolecular nucleophilic addition and elimination. Notably, under the optimized reaction conditions, *N*-methoxy-1*H*-indole-1-carboxamide comfortably reacted with **10.17b** to furnish pentacyclic 2,3′-biindole in 64% yield, which demonstrates the versatile applicability of this methodology in the synthesis of various biologically important motifs (Scheme 10.17).

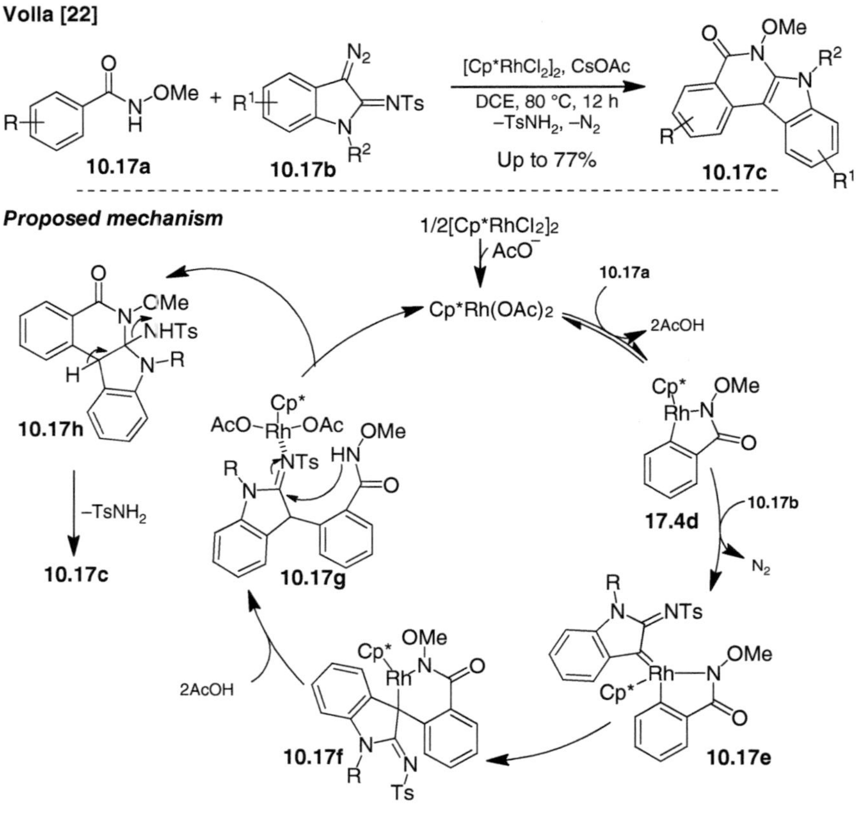

Scheme 10.17 [4 + 2]-Annulation of benzamides and 3-diazoindolin-2-imines. Source: Adapted from Volla [22].

Similarly, Lee and co-workers demonstrated an approach for the construction of indolo-1,2-benzothiazines **10.18c** through rhodium-catalyzed cyclization of sulfoximines **10.18b** with **10.18a** [23]. The catalytic cycle includes C—H activation and subsequent C—C/C—N couplings by extrusion of dinitrogen and *p*-toluenesulfonamide. Moreover, portion-wise addition of 3 equivalent of **10.18a** in three batches in 4 h interval to the reaction mixture improved the yield of the desired product. A wide range of indolo-1,2-benzothiazines was synthesized in yields ranging from 48% to 99% (Scheme 10.18).

In 2023, Muthusamy and co-workers have successfully synthesized thiazolo[4,5-*b*]indol-2-amines from 3-diazoindolin-2-imines and mono-substituted thioureas in

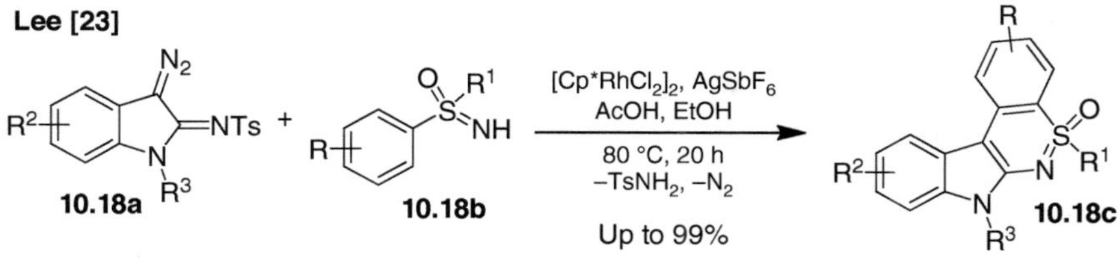

Scheme 10.18 Rhodium-catalyzed reactions of sulfoximines with 3-diazoindolin-2-imines. Source: Adapted from Lee [23].

the presence of boron trifluoride etherate and camphorsulfonic acid by releasing molecular nitrogen and *p*-toluenesulfonamide as by-products [24]. Employment of *N*,*N*′-disubstituted thioamides or cyclic thioamides under optimized reaction conditions furnished only S—H inserted products (Scheme 10.19).

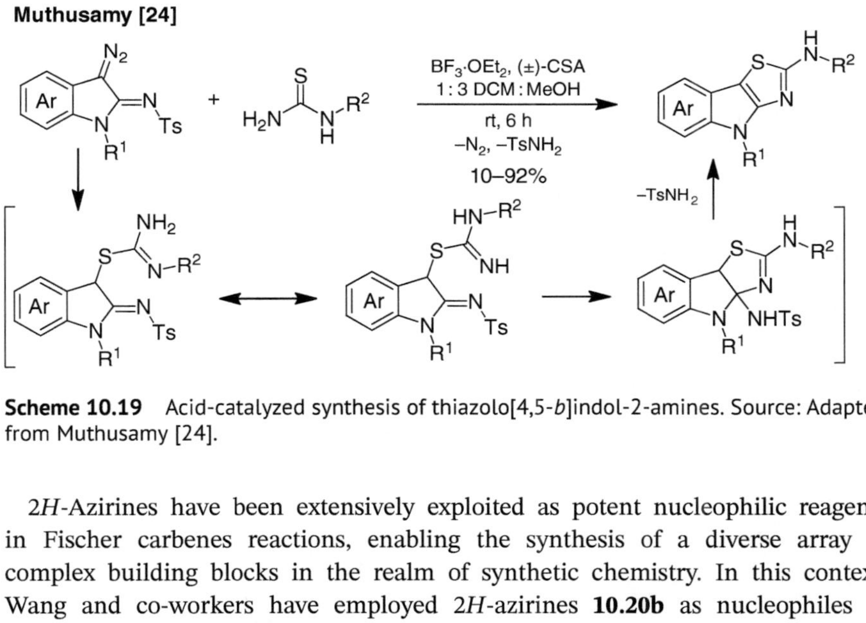

Scheme 10.19 Acid-catalyzed synthesis of thiazolo[4,5-*b*]indol-2-amines. Source: Adapted from Muthusamy [24].

2*H*-Azirines have been extensively exploited as potent nucleophilic reagents in Fischer carbenes reactions, enabling the synthesis of a diverse array of complex building blocks in the realm of synthetic chemistry. In this context, Wang and co-workers have employed 2*H*-azirines **10.20b** as nucleophiles to trap indole embedded α-imino rhodium carbenes. Initially, this reaction produces dihydroindolopyrazine **10.20f**, which can be further transformed into 5*H*-pyrazino[2,3-*b*]indoles **10.20c** through the use of ^{t}BuOK as a base in a convenient one-pot synthesis [25]. According to the proposed mechanism, 2*H*-azirines **10.20b** attack the electrophilic rhodium carbenoid to form a zwitterion intermediate **10.20d**. This intermediate could follow two distinct pathways to reach the product. In the first pathway, ring opening followed by a 6π electrocyclic ring-closing, desulfonylation furnished the product. In the alternative pathway, rearrangement followed by aromatization yielded 5*H*-pyrazino[2,3-*b*]indoles **10.20c**. Additionally, the photoluminescence properties of **10.20c** have been investigated in various forms (Scheme 10.20).

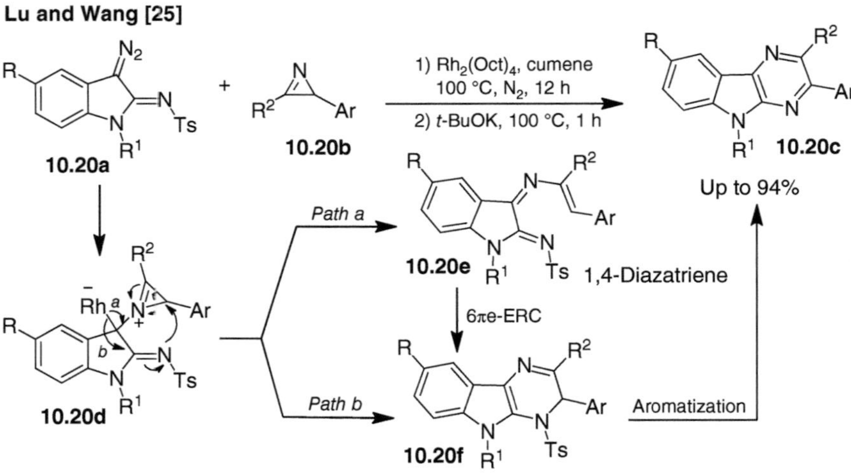

Scheme 10.20 Synthesis of 5*H*-pyrazino[2,3-*b*]indoles. Source: Lu and Wang [25] American Chemical Society.

Similarly, Lee and co-workers also synthesized a variety of indolopyrazines **10.21c** regioselectively from 3-diazoindolin-2-imine **10.21a** and 2*H*-azirine **10.21b** using $Rh_2(Oct)_4$ and Et_3N [26]. Meanwhile, Novikov and co-workers have reported a one-pot synthesis of indolopyrazines **10.21e** from azirines **10.21d** and **10.21a** employing rhodium catalyst in the presence and absence of acid (Scheme 10.21) [27].

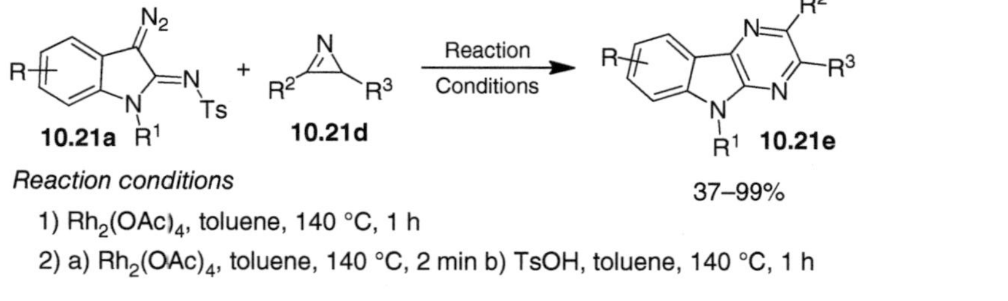

Scheme 10.21 Rhodium-catalyzed reaction of 3-diazoindolin-2-imines with 2*H*-azirines.

On the other hand, the same group ended up with 1,3'-biindoles **10.22c** when 2,2-diaryl-substituted 2*H*-azirines **10.22b** were subjected with **10.22a** under rhodium catalyst [28]. The reaction pathway for this transformation involves the generation of 1,4-diazatriene **10.22e** from **10.22b** and α-imino rhodium carbene **10.22d** and subsequent pseudopericyclic 1,5-electrocyclization followed by prototropic shift to furnish 1,3′-biindoles **10.22c** in good yield (Scheme 10.22).

Novikov [28]

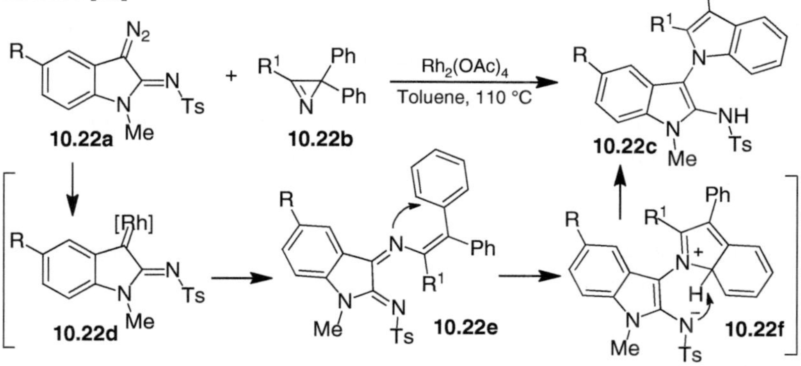

Scheme 10.22 Synthesis of 1,3'-biindoles. Source: Novikov [28]/American Chemical Society.

Copper-catalyzed cyclization of *o*-hydroxyarylenaminones **10.23b** with 3-diazoindolin-2-imines **10.23a** to synthesize 3-indolyl-4*H*-chromen-4-ones **10.23c** as a major product was demonstrated by Lu and Wang [29]. While evaluating the substrate scope of this transformation pyrido[2,3-*b*:6,5-*b*′]diindoles **10.23d** and pyrazino[2,3-*b*:5,6-*b*′]diindole **10.23e** were isolated as byproducts. Moreover, they achieved the synthesis of fluorinated and oxidized derivatives of **10.23c** in excellent yields and regioselectivity employing NFSI and ^{t}BuOOH as reagents (Scheme 10.23).

Lu and Wang [29]

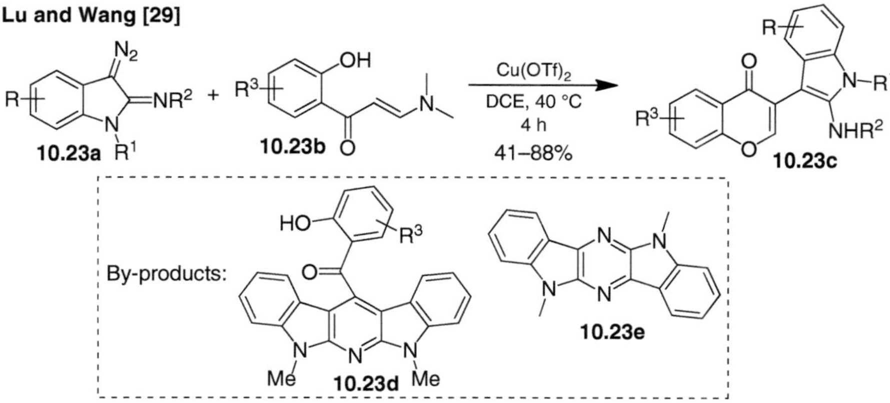

Scheme 10.23 Copper-catalyzed synthesis of 3-indolyl-4*H*-chromen-4-ones. Source: Lu and Wang [29]/American Chemical Society.

Initially, Lu and Wang demonstrated rhodium-catalyzed cyclopropanation of olefins with 3-diazoindolin-2-imine as reported in reference [4].

Subsequently, they expanded their study to investigate the reactivity of 1,3-dienes with 3-diazoindolin-2-imines. Thus, they disclosed rhodium and silver co-catalyzed reactions of differently substituted 1,3-dienes with **10.24a**. Notably, azepino[2,3-*b*]indoles **10.24e** and azepino[2,3-*b*]indol-4(1*H*)-ones **10.24f** were isolated when **10.24a** was subjected with acyclic 1,3-dienes **10.24b**. This transformation occurred through cyclopropanation followed by aza-Cope rearrangement. Conversely, cyclic 1,3-dienes **10.24c**, such as cyclohexadiene and cyclopentadiene when subjected to 3-diazoindolin-2-imine **10.24a**, underwent formal aza-[3 + 2] annulations to furnish hexahydroindolo[2,3-*b*]indoles **10.24d** or cyclopenta[4, 5]-pyrrolo[2,3-*b*]indoles *via* cyclopropanation/1,3-shift [30]. Independently, Lee and co-workers also documented rhodium-catalyzed one-pot synthesis of azepinoindoles **10.24h** from 1,3-dienes and **10.24a** through cyclopropanation followed by aza-Cope rearrangement (Scheme 10.24) [31].

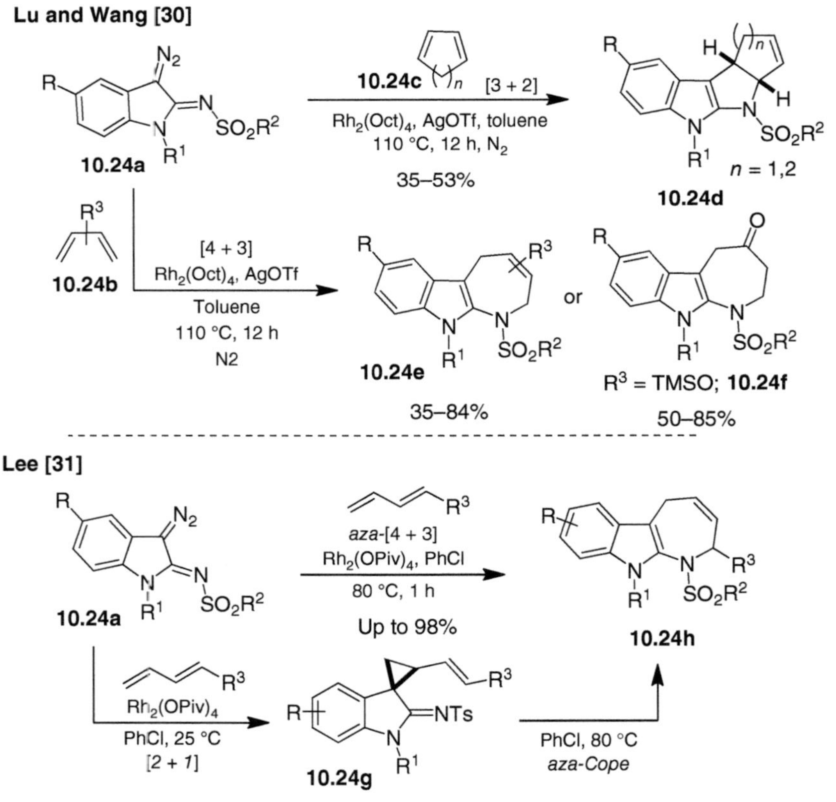

Scheme 10.24 Rhodium catalyzed reactions of 3-diazoindolin-2-imines with 1,3-dienes.

Chen and Pu reported a rhodium-catalyzed denitrogenative [5 + 1]-annulation of benzo[*d*]isoxazole **10.25b** with 3-diazoindolin-2-imines **10.25a** to form spiro[benzo[*e*][1,3]oxazine-2,3-indolin]-2-imines **10.25c** with good functional group tolerance in moderate to good yields [32]. According to the proposed mechanism, an electrophilic α-imino rhodium carbene **10.25d** reacts with the nitrogen atom of benzo[*d*]isoxazole **10.25b** to furnish an adduct **10.25e**. The adduct **10.25e** experiences electronic resonance, facilitating the cleavage of N—O bond and generating an imine intermediate **10.25g**. Subsequently, rearomatization of the aryl moiety generates phenoxy anion **10.25h**, which then cyclizes to give **10.25c** (Scheme 10.25).

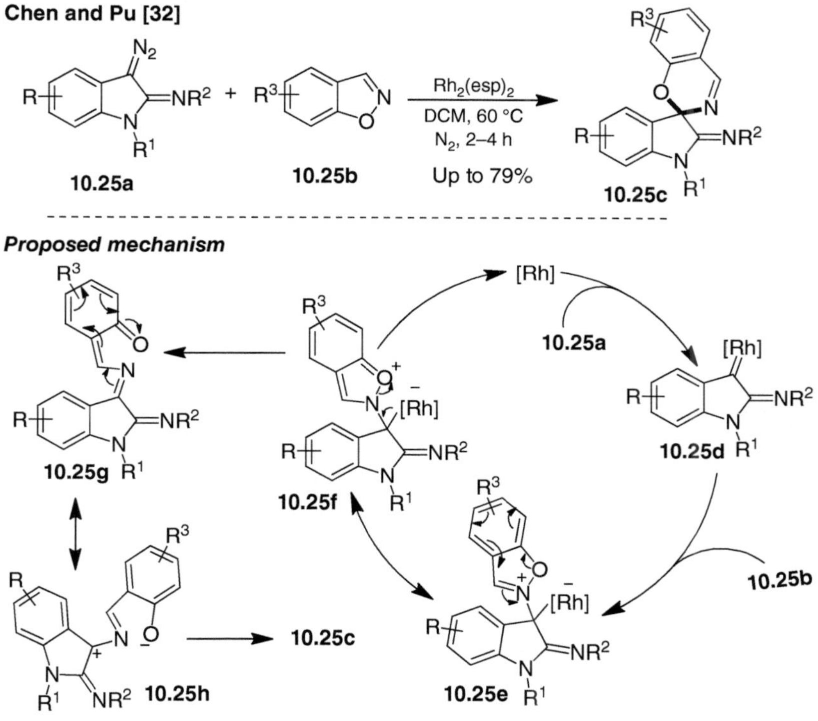

Scheme 10.25 Rh(II)-catalyzed denitrogenative transannulation of benzo[d]isoxazoles. Source: Chen and Pu [32]/John Wiley & Sons.

Soon after, the same group developed rhodium-catalyzed one-pot synthesis of azepino-, azocino-, and azonino-fused diindoles from 3-diazoindolin-2-imines **10.26a** and bromoalkyl-indoles **10.26b** employing $Rh_2(Oct)_4$ and DBU. Under the developed reaction conditions, the synthesis of azepino[2,3-*b*:4,5-*b*′]diindoles, azocino[2,3-*b*:4,5-*b*′]diindoles, and azonino[2,3-*b*:4,5-*b*′]diindoles were achieved in good yield [33]. As previously explained in their earlier work [4], when 3-substituted indoles were exposed to **10.26a** and rhodium catalyst, a cyclopropanation followed by rearrangement results in the formation of 2,3′-biindole **10.26f**. Subsequent cyclization of 2,3′-biindole **10.26f** in the presence of DBU led to the formation of diindoles **10.26c** (Scheme 10.26).

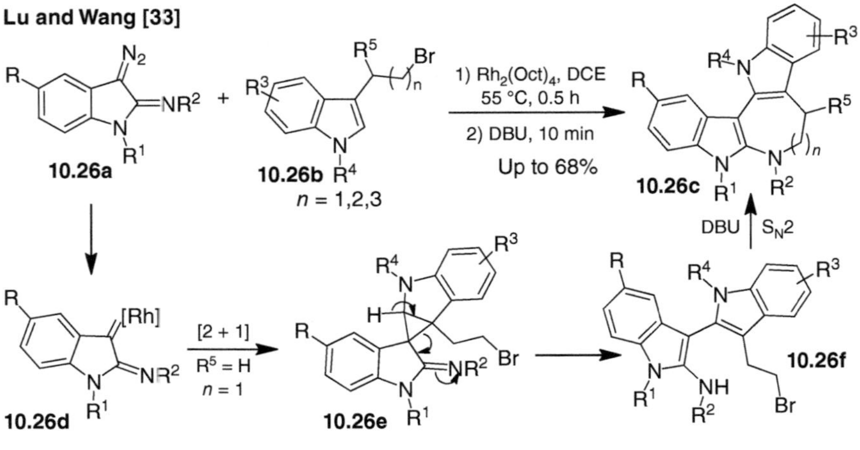

Scheme 10.26 Rhodium-catalyzed synthesis of azepino, azocino, and azonino diindoles. Source: Lu and Wang [33]/American Chemical Society.

Following the cyclopropanation of 3-diazoindolin-2-imines with alkene [4], in 2019, Wang and co-workers introduced an alternative strategy for synthesizing spiro[cyclopropane-1,3′-indolin]-2′-imines **10.27c** [34]. This method entailed a copper-catalyzed reaction of sulfoxonium ylides **10.27b** with 3-diazoindolin-2-imines **10.27a**. The proposed mechanism for this transformation begins with the formation of a copper carbene **10.27d** from the diazo compound **10.27a**, which subsequently interacts with the sulfoxonium ylide **10.27b**, resulting in the formation of alkene intermediate **10.27f**. This alkene intermediate **10.27f** is attacked by another molecule of sulfoxonium ylide **10.27b**, to furnish **10.27c** (Scheme 10.27).

Continuing their research into the functionalization of 3-diazoindolin-2-imines, copper-catalyzed [2+2+1] cycloaddition of diazoimines **10.28a** with hexahydro-1,3,5-triazines **10.28b** have been disclosed for the synthesis of spiro[imidazolidine-4,3′-indolin]-2′-imines **10.28c** in good yields with wide functional group tolerance [35]. Notably, the dual characteristics of hexahydro-1,3,5-triazines **10.28b**, exhibiting nucleophilic reactivity of nitrogen atoms and the capacity to decompose into formaldimines, assisted in this transformation (Scheme 10.28).

The group of Wang, Yu, and Anbarasan have independently disclosed denitrogenative [4+1] annulation reactions of 3-diazoindolin-2-imines with suitable coupling partners to form 3-spiroindoline compounds. In these reactions, 3-diazoindolin-2-imines act as a one-carbon synthon, while other coupling partner serves as a four-atom synthon. For instance, Yu group utilized α-thioxo ketene *N,S*-acetals **10.29b** as four-atom synthon in copper-catalyzed [4+1]-annulation with **10.29a** to afford spiroindoline-fused S-heterocycles **10.29c**. Copper catalysts showed better performance for this transformation than rhodium catalysts (Scheme 10.29) [36].

On the other hand, Lu and Wang successfully synthesized spiro[indoline-3,2-oxazolidin]-2-imines **10.30c** from 2-(phenylamino)ethanols **10.30b** and diazoimines **10.30a** through copper-catalyzed [4+1]-annulation, in good yields

Lu and Wang [34]

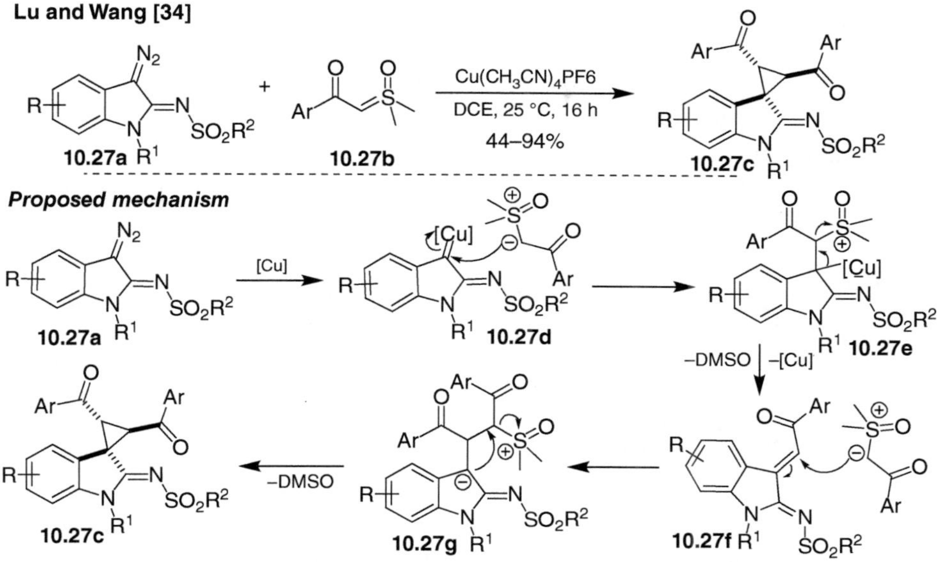

Scheme 10.27 Copper-catalyzed synthesis of spiro[cyclopropane-1,3′-indolin]-2′-imines. Source: Lu and Wang [34]/American Chemical Society.

Lu and Wang [35]

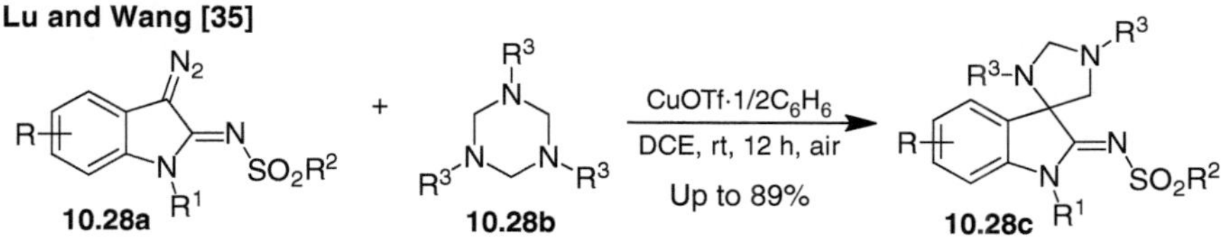

Scheme 10.28 Copper-catalyzed synthesis of spiro[imidazolidine-4,3′-indolin]-2′-imines. Source: Lu and Wang [35]/American Chemical Society.

Yu [36]

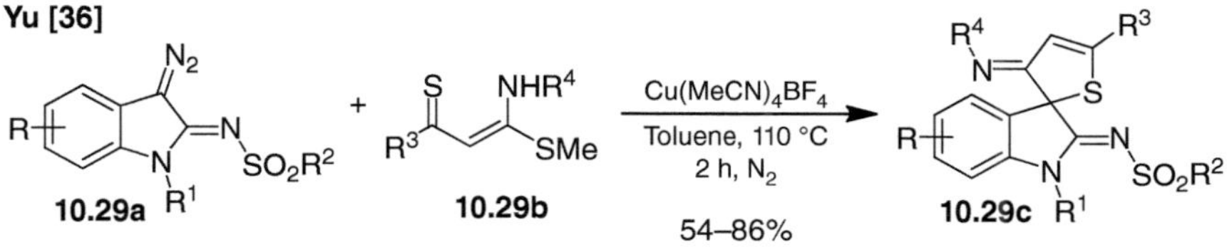

Scheme 10.29 Copper-catalyzed [4 + 1]-annulation of enaminothiones with diazoimines. Source: Adapted from Yu [36].

with high regioselectivity and stereoselectivity [37]. This annulation involves a sequential cascade of N—H insertion and C—O coupling. Additionally, the synthesis of spiro[indoline-3,2-oxazolidin]-2-ones was accomplished through a sequential synthesis of **10.30c**, followed by the hydrolysis with NaOH in one-pot (Scheme 10.30).

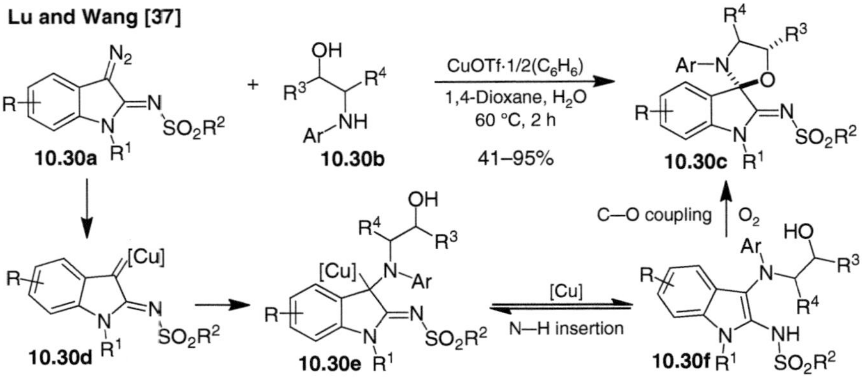

Scheme 10.30 [4 + 1]-Annulation of diazoimines with 2-(arylamino)ethanols. Source: Adapted from Lu and Wang [37].

Meanwhile, Anbarasan and co-workers reported the synthesis of spirobiindolines **10.31c** from diazoimines **10.31b** and *o*-acylanilines **10.31a** under rhodium catalyst [38]. This transformation involves initial *N*-ylide formation followed by cyclization and then dehydration. This method displayed remarkable versatility and generated a variety of spirobiindolines with excellent tolerance for various functional groups and high yields. Furthermore, acid-catalyzed nucleophilic addition of indoles and pyrroles to spirobiindolines **10.31c** furnished tetrasubstituted indolines **10.31e**, which demonstrated the potential utility of this method (Scheme 10.31).

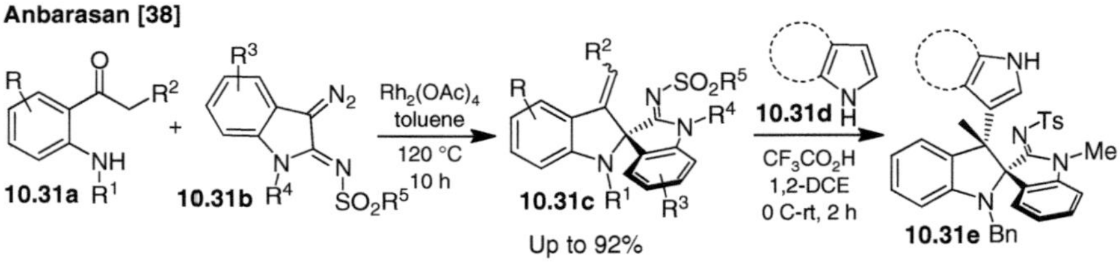

Scheme 10.31 Rh-catalyzed [4 + 1]-annulation of 3-diazoindolin-2-imine with *o*-acylaniline. Source: Adapted from Anbarasan [38].

Shankaraiah and co-workers explored the ruthenium-catalyzed [3 + 3]-annulation of diazoimines **10.32a** and *N,N′*-cyclic azomethine imines **10.32b** for the construction of biologically significant fused α-carbolines **10.32c** [39]. Notably, the directing group present in *N,N′*-cyclic azomethine imines **10.32b** played a key role in facilitating C—H activation, followed by annulation, and aromatization. Additionally, a mechanistic pathway was proposed by tracking intermediates during the catalytic process using ESI-MS. This method enabled the successful synthesis of various substituted indoloisoquinolines in yields ranging from 50% to 97% (Scheme 10.32).

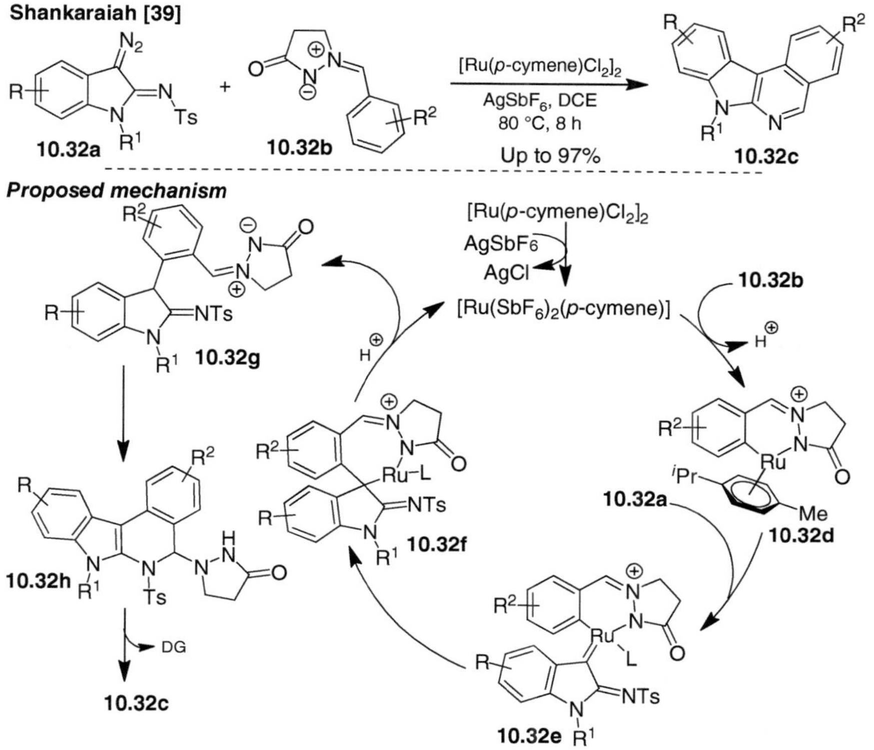

Scheme 10.32 Ru(II)-catalyzed synthesis of fused α-carbolines. Source: Adapted from Shankaraiah [39].

10.5 Miscellaneous Reactions

Lu and Wang observed that the oxygen atom of arylaldehydes is capable to capture the indole embedded α-imino rhodium carbenes. Consequently, treatment of aromatic aldehydes **10.33b** and α,β-unsaturated aldehydes **10.33e** with 3-diazoindolin-2-imines **10.33a** in the presence of $Rh_2(HFB)_4$ and $Rh_2(TFA)_4$ resulted (*E*)-2-aminomethylene-3-oxoindoles **10.33c** and pyranoindoles **10.33f**, respectively [40]. Investigation of substituent effects in this reaction revealed that electron-withdrawing group substituted benzaldehyde afforded the product in higher yields compared to electron-donating group substituted benzaldehydes. Furthermore, heteroaryl aldehydes compatibly furnished the corresponding products in good yields under these reaction conditions. The synthetic utility of the protocol was explained by converting **10.33c** to chromeno[3,2-*b*]indol-11-one skeleton **10.33d** on reaction with NaH (Scheme 10.33).

Three strategies for the arylation of 3-diazoindolin-2-imine were demonstrated by Lu and Wang employing direct C—H bond functionalization as a key step [11–14]. Additionally, the palladium-catalyzed coupling of arylboronic acids **10.34b** with diazoimines **10.34a** afforded arylation product 2-amino-3-arylindoles

Lu and Wang [40]

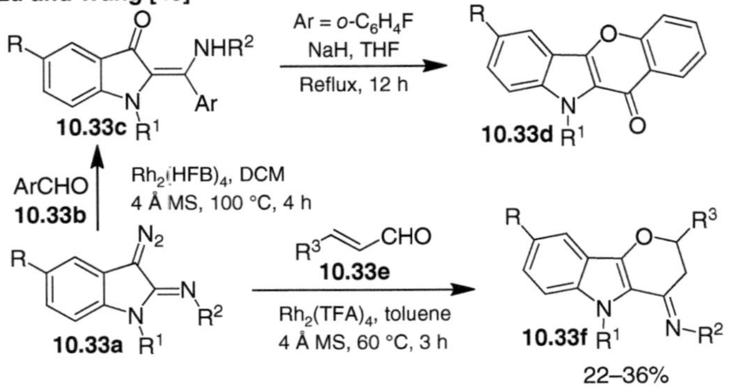

Scheme 10.33 Rhodium-catalyzed reactions of diazoimines with aryl- and acrylaldehydes. Source: Lu and Wang [40]/American Chemical Society.

Lu and Wang [41]

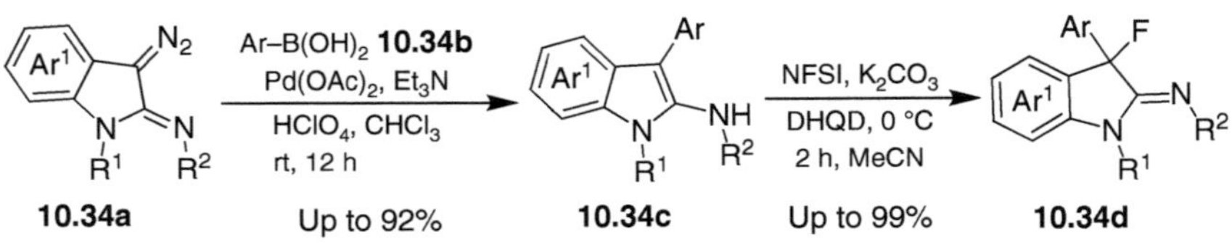

Scheme 10.34 Pd(II)-catalyzed arylation of diazoimines with arylboronic acids. Source: Lu and Wang [41]/American Chemical Society.

10.34c in good yield [41]. These products were further fluorinated to yield 3-aryl-3-fluoroindolin-2-imines **10.34d** using NFSI (Scheme 10.34).

Subsequently, the same group demonstrated $BF_3{\cdot}OEt_2$ catalyzed arylation of 3-diazoindolin-2-imines **10.35a** with arylaldehydes **10.35b** to synthesize 2-amino-3-arylindoles **10.35c**. The possible mechanism for this transformation is as follows: $BF_3{\cdot}OEt_2$ activates the carbonyl carbon, which is then attacked by the nucleophilic carbon of the diazo compound, forming a Tiffeneau–Demjanov rearrangement-like intermediate **10.35d**. Subsequently, aryl migration and concomitant nitrogen extrusion followed by hydrolysis furnished the product **10.35c**. However, reaction of 1-naphthaldehyde **10.35g** and diazo compound **10.35a** in the presence of $BF_3{\cdot}OEt_2$ in DCE solvent resulted *E*-configured alkene **10.35i** *via* $BF_3{\cdot}OEt_2$ promoted (3 + 2)-cycloaddition, followed by retro (3 + 2)-cycloaddition (Scheme 10.35) [42].

Lu and Wang introduced an approach for the synthesis of 3-allyl-2-aminoindoles **10.36c** through a copper-catalyzed reaction between **10.36a** and allyltrimethylsilanes **10.36b** [43]. The reaction goes through the generation of copper carbenoid **10.36d** from **10.36a**, which then reacts with **10.36b** to form a β-silicon stabilized carbocation intermediate **10.36e**. This intermediate can lead to the product **10.36c** through two possible pathways. One pathway involves electrophilic cyclization, resulting [3 + 2]-cycloaddition intermediate **10.36f**, followed by TMS elimination to

Lu and Wang [42]

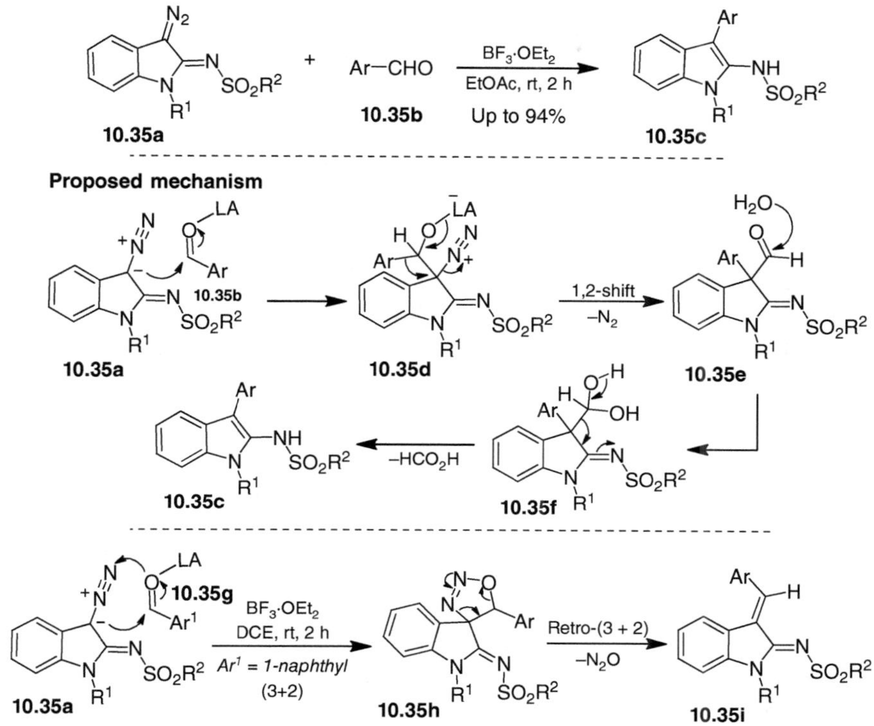

Scheme 10.35 $BF_3 \cdot Et_2O$-catalyzed arylation of 3-diazoindolin-2-imines with arylaldehydes. Source: Lu and Wang [42]/with permission of Elsevier.

yield the desired product **10.36c**. Alternatively, the carbocation intermediate **10.36e** can rearrange to a cyclopropane **10.36g**, which further undergoes TMS elimination to produce the final product **10.36c** (Scheme 10.36).

The incorporation of SCF_3 group into organic molecules enhances bioavailability, transmembrane permeation, and metabolic stability due to the strong electron-withdrawing effect and lipophilicity of SCF_3. Hence, Lu and Wang successfully introduced the SCF_3 group into diazoimines **10.37a** by employing $AgSCF_3$ in the presence of copper catalyst, resulting in the formation of 3-((trifluoromethyl)thio)-2-aminoindoles **10.37b** [44]. Mechanistically, reactive $CuSCF_3$ species, formed from $AgSCF_3$ and CuI, react with the diazo compound **10.37a**, resulting in copper carbene **10.37c** along with nitrogen gas extrusion. Migratory insertion to **10.37d** followed by protonation/aromatization affords the product **10.37b** (Scheme 10.37).

Meanwhile, the same group exemplified gold-catalyzed synthesis of 2-iminoindolin-3-ones **10.38c** from **10.38a** and nitrones **10.38b** [45]. This transformation involves a sequence of reactions, such as oxygen transfer and imine exchange, facilitated by the donation of oxygen and aniline by the nitrones.

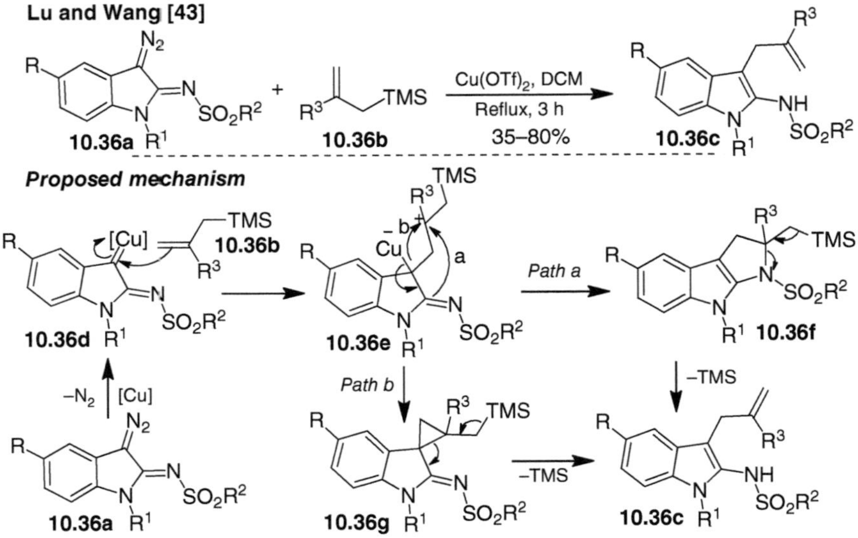

Scheme 10.36 Reaction of diazoimines with allyltrimethylsilanes. Source: Lu and Wang [43]/with permission of Elsevier.

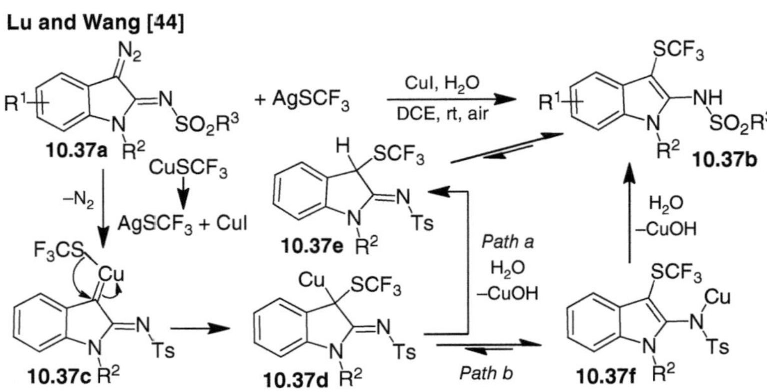

Scheme 10.37 Synthesis of 3-((trifluoromethyl)thio)-2-aminoindoles. Source: Adapted from Lu and Wang [44].

Furthermore, 2-iminoindolin-3-ones **10.38c** were converted into 2-alkynyl-2,3-dihydroquinazolin-4(1*H*)-one **10.38d** by subjecting with silver acetylide, which is *in situ* generated from silver catalyst and terminal alkyne. This transformation occurred due to methanol-initiated ring expansion of 2-iminoindolin-3-one **10.38c**, followed by intramolecular nucleophilic substitution, yielded iminium-type intermediate **10.38k**. The nucleophilic addition of alkynyl silver on to iminium-type intermediate **10.38k** furnished **10.38d** (Scheme 10.38).

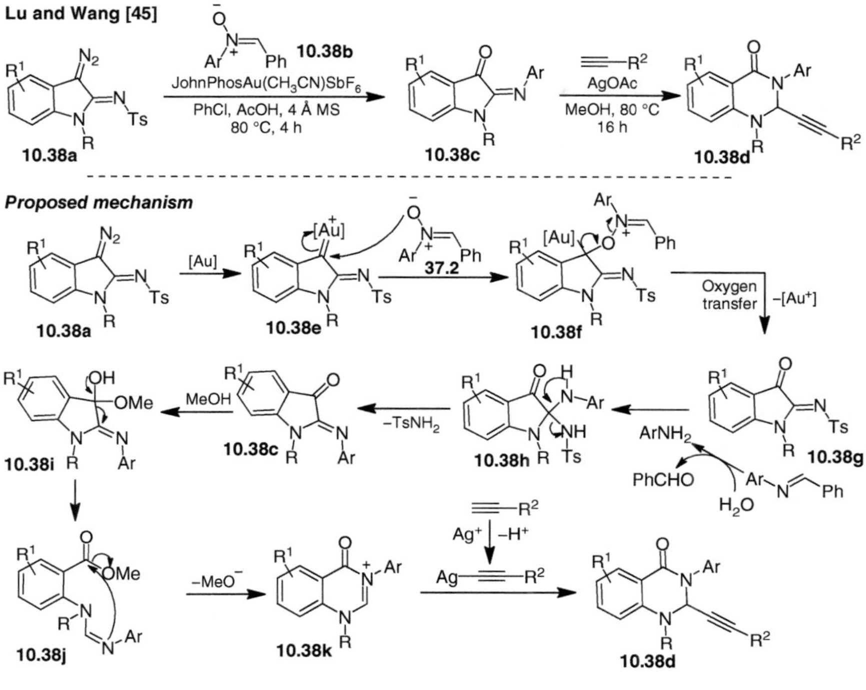

Scheme 10.38 Gold catalyzed synthesis of 2-iminoindolin-3-ones and their conversion. Source: Adapted from Lu and Wang [45].

Wu and Chen disclosed the synthesis of 2,3-diiminoindolines **10.39c** through copper-catalyzed reaction between 3-diazoindolin-2-imines **10.39a** and benzo[*c*] isoxazoles **10.39b** [46]. The proposed mechanism for this transformation involves the initial formation of copper carbene intermediate **10.39d** followed by nucleophilic addition of **10.39b** to generate ylide **10.39e**. Inherent aromatization of the ylide **10.39e** produces zwitter ion **10.39f**, subsequent ring opening results in **10.39c**. Replacing the benzo[*c*]isoxazole with 5-methylisoxazole under the reaction conditions did not afford the corresponding product, demonstrating that the aromatization step is crucial for this transformation. Additionally, Pd-catalyzed hydrogenation of 2,3-diiminoindoles enables an efficient method for the synthesis of 2,3-diaminoindoles (Scheme 10.39).

Lu and Wang have extensively explored the reactivity of 3-diazoindolin-2-imines. During their study, they discovered a novel reactivity of 3-diazoindolin-2-imines with *in situ* generated diazo compound in the presence of copper catalyst [47]. Notably, they accomplished copper-catalyzed stereoselective olefination of **10.40a** with aryl hydrazones **10.40b**. In this transformation, copper played a dual role in oxidizing hydrazones to diazo compounds **10.40f** and cross coupling of two different diazo compounds (Scheme 10.40).

Bao and co-workers reported rhodium-catalyzed synthesis of 2-iminoindolin-nitrones **10.41c** from diazoimines **10.41a** and nitrosoarenes **10.41b** [48]. Mechanistically, the rhodium carbene, generated from **10.41a** and rhodium catalyst,

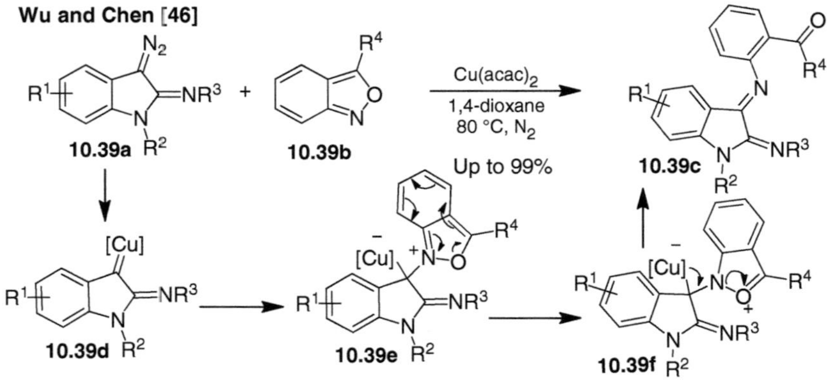

Scheme 10.39 Synthesis of 2,3-diiminoindoles from diazoimines and benzo[c]isoxazoles. Source: Wu and Chen [46]/with permission of Elsevier.

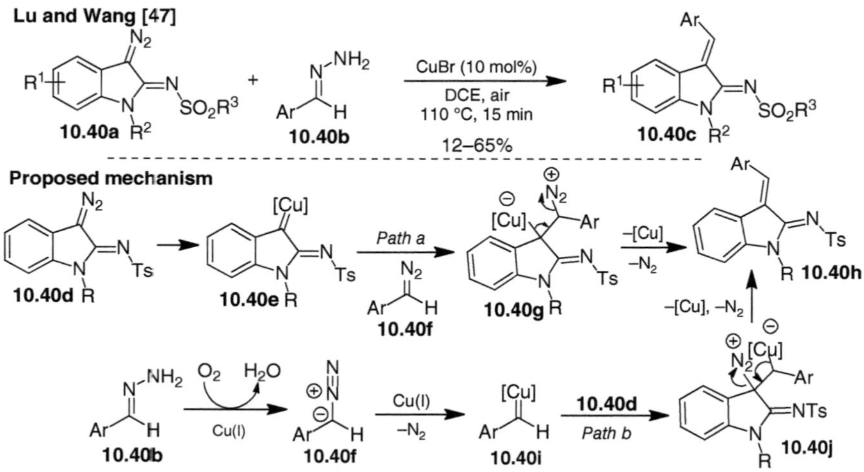

Scheme 10.40 Stereoselective synthesis of (*E*)-3-arylideneindolin-2-imines. Source: Lu and Wang [47]/American Chemical Society.

was attacked by the nitrogen atom of **10.41b** due to the higher nucleophilicity of the nitrogen over the oxygen atom. DFT calculations were employed to elucidate the chemoselectivity of nitrosoarenes **10.41b** with **10.41a** under rhodium catalyst (Scheme 10.41).

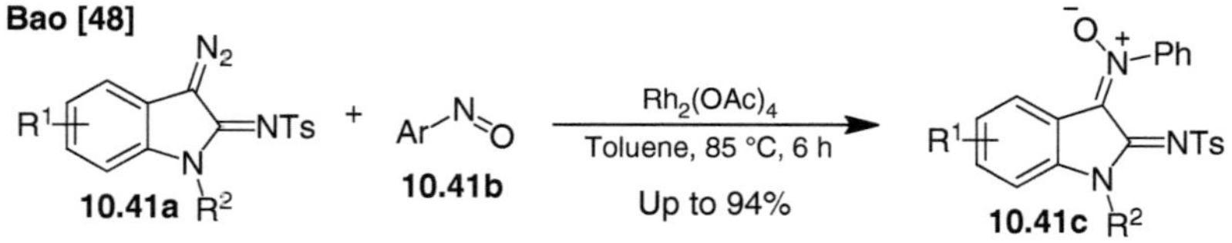

Scheme 10.41 Rhodium catalyzed transformation of diazoimines with nitrosoarenes. Source: Bao [48]/John Wiley & Sons.

10.6 Summary and Outlook

The development of novel organic transformations to synthesize valuable organic molecules is always in constant demand. As a result, acceptor-donor metal carbenoids have gained attention due to their capability to undergo a sequence of cascade reactions with chemoselectivity to generate complex structural motifs. *N*-sulfonyl-1,2,3-triazoles have emerged as efficient surrogates for donor–acceptor metal carbenoids and paved the way for groundbreaking discoveries in the field of organic chemistry. The indole-fused analog, viz. triazoloindole, which exists as 3-diazoindolin-2-imine, generates α-imino metal carbenoids in the presence of various metal catalysts. These carbenoids undergo a series of transformations, including insertion reactions, annulations, cycloaddition, and rearrangement, to yield valuable compounds embedded with the indole structure. This chapter provided a comprehensive account of the documented acid and transition metal-catalyzed denitrogenative transformations of 3-diazoindolin-2-imines, as developed by several research groups. Most of these transformations are rather low-lying fruits and the in-depth potential is yet to be disclosed. Furthermore, the application of 3-diazoindolin-2-imine in asymmetric synthesis has not been studied and is expected in the near future.

References

1 Singh, T.P., and Singh, O.M. (2018). Recent progress in biological activities of indole and indole alkaloids. *Mini-Rev. Med. Chem.* 18: 9–25.

2 Davies, H.M.L., and Denton, J.R. (2009). Application of donor/acceptor-carbenoids to the synthesis of natural products. *Chem. Soc. Rev.* 38: 3061–3071.

3 Akter, M., Rupa, K., and Anbarasan, P. (2022). 1,2,3-Triazole and its analogues: new surrogates for diazo compounds. *Chem. Rev.* 122: 13108–13205.

4 Xing, Y., Sheng, G., Wang, J. et al. (2014). Preparation of triazoloindoles via tandem copper catalysis and their utility as α-imino rhodium carbene precursors. *Org. Lett.* 16: 1244–1247.

5 Sheng, G., Huang, K., Chi, Z. et al. (2014). Preparation of 3-diazoindolin-2-imines via cascade reaction between indoles and sulfonylazides and their extensions to 2,3-diaminoindoles and imidazo[4,5-*b*]indoles. *Org. Lett.* 16: 5096–5099.

6 Sheng, G., Ma, S., Bai, S. et al. (2018). Rhodium-catalyzed reactions of 3-diazoindolin-2-imines with enamines and their extensions towards 5*H*-pyrazino [2,3-*b*] indoles. *Tetrahedron* 74: 2151–2157.

7 Gao, K., Kou, L., Fu, R., and Bao, X. (2020). Synthesis of 2-(3-arylallylidene)-3-oxindoles via dirhodium(II)-catalyzed reaction of 3-diazoindolin-2-imines with 1-aryl-substituted allylic alcohols and computational insights. *Adv. Synth. Catal.* 362: 1292–1297.

8 Wang, L., Wu, Y., Liu, Y. et al. (2017). Copper-catalyzed P—H insertions of α-imino carbenes for the preparation of 3-phosphinoylindoles. *Org. Lett.* 19: 782–785.

9 Cai, M., Liu, K., Kowah, J.A.H. et al. (2021). Mechanism of P—H insertion of α-imino copper carbenes: 1,1-Insertion or 1,3-insertion? *Chem. Phys. Lett.* 785: 139126.

10 Mangali, R., and Muthusamy, S. (2022). BF3·OEt2 catalyzed S—H insertion reactions of α-diazo imidamides and enolizable thioamides under metal-free conditions. *New J. Chem.* 46: 23173–23178.

11 Du, Z., Xing, Y., Lu, P., and Wang, Y. (2015). Copper-catalyzed cascade double C3-indolations of 3-diazoindolin-2-imines with indoles: Convenient access to 3,3-diaryl-2-iminoindoles. *Org. Lett.* 17: 1192–1195.

12 Li, Z., Zhou, X., Lu, P., and Wang, Y. (2016). Preparation of 3-aryl-2-aminoindoles via rhodium-catalyzed coupling reaction between 2-arylpyridines and 3-diazoindolin-2-imines. *J. Org. Chem.* 81: 9433–9437.

13 Li, Z., Xie, J., Wu, L. et al. (2020). Co(III)-catalyzed reaction between 3-diazoindolin-2-imines and 1-pyrimidinylindoles for the synthesis of 2,3′-biindoles. *Tetrahedron* 76: 131371.

14 Qi, J., Wang, Z., Lang, B. et al. (2017). TfOH-catalyzed reaction between 3-diazoindolin-2-imines and electron-rich arenes: access to 3-aryl-2-aminoindoles. *J. Org. Chem.* 82: 12640–12646.

15 Huang, Z., Lin, J., Ma, J. et al. (2022). Borane-catalyzed, HFIP-assisted carbene insertion into internal alkenyl C—H bonds under metal-free conditions. *Adv. Synth. Catal.* 364: 4333–4346.

16 Ma, F., Qian, J., Lu, P., and Wang, Y. (2018). Convenient synthesis of 2-amino-3-(arylthio)indoles via the Rh-catalyzed reaction of 3-diazoindol-2-imines with thioesters. *Org. Biomol. Chem.* 16: 439–443.

17 Wu, L., Li, Z., Lu, P., and Wang, Y. (2018). Copper-catalyzed syntheses of 3-allyl-3-arylthioindolin-2-imines and 3-allenyl-3-arylthioindolin-2-imines from 3-diazoindolin-2-imines. *J. Org. Chem.* 83: 13956–13964.

18 Sheng, G., Huang, K., Ma, S. et al. (2015). Preparation of 3-aryl-2-aminoindoles, 3-allyl-3-amino-2-iminoindolines, and tetrahydro-[1,4]diazepino[2,3-*b*]indoles from 3-diazoindolin-2-imines. *Chem. Commun.* 51: 11056–11059.

19 Zhou, Y., Li, Z., Ma, F. et al. (2019). Palladium-catalyzed synthesis of 3-haloindol-2-amines from 3-diazoindolin-2-imines and alkyl halides. *J. Org. Chem.* 84: 6655–6668.

20 Wang, C., Zhang, H., Lang, B. et al. (2015). Rh-catalyzed reactions of 3-diazoindolin-2-imines: synthesis of pyridoindoles and tetrahydrofuropyrroloindoles. *Org. Lett.* 17: 4412–4415.

21 Wang, H., Li, L., Yu, S. et al. (2016). Rh(III)-catalyzed C—C/C—N coupling of imidates with α-diazo imidamide: synthesis of isoquinoline-fused indoles. *Org. Lett.* 18: 2914–2917.

22 Pal, K., Adak, S., Dey, A., and Volla, C.M.R. (2020). Rh(III)-catalyzed denitrogenative [4+2] annulation of benzamides and 3-diazoindolin-2-imines: expedient access to indolo[2,3-c] isoquinolin-5-ones. *Chem. Asian J.* 15: 1052–1056.

23 Ko, G.H., Son, J.Y., Kim, H. et al. (2017). Synthesis of indolo-1,2-benzothiazines from sulfoximines and 3-diazoindolin-2-imines. *Adv. Synth. Catal.* 359: 3362–3370.

24 Mangli, R., and Muthusamy, S. (2023). Acid-catalyzed one pot synthesis of thiazolo [4,5-*b*]indolamines from α-diazoimidamides and thioureas. *Asian J. Org. Chem.* 12: e202300142.

25 Ding, H., Wang, Z., Bai, S. et al. (2017). Rh-catalyzed conversion of 3-diazoindolin-2-imines to 5*H*-pyrazino[2,3-*b*]indoles with photoluminescent properties. *Org. Lett.* 19: 6514–6517.

26 Baek, Y., Maeng, C., Kim, H., and Lee, P.H. (2018). Regioselective synthesis of indolopyrazines through a sequential rhodium-catalyzed formal [3 + 3] cycloaddition and aromatization reaction of diazoindolinimines with azirines. *J. Org. Chem.* 83: 2349–2360.

27 Ruvinskaya, J.O., Rostovskii, N.V., Filippov, I.P. et al. (2018). A novel approach to 5*H*-pyrazino[2,3-*b*]indoles via annulation of 3-diazoindolin-2-imines with 2*H*-Azirines or 5-alkoxyisoxazoles under Rh(II) catalysis. *Org. Biomol. Chem.* 16: 38–42.

28 Khaidarov, A.R., Rostovskii, N.V., Zolotarev, A.A. et al. (2019). Synthesis of 1-(2-aminovinyl)indoles and 1,3′-biindoles by reaction of 2,2-diaryl-substituted 2*H*-azirines with α-imino Rh(II) carbenoids. *J. Org. Chem.* 84: 3743–3753.

29 Qian, J., Lin, Z., Wang, Z. et al. (2019). Copper-carbene-triggered electrophilic cyclization of *o*-hydroxyarylenaminones with 3-diazoindolin-2-imines: synthesis of 3-indolyl-4*H*-chromen-4-ones and pyrido[2,3-b:6,5-b′]diindoles. *J. Org. Chem.* 84: 6395–6404.

30 Lang, B., Zhu, H., Wang, C. et al. (2017). Rhodium-catalyzed cycloadditions between 3 – diazoindolin-2-imines and 1,3-dienes. *Org. Lett.* 19: 1630–1633.

31 Kim, S., Kim, H., Um, K., and Lee, P.H. (2017). Synthesis of azepinoindoles via rhodium-catalyzed formal aza-[4 + 3] cycloaddition reaction of 3-diazoindolin-2-imines with 1,3-dienes in one-pot. *J. Org. Chem.* 82: 9808–9815.

32 Han, C., Wu, W., Chen, Z., and Pu, S. (2019). Rhodium-catalyzed [5 + 1]-cycloaddition reactions to spiro-benzo[e][1,3]oxazineindoline imines. *Asian J. Org. Chem.* 8: 1385–1389.

33 Sheng, G., Li, Z., Mao, J. et al. (2019). General approach to construct azepino [2,3-*b*:4,5-*b*′]diindoles, azocino[2,3-*b*:4,5-*b*′]diindoles, and azonino[2,3-*b*:4,5-*b*′]diindoles via Rh(II)-catalyzed reactions of 3-diazoindolin-2-imines with 3-(bromoalkyl)indoles. *J. Org. Chem.* 84: 9561–9569.

34 Suleman, M., Li, Z., Lu, P., and Wang, Y. (2019). Copper-catalyzed dimerization of sulfoxonium ylides with 3-diazoindolin-2-imines: a practical and efficient approach to spiro[cyclopropane-1,3′-indolin]-2′-imines. *Eur. J. Org. Chem.* 2019: 4447–4456.

35 Zhou, Y., Ma, F., Lu, P., and Wang, Y. (2019). Preparation of spiro[imidazolidine-4,3′-indolin]-2′-imines via copper(I)-catalyzed formal [2 + 2 + 1] cycloaddition of 3-diazoindolin-2-imines and triazines. *Org. Biomol. Chem.* 17: 8849–8852.

36 Huang, Z., He, Y., Wang, L. et al. (2022). Copper-catalyzed [4 + 1] annulation of enaminothiones with indoline-based diazo compounds. *J. Org. Chem.* 87: 4424–4437.

37 Sheng, G., Ma, S., Bai, S. et al. (2018). A copper-catalyzed reaction of 3-diazoindolin-2-imines with 2-(phenylamino)ethanols: convenient access to spiro[indoline-3,2′-oxazolidin]-2-imines. *Chem. Commun.* 54: 1529–1532.

38 Rupa, K., and Anbarasan, P. (2023). Rhodium catalyzed [4 + 1]-annulation of o-acylanilines with 3-diazoindoline-2-imines. *Org. Lett.* 25: 6357–6362.

39 Valapil, D.G., Mishra, P., Jungare, K., and Shankaraiah, N. (2023). Ru(II)-catalyzed synthesis of indolo[2,3-*c*]isoquinolines via [3 + 3] annulation of *N*,*N*′-cyclic azomethine ylides and 3-diazoindolin-2-imines. *New J. Chem.* 47: 17586–17591.

40 Qian, J., Sheng, G., Huang, K. et al. (2016). α-Amidino rhodium carbenes: key intermediates for the preparation of (*E*)-2-aminomethylene-3-oxoindoles and pyranoindoles. *Org. Lett.* 18: 3682–3685.

41 Ding, H., Bai, S., Lu, P., and Wang, Y. (2017). Preparation of 2-amino-3-arylindoles via Pd-catalyzed coupling between 3-diazoindolin-2-imines and arylboronic acids as well as their extension to 3-aryl-3-fluoroindolin-2-imines. *Org. Lett.* 19: 4604–4607.

42 Huang, K., Lu, P., and Wang, Y. (2019). BF3-promoted reactions between aryl aldehydes and 3-diazoindolin-2-imines: access to 2-amino-3-arylindoles. *Tetrahedron* 75: 3779–3787.

43 Lang, B., Zhou, Y., Lu, P., and Wang, Y. (2019). Copper-catalyzed synthesis of 3-allyl-2-aminoindoles from 3-diazoindolin-2-imines and allyltrimethylsilane. *Tetrahedron* 75: 1597–1604.

44 Lang, B., Suleman, M., Lu, P., and Wang, Y. (2020). Copper(I)-promoted trifluoromethylthiolation of 3-diazoindolin-2-imines with AgSCF3: synthesis of 3-((trifluoromethyl)thio)-2-aminoindoles. *Chin. J. Org. Chem.* 40: 3300–3306.

45 Lin, Z., Qian, J., Lu, P., and Wang, Y. (2020). Syntheses of 2-iminoindolin-3-ones and 2-alkynyl-2,3-dihydroquinazolin-4(1*H*)-ones from 3-diazoindolin-2-imines. *J. Org. Chem.* 85: 11766–11777.

46 Wu, W., and Chen, Z. (2020). Synthesis of 2,3-diiminoindolines and 2,3-diaminoindoles via copper-catalyzed donor–acceptor metallo carbenoid formation and hydrogenation reactions. *Tetrahedron Lett.* 61: 152314.

47 Xie, J., Wu, L., Lu, P., and Wang, Y. (2022). Olefination of 3-diazoindolin-2-imines with hydrazones: an approach toward stereoselective synthesis of (*E*)-3-arylideneindolin-2-imines. *J. Org. Chem.* 87: 10664–10672.

48 Fu, R., Kou, L., Gao, K. et al. (2021). Understanding mechanistic differences between 3-diazoindolin-2-imines and *N*-sulfonyl-1,2,3-triazoles in the Rh2(II)-catalyzed reactions with nitrosoarenes. *Chin. J. Chem.* 39: 1565–1572.

11

Metal-Catalyzed Denitrogenative Transformations of Pyridotriazoles

Deepa Rawat[1,2], Abhisek Joshi[1,2], and Subbarayappa Adimurthy[1]

[1]*Academy of Scientific & Innovative Research (AcSIR), Natural Products & Green Chemistry Division, Ghaziabad, Uttar Pradesh, India*
[2]*CSIR-Central Salt & Marine Chemicals Research Institute, Natural Products & Green Chemistry Division, G. B. Marg, Bhavnagar, Gujarat, India*

11.1 Introduction

In the past decade, much progress has been made in the field of ring-opening reactions of [1,2,3]triazolo[1,5-*a*]pyridines (pyridotriazoles), which involves the carbene surrogates catalyzed by transition metals in the construction of new carbon–carbon and carbon-heteroatom bonds. It was observed that transition metal complexes are effective catalysts for the generation of metal carbenoid intermediate with diazo compounds, which allows it to react as "electrophiles" [1]. It has been demonstrated that pyridotriazoles have the ability to exist in equilibrium with their acyclic diazo form, which, in the presence of metal catalyst, can generate a reactive metal-carbenoid intermediate (Figure 11.1). The catalytic transformations involving such metal carbenes, generated directly from readily available pyridotriazoles have attracted enormous attention as they avoid the use of hazardous diazo compounds as starting substrates for carbene generation [2].

In view of the advances in transition-metal-catalyzed ring-opening of pyridotriazole to construct transannulated N-heterocycles, Rh(II) catalysis has emerged as a versatile and powerful approach in the past few decades [3]. Pioneered works on the rhodium catalysis involve rhodium_carbene intermediates to access complex valuable cyclized products [4, 5]. The uniqueness of rhodium metal species is it can undergo five- [6], six- [7], and seven-membered rhodacycle intermediates [8] and can lead to corresponding five-, six-, and seven-membered ring products. However, eight-membered is quite challenging because eight-membered ring or its intermediates are energetically unstable. Utilizing this tactic to construct seven-membered rings is elusive to achieve higher-membered complex ring

Denitrogenative Transformation of Nitrogen Heterocycles: Synthesis, Reactions and Applications, First Edition. Edited by Pazhamalai Anbarasan.

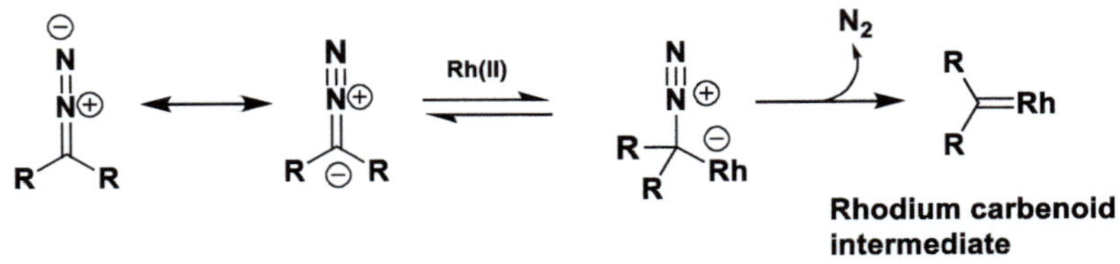

Figure 11.1 Formation of rhodium carbenoid intermediate.

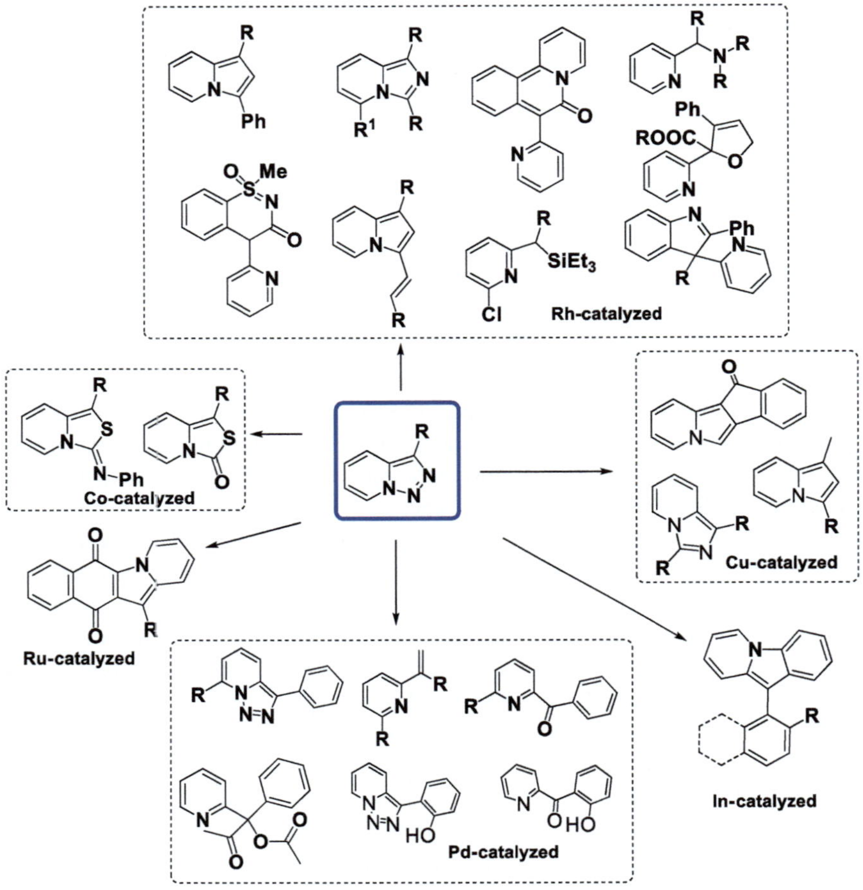

Figure 11.2 Metal-catalyzed transannulation of pyridotriazoles.

systems [9]. In addition to rhodium, other metals such as palladium-, copper-, cobalt-, ruthenium-, and indium-catalyzed transannulation of pyridotriazoles with various annulation partners for the synthesis of diverse heterocycles are discussed (Figure 11.2). The transannulation of pyridotriazoles reports covered in this chapter up to the year 2022 provides the readers with insights into these particular transformations.

11.2 Rhodium-Catalyzed Transannulation of Pyridotriazoles

11.2.1 Rhodium-Catalyzed Transannulation of Pyridotriazoles with Alkyl Silane, Alkynes, and Nitriles

In 1986, Helquist's group employed a methodology that involved rhodium carbenoid intermediate for the construction of five-membered oxygen-containing heterocycles. Using this methodology of annulation of diazocarbonyl compounds with nitriles, a series of oxazoles have been prepared [10]. Padwa et al. [11] have employed the same method for the synthesis of furans from diazo compounds. In 1988 Doyle and co-workers reported the remarkable method for Si—H insertion into diazo compounds through metal carbenoid intermediates. By this approach, they have synthesized different silyl esters and ketones [12] (Figure 11.3).

In 2007, Govergyan's group [13] first confirmed the ring-opening of pyridotriazole and its rhodium carbenoid species, followed by the insertion of Si-H bond from triethyl silane **11.1d** (Scheme 11.1). They used Doyle and co-worker's method for the efficient trapping of Rh carbenoids. Pyridotriazoles **11.1a** having 7-halo substituent undergo extrusion of dinitrogen from the intermediate **11.1b** and generate rhodium carbene complex **11.1c** in the presence of Rh catalyst.

Similarly, to examine this speculation for the annulation of imino diazo compounds with alkynes to form a pyrrole ring, the same group explored [2+1] and [2+3] cycloaddition reactions in the presence of $Rh_2(OAc)_4$ with alkynes and surprisingly they observed the transannulation of pyridotriazole **11.2a** with phenylacetylene **11.2b** and obtained a mixture of cyclopropane **11.2c** and indolizine **11.2d** (Scheme 11.2).

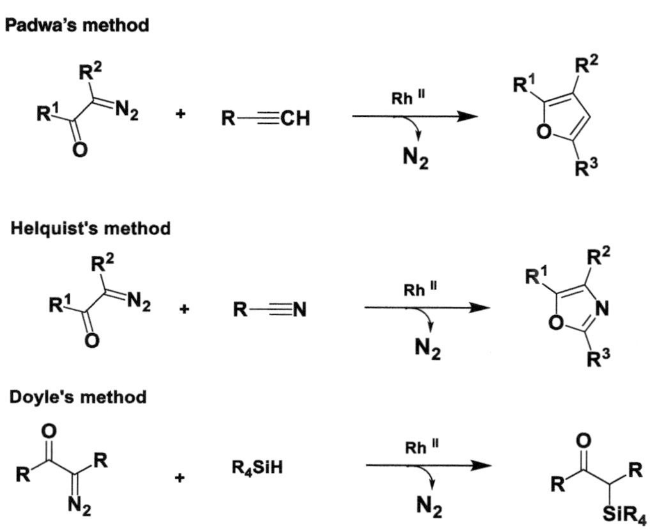

Figure 11.3 Rh-catalyzed annulation and insertion reaction of diazo compounds.

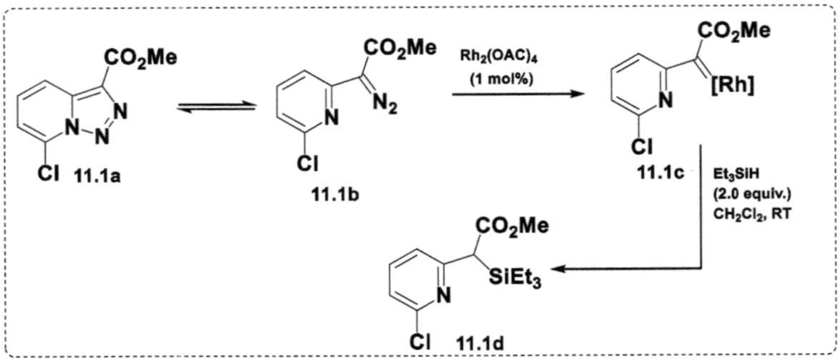

Scheme 11.1 First metallocarbene intermediate by denitrogenation of pyridotriazole.

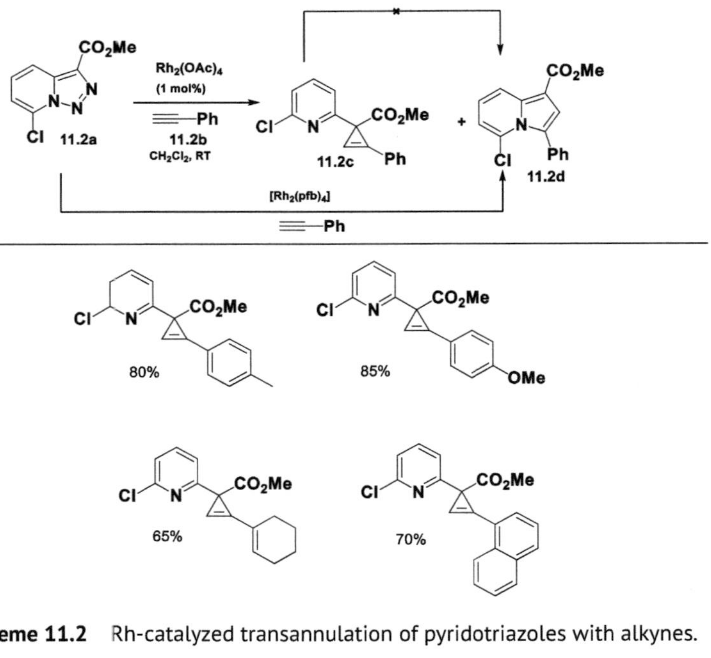

Scheme 11.2 Rh-catalyzed transannulation of pyridotriazoles with alkynes.

Interestingly, the cyclopropane ring does not undergo ring expansion or isomerization to form indolizine. The ratio of the aforementioned products constantly remained throughout the course of the reaction. Later, they improved the yield of indolizine **11.2d** by taking rhodium(II) heptafluorobutyrate as a catalyst. Thus, transannulation of pyridotriazoles with a series of aryl and alkenyl alkynes proceeded with high chemo selectively to produce the indolizines in good yields. All aryl alkynes, such as electron-rich, electron-deficient, and sterically hindered, were equally effective in the reaction. Inspired by the successful formation of a pyrrole ring, which has a nitrogen atom on the bridge, the transannulation of pyridotriazoles with alkynes, they examined the N-fused imidazole ring formation

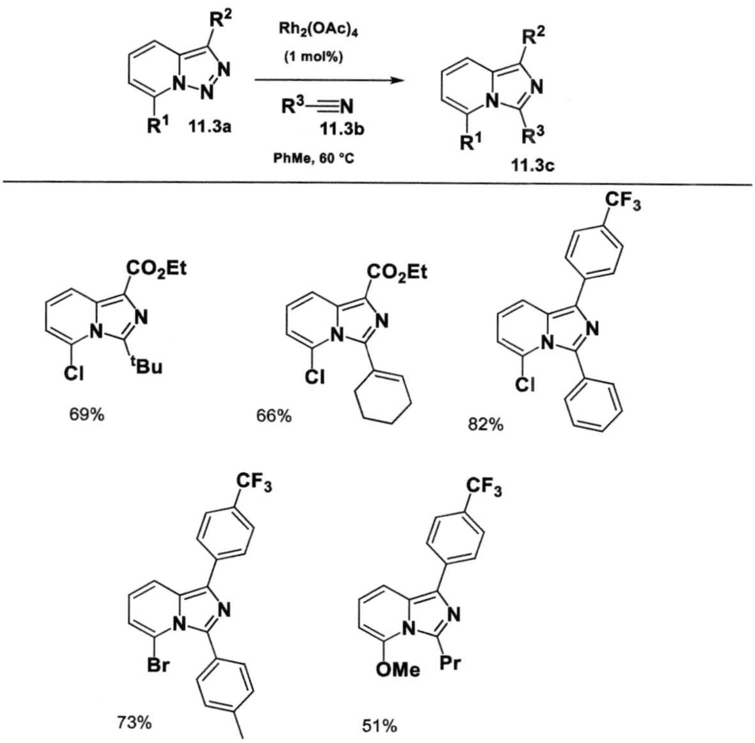

Scheme 11.3 Rh-catalyzed transannulation of pyridotriazoles with nitriles.

by the reaction of pyridotriazoles **11.3a** with nitriles **11.3b** and obtained N-fused imidazopyridines **11.3c** (Scheme 11.3).

Both transformations proceeded the same pathway through *in situ* generation of rhodium carbenoid species in the presence of rhodium catalyst. For transannulation via nitriles, initially nucleophilic attacks on nitrile at the electrophile center of metal carbenoid **B** via either path *a* or path *b*. In path *a*, zwitterion intermediate **C & D** formation was observed, while by path *b*, simple [2 + 2] cycloaddition takes place to give the intermediate **E,** which further undergoes rearrangement to deliver another intermediate **F & G**, which yields the desired product **11.3c** (Scheme 11.4).

In the same year, Gevorgyan and his co-workers reported cyclopropanation of pyridotriazoles **11.5a** with phenylacetylene **11.5b** via [3 + 2] cycloaddition [14] using Rh$_2$(S-DOSP)$_4$ (Tetrakis[(*S*)-(−)-*N*-(*p*-dodecylphenylsulfonyl)prolinato]dirhodium (II)) catalyst. This was a highly efficient synthesis of 1,3- and 1,2-disubstituted N-fused pyrrolohetero-cycles, including indolizines, pyrrolooxazole, and pyrroloisoquinoline, via a novel regiodivergent transition-metal-catalyzed rearrangement of 3-iminocyclopropenes (Scheme 11.5).

They proposed a mechanistic pathway for the novel regiodivergent rearrangement of imino cyclopropenes **6** into fused pyrroloheterocycles **H** and **I**. According to them, the first cyclopropene **A**, undergoes ring-opening in the presence of Rh(I)

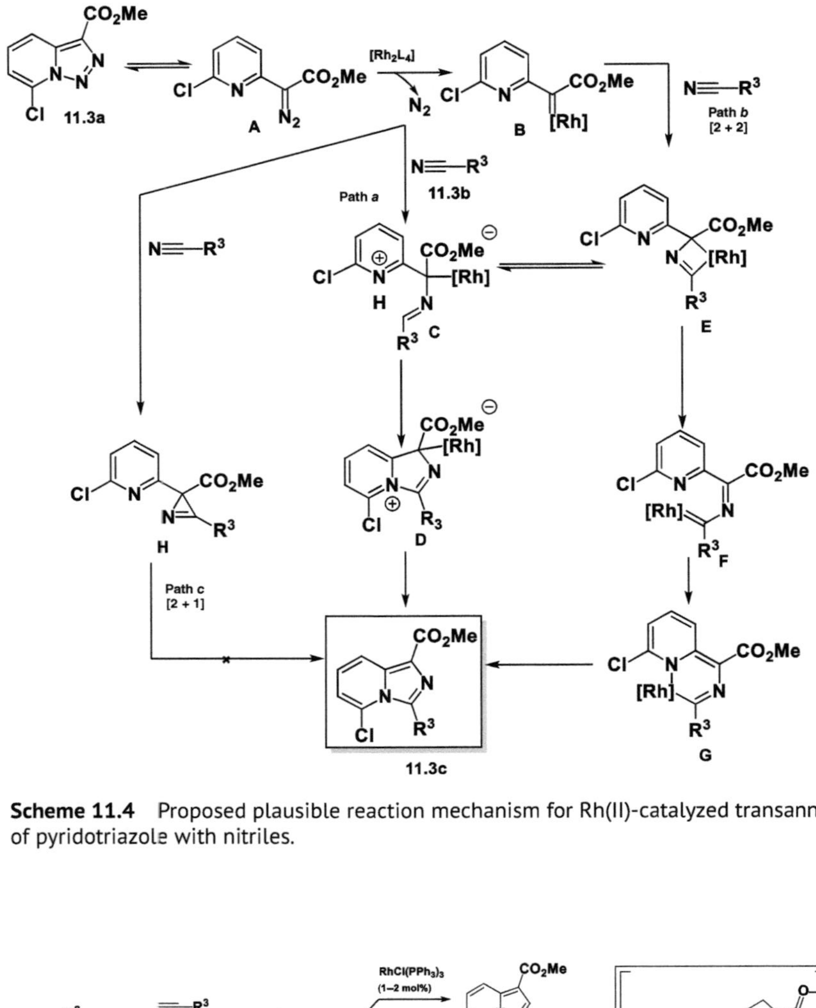

Scheme 11.4 Proposed plausible reaction mechanism for Rh(II)-catalyzed transannulation of pyridotriazole with nitriles.

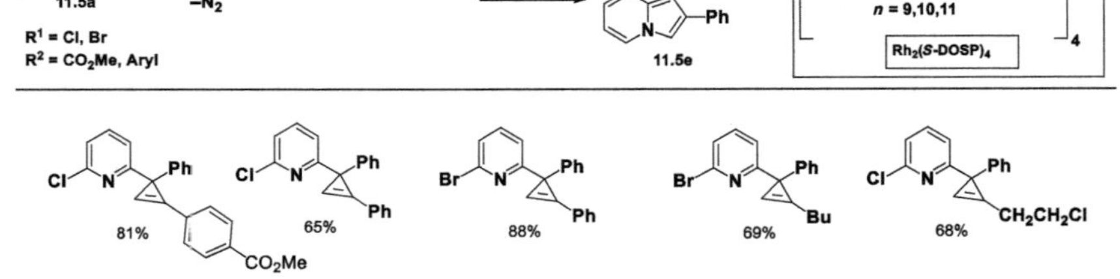

Scheme 11.5 Rh-catalyzed rearrangement of 3-iminocyclopropenes into N-fused pyrroles.

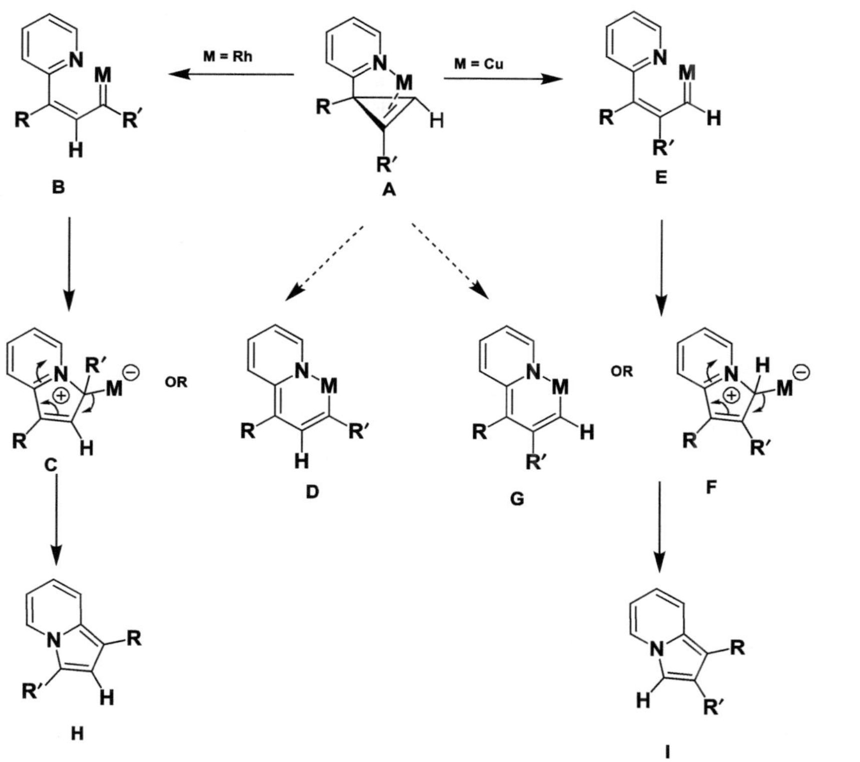

Scheme 11.6 Mechanistic rationale for regiodivergent rearrangement reaction of imino cyclopropenes **A** into fused pyrroloheterocycles.

complex, to produce the most substituted carbenoid derivatives **B**. A nucleophilic attack by nitrogen lone pair on carbenoid center leads to the formation of zwitterion **C**. A subsequent elimination of the metal furnishes regioisomer **D**. In contrast, when Cu(I) catalyst is used, the formation of less-substituted carbenoid **E** occurs. Its cyclization through a zwitterion **F** leads to the product **G** selectively. Alternatively, regioisomers **D** and **G** may arise via a reductive elimination of *aza* metallacycles respectively, which in turn are formed via a 6π-electrocyclization of carbenoids **C** and **F**, or directly from cyclopropane **A**, upon regioselective oxidative addition of the metal (Scheme 11.6).

11.2.2 Rhodium-Catalyzed Transannulation of Pyridotriazoles with Amine and Amides

In continuation of Gevorgyan's group work on applications of diazo-compounds for the synthesis of nitrogen-containing heterocycles, in 2014 they envisioned the reaction of pyridotriazoles with primary amides as a potential route, and synthesized a series of imidazo[1,5-*a*] pyridines (Scheme 11.7) [15]. The methodology involves N—H insertion reactions of pyridotriazoles **11.7a** employing rhodium as a catalyst.

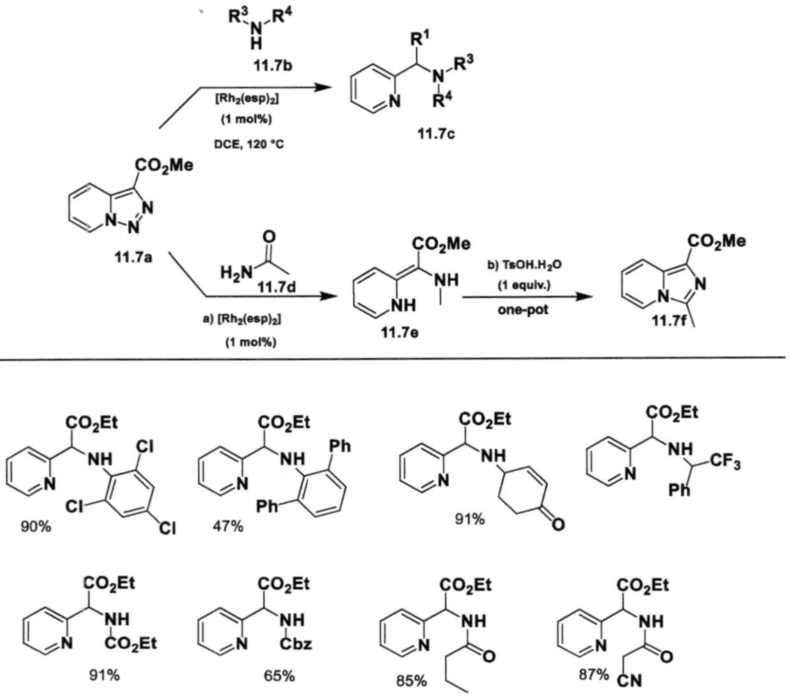

Scheme 11.7 Rhodium-catalyzed NH insertion of pyridyl carbenes.

Different anilines, amines, and enamine as nucleophile were tested using the same methodology, by which various 2-picolylamine **11.7c** derivatives were synthesized. Meanwhile, they also developed a formal one-pot transannulation of pyridotriazoles with primary amides **11.7d**, which proceeds by the rhodium-catalyzed N—H insertion reaction and synthesized imidazo[1,5-*a*]pyridines **11.7f**. The reaction worked efficiently with various alkyl and aryl amides, with both substituted and unsubstituted pyridotriazoles. Notably, the method with amides has a much broader scope compared to the previously developed transannulation reaction of pyridotriazoles with nitriles. Here, the activating group at C-7 position on pyridotriazole is not necessary, and the substituent at C3 is not limited to an electron-withdrawing group. They proposed that transformation with N—H insertion reaction of pyridotriazoles proceeds through the ylide mechanism [16].

11.2.3 Rhodium-Catalyzed [4 + 3] Cycloaddition of Pyridotriazoles with 2-Phenylpyridine and *S*-Aryl Sulfoximines

In 2015, Strassert and Glorius reported the C—H activation of 2-phenylpyridine **11.8b** through ring-opening of pyridotriazoles **11.8a** and employing Rh(III)-as a catalyst for the synthesis of fluorescent moiety **11.8c** (Scheme 11.8). These synthesized molecules were used as a fluorophore for the detection of metal ions [17].

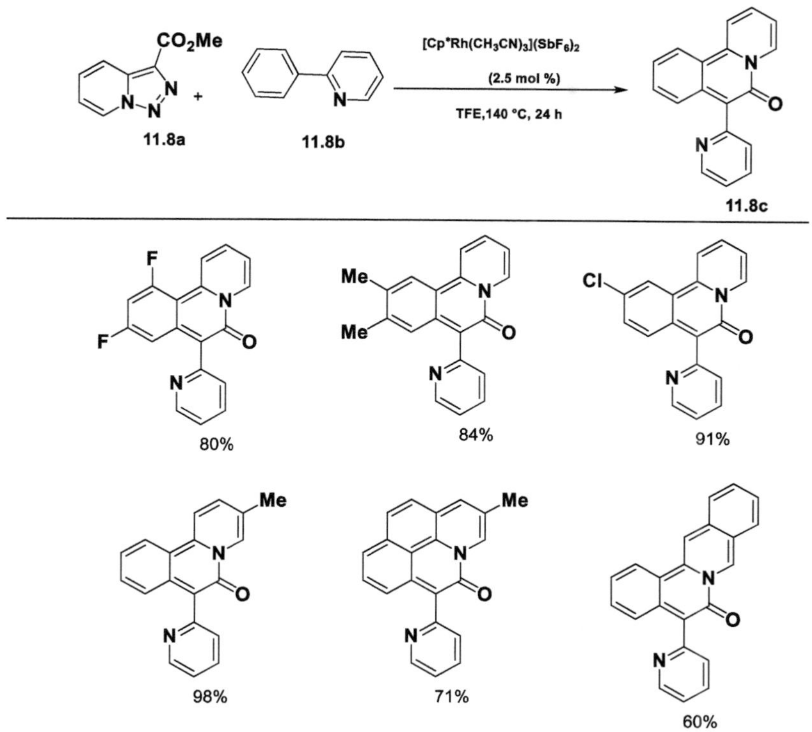

Scheme 11.8 Rh(III)-catalyzed transannulation of pyridotriazole with 2-phenylpyridine.

The 2-phenylpyridine played as a dual role both as directing group in the C—H activation step and acting as a nucleophile in cyclization step.

They proposed a mechanism for Rh(III)-catalyzed C—H insertion in which, initially, the cationic rhodium first coordinates reversibly with 2-phenylpyridine **11.8b** and undergoes a cationic rhodacycle **A** through C—H activation (Scheme 11.9). Further, it reacts with the diazo intermediate **B** generated from pyridotriazole, and with the release of N_2, it delivers carbene intermediate **C**. A pyridyl carbene undergoes migratory insertion and provide the intermediate **D**; it undergoes proto de-rhodation and generates another intermediate **E**. A nucleophilic addition of pyridine takes place on the ester-activated carbonyl group and yields the desired product **11.8c** via cyclic intermediate **F** with the subsequent elimination of methanol and by regenerating the catalyst.

P. H. Lee's group in 2016 developed a rhodium-catalyzed transannulation of pyridotriazoles with *S*-aryl sulfoximines through carbene insertions into aromatic C—H bonds [18]. It involves denitrogenative cyclization followed by the elimination of alcohols for the synthesis of a series of 1,2-benzothiazines **11.10c** bearing pyridyl as well as carbonyl groups. The method involves through N—H/C—H activation of alkyl aryl sulfoximines **11.10b** with broad substrate scope, high functional group tolerance, and good regioselectivity was observed (Scheme 11.10). All C-7 substituted and unsubstituted pyridotriazoles were well tolerated under these conditions.

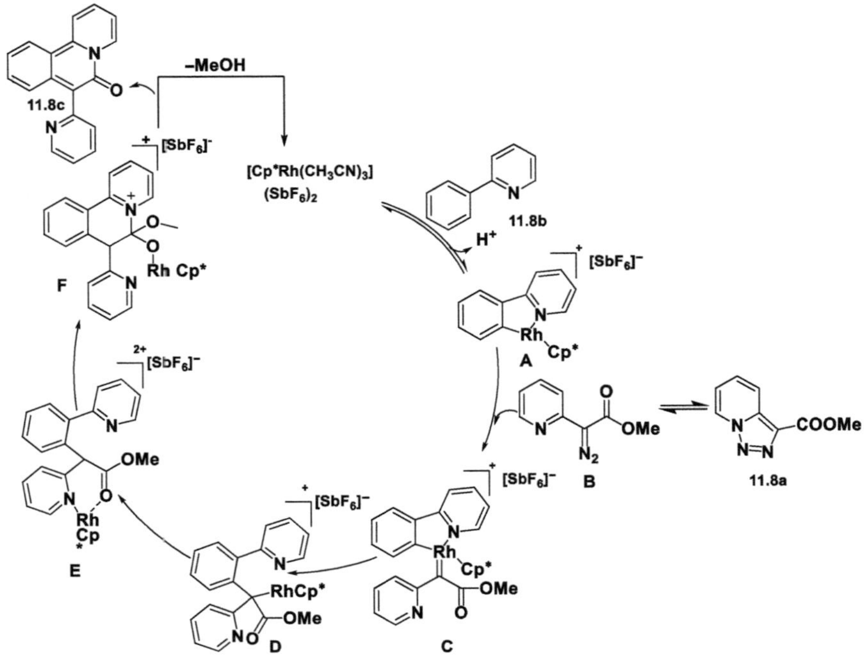

Scheme 11.9 Plausible reaction mechanism transannulation of pyridotriazole with 2-phenylpyridine.

A plausible reaction mechanism was outlined in Scheme 11.11 for the transannulation of pyridotriazole with *S*-aryl sulfoximines. Initially, the catalyst $[Cp^*Rh(MeCN)_3][(SbF_6)]_2$ in the presence of NaOAc, generates acetate-ligated cationic rhodium(III) complex species **A**. Reaction of **A** with sulfoximine **11.11b** results the intermediate **C** by deprotonation. From **C** the rhodacycle intermediate **C** will be formed via electrophilic C—H bond cleavage. On the other side, from the pyridotriazole **11.11a** diazoimine **D** generated *in situ*, which will coordinated with rhodium complex **E** to afford the rhodium carbene intermediate **F**. Migratory insertion of the carbene **F** into the rhodium–carbon bond leads rhodacyclic intermediate **G**. Addition of acid to **G**, generates intermediate **H**. Nucleophilic addition to the activated pyridyl ester and followed by elimination of alcohol give the desired transannulated product **11.11c** via the cyclic intermediate **H**.

11.2.4 Rhodium-Catalyzed Transannulation of Pyridotriazoles with 1,3-Dienes

Lee group in 2017 developed an efficient one-pot synthetic method for the synthesis of functionalized indolizine derivatives through Rh-catalyzed [2 + 1]-cyclopropanation (Scheme 11.12) [19]. The method involves Rh-catalyzed ring expansion and subsequent oxidation using MnO_2 and pyridotriazoles with 1,3-dienes. The authors observed that ring-opening of pyridotriazole with rhodium catalyst generates rhodium carbenoid intermediate, and its reaction with 1,3-dienes through [2 + 1] cycloaddition gives the cyclopropane intermediate.

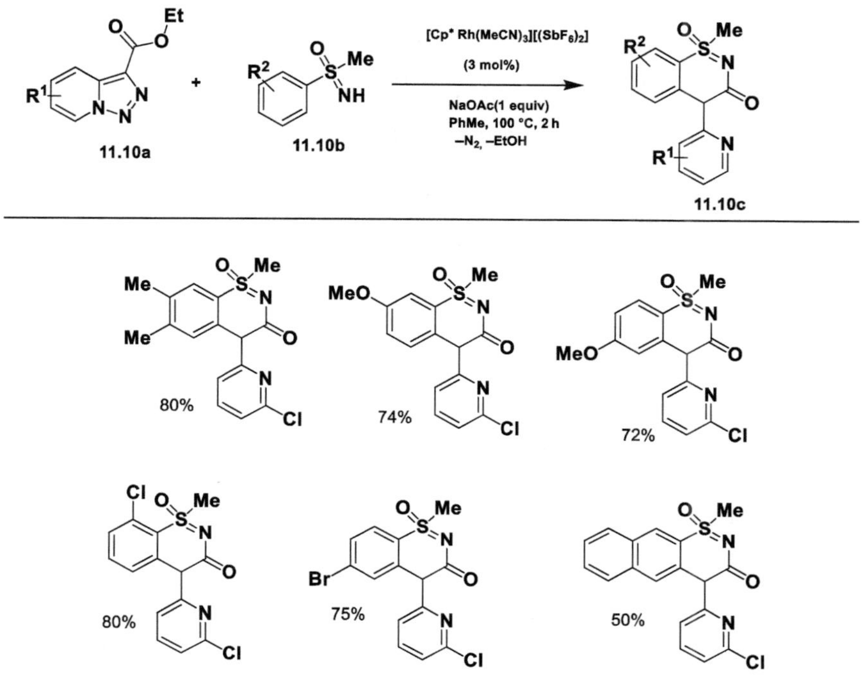

Scheme 11.10 Synthesis of 1,2-benzothiazines by from *S*-aryl Sulfoximines and Pyridotriazoles.

The cyclopropane ring expansion takes place only in the presence of a palladium catalyst and its further cyclization followed by the oxidation with MnO_2 gives the desired product.

11.2.5 Rhodium-Catalyzed Transannulation of Pyridotriazoles with *N*-Phenylbenzimidamides

In 2019, Dong group reported an efficient synthetic method using rhodium-catalyzed highly selective C—H bond activation of *N*-phenylbenzimidamides **11.13b** and through subsequent coupling with pyridotriazoles **11.13a**. They observed excellent chemo selectivity and a series of N-heterocycles having quaternary carbon center **11.13c** with the pyridyl moiety were synthesized (Scheme 11.13) [20]. The reaction tolerates diverse functional groups and conveniently affords various 3*H*-indole derivatives.

A plausible reaction mechanism is shown in Scheme 11.14. Initially, coordination of **11.13b** with the catalyst takes place in the presence of acetate or NTf ion and subsequent cyclometallation affords a rhodacyclic intermediate **A**. The diazoimine **B** will generate *in situ* from the pyridotriazole **11.13a,** after the elimination of N_2 to deliver the rhodium carbene **C** through the migratory insertion, which provides seven-membered rhodacyclic species **D**. Thereafter, the Rh—carbon (alkyl) bond undergoes migratory insertion into the C—N bond to generate another intermediate **E**. Finally, the 3*H*-indole product **11.13c** was obtained from **E** by the elimination of one molecule of NH_3 with regeneration of Rh(III) catalyst.

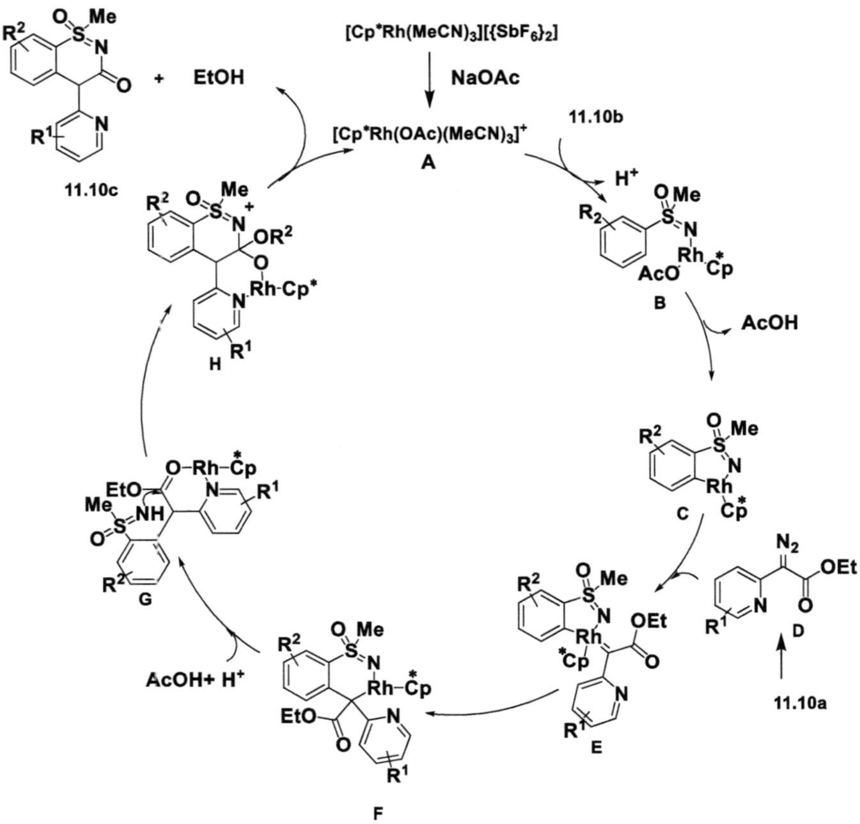

Scheme 11.11 Plausible reaction mechanism transannulation of pyridotriazole with s-aryl sulfoximines.

11.2.6 Rhodium-Catalyzed [4 + 1] Cycloaddition of Pyridotriazoles with Propargyl Alcohols

In the same year 2019, W. Hu and co-workers also reported a transannulation methodology of pyridotriazoles with various propargylic alcohols [21]. A series of 2,5-dihydrofurans were synthesized through [4+1] cycloaddition reaction. Through the present protocol various 2-pyridyl-substituted 2,5-dihydrofuran derivatives accessed in moderate to high yields (Scheme 11.15).

In most of the previously reported transformations, pyridotriazole served as a three-element "N—C—C" synthon in [3+2]-cycloaddition reactions with different nucleophiles such as alkene, alkyne, nitriles, and amides. However, no cyclization reaction with the present type of species as a "one-C" building block was disclosed. This was the first method developed for [4+1] cycloaddition of pyridotriazole (Scheme 11.15).

According to the proposed mechanism, first, the ring-opening of pyridotriazole **11.15a** provides the diazo intermediate **A**, and its reaction with Rh metal delivers rhodium carbenoid intermediate **B** through denitrogenation. The nucleophilic

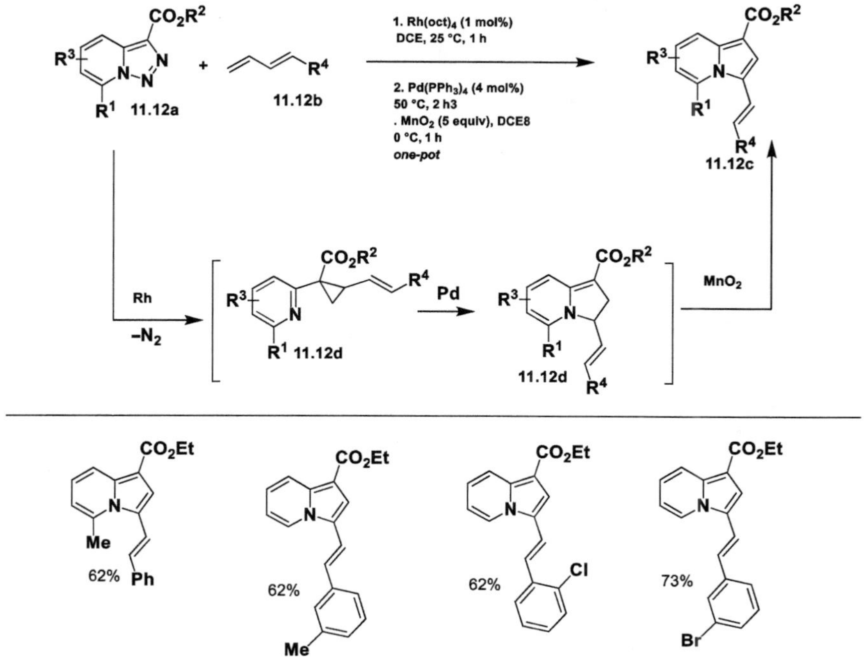

Scheme 11.12 Rhodium-catalyzed [2 + 1]-cyclopropanation, palladium-catalyzed ring expansion, from pyridotriazoles and 1,3-dienes.

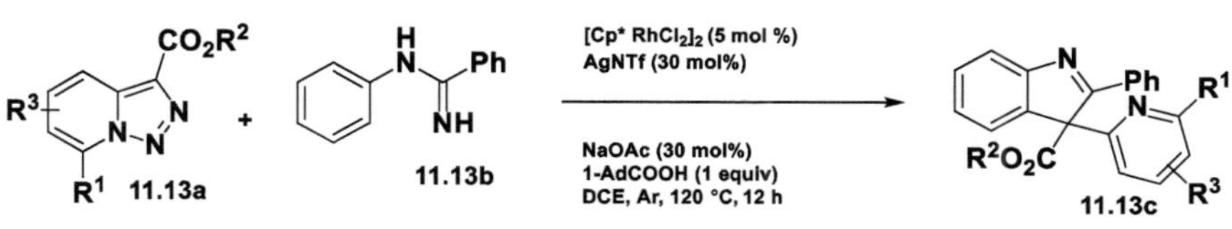

Scheme 11.13 Rh(III)-catalyzed annulation between *N*-phenylbenzimidamides and pyridotriazoles.

attack of propargylic alcohol **11.15b** may provide another intermediate **C**. Through intramolecular [4 + 1] cycloaddition reaction of **C** generates intermediate **D.** Finally, demetallation leads to the final product **11.15c** (Scheme 11.16).

11.3 Palladium-Catalyzed Ring-Opening of [1,2,3]Triazolo[1,5-*a*]pyridines

There are numerous applications of palladium catalysis in the preparation of various agrochemicals, pharmaceuticals, and advanced materials at both laboratory and industrial scale, and hence the palladium catalysis become a powerful tool in organic synthesis [22]. Nobel Prize to Heck, Suzuki, and Negishi underlined the role of palladium catalysis in 2010 for their pioneering work in this particular field [23].

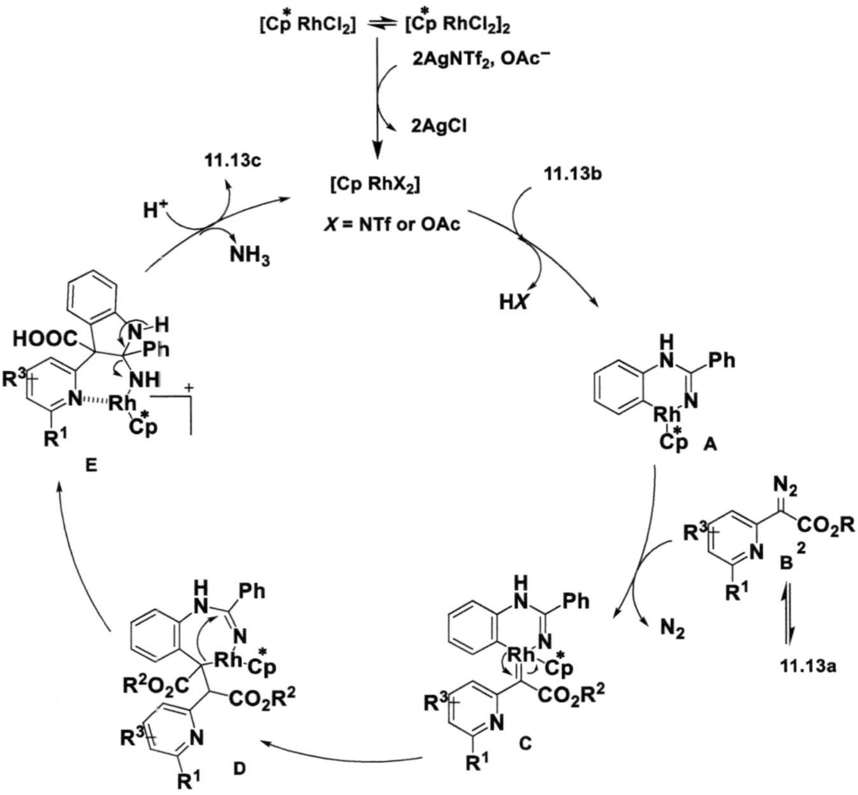

Scheme 11.14 Plausible reaction mechanism of pyridotriazole with *N*-phenylbenzimidamides.

Palladium has its unique role in the catalytic field unlike other transition metals, as it can exist in +2, +3, and +4 oxidation states. Due to its variable oxidation state, numerous methodologies have reported on palladium-catalyzed transformations in the field of coupling reactions, annulations, transannulations, C-H bond functionalization, etc.

Considering the important role of Pd-catalysis in ring-opening of pyridotriazoles, a summary of the main achievements in this area is included here and to make it easier for the reader to find the respective synthetic applications in the following subsections.

11.3.1 Palladium-Catalyzed Ring-Opening of Pyridotriazole with Aryl Halides

In 2016, Hong *et al.* developed the first palladium-catalyzed strategy for the selective (C7) arylation of pyridotriazole with aryl bromides. Further, they have described the synthesis of 6-aryl-2-styrylpyridines, employing methyl substituents at C3 position of pyridotriazole using different bases (Scheme 11.17) [24]. The palladium carbenoid

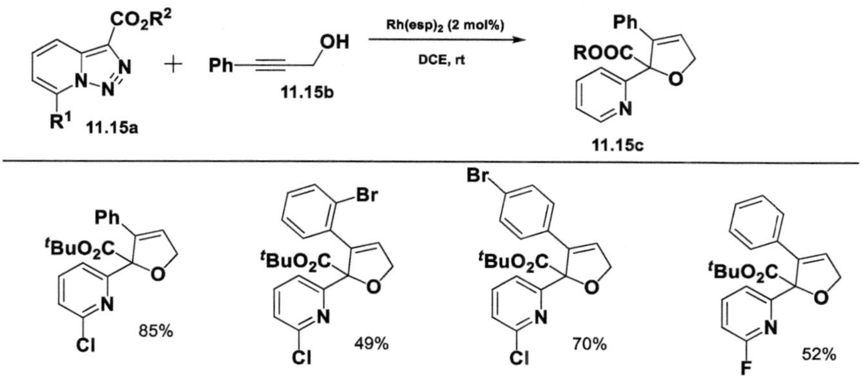

Scheme 11.15 Rhodium (II)-catalyzed formal [4 + 1]-cycloaddition of pyridotriazoles with propargyl alcohols.

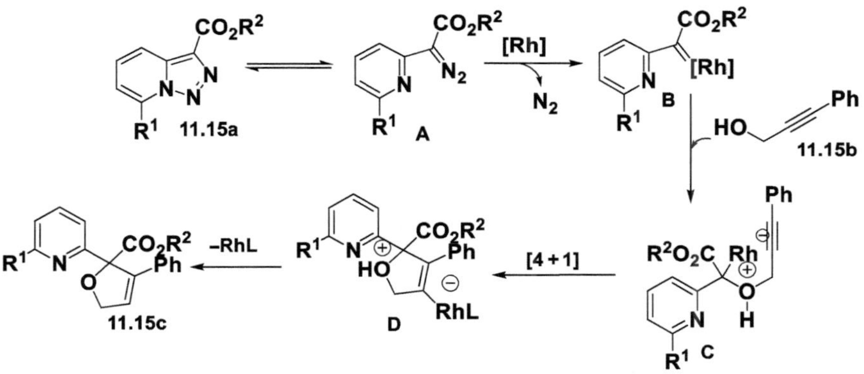

Scheme 11.16 Plausible reaction mechanism of pyridotriazole with propargyl alcohols.

intermediate is generated directly from pyridotriazole **11.17a** and its subsequent reaction with aryl bromides delivers the desired product **11.17d** through denitrogenative pathway.

In 2019, Adimurthy's group also reported the palladium-catalyzed selective arylation of pyridotriazoles at C7 using different aryl halides [25]. They have described the synthesis of 2,6-disubstituted pyridines via palladium-catalyzed aerobic oxidative denitrogenation of pyridotriazoles with aryl halides (Cl, Br, and I) (Scheme 11.18). They performed some mechanistic experiments, which indicated palladium-catalyzed C-6 arylation followed by the ring-opening of pyridotriazole by oxidative denitrogenation leads to the 2,6-disubstituted pyridines. The protocol exceptionally works well with various aryl halides having both electron-rich and deficient groups. The same reaction conditions were extended to regioselective C-3 and C-8 diarylation of several imidazo[1,2-*a*]pyridines. In all the cases, they obtained moderate to good yields of both 2,6-substituted pyridines and polyarylated imidazo[1,2-*a*]pyridines. To validate the feasibility of the process, they have synthesized four products at the gram scale under the optimized conditions.

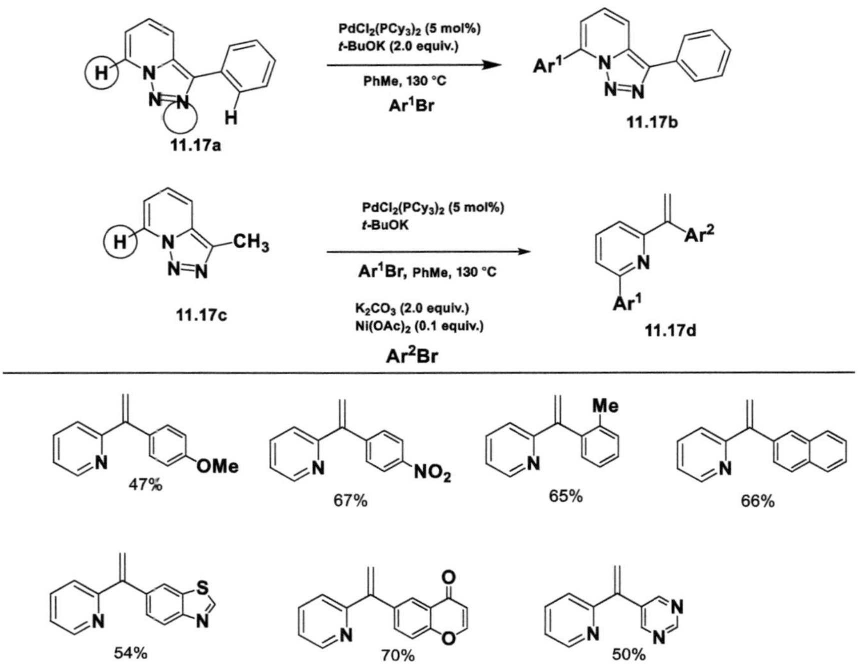

Scheme 11.17 Palladium-catalyzed arylation of [1,2,3]triazolo[1,5-*a*]pyridine.

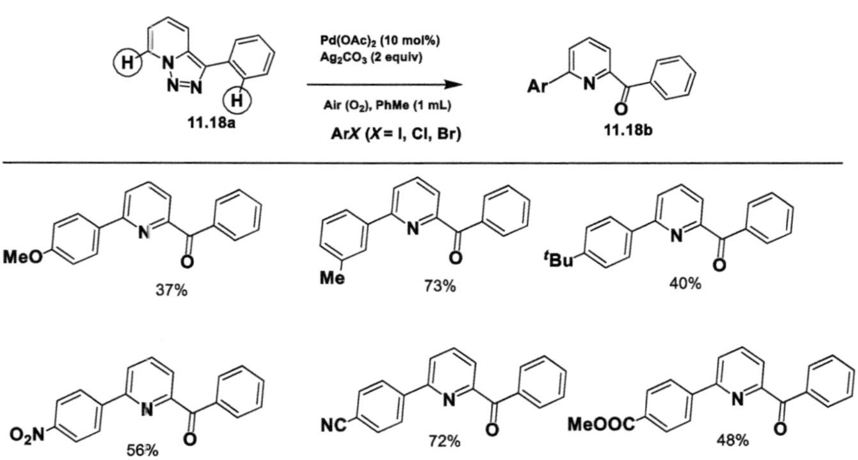

Scheme 11.18 Pd-catalyzed regioselective synthesis of 2,6-disubstituted pyridines through denitrogenation of pyridotriazoles.

A plausible reaction mechanism for the aforementioned transformation has been shown in Scheme 11.19. Initially, palladium coordinated with the nitrogen atom of **11.18a** generates intermediate **A**. Subsequently, oxidative addition of aryliodide generates palladium(IV) complex **B**, followed by the reductive elimination giving rise to

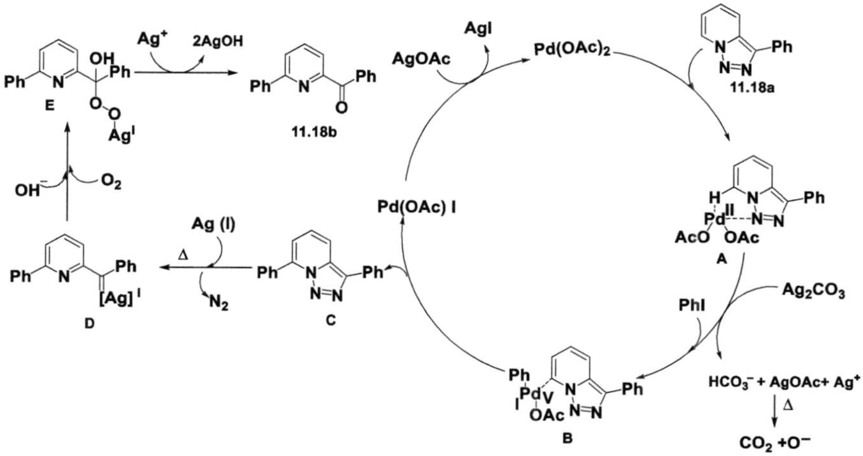

Scheme 11.19 Plausible reaction mechanism for Pd(II) catalyzed arylation of pyridotriazole.

an arylated intermediate **C** and regenerating the active $Pd(OAc)_2$ by the addition of silver acetate. Reaction of **C** with silver salt at 120 °C generates the silver carbene intermediate **D** through denitrogenation of pyridotriazole. Successively, the addition of oxygen to **D** in the presence of base generates the hydroxy organosilver peroxide intermediate **E**. Finally, the elimination of silver(I) species produces the desired product **11.18b** (Scheme 11.19).

11.3.2 Palladium-Catalyzed Ring-Opening of Pyridotriazole with Acetic Anhydride

In the same year, Adimurthy and co-workers have developed a Pd-catalyzed synthesis of 2-oxo-1-phenyl-1-(pyridin-2-yl)propyl acetates **11.20b** from pyridotriazoles **11.20a** and acetic anhydride (Scheme 11.20) [26]. Using acetic anhydride as acylating reagent, an interesting product 2-oxo-1-phenyl-1-(pyridin-2-yl)propyl acetates were obtained through the ring-opening of pyridotriazoles. They also obtained an interesting potential drug intermediates 2-([1,2,3]triazolo[1,5-*a*]pyridin-3-yl)phenol **11.20d** (2-hydroxyphenyl)pyridin-2-yl)methanone **11.20e** through acyloxylation of **11.20a** with carboxylic acids followed by its hydrolysis.

A plausible reaction mechanism for the denitrogenative transannulation of **11.20a** with acetic anhydride is shown in Scheme 11.21. Initially, denitrogenation of **11.20a** in the presence of catalyst $Pd(PPh_3)_4$ may generate the metallocarbene intermediate **A**. The insertion of Ac_2O to the intermediate **A** may generates metalated intermediate **B**. Which upon elimination of $[Pd(PPh_3)_2]$ generates another intermediate **C**. Due to the positive charge on oxygen of Ac_2O, the negative charge present on the central carbon may attack electrophilic carbonyl carbon and deliver the desired product **11.20b**.

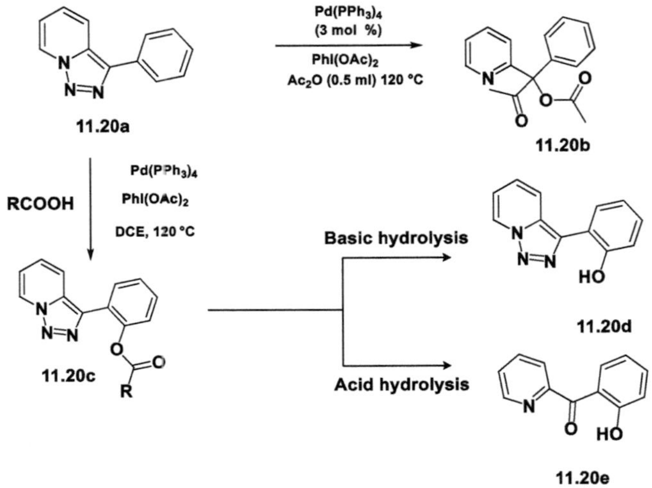

Scheme 11.20 Pd-catalyzed synthesis of 2-oxo-1-phenyl-1-(pyridin-2-yl)propyl acetates from pyridotriazoles.

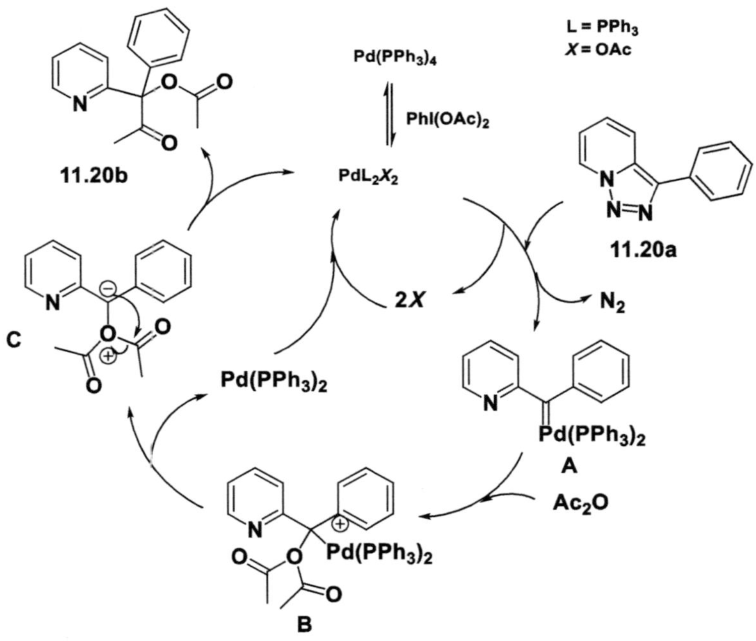

Scheme 11.21 Proposed mechanism for the Pd-catalyzed reaction of pyridotriazoles with acetic anhydride.

11.4 Copper-Catalyzed Denitrogenative Transannulation of Pyridotriazoles

With well understanding of the reaction mechanism of Ullmann reaction, copper-catalyzed reactions have become powerful tools in organic synthesis in the past century and served well for C—C, C—N, C—S, C—O, and other bond formation reactions [27]. Copper catalysis field is more attractive to the synthetic chemists due to several reasons. First of all, copper has its variable Cu^0, Cu^I, Cu^{II}, and Cu^{III} oxidation states, and very cheap compared to palladium and other transition metals and copper is abundantly available on the earth. Furthermore, copper salts generally possess a low toxicity.

11.4.1 Copper-Catalyzed Transannulation of Pyridotriazoles with Alkynes

Gevorgyan's group reported a copper-catalyzed denitrogenation of pyridotriazoles **11.22a** with terminal alkynes **11.22b** under aerobic conditions and obtained indolizine derivatives **11.22c** (Scheme 11.22) [28]. This methodology does not require substitution at the C7 position of pyridotriazoles, does not require electron-deficient pyridotriazoles, and dispenses the use of expensive rhodium(II) catalysts. Employment of this method allowed to efficiently produce indolizines with electronically diverse aryl acetylenes. Aliphatic and heteroaromatic alkynes also reacted well, which was not possible under the previously reported rhodium-catalyzed conditions [13]. Most notably, unsubstituted pyridotriazoles turned out to be the competent substrates for this reaction as well. Representative indolizine derivatives are presented in Scheme 11.22.

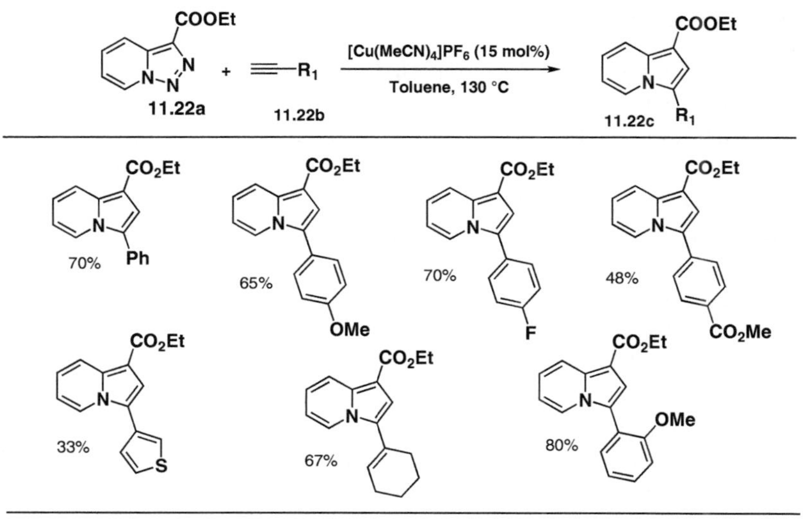

Scheme 11.22 Cu-Catalyzed transannulation with alkynes.

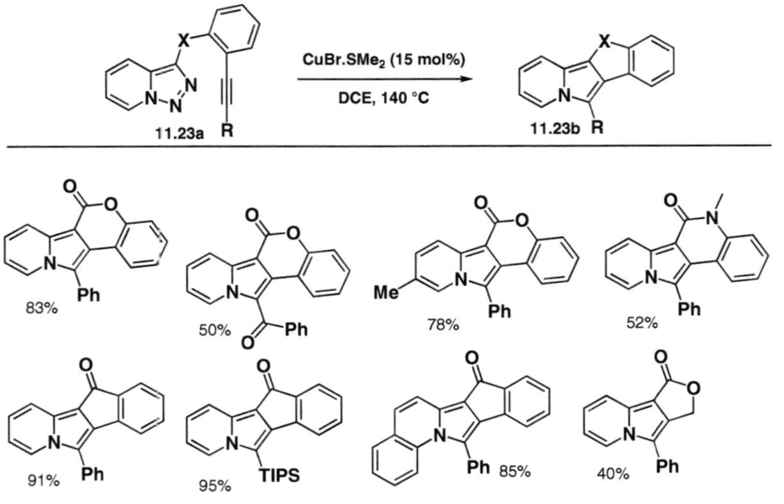

Scheme 11.23 Cu-Catalyzed intramolecular transannulation.

In continuation of Gevorgyan's work on the synthesis of N-heterocyclic structures via transannulation chemistry, the same group reported the assembly of diverse polycyclic frameworks via the copper-catalyzed intramolecular transannulation reactions (Scheme 11.23) [29]. Pyridotriazoles tethered with internal alkynes **11.23a** participated well in this intramolecular transannulation reaction to produce the corresponding tri-, tetra-, and pentacyclic fused indolizine derivatives **11.23b** in a single step with moderate to good yields.

11.4.2 Copper-Catalyzed Transannulation of Pyridotriazoles with Amines and Amino Acids

In 2016, Adimurthy and co-workers reported copper-catalyzed transannulation of pyridotriazoles with benzylamines under aerobic conditions and obtained imidazo[1,5-*a*]pyridines [30]. Pyridotriazoles **11.24a** reacted efficiently with different substituted benzyl amines **11.24b** to produce the corresponding imidazopyridines **11.24c** in high yields (Scheme 11.24). In addition to benzylamines, heteroaromatic amines and cyclic amines reacted well to afford the desired products. The authors observed good functional group tolerance with benzyl amines having electron-donating, -withdrawing, and halogen substituents. However, no reaction was observed with *aza*-heterocyclic or aliphatic amines under the optimized conditions. The representative products are presented in the scheme below (Scheme 11.24).

Authors proposed a probable mechanism for the transannulation of pyridotriazoles with benzyl amines as shown in Scheme 11.25. First, the α-imino diazo compound **A** generates the Cu—carbene complex **B**, which is generated from pyridotriazole **11.24a**. Migratory insertion of benzylamine at the carbene C-atom of **A** forms the intermediate **B**, which further undergoes oxidative dehydrogenation to

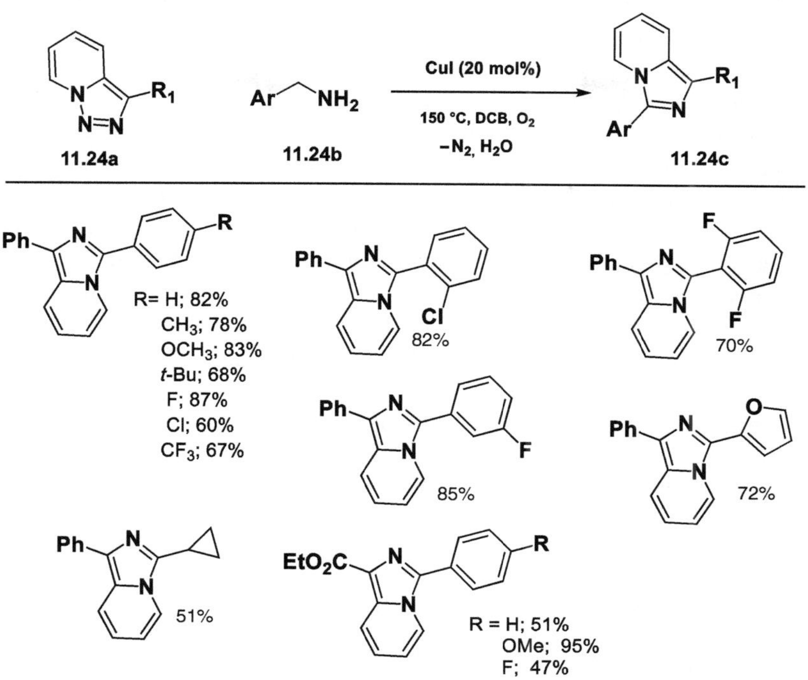

Scheme 11.24 Cu(I)-catalyzed transannulation with benzyl amines.

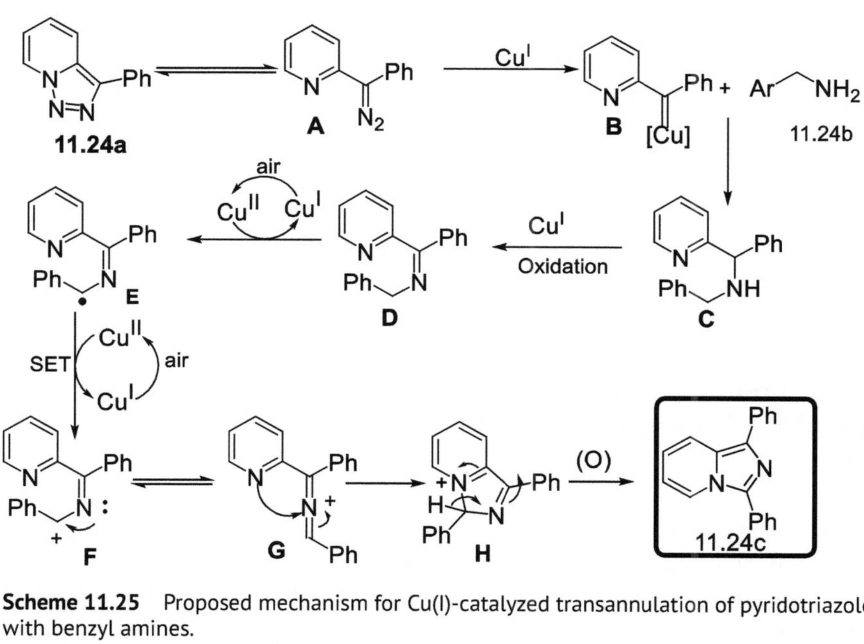

Scheme 11.25 Proposed mechanism for Cu(I)-catalyzed transannulation of pyridotriazoles with benzyl amines.

provide imine intermediate **C**. Subsequently, **C** further undergoes single electron transfer in the process (intermediates **D–G**). Attacking the ring nitrogen lone pair of electrons from **G** leads to a cyclic intermediate **H**. Finally, aromatization of **H** will give the desired product **11.24c** (Scheme 11.25).

α-Amino acids are more readily available and are more stable than other starting materials in nature. Therefore, the decarboxylative reaction of α-amino acids provides a very efficient synthetic method for the synthesis of azaheterocycles [31]. Further, Adimurthy's group has evaluated the strategy on denitrogenative transannulation reaction of pyridotriazoles **11.26a** with some amino acids **11.26b** and their derivatives under copper catalytic conditions [30] and the results are included in Scheme 11.26. Pyridotriazoles react with substituted amino acid derivatives, including phenyl glycine derivatives (4-Cl, 2-Cl, 4-F, and 4-OH), alanine, and leucin, provided the corresponding imidazo[1,5-*a*]pyridines **11.26c** in moderate to excellent yields via decarboxylation followed by transannulation (Scheme 11.26). The representative imidazo[1,5-*a*]pyridines synthesized through decarboxylation of amino acids with pyridotriazoles are given in Scheme 11.26.

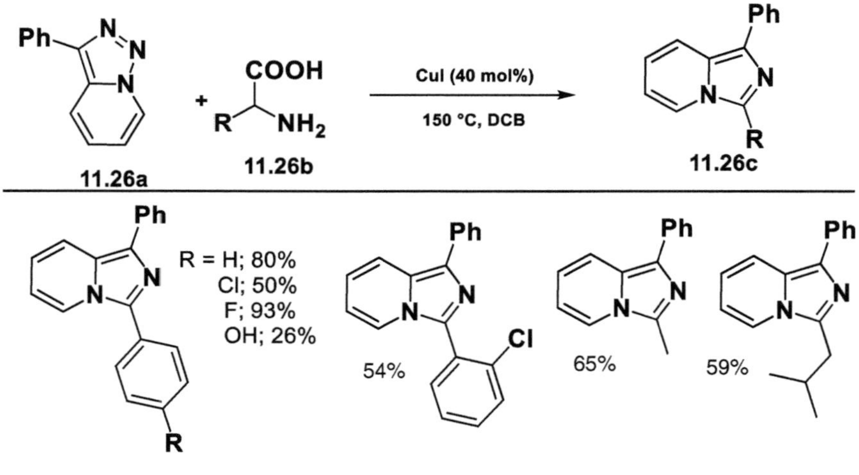

Scheme 11.26 Cu(I)-catalyzed decarboxylative transannulation of pyridotriazoles with α-amino acids.

11.5 Cobalt-Catalyzed Transannulation of Pyridotriazoles with Isothiocyanates and Xanthate Esters

In 2020, Gevorgyan's group successfully developed a cobalt-catalyzed radical transannulation method for pyridotriazoles [32]. The transannulation of pyridotriazole **11.27a** with isothiocyanates **11.27b** and xanthate esters **11.27d** to convert it into two N-fused heterocyclic aromatic systems in a single step (Scheme 11.27). Among these two, one is imino-thiazolopyridines **11.27c** and another is oxo-triazolopyridine

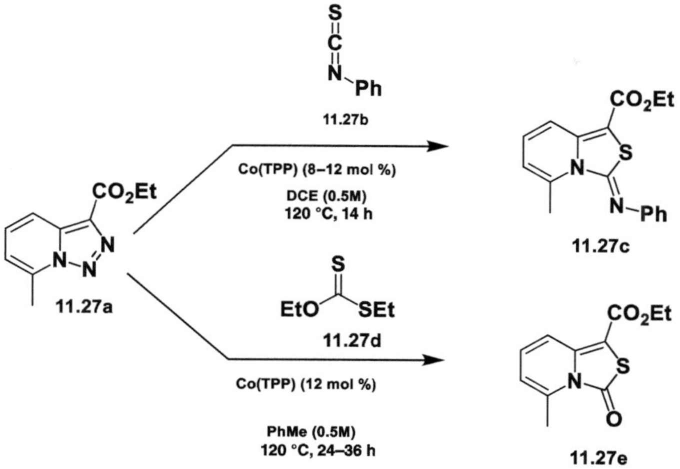

Scheme 11.27 Co-catalyzed transannulation of pyridotriazoles with isothiocyanates and xanthate esters.

11.27e derivatives via radical pathway. They have performed some radical trapping experiments to propose the mechanism for this transannulation, which is slightly different than the earlier-reported ionic transannulations. First, xanthate ester acts as a carbonyl sulfide surrogate. Pyridotriazole **11.27a** exists in equilibrium with its open diazo tautomer **A**. After denitrogenation of **A** and its reaction with cobalt catalyst produces a key α-Co(III)-pyridyl radical intermediate **B**. Subsequently, its reaction with isothiocyanate or carbonyl sulfide generates radical intermediate **C**. This radical intermediate, upon cyclization, leads to imino-thiazolopyridine **11.28c** or oxo-thiazolopyridines**11.28e** and Co(III)-catalyst **D** is regenerated (Scheme 11.28).

11.6 Ruthenium-Catalyzed Transannulation of Pyridotriazoles with Naphthoquinones

In 2022 Adimurthy's group revealed an expeditious ruthenium-catalyzed denitrogenative transannulation of pyridotriazoles **11.29a** with 1,4-naphthoquinones **11.29b** for the synthesis of benzo[f]pyrido[1,2-*a*]indole-6,11-diones **11.29c** [33]. The reaction works very well with 5-hydroxy-naphthalene-1,4-dione, 5,8-dihydroxy-naphthalene-1,4-dione, and 5-methoxynaphthalene-1,4-dione (Scheme 11.29). The variety of pyridotriazole esters such as ethyl, methyl, *n*-butyl, benzyl, and *tert* butyl, including aliphatic cyclic groups (cyclo-pentyl, cyclohexyl, and cyclo-octyl) works smoothly and provide the corresponding transannulated benzo pyrido[1,2-*a*]indoles derivatives in good to excellent yields.

Based on previous literature reports and control experiments, they have proposed a plausible reaction mechanism (Scheme 11.30). Initially, the active Ru(II) species **I** is generated *in situ* in the presence of AcOH. At 120 °C, the denitrogenation of triazole **11.29a** leads to the formation of ruthenium carbenoid intermediate **III**. There may

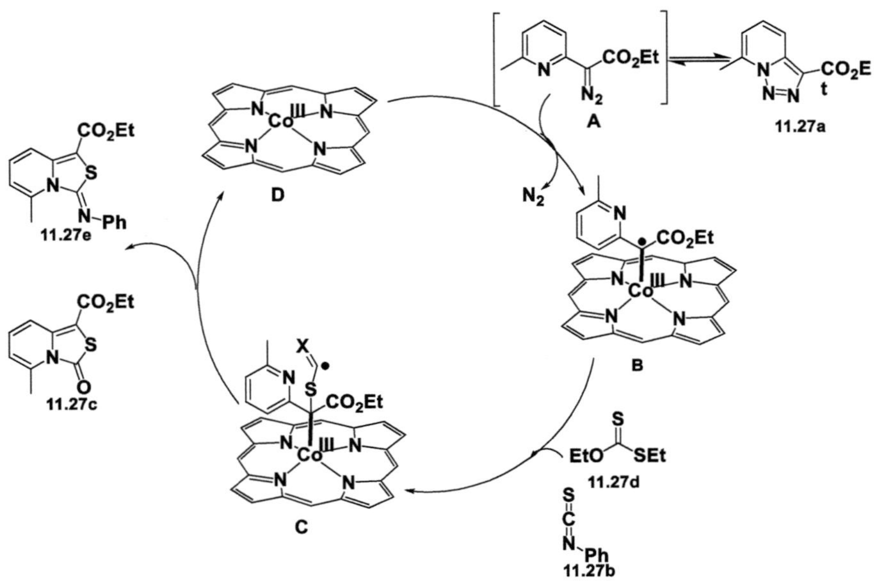

Scheme 11.28 Plausible reaction mechanism for Co-catalyzed tranasannulation of pyridotriazoles.

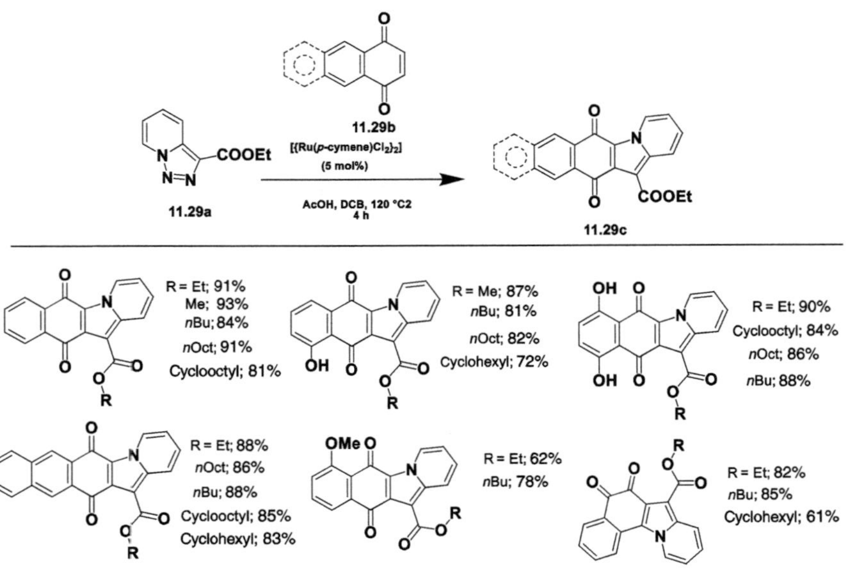

Scheme 11.29 Ru-catalyzed transannulation of pyridotriazole.

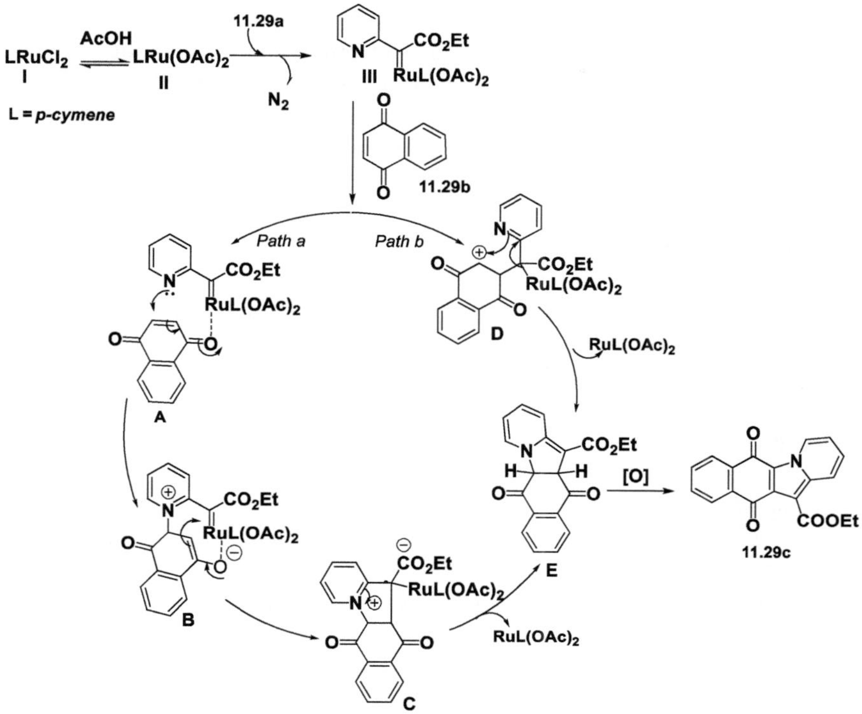

Scheme 11.30 Plausible reaction mechanism for Tr-catalyzed tranasannulation of pyridotriazoles.

be two possibilities to achieve the final product **11.29c** either path (a) or path (b). Through *path a*, **11.29b** coordinates with ruthenium to generate intermediate **A**. Subsequently, N-atom of pyridine attacks at the electrophilic carbon of naphthoquinone to produce intermediate **B**. Intramolecular nucleophilic attack at the metal carbene center generates intermediate **C**, which upon demetallation may produce the intermediate **E**, and its oxidation leads to the final product **11.29c**. On the other hand, if **11.29b** acts as a nucleophile, there might be another possibility to deliver the desired product (*path b*). In this case, **11.29b** attacks at the electrophilic carbene center of species **III** to deliver the intermediate **D**. Its demetallative cyclization may yield the intermediate **E**, which undergoes oxidation to deliver the final product **11.29c**. By path (b), the presence of a carbocation center at the α-position of the carbonyl group of intermediate **D**, makes it highly unstable. Therefore, they suggest that *path a* is more favorable than *path b*.

11.7 Summary and Conclusion

In this chapter, we have described the transannulation of pyridotriazoles with different annulation partners under various catalytic conditions. The synthesis of diverse heterocycles through the transannulation was described in brief along with their mechanistic pathway. A wide range of annulation partners was covered, including the recent advancements, via the ring-opening of pyridotriazoles, which generates metal-carbenoid intermediates that were transformed to a range of heterocycles through the coupling of various nucleophiles with varied conditions.

Acknowledgments

CSIR-CSMCRI Communication No. 156/2023. We are thankful to CSIR-CSMCRI (MLP-0074 and HCP-49) for financial support.

References

1 Pirrung, M.C. and Morehead, A.T. (1996). Saturation kinetics in dirhodium(II) carboxylate-catalyzed decompositions of diazo compounds. *J. Am. Chem. Soc.* 118: 8162.

2 Ma, G., Wei, K.F., Song, M. et al. (2023). Recent advances in transition-metal-catalyzed Büchner reaction of alkynes. *Org. Biomol. Chem.* 21: 5150.

3 Liu, C.X., Yin, S.Y., Zhao, F. et al. (2023). rhodium-catalyzed asymmetric C—H functionalization reactions. *Chem. Rev.* 123: 10079.

4 (a) Liu, C.F., Liu, M., and Dong, L.J. (2019). Iridium(III)-catalyzed tandem annulation synthesis of pyrazolo[1,2-α]cinnolines from pyrazolones and sulfoxonium ylides. *J. Org. Chem.* 84: 409. (b) Shi, X., Wang, R., Zeng, X. et al. (2018). Ruthenium(II)-catalyzed oxidant-free coupling/cyclization of benzimidates and sulfoxonium ylides to form substituted isoquinolines. *Adv. Synth. Catal.* 360: 4049.

5 Wu, X., Xiao, Y., Yu, S.S.J., and Cheng, J. (2019). Rhodium-catalyzed reaction of sulfoxonium ylides and anthranils toward indoloindolones via a (4 + 1) annulation. *Org. Lett.* 21: 6653.

6 Seoane, A., Noelia, C., Noelia, Q. et al. (2014). Rhodium(iii)-catalyzed dearomatizing (3 + 2) annulation of 2-alkenylphenols and alkynes. *J. Am. Chem. Soc.* 136: 7607.

7 (a) Hong, C., Yu, S., Liu, Z., and Zhang, Y. (2021). Rhodium(III)-catalyzed annulation of enamides with sulfoxonium ylides toward isoquinolines. *RSC Adv.* 11: 11490. (b) Wu, S., Zeng, R., Fu, C. et al. (2015). Rhodium-catalyzed C—H functionalization-based approach to eight-membered lactams. *Chem. Sci.* 6: 2275.

8 Seoane, A., Casanova, N., Quiñones, N. et al. (2014). Straightforward assembly of benzoxepines by means of a rhodium(III)-catalyzed C—H functionalization of *o*-vinylphenols. *J. Am. Chem. Soc.* 136: 834.

9 (a) Semakul, N., Jackson, K.E., Pato, R.S., and Rovis, T. (2017). Heptamethylindenyl (ind*) enables diastereoselective benzamidation of cyclopropenes via Rh(III)-catalyzed C—H activation. *Chem. Sci.* 8: 1015. (b) Gilbertson, S.R. and

Brenton, D.B. (2002). Rhodium catalyzed [4 + 2 + 2] cycloaddition and alkyne insertion: a new route to eight-membered rings. *J. Am. Chem. Soc.* 124: 8784.

10 Connell, R., Scavo, F., and Helquist, P. (1986). Rh-catalyzed transannulation of pyridotriazoles with alkynes and nitriles. *Tetrahedron Lett.* 27: 5559.

11 (a) Padwa, A. and Kinder, F.R. (1993). Rhodium(II)-catalyzed cyclization of 2-alkynyl 2-diazo-3-oxobutanoates as a method for synthesizing substituted furans. *J. Org. Chem.* 58 (1): 21. (b) Padwa, A., Kassir, J.M., and Xu, S.L. (1991). Rhodium-catalyzed ring-opening reaction of cyclopropenes. control of regioselectivity by the oxidation state of the metal. *J. Org. Chem.* 56: 6971.

12 Bagheri, V., Doyle, M.P., Taunton, J., and Claxton, E.E. (1988). A new and general synthesis of a-silyl carbonyl compounds by silicon-hydrogen insertion from transition metal-catalyzed reactions of diazo esters and diazo ketones. *J. Org. Chem.* 53: 6158.

13 Chuprakov, S., Hwang, F.W., and Gevorgyan, V. (2007). Rh-catalyzed transannulation of pyridotriazoles with alkynes and nitriles. *Angew. Chem. Int. Ed.* 46: 4757.

14 Chuprakov, S. and Gevorgyan, V. (2007). Regiodivergent metal-catalyzed rearrangement of 3-iminocyclopropenes into n-fused heterocycles. *Org. Lett.* 9: 4463.

15 Shi, Y., Gulevich, A.V., and Gevorgyan, V. (2014). Rhodium-catalyzed NH insertion of pyridyl carbenes derived from pyridotriazoles: a general and efficient approach to 2-picolylamines and imidazo[1,5-*a*]pyridines. *Angew. Chem. Int. Ed.* 53: 14191.

16 (a)Gillingham, D. and Fei, N. (2013). Catalytic *X*—H insertion reactions based on carbenoids. *Chem. Soc. Rev.* 42: 4918. (b) Zhang, Z. and Wang, J. (2008). Recent development of reactions with α-diazocarbonyl compounds as nucleophiles. *Tetrahedron* 64: 6577.

17 Kim, J.H., Gensch, T., Zhao, D. et al. (2015). Rh(III)-catalyzed C—H activation with pyridotriazoles: direct access to fluorophores for metal-ion detection. *Angew. Chem. Int. Ed.* 54: 10975.

18 Jeon, W.H., Son, J.Y., Kim, J.E., and Lee, P.H. (2016). Synthesis of 1,2-benzothiazines by a rhodium-catalyzed domino C—H activation/cyclization/elimination process from s-aryl sulfoximines and pyridotriazoles. *Org. Lett.* 18: 3498.

19 Kim, H., Kim, S., Kim, J. et al. (2017). One-pot synthesis of indolizines via sequential rhodium-catalyzed [2 + 1]-cyclopropanation, palladium-catalyzed ring expansion, and oxidation reactions from pyridotriazoles and 1,3-dienes. *Org. Lett.* 19: 5677.

20 Xu, H. B.; Zhu, Y. Y. And Dong, L. (2019) Specific synthesis of 3*H*-indole derivatives via Rh(III)-catalyzed cascade annulation between *n*-phenylbenzimidamides and pyridotriazoles. *J. Org. Chem.* 84, 16286.

21 Lv, X., Yang, H., Shi, T. et al. (2019). Rhodium(II)-catalyzed formal [4 + 1]-cycloaddition of pyridotriazoles and propargyl alcohols: synthesis of 2,5-dihydrofurans. *Adv. Synth. Catal.* 361: 1265.

22 Wu, X.F., Neumann, H., and Beller, M. (2013). Synthesis of heterocycles via palladium-catalyzed carbonylations. *Chem. Rev.* 113: 1.

23 (a) Wu, X.F., Anbarasan, P., Neumann, H., and Beller, M. (2010). Palladium-catalyzed carbonylative synthesis of quinazolinones from 2-aminobenzamide and aryl bromides. *Angew. Chem. Int. Ed.* 49: 9047. (b) Negishi, E. (2011). Magical power of transition metals: past, present, and future. *Angew. Chem. Int. Ed.* 50: 6738. (c) Suzuki, A. (2011). Cross-coupling reactions of organoboranes: an easy way to construct C—C bonds. *Angew. Chem. Int. Ed.* 50: 6722.

24 Moon, Y., Kwon, S., Kang, D. et al. (2016). Palladium-catalyzed divergent arylation with triazolopyridines: one-pot synthesis of 6-aryl-2-α-styrylpyridines. *Adv. Synth. Catal.* 358: 958.

25 Joshi, A., Semwal, R., Suresh, E., and Adimurthy, S. (2019). Pd-catalyzed regioselective synthesis of 2,6-disubstituted pyridines through denitrogenation of pyridotriazoles and 3,8-diarylation of imidazo[1,2-*a*]pyridines. *Chem. Commun.* 55: 10888.

26 Rawat, D., Kumar, R., and Adimurthy, S. (2019). Pd-catalyzed ortho selective C—H acyloxylation and hydroxylation of pyridotriazoles. *Eur. J. Org. Chem.* 2019: 7874.

27 (a) Hassan, J., Sevignon, M., Gozzi, C. et al. (2002). Aryl–aryl bond formation one century after the discovery of the ullmann reaction. *Chem. Rev.* 102: 1359. (b) Monnier, F. and Taillefer, M. (2009). Catalytic C—C, C—N, and C—O ullmann-type coupling reactions. *Angew. Chem. Int. Ed.* 48: 6954.

28 Helan, V., Gulevich, A.V., and Gevorgyan, V. (2015). Cu-catalyzed transannulation reaction of pyridotriazoles with terminal alkynes under aerobic conditions: efficient synthesis of indolizines. *Chem. Sci.* 6: 1928.

29 Shi, Y. and Gevorgyan, V. (2015). Cu-catalyzed transannulation reaction of pyridotriazoles: general access to fused polycyclic indolizines. *Chem. Commun.* 51: 17166.

30 Joshi, A., Chandra Mohan, D., and Adimurthy, S. (2016). Copper-catalyzed denitrogenative transannulation reaction of pyridotriazoles: synthesis of imidazo[1,5-*a*]pyridines with amines and amino acids. *Org. Lett.* 18: 464.

31 (a) Wang, Q., Zhang, S., Guo, F. et al. (2012). Natural α-amino acids applied in the synthesis of imidazo[1,5-*a*]N-heterocycles under mild conditions. *J. Org. Chem.* 77: 11161. (b) Xu, K., Wang, Z., Zhang, J. et al. (2015). Cobalt-catalyzed decarboxylative acetoxylating of amino acids and aryl acetic acids. *Org. Lett.* 17: 4476.

32 Zhang, Z. and Gevorgyan, V. (2020). Co-catalyzed transannulation of pyridotriazoles with isothiocyanates and xanthate esters. *Org. Lett.* 22: 8500.

33 Rawat, D. and Adimurthy, S. (2022). Transannulation of pyridotriazoles with naphthoquinones and indoles: synthesis of benzo[*f*]pyrido[1,2-*a*]indoles and indolizino[3,2-*b*]indoles. *Adv. Synth. Catal.* 364: 71.

12

Metal-Free Denitrogenative Transformation of Pyridotriazoles

Sharma Happy, Akash Ratwan, and Dongari Yadagiri

Laboratory of Synthesis and Catalysis, Department of Chemistry, Indian Institute of Technology Roorkee, Roorkee 247667, Uttarakhand, India

12.1 Introduction

N-Containing heterocyclic compounds are important moieties for synthesizing diverse pharmacological and biologically important compounds [1]. Pyridotriazoles are an important subclass of N-containing heterocyclic compounds, which have found great recognition in the last few years because of their reactivity and diverse properties [2]. Particularly, pyridotriazoles have been employed as a precursor of carbene or metallocarbene via denitrogenative transformation, whether catalytically or non-catalytically. Pyridotriazole **I** are known to undergo ring-chain isomerization to produce carbene precursor such as α-diazoimine **II** [3] (Figure 12.1). Initial discovery by Gevorgyan group for transannulation of pyridotriazole with nitriles in the presence of rhodium catalyst [4]. From the last decade, numerous novel synthetic methods have been developed for the synthesis of N-containing heterocyclic compounds, which include indolizines, imidazopyridines, pyridoisoquinolinones, and pyridoindoles, as well as a variety of functionalized pyridines, have been developed based on the denitrogenative transformation of pyridotriazoles via transannulation reactions, *X*—H insertion reaction and other transformations [5]. However, most of the cases need to use expensive transition metal catalysts. This chapter mainly discusses the metal-free transformation of pyridotriazoles.

12.2 Acid-Catalyzed Denitrogenative Transformation of Pyridotriazoles

12.2.1 Lewis-Acid Catalyzed Reactions

In 2016, the Adimurthy group showed the transition-metal free transannulation [6] of pyridotriazoles with nitriles catalyzed by Lewis acid such as $BF_3.OEt_2$ for the

Denitrogenative Transformation of Nitrogen Heterocycles: Synthesis, Reactions and Applications, First Edition. Edited by Pazhamalai Anbarasan.

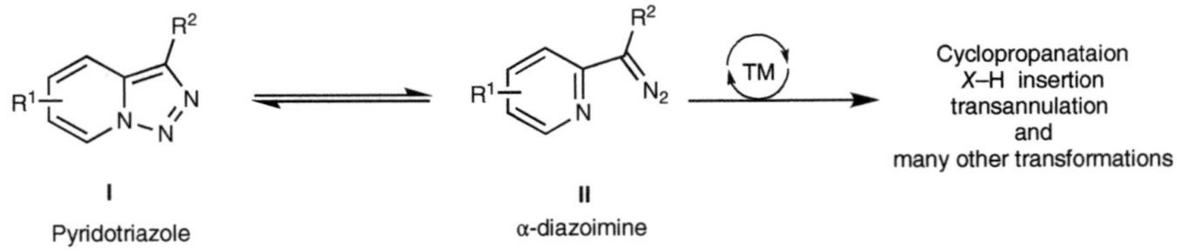

Figure 12.1 Ring-chain isomerism of pyridotriazoles and its catalytic transformations.

synthesis of imidazopyridines (Scheme 12.1). They found that dichlorobenzene (DCB) and dichloroethane (DCE) are the essential solvent mixture for obtaining excellent yield. Pyridotriazole **12.1a** would be in equilibrium with α-diazoimine **12.1d** at a higher temperature and then react with $BF_3{\cdot}OEt_2$ to produce the intermediate **12.1e**, which would undergo two distinct pathways to form **12.1c**. In path *a*, intermediate **12.1e** undergoes a nucleophilic substitution with nitrile derivative

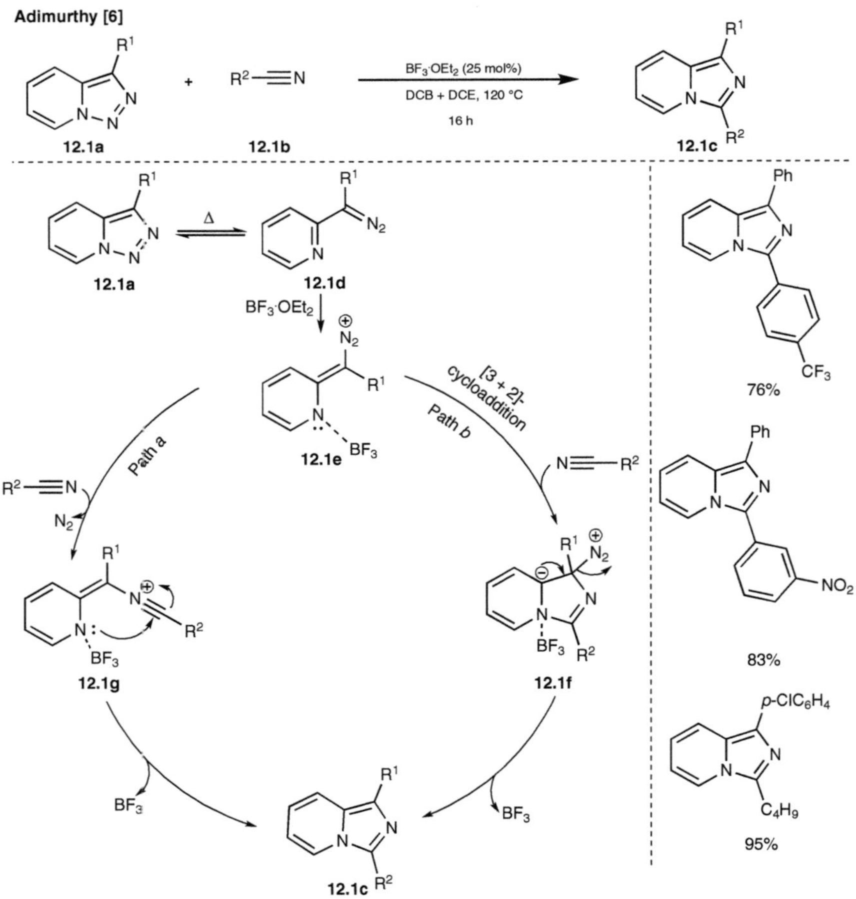

Scheme 12.1 $BF_3{\cdot}OEt_2$-catalyzed transannulation of pyridotriazoles with nitriles. Source: Adimurthy [6]/American Chemical Society.

12.1b to form intermediate **12.1g** via elimination of molecular nitrogen, followed by nucleophilic attack of pyridine nitrogen to produce the imidazopyridines **12.1c** (Scheme 12.1). In path *b*, product **12.1c** would form through [3 + 2]-cycloaddition of **12.1e** with nitrile, giving the **12.1f** by eliminating N_2. Pyridotriazoles are compatible with electron-withdrawing, electron-donating groups containing functional groups, and aromatic, aliphatic nitrile substrate reacted well, giving the corresponding imidazopyridines derivatives **12.1c**.

In 2019, the Adimurthy group showed the synthesis of pyrido[1,2-*a*]indoles in the presence of indium catalyst from the reaction between arenes and pyridotriazoles via transannulation fashion [7]. Indium carbenoid **12.2d** generated from pyridotriazole **12.2a** in the presence of indium triflate $In(OTf)_3$. Pyridinium Aza-Nazarov type intermediate **12.2e** is generated from the insertion of arenes **12.2b** into carbenoid **12.2d** (Scheme 12.2). Bis-cationic pyridinium tertiary carbocation **12.2f** would form

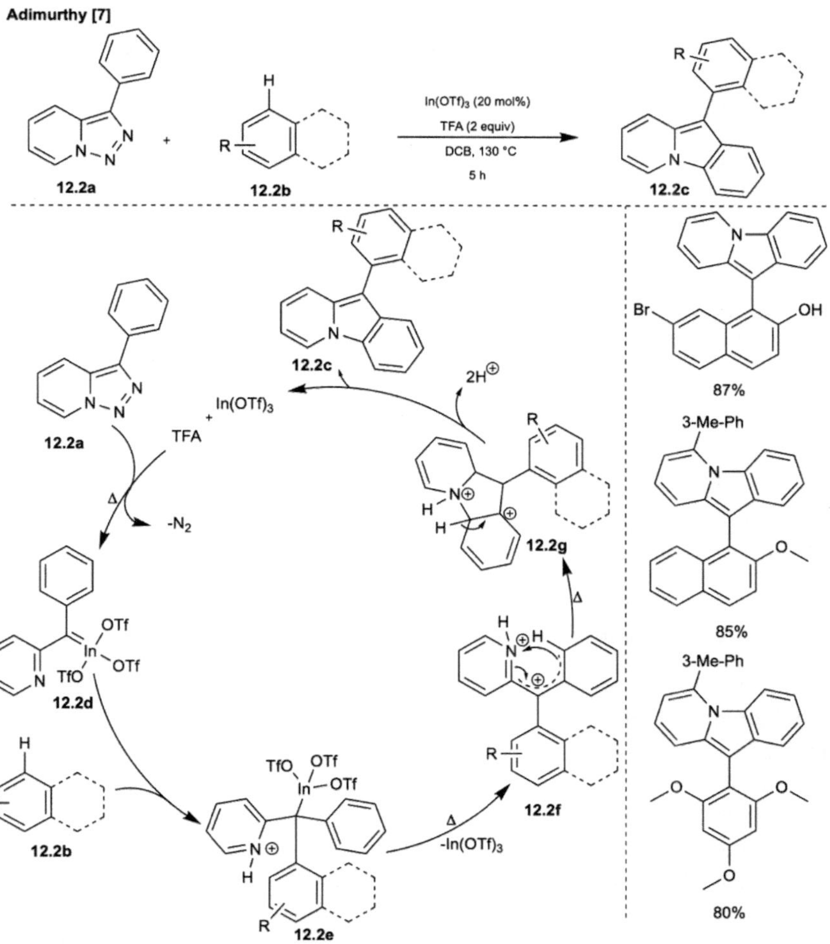

Scheme 12.2 $In(OTf)_3$-catalyzed denitrogenative transannulation of pyridotriazoles. Source: Adimurthy [7]/American Chemical Society.

under acidic conditions upon $In(OTf)_3$ elimination. Intermediate **12.2f** brings the phenyl group close to the pyridine nitrogen due to the high temperature and steric of the two bulky aryl groups, giving the other intermediate **12.2g** via C—H amination. Oxidation of **12.2g** led to the formation of product **12.2c**. The proposed mechanism is well-supported by the DFT calculations.

In 2021, the Adimurthy group extended to isothiocyanates for transannulation with pyridotriazoles [8]. They have studied the synthesis of thiazolo[3,4-*a*]pyridine-3-imines using $BF_3{\cdot}OEt_2$ Lewis acid as a catalyst and, again, the mixture of solvents DCB + DCE combined played a crucial role in the better yield of the product. Various substituted pyridotriazoles **12.3a** and electronically diverse aromatic, aliphatic substituted isothiocyanates **12.3b** reacted very well, gave the corresponding thiazolo[3,4-*a*]pyridine-3-imines derivatives **12.3c** with moderate to good yields (Scheme 12.3).

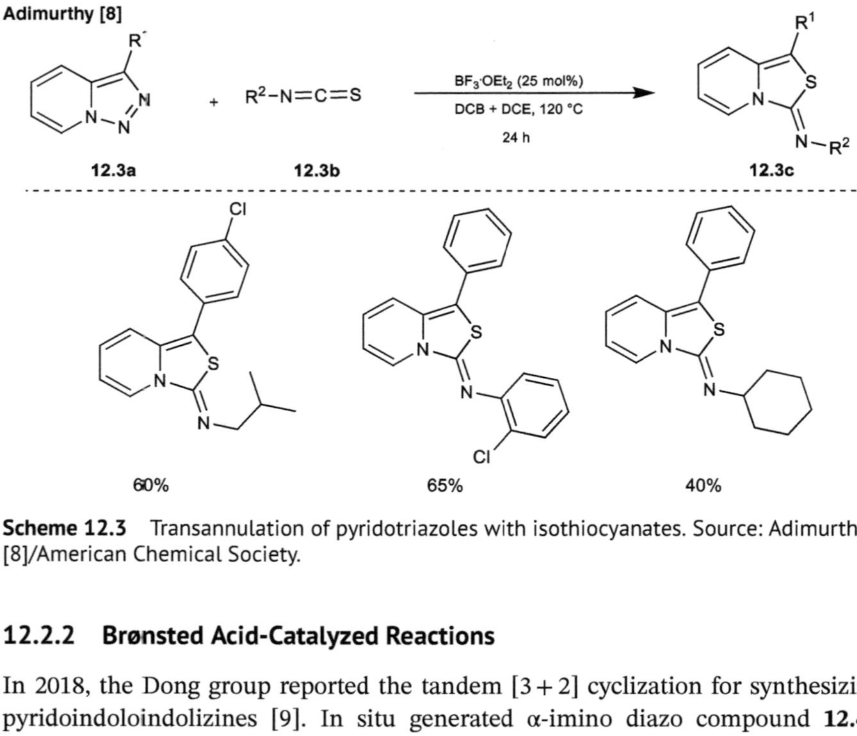

Scheme 12.3 Transannulation of pyridotriazoles with isothiocyanates. Source: Adimurthy [8]/American Chemical Society.

12.2.2 Brønsted Acid-Catalyzed Reactions

In 2018, the Dong group reported the tandem [3 + 2] cyclization for synthesizing pyridoindoloindolizines [9]. In situ generated α-imino diazo compound **12.4d** stabilized by a *p*-toluene sulfonic acid (TsOH) **12.4e** or **12.4f** (Scheme 12.4) and then nucleophile attack of 7-azaindole **12.4b** via elimination of N_2 would form the intermediate **12.4g**, which then undergoes intermolecular nucleophilic cyclization to form intermediate **12.4h** with the elimination of TsOH. Next, it underwent AIBN-mediated oxidation to form the product **12.4c**.

In 2019, the Hamze group developed a metal-free approach for synthesizing pyridylalkylamine **12.5c** derivatives by coupling pyridotriazoles **12.5a** with weakly nucleophilic aniline **12.5b** compounds [10]. TBAB in a catalytic amount and water

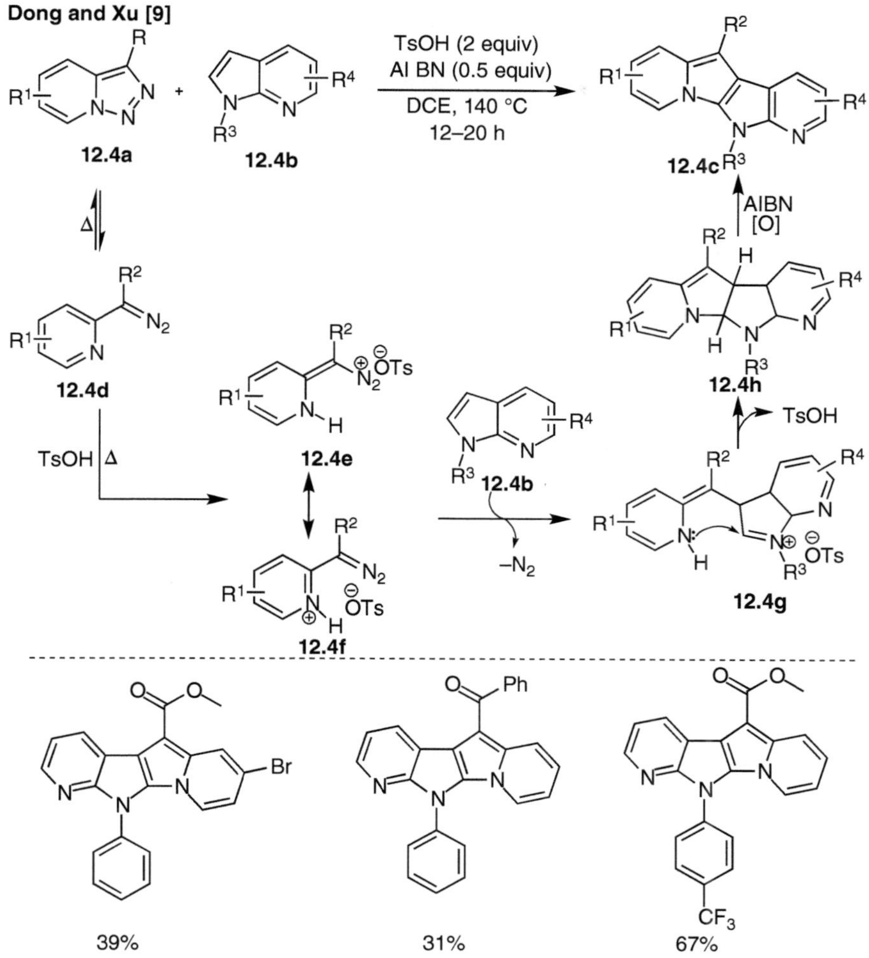

Scheme 12.4 Pyridoindoloindolizines synthesis through Lewis-acid mediated transannulation. Source: Dong and Xu [9]/John Wiley & Sons.

as a co-solvent was used for the success of this transformation, which resulted in the synthesis of pyridylalkylamine derivatives **12.5c**. When pyridotriazole **12.5a** is heated, it undergoes a ring-opening form of diazo compound **12.5d**. In water, the diazo compound **12.5d** would form the intermediate **12.5e** (Scheme 12.5). Next, the nucleophilic bromine of TBAB reacted with the electrophilic carbon of the diazo intermediate **12.5e**, forming the benzyl bromide **12.5f** and subsequent reaction with aniline to form the **12.5g**. Subsequently, this intermediate **12.5g**, upon deprotonation in the presence of tetrabutylammonium hydroxide (TBAH) to form pyridylalkylamine derivatives **12.5c**.

In 2020, the Dehaen group reported acid-mediated denitrogenative transformation for the insertion of C(carbon)-nucleophiles, heteroatom-containing nucleophiles with triazolo[5,1-*a*]isoquinolines to the synthesis of methylene-substituted

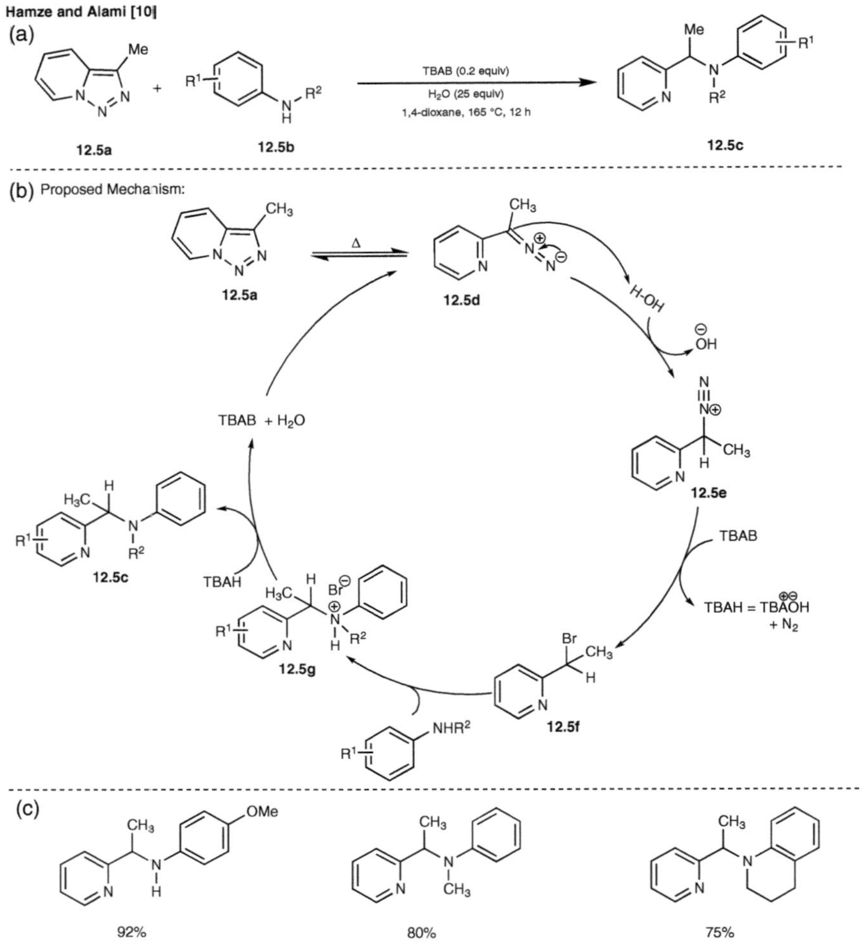

Scheme 12.5 Pyridylalkylamine derivatives from pryridotriazoles. (a) Reaction scheme. (b) Proposed mechanism. (c) Selected examples. Source: Hamze and Alami [10]/John Wiley & Sons.

isoquinoline derivatives [11]. Most C-nucleophiles produced the mixture of regioisomers **12.6d** (Scheme 12.6) in the presence of $HBF_4{\cdot}OEt_2$. In contrast to their nonpolar counterparts, polar nucleophiles [12], for example, heteroatom-containing nucleophiles such as amines, alcohols, and thiols, smoothly reacted in the presence of trifluoroacetic acid under slightly modified conditions to provide the appropriate products in good yields with single regioisomer **12.6c** (Scheme 12.6). This methodology highlights the significance of a methoxy substituent at the isoquinoline moiety, which is crucial for the ring-opening process. Furthermore, the authors also show the synthesis of imidazo[5,1-*a*]isoquinolines via transannulation with nitriles.

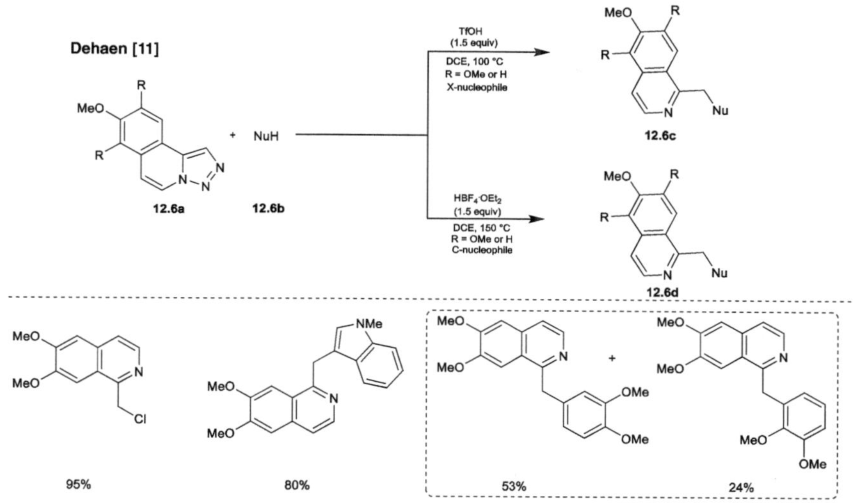

Scheme 12.6 Acid-mediated synthesis of methylene-substituted isoquinolines. Source: Dehaen [11]/American Chemical Society.

12.3 Denitrogenative Transformations of Pyridotriazoles for Forming C–C and C–*X* Bond

12.3.1 Coupling of Boronic Acids with Pyridotriazoles

Shen group in 2019 reported the metal-free cross coupling of arylboronic acids with pyridotriazoles to access the substituted pyridines [13]. The process has straightforward, scaling, simple reaction conditions, good functional tolerance, and a broad substrate scope. Arylboronic acids treated with pyridotriazoles at 100 °C in 1,4-dioxane obtained the cross-coupling product with good yields (Scheme 12.7). Diazo form **12.7d** of pyridotriazole reacts with boronic acid **12.7b** to form 2-picolylboronic acid **12.7f** via a boronate intermediate **12.7e**, and this intermediate **12.7f** then undergoes protodeboronation to give the final product **12.7c** (Path *a*). Since phenylboroxine is also in equilibrium with phenylboronic acid, and in path *b*, boroxine, being the reactive partner, would form the intermediate **12.7g**, followed by 1,2-migration lead to the **12.7h**, and finally, rapid hydrolysis gives the final product **12.7c** (path *b*).

12.3.2 Iodine and Bromine Promoted Transformations of Pyridotriazoles

Recently Li group reported the introduction of the geminal dihalo atoms at the α-position of 2-methylpyridine [14]. They have studied the pyridotriazole **12.8a** with I_2 in acetonitrile as solvent under the nitrogen atmosphere at 80 °C and

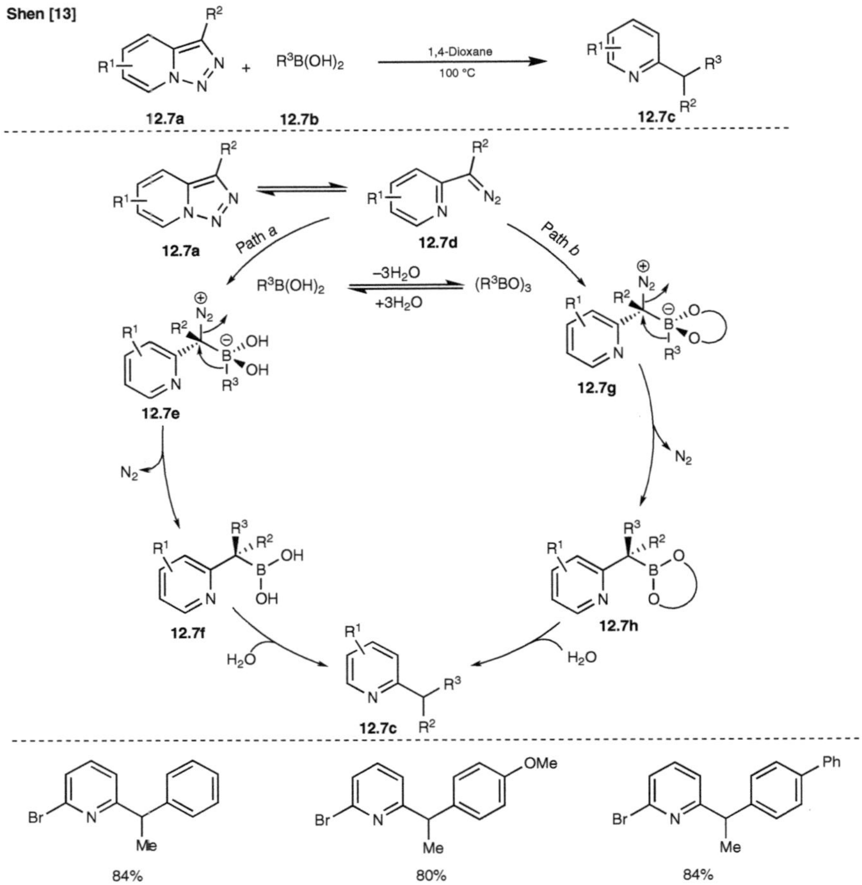

Scheme 12.7 Metal-free cross coupling of arylboronic acids with pyridotriazoles. Source: Shen [13]/American Chemical Society.

isolated the geminal diiodo product **12.8b** in 77% yield. Encouraged by these results, they extended this methodology for synthesizing geminal dibromo compounds **12.8c** in toluene as a solvent within a minute (Scheme 12.8). The in situ generated nucleophilic carbon of diazo compound **12.8d** would attack the halogen atom to give intermediate **12.8e** and the halide anion, which then subsequently displace the nitrogen in an S_N2 fashion, generating the geminal dihalogen-containing methylpyridine derivatives **12.8f** (Scheme 12.8).

12.4 Thermolysis and Photolysis of Pyridotriazoles

In 1978, the Wentrup group reported the thermolysis of pyridotriazoles **12.9a**, which are in equilibrium with the corresponding diazo compound **12.9b**. This diazo

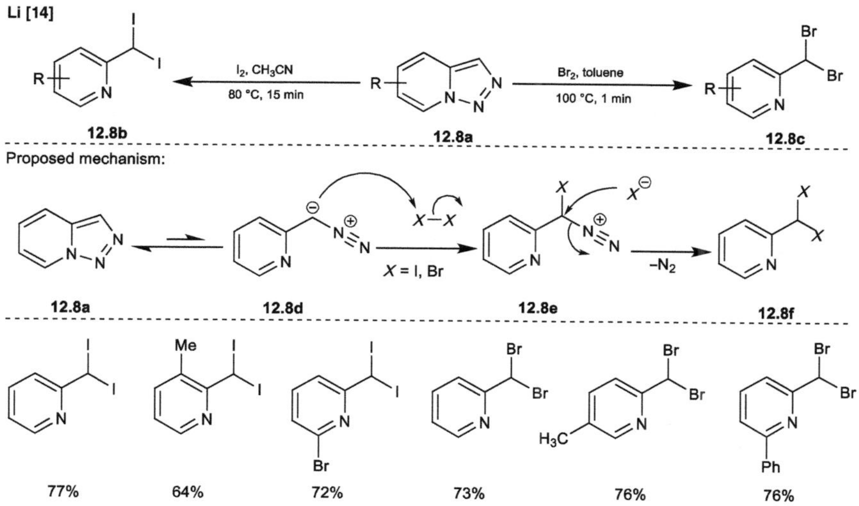

Scheme 12.8 Synthesis of *gem*-di-halo products from pyridotriazoles. Source: Li [14]/John Wiley & Sons.

compound would form carbene, and it was the first report for capturing these carbenes [15]. They performed the reaction of pyridotriazole **12.9a** with fumaronitrile **12.9c** at 150 °C giving the product 3-(2-pyridyl)-1,2-cyclopropanedicarbonitrile **12.9e** (Scheme 12.9) with 32% yield. They suggested that the diazo compound **12.9b** initially would form the 1-pyrazoline **12.9d** through a 1,3-dipolar cycloaddition reaction with fumaronitrile **12.9c**, which then underwent denitrogenation to form the cyclopropane product **12.9e**.

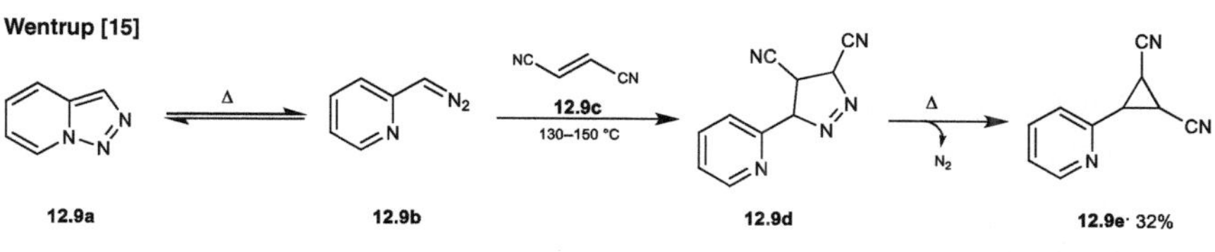

Scheme 12.9 Cyclopropanedicarbonitrile synthesis via thermolysis of pyridotriazoles. Source: Wentrup [15]/John Wiley & Sons.

Based on their previous results, the Wentrup group in 1986 reported the carbene generation from quinoline **12.10a** and isoquinoline-based triazoles **12.10c** upon photo-induced (UV) decomposition (Scheme 12.10a) [16]. Later, in 1993, the Tomioka group showed photo-irradiation of the methyl [1,2,3]triazolo[1,5-*a*] pyridine-3-carboxylate **12.10e** with light having a wavelength greater than 300 nm and observed the prominent band of ketene **12.10f**. They have demonstrated the ketene formation [17] via in situ generated carbene from the pyridotriazole (Scheme 12.10b). The Wentrup group also observed ketene formation from keto and ester-substituted pyridotriazole [18]. At first, photo-irradiation of pyridotriazole

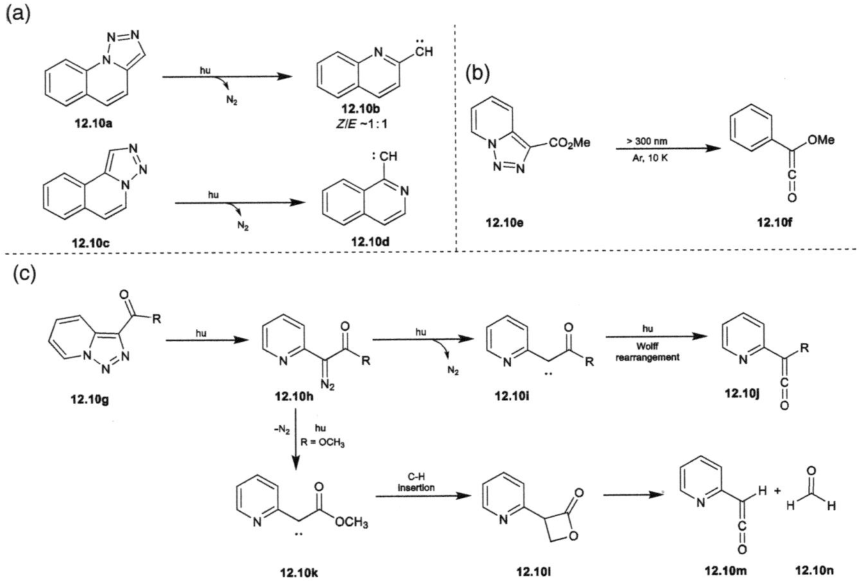

Scheme 12.10 Photolysis of substituted pyridotriazoles. Source: (a) Wentrup [16]/John Wiley & Sons. (b) Komatsu [17]/American Chemical Society. (c) Wentrup [18]/American Chemical Society.

forms a diazo compound, which then undergoes denitrogenation to form carbenes, and these carbenes undergo Wolff rearrangement to form ketene **12.10j**; also observed similar ketene formation from the ester-containing pyridotriazole derivatives. Ketene **12.10m** (Scheme 12.10c) was found to be the major product, along with formaldehyde **12.10n**, and this could be explained by C—H insertion of singlet carbene to tethered methoxy of pyridyl carbene **12.10k** (Scheme 12.10c) followed by fragmentation.

In 1999, the Zimmerman group developed the cyclopropanation of pyridotriazoles **12.11a** using cyclopentadiene **12.11b**. Pyridotriazole, in the presence of UV light, generates a carbene intermediate, which then undergoes cyclopropanation [19] to give a mixture of *exo* and *endo* products (Scheme 12.11a). Later, in 2007, the Tidwell group reported the formation of ylide **12.11f** [20] from 3-pyridyl-substituted pyridotriazoles **12.11e** upon photolysis in CH_3CN as a solvent (Scheme 12.11b).

In 2002, the Wentrup group showed the synthesis of quinolizine-2,4-diones by flash vacuum thermolysis (FVT) of pyridotriazoles [21]. On FVT of pyridotriazoles **12.12a** gives corresponding diazo compound **12.12b** forming carbene **12.12c**. This carbene underwent to generate ketenes predominately in the *S-trans* form of methyl-2-pyridylketenes and phenyl-2-pyridylketenes **12.12d** (Scheme 12.12). Finally, underwent preparative FVT gave a crystalline compound of quinolizine-2,4-diones **12.12e**.

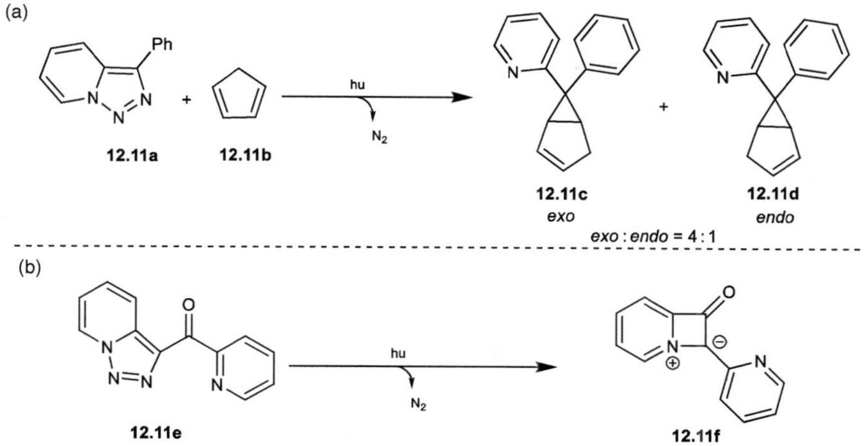

Scheme 12.11 Cyclopropanation and ylide formation from pyridotriazoles. Sources: (a) Zimmerman [19]/American Chemical Society. (b) Tidwell [20]/American Chemical Society.

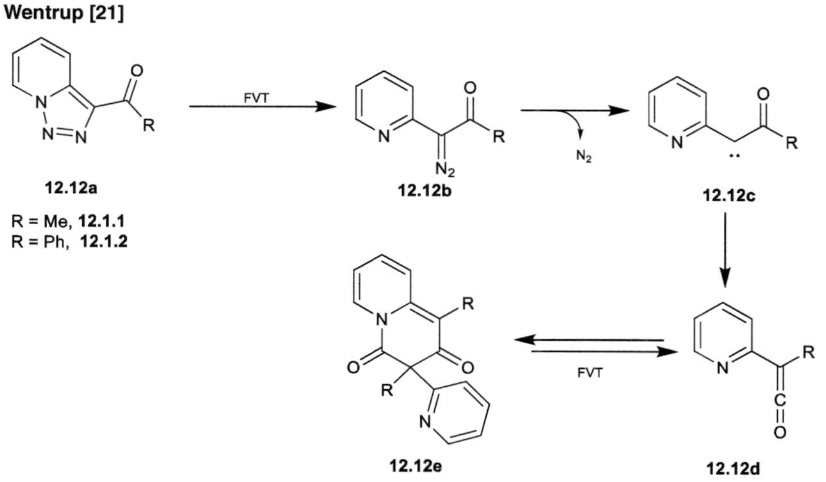

Scheme 12.12 Flash vacuum thermolysis of 3-acyl substituted pyridotriazoles. Source: Wentrup [21]/American Chemical Society.

12.5 Light-Induced Transformations of Pyridotriazoles

In 2019, the Gevorgyan group developed a highly efficient methodology for incorporating arylmethylpyridyl moiety into diverse molecules. This methodology involves light-induced arylation, X—H insertion, and cyclopropanation reaction of pyridyl carbenes generated from pyridotraizoles [22]. Based on UV-absorption studies,

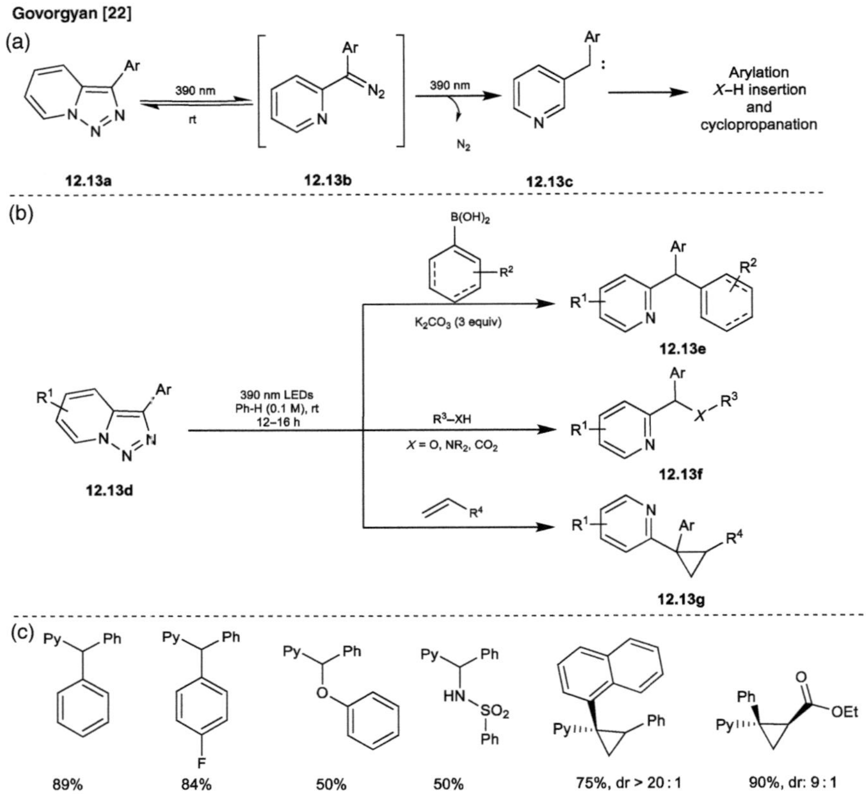

Scheme 12.13 Light-induced arylation, X–H insertion, and cyclopropanation of pyridotriazoles. Source: Govorgyan [22]/The Royal Society of Chemistry.

they have utilized the 390 nm LED for ring-chain tautomerism of the pyridotriazole **12.13a** in equilibrium with its diazo tautomer **12.13b** (Scheme 12.13a), which then forms reactive pyridyl carbene intermediate **12.13c** in the presence of 390 nm LED upon elimination of N_2. Thus, it shows the synthesis of synthetic intermediates (Scheme 12.13a).

They first examined a metal-free arylation reaction of 3-aryl substituted pyridotriazole **12.13d** with aryl/alkenyl boronic acids in the presence of K_2CO_3 as a base under 390 nm LEDs at room temperature and synthesized various substituted diarylpyridylmethane derivatives **12.13e** with excellent yields (Scheme 12.13c). Next, under the standard condition, they performed X—H insertion reaction with pyridyl carbene. They unfold the methodology with O—H insertion with different alcohols, phenol, and carboxylic acids (Scheme 12.13c). They also developed a methodology for N—H insertion with phthalimide and sulfonamides (Scheme 12.13c). Furthermore, when 3-aryl-substituted pyridotriazoles were irradiated in the presence of 390 nm with different alkenes, it resulted in cyclopropanation product via [2+1]-cycloaddition reaction (Scheme 12.13c). However, this approach is only effective with C3-aryl-substituted pyridotriazoles failed other than aryl-substituted pyridotriazole, which includes 3-alkyl-susbtituted-pyridotriazoles such as 3-methyl-pyridotraizoles, which are transparent in this region and do not react under optimized conditions.

12.6 Summary and Outlook

In this book chapter, we explored the metal-free denitrogenative transformations of pyridotriazole. α-Diazoimine, being the reactive partner, is involved in several transannulations, cyclopropanations, and other rearrangement reactions. However, there is still room to expand this chemistry, and new methodologies involving carbene formation from α-diazoimine can be developed. Furthermore, the inherent properties and synthetic potentials of pyridotriazoles in asymmetric chemistry remain unexplored. Since the generation of α-diazoimine from pyridotriazole requires harsh conditions, developing modest techniques for pyridotriazole activation may further unlock the potential of this chemistry.

Acknowledgments

We thank the DST-SERB-CRG (CRG/2022/000442), New Delhi, India, for financial support. H.S. and A.R. sincerely thank the PMRF and UGC for fellowship, respectively.

References

1 (a) Michael, J.P. (2001). *Simple Indolizidine and Quinolizidine Alkaloids. In the Alkaloids: Chemistry and Biology*, vol. 55, 91–258. Academic Press. (b) Michael, J.P. (2002). Quinoline, quinazoline and acridone alkaloids. *Nat. Prod. Rep.* 19: 742–760. (c) Michael, J.P. (2007). Indolizidine and quinolizidine alkaloids. *Nat. Prod. Rep.* 24: 191–222. (d) Michael, J.P. (2008). Indolizidine and quinolizidine alkaloids. *Nat. Prod. Rep.* 25: 139–165. (e) Rogness, D.C., Markina, N.A., Waldo, J.P., and Larock, R.C. (2012). Synthesis of pyrido[1,2-*a*]indole malonates and amines through aryne annulation. *J. Org. Chem.* 77: 2743–2755. (f) Sharma, V. and Kumar, V. (2014). Indolizine: a biologically active moiety. *Med. Chem. Res.* 23: 3593–3606.

2 (a) Jones, G. (2002). The chemistry of the triazolopyridines: an update. *Adv. Heterocycl. Chem.* 83: 1–70. (b) Li, C.G., Xu, G.Q., and Xu, P.F. (2018). Synthesis of quinazolinone derivatives via a visible-light photocatalyzed denitrogenation rearrangement process. *J. Photochem. Photobiol.* 355: 25–31.

3 (a) Regitz, M. (1967). New method of preparative organic chemistry. Transfer of diazo compound. *Angew. Chem. Int. Ed.* 6: 733–749. (b) L'Abbé, G., Godts, F., and Toppet, S. (1985). Diazoimine-triazole equilibrium in fused 1,2,3-triazolo[1,5-*a*]pyrimidines. *J. Chem. Soc. Chem. Commun.* 1985: 589–590.

(c) L'Abbé, G., Godts, F., and Toppet, S. (1986). On the equilibrium position of bis[1,2,3] triazolopyrimidines. *B. Bull. Soc. Chim. Belg.* 95: 679–680. (d) L'Abbé, G. (1990). Molecular rearrangements of 1h-1,2,3-triazoles and 1,2,3-hiadia-zoles. *B. Bull. Soc. Chim. Belg.* 99: 281–291. (e) Aylward, N., Winter, H.W., Eckhardt, U., and Wentrup, C. (2016). Triazoloazine-diazomethylazine valence isomerization [1,2,3]-triazolo[1,5-*a*]pyridines and 2-diazomethylpyridines. *J. Org. Chem.* 81: 667–672.

4 Chuprakov, S., Hwang, F.W., and Gevorgyan, V. (2007). Rh-catalysed transannulation of pyridotriazoles with alkynes and nitriles. *Angew. Chem. Int. Ed.* 46: 4757–4759.

5 (a) Yadagiri, D., Rivas, M., and Gevorgyan, V. (2020). Denitrogenative transformations of pyridotriazoles and related compounds: synthesis of N-containing heterocyclic compounds and beyond. *J. Org. Chem.* 85: 11030–11046. (b) Shen, R., Yin, K., Yu, M., and Li, X. (2023). Progress on pyridotriazoles chemistry: synthesis and application. *Asian J. Org. Chem.* 12: e202200631. (c) Happy, S., Junaid, M., and Yadagiri, D. (2023). Reactivity of quinone methides with carbenes generated from α-diazocarbonyl compounds and related compounds. *Chem. Commun.* 59: 29–42.

6 Joshi, A., Mohan, C.D., and Adimurthy, S. (2016). Lewis acid-catalyzed denitrogenative transannulation of pyridotriazoles with nitriles: synthesis of imidazopyridines. *J. Org. Chem.* 81: 9461–9469.

7 Rawat, D., Ravi, C., Joshi, A. et al. (2019). Indium-catalyzed denitrogenative transannulation of pyridotriazoles: synthesis of pyrido[1,2-*a*]indoles. *Org. Lett.* 21: 2043–2047.

8 Adimurthy, S. (2021). $BF_3 \cdot Et_2O$ catalyzed transannulation of pyridotriazoles with isothiocyanates: synthesis of thiazolo[3,4-*a*]pyridin-3-imines. *New J. Chem.* 45: 20547–20550.

9 Zhang, G.T., Zhang, J., Xu, Y.J., and Dong, L. (2018). Metal-free [3 + 2] tandem cyclization synthesis of unique 11*H*-pyrido[3′,2′ : 4,5]-pyrrolo[3,2-*b*]indolizine from 7-azaindoles and pyridotriazoles. *Eur. J. Org.* 2018: 4197–4201.

10 Lamaa, D., Lin, H.P., Bzeih, T. et al. (2019). A TBAB-catalyzed Csp^3—N bond formation by coupling pyridotriazoles with anilines: a new route to (2-pyridyl)alkylamines. *Eur. J. Org. Chem.* 2019: 2602–2611.

11 Opsomer, T., Van Hoof, M., D'Angelo, A., and Dehaen, W. (2020). 1,2,3-Triazole-mediated synthesis of 1-methyleneisoquinolines: a three-step synthesis of papaverine and analogues. *Org. Lett.* 22: 3596–3600.

12 Mayr, H. and Ofial, A.R. (2008). And it works: nucleophilicity scales for synthesis planning. *Nachr. Chem.* 56: 871–877.

13 Dong, C., Wang, X., Pei, Z., and Shen, R. (2019). Metal-free denitrogenative C-C coupling of pyridotriazoles with boronic acids to afford α-secondary and α-tertiary pyridines. *Org. Lett.* 21: 4148–4152.

14 Tang, T., Chen, J., Xu, Z.F. et al. (2023). Iodine- and bromine-promoted transformations of pyridotriazoles: general access to geminal diiodo and dibromo methylpyridines. *Asian J. Org. Chem.* 12: e202300026.

15 Wentrup, C. (1978). [1,2,3]Triazoloanzine/(diazomrthyl)azine valence tautomers from 5-azinyltetrazoles. *Helv. Chim. Acta.* 61: 1755–1763.

16 Kuzaj, M., Lüerssen, H., and Wentrup, C. (1986). ESR observation of thermally produced triplet nitrenes and photochemically produced triplet cyclohepta-trienylidene. *Angew. Chem. Int. Ed.* 25: 480–482.

17 Tomioka, H., Ichikawa, N., and Komatsu, K. (1993). Photochemistry of 2-(methoxycarbonyl)phenyl azide studied by matrix-isolation spectroscopy. A new slippery energy surface for phenylnitrene. *J. Am. Chem. Soc.* 115: 8621–8626.

18 Kuhn, A., Plüg, C., and Wentrup, C. (2000). Direct observation of a carbene-pyridine ylide by matrix IR spectroscopy. Rearrangements of 2-pyridylacylcarbenes. *J. Am. Chem. Soc.* 122: 1945–1948.

19 Zimmerman, H.E. and Ignatchenko. (1999). A control of the stereochemistry of kinetics protonation: intramolecular proton delivery. *J. Org. Chem.* 64: 6635–6645.

20 Fu, N., Allen, A.D., Kobayashi, S. et al. (2007). Observable azacyclobutenone ylides with antiaromatic character from 2-diazoacetyl-azaaromatics. *J. Am. Chem. Soc.* 129: 6210–6215.

21 Plüg, C., Kuhn, A., and Wentrup, C. (2002). Quinolizine-2,4-diones by reversible dimerisation 2-pyridylketenes. *J. Chem. Soc. Perkin Trans.* 12002: 1366–1368.

22 Zhang, Z., Yadagiri, D., and Gevorgyan, V. (2019). Light-induced metal-free transformations of unactivated pyridotriazoles. *Chem. Sci.* 10: 8399–8404.

13

Recent Development in Denitrogenative Transannulation of Benzotriazoles

Mangal S. Yadav[1], Manoj K. Jaiswal[1], Tanu Gupta[2], and Vinod K. Tiwari[1]

[1] *Department of Chemistry, Institute of Science, Banaras Hindu University, Varanasi, Uttar Pradesh, India*
[2] *Department of Chemistry, Rajendra College, Jai Prakash University, Chapra, Bihar, India*

13.1 Introduction

Heterocyclic compounds are the most prominent class of organic compounds having too much diversity in their structure as well as in synthetic utility and diversely used in the field of medicinal chemistry [1–4]. They have wide industrial applications such as used as corrosion inhibitors, dyestuffs, sanitizers, and in the synthesis of copolymers, and also show diverse biological activities like anticancer, anti-inflammatory, antifungal, antiallergic, antibacterial, anti-HIV, antiviral, and anticonvulsant activities [5–9]. Among the heterocyclic compounds, benzotriazoles represent a prominent class of heterocyclic compounds with diverse biological properties [10–20]. Due to its unique activity and nontoxicity, it is environmentally benign; low cost makes it economically suitable; it has high stability, making it easy to store; and it is a good choice for synthetic utility because of its compatibility with many functionalities, high endurance in a different chemical environment, and its easy introduction and removal in the organic compound (Figure 13.1).

It has been known that benzotriazole has unique reactivity and can be used as a unique synthon for organic transformation in different ways because it can be used as a radical precursor, anion precursor, proton activator, having anion stabilizing property, cation stabilizer, and as a ligand due to the presence of lone pair on nitrogen, helpful in the formation of metal complexes and acts as a good leaving group. In addition, it has been used as a catalyst in several organic reactions, mainly in the coupling reactions. Moreover, its ring cleavage strategy has significant importance for the construction of new carbon–carbon and carbon–heteroatom bonds and facilitates the transannulation reactions for the synthesis of different heterocyclic compounds having different atoms such as nitrogen, sulfur, and oxygen (Figure 13.2).

Due to their unique structural motif, 1,2,3-triazole species readily undergo ring-chain isomerization *via* a Dimroth-type equilibrium to form the diazonium

Denitrogenative Transformation of Nitrogen Heterocycles: Synthesis, Reactions and Applications, First Edition. Edited by Pazhamalai Anbarasan.

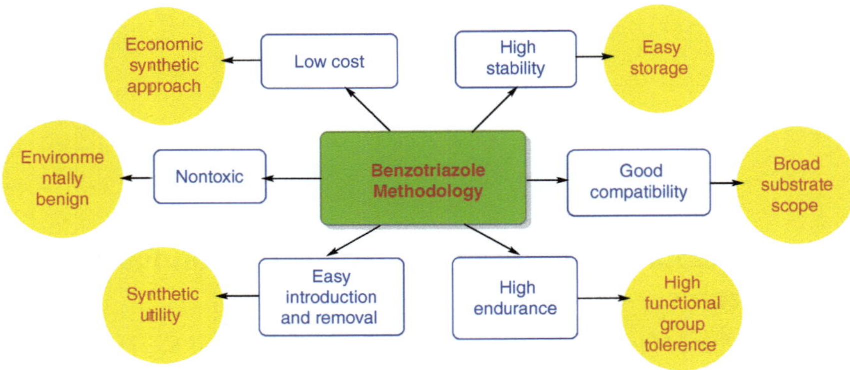

Figure 13.1 Fundamental properties of benzotriazole synthons.

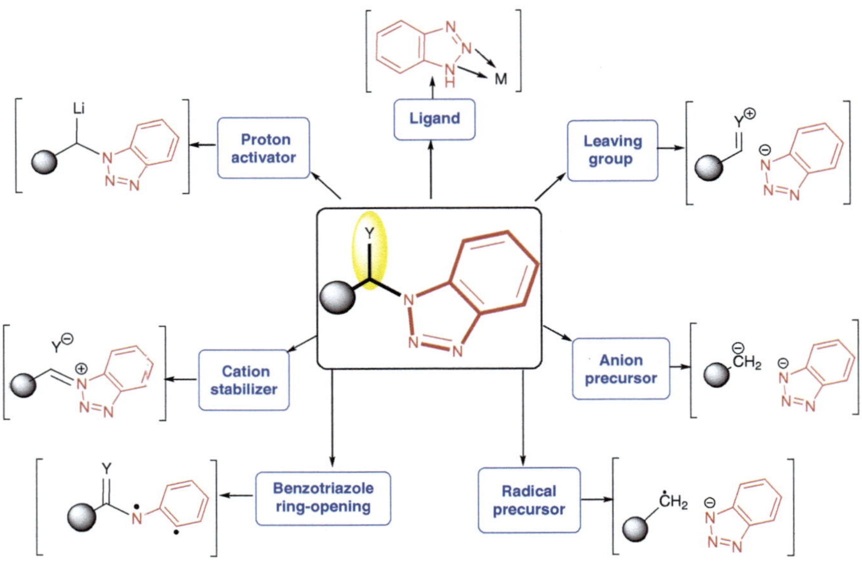

Figure 13.2 Benzotriazole used as a precursor in organic transformation in diverse ways.

or diazo species, while benzotriazoles are more stable and its ring-opening process takes place under harsh condition [21–29], like thermolysis and photolysis and generate *ortho*-amino diazoniums which is used as intermediate for the synthesis of a variety of nitrogen-containing compounds *via* transition metal-catalyzed cross-coupling reactions (Scheme 13.1a) [30–41]. In the past decades, numerous novel reactions have been explored based on this unique reactivity. Among these, much progress has been made in the transition metal-catalyzed diazo reactions of *N*-sulfonyl substituted 1,2,3-triazoles (Scheme 13.1b), which can act as an alternative source of diazo synthon [42–57].

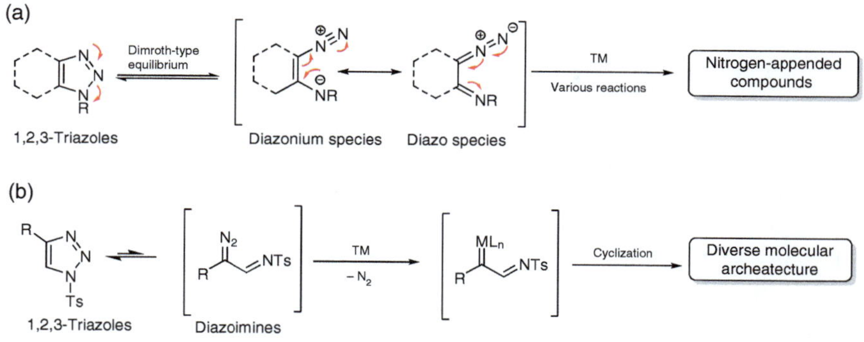

Scheme 13.1 Ring cleavage chemistry of 1,2,3-triazoles and benzotriazole. (a) Dimroth-type equilibrium of 1,2,3-triazoles and diazo/diazonium species used as synthetic equivalents of ortho-amino arenediazoniums. (b) 1,2,3-Triazoles as synthetic equivalents of diazoimines.

In this chapter, we discuss the importance and utility of ring-cleavage of benzotriazole followed by annulation useful for the synthesis of a variety of organic compounds having different heteroatoms like *N*, *O*, and *S*. In 2009, the Nakamura group reported a reaction of *N*-aroylbenzotriazoles with internal alkynes for the synthesis of indole derivatives *via* Pd-catalyzed [3 + 2]-cycloaddition reaction involves a denitrogenative process [58]. This pioneering work became a milestone and benzotriazole derivatives emerged as synthetic precursors of *ortho*-amino arenediazoniums having fascinating chemistry and encouraged the synthesis of a variety of heterocyclic organic compounds from different groups leading to pharmacologically relevant scaffolds **II–XIII**, as shown in Scheme 13.2. A number of transition metal-catalyzed methodologies have been reported in which unsaturated hydrocarbons such as internal and terminal alkynes, 1,3-dienes, allenes, and *N*-allenamides are used as coupling partners with benzotriazoles. These transition metals promote the benzotriazole ring cleavage (BtRC) followed by intermolecular cyclizations leading to the formation of an array of biologically important scaffolds [58–65]. Tiwari's group also reported the synthesis of heterocycles likewise, benzothiazoles and benzoxazoles *via* BtRC followed by cyclization [66–73]. In addition, classical transition-metal catalyzed cross-coupling reactions of benzotriazoles have been reported such as arylation, alkenylation, alkylation, and carbonylation [59, 74–76]. Similar to arenediazonium salts, benzotriazoles can be used as precursors of aryl radical and the radical coupling reactions take place under the initiation of free-radical for the construction of carbon–carbon and carbon–heteroatom bonds [77–89].

Here we focus our attention on the importance and utility of ring-cleavage of benzotriazole followed by annulation useful for the synthesis of a variety of organic compounds. We sincerely hope that this chapter will serve as a useful reference for understanding benzotriazole chemistry in discovering new transformations in organic synthesis.

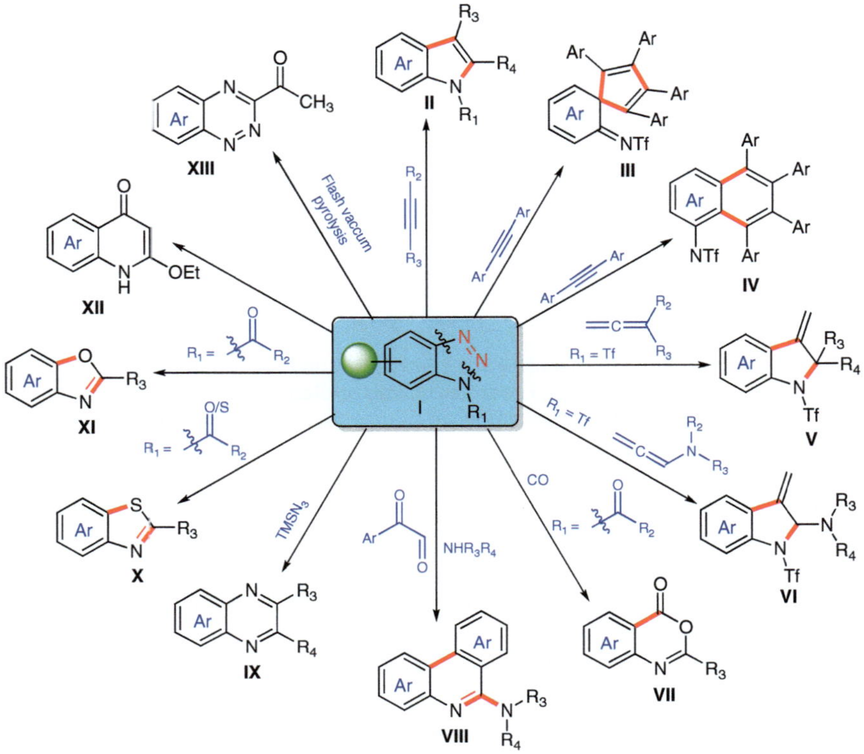

Scheme 13.2 The denitrogenative reactions of benzotriazoles **I** leading to a different biocompatible active heterocyclic compound.

13.2 Common Route for the Synthesis of Benzotriazole Precursors

A number of methodologies have been reported for the synthesis of N1-substituted benzotriazole precursors **I** in good-to-high yields. When *o*-chloro-1,2,3-benzotriazenes **XIV** (obtained from *o*-chloroaniline *via* diazotization followed by coupling with amines) is refluxed with CuI/Cs_2CO_3 in DMF solvent, give the respective benzotriazoles *via* intramolecular *N*-arylation (Scheme 13.3) [90, 91]. Similar expeditious Buchwald–Hartwig-type chemistry has been reported for the synthesis of the same molecule under palladium catalysis [$Pd(OAc)_2$/Cs_2CO_3] [92]. Another method that is furnished for the synthesis of desired *N*1-arylbenzotriazoles **I** through C—H activation of aryl triazines **XV** followed by intramolecular amination under palladium catalysis in the presence of a base [93, 94]. Also cyclocondensation of *o*-(arylamino)arylaminophosphoranes **XVI** in presence of $NaNO_2$ and acetic acid at 0 °C, gives the *N*1-arylbenzotriazoles in good yield within very short time [95]. Alternatively, click chemistry has also been employed for the synthesis of **I** *via* [3 + 2]-cycloaddition benzyne derivatives **XVII** with a variety of organic azides in the presence of CuI/base and give the respective product in high yields [96–98]. Likewise, pyrrolidine catalyzed [3 + 2]-cycloaddition of cyclohexenones **XVIII** with a variety of aryl azides followed by 2,3-dichloro-5,6-dicyano-1,4-benzoquinone (DDQ)-mediated oxidative aromatization in one-pot afforded high yields of *N*1-arylbenzotriazoles **I** [99]. Functionalized NH-benzotriazoles **I** can be synthesized simply and straightforward routes, that is, through copper or palladium-catalyzed NH-arylation of benzotriazole **XIX** [100–105]. Over the past few decades, *N*-acylbenzotriazoles and thioacylbenzotriazoles have been employed as very useful synthons in a number of *N*-, *C*-, *S*-, and *O*-acylation reactions to develop many pharmacologically active molecules.[106] A number of well-established protocols for the *N*-acylation of benzotriazoles are known. As the very first example, a carboxylic acid reacts with thionyl chloride and subsequently reacts with benzotriazole in anhydrous dichloromethane at room temperature to furnish the corresponding *N*-acylbenzotriazoles **XX** in high-to-excellent yields [107]. Other notable examples for the *N*-acylation of benzotriazoles include the reactions of carboxylic acids with NBS/PPh_3/BtH (BtH = benzotriazole) in anhydrous dichloromethane at 0 °C [108], I_2/PPh_3/Et_3N/BtH [109], RSO_2Bt/Et_3N/BtH in THF under refluxing conditions [110, 111], TsCl/Et_3N/DMAP/BtH in dichloromethane [112], 2,4,6-trichloro-1,3,5-triazine (TCT)/BtH/Et_3N in an organic solvent [113], $NaHCO_3$ in an aqueous medium [114], trichloroisocyanuric acid/PPh_3/BtH in dichloromethane [115], *N*-propanephosphonic acid anhydride (T3P@)/pyridine in DMF at room temperature [116], and 2,2′-dipyridylsulfide/Ph_3P/BtH in dichloromethane [117]. *N*-Thioacylbenzotriazoles **XXI** (ROCSBt or RSCSBt)

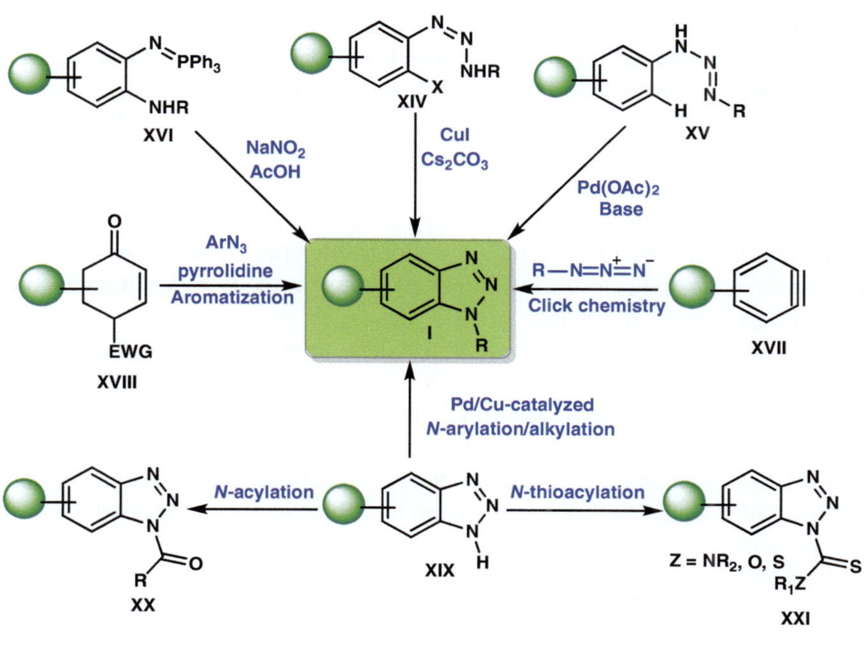

Scheme 13.3 Common synthetic pathway for construction of benzotriazole precursors.

were obtained in excellent yields simply by reacting alcohols or thiols with *bis*(1-benzotriazolyl)methanethione (BtCSBt) in the presence of a suitable base. However, the reactions of secondary amines (R_2NH) with *bis*(1-benzotriazolyl) methanethione furnished the respective *N*-thioacylbenzotriazoles (R_2NCSBt) in excellent yields in 30 minutes, even in the absence of a base (Scheme 13.3).

13.3 Transannulation of Benzotriazole-Appended Scaffolds

Transannulation of benzotriazole ring offers a proficient route for the development of a library of compounds. In this regard, an interesting approach developed by the Nakamura group in 2009 for the synthesis of indole derivatives **13.4c** *via* the reaction of *N*-aroylbenzotriazole **13.4a** with internal alkyne **13.4b** through palladium-catalyzed denitrogenative [3 + 2]-cycloaddition reaction. The reaction was performed under mild conditions to afford the respective product in good yield and easily separable due to the release of nitrogen gas which is formed as a single by-product (Scheme 13.4) [58]. The beauty of this reaction is having broad substrate scope with respect to benzotriazoles and the reaction is purely regioselective on using the unsymmetrical alkyne; the bulkier group comes at the C-2 position of the indole ring and forms as the major product (**13.4cf**: **13.4cg** = 74:26). While the reaction does not proceed with a terminal alkyne. In this reaction, benzotriazoles are efficiently used as *ortho*-amino arene diazoniums, which undergo metal-catalyzed coupling reactions with internal alkyne.

A plausible mechanism has been discussed by the authors. The first step is ring-chain isomerization of benzotriazoles which occurs on heating *via* Dimroth-type equilibrium and forms 2-iminobenzenediazoniums **A**, which undergoes oxidative insertion of Pd(0) and gives Pd(II) intermediate **B** and **B′**, follows insertion of the internal alkyne into the carbon–palladium bond and generates species **C** which finally afforded the respective products *via* reductive elimination with the release of Pd(0) species and completes the catalytic cycle. An obligatory requirement for this reaction is a higher reaction temperature, better facilitated by the presence of a strong electron-withdrawing group on the nitrogen atom for the ring opening of the triazole ring.

Glorius and group in 2017, came with a new approach and performed the reaction of aroylbenzotriazole **13.5a** with terminal alkynes **13.5b** for the synthesis of 2-substituted indole derivatives **13.5c** *via* Ir-catalyzed denitrogenative [3 + 2]-cycloaddition under the irradiation with visible light in moderate to good yields. The reaction has a broad substrate scope with excellent functional group tolerance and is highly regioselective, that is, for achieving the deprotected indoles, a strong withdrawing trifluoromethyl group added on the benzoyl fragment, which provides 2-aryl-substituted indoles without the additional deprotection steps (Scheme 13.5) [59]. The reaction does not proceed with internal alkyne and this scheme shows a complement to the previous scheme [58]. The authors have discussed the plausible mechanism that is based on a radical pathway (Scheme 13.5).

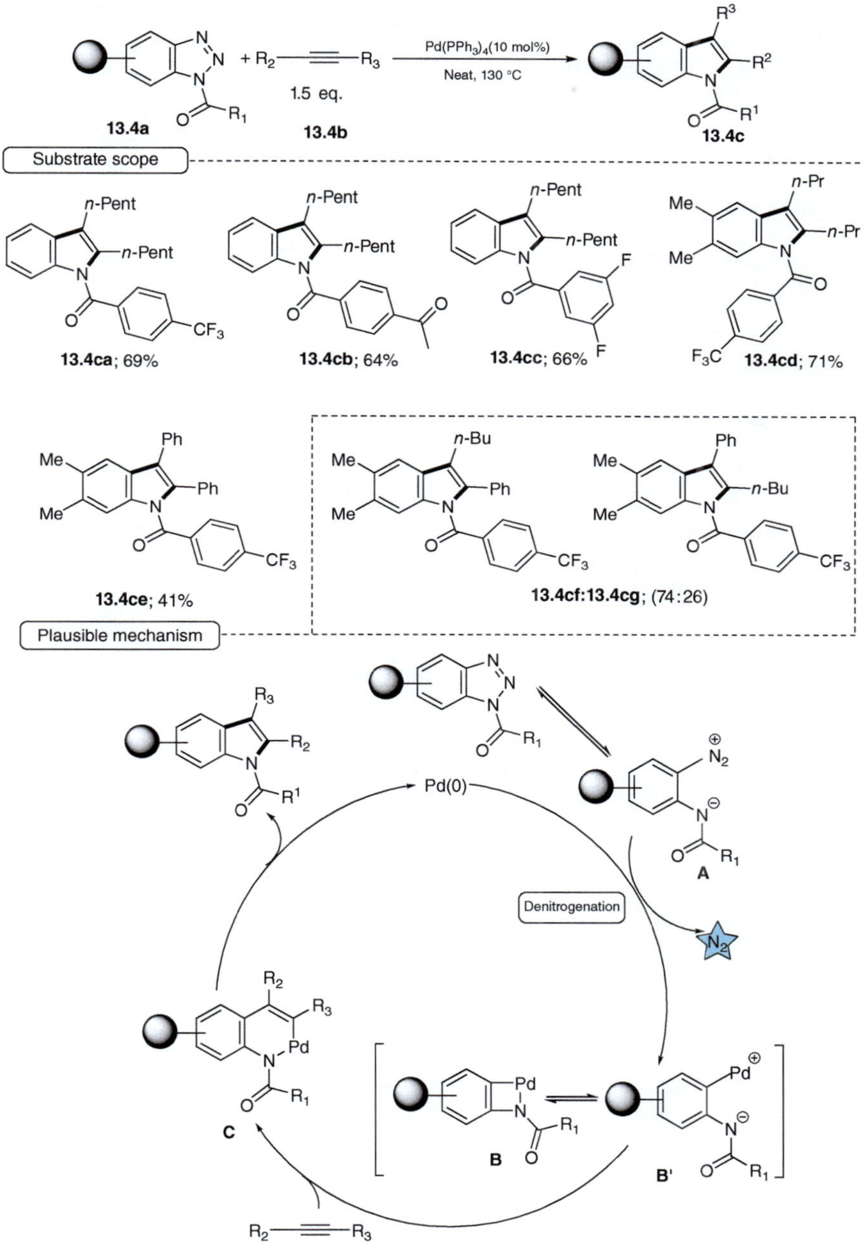

Scheme 13.4 Pd-catalyzed cycloaddition of benzotriazoles with internal alkynes.

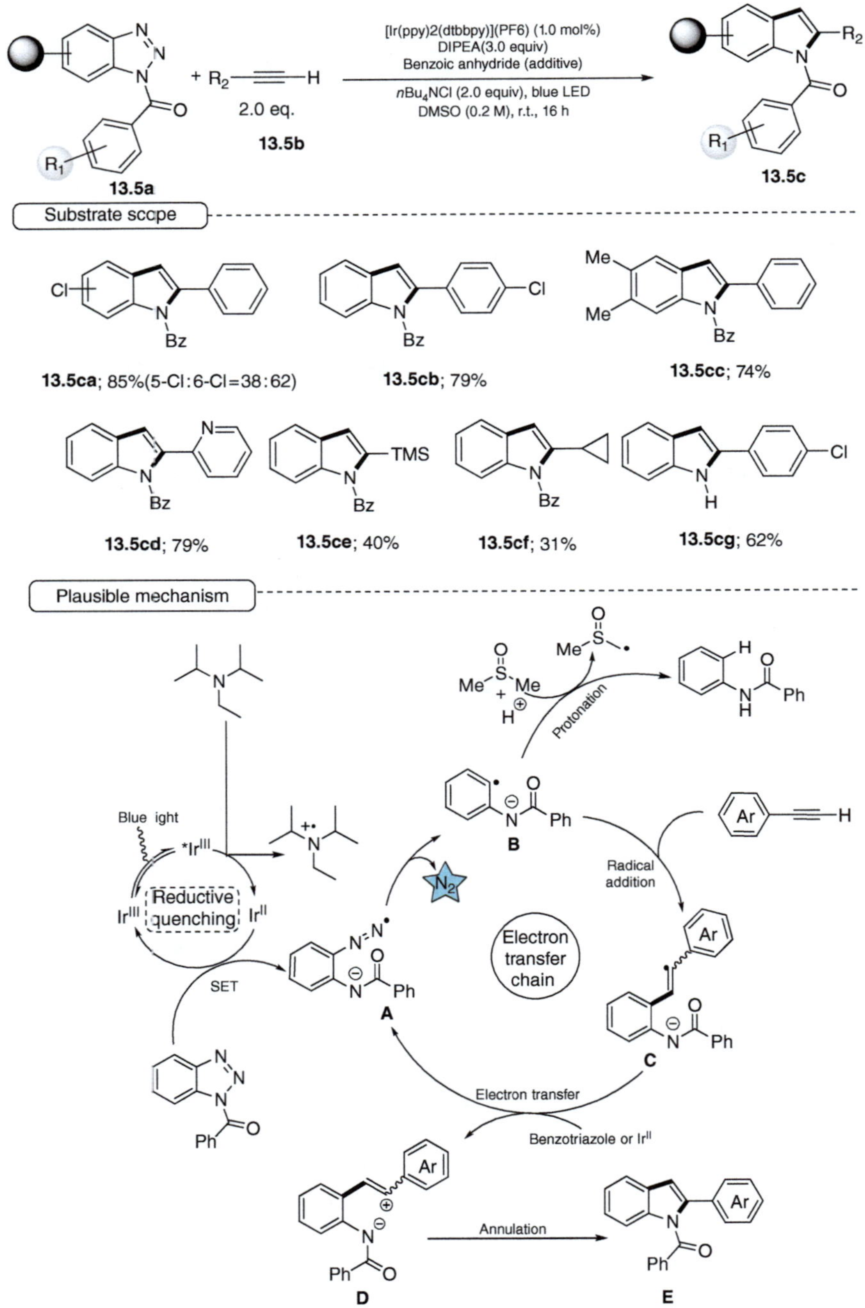

Scheme 13.5 Ir-catalyzed cycloaddition of benzotriazoles with terminal alkynes.

Initially, upon the irradiation of blue light, excited-state Ir(III)* catalyst generates that is reduced by tertiary amine *via* a single-electron transfer process and produces the amine radical cation and Ir(II) species. The interaction of Ir(II) catalyst and benzotriazole produces Ir(III) catalyst and radical anion **A**, which further releases nitrogen gas and generates aryl radical **B**. Further, this aryl radical **B** adds to the triple bond of alkyne and gives the stabilized radical intermediate **C**, which is further oxidized by another molecule of benzotriazole or Ir(II) catalyst and produces the zwitterionic intermediate **D**, undergoes intramolecular cyclization give the respective product.

Further, Tang et al. have discussed reaction of benzotriazole derivatives having triflate group **13.6a** with internal alkynes for the synthesis of 5,6-spiro bicycles **13.6b** and naphthyl amines **13.7a** *via* the Pd-catalyzed novel denitrogenative [2 + 2 + 1] and [2 + 2 + 2] cycloadditions reaction (Schemes 13.6 and 13.7) [60].

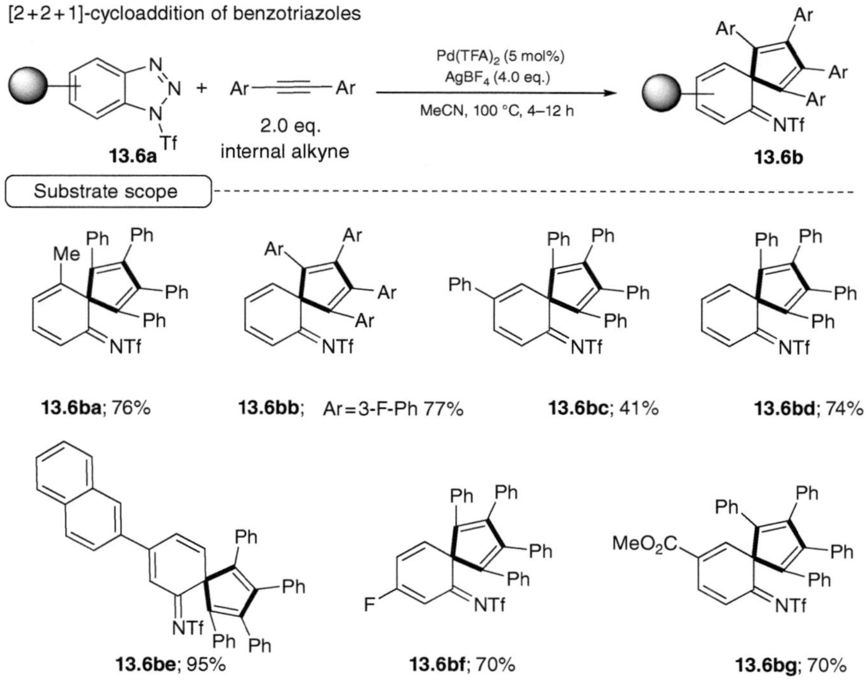

Scheme 13.6 Pd-catalyzed [2 + 2 + 1]-cycloaddition of benzotriazoles with internal alkynes.

It was noticed that, on increasing the temperature, 5,6-spiro bicyclic analogs **13.6b** gradually convert into naphthyl amines **13.7a** *via* a ring-expansion process. In the same article, an alkenylation of benzotriazoles with terminal alkynes to furnish an array of styrene derivatives **13.7b** under a given reaction condition. (Scheme 13.7) [60]. As it has been discussed in previous Schemes 13.4 and 13.5, the indole derivative is not produced by the reaction of benzotriazole with alkyne because of the

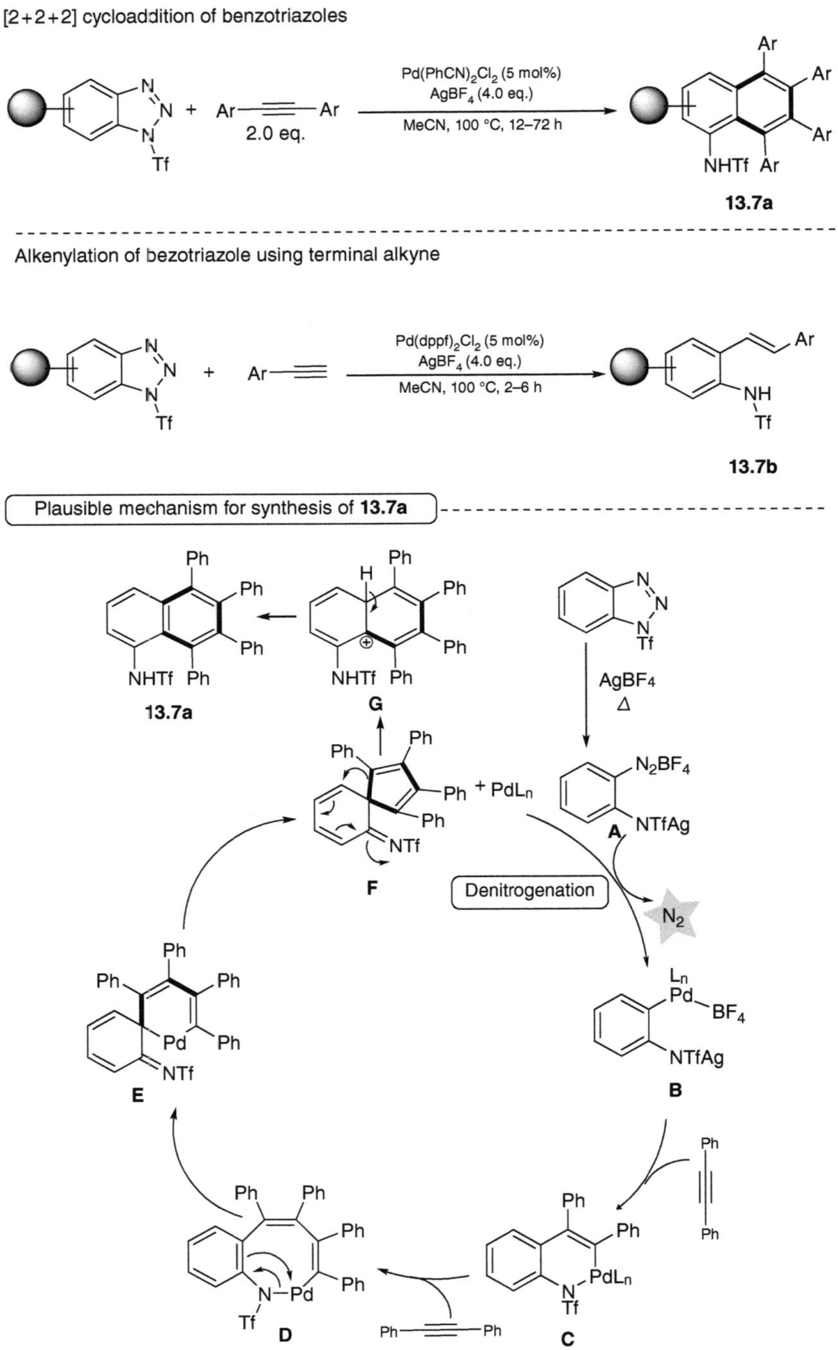

Scheme 13.7 Pd-catalyzed [2 + 2 + 2] cycloaddition of benzotriazoles with internal alkynes.

presence of more electron-deficient substituent (Tf-) on the nitrogen atom of benzotriazole and the use of $AgBF_4$ stabilizing arenediazonium species. The scheme also shows the broad range of substrate scope to benzotriazoles with internal alkynes.

A plausible mechanism has been proposed by the group. They revealed that in assistance with a silver catalyst and at the elevated temperature, the benzotriazole ring opens up and furnished the intermediatory arenediazonium tetrafluoroborate **A**, which undergoes oxidative addition reaction with Pd(0)-catalyst and produces the Pd(II)-complex **B** along with the expulsion of molecular nitrogen. The intermediate **B** follows the insertion of internal alkynes into carbon-palladium and gives palladacycle species **C**. However, species **C** does not undergo reductive elimination and does not produce the respective indole derivative because of the presence of an effective electron-deficient group on the nitrogen atom of benzotriazole. The second molecule of the internal alkyne is assembled in species **C** to produce intermediate **D**, which rearranges itself through 1,3-shift of the metal center and forms six-membered palladacycle species **E**, which further produces 5,6-spiro bicycle **F** *via* reductive elimination. Further on increasing the temperature, the 5,6-spiro bicycle **F** undergoes ring-expansion, leads to the generation of species **G** and finally rearranges itself in such a way as to give the corresponding product **13.7a** that is thermodynamically more stable.

Further, Zhang et al. has discussed the reaction of benzotriazole derivatives **13.6a** with allenes **13.8a** and in the presence of chiral ligand, that is, sulfinamidephosphine-type ligand **12a** produces respective 3-methyleneindoline derivatives **13.8c**. Furthermore, *N*-allenamides **13.9a** and ligand **13.9b** produces corresponding optically active compound 3-methylene indoline derivatives **13.9c**, reaction is highly enantioselective and regioselective, proceeds through Palladium-catalyzed denitrogenative process followed by intermolecular cyclization (Schemes 13.8 and 13.9) [61, 62]. The reaction has a broad substrate scope, progressed viably under mild reaction conditions without base and simply endured at gram scale with excellent yield. This reaction is extremely useful and having good practical value because the product, that is, 3-methyleneindolines can be easily oxidized and gives respective chiral ketone in good yield.

The general mechanistic pathway for the synthesis of developed compounds has been discussed in Scheme 13.9. When reaction mixture was allowed at the appropriate temperature, the benzotriazoles ring form arenediazonium **A** *via* ring-chain isomerization, that is, known as Dimroth-type equilibrium, undergo oxidative addition of Pd(0) catalyst to form intermediate **B** or **B′** accompanied with the expulsion of nitrogen gas, then followed by the insertion of allenes or *N*-allenamides into the carbon-palladium bond gives π-allylpalladium complex **C** or **C′** which finally give respective cyclized products and palladium catalyst regenerate through intramolecular allylic substitution and complete the catalytic cycle.

A report documented by Wu et al. in 2017, and has discussed synthesis of benzoxazinones **13.10c** in assistance of bimetallic catalytic combination of silver and palladium. This catalytic system catalyzed carbonylative activation of benzotriazoles **13.10a** by using carbon monoxide **13.10b** in the place of unsaturated hydrocarbons under mild reaction conditions in moderate to good yields with excellent functional

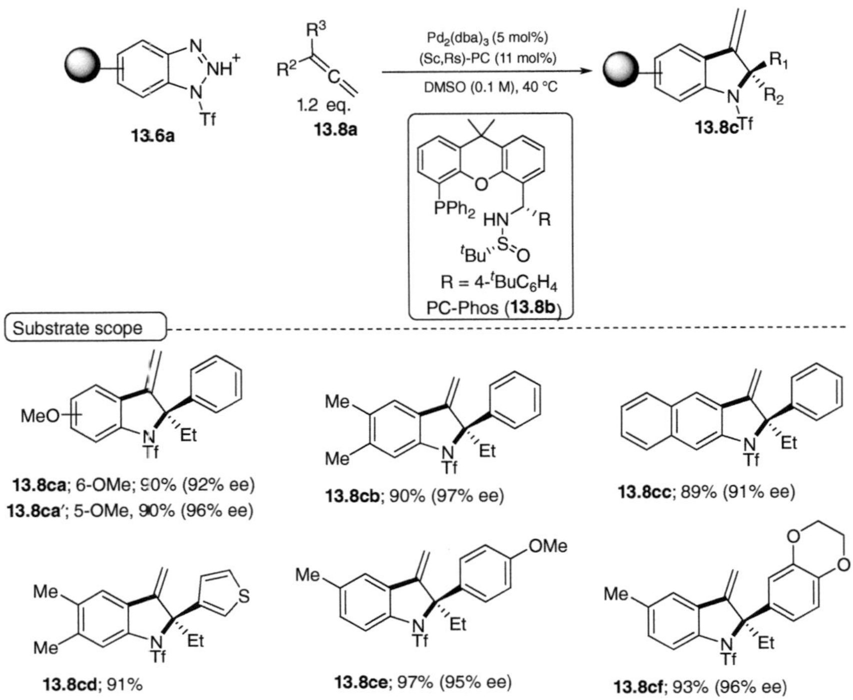

Scheme 13.8 Pd-catalyzed [3 + 2]-cycloaddition of benzotriazoles with allenes.

group tolerance and wide substrate scope (Scheme 13.10) [63]. The plausible mechanism of the reaction has been discussed by the authors (Scheme 13.10). The first step is a temperature-promoted reaction in the presence of silver salt that leads to the formation of arenediazonium intermediate **A**, which is formed from benzotriazoles following the oxidative addition of **A** with Pd(0) catalyst and produces the organopalladium complex **B**, which undergoes insertion reaction with CO to produce a seven-membered intermediate **C**, which finally produces respective product along with the regeneration of active Pd(0) catalyst and complete the catalytic cycle.

A number of methodologies for the synthesis of different heterocyclic compounds have been discussed earlier. In continuation to it, Ahmed et al. first time reported the synthesis of various phenanthridine derivatives **13.11c** through a copper-catalyzed three-component reaction of benzotriazole **13.10a**, 2-oxoaldehydes **13.11a** and secondary amine **13.11b** under very simple reaction condition involve denitrogenative process followed by cyclization reaction (Scheme 13.11) [64]. This protocol is very economical because reactants are very cheap and readily available and shows the wide substrate scope and good functional group tolerance. However, 2-oxoaldehydes, which are *ortho*-substituted, were not suitable for this reaction and benzotriazoles having one substituted group gave a mixture of isomers with compatible regioselectivity. A plausible mechanism has been discussed (Scheme 13.11). The first step is the reaction of 2-oxoaldehydes with secondary amine, which generates 2-oxoiminium ion **A**. Species **A** reacts with benzotriazole in the presence of base

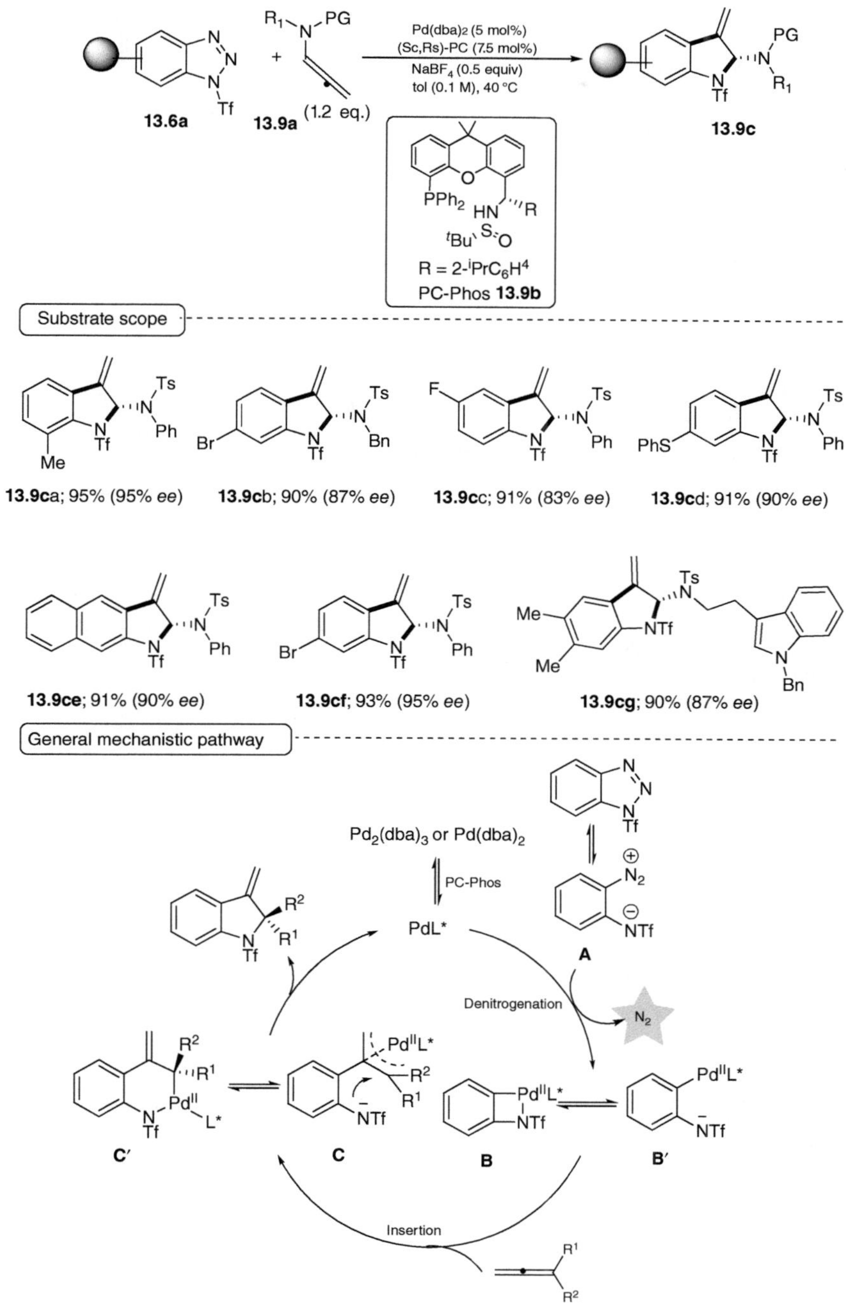

Scheme 13.9 Pd-catalyzed [3 + 2]-cycloaddition of benzotriazoles with *N*-allenamides.

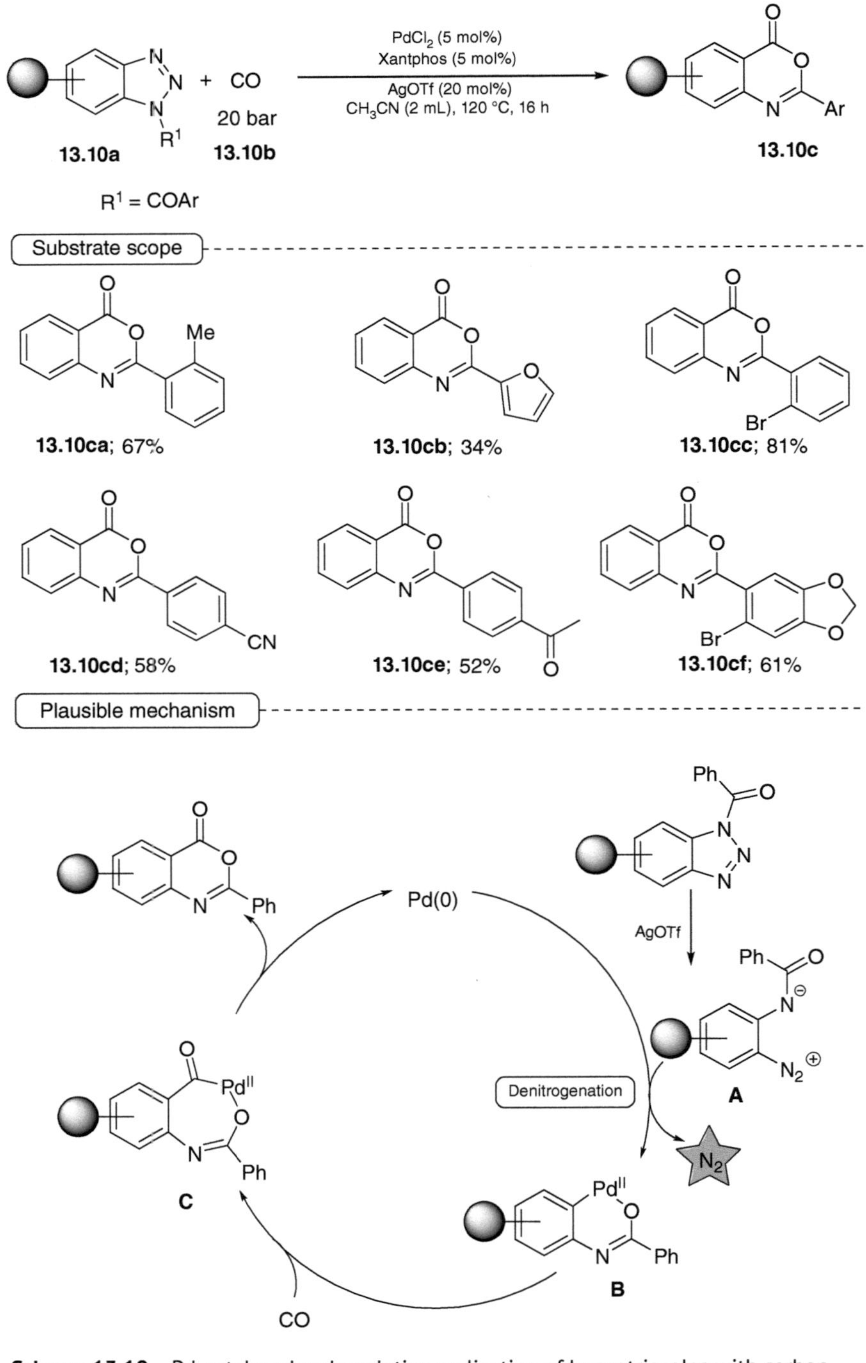

Scheme 13.10 Pd-catalyzed carbonylative cyclization of benzotriazoles with carbon monoxide.

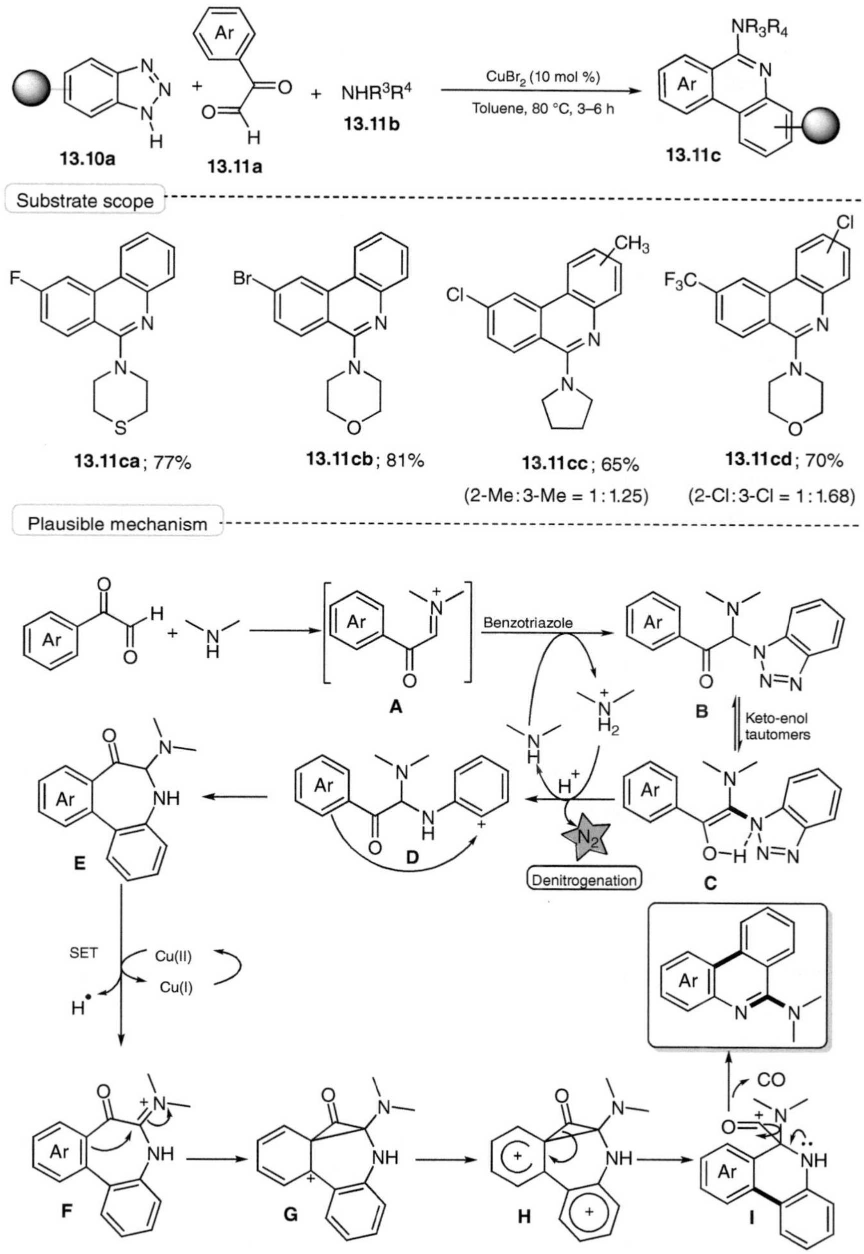

Scheme 13.11 Cu-catalyzed three-component denitrogenative cyclization reaction of benzotriazoles.

form α,α-diamino carbonyl adduct **B**, which undergoes tautomerism and generates its enolic form **C** which undergoes denitrogenation process and produces reactive aryl electrophilic species **D** *via* a five-membered hydrogen-bonding cycle which gives cyclic product **E** through electrophilic substitution reaction and immediately convert into second iminium ion **F** under the oxidative conditions. The sequential formation of intermediate species **G**, **H**, and **I** takes place through rearrangement reaction and finally, the decarbonylation of **I** afforded the respective product.

Free radical cyclization is one of the important methods for the synthesis of a variety of heterocyclic compounds. Recently, Shi et al. discussed the synthesis of quinoxaline **13.12b** *via* the reaction of benzotriazole derivative **13.12a** with azide molecule, that is, $TMSN_3$ in which ring-opening of benzotriazoles take place through the radical addition to *N*-vinyl benzotriazoles **13.12a** under mild conditions (Scheme 13.12) [65]. The reaction is highly chemoselective and nature of product depends on the type of azide radical is generated. The reaction exhibited excellent chemoselectivity by controlling precursors of azide radical for the selective synthesis of quinoxaline or nitrile compounds. This reaction has good substrate scope, that is, the presence of different EDG and EWG on benzotriazole derivative gives the respective product in good to excellent yield with easy handling. A series of substituted quinoxalines could be obtained in moderate to good yields by using as precursors of azide radical. In addition, a wide scope of functional groups on both aromatic and aliphatic *N*-vinylbenzotriazoles was suitable for this reaction. The mechanism has also been discussed showing the synthesis of quinoxaline derivatives is relatively simple *via* the sequential release of nitrogen gas. In this reaction, azide radical is generated *in situ* due to the presence of $PhI(OAc)_2$ along with $TMSN_3$ and is trapped by *N*-vinylbenzotriazoles to afford radical intermediate **A** or **A′**. Subsequently, the ring-opening of triazoles could lead to the formation of aryl radical **B** or **B′**, followed by the release of another molecule of nitrogen gas to produce nitrogen-centered radical **C**, which finally undergoes the oxidative addition reaction and gives the desired product, that is, quinoxaline derivative.

Benzotriazole derivatives **15** have also been used for the synthesis of benzothiazole **24** and benzoxazole derivatives **25** through denitrogenative process followed by intramolecular cyclization reactions. Tiwari's group has recently done lots of work in this field [66–72]. In 2013, the group reported the synthesis of various glycoconjugate benzothiazole derivatives **24** in the presence of tributyltin hydride along with AIBN from protected carbohydrates through the intramolecular cyclization (Scheme 13.13) [66]. The merit of this protocol is the mild reaction condition with short reaction time, simple workup, good to excellent yield, and broad substrate scope with excellent functional group tolerance. The plausible mechanism of radical cyclization for the synthesis of benzothiazole derivative **13.13b** has been discussed. The first step is homolytic cleavage of tributylstannane, which takes place at higher temperatures and in the presence of AIBN. The generation of stannanyl radical takes place, which undergoes addition with thiocarbonyl of benzotriazoles **13.13a** to produce radical intermediate **A**, which rapidly captures hydrogen radical and gives intermediate **B**, where the cleavage of benzotriazole ring occurs through

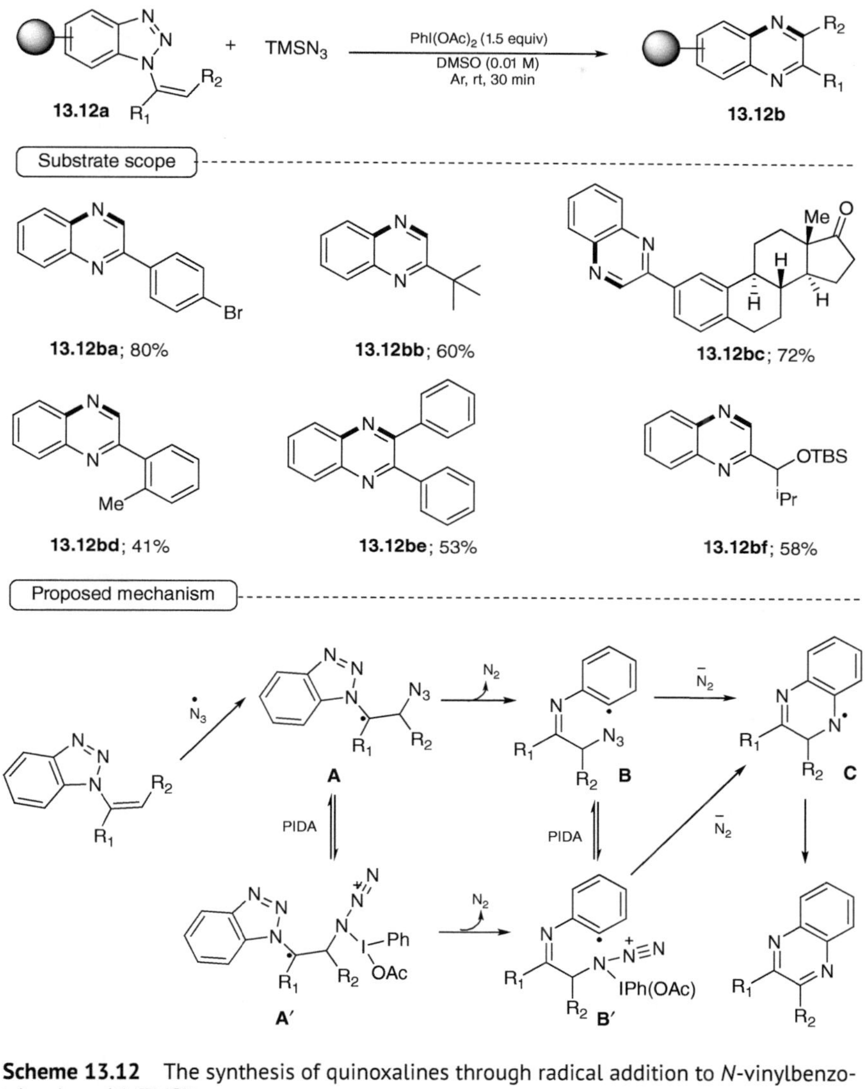

Scheme 13.12 The synthesis of quinoxalines through radical addition to *N*-vinylbenzotriazoles with $TMSN_3$.

the β-scission of N—N bond benzotriazole, led to the formation of biradical intermediate **C**; further, it forms aryl radical intermediate **D** or **E**, that is, resonance stabilized accompanying with the release of nitrogen gas and finally, these intermediate undergo thermally induced oxidation process and give the respective product, that is, benzothiazole **13.13b** which is thermodynamically more stable.

Another methodology is also reported for the synthesis of benzothiazole derivatives in which the reactant was the same as in the earlier, but the first time $(TMS)_3SiH$ in the presence of AIBN is used as a radical reducing agent at the place of toxic stannane which is used for radical cyclization and produce wide

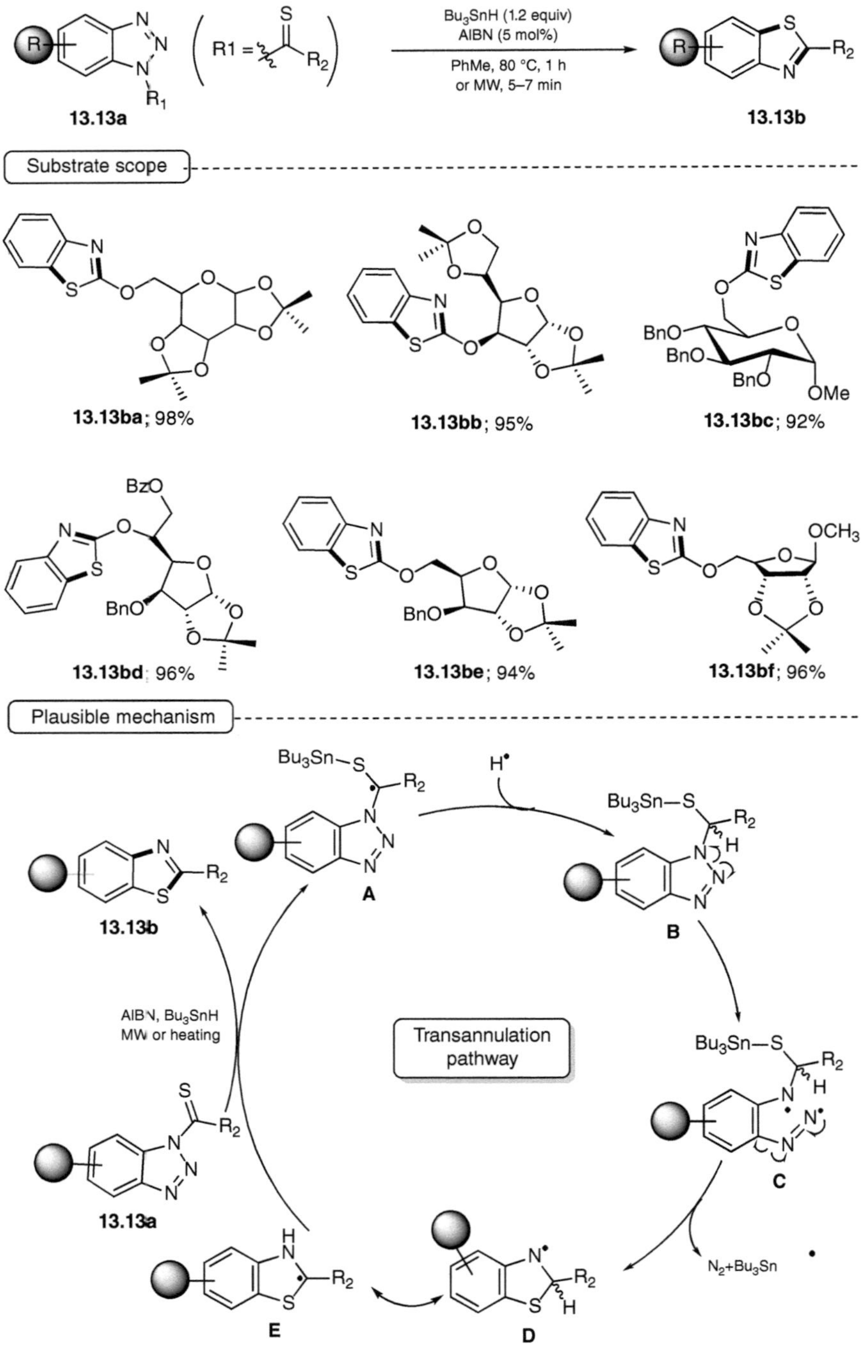

Scheme 13.13 Intramolecular cyclization of benzotriazole derivatives using stannane as a radical reducing agent (Tiwari BtRC route).

variety of 2-N/S/C-substituted benzothiazoles *via* radical catalyzed ring-cleavage of benzotriazole (Scheme 13.14) [67]. The beauty of the reaction is that it can bear a wide variety of functional group tolerance and also can have a broad substrate scope with good-to-excellent yield.

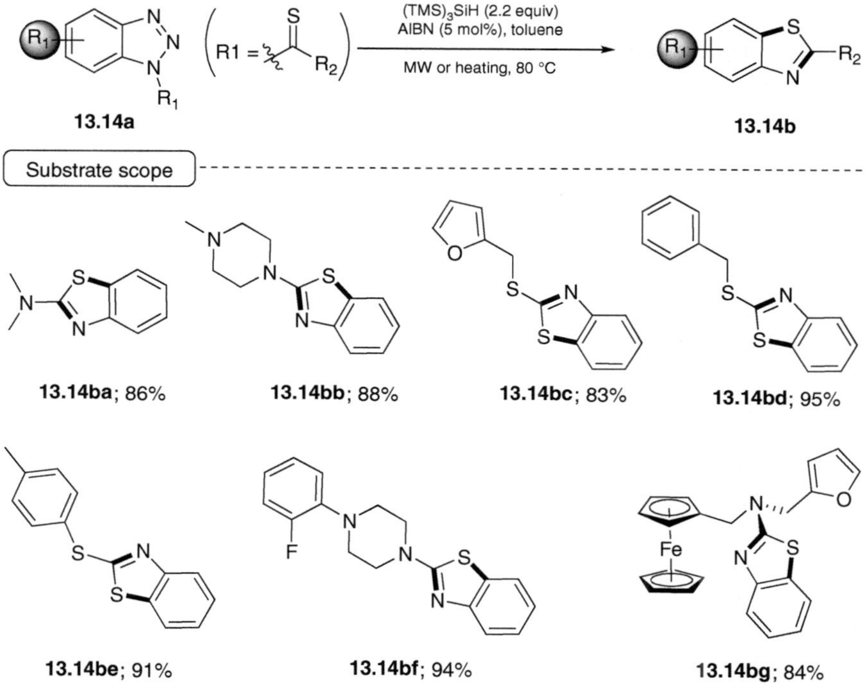

Scheme 13.14 Intramolecular cyclization of benzotriazole derivatives using $(TMS)_3SiH$ as a radical reducing agent (Tiwari BtRC route).

Recently, a green approach is developed for the synthesis of benzothiazoles **13.15b** by using polymethylhydrosiloxane (PMHS) as radical reducing agent at the place of Bu_3SnH and $(TMS)_3SiH$ (Scheme 13.15) [68, 69]. It also shows a wide substrate

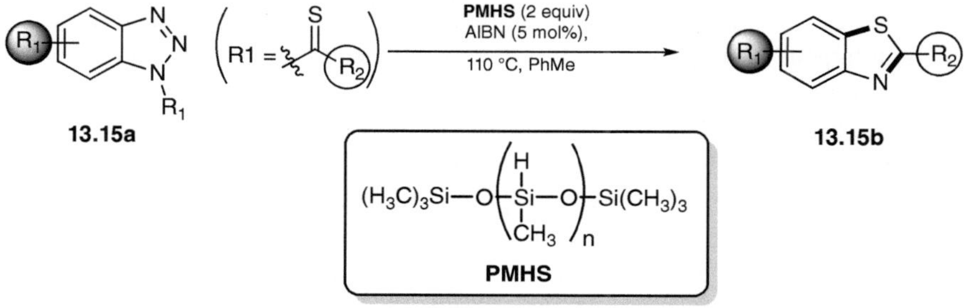

Scheme 13.15 Intramolecular cyclizations of benzotriazole derivatives using industrial waste PMHS as biocompatible radical reducing agent (Tiwari BtRC route).

scope with good compatibility and gives moderate to good yield. This reaction is very economical, cost-effective, and nontoxic because this PMHS is released as a by-product from the silicone industry and is more practical and greener than their previous methods.

In 2017, Tiwari's group also reported the synthesis of benzoxazoles **25** *via* the ring-opening reaction of *N*-acylbenzotriazoles **15** followed by Lewis acid-mediated cyclization. The devised benzotriazole ring cleavage (BtRC) route was established by treating *N*-acylbenzotriqazile with anhydrous aluminum chloride at a higher temperature to give the respective product in moderate-to-good yield (Scheme 13.16) [70]. The reaction is very economical concerning catalyst and substrate, both are inexpensive and easily available and proceed under mild conditions. In addition, the reaction has broad substrate scope, showing both electron-withdrawing and electron-donating groups (EWG & EDG) on the aromatic ring (R_1 and R_2) give the desired product. This BtRC reaction was also performed at the gram scale level under the optimized condition and gave the product in good yield. When the aliphatic *N*-acylbenzotriazoles are used under the same reaction condition, the reaction proceeds *via* the traditional Friedel-Crafts acylation reaction and affords the corresponding ketones in excellent yields. This novel protocol indeed provides simple and practical access to a variety of benzoxazoles. A plausible mechanism has been discussed for the reaction is given in Scheme. In the first step, Lewis-acid $AlCl_3$ coordinates with the oxygen atom of *N*-acylbenzotriazoles and generates intermediate **A**, undergoes charge transfer process, and produces carbocation intermediates **B**, which is resonance stabilized by aromatic ring. Further, this species removes one molecule of nitrogen gas and intermediate **C** is formed in which conjugation is extended and delocalized the positive charge on aryl ring of benzotriazole. Finally, the cyclization takes place and gives the desired cyclized oxazole products.

Cox et al. reported the synthesis of (1*H*)-2-ethoxyquinolin-4-one **13.17b** from ethyl (*E*)-3-(benzotriazol-1-yl)acrylate **13.17a** *via* flash vacuum pyrolysis (FVP) through cleavage of benzotriazole ring. The precursor **13.17a** was prepared from the reaction of ethyl propiolate and benzotriazoles (Scheme 13.17) [118]. The plausible mechanism has been discussed here. On carrying out flash vacuum pyrolysis on precursor molecule **13.17a** at 450 °C, it undergoes benzotriazole ring opening followed by the expulsion of N_2 to form diradical intermediate **A**, which undergoes a hydrogen shift to give another intermediate, that is, ketanimine **B**, which undergoes 1,3-migration of the ethoxy group leads to the formation of ketene intermediate **C**, which undergoes cyclization followed by aromatization and give the desired product (1*H*)-2-ethoxyquinolin-4-one (Scheme 13.17) [118].

Aitken et al. reported the synthesis of 3-acetyl-1,2,4-benzotriazine **13.18b** through the unusual extrusion of trialkylphosphine instead of trialkylphosphine oxide when benzotriazolyl ylides were subjected to flash vacuum pyrolysis. A plausible mechanism has been discussed for the synthesis of triazine derivative in which 1-(chloromethyl) benzotriazole **13.18a** is treated with tri-*n*-butylphosphine which

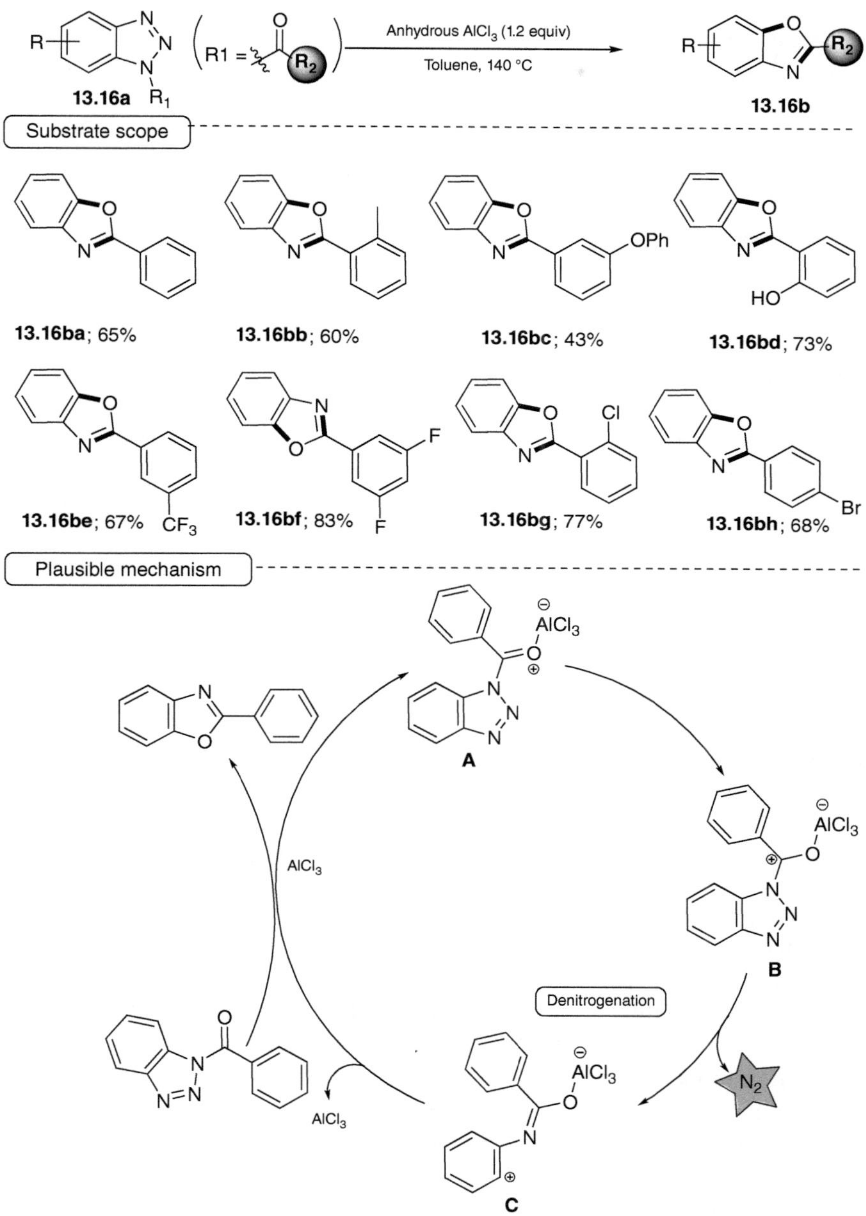

Scheme 13.16 $AlCl_3$-mediated cyclization of *N*-acylbenzotriazoles for the synthesis of benzoxazoles (Tiwari BtRC route).

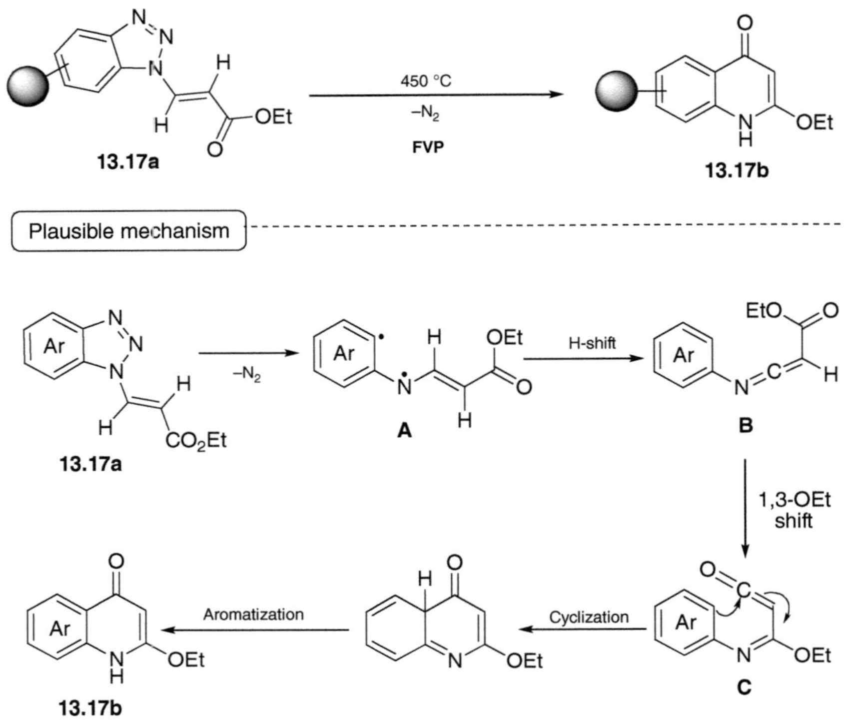

Scheme 13.17 Flash vacuum pyrolysis of 1-benzoylbenzotriazole phenylhydrazones to give (1*H*)-2-ethoxyquinolin-4-one.

generates phosphonium chloride **A**, which reacts with *n*-BuLi and acetylchloride and give another ylide **B**. This ylide **B** undergoes flash vacuum pyrolysis at 450 °C and generates carbene intermediate **C** and release tributylphosphine. The intermediate C rearranges itself in such a manner to give the desired product 3-acetyl-1,2,4-benzotriazine **13.18b** (Scheme 13.18) [119].

Zhong et al. reported the synthesis of 3-Aryl-1,2,4-benzotriazines **13.19c** *via* the cleavage of benzotriazole ring. When authors were trying to prepare homoallylic amine through the reaction of 1,1-bis(benzotriazol-1-yl) methylarenes **13.19a** with allylsamarium bromide **13.19b**, but surprisingly they obtained 3-aryl-1,2,4-benzotriazines **13.19c** in average yields (Scheme 13.19) [120]. The plausible mechanism has been discussed by the author that is based on single electron transfer (SET) mechanism in which compound **13.19a** reacts with allyl samarium bromide **13.19b** and produces intermediate **A** which on radical fragmentation furnish the intermediate **B**, rearrange itself in intermediate **C** which further undergoes cyclization and can afford the desired product 3-Aryl-1,2,4-benzotriazines **13.19c**. On the other hand, if the intermediate **A** abstracts hydrogen without undergoing any radical fragmentation produces 1-benzyl-1*H*-benzo[*d*][1,2,3]triazole **13.19d** (Scheme 13.19).

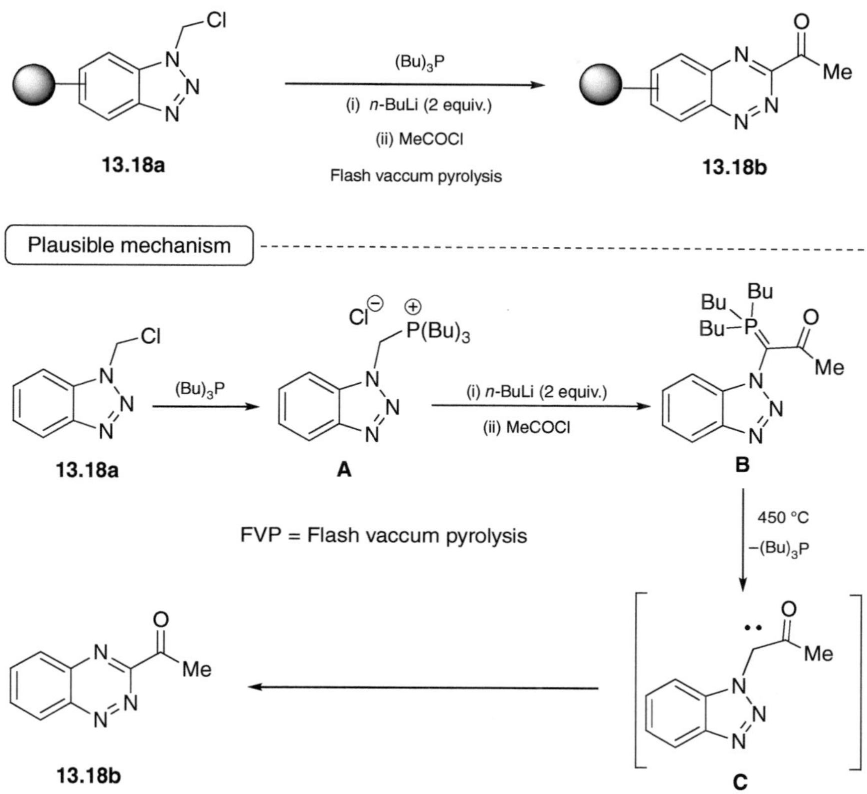

Scheme 13.18 Synthesis of 3-acetyl-1,2,4-benzotriazine through flash vacuum pyrolysis of benzotriazolyl ylides.

Further in 2013, the same group reported the synthesis of 3-Aryl-1,2,4-benzotriazines with enhanced yield by using hexamethylphosphoramide (HMPA) along with allylsamarium bromide. It was observed that a combination of allyl samarium bromide/HMPA suppresses the nucleophilicity and enhances the reducing nature of allylsamarium bromide and hence increases the yield of product (Scheme 13.20) [121].

Number of derivatives of 3-aryl-1,2,4-benzotriazines have been synthesized by Wang et al. in the presence and the absence of HMPA (Schemes 13.19 and 13.20) and yield has been compared and summarized (Scheme 13.21) [120, 121].

Berkett et al. have reported the synthesis of bromo-dihydrobenzofurans **13.22b** from NBS (2 equiv.) mediated oxidative ring opening of aminobenzotriazole derivatives **13.22a**, but derivative of **13.22a** having acetone in the side chain **13.22c** does not undergo cyclization under the similar reaction condition and gives dibromo derivative **13.22d**, when we add on more equivalents of NBS along with 5.0 equivalent of 1-pentene then bromo-dihydrobenzofuran **13.22e** is formed (Scheme 13.22) [122].

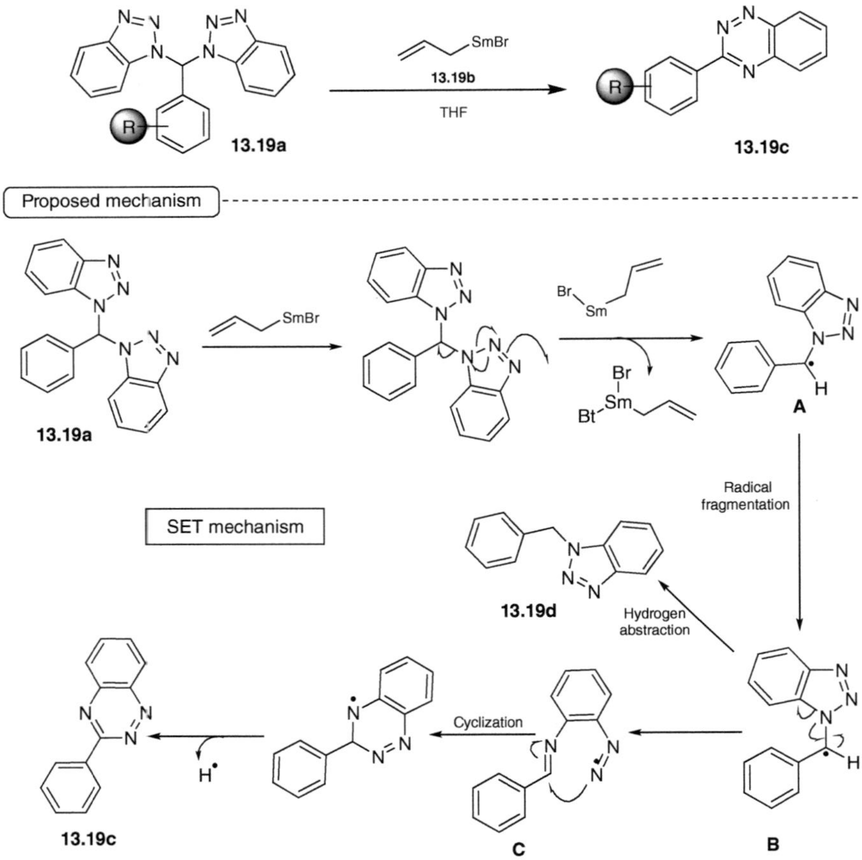

Scheme 13.19 BtRC-mediated synthesis of 3-Aryl-1,2,4-benzotriazines **13.19c** from 1,l-bis(benzotriazol-1-yl) methylarenes **13.19a**.

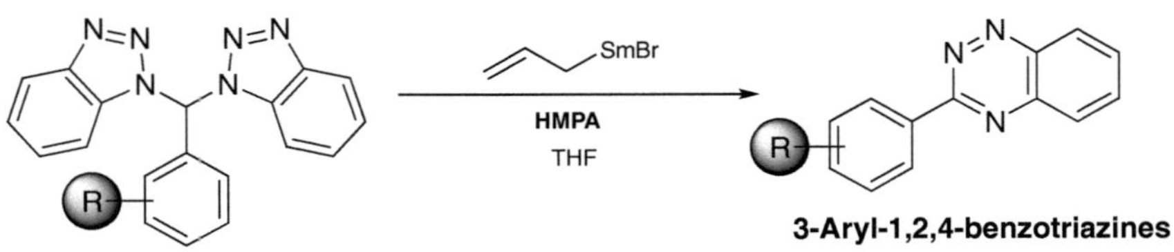

Scheme 13.20 AllylSmBr/HMPA: A potential SET reagent for the synthesis of 3-Aryl-1,2,4-benzotriazines derivative.

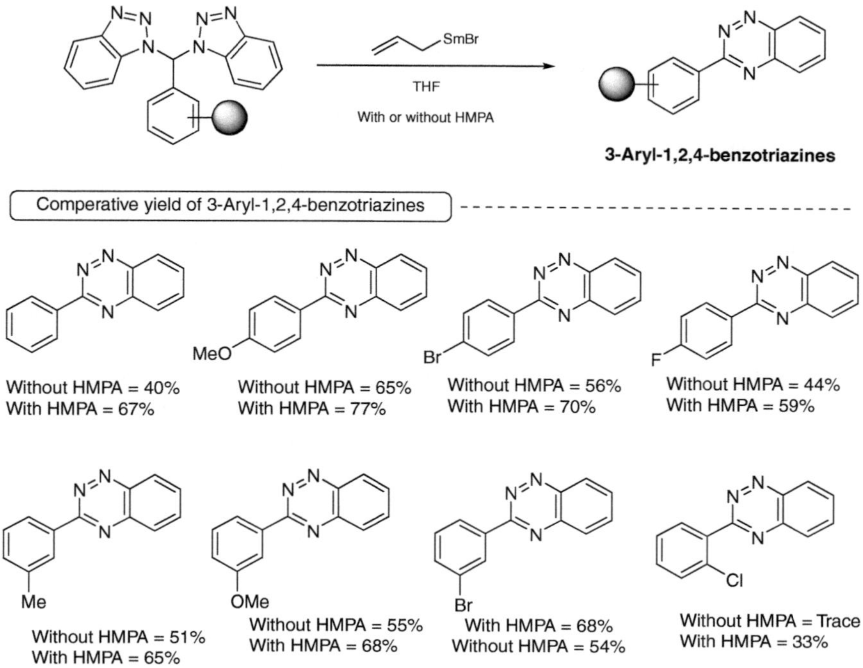

Scheme 13.21 3-Aryl-1,2,4-benzotriazine derivatives obtained without HMPA with HMPA.

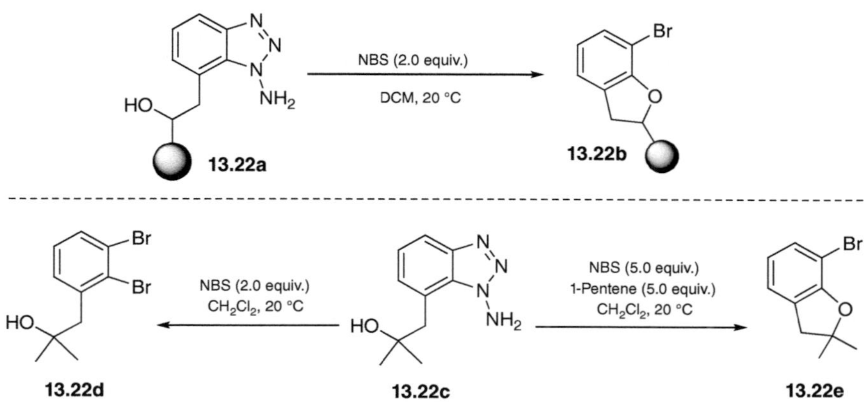

Scheme 13.22 NBS-promoted synthesis of bromo-dihydrobenzofurans derivative.

13.4 Conclusions and Future Perspective

In this chapter, we have focused on the functionalization of benzotriazoles *via* a denitrogenative process and followed by cyclization to produce a variety of heterocyclic compounds. We know very well that benzotriazoles readily undergo Dimroth-type equilibrium in which they form diazonium or diazo species and are used as precursors for the synthesis of new carbon-carbon and carbon-heteroatom bonds. Benzotriazole ring used as a precursor for the synthesis of indole derivative on reaction with the terminal alkyne and internal alkyne in the presence of transition metal. Having a triflate group on benzotriazole ring and reacting with alkyne, produce 5,6-spiro bicycles and naphthyl amine *via* denitrogenative [2+2+1] and [2+2+2] cycloadditions reaction. Also, with allenes and *N*-allenamides, it produces 3-methyleneindoline derivatives. Its application for constructing benzoxazinone derivative through carbonylative activation of benzotriazoles using carbon monoxide is summarized. It is also used for the synthesis of phenanthridine derivatives on reacting with the 2-oxoaldehydes and secondary amine. In addition to it, it is used for the synthesis of quinoxaline on reacting with azide molecule, that is, $TMSN_3$. Benzotriazole also helpful in the synthesis of benzothiazole on reacting with tributyltin hydride along with AIBN, $(TMS)_3SiH$ in the presence of AIBN and polymethylhydrosiloxane (PMHS) as radical reducing agent and also produce benzoxazole derivatives *via* Lewis-acid catalyzed reaction and produce (1*H*)-2-ethoxyquinolin-4-one and 3-acetyl-1,2,4-benzotriazine derivative via flash vacuum pyrolysis (FVP). In this way, we see how benzotriazole moiety has high synthetic value and is used for the synthesis of a variety of organic compounds.

Acknowledgments

The authors sincerely thank the Science Engineering and Research Board (SERB), New Delhi (Grant No. CRG/2021/002776) and Council of Scientific & Industrial Research (CSIR), New Delhi (Scheme No.: 02/(0345)/19/EMR-II) for the funding. MKJ gratefully acknowledge CSIR for the fellowship (SRF). T.G. gratefully acknowledges the Department of Chemistry, Rajendra College, Jai Prakash University, Chapra, and the Indian National Science Academy, New Delhi. VKT thank the Institute of Eminence (IoE)-Banaras Hindu University (BHU) for the incentive grant and MSY sincerely acknowledge IoE-BHU for RJP-PDF.

References

1 Nakamura, I. and Yamamoto, Y. (2004). Transition-metal-catalyzed reactions in heterocyclic synthesis. *Chem. Rev.* 104: 2127–2198.
2 Kim, H., Tung, T.T., and Park, S.B. (2013). Privileged substructure-based diversity-oriented synthesis pathway for diverse pyrimidine-embedded polyheterocycles. *Org. Lett.* 15: 5814–5817.
3 (a) Leitch, D.C., Kayser, L.V., Han, Z.-Y. et al. (2015). A palladium-catalysed multicomponent coupling approach to conjugated poly(1,3-dipoles) and polyheterocycles. *Nat. Commun.* 6: 7411. (b) Yu, J., Singh, A.S., Yan, G. et al. (2020). Recent developments on denitrogenative functionalization of benzotriazoles. *Synthesis* 52: 3781–3800. (c) Kale, R.R., Prasad, V., Mohapatra, P.P.,

and Tiwari, V.K. (2010). ChemInform abstract: recent developments in benzotriazole methodology for construction of pharmacologically important heterocyclic skeletons. *Monatsh. Chem.* 141: 1159–1182.

4 (a) Stępien, M., Gonka, E., Zyła, M., and Sprutta, N. (2017). Recent advances in heterocyclic nanographenes and other polycyclic heteroaromatic compounds. *Chem. Rev.* 117: 3479–3716. (b) Tiwari, V.K., Mishra, B.B., Mishra, K.B. et al. (2016). Cu-catalyzed click reaction in carbohydrate chemistry. *Chem. Rev.* 116: 3086–3240. (c) Agrahari, A.K., Bose, P., Jaiswal, M.K. et al. (2021). Cu(I)-catalyzed click chemistry in glycoscience and their diverse applications. *Chem. Rev.* 12: 7638–7956.

5 Tsuzuki, Y., Tomita, K., Sato, Y. et al. (2004). Synthesis and structure–activity relationships of 3-substituted 1,4-dihydro-4-oxo-1-(2-thiazolyl)-1,8-naphthyridines as novel antitumor agents. *Bioorg. Med. Chem. Lett.* 14: 3189–3193.

6 Srivastava, S.K., Jaggi, M., Singh, A.T. et al. (2007). Anticancer and anti-inflammatory activities of 1,8-naphthyridine-3-carboxamide derivatives. *Bioorg. Med. Chem. Lett.* 17: 6660–6664.

7 Insuasty, B., Becerra, D., Quiroga, J. et al. (2013). Microwave-assisted synthesis of pyrimido[4,5-*b*][1,6]naphthyridin-4(3*H*)-ones with potential antitumor activity. *J. Eur. J. Med. Chem.* 60: 1–9.

8 Bernardino, A.M.R., Castro, H.C., Frugulhetti, I.C.P.P. et al. (2008). SAR of a series of anti-HSV-1 acridone derivatives, and a rational acridone-based design of a new anti-HSV-1 3*H*-benzo[*b*]pyrazolo[3,4-*h*]-1,6-naphthyridine series. *Bioorg. Med. Chem.* 16: 313–321.

9 Zhuang, L., Wai, S.J., Embrey, M.W. et al. (2003). Design and synthesis of 8-hydroxy-[1,6]naphthyridines as novel inhibitors of HIV-1 integrase in vitro and in infected cells. *J. Med. Chem.* 46: 453–456.

10 Kolb, H.C., Finn, M., and Sharpless, K.B. (2001). Click chemistry: diverse chemical function from a few good reactions. *Angew. Chem. Int. Ed.* 40: 2004–2021.

11 Yogeeswari, P., Sriram, D., Ramamoorthy, L. et al. (2002). Anticonvulsant and neurotoxicity evaluation of some 6-chlorobenzothiazolyl-2-thiosemicarbazones. *Eur. J. Med. Chem.* 37: 231–236.

12 Norris, P. (2008). Pyranose N-glycosyl amines: emerging targets with diverse biological potential. *Curr. Top. Med. Chem.* 8: 101–113.

13 De, S.K., Stebbins, J.L., Chen, L.-H. et al. (2009). Design, synthesis, and structure-activity relationship of substrate competitive, selective, and in vivo active triazole and thiadiazole inhibitors of the c-Jun N-terminal kinase. *Med. Chem.* 52: 1943–1952.

14 Massari, S., Daelemans, D., Barreca, M.L. et al. (2010). *J. Med. Chem.* 53: 641.

15 Mishra, B.B. and Tiwari, V.K. (2011). Natural products: an evolving role in future drug discovery. *Eur. J. Med. Chem.* 46: 4769–4807.

16 Tornøe, C.W., Christensen, C., and Meldal, M. (2002). Peptidotriazoles on solid phase: [1,2,3]-triazoles by regiospecific copper(I)-catalyzed 1,3-dipolar cycloadditions of terminal alkynes to azides. *J. Org. Chem.* 67: 3057–3064.

17 Angelo, N.G. and Arora, P.S. (2005). Nonpeptidic foldamers from amino acids: synthesis and characterization of 1,3-substituted triazole oligomers. *J. Am. Chem. Soc.* 127: 17134–17135.

18 Oh, K. and Guan, Z. (2006). A convergent synthesis of new β-turn mimics by click chemistry. *Chem. Commun.* 29: 3069–3071.

19 Caleta, I., Kralj, M., Marjanovic, M. et al. (2009). Novel cyano- and amidino-benzothiazole derivatives: synthesis, antitumor evaluation, and X-ray and quantitative structure–activity relationship (QSAR) analysis. *J. Med. Chem.* 52: 1744–1756.

20 Pedersen, D.S. and Abell, A. (2011). 1,2,3-Triazoles in peptidomimetic chemistry. *Eur. J. Org. Chem.* 2011: 2399–2411.

21 Katritzky, A.R., Ji, F.B., Fan, W.Q. et al. (1992). Novel Dimroth rearrangements of the benzotriazole system: 4-amino-1-(arylsulfonyl)benzotriazoles to 4-[(arylsulfonyl)amino]benzotriazoles. *J. Org. Chem.* 57: 190–195.

22 Katritzky, A.R., Yang, B., and Dalal, N.S. (1998). Novel heteroatom-linked analogues of trityl radicals: diaryl(benzotriazol-1-yl)methyl radical dimers. *J. Org. Chem.* 63: 1467–1472.

23 Katritzky, A.R., Lan, X., Yang, J., and Denisko, O. (1998). Properties and synthetic utility of N-substituted benzotriazoles. *Chem. Rev.* 98: 409–548.

24 Katritzky, A.R., Du, W., Matskawa, Y. et al. (1999). Polycyclic fused phenanthridines: An alternative approach from benzotriazoles. *J. Heterocyclic Chem.* 36: 927–932.

25 Micó, X.I., Ziegler, T., and Subramanian, L.R. (2004). A versatile direct approach to *ortho*-substituted azobenzenes from benzotriazoles. *Angew. Chem. Int. Ed.* 43: 1400–1403.

26 Katritzky, A.R., Khelashvili, L., Le, K.N.B. et al. (2007). Synthesis of *o*-sulfamidotriazobenzenes from 1,1′-sulfonylbis(benzotriazole). *J. Org. Chem.* 72: 5805–5808.

27 Uhde, M., Anwar, M.U., and Ziegler, T. (2008). Reaction of *N*-nonaflyl and *N*-cyano-benzotriazoles with enamines. *Synth. Commun.* 38: 881–888.

28 Kim, T. and Kim, K. (2010). Practical method for synthesis of 2,3-disubstituted indole derivatives promoted by *β*-(benzotriazol-1-yl)allylic *O*-stannyl ketyl radicals. *Tetrahedron Lett.* 51: 868–871.

29 Katritzky, A.R. and Rachwal, S. (2011). Synthesis of heterocycles mediated by benzotriazole. 2. Bicyclic systems. *Chem. Rev.* 111: 7063–7120.

30 Miyaura, N. and Suzuki, A. (1995). Palladium-catalyzed cross-coupling reactions of organoboron compounds. *Chem. Rev.* 95: 2457–2483.

31 Sun, C.-L., Li, B.-J., and Shi, Z.-J. (2011). Direct C—H transformation via iron catalysis. *Chem. Rev.* 111: 1293–1314.

32 Hartwig, J.F. (2012). Borylation and silylation of C–H bonds: a platform for diverse C—H bond functionalizations. *Acc. Chem. Res.* 45: 864–873.

33 Mo, F., Dong, G., Zhang, Y., and Wang, J. (2013). Recent applications of arene diazonium salts in organic synthesis. *Org. Biomol. Chem.* 11: 1582–1593.

34 Xiao, Q., Zhang, Y., and Wang, J. (2013). Diazo compounds and *N*-tosylhydrazones: novel cross-coupling partners in transition-metal-catalyzed reactions. *Acc. Chem. Res.* 46: 236–247.

35 Zhang, F. and Spring, D.R. (2014). Arene C—H functionalisation using a removable/modifiable or a traceless directing group strategy. *Chem. Soc. Rev.* 43: 6906–6919.

36 Cherney, A.H., Kadunce, N.T., and Reisman, S.E. (2015). Enantioselective and enantiospecific transition-metal-catalyzed cross-coupling reactions of organometallic reagents to construct C—C bonds. *Chem. Rev.* 115: 9587–9650.

37 Ruiz-Castillo, P. and Buchwald, S.L. (2016). Applications of palladium-catalyzed C–N cross-coupling reactions. *Chem. Rev.* 116: 12564–12649.

38 Yan, G., Zhang, Y., and Wang, J. (2017). Recent advances in the synthesis of aryl nitrile compounds. *Adv. Synth. Catal.* 359: 4068–4105.

39 Xia, Y., Qiu, D., and Wang, J. (2017). Transition-metal-catalyzed cross-couplings through carbene migratory insertion. *Chem. Rev.* 117: 13810–13889.

40 Mo, F., Qiu, D., Zhang, Y., and Wang, J. (2018). Renaissance of Sandmeyer-type reactions: conversion of aromatic C–N bonds into C–X bonds (X = B, Sn, P, or CF_3). *Acc. Chem. Res.* 51: 496–506.

41 Korch, K.M. and Watson, D.A. (2019). Cross-coupling of heteroatomic electrophiles. *Chem. Rev.* 119: 8192–8228.

42 Chattopadhyay, B. and Gevorgyan, V. (2012). Transition-metal-catalyzed denitrogenative transannulation: converting triazoles into other heterocyclic systems. *Angew. Chem. Int. Ed.* 51: 862–872.

43 Gulevich, A.V. and Gevorgyan, V. (2013). Versatile reactivity of rhodium-iminocarbenes derived from *N*-sulfonyl triazoles. *Angew. Chem. Int. Ed.* 52: 1371–1373.

44 Miura, T., Tanaka, T., Biyajima, T. et al. (2013). One-pot procedure for the introduction of three different bonds onto terminal alkynes through *N*-sulfonyl-1,2,3-triazole intermediates. *Angew. Chem. Int. Ed.* 52: 3883–3886.

45 Parr, B.T., Green, S.A., and Davies, H.M.J. (2013). Rhodium-catalyzed conversion of furans to highly functionalized pyrroles. *J. Am. Chem. Soc.* 135: 4716–4718.

46 Yoo, E.J., Ahlquist, M., Kim, S.H. et al. (2007). Copper-catalyzed synthesis of *N*-sulfonyl-1,2,3-triazoles: controlling selectivity. *Angew. Chem. Int. Ed.* 46: 1730.

47 Yang, J., Zhu, C., Tang, X., and Shi, M. (2014). Rhodium(II)-catalyzed intramolecular annulation of 1-sulfonyl-1,2,3-triazoles with pyrrole and indole rings: facile synthesis of N-bridgehead azepine skeletons. *Angew. Chem. Int. Ed.* 53: 5142–5146.

48 Davies, H.M.L. and Alford, J.S. (2014). Reactions of metallocarbenes derived from *N*-sulfonyl-1,2,3-triazoles. *Chem. Soc. Rev.* 43: 5151–5162.

49 Schultz, E.E., Lindsay, V.N.G., and Sarpong, R. (2014). Expedient synthesis of fused azepine derivatives using a sequential rhodium(II) catalyzed cyclopropanation/1-*aza*-cope rearrangement of dienyltriazoles. *Angew. Chem. Int. Ed.* 53: 9904–9908.

50 Chuprakov, S., Worrell, B.T., Selander, N. et al. (2014). Stereoselective 1,3-insertions of rhodium(II) azavinyl carbenes. *J. Am. Chem. Soc.* 136: 195.

51 Ryu, T., Baek, Y., and Lee, P.H. (2015). Synthesis of pyrazines from rhodium-catalyzed reaction of 2*H*-azirines with *N*-sulfonyl 1,2,3-triazoles. *J. Org. Chem.* 80: 2376–2383.

52 Lei, X., Gao, M., and Tang, Y. (2016). Rh(II)-catalyzed transannulation of *N*-sulfonyl-1,2,3-triazoles with 2,1-benzisoxazoles or 1,2-benzisoxazoles. *Org. Lett.* 18: 4990–4993.

53 Li, Y., Zhang, R., Ali, A. et al. (2017). Synthesis of 6-substituted piperidin-3-ones via Rh(II)-catalyzed transannulation of *N*-sulfonyl-1,2,3-triazoles with electron-rich aromatic nucleophiles. *Org. Lett.* 19: 3087–3090.

54 Song, W., Zheng, N., Li, M. et al. (2018). Regiodivergent rhodium(I)-catalyzed azide-alkyne cycloaddition (RhAAC) to access either fully substituted sulfonyl-1,2,3-triazoles under mild conditions. *Org. Lett.* 20: 6705–6709.

55 Sontakke, G.S., Pal, K., and Volla, C.M.R. (2019). Rh(II)-catalyzed denitrogenative transannulation of *N*-sulfonyl-1,2,3-triazolyl cyclohexadienones for the synthesis of benzofurans and cyclopropa[*cd*]indole-carbaldehydes. *J. Org. Chem.* 84: 12198–12208.

56 Pal, K., Sontakke, G.S., and Volla, C.M.R. (2019). Rh(II)-catalyzed highly diastereoselective cascade transannulation of *n*-sulfonyl-1,2,3-triazoles and vinyl benzoxazinanones. *Org. Lett.* 21: 3716–3720.

57 Liu, Y., Xie, P., Li, J. et al. (2019). Nickel-catalyzed coupling of *N*-sulfonyl-1,2,3-triazole with *H*-phosphine oxides: stereoselective and site-selective synthesis of α-aminovinylphosphoryl derivatives. *Org. Lett.* 21: 4944–4949.

58 Nakamura, I., Nemoto, T., Shiraiwa, N., and Terada, M. (2009). Palladium-catalyzed indolization of *N*-aroylbenzotriazoles with disubstituted alkynes. *Org. Lett.* 11: 1055–1058.

59 Teders, M., Pitzer, L., Buss, S., and Glorius, F. (2017). Regioselective synthesis of 2-substituted indoles from benzotriazoles and alkynes by photoinitiated denitrogenation. *ACS Catal.* 7: 4053–4056.

60 Wang ng, Z., Chen, X., and Tang, Y. (2018). Palladium-catalyzed denitrogenative cycloadditions and alkenylations of benzotriazoles with alkynes. *Org. Chem. Front.* 5: 2815–2819.

61 Wang, Y.H., Li, Y.H., Fan, Y.J. et al. (2017). Palladium-catalyzed denitrogenative functionalizations of benzotriazoles with alkenes and 1,3-dienes. *Chem. Commun.* 53: 11873–11876.

62 Zhang, P.-C., Han, J., and Zhang, J. (2019). Pd/PC-phos-catalyzed enantioselective intermolecular denitrogenative cyclization of benzotriazoles with allenes and N-allenamides. *Angew. Chem. Int. Ed.* 58: 11444–11448.

63 Yin, Z.P., Wang, Z.C., and Wu, X.F. (2017). Silver and palladium cocatalyzed carbonylative activation of benzotriazoles to benzoxazinones under neutral conditions. *Org. Lett.* 19: 6232–6235.

64 Battula, S., Kumar, A., Gupta, A.P., and Ahmed, Q.N. (2015). 2-Oxo-driven N_2 elimination induced decarbonylative cyclization reaction in benzotriazoles to 6-aminophenanthridines. *Org. Lett.* 17: 5562–5565.

65 Su, Y., Petersen, J.L., Gregg, T.L., and Shi, X. (2015). Ambient benzotriazole ring opening through intermolecular radical addition to vinyltriazole. *Org. Lett.* 17: 1208–1211.

66 Kumar, D., Mishra, A., Mishra, B.B. et al. (2013). Synthesis of glycoconjugate benzothiazoles via cleavage of benzotriazole ring. *J. Org. Chem.* 78: 899–909.

67 Kumar, D., Mishra, B.B., and Tiwari, V.K. (2013). Synthesis of 2-*N*/*S*/*C*-substituted benzothiazoles via intramolecular cyclative cleavage of benzotriazole ring. *J. Org. Chem.* 79: 251–266.

68 Yadav, M.S., Singh, A.S., Agrahari, A.K. et al. (2019). Silicon industry waste polymethylhydrosiloxane-mediated benzotriazole ring cleavage: a practical and green synthesis of diverse benzothiazoles. *ACS Omega* 4: 6681–6689.

69 Singh, A.S., Mishra, N., Yadav, M.S., and Tiwari, V.K. (2019). Free radical synthetic protocol for benzothiazoles *via* ring opening of benzotriazole: a two-step organic chemistry experiment for undergraduate and postgraduate students. *J. Heterocyclic Chem.* 56: 275–280.

70 Singh, A.S., Mishra, N., Kumar, D., and Tiwari, V.K. (2017). Lewis-acid-mediated benzotriazole ring cleavage (BtRC) strategy for the synthesis of 2-aryl benzoxazoles from *N*-acylbenzotriazoles. *ACS Omega* 2: 5044–5051.

71 Kumar, D., Singh, A.S., and Tiwari, V.K. (2015). An unprecedented deoxygenation protocol of benzylic alcohols using bis(1-benzotriazolyl)methanethione. *RSC Adv.* 5: 31584–31593.

72 Singh, A.S., Kumar, D., Mishra, N., and Tiwari, V.K. (2017). An unprecedented synthesis of *N*-phenyl amides *via* cleavage of benzotriazole ring under free radical condition. *Chem. Sel.* 2: 224–229.

73 Wang, Y.H., Wu, Y.F., Li, Y.H., and Tang, Y.F. (2017). Denitrogenative Suzuki and carbonylative Suzuki coupling reactions of benzotriazoles with boronic acids. *Chem. Sci.* 8: 3852–3857.

74 Hopkinson, M.N., Gómez-Suárez, A., Teders, M. et al. (2016). Accelerated discovery in photocatalysis using a mechanism-based screening method. *Angew. Chem. Int. Ed.* 55: 4361–4366.

75 Teders, M., Gómez-Suárez, A., Pitzer, L. et al. (2017). Diverse visible-light-promoted functionalizations of benzotriazoles inspired by mechanism-based luminescence screening. *Angew. Chem. Int. Ed.* 56: 902–906.

76 Jian, Y., Chen, M., Huang, B. et al. (2018). Visible-light-induced C(sp^2)–P bond formation by denitrogenative coupling of benzotriazoles with phosphites. *Org. Lett.* 20: 5370–5374.

77 Katritzky, A.R. (2004). Introduction: heterocycles. *Chem. Rev.* 104: 2125–2156.

78 Saracoglu, N. (2007). Functionalization of indole and pyrrole cores via Michael-type additions. *Top. Heterocycl. Chem.* 11: 1–62.

79 Vitaku, E., Smith, D.T., and Njardarson, J.T. (2014). Analysis of the structural diversity, substitution patterns, and frequency of nitrogen heterocycles among U.S. FDA approved pharmaceuticals. *J. Med. Chem.* 57: 10257–10274.

80 Pennington, L.D. and Moustakas, D.T. (2017). The necessary nitrogen atom: a versatile high-impact design element for multiparameter optimization. *J. Med. Chem.* 60: 3552–3579.

81 Jarhad, D.B., Mashelkar, K.K., Kim, H.-R. et al. (2018). Dual-specificity tyrosine phosphorylation-regulated kinase 1A (DYRK1A) inhibitors as potential therapeutics. *J. Med. Chem.* 61: 9791–9810.

82 Garces, A.E. and Stocks, M.J. (2019). Class 1 PI3K clinical candidates and recent inhibitor design strategies: a medicinal chemistry perspective. *J. Med. Chem.* 62: 4815–4850.

83 Wang, S. and Xi, C. (2019). Recent advances in nucleophile-triggered CO_2-incorporated cyclization leading to heterocycles. *Chem. Soc. Rev.* 48: 382–404.

84 Bariwal, J., Voskressensky, L.G., and Van der Eycken, E.V. (2018). Recent advances in spirocyclization of indole derivatives. *Chem. Soc. Rev.* 47: 3831–3848.

85 Lu, L.-Q., Li, T.-R., Wang, Q., and Xiao, W.-J. (2017). Beyond sulfide-centric catalysis: recent advances in the catalytic cyclization reactions of sulfur ylides. *Chem. Soc. Rev.* 46: 4135–4149.

86 Xuan, J. and Studer, A. (2017). Radical cascade cyclization of 1,*n*-enynes and diynes for the synthesis of carbocycles and heterocycles. *Chem. Soc. Rev.* 46: 4329–4346.

87 Xia, Y. and Wang, J. (2017). *N*-Tosylhydrazones: versatile synthons in the construction of cyclic compounds. *Chem. Soc. Rev.* 46: 2306–2362.

88 Chen, J.-R., Hu, X.-Q., Lu, L.-Q., and Xiao, W.-J. (2015). Formal [4+1] annulation reactions in the synthesis of carbocyclic and heterocyclic systems. *Chem. Rev.* 115: 5301–5365.

89 Guo, X.-X., Gu, D.-W., Wu, Z., and Zhang, W. (2015). Copper-catalyzed C—H functionalization reactions: efficient synthesis of heterocycles. *Chem. Rev.* 115: 1622–1651.

90 Kale, R.R., Prasad, V., Hussain, H.A., and Tiwari, V.K. (2010). Facile route for N_1-aryl benzotriazoles from diazoamino arynes via CuI-mediated intramolecular N-arylation. *Tetrahedron Lett.* 51: 5740–5743.

91 Mukhopadhyay, C., Tapaswi, P.K., and Butcher, R.J. (2010). A ligand-free copper(1) catalysed intramolecular N-arylation of diazoaminobenzenes in PEG-water: an expeditious protocol towards regiospecific 1-aryl benzotriazoles. *Org. Biomol. Chem.* 8: 4720–4729.

92 Zhou, J., He, J., Wang, B. et al. (2011). 1,7-Palladium migration via C—H activation, followed by intramolecular amination: regioselective synthesis of benzotriazoles. *J. Am. Chem. Soc.* 133: 6868–6870.

93 Kiran Kumar, R., Ali, M.A., and Punniyamurthy, T. (2011). Pd-catalyzed C—H activation/C–N bond formation: a new route to 1-aryl-1*H*-benzotriazoles. *Org. Lett.* 13: 2102–2105.

94 Zhu, Z., Liu, Q.L., Li, W., and Zhu, Y.M. (2011). *Heterocycles* 83: 2057.

95 Lukasik, E. and Wrobel, Z. (2014). 2-(Arylamino)aryliminophosphoranes as easily available and convenient starting materials in the synthesis of 1,2,3-benzotriazoles. *Synlett* 25: 1987–1990.

96 Shi, F., Waldo, J.P., Chen, Y., and Larock, R.C. (2008). Benzyne click chemistry: synthesis of benzotriazoles from benzynes and azides. *Org. Lett.* 10: 2409–2412.

97 Chandrasekhar, S., Seenaiah, M., Rao, C.L., and Reddy, C.R. (2008). A smooth access to benzotriazoles via azide-benzyne cycloaddition. *Tetrahedron* 64: 11325–11327.

98 Zhang, F. and Moses, J.E. (2009). Benzyne click chemistry with in situ generated aromatic azides. *Org. Lett.* 11: 1587–1590.

99 Ramachary, D.B. and Shashank, A.B. (2013). Organocatalytic triazole formation, followed by oxidative aromatization: regioselective metal-free synthesis of benzotriazoles. *Chem. Eur. J.* 19: 13175–13181.

100 Chen, Z.Y. and Wu, M.J. (2005). Reaction of (*Z*)-1-aryl-3-hexen-1,5-diynes with sodium azide: synthesis of 1-aryl-1*H*-benzotriazoles. *Org. Lett.* 7: 475–477.

101 Mandadapu, A.K., Sharma, S.K., Gupta, S. et al. (2011). Unprecedented Cu-catalyzed coupling of internal 1,3-diynes with azides: one-pot tandem cyclizations involving 1,3-dipolar cycloaddition and carbocyclization furnishing naphthotriazoles. *Org. Lett.* 13: 3162–3165.

102 Ueda, S., Su, M., and Buchwald, S.L. (2011). Highly N^2-selective palladium-catalyzed arylation of 1,2,3-triazoles. *Angew. Chem. Int. Ed.* 50: 8944–8947.

103 Liu, Y., Yan, W., Chen, Y. et al. (2008). Efficient synthesis of *N*-2-Aryl-1,2,3-triazole fluorophores via post-triazole arylation. *Org. Lett.* 10: 5389–5392.

104 Antilla, J.C., Baskin, J.M., Barder, T.E., and Buchwald, S.L. (2004). Copper–diamine-catalyzed *N*-arylation of pyrroles, pyrazoles, indazoles, imidazoles, and triazoles. *J. Org. Chem.* 69: 5578–5587.

105 Kang, S.K., Lee, S.H., and Lee, D. (2000). Copper-catalyzed *N*-arylation of amines with hypervalent iodonium salts. *Synlett* 2000: 1022–1024.

106 Yadav, M.S., Gupta, A., Bose, P. et al. (2023). Synthetic utility of *N*-acylbenzotriazoles. *SynOpen* 7: 430–465.

107 Katritzky, A.R., Suzuki, K., Singh, S.K., and He, H.Y. (2003). Regiospecific C-acylation of pyrroles and indoles using *N*-acylbenzotriazoles. *J. Org. Chem.* 68: 5720.

108 Singh, A.S., Agrahari, A.K., Singh, M. et al. (2017). A new methodology for the synthesis of *N*-acylbenzotriazoles. *ARKIVOC (v)* 2017: 80–87.

109 Katritzky, A.R., Zhang, Y., and Singh, S.K. (2003). Efficient conversion of carboxylic acids into *N*-acylbenzotriazoles. *Synthesis* 2003: 2795–2798.

110 Duangkamol, C., Wangngae, S., Pattarawarapan, M., and Phakhodee, W. (2014). Acyloxyphosphonium versus aminophosphonium intermediates: application tothe synthesis of *N*-acylbenzotriazoles. *Eur. J. Org. Chem.* 32: 7109–7112.

111 Katritzky, A.R., He, H.-Y., and Suzuki, K. (2000). *N*-acylbenzotriazoles: neutral acylating reagents for the preparation of primary, secondary, and tertiary amide. *J. Org. Chem.* 65: 8210–8213.

112 Agha, K.A., Abo-Dya, N.E., Ibrahim, T.S., and Abdel-Aal, E.H. (2016). Efficient synthesis of *N*-acylbenzotriazoles using tosyl chloride: en route to suberoylanilide hydroxamic acid (SAHA). *ARKIVOC (iii)* 2016: 161–170.

113 Wet-Osot, S., Duangkamol, C., Pattarawarapan, M., and Phakhodee, W. (2015). Facile synthesis of *N*-acylbenzotriazoles from carboxylic acids mediated by 2,4,6-trichloro-1,3,5-triazine and triethylamine. *Monatsh. Chem.* 146: 959–963.

114 Wet-Osot, S., Duangkamol, C., and Pattarawarapan, M. (2015). TCT-mediated synthesis of *N*-acylbenzotriazoles in aqueous media. *Tetrahedron Lett.* 56: 6998–7000.

115 Singh, M., Singh, A.S., Mishra, N. et al. (2019). Trichloroisocyanuric acid mediated high-yielding synthesis of *N*-acylbenzotriazoles under mild reaction conditions. *Synthesis* 51: 2183–2190.

116 Laconde, G., Amblard, M., and Martinez, J. (2019). A simple and versatile method to synthesize *N*-acyl-benzotriazoles. *Tetrahedron Lett.* 60: 341–343.

117 Singh, A.S., Agrahari, A.K., Singh, S.K. et al. (2019). An improved synthesis of urea derivatives from *N*-acylbenzotriazole via curtius rearrangement. *Synthesis* 51: 3443–3450.

118 Cox, M., Heidarizadeh, F., and Prager, R.H. (2000). Flash vacuum pyrolysis of N-alkenylbenzotriazoles and N-alkenylisoxazolones. *Aust. J. Chem.* 53: 665–671.

119 Aitken, R.A., Fairhurst, I.M., Ford, A. et al. (1997). Flash vacuum pyrolysis of stabilised phosphorus ylides. Part 11. Unexpected generation and rearrangement of acetyl benzotriazolyl carbine. *J. Chem. Soc. Perkin Trans.* 1: 3107–3112.

120 Zhong, Z., Hong, R., and Wang, X. (2010). Construction of 3-aryl-1,2,4-benzotriazines via unprecedented rearrangement of bis(benzotriazol-1-yl)methylarenes.s. *Tetrahedron Lett.* 51: 6763–6766.

121 Yin, R., Zhou, L., Liu, H. et al. (2013). Reactivity of allylSmBr/HMPA: facile synthesis of 3-aryl-1,2,4-benzotriazines. *Tetrahedron Lett.* 31: 143–148.

122 Birkett, M.A., Knight, D.W., and Mitchell, M.B. (1993). A new approach to dihydrobenzofurans by intramolecular trapping of benzynes by hydroxyl functions. *Tetrahedron Lett.* 34: 6939–6940.

14

Denitrogenative Transannulation of Thiadiazoles

Phil Ho Lee

Department of Chemistry, Kangwon National University, Chuncheon 24341, Republic of Korea

14.1 Introduction

Heterocyclic compounds are one of the most various and abundant moieties in organic compounds and have found diverse applications in bioactive compounds, agrochemicals, and electron materials. The conversion of an easily accessible heterocycle to another functionalized heterocycle provides a strong and step-economical approach to rapidly obtain a heterocyclic library of structurally complex and assorted heterocycles. In this regard, the transannulation reactions of heterovinyl transition-metal carbenes with unsaturated compounds represent some of the most significant methods for heterocyclic compounds. Transition metal-catalyzed transannulation reactions of α-oxo and α-imino carbenes have furnished synthetic methods of structurally various heterocyclic compounds. In 2016, Gevorgyan's group made seminal breakthroughs in producing thiavinyl carbenes and revaling their reactivities [1]. After that, some groups reported that thiavinyl carbenes were able to transannulate with unsaturated compounds, such as alkynes, alkenes, nitriles, and phosphaalkynes, to afford *S*-containing five-membered heterocyclic compounds catalyzed by rhodium catalyst (Scheme 14.1). Currently, one of the most important research fields in thiavinyl carbene chemistry is to find novel transannulating agents and distinctive reactivities of thiavinyl carbenes.

14.2 Rh-Catalyzed Reaction of 1,2,3-Thiadiazoles with Alkynes

14.2.1 Rh-Catalyzed Thiavinyl Carbenes from 1,2,3-Thiadiazoles for the Synthesis of Multisubstituted Thiophenes

In 2016, Gevorgyan and co-workers have reported for the first time that reactive Rh thiavinyl carbenes can be effectively produced from 1,2,3-thiadiazoles [1].

Denitrogenative Transformation of Nitrogen Heterocycles: Synthesis, Reactions and Applications,
First Edition. Edited by Pazhamalai Anbarasan.

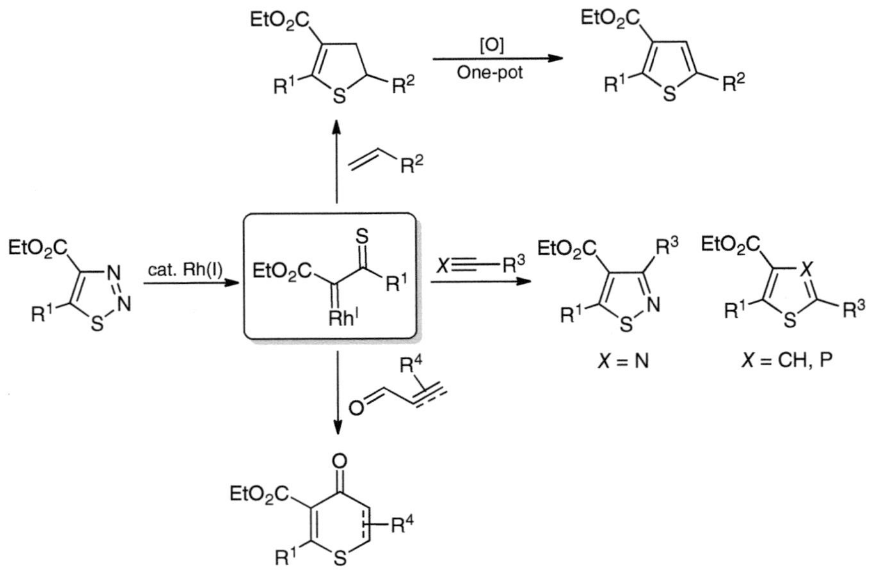

Scheme 14.1 Transannulation reactions of 1,2,3-thiadiazoles with unsaturated compounds.

Rh thiavinyl carbenes, which are useful analogs of the widely known Rh *oxa*- and *aza*-vinyl carbenes, allowed a novel modular method for the synthesis of a multisubstituted thiophene through transannulation of 1,2,3-thiadiazoles with alkynes. This method features a broad scope, high efficiency, and regioselectivity (Scheme 14.2).

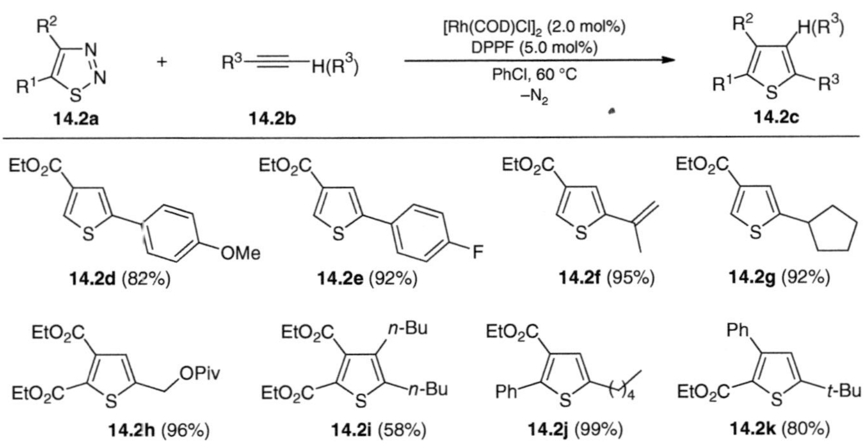

Scheme 14.2 Rh-catalyzed reaction of 1,2,3-thiadiazoles with alkynes. Source: Adapted from Gevorgyan et al. [1].

14.2.2 Rh-Catalyzed Intramolecular Transannulation Reaction of Alkynyl Thiadiazoles for the Synthesis of 5,*n*-fused Thiophenes

In 2017, Lee and co-workers have demonstrated a method for the synthesis of a variety of thiophenes fused with lactams, lactons, or cyclic ethers from Rh-catalyzed intramolecular transannulation reaction of alkynyl thiadiazoles (Scheme 14.3) [2]. This transannulation reaction gives an efficient platform for the construction of a number of 5,*n*-fused thiophenes from readily accessible starting materials with the release of molecular nitrogen.

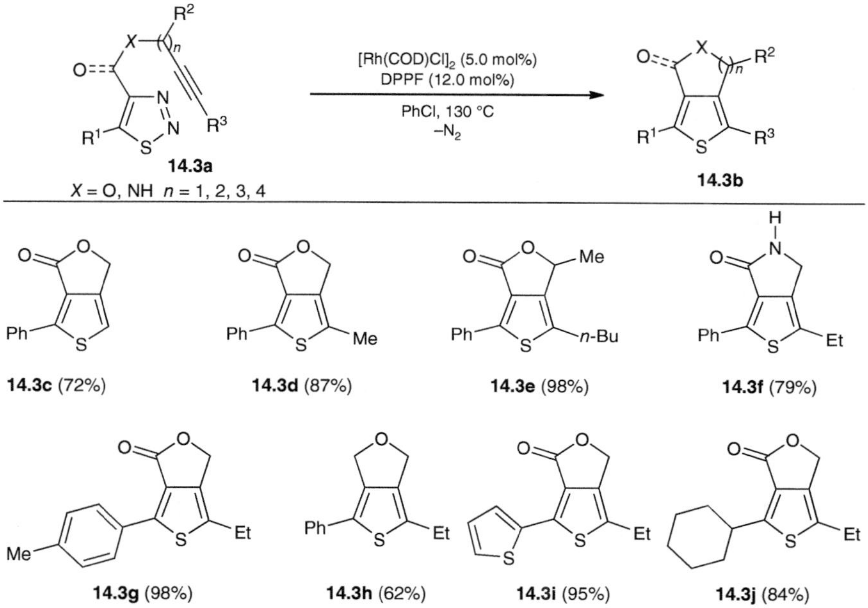

Scheme 14.3 Rh-catalyzed intramolecular transannulation reactions of alkynyl tethered 1,2,3-thiadiazoles. Source: Lee et al. [2]/American Chemical Society.

14.2.3 Rh-Catalyzed Denitrogenative Transformations of 1,2,3-Thiadiazoles for the Synthesis of Multisubstituted Furans and Thiophenes

Recently, Keaveney and co-workers have reported the first Rh-catalyzed denitrogenative transannulation reaction of 1,2,3-thiadiazoles to produce multisubstituted furans (Scheme 14.4) [3]. The cationic $[Rh(COD)_2]BF_4$ catalyst was crucial to this intramolecular transannulation reaction. Experimental and computational investigations disclosed the essential role of electron-donating groups, including dialkylamino and ethoxy groups, at the C5-position in stimulating all key steps of the catalytic cycle. Furthermore, ligand-controlled reactivity with the addition of DPPF ligand was found to prefer the intermolecular transannulation reaction

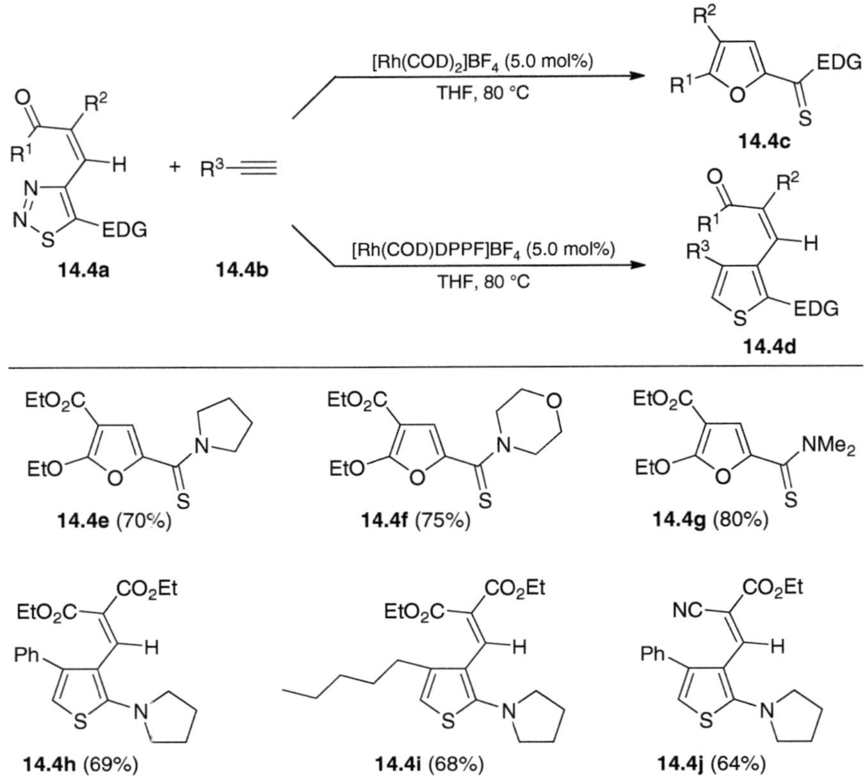

Scheme 14.4 Rh-catalyzed transannulation reactions of 4-vinyl-1,2,3-thiadiazoles. Source: Keaveney et al. [3]/American Chemical Society.

with alkynes over the intramolecular process. Mechanistic investigations proved that this selectivity arose from the formation of a relatively stable cyclometalated Rh(III) intermediate when using DPPF as well as the gradual formation of an inactive $[Rh(DPPF)_2THF]^+$ complex. $[Rh(COD)DPPF]BF_4$-catalyzed intermolecular transannulation reaction of 1,2,3-thiadiazoles with terminal alkynes gave regioselectivity for the unpredicted 2,3,4-substituted thiophene. The developed intermolecular transannulation reaction affords a novel method to regioselectively prepare multifunctionalized thiophenes. These results indicate that the organorhodium species is a four-membered cyclometalated Rh(III) intermediate, not the often suggested α-thiavinyl carbene.

14.2.4 Rh-Catalyzed 1,1-Hydroacylation of Thioacyl Carbenes with Alkynyl Aldehydes and Subsequent Cyclization for the Synthesis of 4*H*-Thiopyran-4-ones

Rh-catalyzed 1,1-hydroacylation of 1,2,3-thiadiazoles with alkynyl aldehydes followed by 6-*endo-dig* cyclization are accessed, producing a wide range of 4*H*-thiopyran-4-ones (Scheme 14.5) [4]. The oxidative addition of Rh(I) to aldehydes is

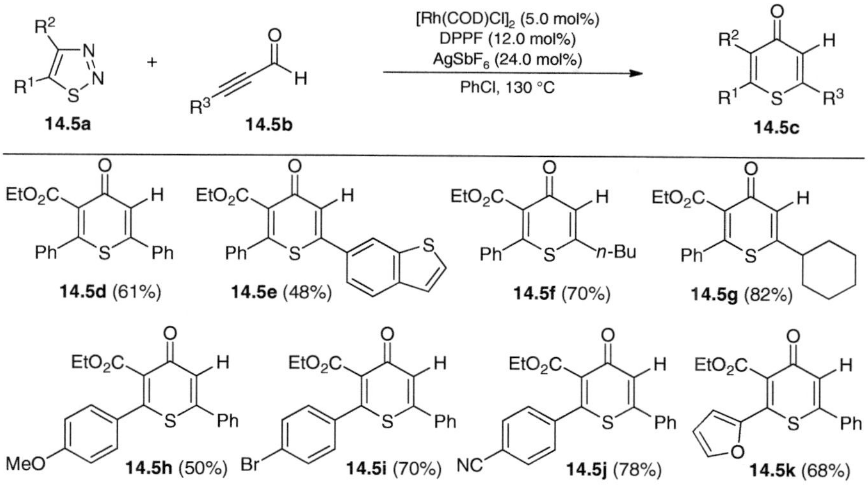

Scheme 14.5 Rh-Catalyzed 1,1-hydroacylation of thioacyl carbenes with alkynyl aldehydes followed by cyclization. Source: Xu and Yang et al. [4]/American Chemical Society.

proposed to be the turnover-limiting step. Applications to estrones provided formal [3 + 3] transannulated product in the structural variation of natural products.

14.3 Rh-Catalyzed Reaction of 1,2,3-Thiadiazoles with Alkenes

14.3.1 Regioselective Synthesis of Dihydrothiophenes and Thiophenes via the Rh-Catalyzed Transannulation of 1,2,3-Thiadiazoles with Alkenes

In 2016, Lee group has developed a method for the regioselective synthesis of a wide range of dihydrothiophenes from Rh-catalyzed transannulation reaction of 1,2,3-thiadiazoles with a variety of alkenes (Scheme 14.6) [5]. When 1-hexene and 1-decene were reacted with 1,2,3-thiadiazole (**14.6a**) in chlorobenzene, the desired dihydrothiophene products (**14.6d** and **14.6e**) were regioselectively obtained in high yields. 4-Phenyl-1-butene was transannulated to provide dihydrothiophene **14.6f** in 95% yield. Trimethylsilyl, ether, and acetoxy groups on aliphatic alkenes (**14.6b**) did not affect the efficiency of the transannulation reaction, producing a variety of dihydrothiophenes (**14.6g**, **14.6i**, and **14.6j**) in good to excellent yields. Bromo and carbonyl groups slightly influenced the transannulation reaction, producing the desired dihydrothiophenes **14.6h** (75%) and **14.6k** (70%). It is noteworthy that a hydroxyl group was tolerable with the present method. Also, the directing effect was not found upon increasing the chain length between the hydroxyl group and the double bond. A number of ω-hydroxy-1-alkenes were smoothly transannulated to furnish the desired dihydrothiophenes (**14.6l–14.6o**) in good to excellent yields ranging from 85% to 95%. No isomeric dihydrothiophenes were observed in any reactions. In the case of aromatic olefins, electronic modification of the substituents

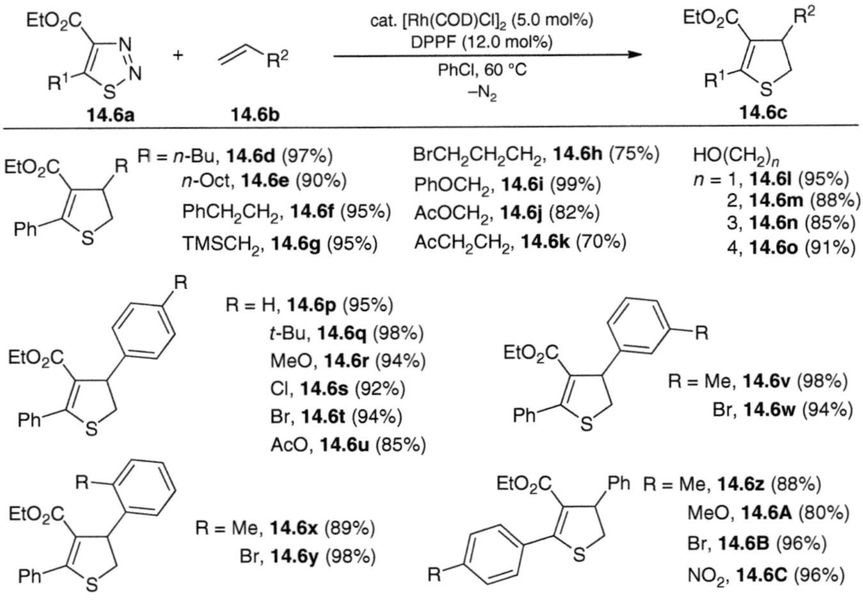

Scheme 14.6 Scope of thiadiazoles. Source: Lee et al. [5]/American Chemical Society.

on the aryl group did not influence the reaction efficiency. Electron-donating groups (Me, *t*-Bu, MeO, and AcO) as well as electron-withdrawing groups (Cl and Br) on the aryl ring did not affect the efficiency of the transannulation reaction, providing the corresponding dihydrothiophenes.

Encouraged by these results, the preparation of multitransannulated products consisting of phenyl and dihydrothiophenyl rings was achieved using the developed reaction (Scheme 14.7). When thiadiazole (**14.7a**) (2.5 equiv) was reacted with 1,4-divinylbenzene (**14.7b**, 1.0 equiv), two-fold transannulation reactions took place, providing the formation of linear pentacyclic compounds consisting of phenyl and dihydrothiophenyl rings (**14.7c**) in 83% yield. A linear pentacyclic compound with different connectivity of the sulfur and carbon bond in two dihydrothiophenyl rings was also synthesized. When 1,4-di(thiadiazolyl)benzene **14.7d** (1.0 equiv) was treated with styrene (4.0 equiv), the desired linear pentacyclic compound **14.7e** consisting of three phenyl and two dihydrothiophenyl rings was obtained in 80% yield.

Next, we tried to the synthesis of thiophenes from Rh-catalyzed transannulation reaction of thiadiazoles with alkenes and subsequent oxidation in one-pot (Scheme 14.8). Rh-catalyzed transannulation reaction of **14.8a** with 1-hexene and 5-bromo-1-pentene followed by oxidation with DDQ afforded the desired thiophenes **14.8d** (90%) and **14.8e** (75%). Gratifyingly, the catalytic transannulation reaction and subsequent oxidation with styrene and 4-*tert*-butyl and 4-chloro-substituted

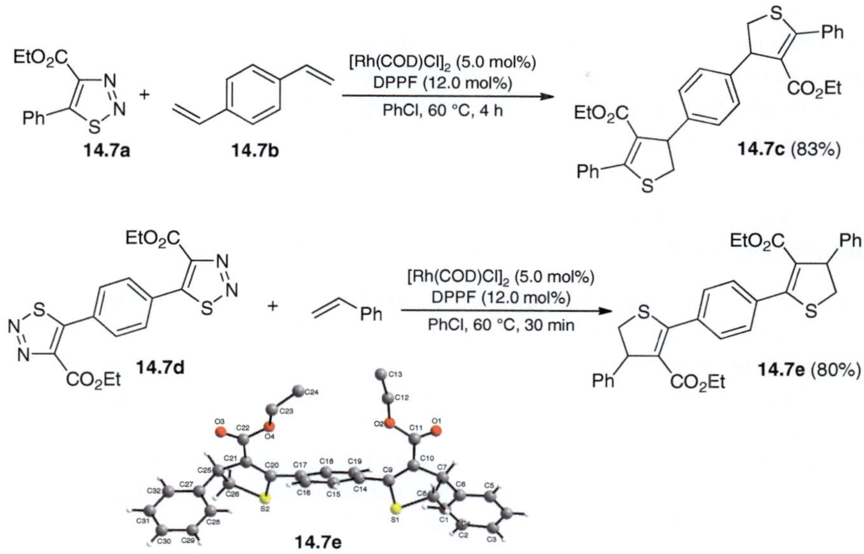

Scheme 14.7 Synthesis of oligomeric compounds. Source: Lee et al. [5]/American Chemical Society.

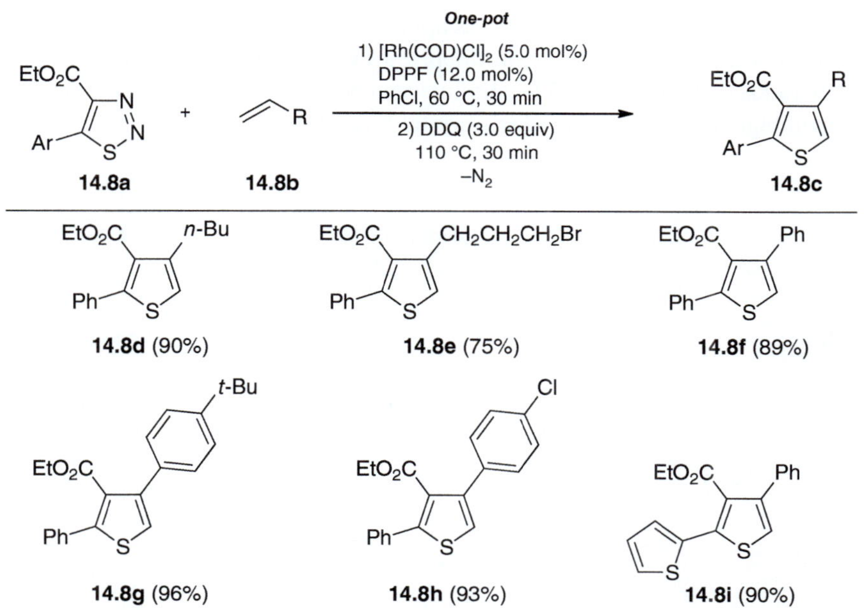

Scheme 14.8 Synthesis of thiophenes from thiadiazoles and alkenes *via* Rh-catalyzed transannulation and subsequent oxidation in one-pot. Source: Lee et al. [5]/American Chemical Society.

styrene smoothly took place to provide the corresponding thiophenes (**14.8f**, **14.8g**, and **14.8h**) in good to excellent yields. Thiophen-2-yl-substituted thiadiazole was transannulated with styrene to give rise to the corresponding thiophene **14.8i** in 90% yield. The present method has the advantages of broad substrate scope, wide functional group compatibility, and high regioselectivity.

14.3.2 Ir-Catalyzed Diastereospecific and Enantioselective (3 + 2) Transannulations of 1,2,3-Thiadiazoles with Norbornenes

A variety of 1,2,3-thiadiazoles (**14.9a**) were good transannulating reaction partners of 1,4-dihydro-1,4-methanonaphthalene (**14.9b**) catalyzed by $[Ir(cod)Cl]_2$ in the presence of DPPF, leading to the formation of transannulated products in excellent yields (Scheme 14.9) [6]. In the case of 5-aryl thiadiazole, electronic modification of the substituents on the aryl group did not influence the reaction efficiency. Electron-withdrawing groups (F, Cl, Br, NO_2, and CF_3) as well as electron-donating methyl group on the aryl ring did not affect the efficiency of the transannulation reaction, affording the corresponding dihydrothiophenes **14.9e**–**14.9j** in high yields. In addition, methylenedioxyphenyl-, furyl-, and methyl-substituted thiadiazoles were smoothly converted to the desired products (**14.9k**–**14.9m**). 5-Unsubstituted thiadiazole underwent a transannulation reaction to provide dihydrothiophene (**14.9n**) in 98% yield.

The asymmetric version of the transannulation reactions was carried out with chiral phosphine ligands. (*R*)-TolBINAP was optimal ligand. With this ligand, different optically active polycyclic 2,3-dihydrothiophenes were obtained with 61–79% *ee* in 80–98% yields (Scheme 14.10). The absolute stereochemistry of the chiral products was unambiguously assigned by single-crystal X-ray diffraction.

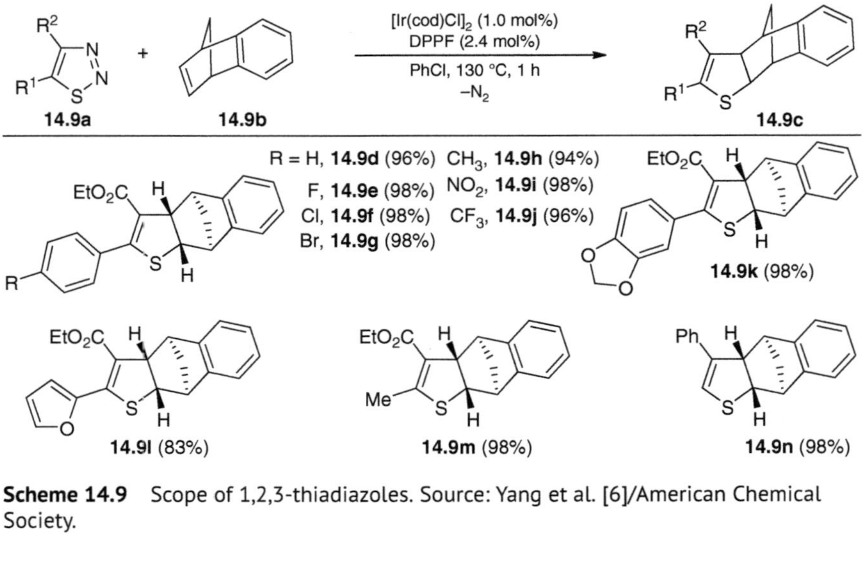

Scheme 14.9 Scope of 1,2,3-thiadiazoles. Source: Yang et al. [6]/American Chemical Society.

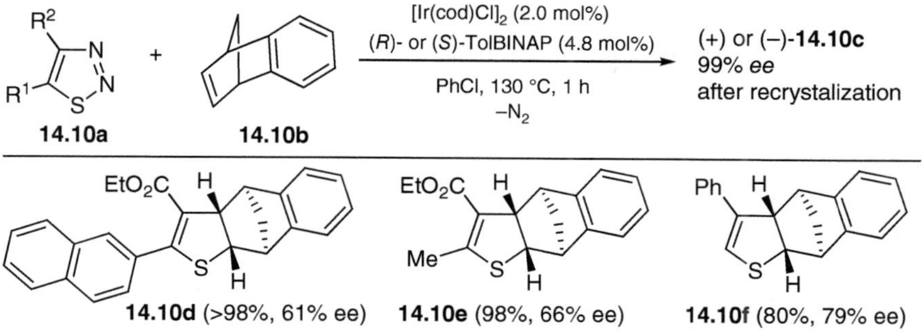

Scheme 14.10 Asymmetric (3 + 2) transannulations of norborenes. Source: Yang et al. [6]/ American Chemical Society.

14.3.3 Rh-Catalyzed 1,1-Hydroacylation of Thioacyl Carbenes with Alkynyl Aldehydes and Subsequent Cyclization

α,β-Enals (**14.11b**) were good substrates for Rh-catalyzed formal (3 + 3) transannulation reactions (Scheme 14.11) [7]. Cinnamaldehydes were converted to 2,3-dihydro-4*H*-thiopyran-4-ones (**14.11d**) in 61% yield. Next, spiro 2,3-dihydro-4*H*-thiopyran-4-ones (**14.11e, 14.11f, 14.11g,** and **14.11h**) were obtained in 42–73% yields. To investigate the β-substituent effect of (*E*)-alk-2-enals, they were treated with thiadiazole under the optimal reaction conditions. Compared with the reaction of acrolein, the presence of β-substituent markedly enabled the transannulation reactions, leading to the formation of **14.11i–14.11l** in good yields ranging from 65% to 83%. In the meantime, the scope of 1,2,3-thiadiazoles was examined by reacting with 3-methylbut-2-enal under the optimal reaction conditions. A number of ethyl

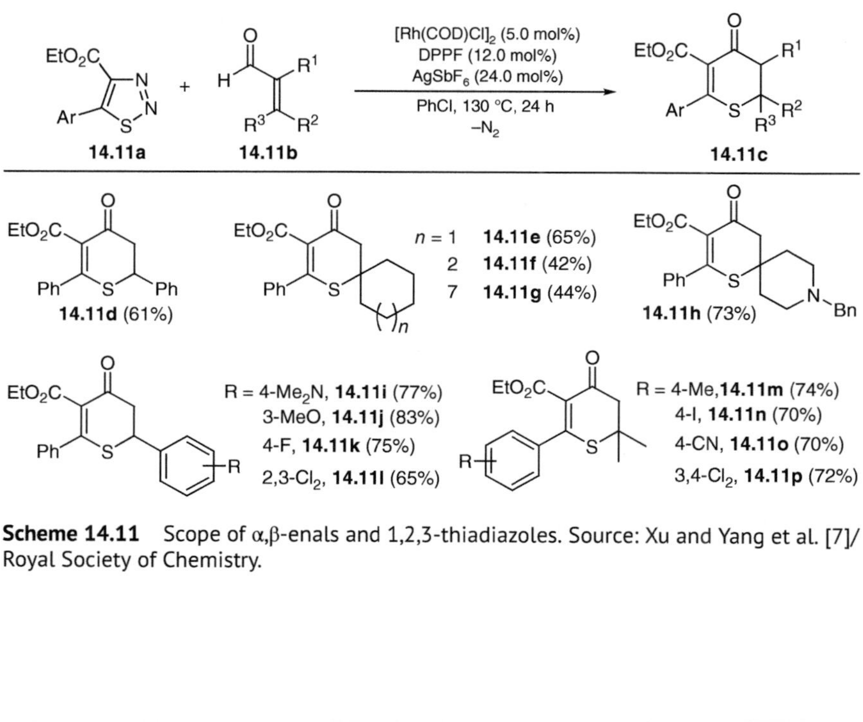

Scheme 14.11 Scope of α,β-enals and 1,2,3-thiadiazoles. Source: Xu and Yang et al. [7]/ Royal Society of Chemistry.

5-aryl-1,2,3-thiadiazole-4-carboxylates underwent the transannulation reaction to furnish the corresponding 2,3-dihydrothiopyran-4-ones (**14.11m–14.11p**, 70–74%). Functional groups such as halogen and cyano are tolerable.

14.4 Rh-Catalyzed Reaction of 1,2,3-Thiadiazoles with Carbonyl Compounds

14.4.1 Ag-Mediated Annulation of 5*H*-1,2,3-Thiadiazoles with 1,3-Dicarbonyls for the Synthesis of Polysubstituted Furans

In 2022, Huang and co-workers investigated the scope and limitations of 5*H*-1,2,3-thiadiazoles (Scheme 14.12) [8]. Electron-withdrawing as well as electron-donating groups at 4-position of the aryl ring of 4-aryl-1,2,3-thiadiazoles were smoothly converted to the desired transannulated products in moderate to good yields. Methoxycarbonyl, nitrile, halide, and double/triple bond on the aryl ring of 4-aryl-1,2,3-thiadiazoles were compatible (**14.12e**, **14.12g**, **14.12i**, **14.12j**, and **14.12k**). Substituents at 2- or 3-position of the phenyl ring had little effect on these transformations. Also, polysubstituents on phenyl rings were used for the preparation of furans. However, alkyl-substituted thiadiazoles were converted in low yield. Cyclohexanedione was transannulated to the corresponding furan **14.12l** in 40% yield.

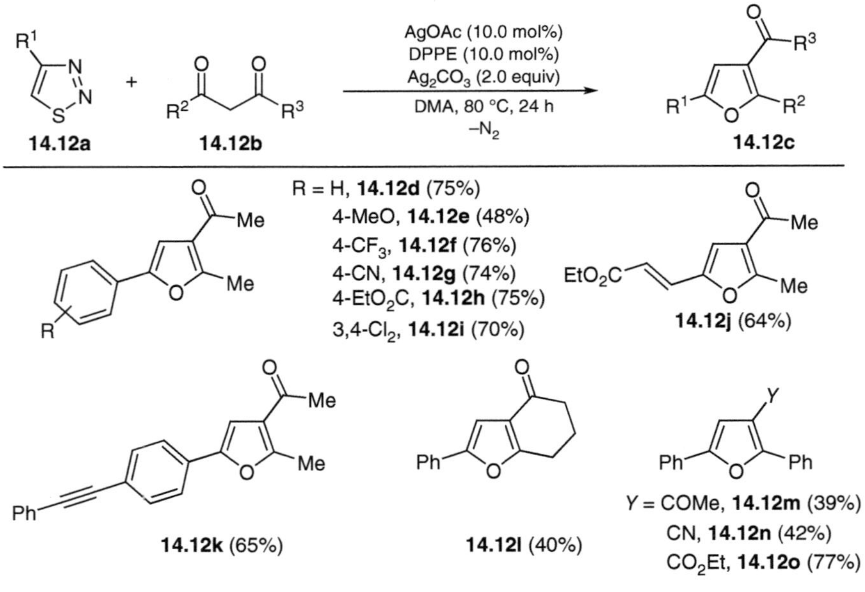

Scheme 14.12 Scope of 5*H*-1,2,3-thiadiazoles and 1,3-dicarbonyls. Source: Huang et al. [8]/Royal Society of Chemistry.

14.5 Rh-Catalyzed Reaction of 1,2,3-Thiadiazoles with Nitriles

14.5.1 Rh-Catalyzed Transannulation of 1,2,3-Thiadiazoles with Nitriles for the Synthesis of Isothiazole

A synthetic approach for a number of isothiazoles was developed by Rh-catalyzed transannulation of 1,2,3-thiadiazoles with alkyl, aryl, and heteroaryl nitriles, which proceeds through an α-thiavinyl Rh-carbenoid intermediate (Scheme 14.13) [9]. Ethyl 5-phenyl-1,2,3-thiadiazole-4-carboxylate (**14.13a**) was treated with a variety of aromatic and aliphatic nitriles (**14.13b**) to prove the scope and efficiency of Rh-catalyzed transannulation reaction. Although 2-fluorobenzonitrile was converted to the corresponding isothiazole in 57% yield due to steric hindrance, 4-fluorobenzonitrile was transannulated to the desired isothiazole (**14.13e**) in 99% yield. Transannulation with 3-chlorobenzonitrile was workable, producing the corresponding isothiazole (**14.13f**) in excellent yields. Cyclization with 4-bromobenzonitrile and 3-iodobenzonitrile afforded the isothiazoles **14.13g** and **14.13h**, respectively, in quantitative yields. Transannulation with a benzonitrile having a labile acetyl group proceeded to provide **14.13i** in 98% yield. The mono-isothiazole **14.13j** was selectively produced in 95% yield from

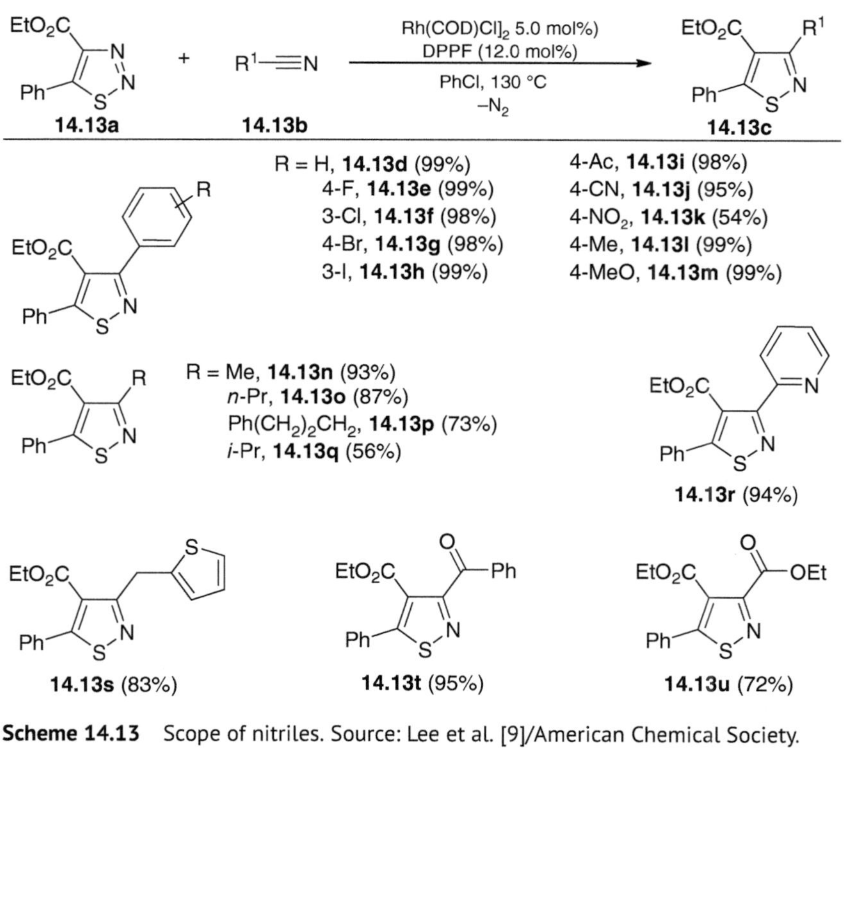

Scheme 14.13 Scope of nitriles. Source: Lee et al. [9]/American Chemical Society.

1,4-dicyanobenzene under the optimized reaction conditions. The benzonitrile with the electron-withdrawing 4-nitro group yielded the desired product **14.13k** in 54% yield. 4-Methylbenzonitrile and 4-methoxybenzonitrile were transannulated to **14.13l** and **14.13m**, respectively, in quantitative yields. Transannulation of **14.13a** with aliphatic nitriles was performed under the optimized conditions. Transannulation with acetonitrile, butyronitrile, and 4-phenylbutyronitrile led to the corresponding products **14.13n**, **14.13o**, and **14.13p**, respectively, in good to excellent yields (73–93%). However, isobutyronitrile was less reactive due to steric effects. Heterocyclic nitriles, such as 2-cyanopyridine and 2-thiopheneacetonitrile, reacted with **14.13a** to give **14.13r** (94%) and **14.13s** (83%). Furthermore, benzoyl cyanide and ethyl cyanoformate were converted to the desired products **14.13t** (95%) and **14.13u** (72%), respectively, *via* the Rh-catalyzed transannulation reaction.

Based on these results, a wide range of thiadiazoles were transannulated with benzonitrile, 4-bromobenzonitrile, and 1,4-dicyanobenzene (Scheme 14.14). Ethyl

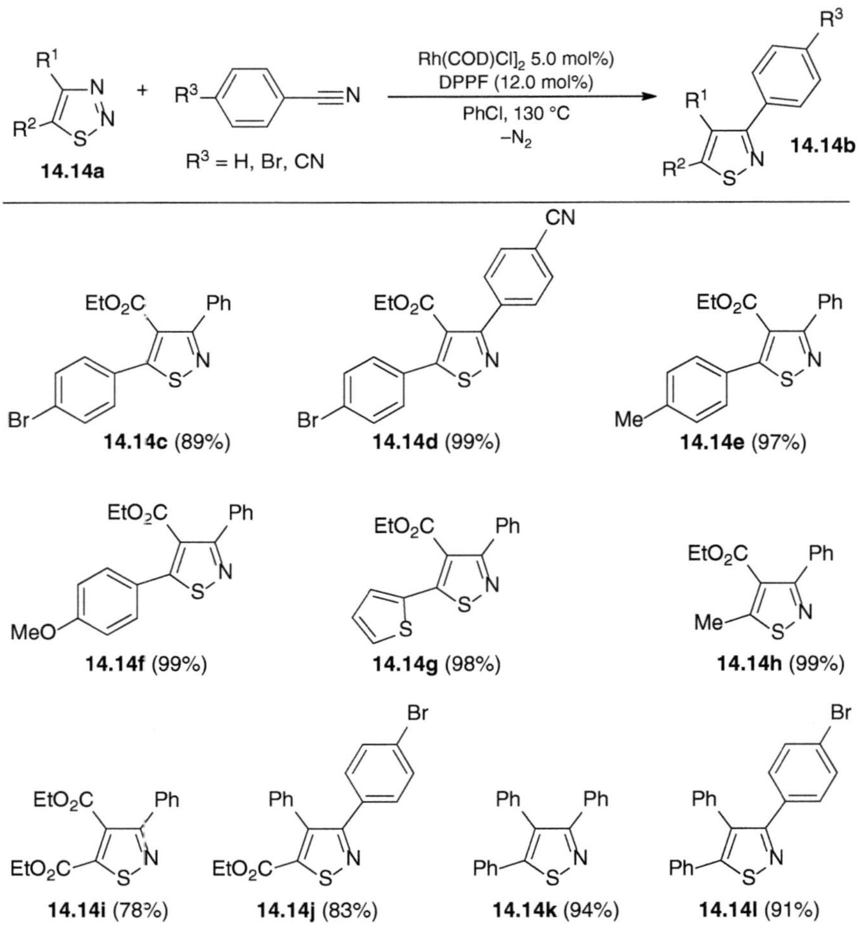

Scheme 14.14 Scope of thiadiazoles. Source: Lee et al. [9]/American Chemical Society.

5-aryl-1,2,3-thiadiazole-4-carboxylates with an electron-withdrawing bromo group or electron-donating methyl or methoxy group on the aryl ring were efficiently reacted with benzonitrile, affording the desired isothiazoles **14.14c–14.14f** in good to excellent yields (89–99%). Thiadiazoles with 2-thienyl and methyl group were converted to the desired isothiazole **14.14g** and **14.14h** in excellent yields. Diethyl 1,2,3-thiadiazole-4,5-dicarboxylate was transannulated by the developed reaction to give the isothiazole **14.14i** in 78% yield. When the thiadiazole was reacted with 4-bromobenzonitrile, the corresponding isothiazole **14.14j** was obtained in 83% yield. When 4,5-diphenyl-1,2,3-thiadiazole was reacted with benzonitrile and 4-bromobenzonitrile, the corresponding isothiazoles **14.14k** and **14.14l** were produced in 94% and 91% yields, respectively.

Oligomeric arylene compounds consisting of benzene and isothiazole rings were accessed using the developed reaction. For example, the isothiazole produced by the transannulation of ethyl 5-phenyl-1,2,3-thiadiazole-4-carboxylate with 1,4-dicyanobenzene was further reacted with **14.15a** to give the (benzene/isothiazole) pentaoligomer **14.15b** in 82% yield under the modified optimum conditions (Scheme 14.15). Pentaoligomers with a different connectivity of nitrogen and sulfur bonds in two isothiazole rings were also synthesized. When 1,4-di(thiadiazolyl)benzene was reacted with 3-iodobenzonitrile, the desired penta(arylene) **14.15f** consisting of three benzene and two isothiazole rings was obtained in 68% yield.

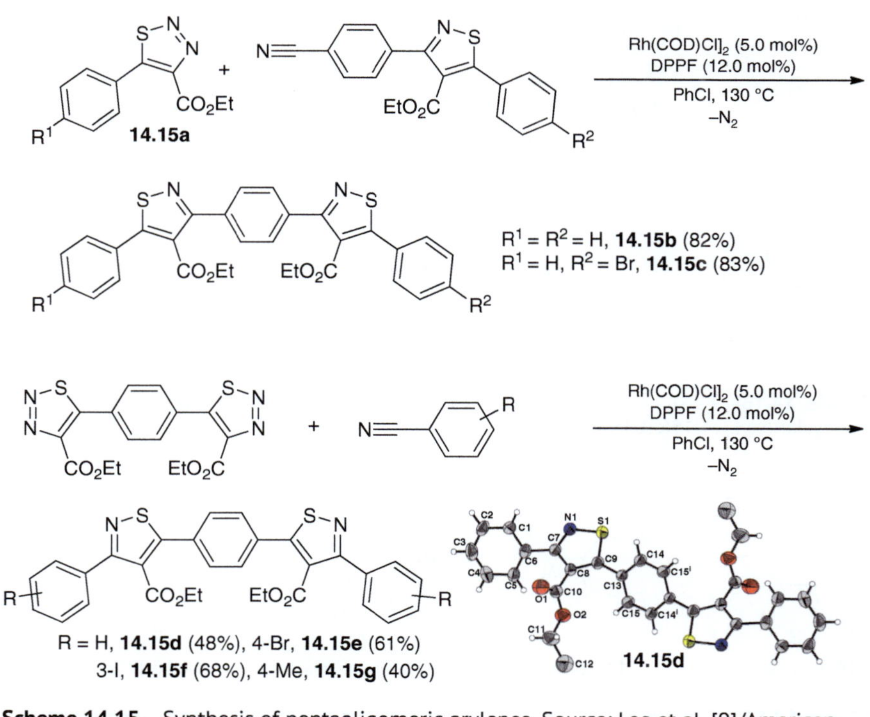

Scheme 14.15 Synthesis of pentaoligomeric arylenes. Source: Lee et al. [9]/American Chemical Society.

14.5.2 Synthesis of Bicyclic Isothiazoles *via* Intramolecular Rh-Catalyzed Transannulation of Cyanothiadiazoles

In recent years, many metal-catalyzed transannulation reactions *via* metal-carbenoid intermediates from *N*-sulfonyl-1,2,3-triazoles and pyridotriazoles have been reported. More recently, Gevorgyan and co-workers described Rh-catalyzed transannulation of 1,2,3-thiadiazoles with alkynes for the synthesis of thiophenes *via* an α-thiavinyl carbene as a 1,3-dipole equivalent [1]. In addition, Lee group developed the intermolecular Rh-catalyzed transannulation of 1,2,3-thiadiazoles with nitriles and alkenes to produce 4-substituted isothiazoles and dihydrothiophenes, respectively [5, 9]. These results are in obvious contrast with earlier findings of Gevogyan and co-workers from the point of view of regioselectivity. Stimulated by these results and interested in the latent uses of bicyclic isothiazoles in medicinal chemistry, electronic material chemistry, and pharmaceuticals, Lee group reported an efficient intramolecular Rh-catalyzed transannulation of cyanothiadiazoles containing an ester, amide, or ether as a linker, leading to the straightforward preparation of bicyclic isothiazoles (Scheme 14.16) [10].

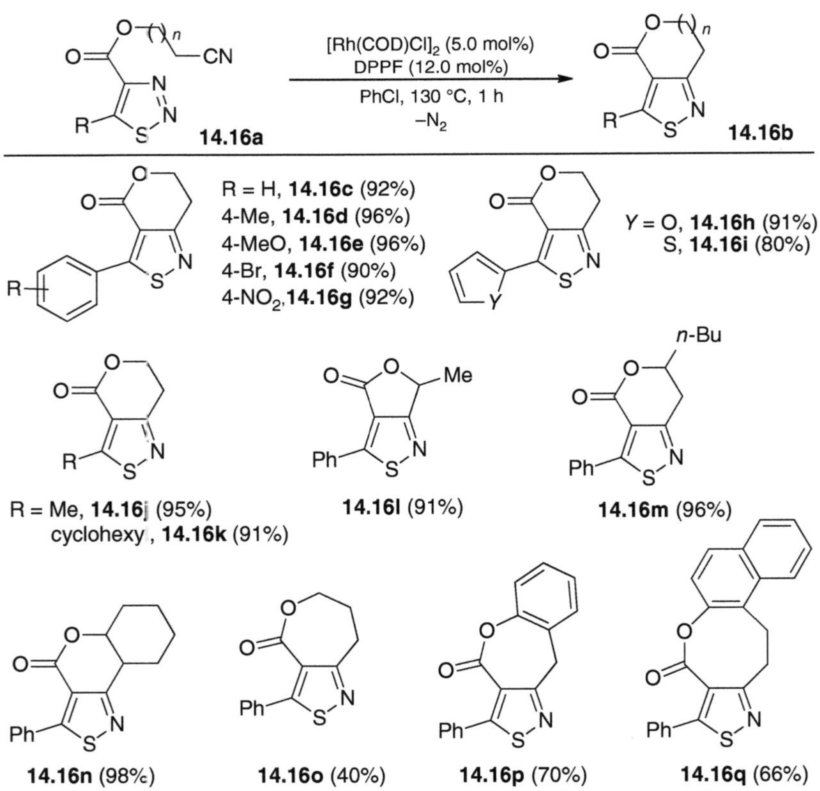

Scheme 14.16 Substrate scope of thiadiazoles bearing a cyanoalkoxycarbonyl groups. Source: Lee et al. [10]/American Chemical Society.

The scope and limitation of Rh-catalyzed intramolecular transannulation of a number of cyanothiadiazoles (**14.16a**) were examined. The electronic modification of the substituents on the aryl group of 2-cyanoethyl 5-aryl-1,2,3-thiadiazole-4-carboxylate did not influence the reaction efficiency. Electron-donating methyl and methoxy groups afforded the desired bicyclic isothiazoles in 96% yields. The conditions of Rh-catalyzed intramolecular transannulation reaction were tolerable with bromo and nitro groups, producing **14.16f** and **14.16g** in 90% and 92% yields, respectively. 2-Furyl-substituted cyanothiadiazole tolerated the optimized conditions and yielded the corresponding 2-furyl-substituted bicyclic isothiazole **14.16h** in 91% yield. 2-Thiophenyl-substituted cyanothiadiazoles afforded the desired bicyclic isothiazoles (**14.16i**) in 80% yield. The present method worked equally well even with methyl- and cyclohexyl-substituted cyanothiadiazoles, resulting in the facile construction of bicyclic isothiazoles in high yields. Encouraged by these results, the effect of the chain length and substituents on the cyanoalkyl group on the Rh-catalyzed transannulation was investigated. Thiadiazoles with cyanomethyl groups prepared from 2-hydroxypropanenitrile underwent Rh-catalyzed transannulation, affording the desired bicyclic thiazole (**14.16l**) in 91% yield. Thiadiazole obtained from the esterification reaction of 5-phenyl isothiazole-4-carboxylic acid with 3-hydroxyheptanenitrile was compatible with the reaction conditions to afford the desired isothiazole (**14.16m**) in 96% yield. The present method worked equally well with thiadiazole accessed from 2-hydroxycyclohexanecarbonitrile, leading to the production of tricyclic isothiazole **14.16n** in 98% yield. Next, Rh-catalyzed transannulation of cyanoalkyl 5-phenyl-1,2,3-thiodiazolyl-4-carboxylates with an enlarged alkyl chain was examined. Although thiadiazole obtained from 4-hydroxybutanenitrile was less reactive, the desired bicyclic thiazole was produced in an acceptable yield. When thiadiazole accessed from 2-(2-hydroxyphenyl)acetonitrile was reacted with the Rh catalyst, tricyclic isothiazole (**14.16p**) was obtained in 70% yield. Also, thiadiazole having 3-(2-hydroxynaphthalen-1-yl)propanenitrile moiety was transannulated to give tetracyclic isothiazole (**14.16q**) in 66% yield.

14.5.3 Rh-Catalyzed Highly Chemoselective and Diastereoselective (3 + 2) Transannulations of 1,2,3-Thiadiazoles with Cyanoepoxides

A diastereoselective and chemoselective (3 + 2) transannulation of 1,2,3-thiadiazoles with cyanoepoxides has been demonstrated (Scheme 14.17) [11]. Although the use of sterically smaller DPPF allows the synthesis of *trans*-products from *cis*- or *trans*-cyanoepoxides up to 95% yields and 100 : 0 *dr*., the use of sterically larger DPEPhos allows the preparation of *cis*-epoxyl isothiazoles from *cis*-cyanoepoxides up to 95% yields and 100 : 0 *dr*. Bisphosphines and epoxides play dual roles. Bisphosphines serve not only as ligands, but also as catalysts to catalyze the isomerization of *cis*-epoxides to *trans*-isomers. The diastereoselectivity is controlled by the kinetic competition between the direct transannulation of *cis*-cyanoepoxides and the bisphosphine-catalyzed isomerization of *cis*-products. The epoxy group's large steric hindrance guarantees excellent chemoselectivity toward the (3 + 2) annulation,

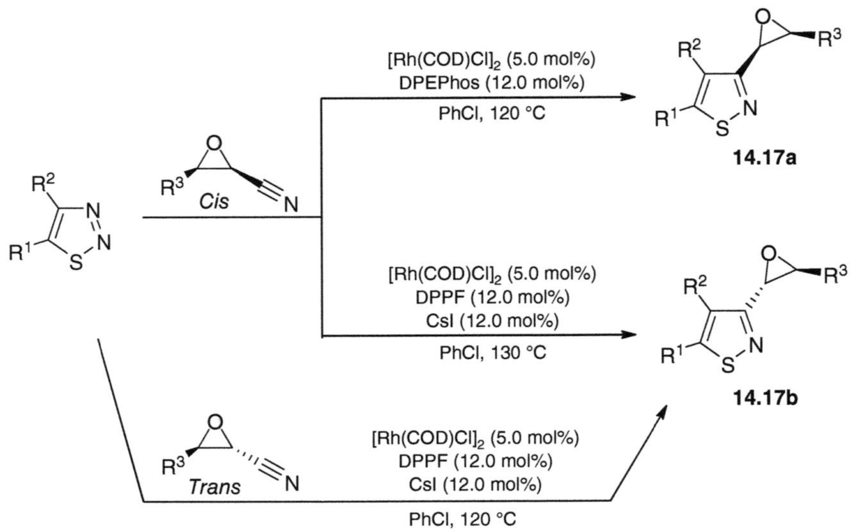

Scheme 14.17 Reactions of thioacyl carbenes with cyanoepoxides. Source: Yang et al. [11]/Royal Society of Chemistry.

and its electron-withdrawing ability significantly improves the reactivity of the adjacent cyano group. To address the controversy over the organorhodium intermediates, Prof. Yang and co-workers suggest the resonation between Lee's umpolung 1,3-dipoles, Bao's cyclometalated Rh(III) complexes, and thioacyl-coordinated Rh(I) carbenes.

14.5.4 Transannulaction of 1,2,3-Thiadiazoles and Alk-2-enenitriles for the Synthesis of 3-(Alk-1-enyl)isothiazoles

1,2,3-Thiadiazoles are precursors of thiovinyl carbenes and thioketenes and show various reactions with unsaturated compounds. Rh-catalyzed reactions of 1,2,3-thiadiazoles and alk-2-enenitriles such as alkyl, aryl, and heteroaryl acrylonitriles have been examined, demonstrating chemospecific reactivity to afford 3-vinylisothiazoles *via* the (3 + 2) transannulation between the cyano group of alk-2-enenitriles and *in situ* generated thioacylcarbenes (Scheme 14.18) [12]. With the optimal conditions in hand, the scope and generality of alk-2-enenitriles were examined as shown in Scheme 14.18. A variety of *para*-substituted cinnamonitriles were first investigated, including those with electron-donating groups such as methoxy, methyl, phenyl, and methylthio and those with electron-withdrawing groups such as bromo, trifluoromethyl, and nitro, producing the desired products **14.18e–14.18k** in moderate to good yields. Compared with other cinnamonitriles, cinnamonitriles with methylthio, bromo, and strong electron-deficient trifluoromethyl and nitro provided the corresponding isothiazole derivatives in high yields. Heteroaryl derived 3-(thiophen-2-yl) acrylonitrile worked well, affording the desired product **14.18l** in 73% yield. For aliphatic (*E*)-but-2-enenitrile, it gave the desired products **14.18m** in very low yields. Diphenylacrylonitrile also worked well, producing the desired product

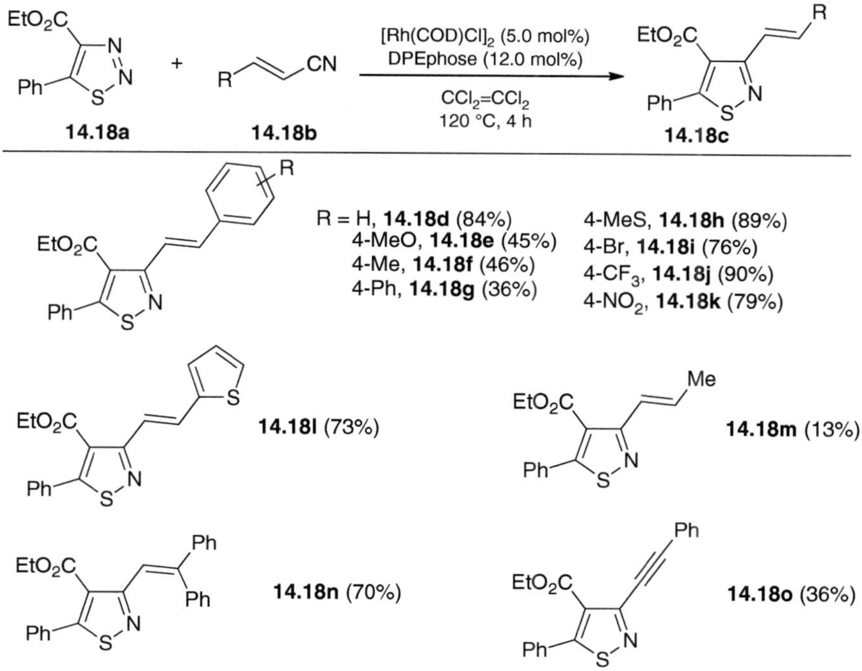

Scheme 14.18 Reaction with alk-2-enenitriles and alk-2-ynenitrile. Source: Xu et al. [12]/John Wiley & Sons.

14.18n in 70% yield. Besides, 3-phenylpropiolonitrile was examined, providing the corresponding 3-(phenylethynyl)isothiazole derivative **14.18o** in 36%.

14.6 Rh-Catalyzed Reaction of 1,2,3-Thiadiazoles with Imines

14.6.1 (2 + 2 + 2) Annulations of Thioketenes and Imines

(2 + 2 + 2) Annulations by using cyclic imines are demonstrated, and further supported by theoretical computations [13]. The zwitterionic intermediates *in situ* generated from thioketenes and imines undergo a stepwise nucleophilic *endo*-addition/*Si*-attack pathway with a second molecule of imine, giving (2,4)-*cis*-(4,5)-*cis*-(2 + 2 + 2) annulated products, which completely epimerize into (2,4)-*cis*-(4,5)-*trans*-annulated products in the presence of bases. Selectivity of the thio-Staudinger cycloadditions depends on the substituents of thioketenes and imines. Generally, when reacting with *N*-alkyl linear imines, monoalkyl, and disubstituted thioketenes prefer (2 + 2) pathway. However, when reacting with cyclic imines, except for *tert*-alkylthioketenes that favor (2 + 2) pathway, they prefer (2 + 2 + 2) pathways. The reactivities of Staudinger, thio-Staudinger, and sulfa-Staudinger intermediates are also compared. All three intermediates can be subjected to conrotatory ring closures to produce four-membered ring

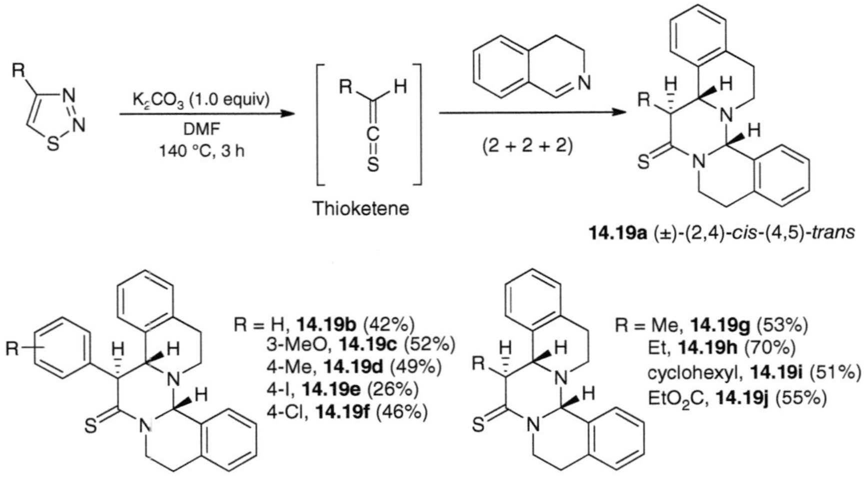

Scheme 14.19 Scope of (2 + 2 + 2) annulations. Source: Xu et al. [13]/Royal Society of Chemistry.

products. When reacting with cyclic imines, the Staudinger intermediates prefer the hetero-Diels–Alder cycloadditions, while the thio- and sulfa-Staudinger intermediates favor stepwise nucleophilic additions.

The experimental and stereochemical results are summarized in Scheme 14.19. The stereochemistry of the (2 + 2 + 2) annulated products affords significant clues to the mechanistic rationale. The C2 and C4 stereochemistry was judged by the NOE determination and by the XRD analysis of product **14.19i**. The obtained (2 + 2 + 2) annulated products exhibited specific stereochemistry. For arylthioketenes, regardless of the electron-donating or electron-withdrawing groups on the aryl rings, all the products (**14.19c–14.19f**) possessed (2,4)-*cis*-(4,5)-*trans*-configurations and were obtained in 26–52% yields. For alkylthioketenes, including primary methyl and ethylthioketenes and secondary cyclohexylthioketenes, and electron-deficient ethoxycarbonylthioketene, the desired products (**14.19g–14.19j**) were isolated in 51–70% yields, and they all displayed the same (2,4)-*cis*-(4,5)-*trans*-configurations.

14.7 Rh-Catalyzed Reaction of 1,2,3-Thiadiazoles with Phosphaalkynes

14.7.1 Rh-Catalyzed Cyclization Reactions of Thiadiazoles with Phosphaalkynes for the Synthesis of 1,3-Thiaphospholes

Rh-catalyzed synthetic method was reported for the synthesis of 1,3-thiaphospholes [14]. Cyclization reactions of a variety of 1,2,3-thiadiazoles with phosphaalkynes in the presence of a rhodium precursor and a dcypp ligand provide the corresponding 1,3-thiaphospholes *via* denitrogenation of the 1,2,3-thiadiazoles. Under the optimized reaction conditions, reactions of a number of 1,2,3-thiadiazole derivatives were investigated (Scheme 14.20). 1,3-Thiaphospholes having methyl

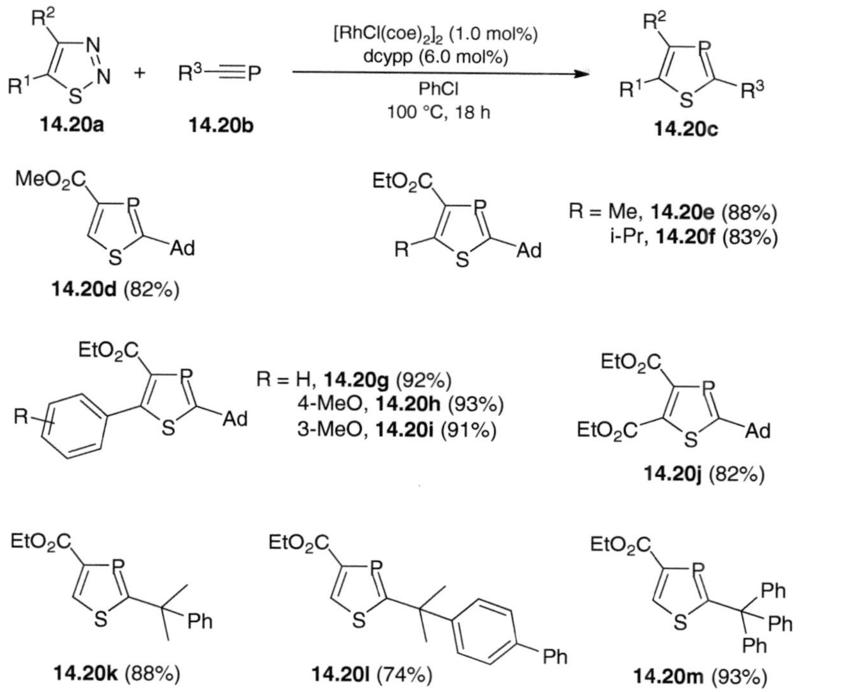

Scheme 14.20 Reaction of 1,2,3-thiadiazoles with phosphaalkynes. Source: Nishibayashi et al. [14]/John Wiley & Sons.

and ethyl ester were produced in good yields. The introduction of isopropyl, phenyl, 4-methoxyphenyl, and 3-methoxyphenyl group at the 5-position of 1,3-thiaphosphole was successful in providing the corresponding products in high yields. In addition, we found that a 1,3-thiaphosphole having two ester groups at the 4- and 5-positions (**14.20j**) was produced in 82% yield. Moreover, the introduction of 2-phenylprop-2-yl group into the phosphaalkyne was applicable to provide the corresponding product (**14.20k**) in 88% yield. Biphenyl moiety was compatible with this reaction system to form **14.20l** in 74% yield. Noteworthy is that a highly sterically hindered phosphaalkyne having a trityl group proceeds smoothly in the present reaction conditions, where tri-phenylmethylphosphaethyne was transformed into the desired product (**14.20m**) in 93% yield.

14.8 Rh-Catalyzed Reaction of 1,2,3-Thiadiazoles with Aziridines

14.8.1 Synthesis of Thiazolines *via* the Thioketene-Induced Ring Expansion of Aziridines

In 2022, Xu and co-worker investigated the substrate scopes of 1,2,3-thiadiazoles and 2-arylaziridines **14.21a** (Scheme 14.21) [15]. The reactions of different

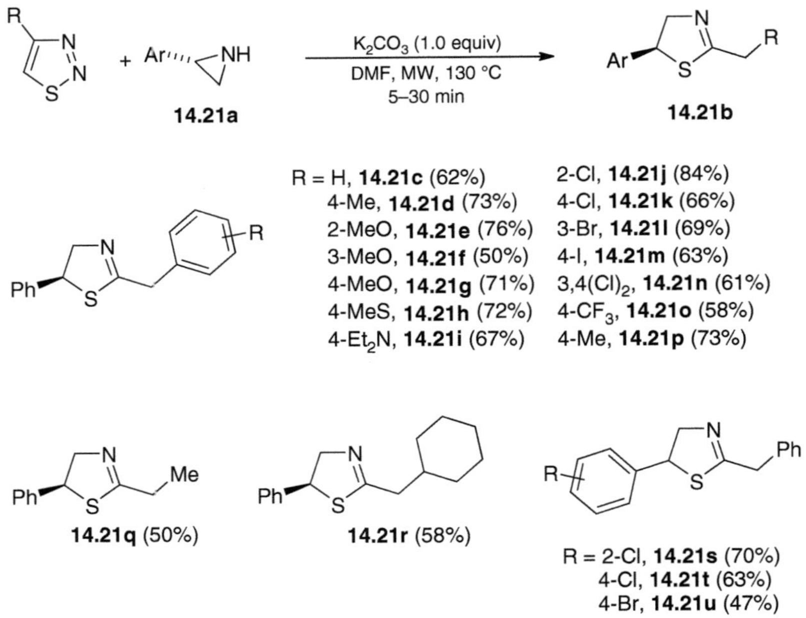

Scheme 14.21 Scope of 1,2,3-thiadiazoles and 2-arylaziridines. Source: Xu et al. [15]/ Royal Society of Chemistry.

1,2,3-thiadiazoles and 2-phenylaziridine provided the desired thiazolines in various yields without steric effect for 4-aryl-1,2,3-thiadiazoles. However, bulky 2-*tert*-butyl-1,2,3-thiadiazole afforded the corresponding thiazoline in a low yield of 21%, showing the effect of steric hindrance. In General, the electron-rich thiadiazoles resulted in much better yields than the electron-deficient ones. Secondly, different 2-arylaziridines were converted to the desired thiazolines in moderate to good yields.

14.9 Conclusions

In recent years, many metal-catalyzed transannulation reactions *via* metal-carbenoid intermediates from *N*-sulfonyl-1,2,3-triazoles and pyridotriazoles have been reported. More recently, Gevorgyan and co-workers described the Rh-catalyzed transannulation of 1,2,3-thiadiazoles with alkynes for the synthesis of thiophenes *via* an α-thiavinyl carbene as a 1,3-dipole equivalent. In addition, Rh-catalyzed transannution reactions of 1,2,3-thiadiazoles with alkynes, alkenes, carbonyl compounds, nitriles, imines, phosphaalkynes, and aziridines provided a wide range of *S*- and *O*-containing five-membered heterocyclic compounds. At the present time, one of the most important research fields in the thiavinyl carbene chemistry is to find novel transannulating agents and distinctive reactivities of thiavinyl carbenes. These methods are very powerful in synthesizing a variety of heterocyclic compounds and therefore is expected to be widely used in the future.

References

1 Gevorgyan, V. and Kurandina, D. (2016). Rh-catalyzed thiavinyl carbenes from 1,2,3-thiadiazoles enable modular synthesis of multisubstituted thiophenes. *Org. Lett.* 18: 1804–1807.

2 Kim, J.E., Lee, J., Yun, H. et al. (2017). Rhodium-catalyzed intramolecular transannulation reaction of alkynyl thiadiazoles enabled 5,*n*-fused thiophenes. *J. Org. Chem.* 82: 1437–1447.

3 Tokareva, M.A., Pernik, I., Lewis, W. et al. (2019). Rh(I)-catalyzed denitrogenative transformations of 1,2,3-thiadiazoles: ligand-controlled product selectivity and the structure of the key organorhodium intermediate revealed. *ACS Catal.* 12: 5574–5584.

4 Zhou, B., Wu, Q., Dong, Z. et al. (2019). Rh-catalyzed 1,1-hydroacylation of thioacyl carbenes with alkynyl aldehydes and subsequent cyclization. *Org. Lett.* 21: 3594–3599.

5 Son, J.-Y., Kim, J., Han, S.H. et al. (2016). Regioselective synthesis of dihydrothiophenes and thiophenes via the rh-catalyzed transannulation of 1,2,3-thiadiazoles with alkenes. *Org. Lett.* 18: 5408–5411.

6 Chen, C., Fang, S., Dong, Z. et al. (2022). Catalytic diastereospecific and enantioselective (3 + 2) transannulations of 1,2,3-thiadiazoles with strained norbornene derivatives. *Org. Lett.* 24: 2110–2114.

7 Wu, Q., Dong, Q., Xu, J., and Yang, Z. (2021). Sulfur-controlled and rhodium-catalyzed formal (3 + 3) transannulation of thioacyl carbenes with alk-2-enals and mechanistic insights. *Org. Biomol. Chem.* 19: 3173–3180.

8 Feng, M., Jiang, M., Jiang, H., and Huang, L. (2022). Silver-mediated annulation between 5-*H*-1,2,3-thiadiazoles and 1,3-dicarbonyl compounds to construct polysubstituted furans. *Org. Chem. Front.* 9: 5864–5871.

9 Seo, B., Kim, Y.G., and Lee, P.H. (2016). Synthesis of isothiazole via the rhodium-catalyzed transannulation of 1,2,3-thiadiazoles with nitriles. *Org. Lett.* 18: 5050–5053.

10 Seo, B., Kim, H., Kim, Y.G. et al. (2017). Synthesis of bicyclic isothiazoles through an intramolecular rhodium-catalyzed transannulation of cyanothiadiazoles. *J. Org. Chem.* 82: 10574–10582.

11 Dong, Z., Chen, C., Wang, J. et al. (2022). Dual roles of bisphosphines and epoxides: Rh-catalyzed highly chemoselective and diastereoselective (3 + 2) transannulations of 1,2,3-thiadiazoles with cyanoepoxides. *Org. Chem. Front.* 8: 6687–6698.

12 Wu, Q., Chen, N., and Xu, J. (2022). Chemoselectivity in the transannulaction of 1,2,3-thiadiazoles and alk-2-enenitriles: specific synthesis of 3-(alk-1-enyl)isothiazoles. *ChemistrySelect* 7: e202103943.

13 He, W., Zhuang, J., Du, H. et al. (2017). Stereochemistry and mechanistic insights in the $[2^t + 2^i + 2^i]$ annulations of thioketenes and imines. *Org. Biomol. Chem.* 15: 9424–9432.

14 Liang, W., Nakajima, K., and Nishibayashi, Y. (2020). Rhodium-catalyzed cyclization reactions of thiadiazoles with phosphaalkynes to prepare 1,3-thiaphospholes. *Eur. J. Org. Chem.* 2020: 3879–3882.

15 Wu, Q. and Xu, J. (2022). Regio- and stereoselective synthesis of thiazoline derivatives via the thioketene-induced ring expansion of aziridines. *Chem. Commun.* 58: 2714–2714.

15

Denitrogenative Transformations of Benzotriazinones and Related Compounds

Ramaraju Korivi, Popuri Sureshbabu, Dhanip Kumar Sharma, and Subramaniyan Mannathan

Department of Chemistry, SRM University-AP, Amaravati, Andhra Pradesh, India

15.1 Introduction

1,2,3-Benzotriazin-4(3*H*)-one, an important class of 1,2,3-triazoles, has attracted considerable attention because of its biological activities, including antimicrobial [1], anti-inflammatory [2] antidepressant [3], anticancer [4, 5], antiulcer [6] antidiarrheal [7], and anesthetics [8]. It has also been used as an excellent reagent in peptide synthesis (C—N bond formation) [9]. In recent years, 1,2,3-benzotriazin-4(3*H*)-ones have been well explored as potential precursors in various denitrogenative annulation and cross-coupling reactions to prepare heterocycles and *ortho*-substituted benzamide derivatives.

15.2 Metal-Catalyzed Denitrogenative Transformations

Transition-metal catalyzed annulations and cross-coupling reactions are arguably one of the efficient protocols for generating cyclic compounds and forming various carbon–carbon, carbon–hetero bonds in organic synthesis [10–12]. In this context, transition-metal complexes based on Rhodium (Rh), Ruthenium (Ru), Iridium (Ir), Palladium (Pd), Cobalt (Co), Nickel (Ni), and Iron (Fe) have been successfully employed to achieve such transformations effectively [13–16]. Aryl (pseudo) halides are one of the effective and frequently employed coupling partners in various cross-coupling and annulation reactions [17]. Despite their effectiveness, these precursors generate a stoichiometric amount of (pseudo) halides as waste at the end of the reaction. Consequently, several alternative strategies have been developed to address these issues [18, 19]. Among them, denitrogenative reactions of 1,2,3-triazoles have emerged as a useful strategy for performing various transannulation and cross-coupling reactions with the extrusion of nitrogen molecules as by-products [20–22]. In this context, metal-catalyzed denitrogenative annulation and cross-coupling reactions of 1,2,3-benzotriazin-4(3*H*)-ones were studied. Unlike

Denitrogenative Transformation of Nitrogen Heterocycles: Synthesis, Reactions and Applications, First Edition. Edited by Pazhamalai Anbarasan.

other 1,2,3-triazoles, 1,2,3-benzotriazin-4(3*H*)-ones undergo oxidative addition with low valent metal complex catalyst, forming an *aza*-metallacycle (**15.1A**) as the key intermediate (Scheme 15.1). Capturing this key intermediate with a suitable coupling partner may lead to annulated or cross-coupling products.

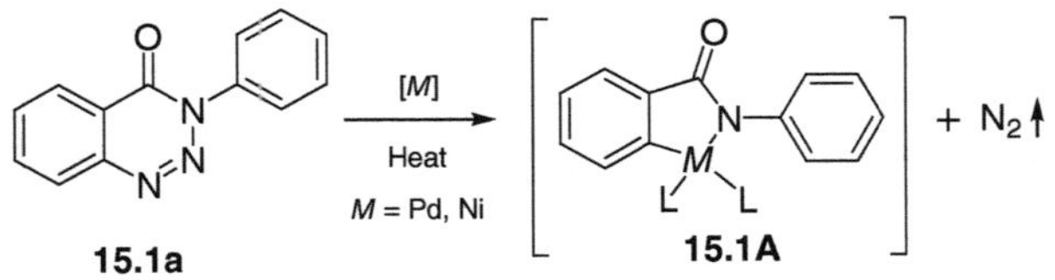

Scheme 15.1 Generation of reactive intermediates upon denitrogenation of 1,2,3-benzotriazin-4(3*H*)-ones.

15.3 Nickel and Palladium-Catalyzed Denitrogenative Transannulation Reactions

15.3.1 Synthesis of Isoquinolones and Related Derivatives

In 2008, Murakami group demonstrated an efficient method for the synthesis of 1(2*H*)-isoquinolones (**15.2c**) by nickel-catalyzed denitrogenative alkyne insertion of 1,2,3-benzotriazin-4(3*H*)-ones (**15.2a**) (Scheme 15.2) [23]. The reaction is compatible with various *N*-substituted 1,2,3-benzotriazin-4(3*H*)-ones and symmetrical alkynes (**15.2b**). In the case of unsymmetrical alkynes, the reaction yields different regioisomers depending on the ligand used. The reaction proceeds via a five-membered nickelacycle (**15.2A**) through denitrogenative activation

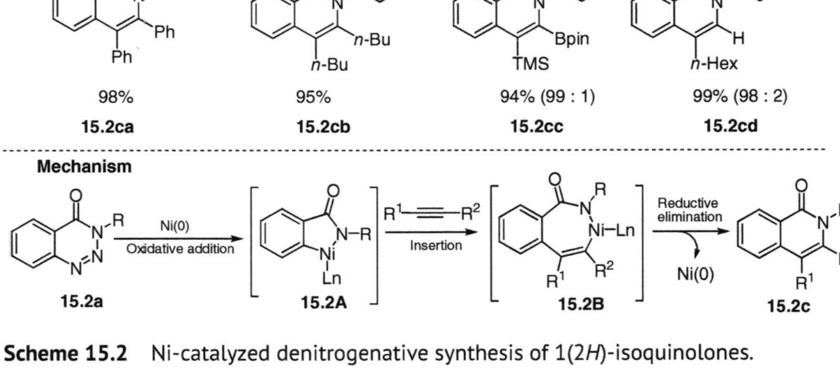

Scheme 15.2 Ni-catalyzed denitrogenative synthesis of 1(2*H*)-isoquinolones.

of the 1,2,3-benzotriazin-4(3*H*)-one followed by insertion of alkynes to form the intermediate (**15.2B**). Reductive elimination ultimately leads to the desired product (**15.2c**).

Subsequently, the same group reported a regio- and enantioselective synthesis of 3,4-dihydroisoquinolin-1(2*H*)-ones (**15.3c**) and (**15.3d**) from 1,2,3-benzotriazin-4(3*H*)-ones using allenes (**15.3b**) under nickel catalysis (Scheme 15.3) [24]. In this reaction, the PMe_3 ligand is employed for the nonasymmetric version, while *i*-Pr-FOXAP ligand is utilized for the enantioselective annulation reaction. This method demonstrates compatibility with aliphatic and aromatic terminal allenes, bearing various functional groups, resulting in the formation of desired products (**15.3c**) in good yields with high selectivity. It is important to mention that when *tert*-butyl and trialkylsilyl substituted allenes are involved, (**15.3d**) is preferably achieved due to the steric hindrance created by the bulky substituent at the metal center during the reductive elimination process.

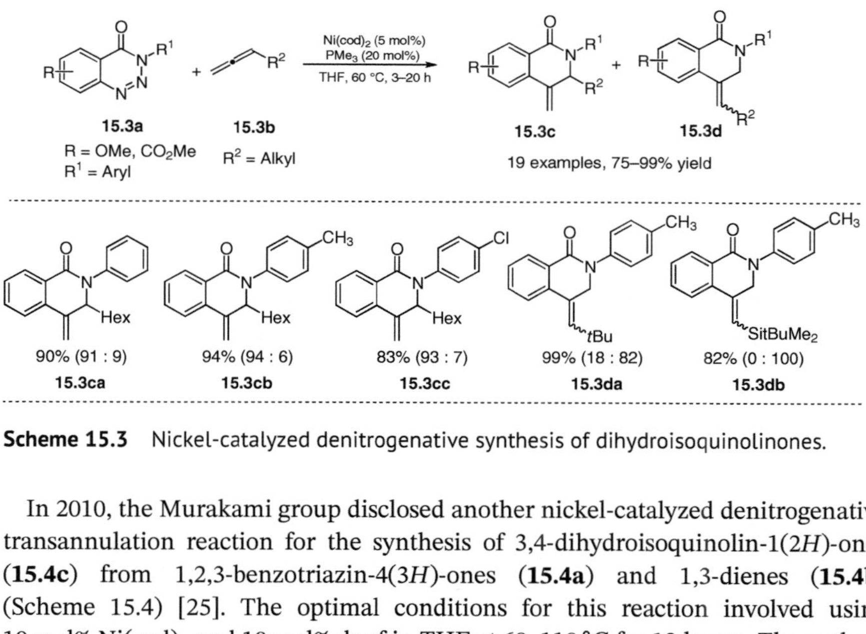

Scheme 15.3 Nickel-catalyzed denitrogenative synthesis of dihydroisoquinolinones.

In 2010, the Murakami group disclosed another nickel-catalyzed denitrogenative transannulation reaction for the synthesis of 3,4-dihydroisoquinolin-1(2*H*)-ones (**15.4c**) from 1,2,3-benzotriazin-4(3*H*)-ones (**15.4a**) and 1,3-dienes (**15.4b**) (Scheme 15.4) [25]. The optimal conditions for this reaction involved using 10 mol% $Ni(cod)_2$ and 10 mol% dppf in THF at 60–110 °C for 12 hours. The authors proposed a similar reaction mechanism, which begins with the formation of an

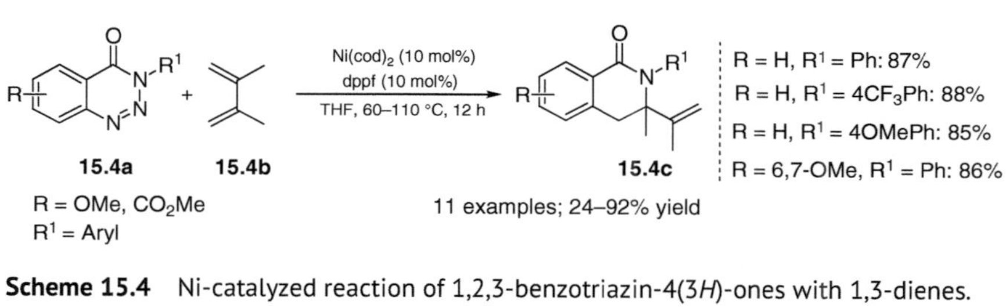

Scheme 15.4 Ni-catalyzed reaction of 1,2,3-benzotriazin-4(3*H*)-ones with 1,3-dienes.

azanickelacycle as the key intermediate, followed by 1,3-diene insertion to form (**15.5A**), which is in equilibrium with zwitterionic π-allylnickel species (**15.5B**). Subsequent intramolecular recombination of (**15.5B**) results in the formation of product (**15.4c** and **15.5d**). On the other hand, intermediate (**15.5A**) can undergo β-hydride elimination and reductive elimination, leading to the formation of (**15.5c**) and the generation of nickel(0) (Scheme 15.5).

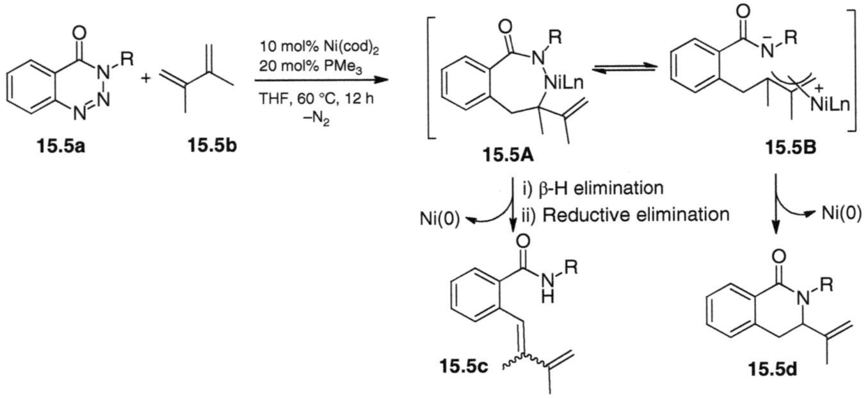

Scheme 15.5 Ni-catalyzed reaction pathway of 1,2,3-benzotriazin-4(3*H*)-ones with 1,3-dienes.

Nevertheless, when the reaction was performed with methyl acrylate (**15.6b**) as the alkene partner using P(*n*-Bu)$_3$ as a ligand, it yielded 3,4-dihydroisoquinolin-1(2*H*)-one product (**15.6c**) in high yield with high regioselectivity (Scheme 15.6a) [25]. Furthermore, various electron-deficient alkenes (**15.6b′**) participated in the annulation reaction, providing the corresponding products (**15.6c′**) in 11–99% yield, respectively (Scheme 15.6b).

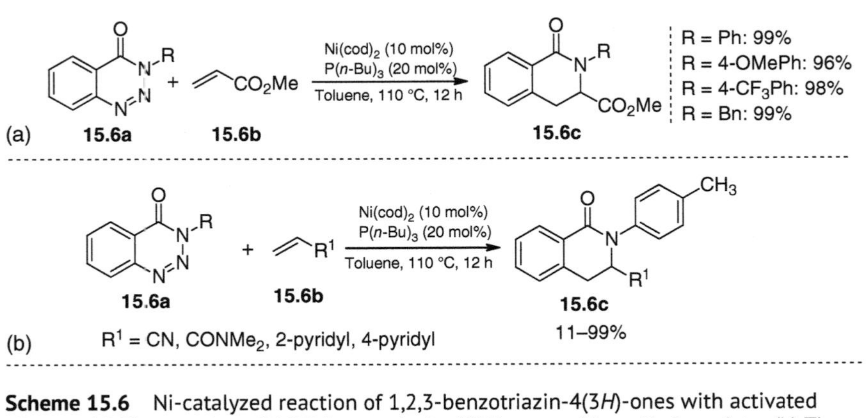

Scheme 15.6 Ni-catalyzed reaction of 1,2,3-benzotriazin-4(3*H*)-ones with activated alkenes. (a) The reaction of 1,2,3-benzotriazin-4(3*H*)-ones with methyl acrylate; (b) The reaction of 1,2,3-benzotriazin-4(3*H*)-ones with other activated alkenes.

In 2015, Liu and his co-workers demonstrated an asymmetric Ni/bis(oxazoline)-catalyzed denitrogenative transannulation approach to produce axially chiral isoquinolones (**15.7c**) with excellent enantioselectivity, reaching up to 98% *ee* (Scheme 15.7) [26]. The use of bulky internal alkynes bearing a naphthalene ring and the chiral bis(oxazoline) ligand (*S,S*-**L1**) is crucial for the success of this reaction. Density functional theory (DFT) calculations were performed to model intermediate geometries and characterize transition states, providing insights into the observed atroposelectivity. In 2016, the research group led by Guo and Liu conducted comprehensive studies using DFT for the Ni-catalyzed denitrogenative transannulation reaction to produce axially chiral isoquinolones [27].

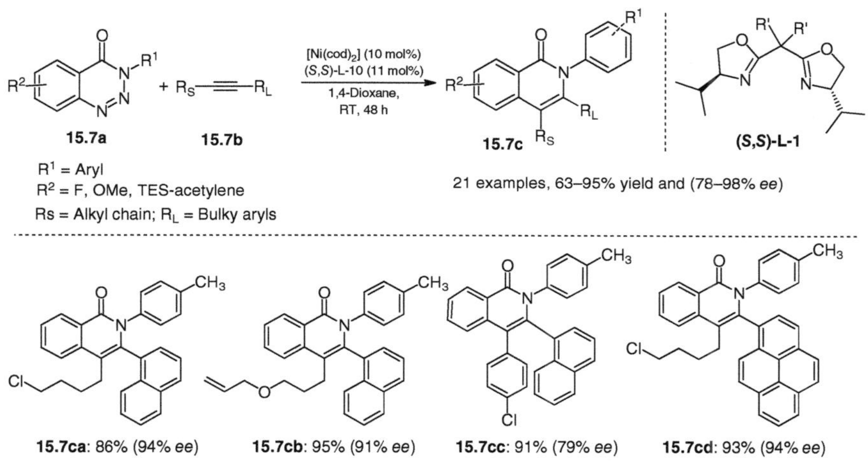

Scheme 15.7 Ni-catalyzed denitrogenative transannulation of 1,2,3-benzotriazin-4(3*H*)-ones and bulkier internal alkynes.

15.3.2 Synthesis of Phenanthridinones

Cheng *et al.* described a Ni-catalyzed denitrogenative annulation of 1,2,3-benzotriazin-4(3*H*)-ones (**15.8a**) and arynes (**15.8b**) to prepare medicinally valuable phenanthridinones (**15.8c**) (Scheme 15.8) [28]. A wide range of 1,2,3-benzotriazin-4(3*H*)-ones (**15.8a**) and benzynes (**15.8b**) actively participated in the reaction, resulting in various substituted phenanthridinones in good to high yields.

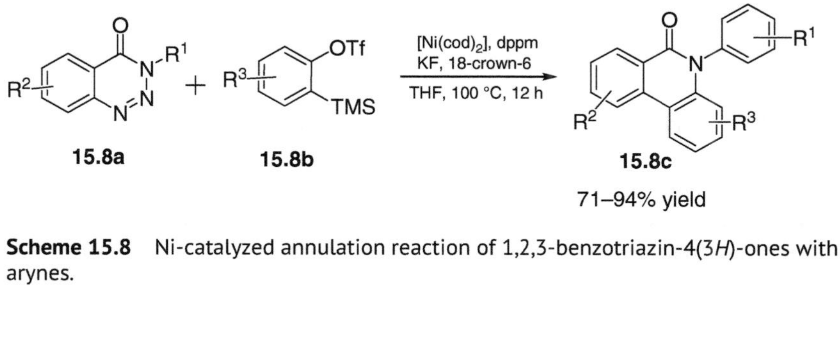

Scheme 15.8 Ni-catalyzed annulation reaction of 1,2,3-benzotriazin-4(3*H*)-ones with arynes.

Notably, the bidentate phosphine ligand, dppm, combined with 18-crown-6/KF, proved crucial for this reaction's success.

15.3.3 Synthesis of Isoindolin-1-ones and Related Derivatives

In 2011, Murakami and co-workers reported a Pd-catalyzed denitrogenative synthesis of 3-(imino)isoindolin-1-ones (**15.9c**) from 1,2,3-benzotriazin-4(3*H*)-ones (**15.9a**) and isocyanides (**15.9b**) (Scheme 15.9) [29]. The reaction exhibits tolerance to various functional groups present on the 1,2,3-benzotriazin-4(3*H*)-ones and isocyanides substrates. The mechanistic pathway closely parallels nickel-catalyzed annulation reactions involving 1,2,3-benzotriazin-4(3*H*)-ones and alkynes. The process initiates with the oxidative addition of a 1,2,3-benzotriazin-4(3*H*)-ones to Pd(0), generating a five-membered azapalladacycle (**15.9A**), followed by the insertion of isocyanides to Pd-C bond yields the desired product and regeneration of the active palladium(0) catalyst for the further cycle.

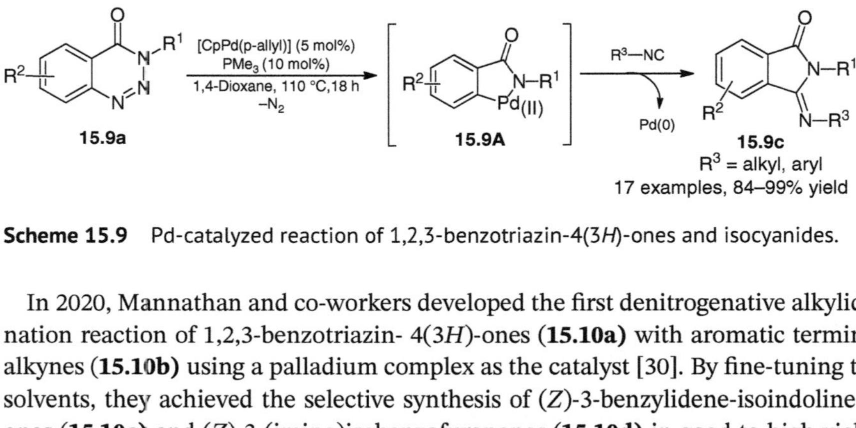

Scheme 15.9 Pd-catalyzed reaction of 1,2,3-benzotriazin-4(3*H*)-ones and isocyanides.

In 2020, Mannathan and co-workers developed the first denitrogenative alkylidenation reaction of 1,2,3-benzotriazin- 4(3*H*)-ones (**15.10a**) with aromatic terminal alkynes (**15.10b**) using a palladium complex as the catalyst [30]. By fine-tuning the solvents, they achieved the selective synthesis of (*Z*)-3-benzylidene-isoindoline-1-ones (**15.10c**) and (*Z*)-3-(imino)isobenzofuranones (**15.10d**) in good to high yields (Scheme 15.10). On the other hand, the reaction of (**15.11a**) with aliphatic terminal alkynes (**15.11b**) furnished *ortho*-alkynylated benzamides (**15.11c**) and (**15.11d**) (Scheme 15.11).

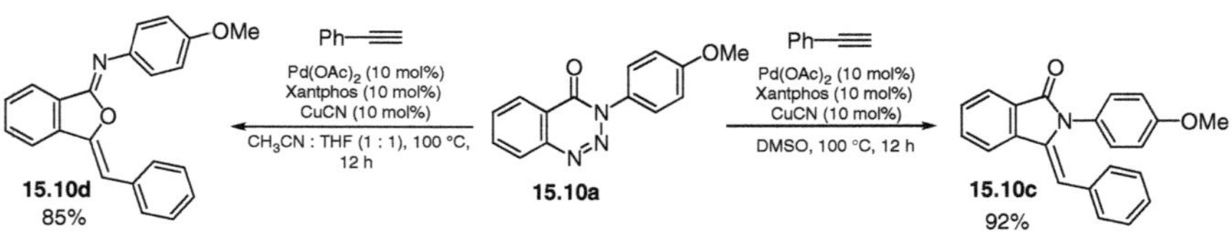

Scheme 15.10 Pd-catalyzed annulation reaction of 1,2,3-benzotriazin-4(3*H*)-ones and aromatic alkynes.

The proposed mechanism is illustrated in Scheme 15.12. The reactions proceed by forming five-membered azapalladacyclic intermediate (**15.12A**). Intermediate (**15.12A**) then coordinates with aromatic alkyne to form intermediate (**15.12B**),

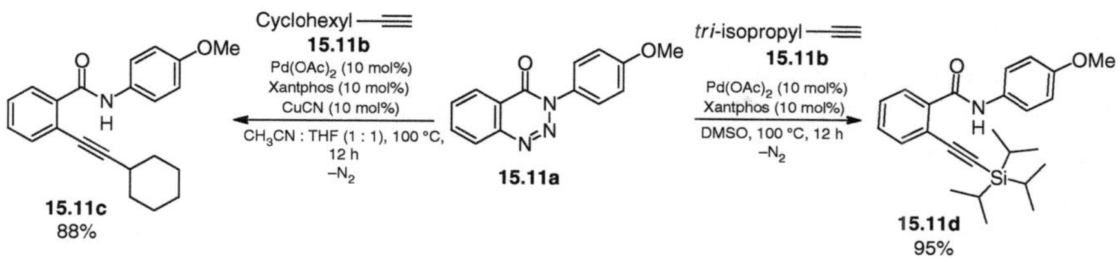

Scheme 15.11 Pd-catalyzed coupling reaction of 1,2,3-benzotriazin-4(3*H*)-ones and aliphatic alkynes.

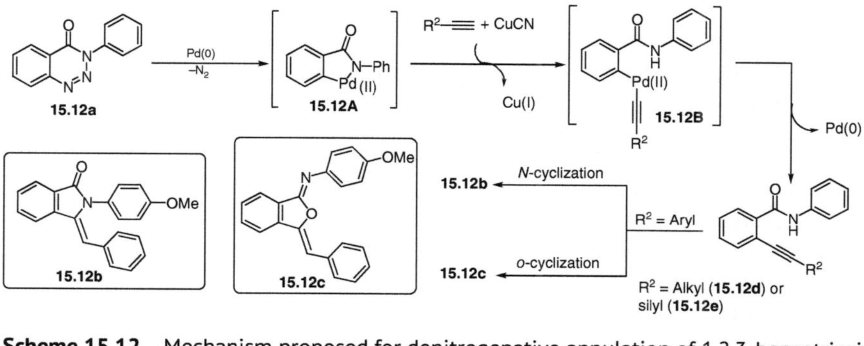

Scheme 15.12 Mechanism proposed for denitrogenative annulation of 1,2,3-benzotriazin-4(3*H*)-ones with terminal alkynes.

which subsequently undergoes reductive elimination to yield the coupled product (**15.12b–15.12e**) (Scheme 15.12). The authors found that copper cyanide (CuCN) is a vital additive, as it not only assists in the transmetallation step but also facilitates the cycloisomerization process, leading to improved product yields.

Very recently, the research group of Mannathan demonstrated Pd(0)-catalyzed denitrogenative transannulation method for the preparation of diverse 3-substituted isocoumarin-1-imine frameworks (**15.13c**) using 1,2,3-benzotriazin-4(3*H*)-ones (**15.13a**) and terminal alkynes (**15.13b**) (Scheme 15.13) [31]. The developed protocol exhibits high regioselectivity, tolerates a wide range of functional groups and

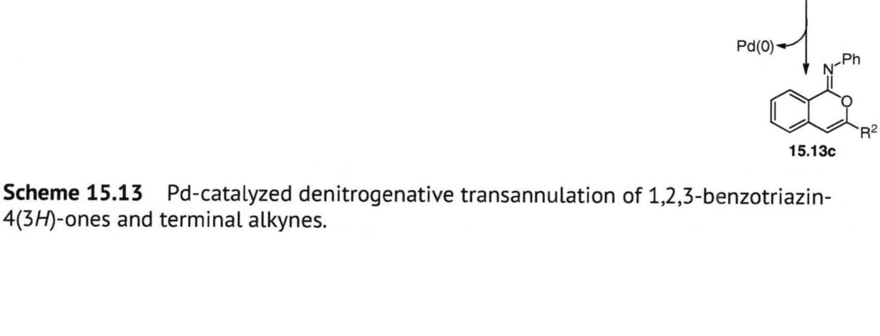

Scheme 15.13 Pd-catalyzed denitrogenative transannulation of 1,2,3-benzotriazin-4(3*H*)-ones and terminal alkynes.

provides corresponding imino-isocoumarin derivatives in good to high yields. The reaction is believed to proceed via a five-membered palladacycle intermediate (**15.13B**), followed by regioselective alkyne insertion to generate (**15.13C**), and subsequent reductive elimination gives the desired product (**15.13c**) (Scheme 15.13). Furthermore, the application of this method was showcased through the one-pot synthesis of biologically significant 3-substituted isocoumarin scaffolds (**15.13c**).

15.4 Nickel-Catalyzed Denitrogenative Cross-Coupling Reactions

15.4.1 Synthesis of *ortho*-Arylated, Alkylated, and Alkenylated Benzamides

In 2018, Mannathan and co-workers developed Ni-catalyzed denitrogenative cross-coupling approach to construct *ortho*-arylated and alkenylated benzamides (**15.14c**) (Scheme 15.14) [32]. The reaction tolerated diverse *N*-aryl and *N*-alkyl substituted 1,2,3-benzotriazin-4(3*H*)-ones (**15.14a**), and successfully applied to a broad range of aryl and alkenyl boronic acids (**15.14b**) to furnish the desired products (**15.14c**) in good yields. The catalytic cycle is believed to proceed through a five-membered azanickelacycle intermediate (**15.15A**) with an extrusion of molecular N_2. Then, sodium carbonate-assisted transmetalation with organoboronic acids and consecutive reductive elimination step delivered the desired *ortho*-arylated benzamides (Scheme 15.15). Synthetic methodology was applied in synthesizing fluorenones and benzylamines, important scaffolds in naturally occurring bioactive molecules [33].

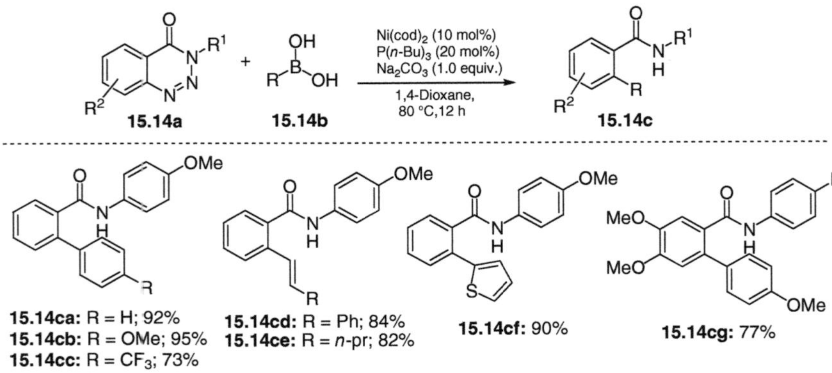

Scheme 15.14 Reaction of 1,2,3-benzotriazin-4(3*H*)-ones and organoboronic acids under Ni(0)-catalysis.

Subsequently, Cheng and co-workers also described the nickel-catalyzed synthesis of *ortho*-substituted benzamides (**15.16c**) from 1,2,3-benzotriazin-4(3*H*)-ones (**15.16a**) and organoboronic acids (**15.16b**) (Scheme 15.16) [34]. In this protocol,

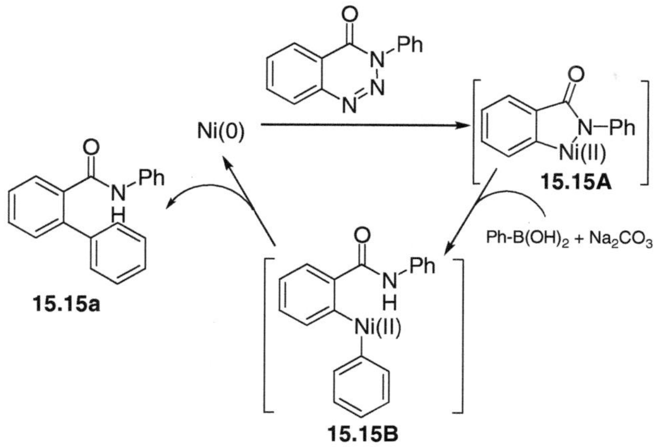

Scheme 15.15 Proposed mechanism for the Ni(0)-catalyzed denitrogenative cross-coupling reaction.

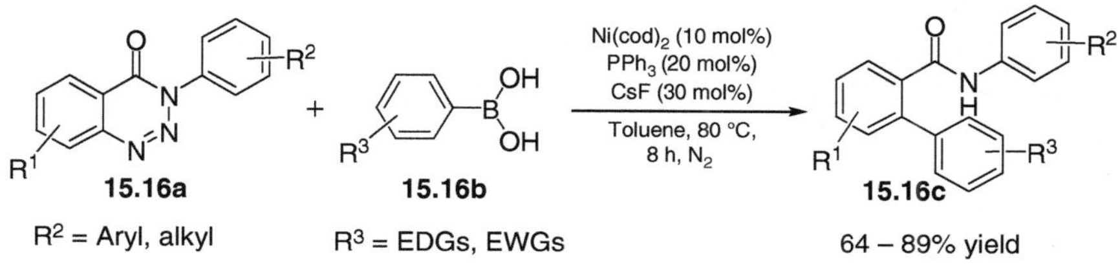

Scheme 15.16 Ni(0)-catalyzed denitrogenative *ortho*-substitutions of 1,2,3-benzotriazin-4(3*H*)-ones with boronic acids.

various substituted 1,2,3-benzotriazin-4(3*H*)-ones and boronic acids underwent denitrogenative coupling to give the desired *o*-substituted benzamides in good to high yields. Based on the competitive experiments, electron-rich boronic acid displayed faster reaction efficacy, which attributed to the result that transmetalation steps slightly faster for the same than electron-deficient counterparts. The broad synthetic utility of the *o*-arylation reaction is further established by the production of densely functionalized bioactive drug molecules, for example, losartan and irbesartan, which are used in treating high blood pressure [35].

Zou's research group demonstrated an efficient denitrogenative cross-coupling reaction between 1,2,3-benzotriazin-4(3*H*)-ones (**15.17a**) and alkyl halides (**15.17b**) using nickel-catalysis (Scheme 15.17) [36]. The reaction requires 10 mol% Ni-catalyst, 2.0 equiv. manganese, 10 mol% potassium iodide and 25 mol% TMSCl in DMA solvent from 0 to 100 °C. This alkylation method's scope was quite broad, and the desired products were formed in good to high yields. The reaction pathway began with forming five-membered nickelacycle (**15.18A**) via oxidative addition of Ni(0) into 1,2,3-benzotriazin-4(3*H*)-ones (**15.18a**) upon expulsing gaseous nitrogen. The reaction involves single-electron oxidation of compound (**15.18A**), yielding Ni(III) alkyl azanickelacycle (**15.18B**), facilitated by an alkyl radical derived from

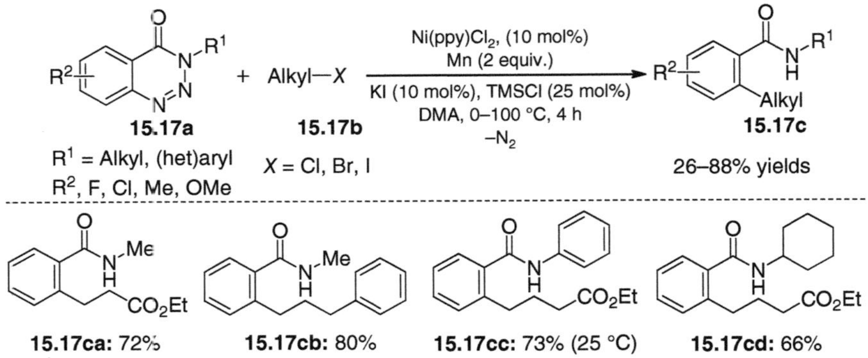

Scheme 15.17 Ni-catalyzed cross-electrophile coupling reactions for the synthesis of *o*-alkylated benzylamides.

the alkyl halide. Subsequently, an intramolecular reductive elimination resulted in generating the aryl/alkyl bond within Ni(I) intermediate (**15.18C**). The process advances with a secondary single-electron oxidation of Ni(I), leading to the generation of Ni(II) species (**15.18D**). The final step encompasses the reduction of Ni(II) species (**15.18D**) through Mn(0) activated by TMSCl, ultimately regenerating the catalytically active Ni(0) species (Scheme 15.18). In this catalytic cycle, nickel followed sequence Ni(II → III → I → II → 0), which was previously devised by the Weix group [37].

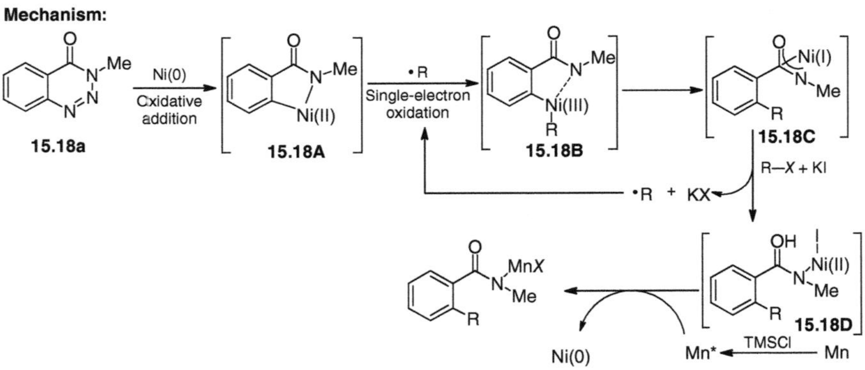

Scheme 15.18 Ni-catalyzed and manganese-assisted reaction mechanism of 1,2,3-benzotriazin-4(3*H*)-ones and alkyl halides.

J. Mao and P. J. Walsh research groups combinedly developed Ni-catalyzed cross-electrophile coupling of 1,2,3-benzotriazin-4(3*H*)-ones (**15.19a**) with aryl bromides (**15.19b**) to access a diverse array of *ortho*-arylated benzamide derivatives (**15.19c**) (Scheme 15.19) [38]. The optimized conditions for this cross-coupling reaction were 10 mol% $NiBr_2$.glyme, bipy (15 mol%), Zn powder (2 equiv.), and DMA (0.5 M) at 80 °C for 12 hours. The scope of the methodology was investigated by

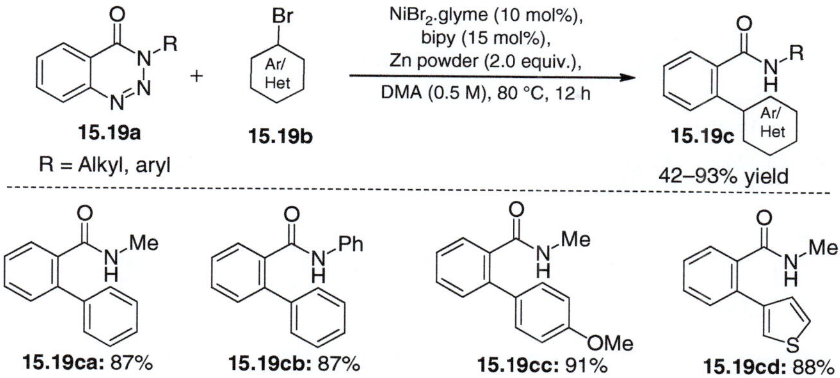

Scheme 15.19 Ni-catalyzed cross-electrophile coupling reactions for the synthesis of *o*-arylated benzylamides.

varying the substitutions on both the coupling partners and a broad range of functional groups was very well tolerated. The reaction pathway began with forming five-membered nickelacycle (**15.20A**) facilitated by oxidative addition of Ni(0) into 1,2,3-benzotriazin-4(3*H*)-ones (**15.20a**). The five-membered *aza*-nickelacycle intermediate (**15.20A**) is proposed to be reduced to a LnNi(I) species (**15.20B**) that undergoes oxidative addition of the aryl bromide (**15.20b**), resulting in intermediate (**15.20C**). The intermediate (**15.20C**) subsequently undergoes reductive elimination to afford *ortho*-arylated benzamide derivative (**15.20c**) and an LnNi(I)–Br species (**15.20D**). The final step encompasses the conversion of intermediate (**15.20D**) to LnNi(0) by Zn powder to close the catalytic cycle (Scheme 15.20).

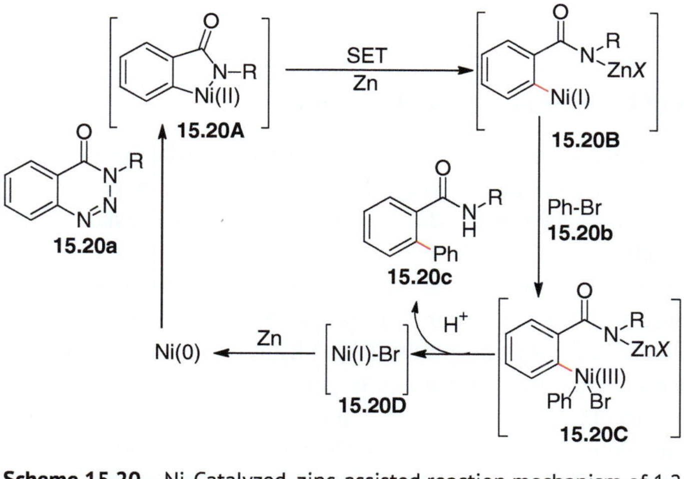

Scheme 15.20 Ni-Catalyzed, zinc-assisted reaction mechanism of 1,2,3-benzotriazin-4(3*H*)-ones and aryl bromides.

In a recent development, Zou and colleagues reported a novel approach involving mechanoredox, nickel cocatalyzed manganese-mediated cross-electrophile

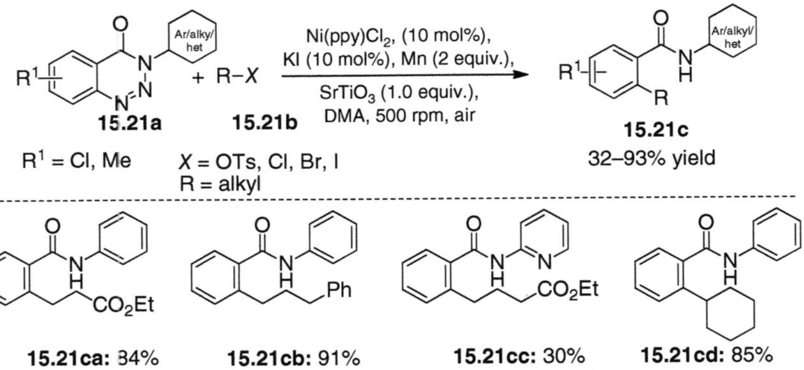

Scheme 15.21 Ni-catalyzed cross-electrophile coupling reactions for the synthesis of *o*-alkylated benzylamides.

coupling for synthesizing *ortho*-functionalized benzamides (**15.21c**) (Scheme 15.21) [39]. This method was developed by a liquid-assisting grinding strategy where the manganese powders and strontium titanate act as a reductant and a cocatalyst. The *ortho*-alkylation reaction was optimized with various transition metal catalysts, activators, metal reductants, and additives in various solvents under an air atmosphere. The final optimum conditions are 10 mol% Ni(bpy)Cl$_2$ cocatalyst, 10 mol% KI, 2.0 equiv. Mn powders as reductant, SrTiO$_3$ (1.0 equiv.) an additive in DMA (η = 1) using 16 stainless-steel balls (ø = 7 mm) in 50 ml jar at 500 rpm, in air. The scope of the methodology is very general, tolerating diverse substitutions on both 1,2,3-benzotriazin-4(3*H*)-ones (**15.21a**) and alkyl(pseudo)halides (**15.21b**), delivering the corresponding *ortho*-alkylated benzamides (**15.21c**) in good to high yields.

15.4.2 Synthesis of Functionalized β-Aryl Ketones

In 2020, Li and his co-workers described a straightforward procedure for synthesizing β-(*o*-amido) aryl ketones (**15.22c**) through Ni(0)-catalyzed C-C bond-forming reactions between 1,2,3-benzotriazin-4(3*H*)-ones (**15.22a**) and unmasked cyclopropanols (**15.22b**) (Scheme 15.22) [40].

This report employs cyclopropanols as efficient coupling partners delivering corresponding β-functionalized ketones in high yields. Notably, the combination of Ni(cod)$_2$ and monodentate alkylphosphine (PMe$_3$) ligand was necessary for the success of this reaction. This protocol was quite general, and a variety of 1,2,3-benzotriazin-4(3*H*)-ones (**15.22a**) and cyclopropanols (**15.22b**) participated in the coupling reactions and delivered the product (**15.22c**) with 49–93% yield.

As depicted in Scheme 15.22, the reaction starts via forming nickelacycle intermediate (**15.22A**), followed by reaction with cyclopropanol (**15.22b**) to generate intermediate (**15.22B**). Subsequently, intermediate (**15.22B**) experiences β-carbon elimination leading to intermediate (**15.22C**), which then undergoes reductive elimination and affords the desired product (**15.22c**). Alternatively, (**15.22C**) can

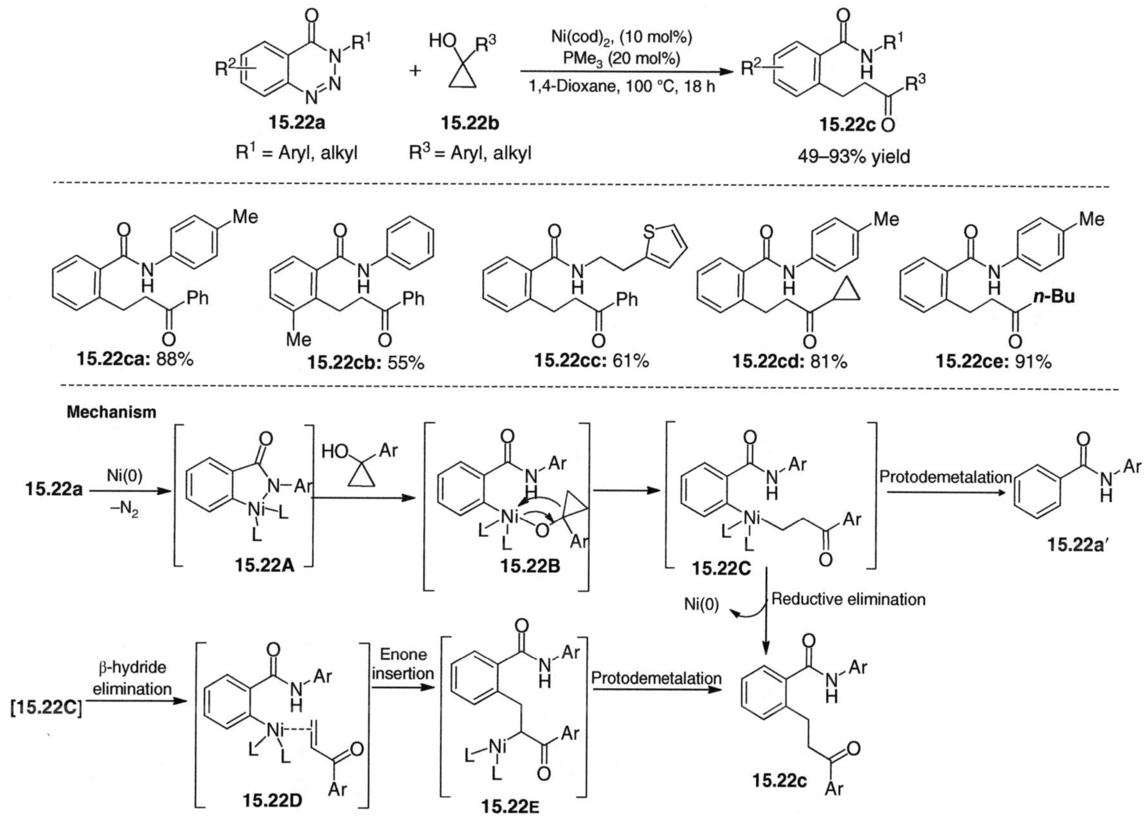

Scheme 15.22 Ni(0)-catalyzed denitrogenative coupling of 1,2,3-benzotriazin-4(3*H*)-ones with cyclopropanols.

undergo β-hydride elimination, leading to the formation of intermediate (**15.22D**). Then, intermediate **15.22**D transforms into (**15.22E**) via enone insertion, resulting in the generation of product (**15.22c**). Meanwhile, Intermediate (**15.22C**) may undergo protodemetalation to furnish the benzamide by-product (**15.22a′**).

15.5 Pd-Catalyzed Cross-Coupling Reactions

15.5.1 Synthesis of *ortho*-Arylated and Alkenylated Benzamides

In 2021, Mannathan and co-workers developed the first denitrogenative Suzuki–Miyaura type cross-coupling of 1,2,3-benzotriazin-4(3*H*)-ones (**15.23a**) with organoboronic acids (**15.23b**) using $Pd(OAc)_2$ (10 mol%), Xantphos (10 mol%), and Na_2CO_3 (1 equiv.) affording the *ortho*-arylated and alkenylate benzamides (**15.23c**) in high yields (Scheme 15.23) [41]. Notably, this protocol demonstrated tolerance to organoboronic acids with electron-deficient substitutions compared to the previously reported Ni-catalyst system. DFT studies supported the necessity of a relatively higher reaction temperature in this reaction compared to the Ni-catalyzed denitrogenative reaction of 1,2,3-benzotriazin-4(3*H*)-ones.

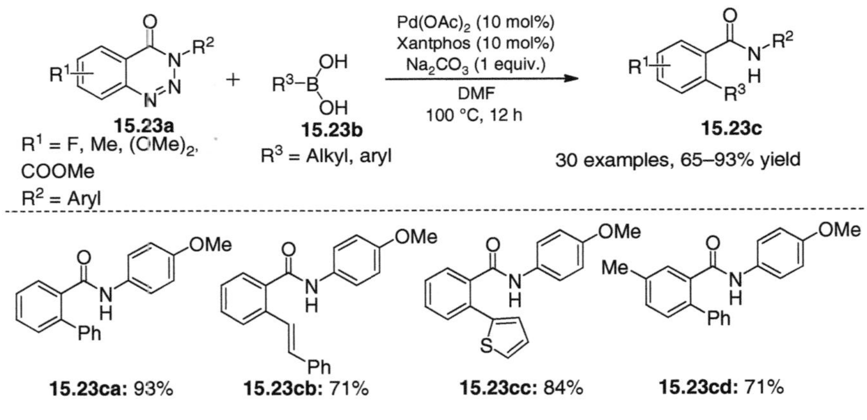

Scheme 15.23 Pd-catalyzed cross-coupling reaction for the synthesis of *o*-functionalized benzylamides.

15.5.2 Synthesis of *ortho*-Methylated Benzamides

In 2023, Li et al. described a straightforward denitrogenative cross-coupling method catalyzed by palladium/xantphos for the synthesis of *ortho*-methylated benzamides (**15.24c**) from 1,2,3-benzotriazin-4(3*H*)-ones (**15.24a**) and DABAL-Me_3 (**15.24b**) (Scheme 15.24) [42]. The versatility of this Pd-catalyzed *ortho*-methylation reaction was explored by varying substituents on the *N*-aryl groups and 1,2,3-benzotriazin-4(3*H*)-one (**15.24a**) backbone, yielding the desired products in moderate to good yields. Under similar reaction conditions, various 1,2,3-benzotriazin-4(3*H*)-ones smoothly reacted with trialkyl aluminum reagents, such as $AlEt_3$ and $Al(Octyl)_3$, to produce *ortho*-ethylated- and octylated benzamides in yields ranging from 27% to 71%.

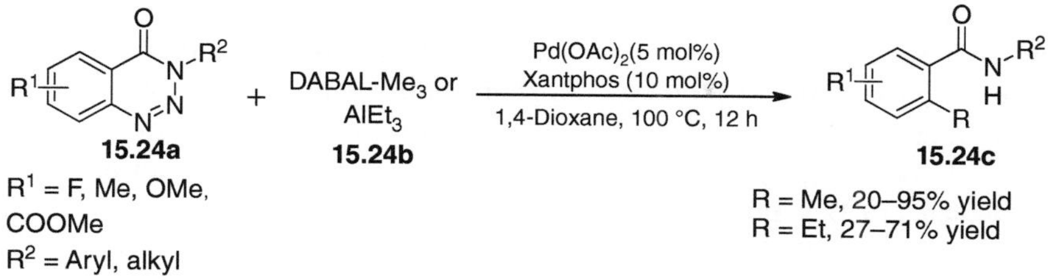

Scheme 15.24 Pd-catalyzed cross-coupling reaction for the synthesis of *o*-alkylated benzylamides.

15.6 Visible-Light-Mediated Denitrogenative Reactions

Photoredox catalysis has emerged as an attractive approach for synthesizing valuable compounds in recent decades [43]. In these reactions, a photoexcited catalyst (organocatalysts, metals such as Ru and Ir) undergoes a single electron transfer (SET) process, which is responsible for generating radicals or radical ions through the oxidation or reduction of the starting materials. In this regard, 1,2,3-benzotriazin-4(3*H*)-ones (**15.25a**) also generate a radical anion intermediate via the SET process, as shown in Scheme 15.25. In this section, the reaction of the radical anion (**15.25A**) intermediate is discussed in detail.

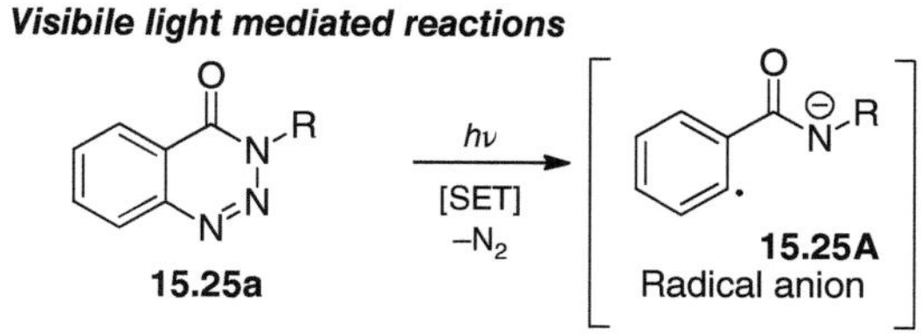

Scheme 15.25 Generation of radical ion intermediate from 1,2,3-benzotriazin-4(3*H*)-ones in the presence of visible light.

15.6.1 Synthesis of Isoquinolones

In 2015, Wang et al. developed a visible light-mediated isoquinolones synthesis from 1,2,3-benzotriazin-4(3*H*)-ones (**15.26a**) and aromatic terminal alkynes (**15.26b**) (Scheme 15.26) [44]. The reaction was performed with 1 mol% of *fac*-Ir(ppy)$_3$ as the

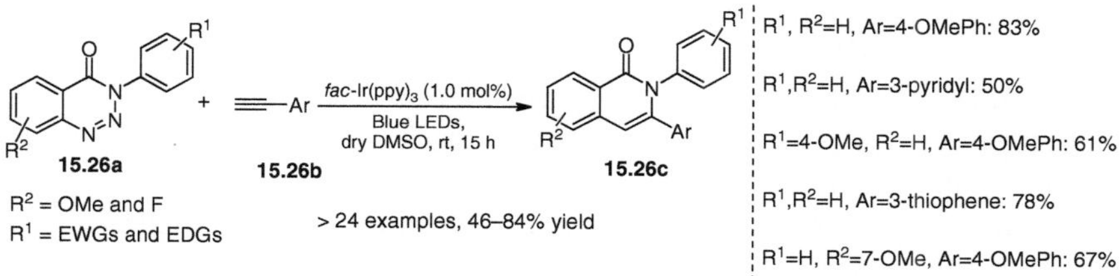

Scheme 15.26 Visible-light mediated reaction of 1,2,3-benzotriazin-4(3*H*)-ones and terminal alkynes.

photocatalyst in DMSO solvent under blue light irradiation. The authors demonstrated a radial clock experiment, which supported the hypothesis that the reaction proceeded through the SET process. Further, control experiments supported the formation of a carbon-centered radical instead of nitrogen-centered radical from 1,2,3-benzotriazin-4(3*H*)-one. Mechanistically, the generation of radical anion intermediate (**15.27A**) initiated the reaction cycle. Further steps, including radical alkyne insertion, cyclization, and successive oxidation, completed the reaction cycle and yielded the desired isoquinolone (**15.27b**) (Scheme 15.27).

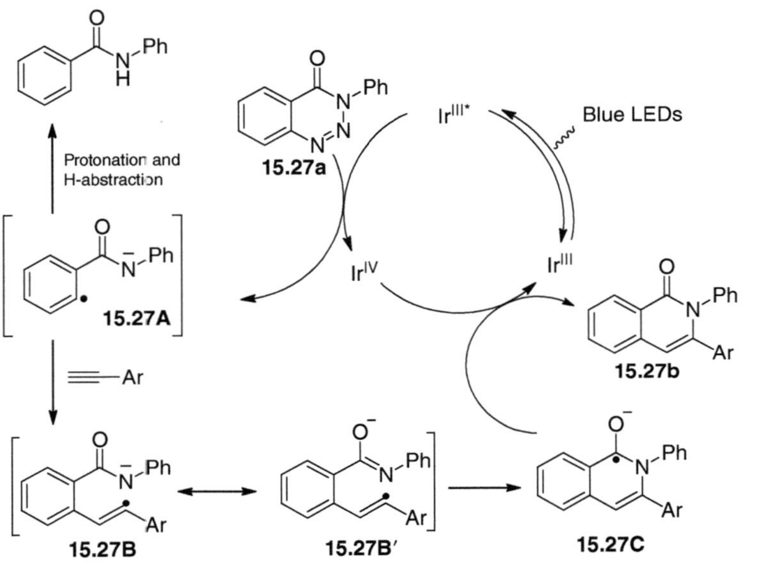

Scheme 15.27 Mechanistic pathway for the synthesis of isoquinolones.

15.6.2 Synthesis of *ortho*-Phosphorylated Aryl Amides

In 2021, Zhao and co-workers documented a mild photo-mediated denitrogenative phosphorylation of 1,2,3-benzotriazin-4(3*H*)-ones (**15.28a**) using arylphosphonates, phosphinates, and phosphine oxides (Scheme 15.28) [45]. The coupling reaction proceeds with 5 mol% Eosin Y and 2 equiv. of DIPEA in combination with CH_3CN/H_2O solvent system in the presence of blue light or sunlight at room temperature. Significantly, this clean approach does not require any transition metals or ligands and has a broad substrate scope irrespective of the electronic nature of substitutions on both coupling partners.

Moreover, B_2Pin_2 was also examined as a radical acceptor in the radical-coupling process and afforded the corresponding *ortho*-borylated (**15.29c**) products in 65–70% yields (Scheme 15.29). Nevertheless, 1,2,3-benzotriazin-4(3*H*)-one (**15.29a**) having an aliphatic chain instead of aryl substitutions at the nitrogen atom failed to produce the respective *o*-phosphorylated products. Notably, the *N*-aryl group was necessary to stabilize the key intermediate, and the electronic and steric factors of the same had an insignificant effect on the reactivity.

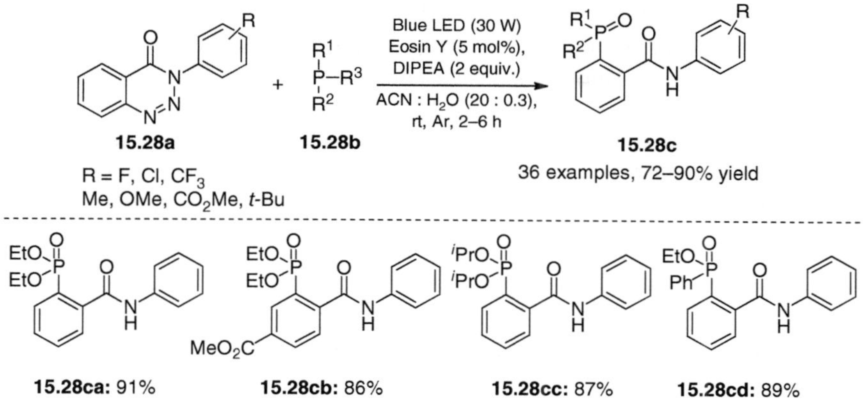

Scheme 15.28 Visible-light induced phosphorylation of 1,2,3-benzotriazin-4(3*H*)-ones.

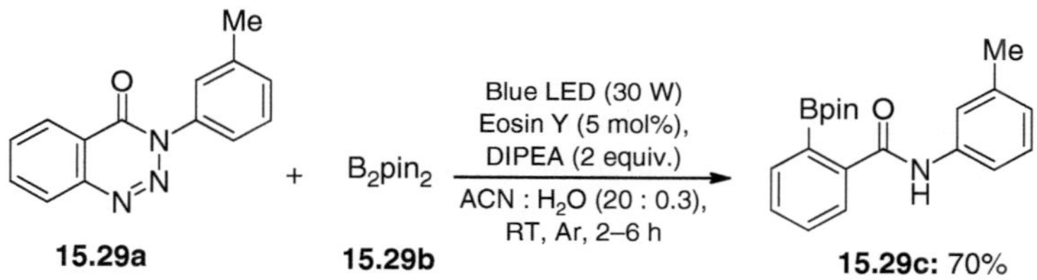

Scheme 15.29 Visible-light induced borylation of 1,2,3-benzotriazin-4(3*H*)-ones with diboron.

15.6.3 Synthesis of *ortho*-Selenylated Benzamides

In 2022, Wang and co-workers demonstrated an efficient method for the synthesis of *ortho*-selenylated benzamides (**15.30c**) in the presence of Iridium photocatalyst and blue light (Scheme 15.30) [46]. The hetero-coupling reaction proceeded smoothly

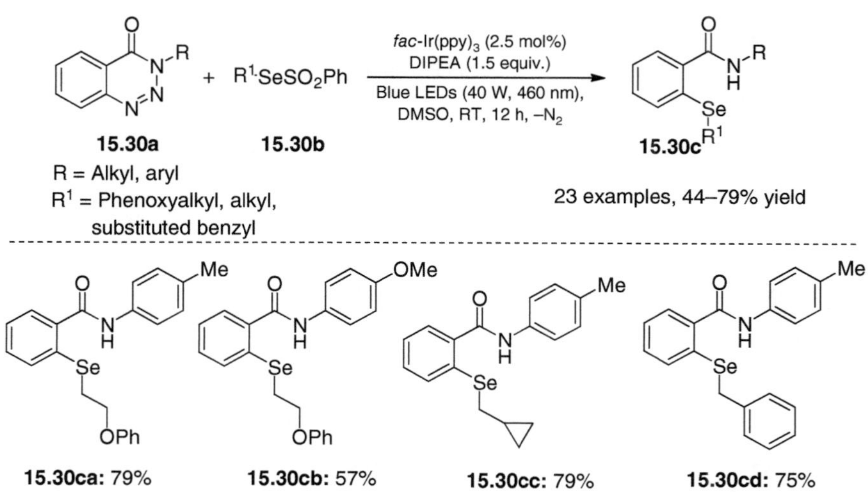

Scheme 15.30 Visible-light-promoted denitrogenative *ortho*-selenylation reaction of 1,2,3-benzotriazin-4(3*H*)-ones.

with 2.5 mol% *fac*-Ir(ppy)$_3$ and 1.5 equiv. of DIPEA in DMSO solvent, under 460 nm blue light at room temperature. The reaction conditions were carefully optimized to achieve good yields. Among the tested photocatalysts – *fac*-Ir(ppy)$_3$, 4-CzIPN, Ru(pbz)$_2$(PF$_6$)$_2$, Eosin Y, and Rose Bengal – *fac*-Ir(ppy)$_3$ proved to be most effective for the reaction. This coupling reaction is scalable and displayed remarkable tolerance toward various functional groups. The synthetic potential of this methodology was evidenced by its application in synthesizing a series of ebselen and its analogs (**15.30c**).

15.7 Metal-free Denitrogenative Transformations

15.7.1 Synthesis of *N*-Isoxazolyl-2-iodobenzamides and 4*H*-Tetrazolo[1,5-*a*] [1,4]-benzodiazepine-6-ones

Raffa's group reported the synthesis of *N*-isoxazolyl-2-iodobenzamides (**15.31b**) from the corresponding 1,2,3-benzotriazin-4(3*H*)-ones (**15.31a**). The reaction was accomplished by subjecting the 1,2,3-benzotriazin-4(3*H*)-ones to reflux conditions in a mixture of acetic acid and potassium iodide for one hour, as depicted in Scheme 15.31 [47, 48]. The synthesized iodo derivatives were tested for antifungal activity and showed comparability to the benodanil reference drug [45]. Subsequently, Kishore et al. presented a unique method for synthesizing 4*H*-tetrazolo [1,5-*a*][1,4]benzodiazepine-6-ones (**15.32b**) from 1,2,3-benzotriazin-4(3*H*)-ones. During the reactions, they observed the formation of 2-azidobenzamide intermediate (**15.32A**) (Scheme 15.32).

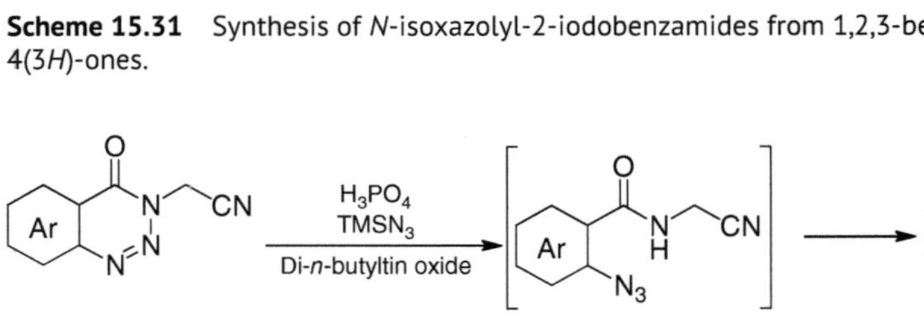

Scheme 15.31 Synthesis of *N*-isoxazolyl-2-iodobenzamides from 1,2,3-benzotriazin-4(3*H*)-ones.

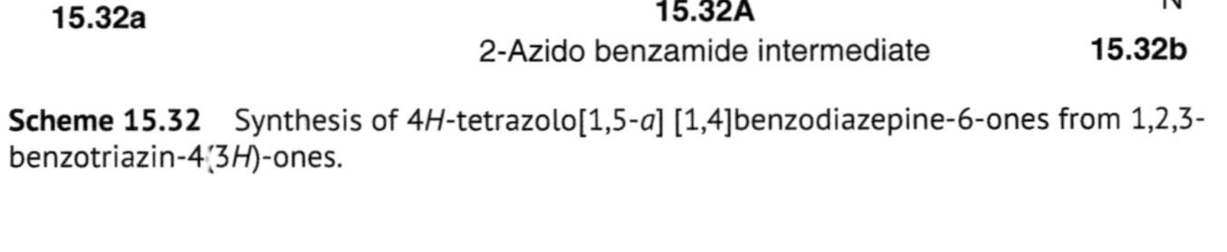

Scheme 15.32 Synthesis of 4*H*-tetrazolo[1,5-*a*] [1,4]benzodiazepine-6-ones from 1,2,3-benzotriazin-4(3*H*)-ones.

15.7.2 Synthesis of 2-Hydroxy-*N*-substituted Benzamides

In 2022, Mannathan' s group reported a metal-free acid mediated denitrogenative synthesis of *N*-aryl- and alkyl-substituted salicylamides, as depicted in (Scheme 15.33) [49]. The desired *ortho*-hydroxylated benzamides (**15.33b**) were obtained in good to high yields by treating 1,2,3-benzotriazin-4(3*H*)-ones (**15.33a**) with 5.0 equiv. of trifluoroacetic acid in toluene at 80 °C for one hour. The reaction is operationally simple and exhibits a broad substrate scope with high functional group tolerance. Mechanistically, the reaction proceeds through a benzene diazonium intermediate (**15.33A**), which then undergoes electrophilic displacement with 2,2,2-trifluoroacetate ion to deliver the *ortho*-trifluoro acetoxylated benzamide (**15.33B**). Subsequent hydrolysis of intermediate (**B**) bestowed the desired *N*-substituted salicylamides (**15.33b**).

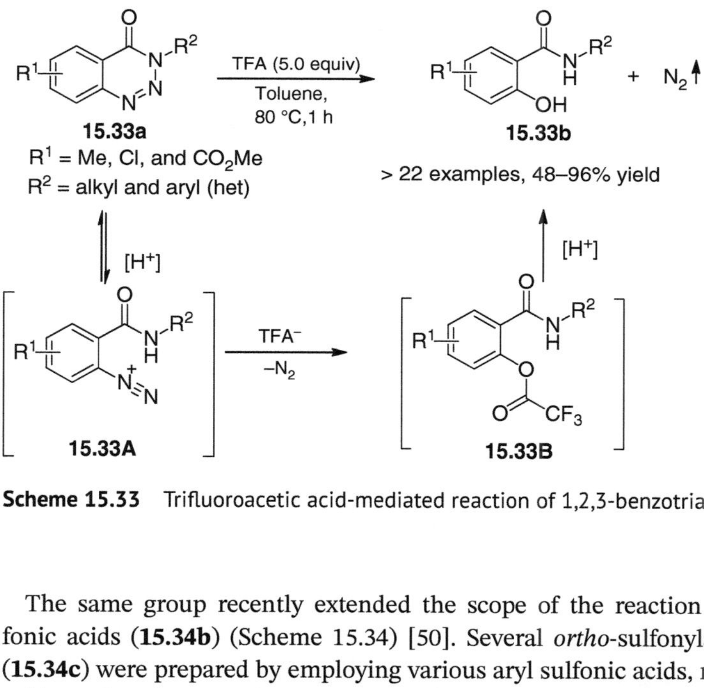

Scheme 15.33 Trifluoroacetic acid-mediated reaction of 1,2,3-benzotriazin-4(3*H*)-ones.

The same group recently extended the scope of the reaction with organosulfonic acids (**15.34b**) (Scheme 15.34) [50]. Several *ortho*-sulfonylated benzamides (**15.34c**) were prepared by employing various aryl sulfonic acids, methane sulfonic acids and trifluoromethane sulfonic acids. The practicality of the reaction is also demonstrated by performing a gram-scale reaction toward the synthesis of 2-(phenylcarbamoyl)phenyl 4-methylbenzenesulfonate (**15.34c**).

15.7.3 Synthesis of 1,2-Benzodithiol-3-ones

Recently, Yao Zhou and co-workers described an effective method for synthesizing 1,2-benzodithiol-3-ones (**15.35b**) via denitrogenative heterotransannulation of benzotriazinones and sodium sulfide in an acidic condition without the need of any

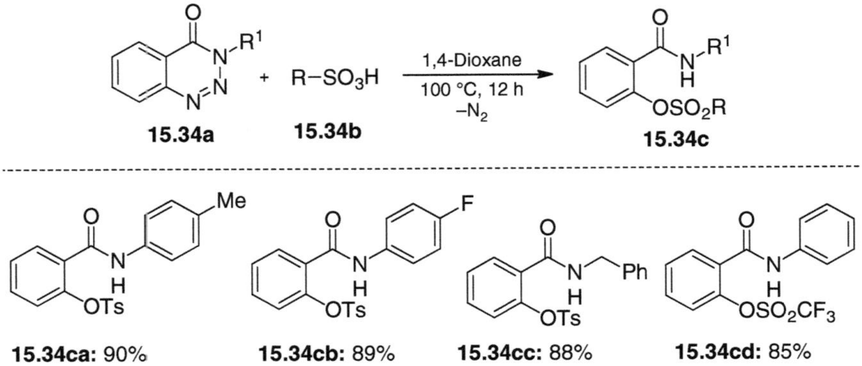

Scheme 15.34 Reaction of 1,2,3-benzotriazin-4(3*H*)-ones with organosulfonic acids.

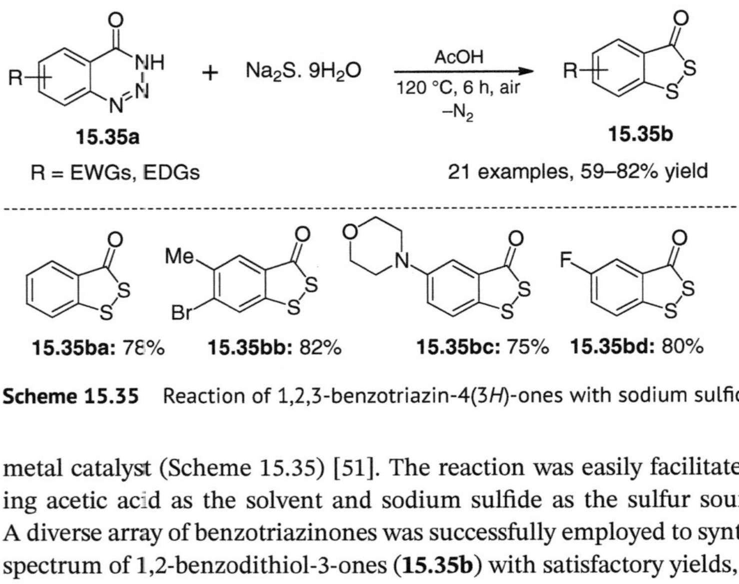

Scheme 15.35 Reaction of 1,2,3-benzotriazin-4(3*H*)-ones with sodium sulfide.

metal catalyst (Scheme 15.35) [51]. The reaction was easily facilitated by employing acetic acid as the solvent and sodium sulfide as the sulfur source at 120 °C. A diverse array of benzotriazinones was successfully employed to synthesize a wide spectrum of 1,2-benzodithiol-3-ones (**15.35b**) with satisfactory yields, ranging from 59% to 82%.

15.8 Thermolysis

Smalley and co-workers investigated the thermolysis reaction of 3-phenyl-1,2,3-benzotriazin-4(3*H*)-one (**15.36a**) (Scheme 15.36) [52, 53]. During the studies, they observed the predominant formation of acridones, alongside a minor amount of phenanthridones (**15.36b**), at 320 °C in the absence of solvents. Conversely, thermolysis of 3-(β-styryl) benzotriazinone (**15.36d**) at 300 °C in paraffin oil resulted in a higher yield of 3-phenylquinolin-4(1*H*)-one (**15.36e**). Mechanistically, the reaction progresses through a zwitterion intermediate (**15.36A**), which can undergo ring closure directly to yield phenanthridinone, or it may proceed toward the formation of acridone (**15.36c**) via a benzazetinone intermediate (**15.36B**).

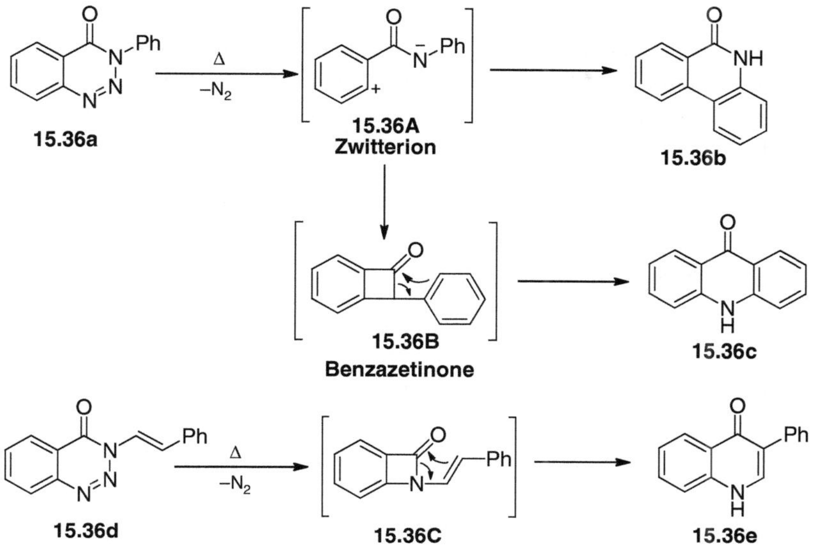

Scheme 15.36 Thermolysis of 1,2,3-benzotriazin-4(3*H*)-ones.

15.9 Lewis-Acid-Mediated Denitrogenation Reactions

In 2022, Long and co-workers reported denitrogenative cleavage of 1,2,3-benzotriazin-4(3*H*)-ones (**15.37a**) in the presence of aluminum halides (Scheme 15.37) [54]. *N*-alkyl substituted benzotriazinones were treated with excess AlI_3 in acetonitrile

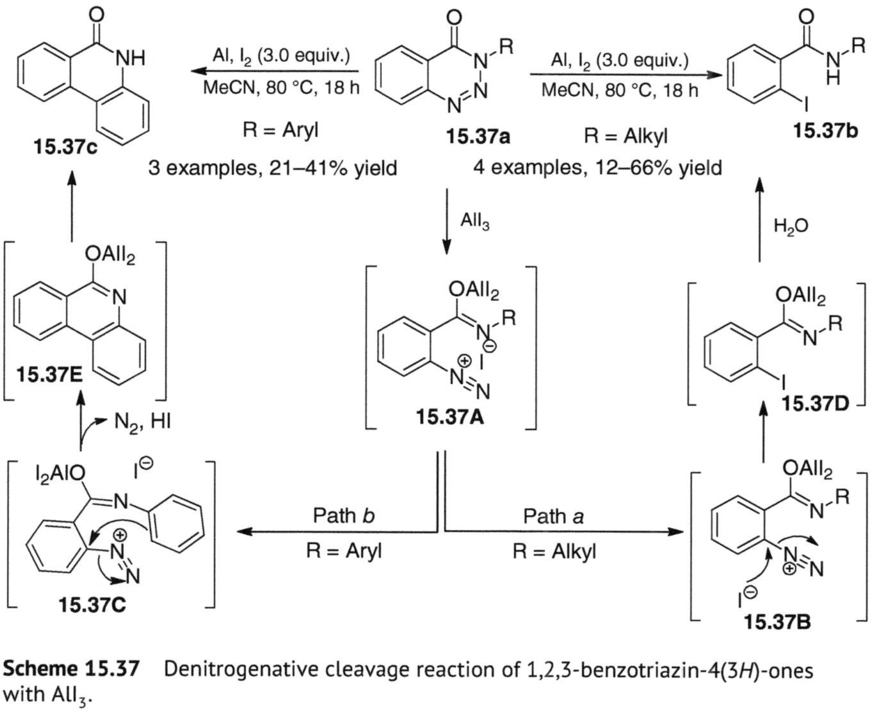

Scheme 15.37 Denitrogenative cleavage reaction of 1,2,3-benzotriazin-4(3*H*)-ones with AlI_3.

at 80 °C, furnished 2-iodo *N*-alkylbenazmides (**15.37b**) in moderate to good yield. Under the same reaction conditions *N*-aryl substituted benzotriazinones provided the 1-phenanthridinones (**15.37c**) in lower yields (Scheme 15.37). On the other hand, benzotriazinones subjected to the cleavage reaction in the presence of $AlCl_3$ delivered the quinazolinone compound (**15.38b**) in 78% yield (Scheme 15.38).

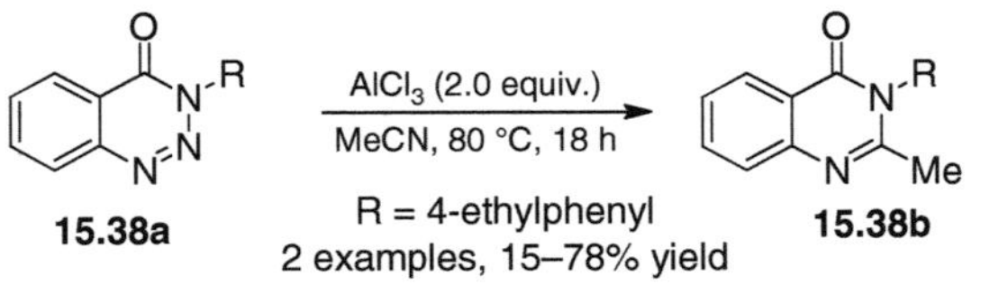

Scheme 15.38 Denitrogenative cleavage reaction of 1,2,3-benzotriazin-4(3*H*)-ones with $AlCl_3$.

15.10 Electrochemical Denitrogenative Transformations

Very recently, in 2023, Chen research group demonstrated an eco-friendly approach for electrochemical denitrogenative transannulation method for the preparation of diverse isoquinolin-1(2*H*)-one (**15.39c**) frameworks using 1,2,3-benzotriazin-4(3*H*)-ones (**15.39a**) and terminal alkynes (**15.39b**) (Scheme 15.39) [55]. During the optimization, different electrolytes (Bu_4NPF_6, *n*-Bu_4NClO_4, *n*-Bu_4NOAc, *n*-Bu_4NBF_4, KPF_6), solvents (DMF, MeCN, 1,4-dioxane, HFIP, THF, and DMSO : H_2O = 8 : 1) and electrodes [Fe(+)|F(−), Pt(+)|Fe(−) and Fe(+)|C(−) and C(+)|Fe(−)] were tested for the annulation reaction and Fe (+)|F(−) electrode, 2.0 equiv. *n*-Bu_4NBF_4

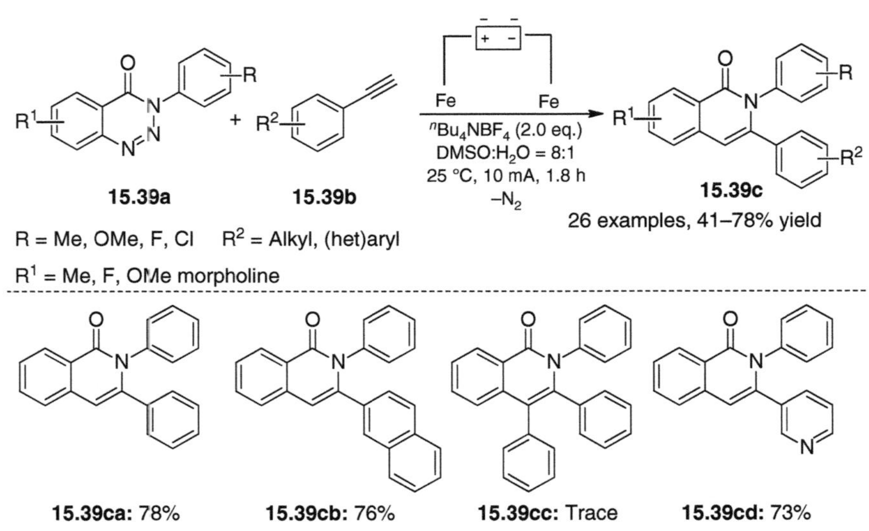

Scheme 15.39 Electrochemical transannulation of 1,2,3-benzotriazin-4(3*H*)-ones and alkynes.

as electrolyte, DMSO : H_2O = 8 : 1 solvent at 25 °C for 1.8 hour provided the isoquinolinone products (**15.39c**) in good yield. The developed protocol does not require metal-catalyst or external reductant and provides corresponding isoquinolinone derivatives (**15.39c**) in good to high yields. The scope of the methodology was investigated by varying the substitutions on both the coupling partners and a broad range of functional groups was very well tolerated.

15.11 Related Compounds

Benzothiatriazine-1,1(2*H*)-dioxides are privileged heterocyclic scaffolds in medicinal chemistry and exhibit a wide range of pharmaceutical properties. For example, benzothiatriazine-1,1(2*H*)-dioxide derivatives act as diuretic agents and nematicidal agents. In recent years, benzothiatriazine-1,1(2*H*)-dioxide has also been explored as potential precursors in various denitrogenative annulation reactions to prepare biologically relevant heterocycles frameworks [29, 56–59].

15.11.1 Synthesis of 3-Iminothiaisoindoline-1,1-dioxides

In 2010, Murakami et al. disclosed Pd-catalyzed denitrogenative annulation reaction of 1,2,3,4-benzothiatriazine-1,1(2*H*)-dioxides (**15.40a**) with isocyanides (**15.40b**) (Scheme 15.40) [29]. This reaction provided a new synthetic route for preparing 3-iminothiaisoindoline-1,1-dioxides (**15.40c**), which are well known for their biological activity.

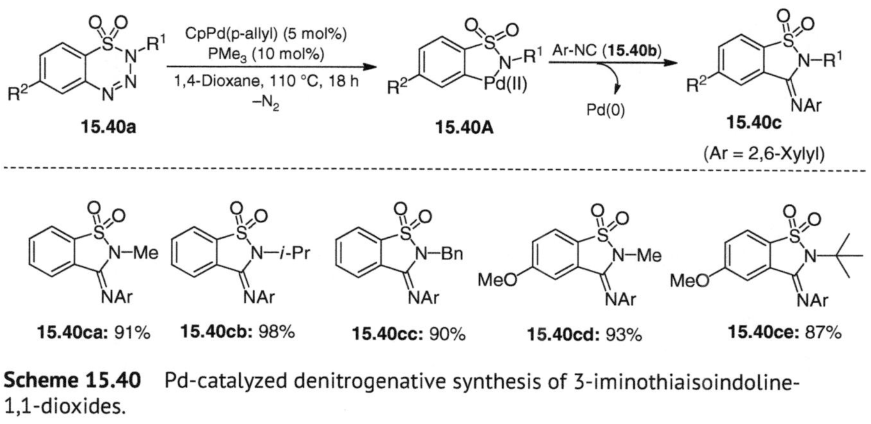

Scheme 15.40 Pd-catalyzed denitrogenative synthesis of 3-iminothiaisoindoline-1,1-dioxides.

15.11.2 Synthesis of 3,4-Dihydro-1,2-benzothiazine-1,1(2*H*)-dioxides

In 2010, Murakami et al. reported a Ni(0)-catalyzed denitrogenative regio- and enantioselective synthesis of substituted 3,4-dihydro-1,2-benzothiazine-1,1(2*H*)-dioxides (**15.41c, 15.41d**) from 1,2,3,4-benzothiatriazine-1,1(2*H*)-dioxides (**15.41a**) and allenes (**15.41b**) (Scheme 15.41) [56]. The reactions were performed using 10 mol%

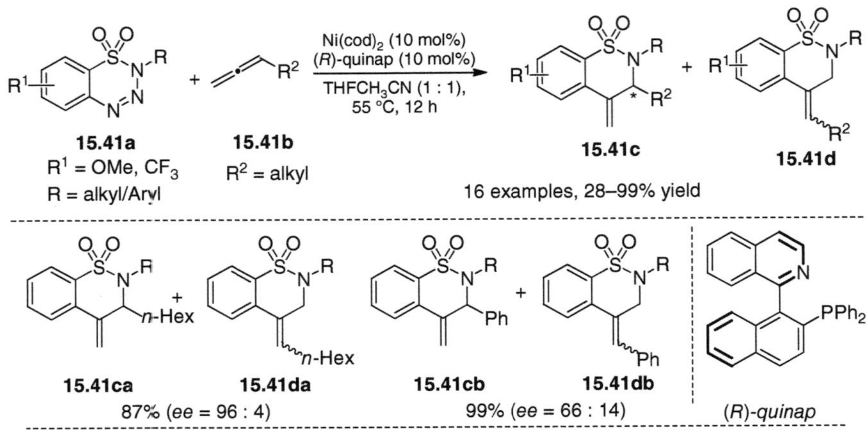

Scheme 15.41 Ni-catalyzed denitrogenative synthesis of 3,4-dihydro-1,2-benzothiazine-1,1(2*H*)-dioxides.

$Ni(cod)_2$ and 10 mol% (*R*)-quinap in THF/CH_3CN (1 : 1) binary solvent system at 55 °C for 12 hours. The scope of substrates is investigated by varying the substituents on 1,2,3-benzothiatriazine-1,1(2*H*)-dioxides (**15.41a**) and allenes (**15.41b**) coupling partners, providing the corresponding products in moderate to good yields with high regio- and enantioselectivities.

15.11.3 Synthesis of Biaryl Sultams

In 2016, Yu research group demonstrated an interesting approach for the synthesis of biaryl sultams (**15.42c**) using visible-light promoted denitrogenative cyclization of *N*-aryl-1,2,3,4-benzothiatriazine-1,1(2*H*)-dioxides (**15.42a**) (Scheme 15.42) [57]. The cyclization reaction proceeds with 1 mol% of $Ru(bpy)_3Cl_2$ in dry DMSO solvent in the presence of blue light at room temperature. The reaction is scalable, and high functional group tolerance helped to establish a good substrate scope of

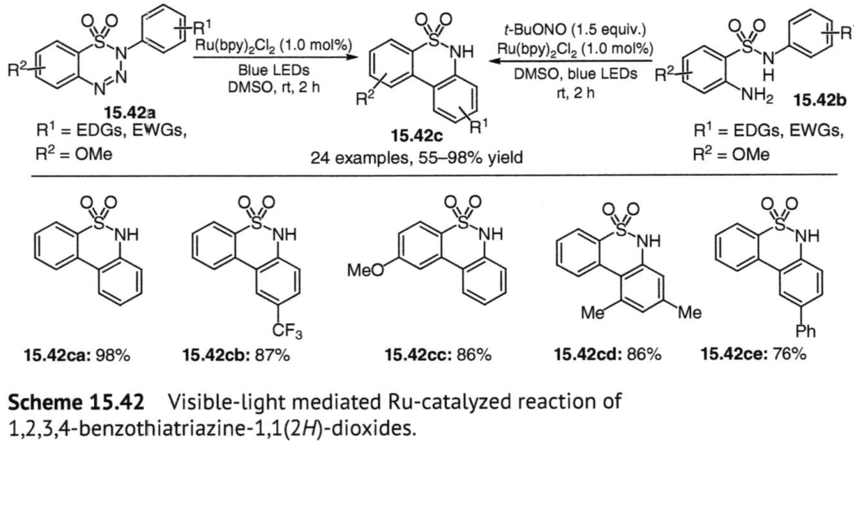

Scheme 15.42 Visible-light mediated Ru-catalyzed reaction of 1,2,3,4-benzothiatriazine-1,1(2*H*)-dioxides.

the reaction with a wide variety of *N*-aryl-1,2,3,4-benzothiatriazine-1,1(2*H*)-dioxides (**15.42a**). The authors also demonstrated one-pot synthesis of biaryl sultams (**15.42c**) from 2-amino-*N*-phenylbenzenesulfonamides (**15.42b**) in the presence TBN sequential diazotization and denitrogenative cyclization process with moderate to good yields.

The mechanism for the synthesis of biaryl sultams was depicted in Scheme 15.43. Photocatalyst Ru(II) initially goes to an excited state Ru(II*) upon irradiation under blue light. The excited state RuII* reacts with alkyne and generates the Ru(III) and radical anion (**15.43A**). Then (**15.43A**) undergoes intramolecular homolytic aromatic substitution process to produce the intermediate radical anion (**15.43B**). Next, (**15.43B**) is oxidized by Ru(III) complex to give intermediate cation (**15.43C**) with the regeneration of the Ru(II) species. Subsequently, the desired biaryl sultam (**15.43b**) is formed after the loss of the proton.

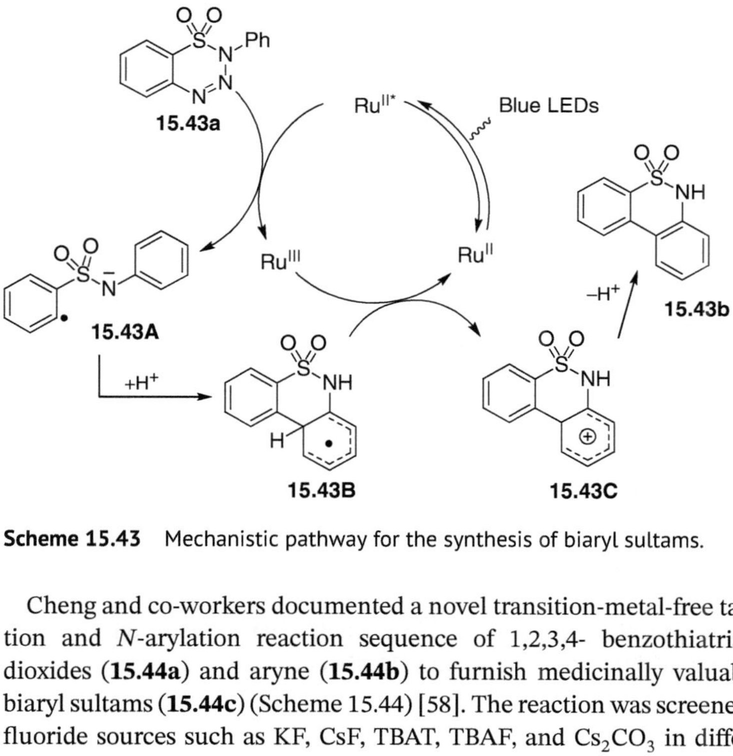

Scheme 15.43 Mechanistic pathway for the synthesis of biaryl sultams.

Cheng and co-workers documented a novel transition-metal-free tandem cyclization and *N*-arylation reaction sequence of 1,2,3,4- benzothiatriazine-1,1(2*H*)-dioxides (**15.44a**) and aryne (**15.44b**) to furnish medicinally valuable *N*-arylated biaryl sultams (**15.44c**) (Scheme 15.44) [58]. The reaction was screened with various fluoride sources such as KF, CsF, TBAT, TBAF, and Cs_2CO_3 in different solvents. Among them 3.0 equiv. KF and 2.0 equiv. 18-crown-6 in THF at 80 °C after eight hours provided the desired product in excellent yield. This presented approach exhibited broad substrate scope across a variety of 1,2,3,4-benzothiatriazine-1,1(2*H*)-dioxides (**15.44a**) and benzyne precursors (**15.44b**); however, the reaction efficiency was hardly affected by the electronics of substituents.

In continuation, the same group (Cheng et al.) also disclosed Ni-catalyzed denitrogenative annulation reaction of 1,2,3,4-benzothiatriazine-1,1(2*H*)-dioxides (**15.45a**) and aryne (**15.45b**) to access *N*-arylated biaryl sultams (**15.45c**) (Scheme 15.45) [59].

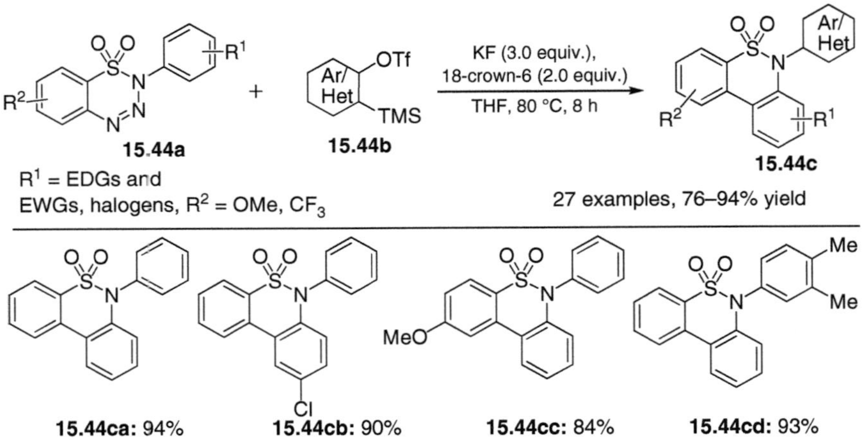

Scheme 15.44 Metal-free reaction of 1,2,3,4-benzothiatriazine-1,1(2*H*)-dioxides with arynes.

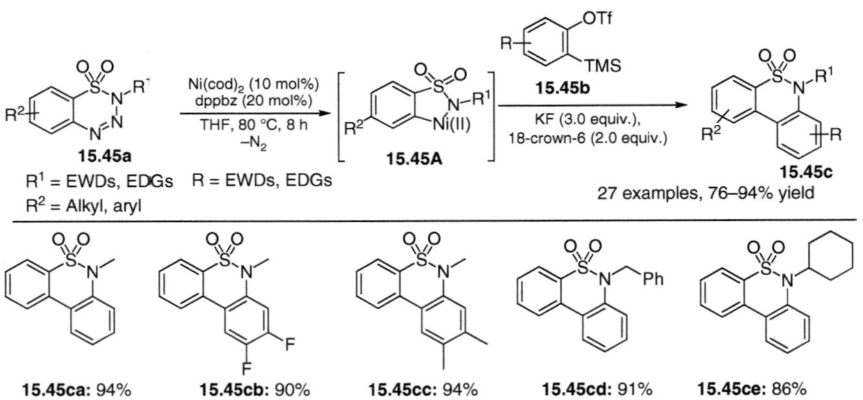

Scheme 15.45 Nickel-catalyzed denitrogenative reaction of 1,2,3,4-benzothiatriazine-1,1(2*H*)-dioxides with arynes.

The scope of the protocol studied with a wide diversity of substituents on both coupling partners provided the desired products (**15.45c**) in good yields.

15.12 Conclusions and Outlook

In conclusion, 1,2,3-benzotriazin-4(3*H*)-ones have witnessed significant developments and advancements in metal-catalyzed, metal-free, visible-light mediated, thermolytic, and electrochemical denitrogenative transformations. These transformations offer a versatile platform for the synthesis of various valuable compounds, ranging from isoquinolones, phenanthridinones, and isoindolin-1-ones to *ortho*-substituted benzamides, including functionalized β-aryl ketones and *ortho*-alkylated benzamides. The versatility and broad substrate scope of 1,2,3-benzotriazin-4(3*H*)-ones in various types of reactions under mild reaction conditions have showcased their ability as successful electrophilic coupling partners. In a similar manner, 1,2,3,4-benzothiatriazine-1,1(2*H*)-dioxides also participated in annulation reactions to prepare biaryl sultams, 3,4-dihydro-1,2-benzothiazine-1,1(2*H*)-dioxides and 3-iminothiaisoindoline-1,1-dioxide. In future, we can anticipate the emergence of numerous denitrogenative coupling/annulation via photocatalytic, electrochemical, and C—H functionalization routes for synthesizing heterocyclic frameworks using 1,2,3-benzotriazin-4(3*H*)-ones. These developments hold promise for expanding synthetic chemists' toolkit and advancing the heterocyclic compound synthesis field.

References

1 Sammaiah, G. and Sarangapani, M. (2008). Synthesis and characterization of biologically significant 3-(2-oxo-1,2,dihydroindol-3-ylideneamino)-3*H*benzo[d][1,2,3] triazin-4-ones. *Asian J. Chem.* 20: 282–286.

2 Daidone, G., Plescia, S., Raffa, D. et al. (1990). Researches on antiinflammatory agents. studies on some new 3-(pyrazol-5-yl)-1,2,3-benzotriazin-4(3*H*)-ones and -quinazolin-4(3*H*)-ones. *Farmaco* 45: 391–398.

3 Fiorino, F., Severino, B., De Angelis, F. et al. (2008). Synthesis and in-vitro pharmacological evaluation of new 5-HT1A receptor ligands containing a 1,2,3-benzotriazin-4(3*H*)-one nucleus. *Arch. Pharm.* 341: 20–27.

4 El Rayes, S.M., Ali, I.A.I., Fathalla, W., and Mahmoud, M.A.A. (2020). Synthesis and biological activities of some new 1,2,3-benzotriazin-4(3*H*)-one derivatives based on molecular docking; promising HepG2 liver carcinoma inhibitors. *ACS Omega* 5 (12): 6781–6791.

5 Raffa, D., Daidone, G., Maggio, B. et al. (1999). Synthesis and antiproliferative activity of novel 3-(indazol-3-yl)-quinazolin-4(3*H*)-one and 3-(indazol-3-yl)-benzotriazin-4(3*H*)-one derivatives. *Arch. Pharm.* 332: 317–320.

6 Takai, H., Obase, H., Nakamizo, N. et al. (1985). Synthesis and pharmacological evaluation of piperidine derivatives with various heterocyclic rings at the 4-position. *Chem. Pharm. Bull.* 33: 1104–1115.

7 Fiorino, F., Caliendo, G., Perissutti, E. et al. (2005). Synthesis by microwave irradiation and antidiarrhoeal activity of 1,2,3-benzotriazin-4(3*H*)-one and saccharine derivatives. *Arch. Pharm.* 338: 548–555.

8 Caliendo, G., Fiorino, F., Grieco, P. et al. (1999). Preparation and local anaesthetic activity of 1,2,3-benzotriazin-4(3*H*)-one and benzoyltriazole derivatives. *Eur. J. Med. Chem.* 34: 1043–1051.

9 Katritzky, A.R., Suzuki, K., and Singh, S.K. (2004). *N*-Acylation in combinatorial chemistry. *ARKIVOC, i* 2004: 12–35.

10 Hummel, J.R., Boerth, J.A., and Ellman, J.A. (2017). Transition-metal-catalyzed C—H bond addition to carbonyls, imines, and related polarised π bonds. *Chem. Rev.* 117: 9163–9227.

11 de Meijere, A. and Diederich, F. (ed.) (2004). *Metal-Catalyzed Cross-Coupling Reactions. 2nd Completely Revised and Enlarged*, vol. 2. Weinheim: Wiley-VCH.

12 Diederich, F. and Stang, P.J. (ed.) (1998). *Metal-Catalyzed Cross-Coupling Reactions*. Weinheim: Wiley-VCH.

13 Ritleng, V., Sirlin, C., and Pfeffer, M. (2002). Ru-, Rh-, and Pd-catalyzed C—C bond formation involving C—H activation and addition on unsaturated substrates: reactions and mechanistic aspects. *Chem. Rev.* 102: 1731–1769.

14 Sanjeev, K., Saiprasad, N., Srilekha, O. et al. (2021). Recent advances in Rh(III)/Ir(III)-catalyzed C—H functionalization/annulation via carbene migratory insertion. *Org. Biomol. Chem.* 19: 1438–1458.

15 Cahiez, G. and Moyeux, A. (2010). Cobalt-catalysed cross-coupling reactions. *Chem. Rev.* 110: 1435–1462.

16 Jana, R., Pathak, T.P., and Sigman, M.S. (2011). Advances in transition metal (Pd, Ni, Fe)-catalyzed cross-coupling reactions using alkyl-organometallics as reaction partners. *Chem. Rev.* 111: 1417–1492.

17 Ayogu, J.I. and Onoabedje, E.A. (2019). Recent advances in transition metal-catalyzed cross-coupling of (hetero)aryl halides and analogues under ligand-free conditions. *Catal. Sci. Technol.* 9: 5233–5255.

18 Yang, Y., Lan, J., and You, J. (2017). Oxidative C—H/C—H coupling reactions between two (hetero)arenes. *Chem. Rev.* 117: 8787–8863.

19 Miyaura, N. and Suzuki, A. (1995). Palladium-catalyzed cross-coupling reactions of organoboron compounds. *Chem. Rev.* 95: 2457–2483.

20 Chattopadhyay, B. and Gevorgyan, V. (2012). Transition-metal-catalyzed denitrogenative transannulation: converting triazoles into other heterocyclic systems. *Angew. Chem., Int. Ed.* 55: 862–872.

21 Akter, M., Rupa, K., and Anbarasan, P. (2022). 1,2,3-Triazole and its analogues: new surrogates for diazo compounds. *Chem. Rev.* 122 (15): 13108–13205.

22 Das, S.K., Roy, S., and Chattopadhyay, B. (2023). Transition-metal-catalyzed denitrogenative annulation to access high-valued N-heterocycles. *Angew. Chem., Int. Ed.* 62: e202210912.

23 Miura, T., Yamauchi, M., and Murakami, M. (2008). Synthesis of 1(2*H*)-isoquinolones by the nickel-catalyzed denitrogenative alkyne insertion of 1,2,3-benzotriazin-4(3*H*)-ones. *Org. Lett.* 10: 3085–3088.

24 Yamauchi, M., Morimoto, M., Miura, T., and Murakami, M. (2010). Enantioselective synthesis of 3,4-dihydroisoquinolin-1(2*H*)-ones by nickel-catalyzed denitrogenative annulation of 1,2,3-benzotriazin-4(3*H*)-ones with allenes. *J. Am. Chem. Soc.* 132: 54–55.

25 Miura, T., Morimoto, M., Yamauchi, M., and Murakami, M. (2010). nickel-catalyzed denitrogenative annulation reactions of 1,2,3-benzotriazin-4(3*H*)-ones with 1,3-dienes and alkenes. *J. Org. Chem.* 75: 5359–5362.

26 Fang, Z.-J., Zheng, S.-C., Guo, Z. et al. (2015). Asymmetric synthesis of axially chiral isoquinolones: nickel-catalyzed denitrogenative transannulation. *Angew. Chem.* 127: 9664–9668.

27 Wang, N., Zheng, S.-C., Zhang, L.-L. et al. (2016). Nickel(0)-catalyzed denitrogenative transannulation of 1,2,3-benzotriazin-4(3*H*)-ones with alkynes:

mechanistic insights of chemical, reactivity and regio- and enantioselectivity from density functional theory and experiment. *ACS Catal.* 6: 3496–3505.

28 Thorat, V.H., Upadhyay, N.S., Murakami, M., and Cheng, C.-H. (2018). Nickel-catalyzed denitrogenative annulation of 1,2,3-benzotriazin-4-(3*H*)-ones with benzynes for construction of phenanthridinone scaffolds. *Adv. Synth. Catal.* 360: 284–289.

29 Miura, T., Nishida, Y., Morimoto, M. et al. (2011). Palladium-catalyzed denitrogenation reaction of 1,2,3-benzotriazin-4(3*H*)-ones incorporating isocyanides. *Org. Lett.* 13: 1429–1431.

30 Balakrishnan, M.H. and Mannathan, S. (2020). Palladium/copper-catalyzed denitrogenative alkylidenation and *ortho*-alkynylation reaction of 1,2,3-benzotriazin-4(3*H*)-ones. *Org. Lett.* 22: 542–546.

31 Hari Balakrishnan, M., Sureshbabu, P., Korivi, R., and Mannathan, S. (2023). Regioselective synthesis of 3-substituted isocoumarin-1-imines via palladium-catalyzed denitrogenative transannulation of 1,2,3-benzotriazin-4(3*H*)-ones and terminal alkynes. *Chem Asian J.* 18: e202300726.

32 Balakrishnan, M.H., Sathriyan, K., and Mannathan, S. (2018). Nickel-catalyzed denitrogenative cross-coupling reaction of 1,2,3-benzotriazin-4(3*H*)-ones with organoboronic acids: an easy access to *ortho*-arylated and alkenylated benzamides. *Org. Lett.* 20: 3815–3818.

33 Micovic, V. and Mihailovic, M. (1953). The reduction of acid amides with lithium aluminum hydride. *J. Org. Chem.* 18: 1190.

34 Thorat, V.H., Upadhyay, N.S., and Cheng, C.-H. (2018). Nickel-catalyzed denitrogenative *ortho*-arylation of 1,2,3-benzotriazin-4(3*H*)-ones with organic boronic acids: an efficient route to losartan and irbesartan drug molecules. *Adv. Synth. Catal.* 360: 4784–4789.

35 Jiménez-Somarribas, A. and Williams, R.M. (2013). Synthetic studies on lemonomycin: construction of the tetracyclic core. *Tetrahedron* 69: 7505–7512.

36 Wang, F., Tong, Y., and Zou, G. (2022). Nickel-catalyzed manganese-assisted denitrogenative cross electrophile-coupling of 1,2,3-benzotriazin-4(3*H*)-ones with alkyl halides for *ortho*-alkylated benzamides. *Org. Lett.* 24: 5741–5745.

37 Weix, D.J. (2015). Methods and mechanisms for cross-electrophile coupling of Csp(2) halides with alkyl electrophiles. *Acc. Chem. Res.* 48: 1767–1775.

38 Lin, T., Wang, Y.-E., Cui, N. et al. (2022). Nickel-catalyzed cross-electrophile coupling of 1,2,3-benzotriazin-4(3*H*)-ones with aryl bromides. *J. Org. Chem.* 87: 16567–16577.

39 Wang, H., Ding, W., and Zou, G. (2023). Mechanoredox/nickel co-catalyzed cross electrophile coupling of 1,2,3-benzotriazin-4(3*H*)-ones with alkyl (pseudo)halides. *J. Org. Chem.* 88: 12891–12901.

40 Li, J., Zheng, Y., Huang, M., and Li, W. (2020). Ni-catalyzed denitrogenative cross-coupling of 1,2,3-benzotriazin-4(3*H*)-ones and cyclopropanols: an easy access to functionalized β-aryl ketones. *Org. Lett.* 22: 5020–5024.

41 Balakrishnan, M.H., Kanagaraj, M., Sankar, V. et al. (2021). Synthesis of *ortho*-arylated and alkenylated benzamides by palladium-catalyzed denitrogenative cross-coupling reactions of 1,2,3-benzotriazin-4(3*H*)-ones with organoboronic acids. *New J. Chem.* 45: 17190.

42 Li, S., Li, Y., Zhu, R. et al. (2023). Synthesis of *ortho*-methylated benzamides *via* palladium-catalyzed denitrogenative cross-coupling reaction of 1,2,3-benzotriazin-4(3*H*)-ones with DABAL-Me_3. *Org. Lett.* 25 (29): 5443–5447.

43 Romero, N.A. and Nicewicz, D.A. (2016). Organic photoredox catalysis. *Chem. Rev.* 116: 10075–10166.

44 Wang, H. and Yu, S. (2015). Synthesis of isoquinolones using visible-light-promoted denitrogenative alkyne insertion of 1,2,3-benzotriazin-4(3*H*)-ones. *Org. Lett.* 17: 4272–4275.

45 Chen, F., Hu, S., Li, S. et al. (2021). Visible-light-induced denitrogenative phosphorylation of 1,2,3-benzotriazin-4(3*H*)-ones: a metal- and additive-free method for accessing orthophosphorylated benzamide derivatives. *Green Chem.* 23: 296–301.

46 Liu, B.-X., Wang, F., Chen, F. et al. (2022). Visible-light-promoted denitrogenative *ortho*-selenylation reaction of 1,2,3-benzotriazin-4(3*H*)-ones: synthesis of *ortho*-selenylated benzamides and ebselen analogs. *Org. Chem. Front.* 9: 2418–2423.

47 Raffa, D., Daidone, G., Maggio, B. et al. (1999). Synthesis and antifungal activity of new N-isoxazolyl-2-iodobenzamides. *II Farmaco* 54: 90–94.

48 Gupta, R., Sirohi, R., and Kishore, D. (2006). A novel approach to the synthesis of 4*H* tetrazolo[1,5-*a*] [1,4]benzodiazepine-6-ones from 1,2,3-benztriazin-4-(3*H*)-ones. *Indian J. Chem. Sect. B: Org. Chem. Incl. Med. Chem.* 45: 2147–2159.

49 Madasamy, K., Balakrishnan, M.H., Korivi, R., and Mannathan, S. (2022). Trifluoroacetic acid-mediated denitrogenative *ortho*-hydroxylation of 1,2,3-benzotriazin-4(3*H*)-ones: a metal-free approach. *J. Org. Chem.* 87: 8752–8756.

50 Korivi, R., Madasamy, K., Sureshbabu, P., and Mannathan, S. (2023). Convenient synthesis of salicylanilide sulfonates from 1,2,3-benzotriazin-4(3*H*)-ones and organosulfonic acids *via* denitrogenative cross-coupling. *ACS Omega* 8: 18306–18311.

51 Zhou, Y., Zhang, B., Dong, J. et al. (2022). Assembly of benzo[*c*][1,2]dithiol-3-ones *via* acid-promoted denitrogenative transannulation of 1,2,3-benzotriazin-4(3*H*)-ones. *Org. Lett.* 24: 9012–9016.

52 Paterson, T.M., Smalley, R.K., Suschitzky, H., and Barker, A.J. (1980). 1,2,3-Benzotriazin-4(3*H*)-ones and related systems. Part 6.[1] Thermal and photolytic decomposition of 3-arylideneamino- and 3-imidoyl- and 3-heteroaryl-1,2,3-benzotriazin-4-ones. *J. Chem. Soc. Perkin Trans. I* 633: 1–6.

53 Barker, A.J., Paterson, T.M., and Smalley, R.K. (1979). 1,2,3-Benzotriazin-4(3*H*)-ones and related systems. Part 5.[1] Thermolysis of 3-aryl- and 3-alkenyl-1,2,3-benzotriazin-4(3*H*)-ones. *J. Chem. Soc. Perkin Trans.* 4: 7–12.

54 Dong, B., Liu, Y., Yang, P. et al. (2022). Denitrogenative cleavage of benzotriazoles and 1,2,3-benzotriazin-4(3*H*)-ones, and selective *N*-desulfonylation of benzotriazoles by aluminum halides. *Tetrahedron Lett.* 103: 53965.

55 Chen, J., Xiao, L., and Qi, L. (2023). Electrochemical annulation of 1,2,3-benzotriazin-4(3*H*)-ones with alkynes to access isoquinolin-1(2*H*)-ones. *Org. Biomol. Chem.* 21: 7295–7299.

56 Miura, T., Yamauchi, M., Kosaka, A., and Murakami, M. (2010). Nickel-catalyzed regio- and enantioselective annulation reactions of 1,2,3,4-benzothiatriazine-1,1(2*H*)-dioxides with allenes. *Angew. Chem. Int. Ed.* 49: 4955–4957.

57 Han, Y., Wang, H., and Yu, S. (2016). Synthesis of biaryl sultams using visible-light-promoted denitrogenative cyclization of 1,2,3,4-benzothiatriazine-1,1-dioxides. *Org. Chem. Front.* 3: 953–956.

58 Thorat, V.H., Hsieh, J.-C., and Cheng, C.-H. (2020). Transition-metal-free tandem cyclization/*N*-arylation reaction: a method to access biaryl sultam derivatives via a diradical pathway. *Org. Lett.* 22: 6623–6627.

59 Thorat, V.H., Tsai, Y.-L., Huang, Y.-R. et al. (2022). Nickel-catalyzed denitrogenative cyclization of 1,2,3,4-benzothiatriazin-1,1(2*H*)-dioxides with arynes to synthesize biaryl sultams. *Org. Lett.* 24: 2915–2920.

16

Denitrogenative Transformations of 3-Aminoindazoles

Xin Li[1] *and Qiuling Song*[2]

[1] *Institute of Next Generation Matter Transformation, College of Materials Science and Engineering, Huaqiao University, Xiamen, China*
[2] *Key Laboratory of Molecule Synthesis and Function Discovery, Fujian Province University, College of Chemistry, Fuzhou University, Fuzhou, China*

16.1 Introduction

Polynitrogen heterocycles are privileged core structural motifs that widely exist in a broad spectrum of pharmaceutical and material molecules. Moreover, they are regarded as significant building blocks in synthetic chemistry, for example, triazoles are extensively employed as precursors to generate α-imino metal carbenoid species, 1,2,4,5-tetrazines play an important role in click chemistry and bioorthogonal chemistry, imidazole and DBU are the most commonly used organic bases.

In recent years, the denitrogenative transformations of polynitrogen heterocycles have emerged as one of the most powerful and reliable tools for constructing structurally complex molecules because they can easily afford a series of synthetically useful metal carbenoids or radical species. Besides triazoles, whose denitrogenation generates α-imino metal carbenoid species, the denitrogenative ring-opening of benzotriazoles, pyridotriazoles, benzotriazinones, and tetrazoles provide a diverse range of synthetically useful compounds and thus have been well documented.

Among the various polynitrogen heterocycle candidates for denitrogenative transformations, 3-aminoindazoles is regarded as a special molecule, because during the denitrogenative ring-opening process an aromatic nitrile moiety could be furnished concomitantly. This is of great synthetic value for the cyano group can be handily transformed into a diverse array of functional groups, thus immensely expanding the molecule complicacy [1]. Moreover, nitrile-containing compounds are pervasively embedded in pharmaceuticals, natural products, herbicides, and agrochemicals [2]. Therefore, the denitrogenative transformations of 3-aminoindazoles are an ideal process and deserve extensive exploration (Scheme 16.1).

Despite its versatility and importance, the denitrogenative ring-opening of 3-aminoindazoles is less reported in contrast to other polynitrogen heterocycles.

Denitrogenative Transformation of Nitrogen Heterocycles: Synthesis, Reactions and Applications, First Edition. Edited by Pazhamalai Anbarasan.

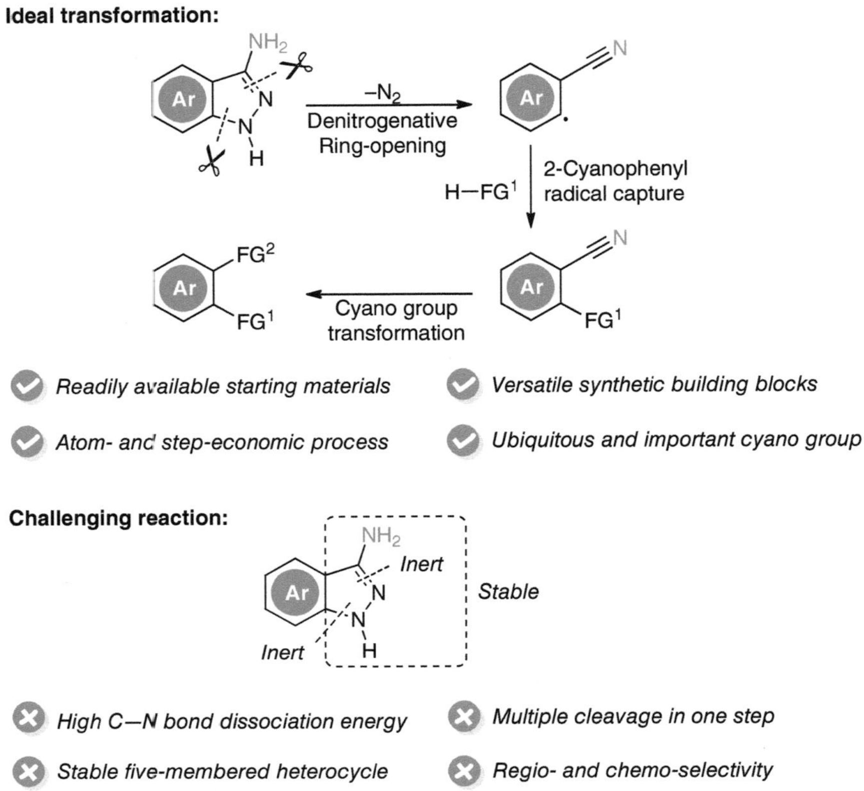

Scheme 16.1 Synthetic value and challenge of denitrogenative transformations of 3-aminoindazoles

Attributing to the high C—N bond dissociation energy, multiple cleavages of unactivated C—N bonds under one set of conditions are challenging. Besides, the direct transformation of relatively stable five-membered heterocycles through selective cleavage of two C—N bonds in one step is still far from being achieved (Scheme 16.1). Consequently, although remarkable progress has been achieved in the last decades in transition-metal-catalyzed selective cleavage of C—N bonds, there is still a great demand for the development of practical strategies for the multifarious C—N bonds cleavage and denitrogenative ring-opening of 3-aminoindazoles.

Only in recent years have chemists addressed the reactivity and selectivity issues of this reaction. By judicious employment of efficient catalytic systems and ingenious design of the reaction process, a diverse array of structurally novel compounds, some of which are difficult to access through traditional synthetic approaches, were thus obtained in good yields. This chapter summarizes and discusses the recent advances in denitrogenative transformations of 3-aminoindazoles, which are organized according to the reaction categories.

16.2 Coupling with Conjugated Unsaturated Systems

16.2.1 Coupling with Heterocycles and Arenes

The first example of denitrogenative ring-opening of 3-aminoindazoles was reported by Song et al. in 2018. In this reaction, an unprecedented oxidative cleavage of two C—N bonds of 3-aminoindazoles occurred under an oxidative condition, thus releasing one equiv. of nitrogen and generating an aryl radical, which subsequently engaged in a C—H arylation with a series of heterocycles and arenes. Mechanistically, the 3-aminoindazole substrates are first oxidized and abstracted a hydrogen atom by a *tert*-butoxyl radical, thus resulting in an intramolecular rearrangement to extrude one molecular nitrogen and form a 2-cyanophenyl radical. The latter then attacks heterocycles or arenes to furnish a radical addition intermediate, which is further oxidized to carbocation species and deprotonated to afford the final coupling products (Scheme 16.2) [3].

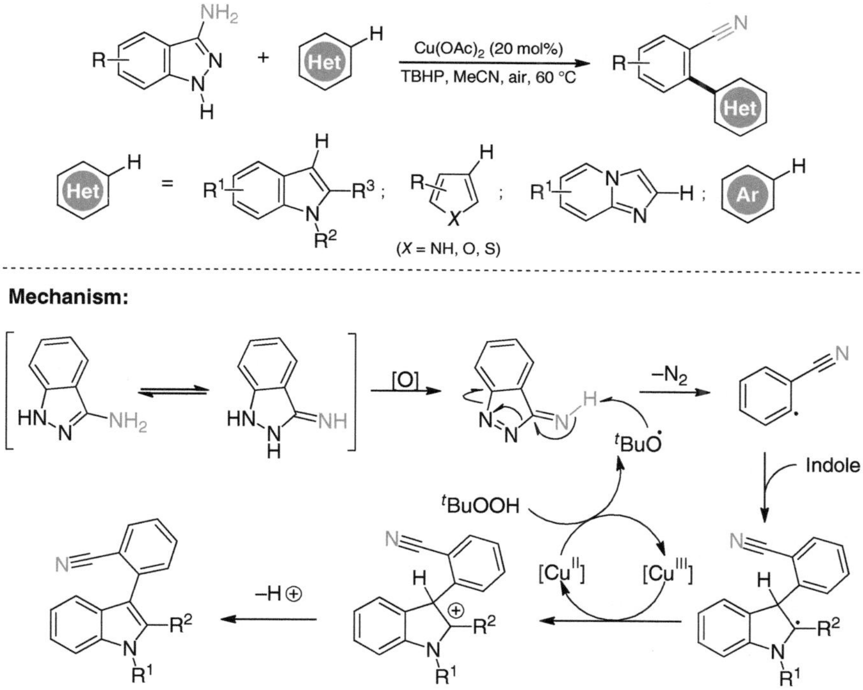

Scheme 16.2 Denitrogenative ring-opening of 3-aminoindazoles for the synthesis of aromatic nitrile-containing (hetero)arenes. Source: Zhou et al. [3]/American Chemical Society.

In the previous work, the generated 2-cyanophenyl radicals were intermolecularly captured by heterocycles or arenes. Instead, if the radical capture occurred intramolecularly, extended π-systems-derived polycyclic aromatic compounds would be obtained, which are highly significant and have comprehensive applications in interdisciplinary research fields. To this end, Song et al. developed an oxidative denitrogenation/intramolecular radical cascade, which was referred to as "aromatic metamorphosis," to construct a series of nitrile-containing polycyclic aromatic compounds in decent yields with good functional group compatibility. DFT calculations not only further consolidated the mechanism proposed earlier by the same authors, but also provided clearer details with quantitative energy variation throughout the whole process (Scheme 16.3) [4].

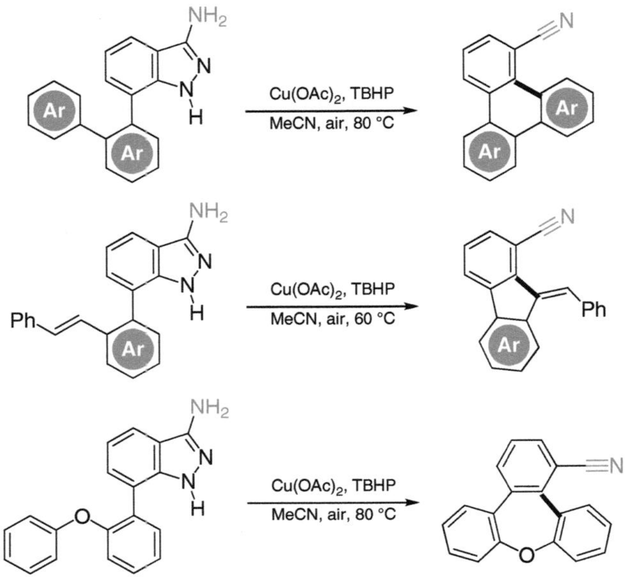

Scheme 16.3 Denitrogenative ring-opening of 3-aminoindazoles for the synthesis of nitrile-containing triphenylenes and fluorenes. Source: Zhou et al. [3]/American Chemical Society.

Employing other kinds of heterocyclic compounds to react with 3-aminoindazoles was disclosed by Yuan et al. in 2023. The authors found that quinoxalin-2(1*H*)-ones, a class of vital structural motifs in natural products, pharmaceuticals, and bioactive molecules, are good candidates to capture the 2-cyanophenyl radicals, thus affording the C3-cyanoarylated products in good yields with a broad substrate scope. Condition screening showed that *tert*-butyl peroxybenzoate (TBPB) performed better than TBHP, while $Cu(OAc)_2$ and acetonitrile were still the optimal catalysts and solvents (Scheme 16.4) [5].

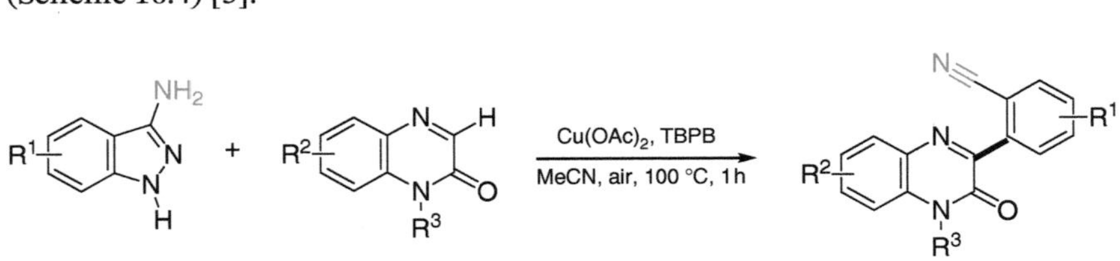

Scheme 16.4 C3-cyanoarylation of quinoxalin-2(1*H*)-ones via denitrogenative ring-opening of 3-aminoindazoles. Source: Mou et al. [5]/Royal Society of Chemistry.

16.2.2 Coupling with Ketene Dithioacetals

The scope of coupling partners can be further expanded, for example, Song et al. found that ketene dithioacetals are ideal substrates in Cu-catalyzed denitrogenative radical addition of 3-aminoindazoles. Owing to the high nucleophilicity of the olefinic C—H bonds in ketene dithioacetals, their C—H arylation proceeded smoothly by employing 3-aminoindazoles as cyanoarylating reagents, thus affording a diverse array of cyanoarylated dithioacetals in decent yields. Interestingly, the outcome of the reaction can be readily controlled by the employed oxidant, that is, adding 2.0 equiv. of TBHP into the reaction system resulted in the denitrogenation ring-opening process to obtain the olefinic C—H cyanoarylated products. Alternatively, in the absence of TBHP, Lewis-acid promoted [3 + 3] annulation reaction occurred predominantly, delivering various *S*-containing pyrimido[1,2-*b*]-indazole derivatives in good yields (Scheme 16.5) [6].

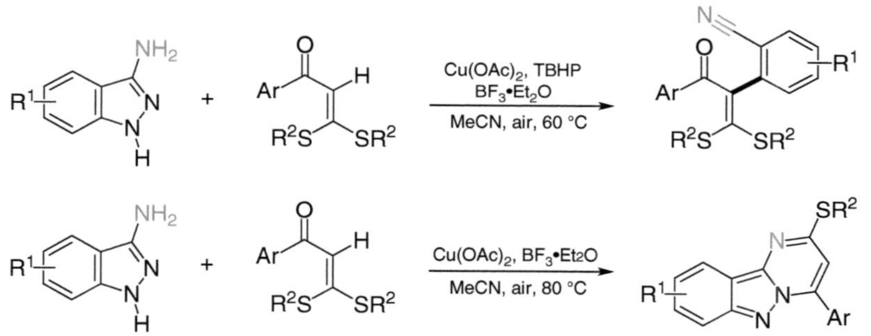

Scheme 16.5 Olefinic C−H cyanoarylation of ketene dithioacetals via denitrogenative ring-opening of 3-aminoindazoles. Source: Zhou et al. [6]/Royal Society of Chemistry.

16.2.3 Coupling with Enamines

Enamines are versatile building blocks in synthetic chemistry. Owing to the π-donating capacity of the nitrogen atom, the C—C double bonds in enamines are relatively electron-rich and thus making themselves good radical acceptors. Song et al. found that by merging their developed denitrogenative ring-opening of 3-aminoindazoles with enamine chemistry, a diverse array of cyanoarylated enamines could be synthesized in decent yields with a wide substrate scope. Of note, the high reactivity of enamine substrates enabled the reaction to proceed at room temperature. Mechanistically, this reaction also proceeds via a radical pathway, and 2-cyanophenyl radical was proven to be the key intermediate of the reaction (Scheme 16.6) [7].

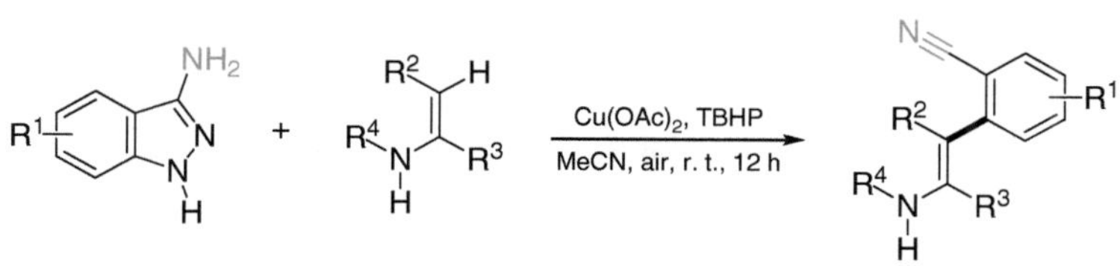

Scheme 16.6 C−H cyanoarylation of Enamines via denitrogenative ring-opening of 3-aminoindazoles. Source: Zhou et al. [7]/Royal Society of Chemistry.

16.2.4 Cascade Radical Processes

Besides simple heterocycles, arenes, and enamines, other conjugated unsaturated systems are also capable of capturing the 2-cyanophenyl radicals. Accordingly, more structurally complicated products could be obtained. For example, in 2021 Liu et al. reported a Cu-catalyzed oxidative dual arylation of C—C double bonds in acrylamides with 3-aminoindazoles. Mechanistically, this protocol involves a radical sequence of 3-aminoindazole oxidation, two C—N bonds cleavage, cyanoaryl radical addition, and intramolecular cyclization. By virtue of this highly efficient method, a wide range of cyanoarylated oxindoles, which are vital structural motifs in natural products, pharmaceuticals, and bioactive molecules, were obtained from simple starting materials (Scheme 16.7) [8].

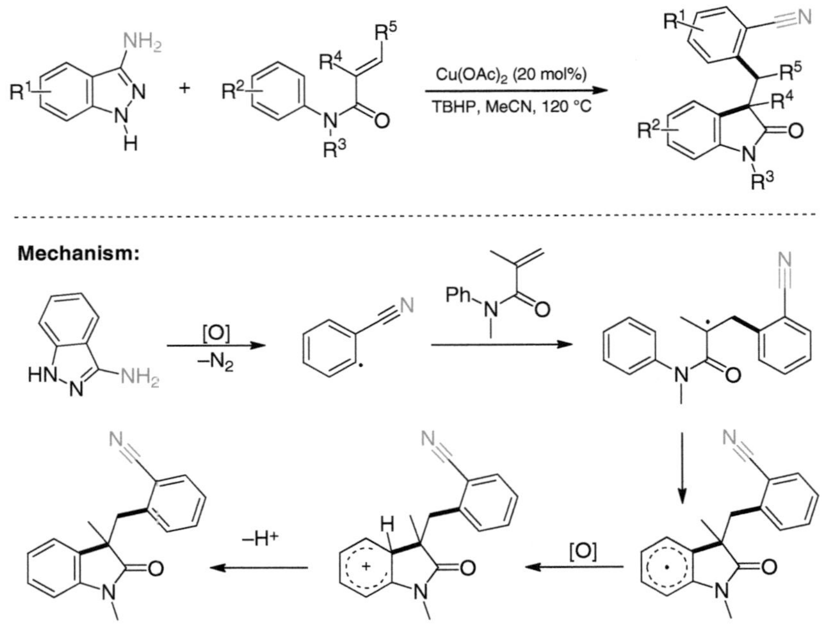

Scheme 16.7 Denitrogenation of 3-aminoindazoles for the synthesis cyanoarylated oxindoles. Source: Wang et al. [8]/American Chemical Society.

Soon after, a similar radical cascade was reported by the same group, which employed *N*-methacryloyl-substituted 2-arylbenzoimidazoles or 2-arylindoles as substrates to afford isoquinolinones, a class of nitrogen-containing heterocycles existed in many bioactive natural products and pharmaceuticals. The synthetic value of this protocol was highlighted by introducing two functional groups into the C—C double bond at the same time and increasing the structural complicacy in an atom- and step-economic manner (Scheme 16.8) [9].

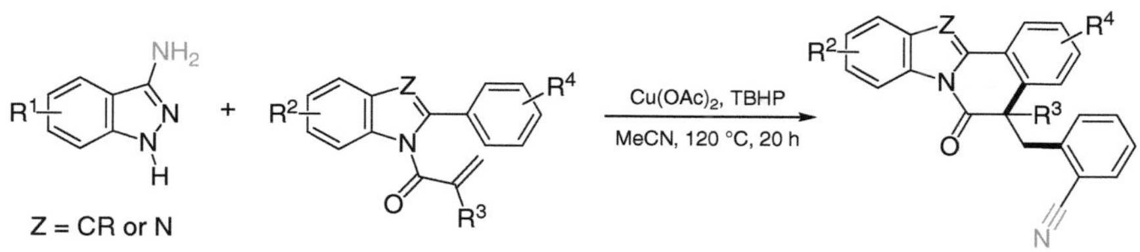

Scheme 16.8 Denitrogenation of 3-aminoindazoles for the synthesis of isoquinolinones. Source: Guo et al. [9]/Royal Society of Chemistry.

16.3 Coupling with Thiols and Diselenides

Besides being captured by conjugated unsaturated systems to form C—C bonds, the *in situ* formed 2-cyanophenyl radical could also couple with other species such as thiols to construct C—S bonds. Aryl sulfides are ubiquitous in pharmaceutical molecules and materials; however, previously reported synthetic methods depended heavily on the cross-coupling of thiols or their surrogates with aryl halides, pseudo halides, Grignard reagents, diazonium salts, boronic acids, triflates, etc. The radical coupling between thiols and 2-cyanophenyl radicals, which could concomitantly introduce an additional functionality in a single step to increase the molecules' complicacy, has never been realized. To address this issue, Song et al. developed an oxidative denitrogenative radical coupling of 3-aminoindazoles employing commercially available thiols and diselenides as coupling partners. A diverse array of 1,2-thiobenzonitriles were thus synthesized in good yields with broad substrate scope, even structurally complicated cysteine-containing peptide was also compatible well with the reaction condition, affording the *S*-arylcysteine product in 62% yield. Mechanistically, 2-cyanophenyl radical is generated from the oxidative ring-opening of 3-aminoindazoles, while thiyl radical is produced from thiol with the help of a *tert*-butoxyl radical. These two radicals subsequently form a radical pair by coordinating with Cu(I) species, forge a C—S bond through the coupling of the radical intermediates, and deliver the desired product (Scheme 16.9) [10].

16.4 Denitrogenative Transannulation of 3-Aminoindazoles

In the above-discussed reactions, the *in situ* generated 2-cyanophenyl radical was captured by different radical acceptors and the cyano group was preserved after the reaction. However, the cyano group is an unsaturated group; it can also accept a radical under appropriate conditions to afford radical addition products. For example, in the reaction of 3-aminoindazole and *N*-(2-methylallyl)-*N*-phenylacetamide, Cheng et al. surprisedly found that the predicated 2-cyanophenyl substituted indoline was not detected. Instead, 2,2-disubstituted indanone was isolated in 55% yield. To explain this unusual result, the authors proposed a cascade denitrogenative

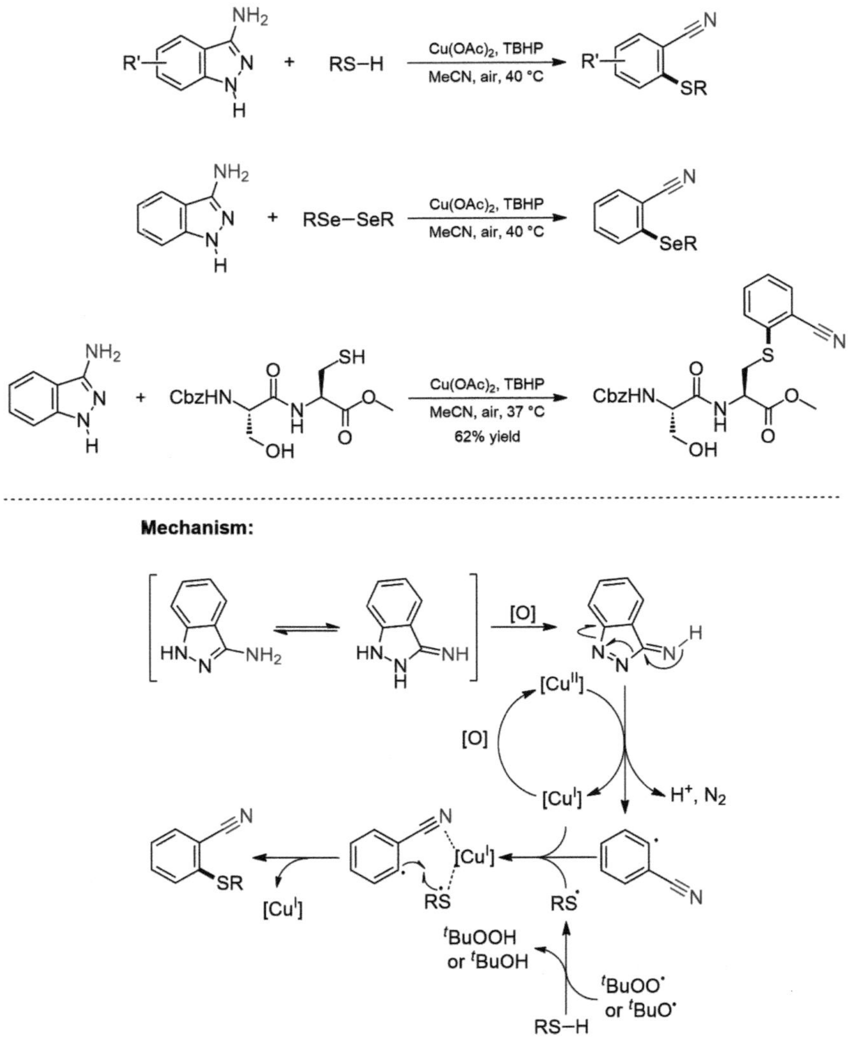

Scheme 16.9 Denitrogenative radical coupling reaction for the synthesis of 1,2-thiobenzonitriles. Source: Zhou et al. [10]/Royal Society of Chemistry.

transannulation/hydrolyzation process, wherein the 2-cyanophenyl radical is captured by alkene substrate to generate another carbon-centered radical intermediate. Subsequent intermolecular radical addition of the alkyl radical to the Cu(III)-coordinated cyano group gives an oxime intermediate, which is hydrolyzed to afford the final indanone product. This reaction is of synthetic value, for it developed an alternative reaction pattern of 2-cyanophenyl radical. By virtue of this reaction, a series of 2,2-disubstituted indanones, which are key structural motifs in various natural products and pharmaceutical compounds that exhibit a variety of biological activities, were obtained in good yields (Scheme 16.10) [11].

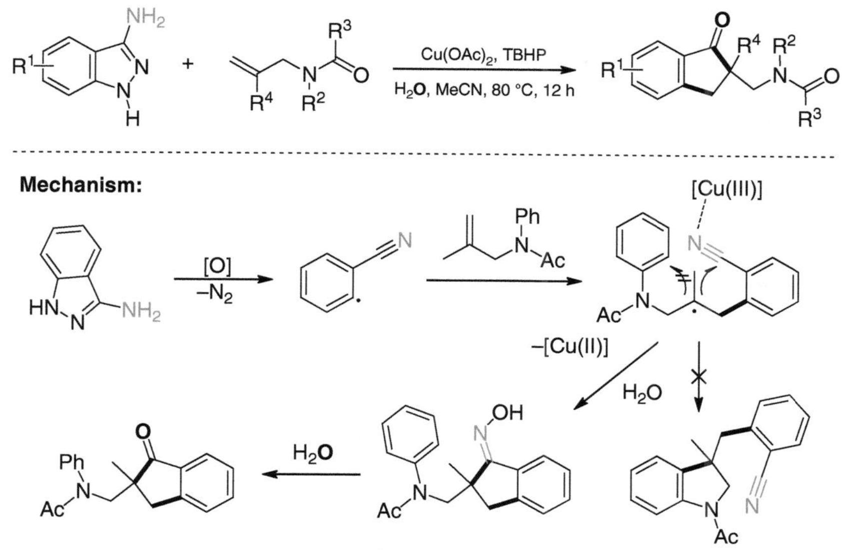

Scheme 16.10 Cascade denitrogenative transannulation/hydrolyzation of 3-aminoindazoles for the synthesis of 2,2-disubstituted indanones. Source: Teng et al. [11]/American Chemical Society.

Nitriles are versatile building blocks that can be handily transformed into a diverse array of functional groups. Therefore, the cyano group generated from the oxidative denitrogenation of 3-aminoindazoles could react with appropriate functionalities existing in the reaction system to afford other valuable products. For example, Song et al. developed an "extrude-and-sew" strategy for the synthesis of diverse functionalized 1-aminoisoquinoline and 3-aminobenzothiophene derivatives. In this reaction, the cyanoarylated products were first generated as key intermediates, which then engaged in the base-promoted intramolecular nucleophilic addition of the cyano groups to furnish 3-aminobenzothiophene and 1-aminoisoquinoline derivatives as final products (Scheme 16.11) [12].

16.5 N—N Bond Cleavage of 3-Aminoindazoles

Above-discussed reactions all involvethe C—N bond cleavage, which extrude N1 and N2 atoms and generate phenyl radicals, keeping the N3 atom in the final product (Scheme 16.12, equation 1). However, N—N bond cleavage, a more challenging disconnection strategy, has never been applied to the oxidative ring-opening chemistry of 3-aminoindazoles until a seminal work was reported by Song et al. in 2018. In this work, an unprecedent N—N bond cleavage occurred under the oxidative conditions, leaving the N1 atom as an amino group in the final product (Scheme 16.12, equation 2). Specifically, employing NBS as a bromine source and ceric ammonium nitrate (CAN) as oxidant, a variety of 2-aminobenzoates were

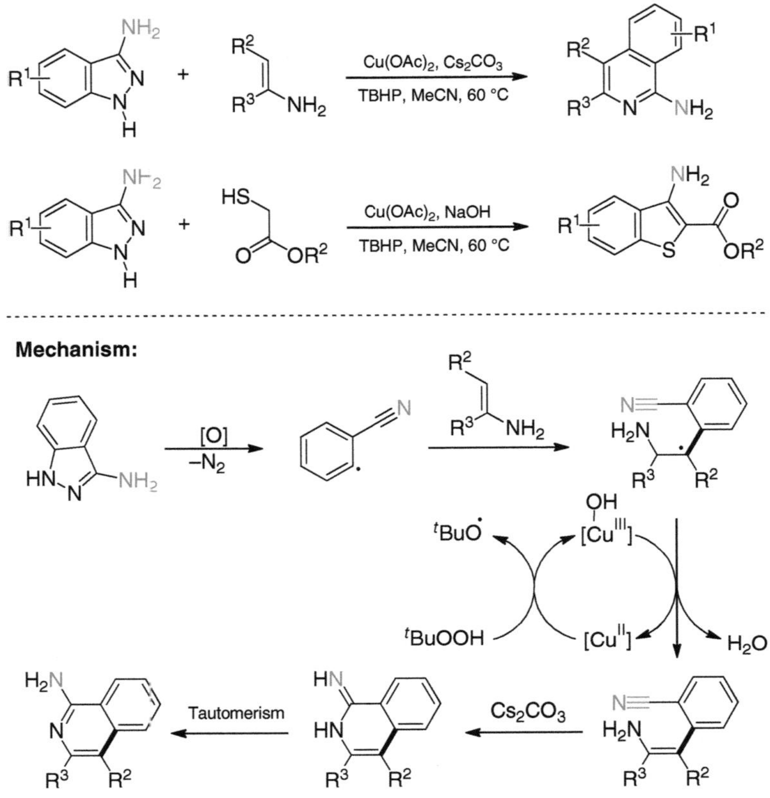

Scheme 16.11 Denitrogenative transannulation of 3-aminoindazoles for the synthesis of 1-aminoisoquinolines and 3-aminobenzothiophenes. Source: Zhou et al. [12]/American Chemical Society.

obtained in good yields at room temperature. Furthermore, the authors found that mono- and dual-brominated aminobenzoates could be controllably achieved by employing the appropriate oxidant and bromine source, thus enriching the synthetic value of the reaction. Although the mechanism was unclear at that stage, the authors still gave a plausible explanation for the outcome of the reaction based on the preliminary mechanistic experiments. The free amino group in 3-aminoindazole is first oxidized into a nitroso group; the resulting intermediate then undergoes a fast [1, 5]-*H* shift to generate an oxime. The latter reacts with ROH to form a six-membered intermediate, which collapses with the assistance of water and NBS. Meanwhile, the N—N bond is disconnected to generate a free 1-amino group and an N—Br amide group. Then, the N—Br bond undergoes a homolytic cleavage to form a dimerization intermediate. After a 1,1-elimination/nitrene decomposition process, the final product 2-aminobenzoate is released along with one molecule of nitrogen (Scheme 16.12) [13].

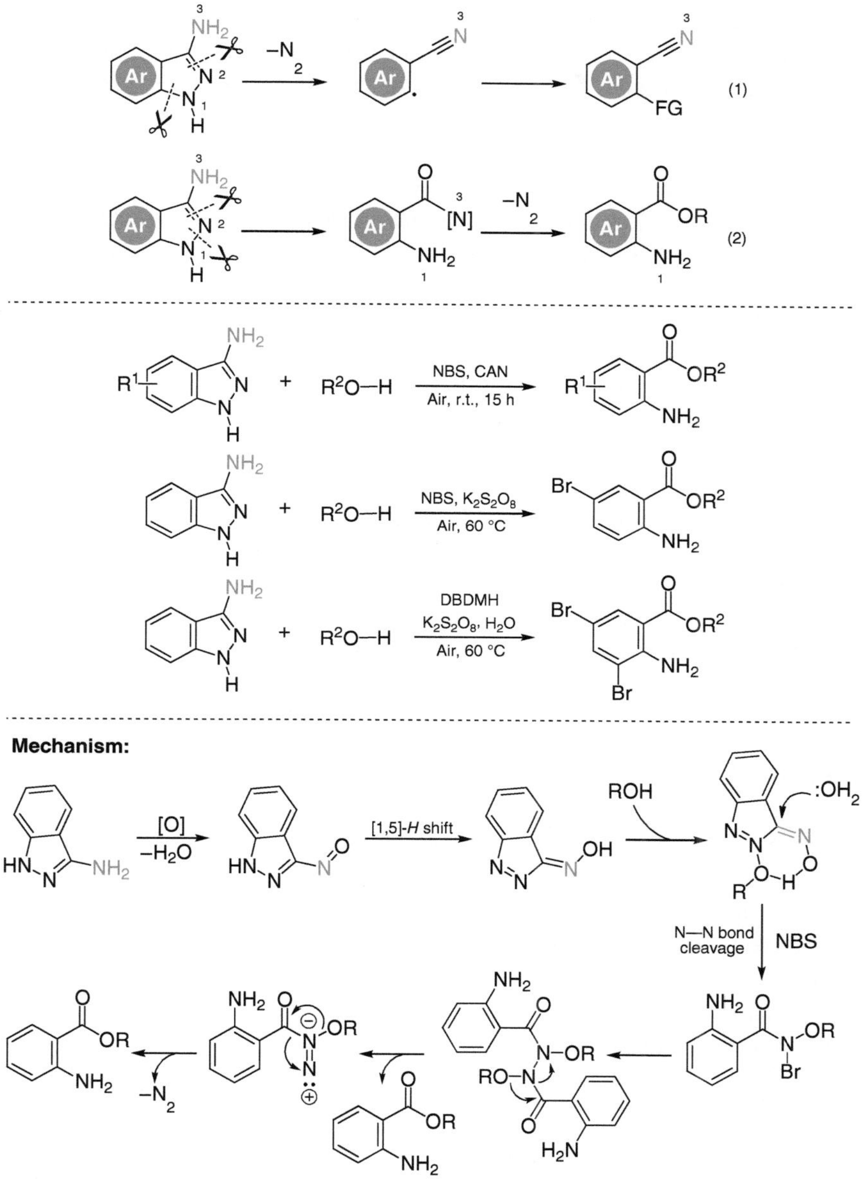

Scheme 16.12 Oxidative N–N bond cleavage of 3-aminoindazoles for the synthesis of 2-aminobenzoates. Source: Zhou and Song [13]/Royal Society of Chemistry.

16.6 Conclusion and Outlook

Despite the high C—N bond dissociation energy, relatively stable five-membered heterocycles, and challenging multiple cleavage of two unactivated C—N bonds in one step, the denitrogenative transformations of 3-aminoindazoles is still regarded as an ideal and synthetically valuable reaction, because it can create various structurally complex molecules bearing a versatile and ubiquitous cyano group in an atom- and step-economic manner. Only in recent years have chemists addressed the reactivity and selectivity issues of this reaction. By judicious choice of the catalysts, coupling reagents and oxidants, as well as the ingenious design of the reaction process, several novel transformations have been disclosed, which include coupling with heterocycles or arenes, coupling with ketene dithioacetals, coupling with enamines, coupling with thiols and diselenides, denitrogenative transannulation, and N—N bond cleavage of 3-aminoindazoles. A diverse array of structurally novel compounds, some of which are difficult to access through traditional synthetic approaches, were thus obtained in good yields.

However, present reaction conditions still have some limitations, for example, the relatively strong oxidative conditions lead to a lack of functional group compatibility, especially for oxidant-sensitive motifs such as alcohols, aldehydes, and anilines. This will become a tricky problem when dealing with highly functionalized substrates. Secondly, the synthetic utilities of this reaction need to be explored, especially in the synthesis or structural modifications of natural products and pharmaceutical molecules. Finally, the present protocol heavily relies on the copper/TBHP oxidation system; new strategies that could generate the 2-cyanophenyl radicals through a milder, more efficient, and more selective pathway are expected.

References

1 Patai, S., and Rappoport, Z. (1983). *Preparation and Synthetic Applications of Cyano Compounds*. New York: Wiley-VCH.

2 Kleemann, A., Engel, J., Kutscher, B., and Reichert, D. (2001). *Pharmaceutical Substance: Synthesis, Patents, Applications*, 4e. Stuttgart: Georg Thieme.

3 Zhou, Y., Deng, S., Mai, S., and Song, Q. (2018). Cu-catalyzed denitrogenative ring-opening of 3-aminoindazoles for the synthesis of aromatic nitrile-containing (hetero)arenes. *Org. Lett.* 20: 6161–6165.

4 Zhou, Y., Lin, L., Wang, Y. et al. (2019). Cu-catalyzed aromatic metamorphosis of 3-aminoindazoles. *Org. Lett.* 21: 7630–7634.

5 Mou, C.-X., Yuan, J.-W., Hu, Q. et al. (2023). Copper-catalyzed oxidative direct C3-cyanoarylation of quinoxalin-2(1*H*)-ones via denitrogenative ring-opening of 3-aminoindazoles. *New J. Chem.* 47: 3783–3792.

6 Zhou, Y., Lou, Y., Wang, Y., and Song, Q. (2019). Oxidant-controlled divergent transformations of 3-aminoindazoles for the synthesis of pyrimido[1,2-*b*]-indazoles and aromatic nitrile-derived dithioacetals. *Org. Chem. Front.* 6: 3355–3359.

7 Zhou, Y., Wang, Y., Song, Z. et al. (2020). Cu-catalyzed C—N bond cleavage of 3-aminoindazoles for the C—H arylation of enamines. *Org. Chem. Front.* 7: 25–29.

8 Wang, Q.-L., Zhou, Q., Liao, J. et al. (2021). Cu-catalyzed oxidative dual arylation of active alkenes: preparation of cyanoarylated oxindoles through denitrogenation of 3-aminoindazoles. *J. Org. Chem.* 86: 2866–2875.

9 Guo, Y., Huang, P.-F., Xiong, B.-Q. et al. (2022). Cu-catalyzed oxidative denitrogenation of 3-aminoindazoles for the synthesis of isoquinolinones. *Org. Biomol. Chem.* 20: 6844–6853.

10 Zhou, Y., Wang, Y., Lou, Y., and Song, Q. (2019). Facile synthesis of 1,2-thiobenzonitriles via Cu-catalyzed denitrogenative radical coupling reaction. *Chem. Commun.* 55: 10265–10268.

11 Teng, J., Sun, S., Yu, J.-T., and Cheng, J. (2019). Copper-catalyzed cascade denitrogenative transannulation/hydrolyzation of 3-aminoindazoles toward 2,2-disubstituted indanones. *J. Org. Chem.* 84: 15669–15676.

12 Zhou, Y., Wang, Y., Lou, Y., and Song, Q. (2019). Cu-catalyzed denitrogenative transannulation of 3-aminoindazoles to assemble 1-aminoisoquinolines and 3-aminobenzothiophenes. *Org. Lett.* 21: 8869–8873.

13 Zhou, Y., and Song, Q. (2018). Oxidative ring-opening of 3-aminoindazoles for the synthesis of 2-aminobenzoates. *Org. Chem. Front.* 5: 3245–3249.

17

Denitrogenative Transformations of Tetrazoles and Pyridotetrazoles

Sandip Kumar Das and Buddhadeb Chattopadhyay

Department of Biological and Synthetic Chemistry, Centre of Biomedical Research, SGPGIMS Campus, Lucknow, Uttar Pradesh, India

17.1 Introduction

Over the years, the exploration of novel strategies for making high-valued N-containing molecules has become a captivating research area, primarily owing to their wide-ranging applications in the fields of medicinal chemistry, drug discovery, natural product synthesis, agrochemicals, and beyond. In this regard, metal-catalyzed denitrogenative transformation stands out as a compelling alternative, offering a step- and atom-efficiency process from easily accessible feedstocks for producing high-valued N-scaffolds. Besides, this method significantly decreases the production of byproducts. The denitrogenative transformation of 1,2,3-triazole and 1,2,3,4-tetrazole systems has been documented in literature [1]. The primary principle behind this strategy involves the generation of either a metal carbene or a nitrene from the respective precursor in the presence of a metal catalyst, achieved through dinitrogen evolution. Subsequently, these reactive intermediates are captured by other reacting partners either intermolecularly or intramolecularly. This process yields numerous essential heterocycles, including molecules of medicinal significance and natural products. This chapter primarily concentrates on the metal-catalyzed denitrogenative transformation of 1,2,3,4-tetrazoles, developed by Chattopadhyay and his co-worker.

In the history of denitrogenative transformation of 1,2,3,4-tetrazole, the pioneering work began with Huisgen and von Fraunberg in 1969 [2]. Their research focused on exploring the reactivity of 1,2,3,4-tetrazole and revealing that Cu powder had the capability to facilitate an annulation reaction from the corresponding tetrazole at reaction temperature (Figure 17.1). While they proposed the formation of these compounds through a Cu-nitrene intermediate, no experimental evidence was provided to substantiate this assertion. Later in 1976, the Wentrup group [3] reported denitrogenative thermal and photochemical nitrene–nitrene rearrangement reactions of tetrazole at very high temperature (>500 °C) under flash vacuum pyrolysis to synthesize nitrogen molecules (Figure 17.1). Despite these early discoveries, no attempts have been seen to use tetrazole as a nitrene source in chemical synthesis.

Denitrogenative Transformation of Nitrogen Heterocycles: Synthesis, Reactions and Applications, First Edition. Edited by Pazhamalai Anbarasan.

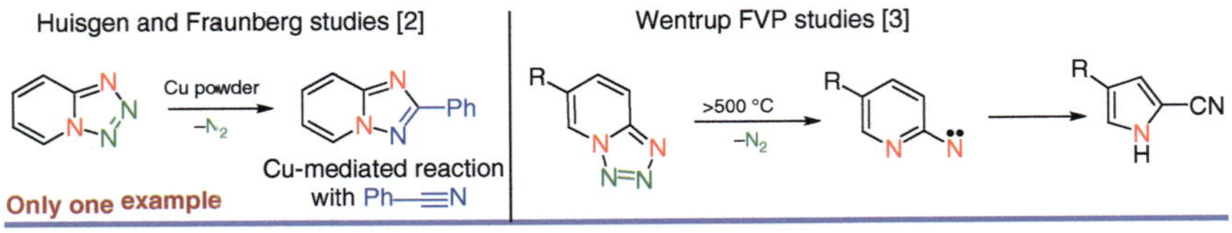

Figure 17.1 Pioneering studies of pyridotetrazole.

This is because of the extreme inertness of this system, which usually remains in closed tetrazolic form rather than the open azide form. In response to these challenges, the Chattopadhyay research group has built a plan to reveal the synthetic potential of 1,2,3,4-tetrazole as versatile aminating precursors to accomplish C—N bond-forming reactions inter and intramolecularly, paving the way for the discovery of smart technologies that would be extremely welcoming in the fields of medicinal chemistry, biochemistry, pharmaceutical chemistry, and related areas. In this chapter, the authors aim to overview the exceptional reactivity of 1,2,3,4-tetrazole with mechanistic insights toward various denitrogenative C—N bond-forming reactions in the presence of transition metal catalysts.

17.2 Rh-Catalyzed Denitrogenative Annulation of Monocyclic Tetrazoles

In 2018, Murakami and co-workers reported rhodium-catalyzed denitrogenation of monocyclic tetrazoles with various styrenes to prepare a number of pyrazolines with high degree of enantioselectivity [4]. In this study, they described that *in situ* generated *N*-sulfonyl tetrazole in the presence of chiral Rh-carboxylate first generates α-azo Rh(II)-carbenoid species.

The carbenoid species then readily undergo cycloaddition reaction with styrene's, leading to the formation of chiral 3,5-diaryl-2-pyrazolines.

After evaluating numerous rhodium catalysts, the authors found that the $Rh_2[(S)\text{-TCPTTL}]_4$ catalyst in the presence of a nonpolar ethyl cyclohexane solvent can provide the desired annulated products with high isolated yield (84%) and enantiomeric excess (96%). Notably, while under the optimized conditions, various aromatic styrene's perform well, aliphatic styrene's exhibit limited reactivity. This is due to the decomposition of corresponding nitrile imine, which infrequently reacts with alkenes. In the mechanism (depicted in Figure 17.2), the annulation reaction begins with the sulfonylation of substrate tetrazole in the presence of Tf_2O, resulting in a mixture of two isomers, **(17.2a)** (*N*1-triflylated) and **(17.2b)** (*N*2-triflylated), which exist in equilibrium. Subsequently, the **(17.2b)** isomer participates in ring chain isomerization to furnish a stable α-diazo diazene **(17.2c)**. Then, with the assistance of a dirhodium catalyst, Rh(II)-carbenoid species **(17.2d)** is generated from diazene intermediate **(17.2c)** through dinitrogen evolution. Following this, an asymmetric

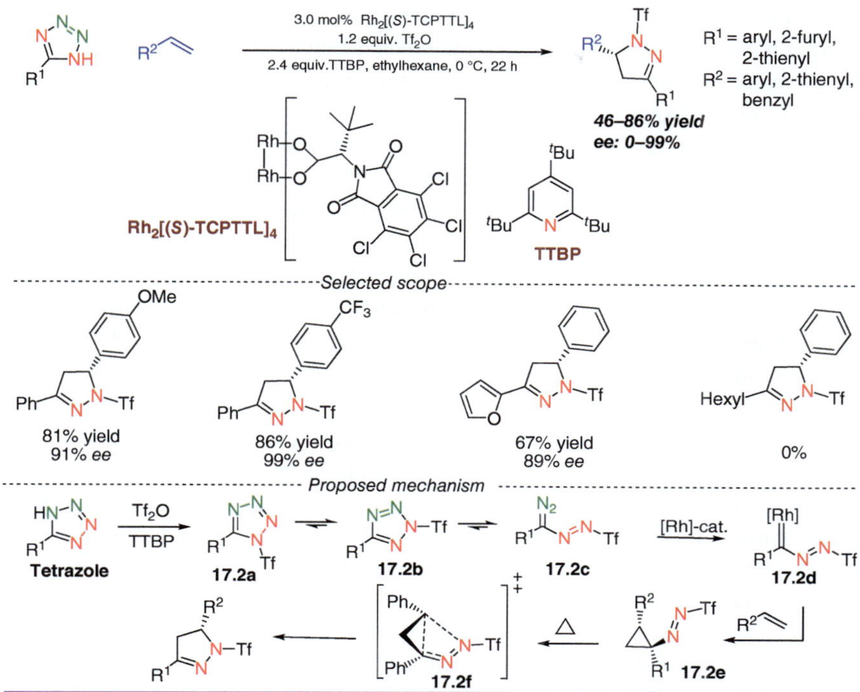

Figure 17.2 Denitrogenative annulation of monocyclic tetrazoles through metal carbene.

cyclopropanation reaction in between Rh^{II}-carbenoid species and styrene stereoselectivity yields cyclopropyldiazene (**17.2e**). Finally, ring expansion of cyclopropane intermediate (**17.2e**) leads to the formation of desired 2-pyrazoline.

17.3 Ir-Catalyzed Denitrogenative C(sp^2)–H Amination of Pyridotetrazoles

Among various electron-rich nitrogen-containing molecular scaffolds, pyridotetrazole occupies a special attention because of its unique structural properties. For example, upon treatment of metal catalysts, it can potentially generate a metal–nitrene intermediate along with an additional pyridyl unit. Thus, the idea was to generate a productive metal–nitrene intermediate for the development of new concepts/new methods to make C—N bonds. The other reason to work on this molecular framework is the presence of a one-nitrogen atom that can completely change the chemical, physical, and biological properties of a molecule. One-N atom can also make the synthetic method extremely challenging. For example, the drastic change can be seen between (i) benzene and pyridine, the difference is only one nitrogen atom, but their properties are completely different; (ii) likewise, indoline and *aza*-indoline (again the difference between is the presence of one nitrogen atom that make these two molecules completely different); (iii) and if we go for the more complex molecular system, penicillin and ampicillin are the best examples. As a result, pyridotetrazole can be used as a new type of aminating agent that can be used to make new molecules to identify new molecular properties related to drug discovery and medicinal chemistry program.

Similar to 1,2,3-triazole, pyridotetrazoles exist in equilibrium with their open azide form (Figure 17.3, i). Literature data suggests that tetrazole-azide isomerization is influenced by the reaction conditions especially by substituents, solvent, and temperature [5]. Unlike the metal-catalyzed denitrogenative transformation of 1,2,3-triaozle, achieving a denitrogenative transformation of pyridotetrazole remains an immense challenge due to the extreme inertness of the tetrazole moiety. Like 1,2,3-triazoles, the key factors for the advance of denitrogenative transformation of 1,2,3,4-tetrazoles should be (i) the ring opening of 1,2,3,4-tetrazoles in the presence of metal catalyst followed by N_2 elimination and (ii) subsequent diverse transformations, enabling the rapid synthesis of various heterocycles in a single step. Accepting all these challenges, in 2018, a denitrogenative amination reaction of pyridotetrazoles was first reported by Chattopadhyay et al. [6].

In this study, the *N*-pyridyl metal nitrene was captured from pyridotetrazole using cp*Ir^{III} cation, enabling the access to several privileged scaffolds such as *aza*-carbazoles, *aza*-indoles, and other high-valued nitrogen heterocycles. This catalytic annulation reaction is known for its broad substrate scope and high tolerance for various functional groups. It's worth noting that electronically and sterically diverse tetrazoles are well-suited to the reaction conditions. Interestingly, in

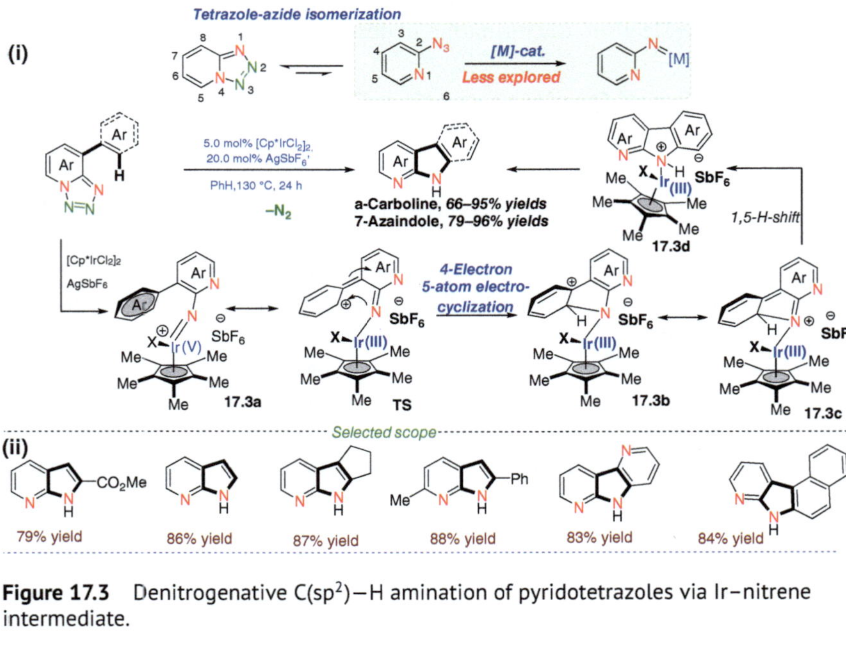

Figure 17.3 Denitrogenative C(sp^2)−H amination of pyridotetrazoles via Ir−nitrene intermediate.

addition to the aryl substituents at the C-8 position of tetrazole, substrates bearing an alkene group at the C-8 position also undergo productive reactions, leading to the formation of 7-azaindole derivatives.

It has been discovered that the Ir-catalyzed annulation reaction proceeds through an electrocyclization pathway. In the electrocyclization mechanism (depicted in Figure 17.3, i), the first step involves the formation of a sterically favored iridium nitrene intermediate (**17.3.a**) from tetrazole, facilitated by the in situ generated Cp*Ir^{III} cation. After that, favored TS participates in a 4-electron 5-atom electrocyclization to create an intermediate with a C—N bond (**17.3b**). Next, resonating structure (**17.3c**) of (**17.3b**) undergoes 1,5-H shift to form the intermediate (**17.3d**). Finally, the desired aminated product is generated from intermediate (**17.3d**), completing the catalytic cycle.

To support this electrocyclization pathway, a series of mechanistic studies were conducted (Figure 17.4a, b). For instance, amination with a nonconjugated alkene (Figure 17.4a, i), conjugated Z-alkene (Figure 17.4a, ii) as well as a kinetic isotope effect (KIE) experiment (intermolecular and intramolecular, Figure 17.4a, iii) which ruled out the possibility of the C—H activation and the insertion mechanism. Subsequently, a nonconjugated tetrazole (Figure 17.4a, iv) was used that resulted in 0% product (**17.4h**) conversion, indicating that continuous conjugation is prerequisite for this amination. Furthermore, in a competition experiment (Figure 17.4b) using *meta*- and *para*-substituted methoxy tetrazoles (**17.4i and17.4j**) in the same vessel, there was a 6% conversion of the *meta*-substituted tetrazole and a 38% conversion of the *para*-OMe tetrazole. This provided evidence for the participation of electrocyclization mechanism.

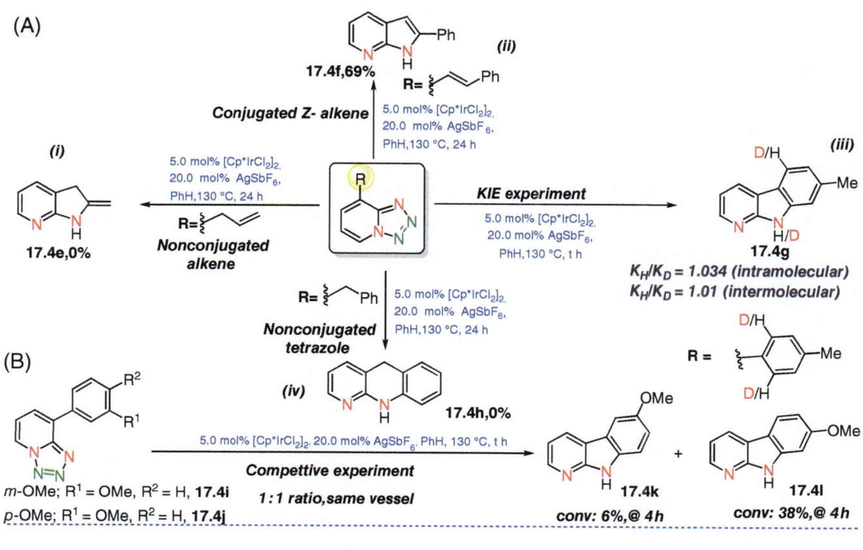

Figure 17.4 Mechanistic studies for supporting electrocyclization pathway. (a) Amination with conjugated and non-conjugated alkene. KIE experiment intermolecularly and intramolecularly. Amination with non-conjugated tetrazole. (b) Competitive experiment using *meta*- and *para*-substituted tetrazole.

17.4 Intermolecular Denitrogenative Annulation of Pyridotetrazoles via Mn-Nitrene Intermediate

As a part of metal nitrene studies, in 2021, Chattopadhyay et al. discovered an Mn-porphyrin-based catalytic system to achieve denitrogenative annulation reaction between tetrazole and nitrile [7]. It was observed that CuI is essential as a cocatalyst for the successful transformation.

Notably, the annulation reactions were performed at two different conditions (as mentioned in Figure 17.5), such as at 100 °C temperature in the presence of lower amount of catalyst loadings (5 mol%) and at 80 °C temperature using (20 mol%) catalyst loadings. In both the conditions, comparable yields can be obtained. For example, various aromatic nitriles and substituted tetrazoles were treated using two sets of reaction conditions (Figure 17.5, conditions A and conditions B) to get the desired 1,2,4-triazolo pyridine products in good to excellent yields.

To get a detailed understanding of the Mn-catalyzed reaction mechanism, a series of control experiments were conducted. While the complete reaction mechanism profile remains unclear, they conducted several mechanistic studies to gain insights. To demonstrate the formation of a Mn-nitrene intermediate in the reaction profile,

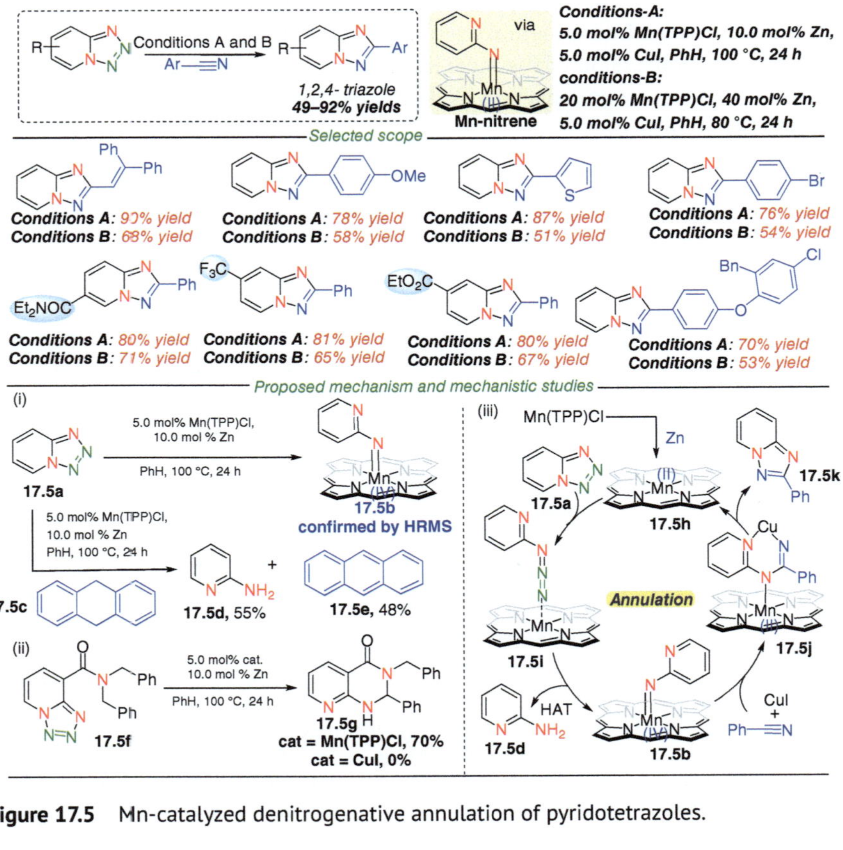

Figure 17.5 Mn-catalyzed denitrogenative annulation of pyridotetrazoles.

several important experiments were performed first, such as at first, an experiment was carried out between 1,2,3,4-tetrazole (**17.5a**) and 9,10-dihydroanthracene (**17.5c**) using the employed reaction conditions (Figure 17.5, i). Examination of the crude reaction mixture revealed 48% anthracene (**17.5e**) and 55% reduced 2-amino pyridine (**17.5d**), which supported the possibility of the formation of a Mn-nitrene intermediate (**17.5b**). Notably, the mass of corresponding Mn-nitrene intermediate (**17.5b**) was detected through high-resolution mass spectroscopy. To provide additional support for the generation of the Mn-nitrene intermediate, a C8-substituted tetrazole bearing a dibenzyl amide unit (**17.5f**) was designed (Figure 17.5, ii). It was hypothesized that if the Mn–nitrene is responsible for this annulation reaction, the corresponding C(sp^3)—H aminated product (**17.5g**) should be observed. Surprisingly, the corresponding aminated product (**17.5g**) in 70% yield was obtained. However, when the same experiment was conducted using CuI as a catalyst, it resulted in no reaction.

Based on the experimental evidence, a mechanism was proposed (Figure 17.5, iii). In the initial step, the in situ generated Mn^{II}TPP (**17.5h**) coordinates with the terminal nitrogen of the tetrazole, leading to the formation of an azide-bound Mn–porphyrin complex (**17.5i**). Subsequently, from complex (**17.5i**), nitrogen evolution occurs, resulting in the formation of manganese-nitrene intermediate (**17.5b**). In the presence of benzonitrile and CuI, this manganese-nitrene intermediate reacts to yield the amidine complex (**17.5j**). Finally, amidine complex undergoes cyclization to produce the annulated product (**17.5k**).

17.5 Denitrogenative Transformation of Pyridotetrazoles via Metalloradical Mechanism

17.5.1 Intermolecular Denitrogenative Annulation of Pyridotetrazoles

Over the last few decades, the concept of metalloradical activation for radical cyclization has attracted growing attention in the field of organic synthesis for its distinct reactivity compared to the traditional ionic reactions. The major advantages of radical reactions over the ionic reactions include (i) radical reaction proceeds with high reactivity, (ii) they are compatible under mild and neutral conditions with a broad spectrum of solvents, and (iii) they exhibit a broad range of functional group tolerance. The research groups of Zhang and de Bruin and Chattopadhyay have indeed made significant contributions to the field of synthetic organic chemistry by developing various transformations based on the Co(II)-metalloradical chemistry. These transformations include radical C—C bond formation, radical C—H amination, aziridation, and many more [8]. However, metalloradical activation of 1,2,3,4-tetrazole is a challenging task due to the extreme inertness of the tetrazole compounds.

Intermolecular denitrogenative annulation of 1,2,3,4-tetrazoles with terminal alkynes is quite challenging due to the competing 1,3-dipolar cycloaddition reaction (click reaction). Therefore, to perform a selective annulation of 1,2,3,4-tetrazoles with terminal alkynes overcoming the conventional click reaction, one would need to carefully design the reaction conditions and potentially employ a metal catalyst or other strategies to direct the reaction toward the desired denitrogenative annulation.

Accepting this challenge, in 2018, Chattopadhyay and co-workers discovered a Fe^{II}-based metalloradical catalytic system that successfully leads the reaction between pyridotetrazole and alkynes toward desired denitrogenative annulation (Figure 17.6). The developed catalytic system needs 5.0 mol% Fe(TPP)Cl, 10 mol% Zn-dust (reducing agent), PhH solvent, and heating to 100 °C [9]. With the help of optimized annulation conditions, the authors examined numerous substituted tetrazoles and terminal alkynes, enabling the synthesis of important imidazopyridines in good to excellent yields. To elucidate the formation of the annulated product, the authors proposed a radical activation mechanism. In the first step of this mechanism, Fe^{III}(TPP)Cl is reduced to the active catalyst Fe^{II}TPP (**17.6a**) in the presence of Zn-dust. Then, the active catalyst Fe^{II}TPP facilitates the generation of a nitrene radical from the corresponding tetrazole through nitrogen evolution. Afterward, nitrene radical intermediate (**17.6c**) is trapped by the terminal alkyne, resulting in the formation of the γ-Fe^{III}-vinyl radical intermediate (**17.6d**). Subsequently, through radical addition, a new β-Fe^{III}-alkyl radical intermediate (**17.6e**) is generated, which undergoes β-radical scission to break the N—Fe bond and form a C=N bond, ultimately yielding the product and completing the catalytic cycle. The presence of nitrene radical intermediate in the reaction medium was

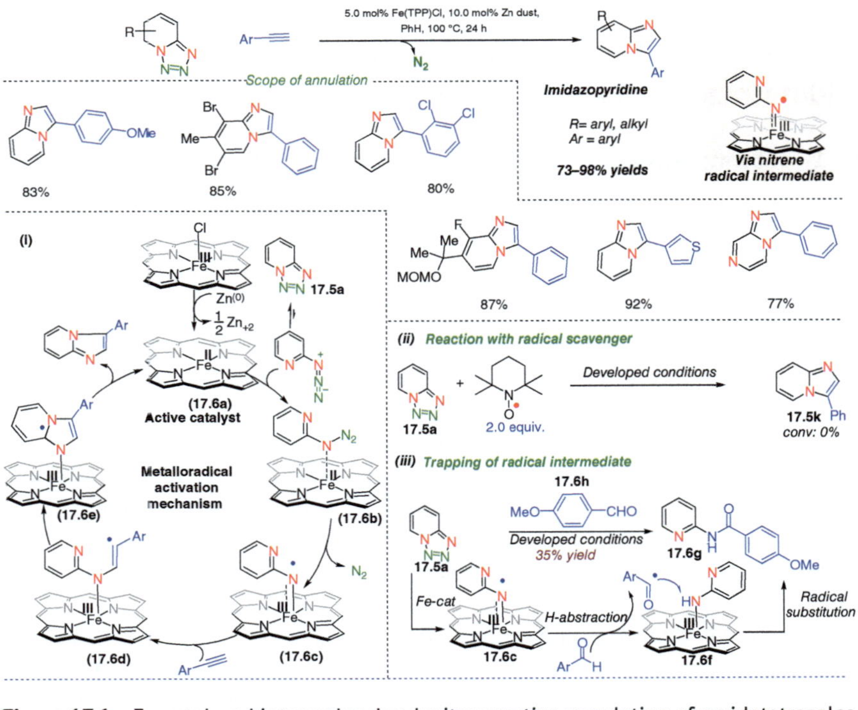

Figure 17.6 Fe-catalyzed intermolecular denitrogenative annulation of pyridotetrazoles with diverse terminal alkynes.

confirmed through various control experiments: (i) radical-scavenging experiment using TEMPO showed no reaction (Figure 17.6, ii). (ii) trapping of the radical intermediate using benzaldehyde (**17.6h**) resulted in the isolation of the trapped amide product (**17.6f**) in 35% yield, indicating the presence of the nitrene radical in the reaction medium (Figure 17.6, iii) and (iii) EPR spectra confirmed the presence of an organic radical with a *g*-value of 1.97 in the reaction medium.

17.5.2 Intramolecular Denitrogenative C(sp^3)—H Amination of Pyridotetrazoles

Compared to the well-established C(sp^2)—H bond aminations, unreactive aliphatic C(sp^3)—H bond amination presents a significant challenge. The primary obstacles in activating these inert C(sp^3)—H bonds are as follows: (i) high BDE (93–105 kcal/mol), (ii) the interaction between the metal d_π orbital (HOMO) and the high energy σ*orbital (LUMO) of the C(sp^3)—H bond is energetically less favorable due to the lack of π-assistance from the C(sp^3)—H bond, and (iii) selectivity issues. Bearing all these challenges in mind, in early 2020, Chattopadhyay et al. discovered an iron-based catalytic system capable of intramolecularly aminating primary, secondary, and tertiary C—H bonds to access a variety of complex nitrogen heterocycles with high levels of selectivity and functional group tolerance (Figure 17.7) [10].

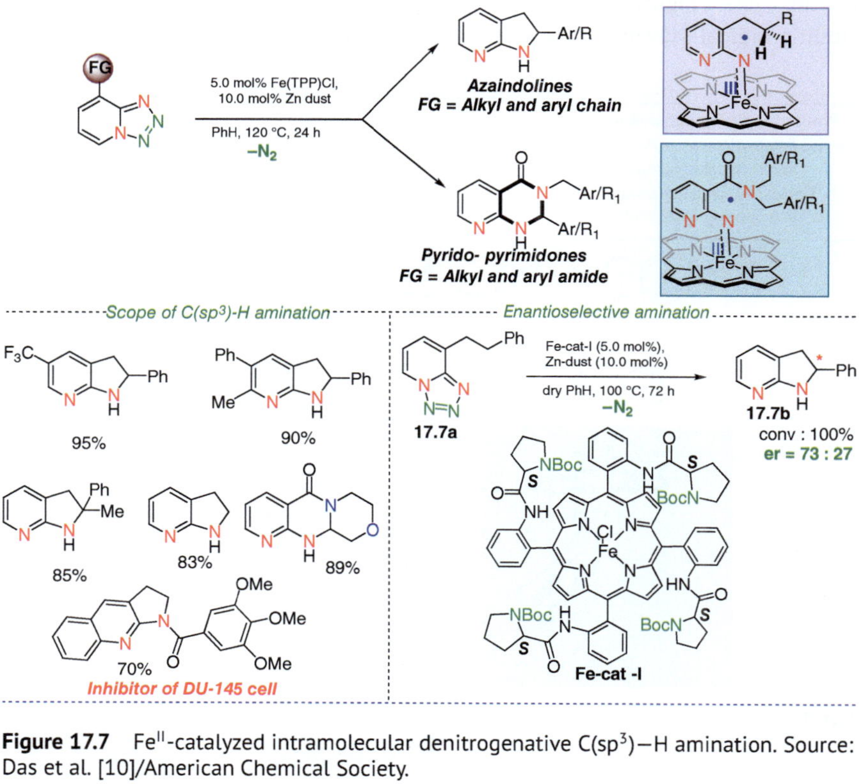

Figure 17.7 Fe^{II}-catalyzed intramolecular denitrogenative C(sp^3)—H amination. Source: Das et al. [10]/American Chemical Society.

This denitrogenative amination method offers access to two important N-heterocycles, 7-azaindoline, and pyrido-pyrimidones (Figure 17.7). These compounds are challenging to produce through conventional methods previously reported, primarily because of the presence of an extra nitrogen atom. Following systematic screening, the authors discovered that with minor adjustments to their previously established Fe-based reaction conditions, the desired C(sp^3)—H aminated products can be obtained quantitatively. The versatility of this amination reaction was found to be noteworthy. It allowed for the amination of various electronically diverse C—H bonds, as well as tetrazoles containing an amide group with different C—H bonds at the C8-position, enabling the synthesis of valuable N-heterocycles with varying ring sizes. Additionally, the authors explored the applicability of this method for the enantioselective C(sp^3)—H amination using a range of chiral catalysts. The investigations revealed that the use of **Fe-cat-I** (α, β, α, β) led to the quantitative formation of the aminated product (**17.7b**) with a 73 : 27 enantiomeric ratio, suggesting the potential for enhancing enantioselectivity through the design of appropriate chiral catalysts.

Detailed experimental findings revealed that this catalytic amination operates through a metalloradical activation mechanism (Figure 17.8). In the initial stage of this mechanism, intramolecular H-atom abstraction takes place via the nitrene radical (**17.8e**). Subsequent radical substitution from the relevant intermediate (**17.8f**) leads to the formation of the desired aminated product, completing the catalytic cycle (Figure 17.8). Notably, a series of mechanistic investigations, including radical-scavenging experiments, intermolecular and intramolecular KIE

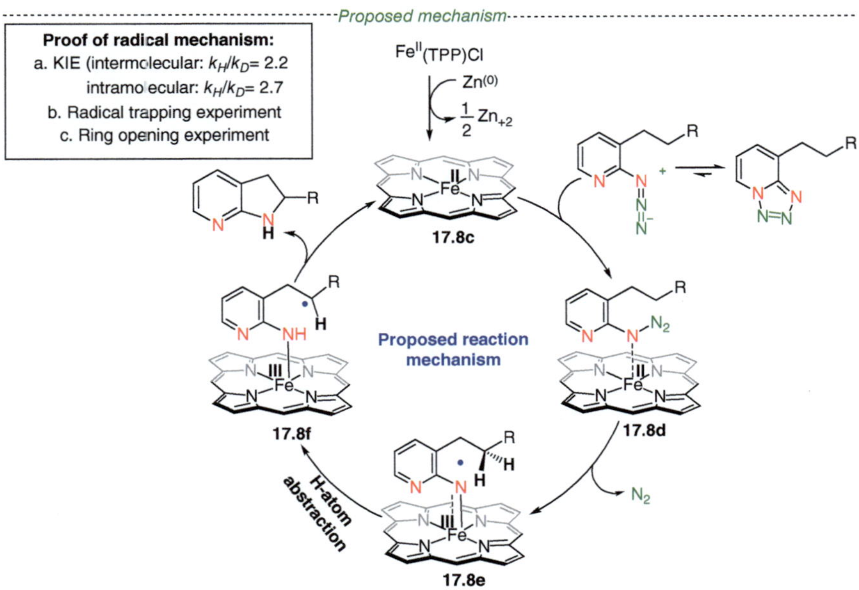

Figure 17.8 Mechanism of intramolecular denitrogenative C(sp^3)—H amination.

measurements, as well as ring-opening experiments, provided valuable insights. These studies suggested that the C—H amination likely proceeds through a radical mechanism, and it is probable that C—H bond cleavage occurs during the turnover-limiting step.

17.5.3 Iron-Catalyzed Intramolecular Denitrogenative Radical Rearrangement of Pyridotetrazoles

In 2021, Chattopadhyay and co-worker further extended the scope of the denitrogenative transformation reaction in the direction of the denitrogenative rearrangement over the C(sp^3)—H bond amination (Figure 17.9) [11]. The authors reported that while the rearrangement chemistry of diazo precursors and organic azides had been extensively explored previously by Murakami [12], Fokin [13], Driver [14], and Shi [15] using Rh-catalysis, the rearrangement of 1,2,3,4-tetrazoles

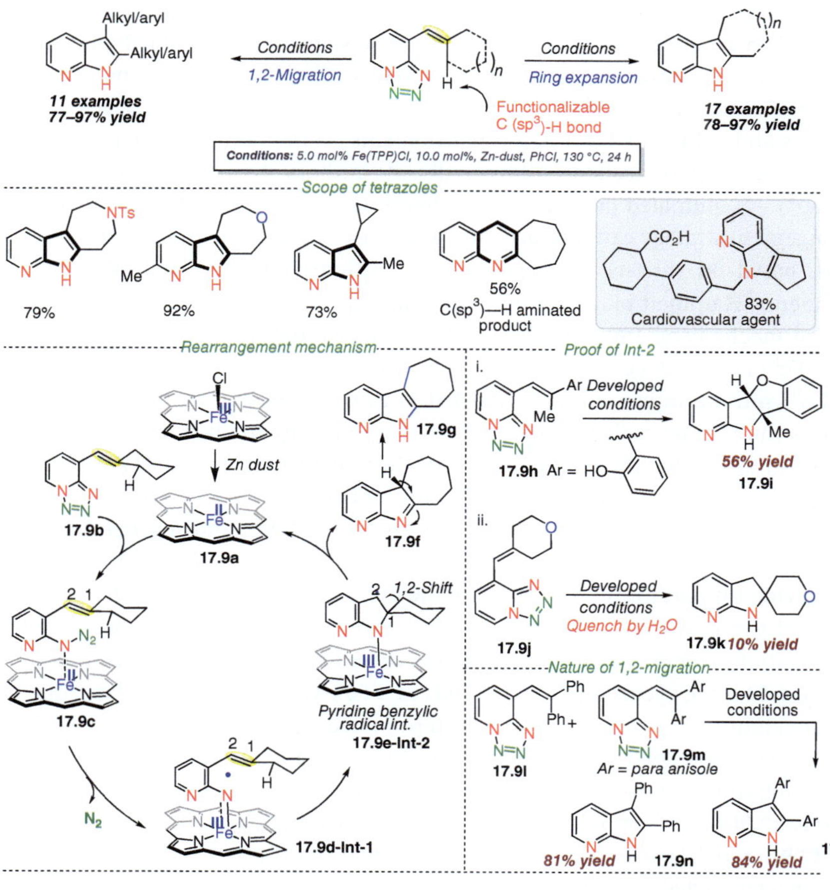

Figure 17.9 Fe^{II}-catalyzed intramolecular rearrangement over C(sp^3)—H amination.

in the presence of base metal complexes remained uncharted territory. In their quest to devise an appropriate method, the authors devised a substrate featuring a disubstituted alkene, as well as a functionalizable C(sp^3)—H bond (e.g., methylene cyclohexane) substituted at the C8 position of the tetrazole. Within this substrate, the potential for molecular rearrangement exists through either a ring expansion or a 1,2-migration.

To facilitate the rearrangement over the competitive C(sp^3)—H bond amination, Chattopadhyay et al. devised a strategy that consisted of the following elements: (i) tetrazole should have a flexible methylene cycloalkane (like cyclohexane) unit; (ii) the cyclohexane segment should maintain a stable chair conformation to ensure that the functionalizable C(sp^3)—H bonds remain sufficiently distant from the reactive radical center, making them inaccessible for C(sp^3)—H amination; and (iii) due to geometric constraints, the pendant C=C bond should be in close proximity to the reactive radical center. After the reaction, this arrangement leads to the formation of rearrangement products rather than C(sp^3)—H aminated products.

To ensure the success of their designed plan, the authors conducted experiments under various conditions and ultimately discovered that employing 5.0 mol% of Fe(TPP)Cl in conjunction with a 10 mol% Zn-dust reductant in PhCl solvent at 130 °C yielded the desired rearrangement product. The developed method proved to be delightfully controllable, exhibiting compatibility with a broad spectrum of substrates and delivering both ring expansion and 1,2-migration products in yields ranging from good to excellent. Additionally, the authors observed the formation of the C(sp^3)—H aminated product 1,8-napthyridine, yielding 56%. Remarkably, this rearrangement strategy can produce one cardiovascular agent with 83% yield.

Next, based on various mechanistic investigations, including radical-trapping experiments, Hammett plots, and ring-opening experiments, the authors proposed a radical mechanism. Initially, Fe^{III}(TPP)Cl is converted into the active catalytic species Fe^{II}(TPP)(**17.9a**). Then, the metal-bound tetrazole complex (**17.9c**) delivers an α-Fe^{III} nitrene intermediate (**17.9d-Int-I**) through N_2 elimination. Following this step, the nitrene radical (**17.9d-Int-1**) is captured by the pendent C=C bond, resulting in the formation of a pyridine benzylic radical intermediate (**17.9e-Int-2**). Subsequently, a 1,2-migration, leading to intermediate (**17.9f**), is succeeded by a 1,3-H shift, ultimately yielding the desired product (**17.9g**), and regenerating the catalyst.

To provide support for the formation of the pyridine benzylic radical, the authors conducted two crucial experiments. The first experiment involved a cascade radical cyclization, where H-atom abstraction occurred from an *ortho*-substituted OH group, yielding the cyclized product (**17.9i**) in a 56% yield. The second experiment was a stoichiometric reaction, which was terminated with water, leading to the formation of a spirocyclized product (**17.9k**) in a 10% yield. Finally, to ascertain whether the 1,2-migration step is concerted or occurs stepwise, a crossover experiment with substrate **17.9l** and **17.9m** was performed that resulted in two distinct types of products (**17.9n** and **17.o**). This experiment confirmed that the 1,2-migration process occurs in a concerted fashion.

17.5.4 Iron-Catalyzed Intramolecular Denitrogenative Radical $C(sp^2)$—H Amination of Pyridotetrazoles

Prompted by the findings in the Fe-porphyrin-catalyzed $C(sp^3)$—H bond amination study [10], Chattopadhyay and co-worker next explored the potential of this catalytic system to activate $C(sp^2)$—H bonds. It was envisioned that if successful, this would mark the inaugural instance of metalloporphyrin-catalyzed intramolecular $C(sp^2)$—H amination of 1,2,3,4-tetrazole. Moreover, this approach holds several advantages over the previously established Ir-catalyzed reaction, including cost-effectiveness, nontoxicity, and biocompatibility of the catalyst. Furthermore, it would involve a distinctly different mechanistic pathway compared to the Ir-catalyzed electrocyclization mechanism.

Thus, driven by this anticipation, the authors introduced the concept of a Fe-catalyzed radical activation of 1,2,3,4-tetrazole and its application in the catalytic intramolecular denitrogenative annulation reactions. This study demonstrates that the annulation can occur through $C(sp^2)$—H amination when the aryl group lacks functionalizable $C(sp^3)$—H bonds. However, remarkably, the reaction proceeds through $C(sp^3)$—H amination with exceptional regioselectivity despite the presence of functionalizable $C(sp^2)$—H bonds (see Figure 17.10). In summary, a single catalytic system exhibits a dual role in promoting $C(sp^2)$—H and $C(sp^3)$—H amination reactions with remarkable selectivity, leading to the synthesis of various valuable N-heterocyclic scaffolds in high yields [16].

With the help of established reaction conditions, the authors made the choice to demonstrate the practical applicability of this amination reaction. As a result, they linked numerous pharmaceutical and medically significant compounds containing aromatic bromide or phenolic groups by initially converting them into their corresponding triflate derivatives for coupling with tetrazoles. Subsequently, they conducted the denitrogenative amination reaction (see Figure 17.10). Their substrate scope revealed the remarkable compatibility of a wide array of medically important and complex molecules with the reaction conditions they have developed. Tetrazoles bearing $C(sp^3)$—H bond present in the *ortho* position of arene ring along with a $C(sp^2)$—H bond yielded only $C(sp^3)$—H aminated product with excellent yield.

Mechanistic investigations, encompassing both experimental and computational approaches, revealed that both the $C(sp^2)$—H and $C(sp^3)$—H amination reactions share a common metalloradical activation mechanism. In the $C(sp^2)$—H mechanism, the open form of 1,2,3,4-tetrazole (**17.10b**) initially forms a weak coordination with the active catalyst, resulting in the formation of a catalyst-substrate complex denoted as (**17.10c**). Subsequent elimination of N_2 from intermediate (**17.10c**) leads to the formation of the Fe^{III}-nitrene intermediate (**17.10d**). In the following step, a radical addition to the adjacent aryl C=C bond forms the desired C—N bond, concomitantly reducing Fe^{III} to Fe^{II}, thereby generating intermediate (**17.10e**). It is likely that the intermediate (**17.10e**) involves weak coordination of the hydropyridoindole to the Fe center of the catalyst. In the subsequent step, the product precursor dissociates from the active catalyst prior to a 1,5-H shift, resulting in the formation of the final product (**17.10g**).

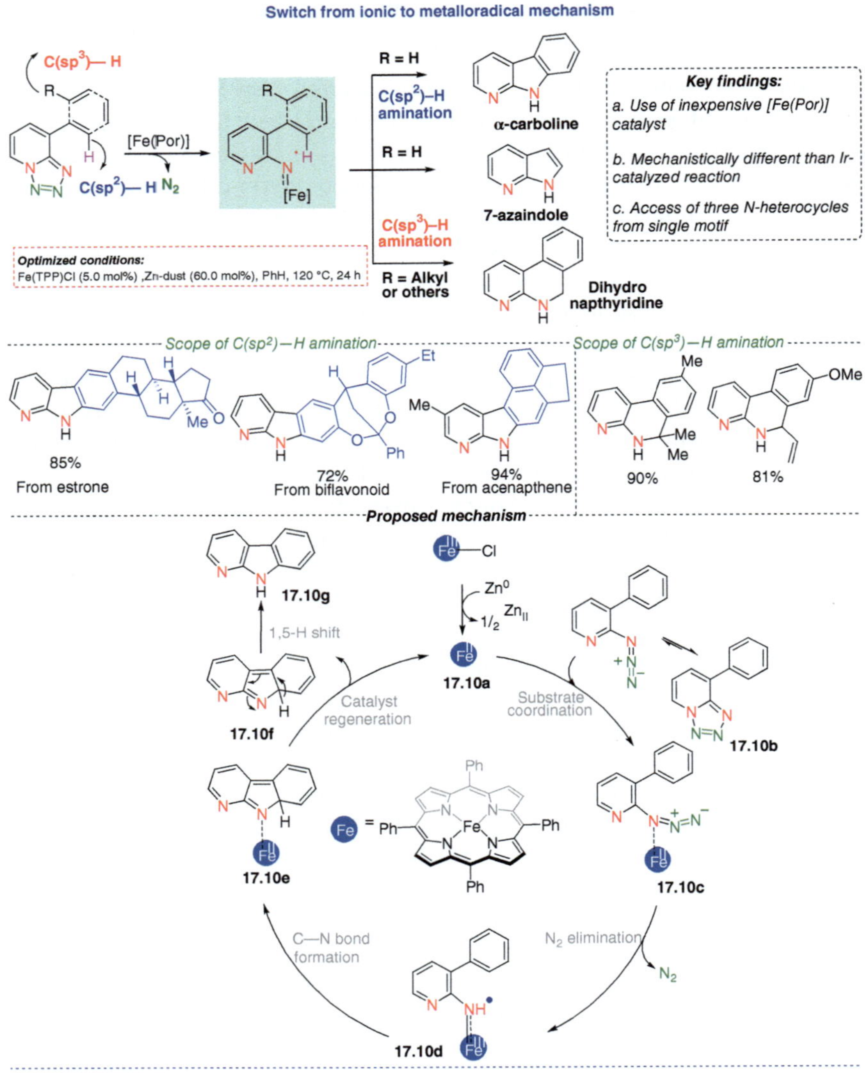

Figure 17.10 Intramolecular radical C(sp^2)—H amination. Source: Das et al. [10]/American Chemical Society.

Markedly, the most favorable pathway for C(sp^2)—H amination was found to proceed through a triplet spin state, whereas the quintet spin state was preferred for the competitive C(sp^3)—H amination. The transition between the triplet and quintet spin states, known as the intersystem crossing (ISC), was identified as more likely to occur before the formation of the nitrene radical intermediate in the case of benzylic C(sp^3)—H amination. Application of the energetic span model to the Gibbs free energy profile indicated that the formation of the nitrene radical constituted the rate-determining step of the catalytic cycle, with C(sp^3)—H amination being favored by 4.1 kcal/mol over C(sp^2)—H amination.

17.5.5 Iron-Catalyzed Intermolecular Benzylic C(sp^3)—H Amination

In a recent report, Chattopadhyay et al. demonstrated the effectiveness of tetrazole as an excellent nitrene precursor for both the intramolecular C(sp^2)—H amination through iridium-catalyzed electrocyclization [6] and C(sp^3)—H amination via a radical mechanism with iron catalysis [10]. Inspired by these breakthroughs, the authors introduced a new method for the intermolecular amination reactions using tetrazole as the nitrene precursor, with the goal of synthesizing benzylamines bearing 2-pyridyl substituents. In their quest to identify suitable reaction conditions for the C(sp^3)—H amination, the authors initiated their investigations using basic feedstock ethyl benzene and 5-amide tetrazole, applying previously established protocols. Unfortunately, their initial catalyst systems, which had proven effective for the intramolecular C(sp^3)—H amination, proved ineffective for achieving intermolecular C(sp^3)—H amination. This difficulty may be attributed to the inherent challenges associated with the intermolecular C(sp^3)—H bond amination. However, after conducting an extensive screening of various catalytic systems, they discovered that an iron porphyrin catalytic system, featuring both electron-donating mesityl and 2,6-dimethoxyphenyl units, successfully provided the anticipated benzylic amination product in high yield [17].

The authors highlighted several novel findings in their work: (i) while existing literature reports for the intermolecular amination at the more challenging secondary C—H bonds of gaseous hydrocarbons using base-metal catalyst systems like cobalt or nickel, intermolecular benzylic C(sp^3)—H amination using 1,2,3,4-tetrazole as a nitrene precursor via iron catalysis remains an underexplored area. Moreover, this work distinguishes itself from many other amination reactions where the nitrogen atom is typically protected and requires removal for further product development; (ii) this method selectively aminates 2^0 benzylic C(sp^3)—H bond over the weaker 3° and sterically accessible 1° benzylic C(sp^3)—H bonds; (iii) the direct incorporation of 2-aminopyridine into benzylic and heterobenzylic positions is significant, given that this motif is prevalent in many bioactive compounds; (iv) the reaction conditions are straightforward, necessitating only 5 mol% of the widely available iron catalyst; and (v) the versatility of this reaction is extensive with regards to benzylic substrate and tetrazole precursor. Under the optimal reaction conditions, the authors observed that several electronically diverse benzylic C(sp^3)—H bonds underwent intermolecular amination, resulting in high yields of the aminated product. The extensive scope of substituted and heterocyclic tetrazoles tested in this amination reaction under the developed conditions proved to be remarkable. Chattopadhyay group also performed several reactions to underscore the practical applications of their intermolecular amination method: (i) they efficiently synthesized a potential drug molecule, Dapoxetine (**17.11e**), known for inhibiting serotonin transport, from a simple starting material (**17.11c**), achieving good yields; (ii) a large-scale synthesis (3.5 mmol) with an impressive yield (71%) was also performed. The authors also illustrated the feasibility of removing the pyridyl group from the aminated product (**17.11a**) through a well-established procedure, resulting in the formation of the secondary amine **17.11b**.

To gain a comprehensive understanding of the C(sp^3)—H amination reaction, Chattopadhyay and co-worker have advanced a radical mechanism, which gains substantial support from a series of pertinent mechanistic investigations: (i) KIE experiments, both in intermolecular and intramolecular fashion, (ii) radical trapping experiments employing TEMPO, and (iii) Hammett plot analysis. Relying on these experimental findings, the authors have advanced the notion of a radical mechanism, as elucidated in Figure 17.11. Furthermore, based on the KIE values obtained in the amination reaction, Chattopadhyay et al. have posited that the rate-limiting step for intermolecular C(sp^3)—H benzylic amination likely involves the cleavage of the C—H bond.

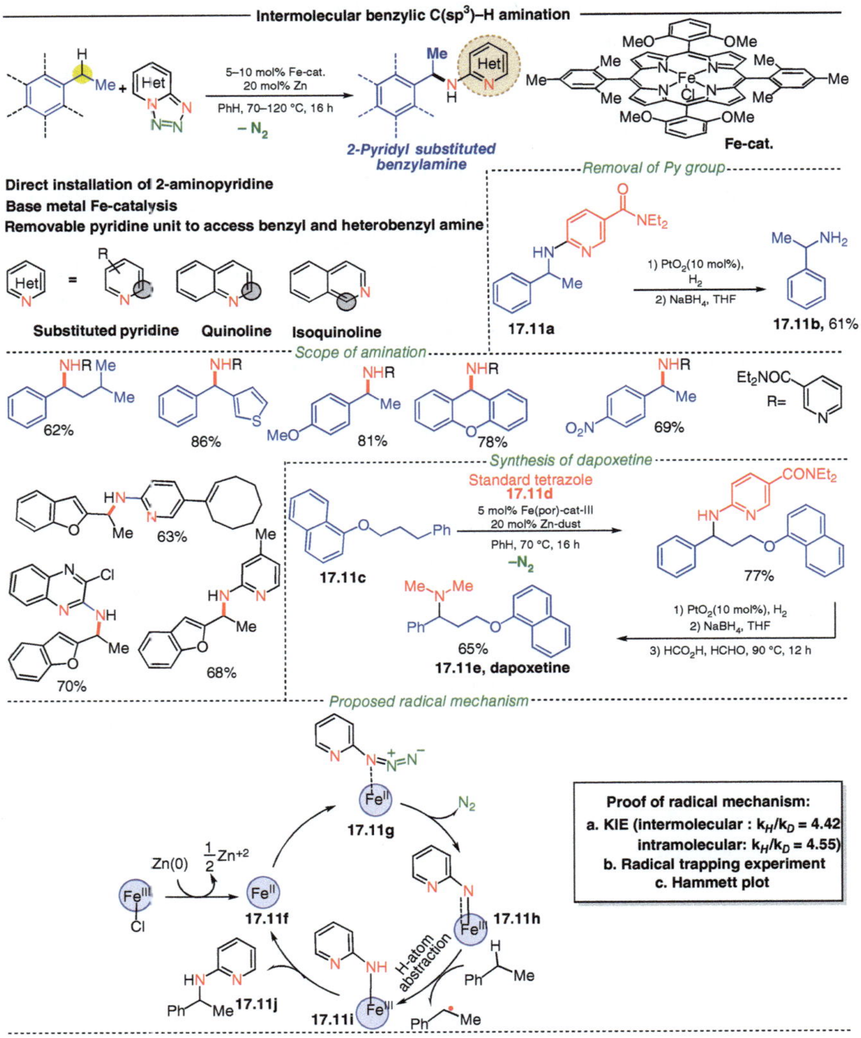

Figure 17.11 Intermolecular radical C(sp^3)−H amination.

17.5.6 Iron-Catalyzed Intermolecular Denitrogenative C—N Cross-Coupling Reactions

In a recent publication, Chattopadhyay and co-worker have detailed a novel approach for facilitating intermolecular C—N cross-coupling reactions between a wide range of tetrazoles and diverse boronic acids (Figure 17.12) [18]. The study highlights that this method operates through a radical activation mechanism previously unexplored in traditional C—N cross-coupling reactions. This C—N bond formation reaction is supported by a combination of experimental studies and density functional theory calculations. The authors demonstrated the versatility of

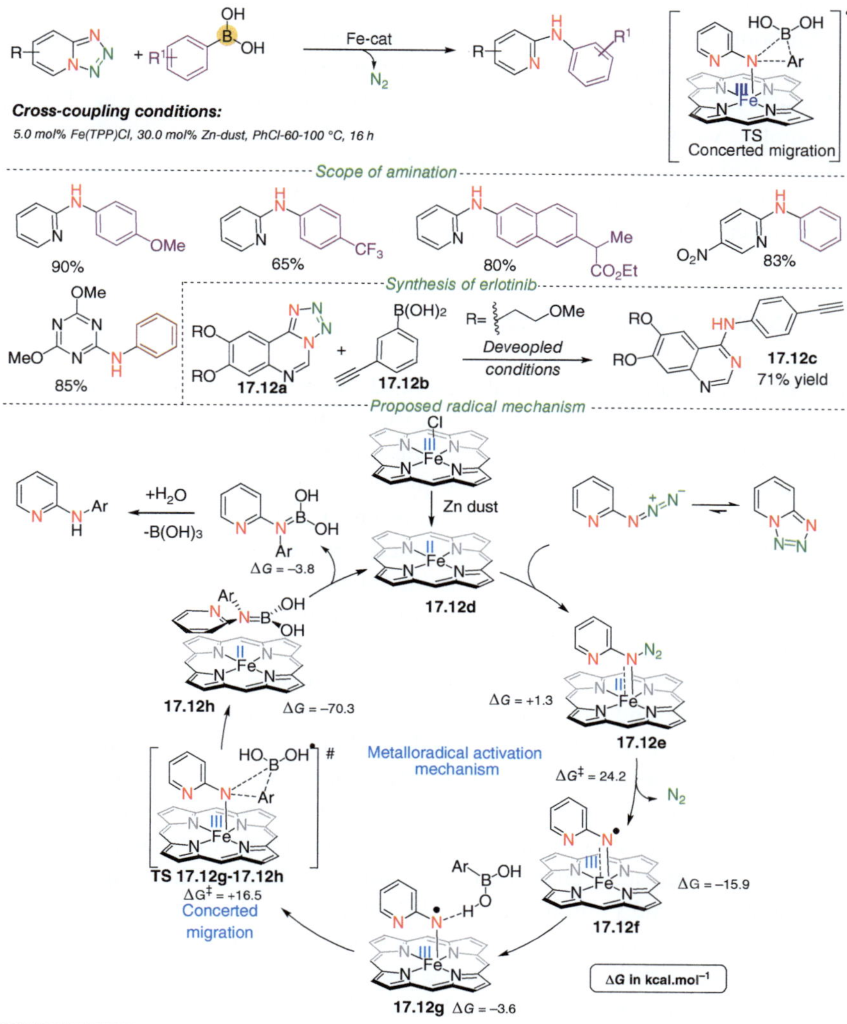

Figure 17.12 Intermolecular radical C—N cross-coupling reaction. Source: Das et al. [18]/American Chemical Society.

their C—N cross-coupling reaction by successfully utilizing a wide range of tetrazoles and a variety of boronic acids. In addition to tetrazoles, numerous aromatic and aliphatic azides also displayed remarkable efficiency in this amination process. The method has been applied to several late-stage C—N cross-coupling reactions involving bioactive scaffolds, as well as enabling a concise synthesis of a drug candidate Erlotinib **17.12c** from corresponding tetrazole (**17.12a**) and boronic acid (**17.12b**).

In the mechanism, the authors disclosed that the reaction of the Fe^{III}-nitrene with phenylboronic acid starts with the formation of a hydrogen bond between the OH group on boron and the lone pair of the nitrene (**17.12g**). Progressing further, the reaction passes through a transition state, known as **3TS 17.12g–17.12h** with an activation barrier of $\Delta G\ddagger = 16.5\,\text{kcal mol}^{-1}$ upon formation of the boron nitrogen bond and concerted migration of the boron aryl group moving toward the nitrene. Subsequently, the formation of the B=N double bond is exergonic ($\Delta G° = -70.3\,\text{kcal mol}^{-1}$), resulting in the formation of a complex (**17.12h**). Within this complex, weak π–π interactions are observed between [Fe(II)Por] and the boramidic acid (B=N = 1.470 Å). The complex is subsequently hydrolyzed to yield the desired reaction product. Simultaneously, [Fe(II)Por] (represented as **17.12d**) reenters the catalytic cycle for further reactions.

17.6 Summary and Outlook

In this chapter, Chattopadhyay et al. described the remarkable reactivity of pyridotetrazole as a nitrene precursor utilizing various metal catalysts to swiftly build a wide array of nitrogen-containing scaffolds. Since pyridotetrazole has one vital pyridyl unit, one can simply access the pyridine-bearing biologically significant heterocyclic cores in one step without compromising efficiency. With the help of metal catalysts, the authors studied the distinctive reactivity of pyridotetrazoles, for example, (i) intramolecular $C(sp^2)$—H amination through Ir–nitrene intermediate, (ii) intermolecular annulation by Fe-based radical mechanism, (iii) radical denitrogenative $C(sp^3)$—H amination (inter and intra), (iv) Mn-based dual reactivity of tetrazole, (iv) iron-catalyzed rearrangement reaction over more feasible $C(sp^3)$—H amination reaction, and (v) iron-catalyzed intermolecular denitrogenative C—N cross-coupling reactions. These developed catalytic systems enable access to a broad spectrum of bioactive and synthetically valuable nitrogen-holding scaffolds. However, some challenges persist to stimulate future research directions: (i) insight into effective catalytic systems with high regio-, chemoselectivity, and wide substrate scope, (ii) enhancing enantioselective amination through the design of more resilient and eco-friendly chiral catalysts, (iii) exploring application in the synthesis and late-stage functionalization of natural and biologically active compounds, and (iv) devising industrially robust catalytic systems capable of achieving suitably high turnover numbers. The authors believe that this book chapter will offer valuable insights into the forthcoming growth of denitrogenative reactions and will serve as a catalyst for further advancements in this expanding and consequential field.

References

1 (a) Das, S.K., Roy, S., and Chattopadhyay, B. (2023). Transition-metal-catalyzed denitrogenative annulation to access high-valued N-heterocycles. *Angew. Chem. Int. Ed.* 62: e202210912. (b) Akter, M., Rupa, K., and Anbarasan, P. (2022). 1,2,3-Triazole and its analogues: new surrogates for diazo compounds. *Chem. Rev.* 122: 13108–13205. (c) Roy, S., Das, S.K., Khatua, H. et al. (2021). Road map for the construction of high-valued N-heterocycles via denitrogenative annulation. *Acc. Chem. Res.* 54: 4395–4409. (d) Chattopadhyay, B. and Gevorgyan, V. (2012). Transition-metal-catalyzed denitrogenative transannulation: converting triazoles into other heterocyclic systems. *Angew. Chem. Int. Ed.* 51: 862–872.

2 (a) von Fraunberg, K. and Huisgen, R. (1969). Kupferkatalysierte zerfallsreaktionen von imidaziden; reaktionen von kupfer-azenen. *Tetrahedron Lett.* 10: 2599–2602. (b) Huisgen, R., von Fraunberg, K., and Sturm, H.J. (1969). Cycloadditionen von pyridyl- und pyrimidyl-aziden. *Tetrahedron Lett.* 10: 2589–2594. (c) Huisgen, R. and von Fraunberg, K. (1969). Zur thermolyse des 2-azido-4.6-dimethyl-pyrimidins; reaktionen des azens. *Tetrahedron Lett.* 10: 2595–2598.

3 Harder, R. and Wentrup, C. (1976). Pyridylnitrenes. Mechanism of ring contraction to cyanopyrroles. *J. Am. Chem. Soc.* 98: 1259–1260.

4 Nakamuro, T., Hagiwara, K., Miura, T., and Murakami, M. (2018). Enantioselective denitrogenative annulation of 1*H*-tetrazoles with styrenes catalyzed by rhodium. *Angew. Chem. Int. Ed.* 57: 5497–5500.

5 (a) Boyer, J.H. and Miller, E.J. (1959). The tetrazole-azidoazomethine equilibrium. *J. Am. Chem. Soc.* 81: 4671–4673. (b) Chattopadhyay, B., Rivera-Vera, C.I., Chuprakov, S., and Gevorgyan, V. (2010). Fused tetrazoles as azide surrogates in click reaction: efficient synthesis of N-heterocycle-substituted 1,2,3-triazoles. *Org. Lett.* 12: 2166–2169.

6 Das, S.K., Roy, S., Khatua, H., and Chattopadhyay, B. (2018). Ir-catalyzed intramolecular transannulation/C(sp^2)—H amination of 1,2,3,4-tetrazoles by electrocyclization. *J. Am. Chem. Soc.* 140: 8429–8433.

7 Khatua, H., Das, S.K., Roy, S., and Chattopadhyay, B. (2021). Dual reactivity of 1,2,3,4-tetrazole: manganese-catalyzed click reaction and denitrogenative annulation. *Angew. Chem. Int. Ed.* 60: 304–312.

8 (a) Lu, H., Jiang, H., Hu, Y. et al. (2011). Chemoselective intramolecular allylic C—H amination versus C=C aziridination through Co(II)-based metalloradical catalysis. *Chem. Sci.* 2: 2361–2366. (b) Kuijpers, P.F., Tiekink, M.J., Breukelaar, W.B. et al. (2017). Cobalt-porphyrin-catalysed intramolecular ring-closing C—H amination of aliphatic azides: a nitrene-radical approach to saturated heterocycles. *Chem. Eur. J.* 23: 7945–7952. (c) Jiang, H., Lang, K., Lu, H. et al. (2016). Intramolecular radical aziridination of allylic sulfamoyl azides by cobalt(II)-based metalloradical catalysis: effective construction of strained heterobicyclic structures. *Angew. Chem. Int. Ed.* 55: 11604–11608. (d) Lu, H.-J. and Zhang, X.P. (2011). Catalytic C—H functionalization by metalloporphyrins: recent developments and future directions. *Chem. Soc. Rev.* 40: 1899–1909.

(e) Wang, Y., Wen, X., Cui, X. et al. (2017). Asymmetric radical cyclopropanation of alkenes with in situ-generated donor-substituted diazo reagents via Co(II)-based metalloradical catalysis. *J. Am. Chem. Soc.* 139: 1049–1052. (f) Lang, K., Torker, S., Wojtas, L., and Zhang, X.P. (2019). Asymmetric induction and enantiodivergence in catalytic radical C—H amination via enantiodifferentiative H-atom abstraction and stereoretentive radical substitution. *J. Am. Chem. Soc.* 141: 12388–12396. (g) Zhou, M., Lankelma, M., van der Vlugt, J.I., and de Bruin, B. (2020). Catalytic synthesis of 8-membered ring compounds via cobalt(III)-carbene radicals. *Angew. Chem. Int. Ed.* 59: 11073–11079. (h) Li, C., Lang, K., Lu, H. et al. (2018). Catalytic radical process for enantioselective amination of C(sp^3)—H bonds. *Angew. Chem. Int. Ed.* 57: 16837–16841. (i) Roy, S., Das, S.K., and Chattopadhyay, B. (2018). Cobalt(II)-based metalloradical activation of 2-(diazomethyl)pyridines for radical transannulation and cyclopropanation. *Angew. Chem. Int. Ed.* 57: 2238–2243.

9 Roy, S., Khatua, H., Das, S.K., and Chattopadhyay, B. (2019). Iron(II)-based metalloradical activation: switch from traditional click chemistry to denitrogenative annulation. *Angew. Chem. Int. Ed.* 58: 11439–11443.

10 Das, S.K., Roy, S., Khatua, H., and Chattopadhyay, B. (2020). Iron-catalyzed amination of strong aliphatic C(sp^3)—H bonds. *J. Am. Chem. Soc.* 142: 16211–16217.

11 Roy, S., Das, S.K., Khatua, H. et al. (2021). Iron-catalyzed radical activation mechanism for denitrogenative rearrangement over C(sp^3)—H amination. *Angew. Chem. Int. Ed.* 60: 8772–8780.

12 Miura, T., Funakoshi, Y., Morimoto, M. et al. (2012). Synthesis of enaminones by rhodium-catalyzed denitrogenative rearrangement of 1-(*N*-sulfonyl-1,2,3-triazol-4-yl)alkanols. *J. Am. Chem. Soc.* 134: 17440–17443.

13 Selander, N., Worrell, B.T., and Fokin, V.V. (2012). Ring expansion and rearrangements of rhodium(II) azavinyl carbenes. *Angew. Chem. Int. Ed.* 51: 13054–13057.

14 Sun, K., Liu, S., Bec, P.M., and Driver, T.G. (2011). Rhodium-catalyzed synthesis of 2,3-disubstituted indoles from β,β-disubstituted stryryl azides. *Angew. Chem. Int. Ed.* 50: 1702–1706.

15 Chen, K., Zhu, Z.-Z., Zhang, Y.-S. et al. (2014). Rhodium(II)-catalyzed intramolecular cycloisomerizations of methylenecyclopropanes with *N*-sulfonyl 1,2,3-triazoles. *Angew. Chem. Int. Ed.* 53: 6645–6649.

16 Das, S.K., Das, S., Ghosh, S. et al. (2022). An iron(II)-based metalloradical system for intramolecular amination of C(sp^2)—H and C(sp^3)—H bonds: synthetic applications and mechanistic studies. *Chem. Sci.* 13: 11817–11828.

17 Khatua, H., Das, S., Patra, S. et al. (2022). Iron-catalyzed intermolecular amination of benzylic C(sp^3)—H bonds. *J. Am. Chem. Soc.* 144: 21858–21866.

18 Das, S., Ehlers, A.W., Patra, S. et al. (2023). Iron-catalyzed intermolecular C—N cross-coupling reactions via radical activation mechanism. *J. Am. Chem. Soc.* 145: 14599–14607.

18

Radical Denitrogenative Transformations of Polynitrogen Heterocycles

Sanjeev Kumar and Vinaykumar Kanchupalli

Department of Chemical Sciences, National Institute of Pharmaceutical Education and Research (NIPER), Hyderabad, Telangana India

18.1 Introduction

In organic synthesis, free radicals have significance importance, due to high chemical reactivity and unique stabilities [1]. Hence, chemical synthesis based on free radical reactions has always been of great interest, and recent eras have witnessed the exploration and expansion of the pool of radical-based chemistry throughout the evolution of modern organic synthesis [2]. Although numerous methods exist for free radical reactions, there is a need for novel, competent, and efficient strategies from readily available precursors that are always of central importance.

On the other hand, denitrogenative transformations have been one of the most attractive strategies for forming several synthetically eminent chemical structures, and they are present in various pharmaceuticals, fine chemicals, natural products, etc. [3]. Among them, poly-nitrogen heterocycles-derived denitrogenative protocols have received significant interest due to their broader scope of the formation of diverse C—C, C—X (X = N, O, S, etc.) bonds and construct of high-valued nitrogen heterocycles [4]. The success of these exceptional reactions includes easily accessible starting materials, highly efficient, step-economical, and the only byproduct is environmentally benign nitrogen gas. For the first time, in 1969, Huisgen and von Fraunberg disclosed transannulation of 1,2,3,4-tetrazole with benzonitrile via a denitrogenative annulation pathway at higher temperatures (Scheme 18.1ai) [5]. Further, Wentrup has demonstrated a flash vacuum pyrolysis of tetrazoles to build the corresponding heterocycles (Scheme 18.1aii) [6]. Unfortunately, this chemistry has never been in considerable attention till 2007. After a long time, Geovrgyan and Fokin groups showed the utility of denitrogenative transformations of pyridotriazoles and triazoles to the synthesis of diverse nitrogen heterocycles under mild and catalytic conditions [7]. Since then, this chemistry has become an attractive field; simultaneously, various poly-nitrogen heterocycles were introduced as a starting material for denitrogenative transformations, which has shown a great impact in modern organic

Denitrogenative Transformation of Nitrogen Heterocycles: Synthesis, Reactions and Applications, First Edition. Edited by Pazhamalai Anbarasan.

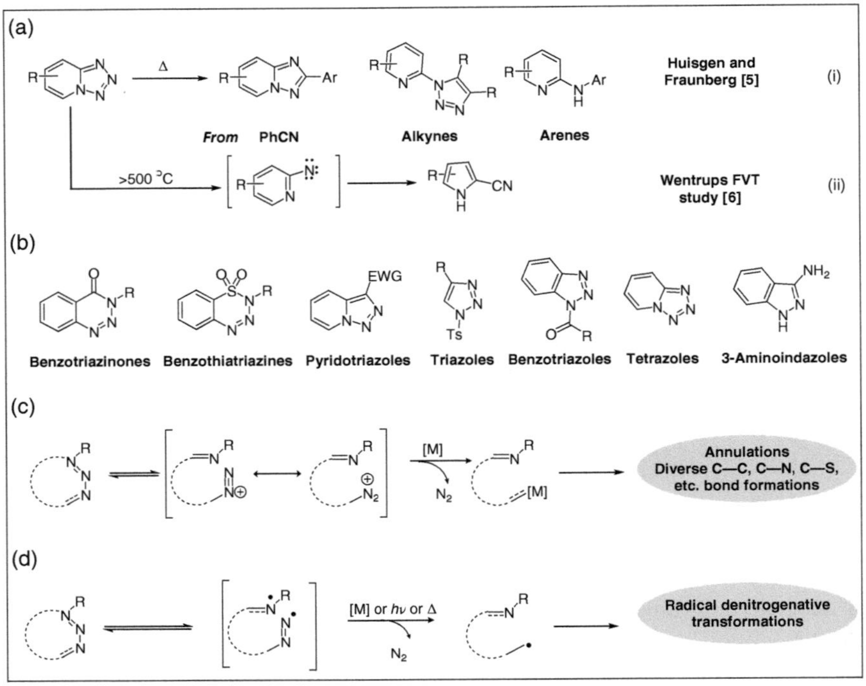

Scheme 18.1 Denitrogenative transformations of poly-nitrogen heterocycles. (a) Classical ring opening reactions of poly-nitrogen heterocycles. (b) Diverse precursors for denitrogenative reactions. Sources: (a) Huigen and Fraunberg [5]/with permission of Elsevier. (b) Wentrups [6]/American Chemical Society. (c) Denitrogenative transformations via ionic mechanism: well established. (d) Denitrogenative transformations via radical mechanism: underdeveloped.

chemistry (Scheme 18.1b). Since their discovery, most of these reactions proceed through an ionic mechanistic pathway, such as ring opening of poly-nitrogen scaffold followed by nitrogen extrusion to form carbene species/metal-carbenoid intermediate, and subsequent ring closure annulations or other synthetic transformations with suitable reaction partners (Scheme 18.1c). Besides that, radical denitrogenative transformations of poly-nitrogen heterocycles have become a hot topic in recent years (Scheme 18.1d). In this chapter, we are going to focus specifically on the radical denitrogenative transformations and their applications.

The radical denitrogenative transformations of poly-nitrogen heterocycles could be classified into the following five types: (i) radical denitrogenation of benzotriazinones and benzothiatriazines, (ii) radical denitrogenation of triazoles and benzotriazoles, (iii) radical denitrogenation of pyridotriazoles, (iv) radical denitrogenation of tetrazoles, and (v) radical denitrogenation of substituted 3-aminoindazoles. These radical methods have diverse advantages, including reaction performed under mild reaction conditions, expanding the reaction diversity by employing different types of coupling precursors and providing some remarkable virtues for assembling various carocyclizations/heterocyclizations that are otherwise challenging to construct motifs.

18.2 Radical Denitrogenation of Benzotriazinones and Benzothiatriazines

Benzotriazin-4(3*H*)-ones and benzothiatriazine-1,1(2*H*)-dioxides are valuable heterocyclic moieties in medicinal chemistry and are key components of a broad spectrum of pharmaceutically relevant compounds [8]. For example, benzotriazin-4(3*H*)-one derivatives are widely used as a potent anesthetic, antitumor, antidepressant, and so on. Similarly, benzothiatriazine-1,1(2*H*)-dioxide analogs are well known for the diuretic and nematocidal agents [9]. Apart from their biological importance, these benzannulated heterocycles also served as useful synthetic precursors in various denitrogenative annulation reactions because they contain a substantial quantity of nitrogen atoms within the ring, which are flexible to proceed for the expulsion of nitrogen gas during the reaction. Based on this concept, the groups of Murakami [10], Liu [11], Nakamura [12], and Cheng [13] demonstrated transition metal-catalyzed denitrogenative annulations with various coupling partners such as olefins, alkynes, allenes and isocyanides, etc. Additionally, in 2015, a successful breakthrough was achieved by Yu and co-workers [14]. With the assistance of a photocatalyst, the group performed a visible-light-mediated regioselective denitrogenative insertion of terminal alkynes in 1,2,3-benzotriazin-4(3*H*)-ones (Scheme 18.2). Notably, the reaction proceeded through a novel radical denitrogenative annulation pathway, and the literature survey revealed that this was the first example of radical denitrogenation transformation of polynitrogen

Yu [14]

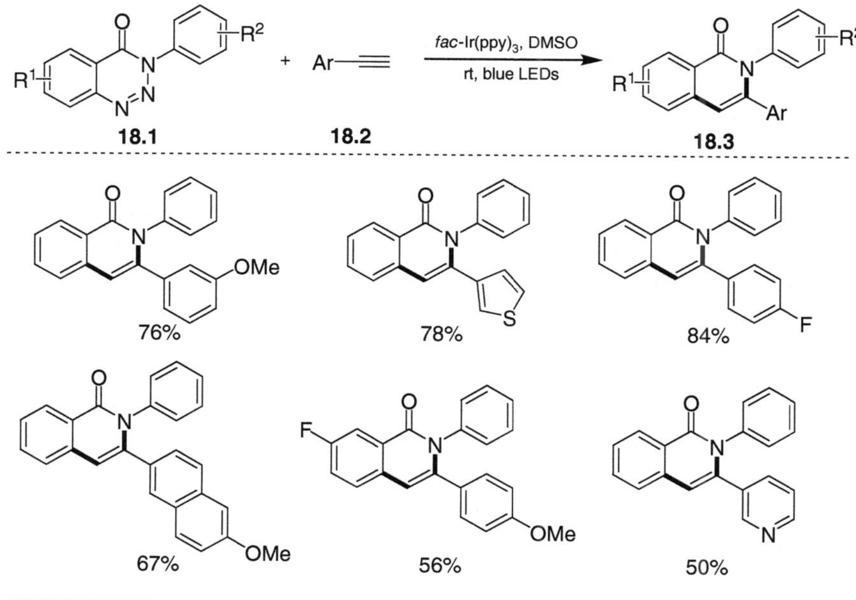

Scheme 18.2 Synthesis of isoquinolones using visible-light-promoted denitrogenative transannulations. Source: Yu [14]/American Chemical Society.

heterocycles. This transannulation strategy showed good compatibility with diverse functionalized 1,2,3-benzotriazinones and alkynes and could deliver various substituted isoquinolones in satisfactory isolated yields under mild reaction conditions.

To support the reaction mechanism, a radical trapping experiment was conducted with TEMPO (2,2,6,6-tetramethylpiperidin-1-oxyl), and it completely suppressed the formation of the desired product (**18.3**) and furnished the radical trapped product, which further detected in HRMS (high-resolution mass spectrometry) studies (Scheme 18.3a). Meanwhile, a radical clock experiment was conducted under the standard conditions by using allyloxy substituted 1,2,3-benzotriazinone (**18.4**) without the addition of alkyne precursor and observed the corresponding 2,3-dihydrobenzofuran derivative (**18.5**) in low yield (Scheme 18.3b). It indicates that the denitrogentation protocol proceeds via an aryl radical rather than a nitrogen-centered radical. Delightfully, aryl radical species are important intermediates for the introduction of aryl moieties in organic reactions. Which is generally derived from the decarboxylation of benzoic acids or deamination of aromatic amines [15]. Hence, this method is a complementary approach for the pioneering reports. The proposed mechanism is depicted in Scheme 18.3c. First, the Ir(III) photocatalyst being irradiated by a blue LED, which generates the excited state Ir(III)*, and it is oxidatively quenched by benzotriazin-4(3*H*)-one (**18.1**) to give the aryl radical anion (**I**) and Ir(IV). The radical intermediate **I** insertion to the alkyne (**18.2**) to forms the intermediate **II**, which subsequently undergoes an intramolecular addition to give intermediate **IV**. Eventually, radical anion (**IV**) is oxidized by Ir(IV) to yield the desired isoquinolone (**18.3**) with regeneration of the Ir(III) catalyst.

Further, the Yu group illustrated an intramolecular denitrogenative radical cyclization of 1,2,3,4-benzothiatriazine-1,1-dioxides to the synthesis of biaryl sultams under photocatalytic-assisted visible-light conditions (Scheme 18.4a) [16]. This efficient and eco-friendly synthetic protocol showed a broad functional group tolerance and moderate to excellent yields. The proposed mechanism involves a formation of excited state Ru(II)* by irradiation of Ru(II) under blue LEDs. The excited state of Ru(II)* is further oxidatively quenched by compound **18.7**, with concomitant generation of the radical anion **I** and Ru(III) complex. Subsequent intramolecular homolytic aromatic substitution (HAS) of **I** to form the radical intermediate **II**. Later, intermediate **II** is oxidized by Ru(III) to furnish cationic intermediate **III**, along with the regeneration of the photocatalyst Ru(II). The cationic intermediate further undergoes deprotonation to provide the sultam derivatives (Scheme 18.4b).

Recently, Cheng group developed a transition-metal-free tandem cyclization/*N*-arylation of 1,2,3,4-benzothiatriazine-1,1-dioxides and *ortho*-(trimethylsilyl)phenyl triflates. It is a two-step, one-pot reaction. The first step involves intramolecular homolytic cyclization of 1,2,3,4-benzothiatriazine-1,1-dioxides, and the second step follows the *in situ* generation of benzyne intermediate from the (trimethylsilyl)phenyl triflates (**18.9**) and subsequent insertion to access diversely functionalized biaryl sultam derivatives (Scheme 18.5) [17]. The control experiments revealed that, when the reaction conducted under the standard condition by using compound **18.7** in the absence of benzyne precursor, delivered the product **18.8a** in a very good yield. In addition, since there was no aryne precursor participating in the reaction, the

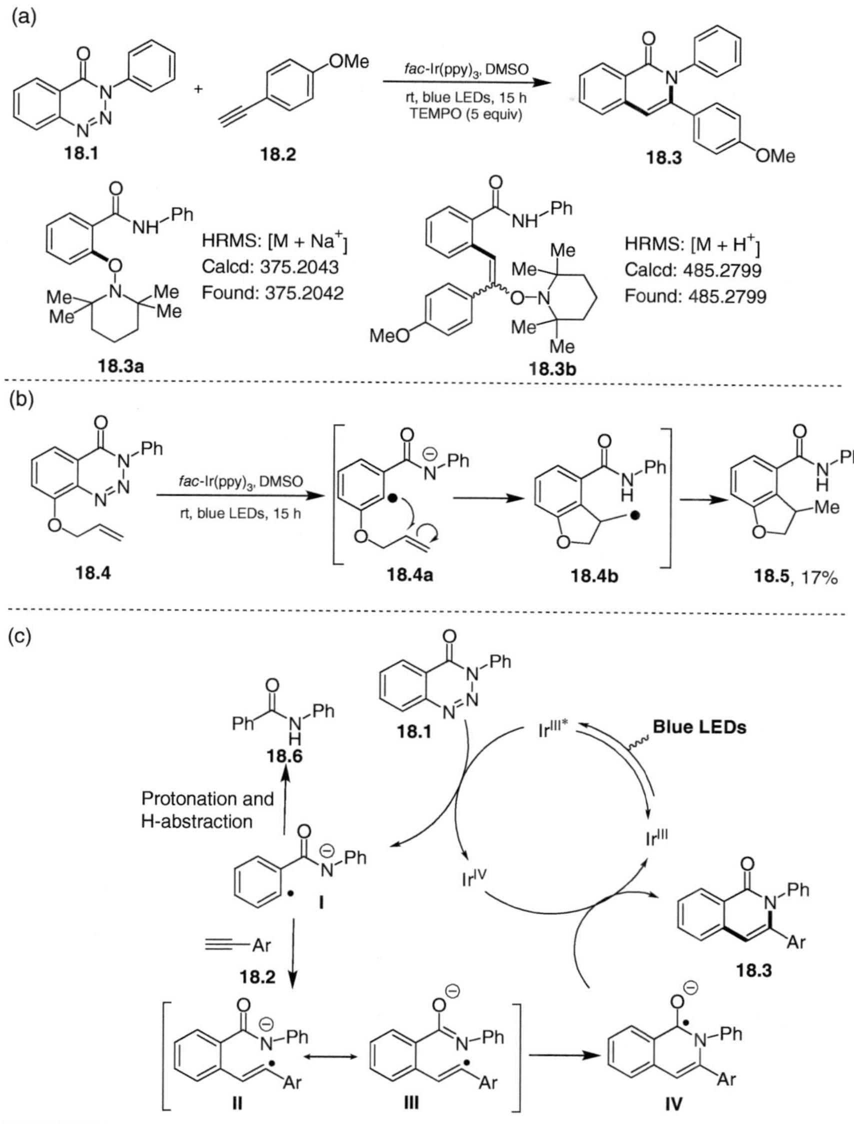

Scheme 18.3 Control experiments and proposed mechanism. (a) Radical trapping experiment with TEMPO. (b) Radical clock experiments. (c) Proposed mechanism.

reaction performed in the absence of the fluoride source observed the same product (**18.8a**) with almost similar yield. Moreover, large amounts of the radical trapping reagents such as BHT and TEMPO were loaded into the reaction, which influenced the reaction yields (Scheme 18.5b, c). The mechanistic investigations confirmed that the reaction proceeds through a diradical intermediate and is formed by thermal activation without any assistance of a photocatalyst or transition-metal complex.

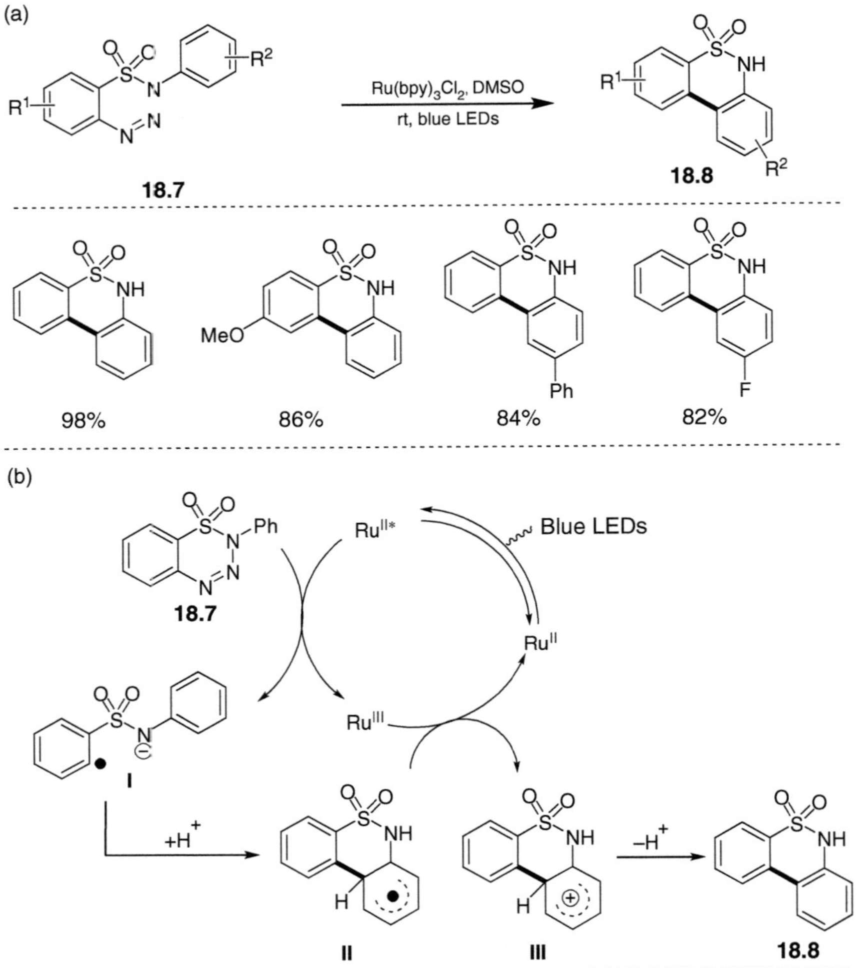

Scheme 18.4 Synthesis of biaryl sultams and plausible mechanism. (a) Synthesis of biaryl sultams. (b) Proposed mechanisms. Source: Yu [16]/American Chemical Society.

Apart from the cyclization products, in 2020, the metal-free, visible-light-induced synthesis of *ortho*-phosphorylated benzamides (**18.11**) was successfully achieved by Zhao and co-workers. Wherein, organic dye Eosin-Y acts as photoredox catalyst and sunlight or a blue LED as a light source. A wide variety of aryl-phosphonates, aryl-phosphinates, and aryl-phosphine oxides can be readily obtained with high regioselectivities under mild reaction conditions (Scheme 18.6) [18].

Shortly after this report, Wang et al. disclosed a visible-light-promoted denitrogenative free radical cross-coupling reaction between 1,2,3-benzotriazin-4(3*H*)-ones (**18.1**) and selenosulfonates (**18.12**) under oxidant-free catalysis conditions. This method features environmentally friendly reaction conditions, broad substrate scope, and excellent functional group compatibility. The developed method was

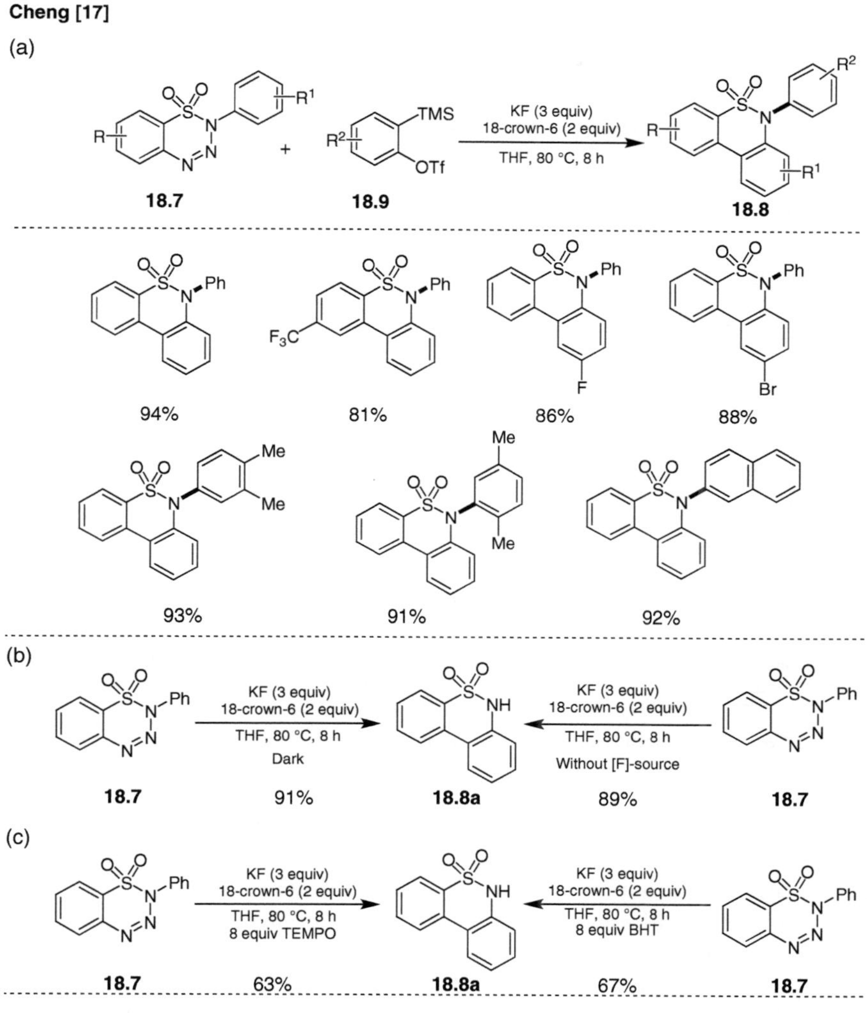

Scheme 18.5 Synthesis of biaryl sultams and control experiments. (a) Synthesis of Biaryl sultams. Source: Cheng [17]/American Chemical Society. (b) Control experiments. (c) Radical scavenger reactions.

broadening the strategies of C—Se bond formation in the field of organic chemistry (Scheme 18.7) [19]. In addition, when the *ortho*-selenylated derivatives exposed to an acid, they provided a slow release of highly potent biological active ebselen motifs in good yields (Scheme 18.7b). The addition of stoichiometric amounts of the radical scavengers such as TEMPO or BHT and 1,1-diphenylethylene to the reaction significantly suppressed the formation of the desired product, which indicated that a radical mechanism might be involved in this transformation (Scheme 18.7c).

Based on control experiments, a suitable mechanism has been proposed, which is depicted in Scheme 18.8. The highly reactive photoredox catalyst (*IrIII) generated by blue light irradiation, which subsequently oxidatively quenched by

Denitrogenative phosphorylation of benzotriazinones: Zhao [18]

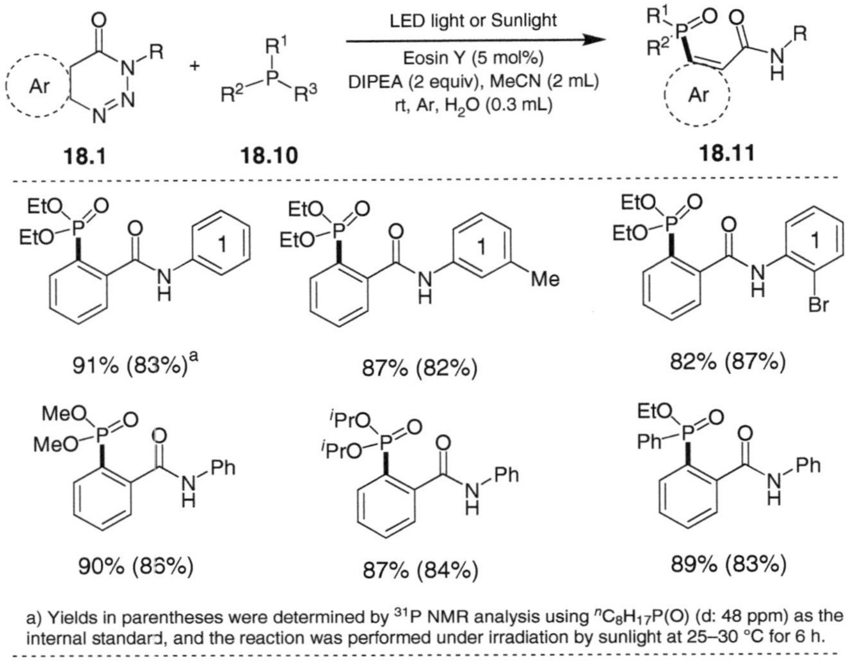

Scheme 18.6 Visible-light-induced denitrogenative phosphorylation of benzotriazinones. Source: Zhao [18]/Royal Society of Chemistry.

1,2,3-benzotriazin-4(3*H*)-ones (**18.1**) to generation of Ir(IV) and radical anion-**I**. At the same time, photoredox catalyst Ir(IV) oxidizes from the DIPEA to regenerate the photocatalyst at its ground state. Simultaneously, the radical anion intermediate-**I** reacts with selenosulfonates (**18.12**) to furnish intermediate-**II**, and a sulfone radical intermediate-**III**. Further, protonation of intermediate-**II** to yield the desired selenated product (**18.13**). The sulfone radical (**III**) is reduced by DIPEA radical to afford an iminium ion (**VII**), and along with sulfone anionic intermediate (**VIII**). Alternatively, homolysis of selenosulfonates (**18.12**) produces the selenium radical (**IV**), which undergoes the homocoupling to form a diselenium intermediate-**V**. Further, the diselenium intermediate-**V** interacts with compound **18.1** and to form anionic intermediate-**II**, along with selenium radical (**IV**).

18.3 Radical Denitrogenation of Triazoles and Benzotriazoles

Triazole scaffolds are one of the significant heterocycles that exhibit broad biological activities and are common motifs in pesticides, pharmaceuticals, and synthetic drugs [20]. Given the importance of the triazoles, their synthesis has attracted much attention. Specifically, 1,3-dipolar cycloaddition or also known as "click

Wang [19]

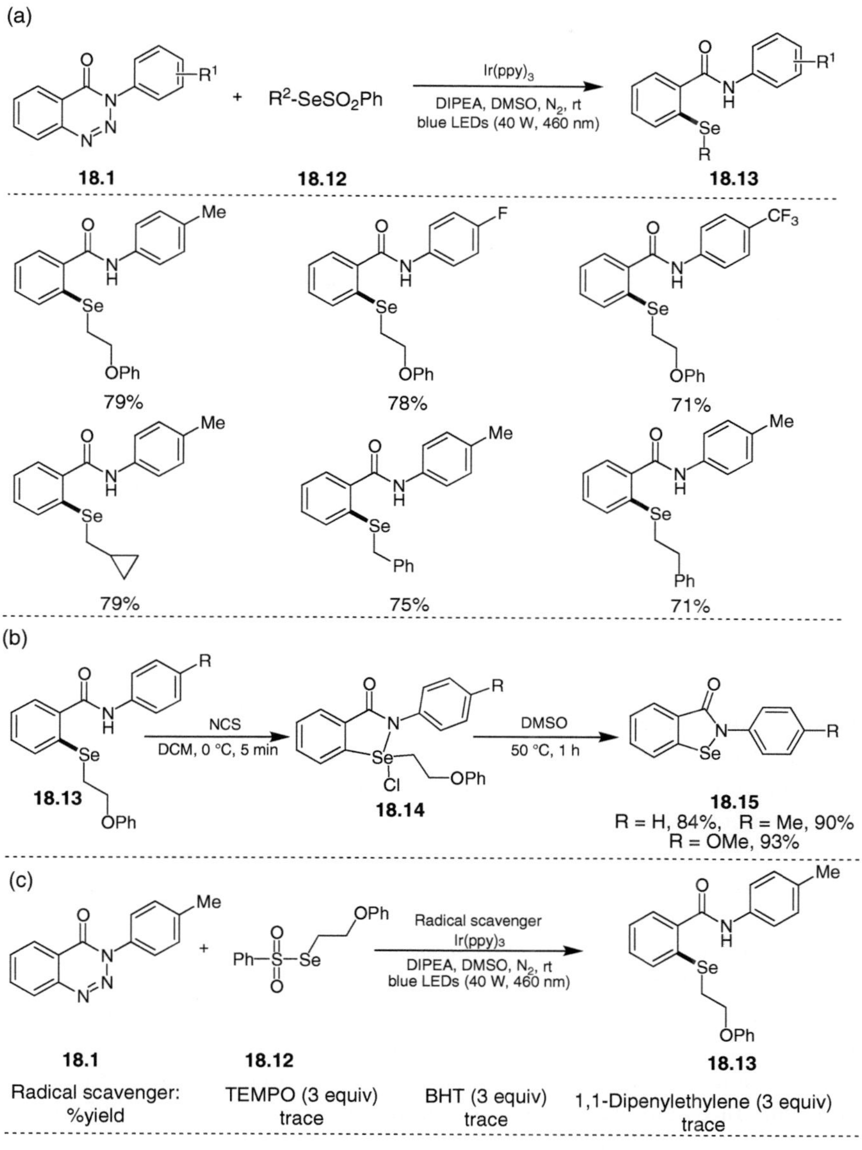

Scheme 18.7 Visible-light-promoted denitrogenative *ortho*-selenylation, applications and control experiments. (a) Visible-light-promoted denitrogenative orthoselenylation reaction of benzotriazinones. Source: Wang [19]/Royal Society of Chemistry. (b) Synthetic applications (epselen derivatives). (c) Control experiments.

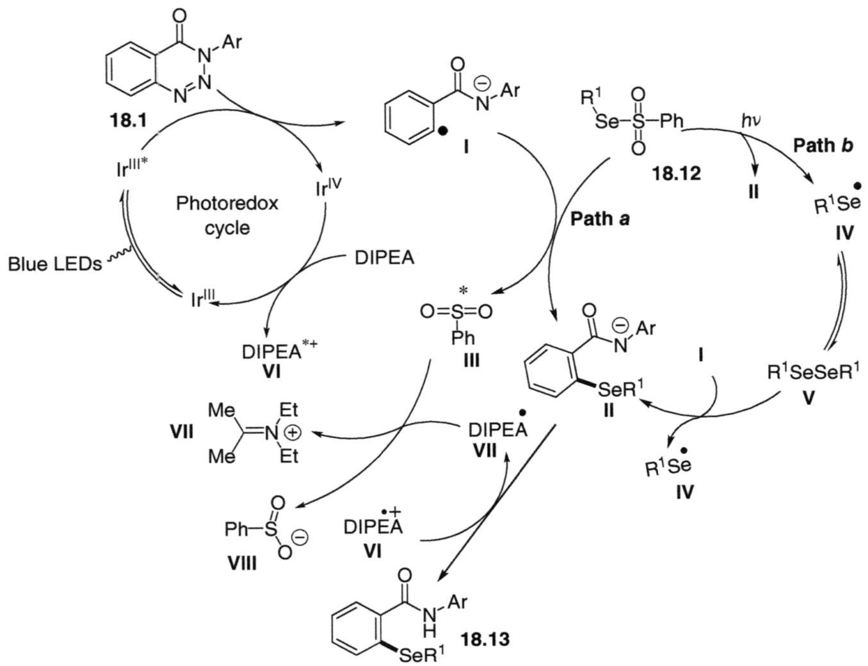

Scheme 18.8 Plausible mechanism.

reaction" between the azide and terminal alkyne is the most popular method for producing triazole moiety [21]. In view of the important application of triazole is "ring-opening" followed by the extrusion of nitrogen gas, and participation in various synthetic transformations are highly attractive approaches, which were developed by Fokin and others [7]. Recently, radical denitrogenative protocols of triazoles have also captured a significant interest. For the first time, Fokin and co-workers discovered a cascade transformation of triazoles (**18.16**) for the practical access to α-oxygenated amidines (**18.17**), which are difficult to synthesize a class of α-oxygenated carboxylic acid derivatives (Scheme 18.9) [22]. The reaction utilized the most inexpensive molecular oxygen as an oxidant, furnished the required products with excellent regioselectivities, and demonstrated high functional group tolerance. Notably, the radical trapping reactions resulted in the formation of compounds **18.18** and **18.19**, which described that the reaction proceeded through a radical mechanism (Scheme 18.9b).

Based on quenching experiments, the authors proposed a tentative mechanism, which is shown in Scheme 18.10. Initially, the lithiated triazole (**A**) was generated by deprotonation of *N*-sulfonyl triazole (**18.16**) with LiHMDS, which subsequently undergoes a visible bubbling of the reaction mixture, indicated a loss of N_2, and affords an amphoteric lithium ketenimine Intermediate-**B**. Further, the Intermediate-**B** is rapidly trapped by the electrophilic iodine, resulting in the iodoketenimine intermediate-**C**, which subsequently reacts with nucleophilic amines and furnishes the iodoamidine product-**D**. Later, autoxidation of the iodoamidine intermediate (**D**) by exposure to atmospheric oxygen generates an

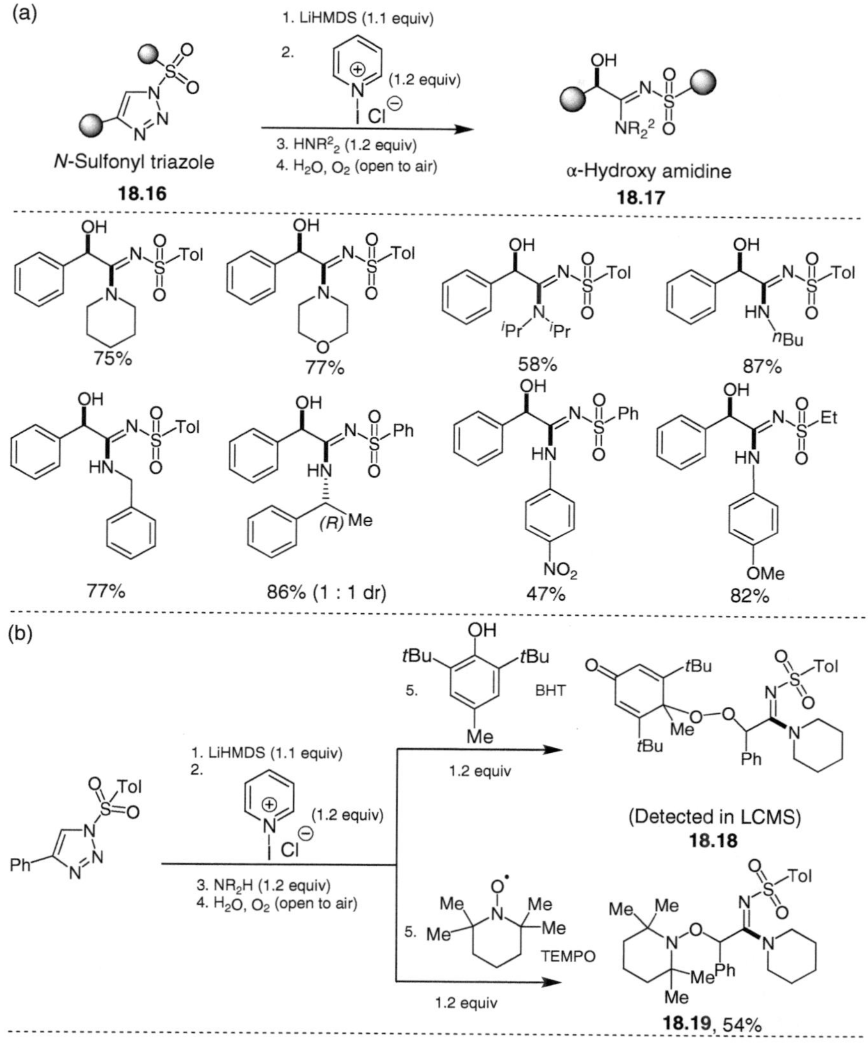

Scheme 18.9 Synthesis of α-oxygenated amidines from triazoles and control experiments. (a) Synthesis of α-oxygenated amidines: Fokin [22]. (b) Radical trapping reactions. Source: Fokin [22]/American Chemical Society.

α-carbon centered radical (**E**), which rapidly reacts with triplet oxygen, forming a peroxy radical **F**. The intermediate-**F** converts hydroperoxy intermediate **G** in the presence of H_2O and I_2, after which the iodide reduces intermediate-**G** to deliver the desired hydroxylated product (**17**). In the absence of H_2O, a 1 : 1 mixture of hydroxy (**I**) and oxo amidines (**J**) is observed via dimerized intermediate-**H** followed by Russell fragmentation.

Similarly, the ring-opening reactions of benzotriazole derivatives emerged as an attractive way to generate a wide variety of nitrogen heterocycles such as indoles, benzoazines, etc. In 2013, the Tiwari group developed a concise and efficacious

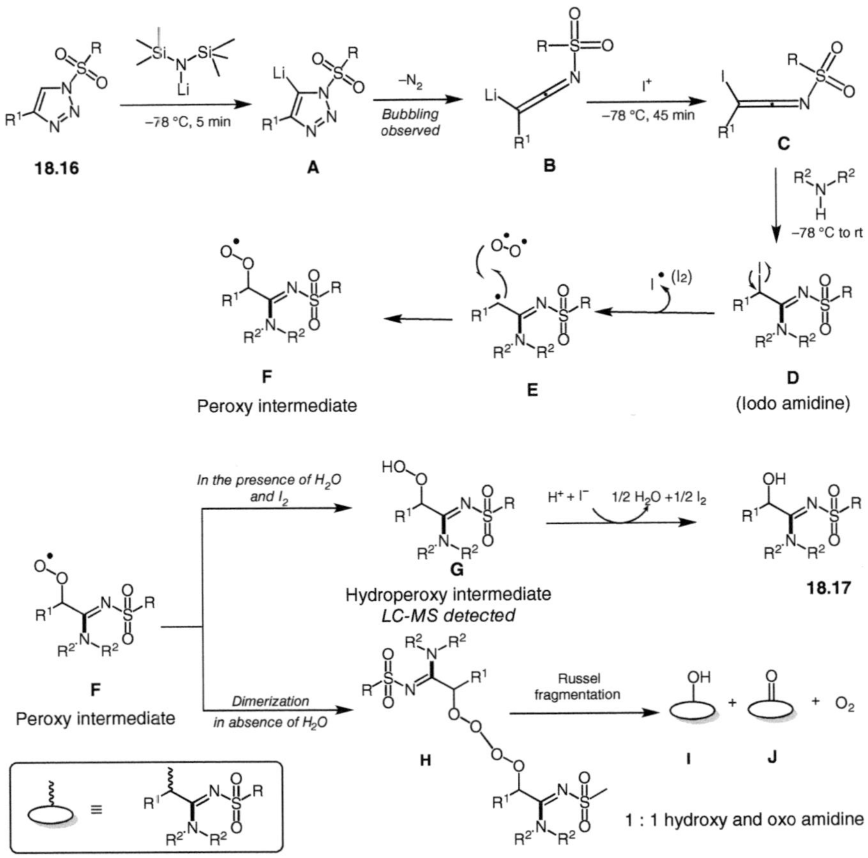

Scheme 18.10 Plausible mechanism.

method for the synthesis of substituted benzothiazoles from benzotriazole (**18.20**) in a one-pot manner. Typically, the reaction provides an easy access to a diverse range of 2-N/S/C-substituted benzothiazoles under mild conditions by using silanes or stannane reagents under heating or microwave irradiated a free radical β-scission of the N—N bond (Scheme 18.11a and Scheme 18.11b) [23]. The same protocol was also utilized for the synthesis of glycoconjugate benzothiazoles (**18.23**) from carbohydrate-conjugated benzotriazole derivatives (Scheme 18.11c) [24].

The author proposed a denitrogenative radical cyclization mechanism for the reaction (Scheme 18.12). The homolytic cleavage of tributyl-stannane initiates a radical reaction, which is propagated by the subsequent addition of stannanyl radical to sulfur of thiocarbonyl group of benzotriazole (**18.20**), and gives the intermediate **A**. Then, a rapid addition of a hydrogen radical to intermediate **A**, affords an intermediate **B**. The intermediate **B** further undergoes a β-scission via a homolytic cleavage of benzotriazole ring, followed by denitrogenation, which affords a biradical intermediate **C**. The intermediate **C** further undergoes an intramolecular cyclization, producing resonance-stabilized nitrogen-centered

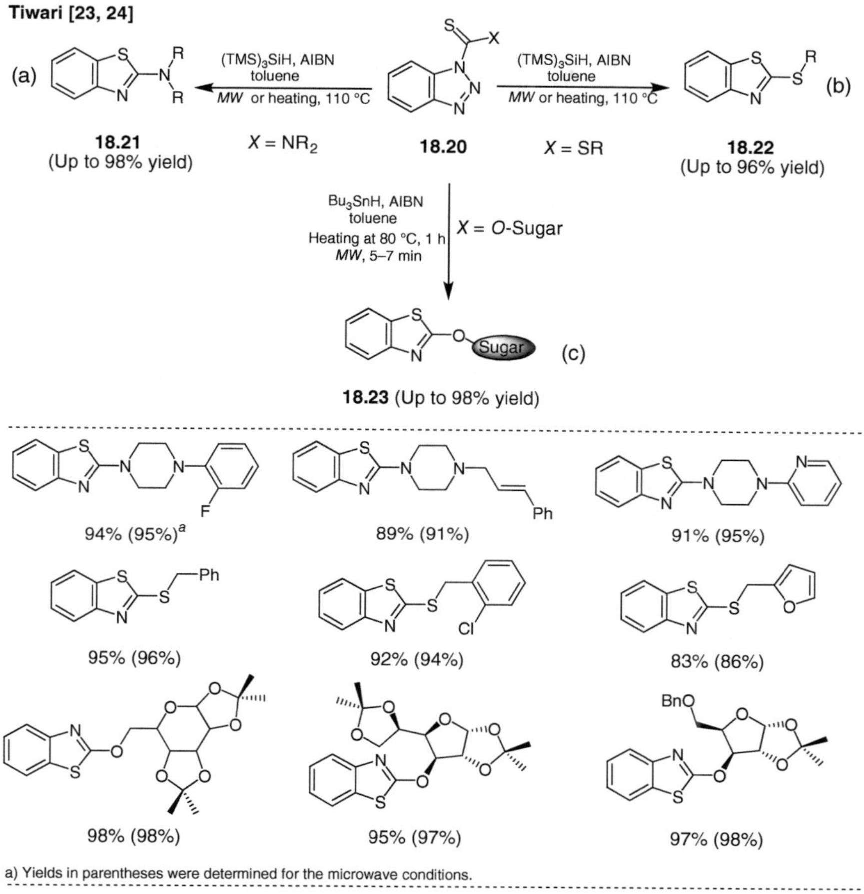

Scheme 18.11 Synthesis of α-oxygenated amidines from triazoles and control experiments. Source: Tiwari [23, 24].

radical **D** or **E**. Finally, thermally induced oxidation process leads to aromatization and furnishes the desired product (**18.23**).

Shi et al. reported an efficient intermolecular radical addition to vinylbenzotriazoles (**18.24**) for the formation of quinoxaline (**18.25**) and imino nitrile (**18.26**) via a radical denitrogenative pathway. It is an alternative approach to triazole ring-opening reactions. Delightfully, good yields and a broad scope of functional group tolerance were observed. Interestingly, the reaction exclusively provided the imino nitrile as a significant product under base conditions, such as sodium azide (NaN_3) as an azide source. In the case of trimethyl silyl azide ($TMSN_3$) as an azide source, the reaction afforded the quinoxaline derivatives in high yields (Scheme 18.13) [25].

The author proposed a suitable reaction mechanism for both products (Scheme 18.14). The formation of quinoxaline (**18.25**) is a very straightforward protocol (**path-*A***). Radical addition of vinyl triazole (**18.24**) to give intermediate **A**,

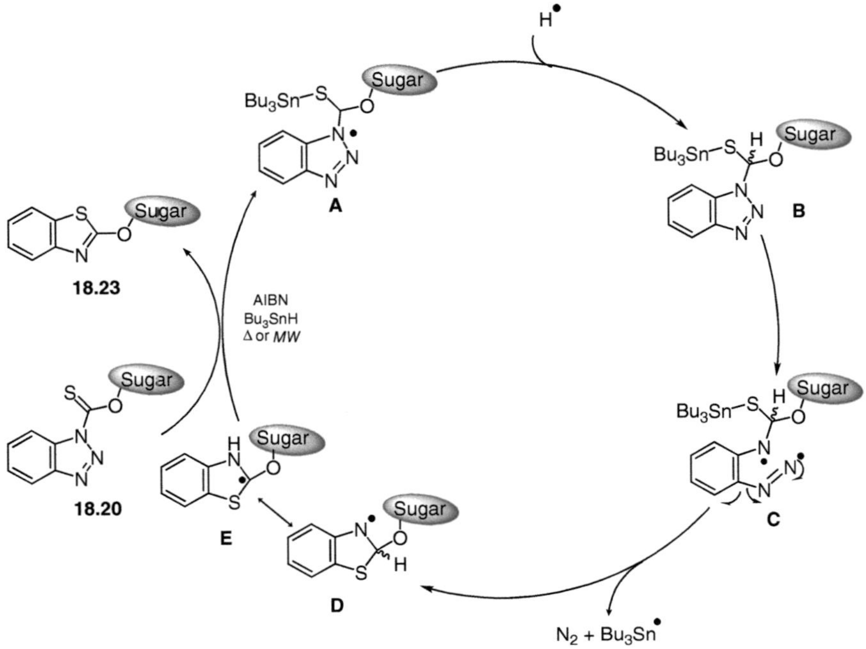

Scheme 18.12 Postulated mechanism for the formation of benzothiazoles.

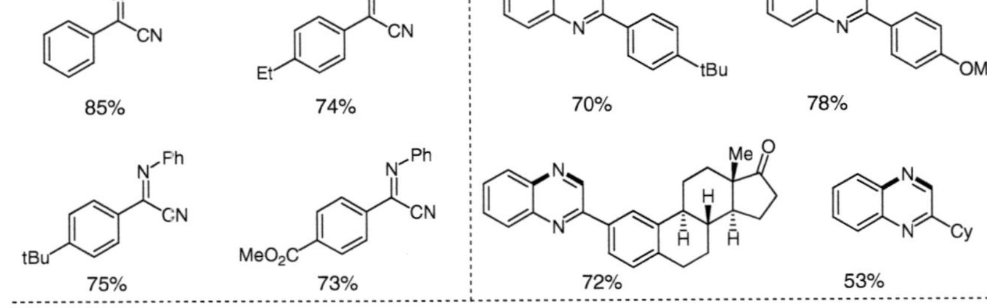

Scheme 18.13 Intermolecular radical addition to vinyltriazoles. Source: Shi [25]/American Chemical Society.

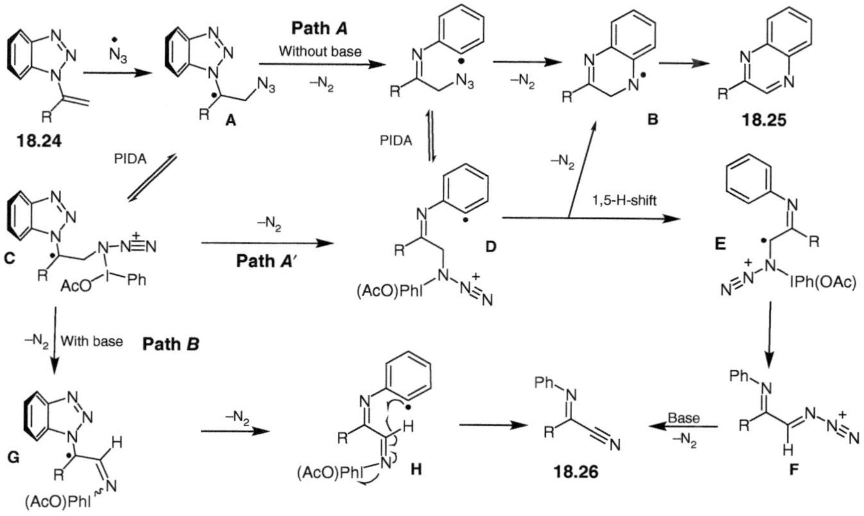

Scheme 18.14 Possible mechanism.

which subsequently undergoes sequential denitrogenation, provides the nitrogen radical intermediate **B**. The intermediate **B** further undergoes aromatization to furnish the desired product **18.25**. Alternatively, intermediate **A** coordinates with PIDA and provides the intermediate **C**, followed by sequential extrusion of two molecular nitrogen gas, which would give intermediate **B** subsequent aromatization, preferably providing the desired cyclized product **18.25**. However, the formation of iminonitrile (**18.26**) is unclear. The intermediate **C** follows two pathways. **Path *A′*** would proceed through a ring opening followed by the expulsion of nitrogen to give arene radical intermediate **D**. Upon participating in the 1,5-H shift and eliminating nitrogen gas, it favors the formation of nitrile product (**18.26**). Alternatively, in path ***B***, the intermediate **C** will also lose the nitrogen gas to form Intermediate **G**, which undergoes ring opening followed by a 1,5-H shift, favors the nitrile derivatives (**18.26**).

In 2016, Glorius et al. achieved diverse *ortho*-functionalized *N*-arylbenzamides such as *ortho*-borylated, -thiolated, and -alkylated *N*-arylbenzamide derivatives from 1-(benzoyl)benzotriazoles (**18.27**) via visible-light-mediated denitrogenative radical cascade pathway (Scheme 18.15) [26]. This method tolerates a broad range of functional groups and produces high yields in a one-pot manner. Moreover, quantum yield measurements and Stern–Volmer analysis supported the fact that the reactions proceed via photocatalytically initiated radical chain mechanisms. A plausible mechanism for the photocatalytic *ortho*-functionalized *N*-arylbenzamides, is shown in Scheme 18.16. Initially, highly reactive photoredox catalyst (*IrIII) generated by visible-light irradiation of photocatalyst Ir(III), which undergoes single-electron transfer with the reductive quencher DIPEA (**I**), resulting in the highly reducing Ir(II) species along with an amine radical cation (**II**). Then, benzotriazole (**18.27**) acquires an electron from this Ir(II)-complex to form the highly

Glorius [26]

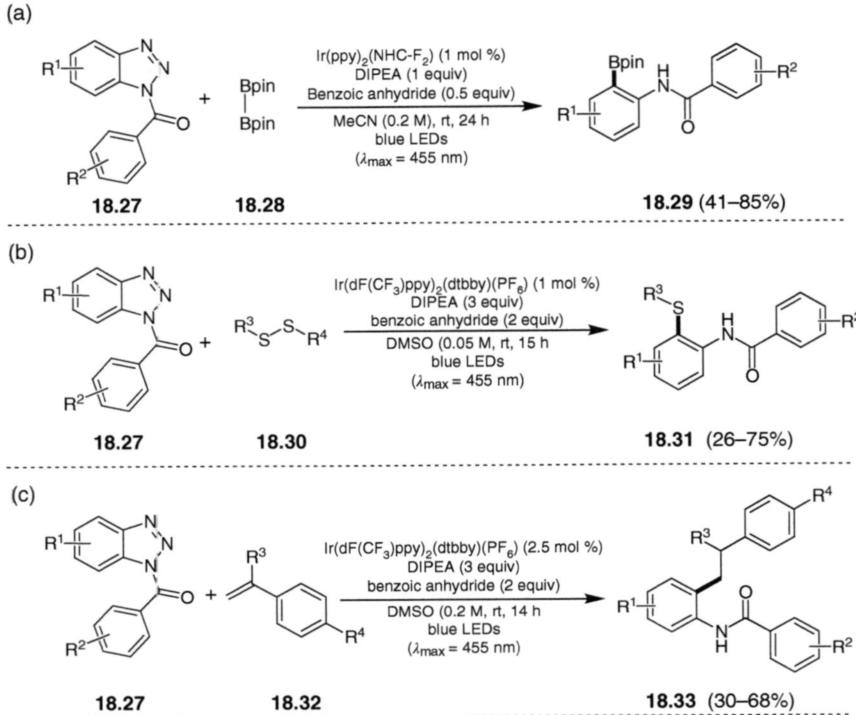

Scheme 18.15 Visible-light-mediated *ortho*-functionalizations. (a) Borylations. (b) Thiolations. (c) Alkylations. Source: Glorius [26]/American Chemical Society.

reactive intermediate **III**, subsequently regenerates the active photocatalyst Ir(III). The intermediate **III** undergoes an extruded nitrogen to furnish the anionic radical intermediate **IV**. Then, the anionic radical (**IV**) reacts with styrene (**18.32**) and provides the stabilized radical-**V**, which subsequently produces the desired product (**18.33**) by hydrogen-atom abstraction from another DIPEA molecule (**I**) or from the radical cation (**II**). The aryl radical anion (**IV**) could also abstract a hydrogen atom directly from compound **II** or solvent to yield the byproduct *N*-phenylbenzamide (**18.34**).

Using the same strategy, the Glorius group further explored the synthesis of biologically important 2-substituted indoles (**18.35**) via visible-light-mediated denitrogenative cascade cyclization strategy between 1-(benzoyl)benzotriazoles (**18.27**) and terminal alkynes (**18.2**). This synthetic protocol displays a high functional group tolerance, a broad substrate scope, and mild reaction conditions. Interestingly, substrate investigation revealed that introducing a strongly electron-withdrawing trifluoromethyl group on the benzoyl fragment directly delivers the corresponding free *NH*-indoles (Scheme 18.17b) [27]. The proposed mechanism is depicted in Scheme 18.18, which follows the same process described in Scheme 18.16. The key

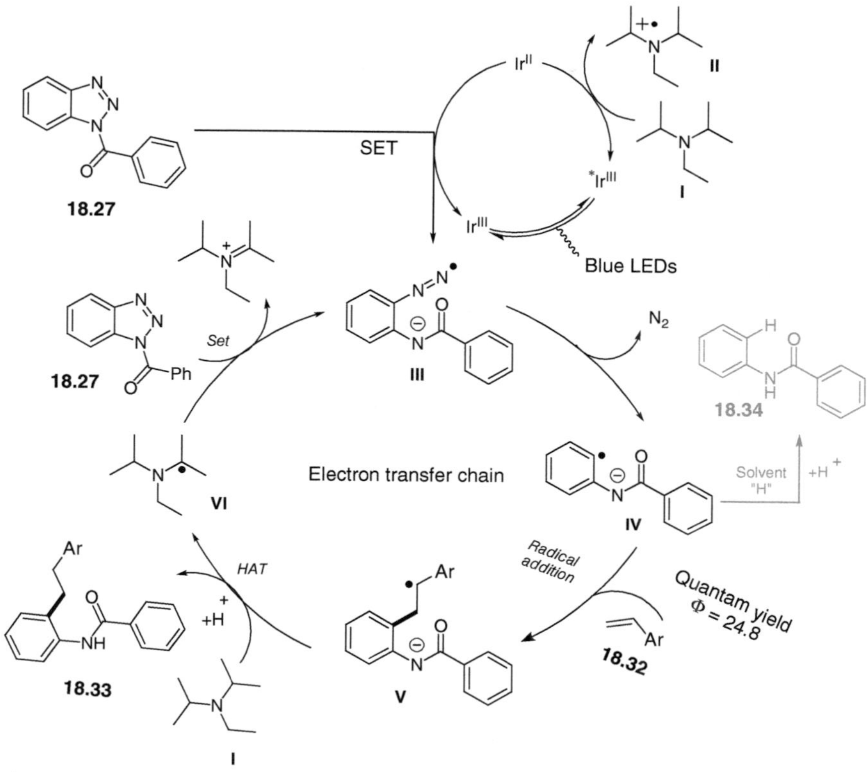

Scheme 18.16 Plausible mechanism for the alkylation of benzotriazole.

steps involved in the following mechanism is visible-light irradiation of photocatalyst to generate excited state Ir(III)/single electron transfer (SET) to the reductive quencher DIPEA/aryl radical formation (**IV**) via interaction between Ir(II) and benzotriazole **18.27** via SET/nitrogen extrusion/alkyne addition. Finally, oxidized by another molecule of benzotriazole **18.27** or Ir(II), followed by annulation, affords the desired indole scaffold (**18.35**).

In 2018, Xia group disclosed a visible-light mediated denitrogenative phosphorylation reaction between benzotriazoles (**18.27**) and phosphites (**18.10**), and this strategy affords diverse substituted aryl phosphonates (**18.37**) under mild conditions. The following reaction proceeds through a photoredox catalytic process (Scheme 18.19) [28].

Very recently, Li and co-workers developed an efficient method of visible-light-induced radical denitrogenative cascade cyclization of vinyl benzotriazoles (**18.24**) with alkyl halides (**18.38**) to substituted phenanthridines (**18.39**). The reaction was performed under mild conditions and provided diverse functionalized phenanthridine analogs in moderate to high yields (Scheme 18.20) [29]. The reaction was well tolerated for diverse functional groups with respect to benzotriazoles and alkyl halides. It is worth noting that the product formation was completely

Glorius [27]

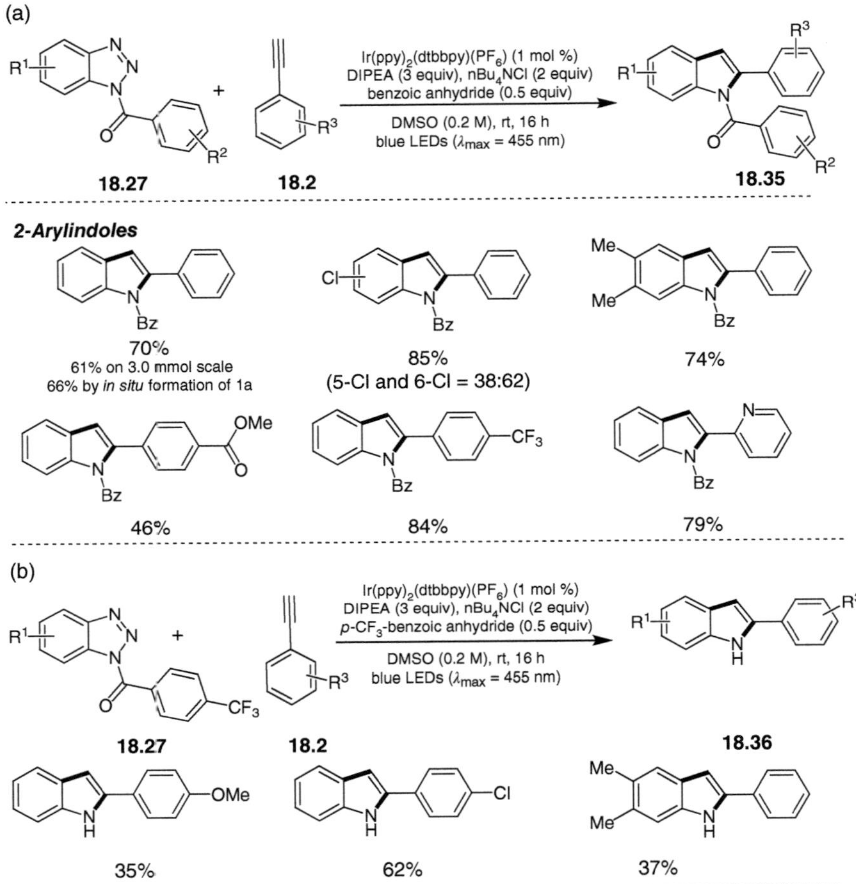

Scheme 18.17 Synthesis of 2-substituted indoles. (a) Synthesis of *N*-benzoyl indoles. (b) Synthesis of N-*H* indoles. Source: Glorius [27]/John Wiley & Sons.

suppressed in the presence of TEMPO, which declared that the reaction proceeds through a radical mechanism (Scheme 18.20b). Moreover, the classic radical clock experiments revealed that the ATRA product **18.42** obtained 50% and 70% of yield with and without DMAP reagent (Scheme 18.20c). These experiments indicated the DMAP is not involved in radical generation steps.

According to mechanistic results and literature precedence, the author proposed a suitable mechanism, which is shown in Scheme 18.21. Initially, the blue light irradiation on alkyl halide (R-Br) (**18.38**) would generate the alkyl radical **A**, which immediately underwent an intermolecular radical addition to **18.24**, affording the radical intermediate **B**. The intermediate-**B** could trap with bromo radical (**Br·**) to form intermediate **C** and proceed through a subsequent elimination with the

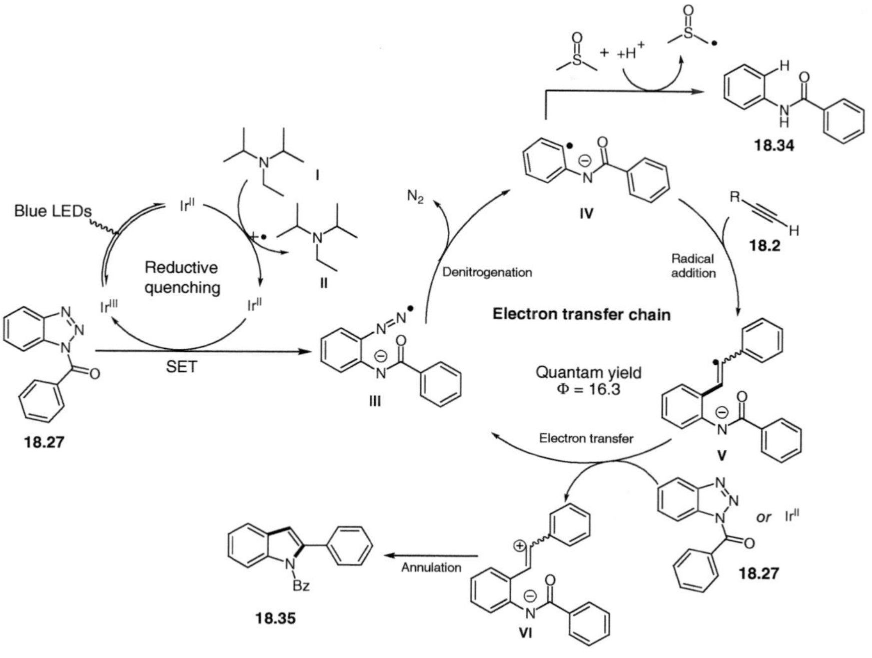

Scheme 18.18 Proposed mechanism.

Xia [28]

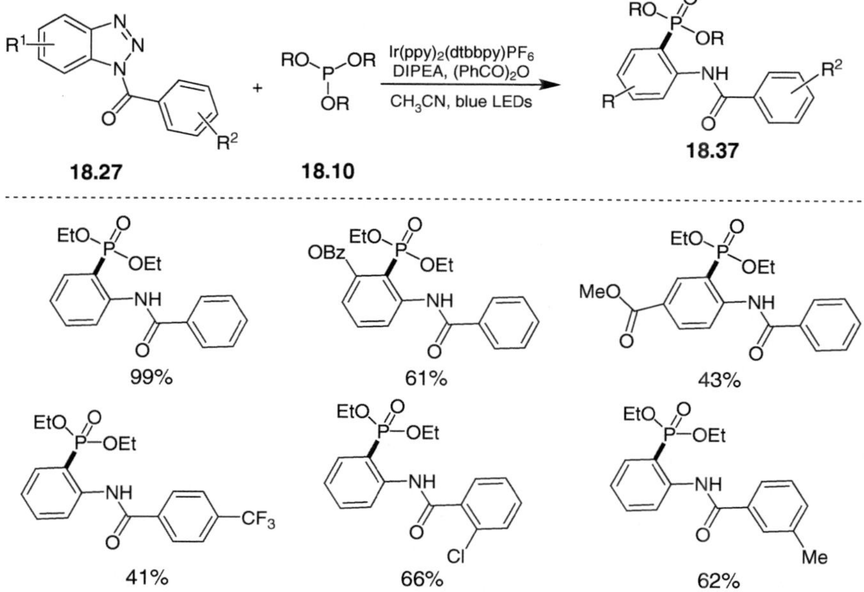

Scheme 18.19 Visible-light-induced C(sp^2)–P bond formation by denitrogenative coupling reactions. Source: Xia [28]/American Chemical Society.

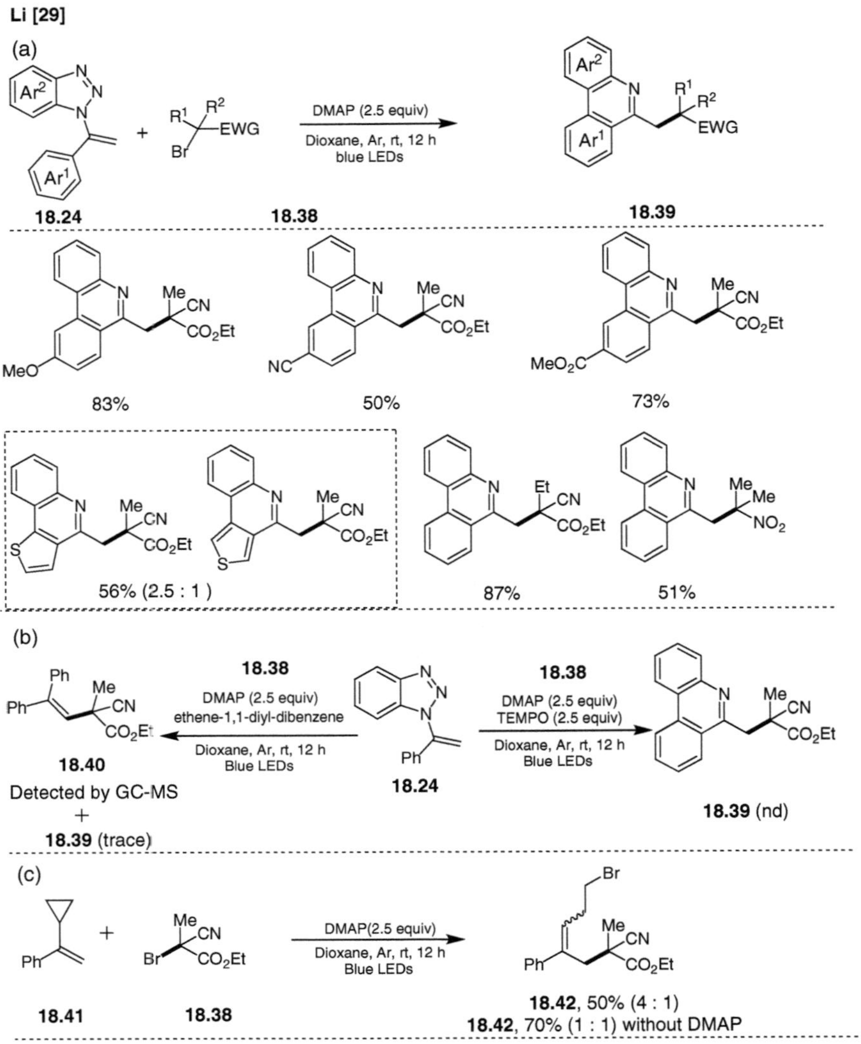

Scheme 18.20 Synthesis of phenanthridines and control experiments. (a) Synthesis of phenanthridines. (b) Radical trapping experiment. (c) Radical clock experiment. Source: Li [29]/American Chemical Society.

help of base, providing the vinyl benzotriazole derivatives (**18.43**). Alternatively, the intermediate **B** would undergo a radical ring-opening followed by denitrogenation to form radical anion intermediate **D** (**path *A***). The intermediate **D** further participates in an intramolecular radical cyclization process to form radical **E** and immediately traps with bromine radical, followed by dehydrohalogenation furnishes the desired product **18.39**. Alternatively, aryl radical anion (**D**) could also abstract a hydrogen atom from solvent or trap directly bromo radical, which could yield the corresponding imine byproducts (**18.44** or **18.45**).

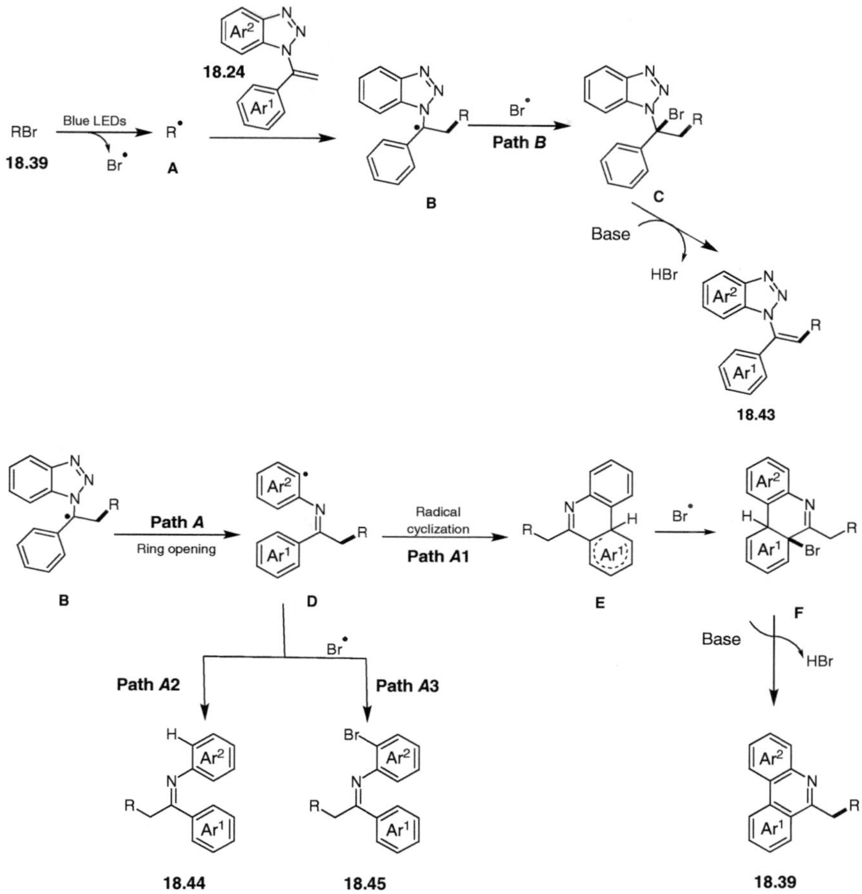

Scheme 18.21 Proposed mechanism.

18.4 Radical Denitrogenation of Pyridotriazoles

Pyridotriazoles are important building blocks, which are frequently found in many pharmaceutical and material science fields [30]. Additionally, having the triazole moiety in the pyridotriazoles, it is recognized as a suitable precursor for the diverse denitrogenative tranannulations via a ring opening followed by a nitrogen extrusion strategy [31]. In 2016, Adimurthy et al. demonstrated a denitrogenative transannulation of pyridotriazoles (**18.46**) with benzylamines (**18.47**) or amino acids (decarboxylation) (**18.48**) for the synthesis of imidazo[1,5-*a*]pyridines (**18.49**) by using copper-catalyzed aerobic oxidation conditions. In this system, molecular oxygen (O_2) from air is used as the sole oxidant. Under similar conditions, diverse benzyl amines and amino acid derivatives were well tolerated and provided the desired imidazopyridine analogs in moderate to excellent yields. Overall, this synthetic strategy displays a broad substrate scope with a high functional group tolerance

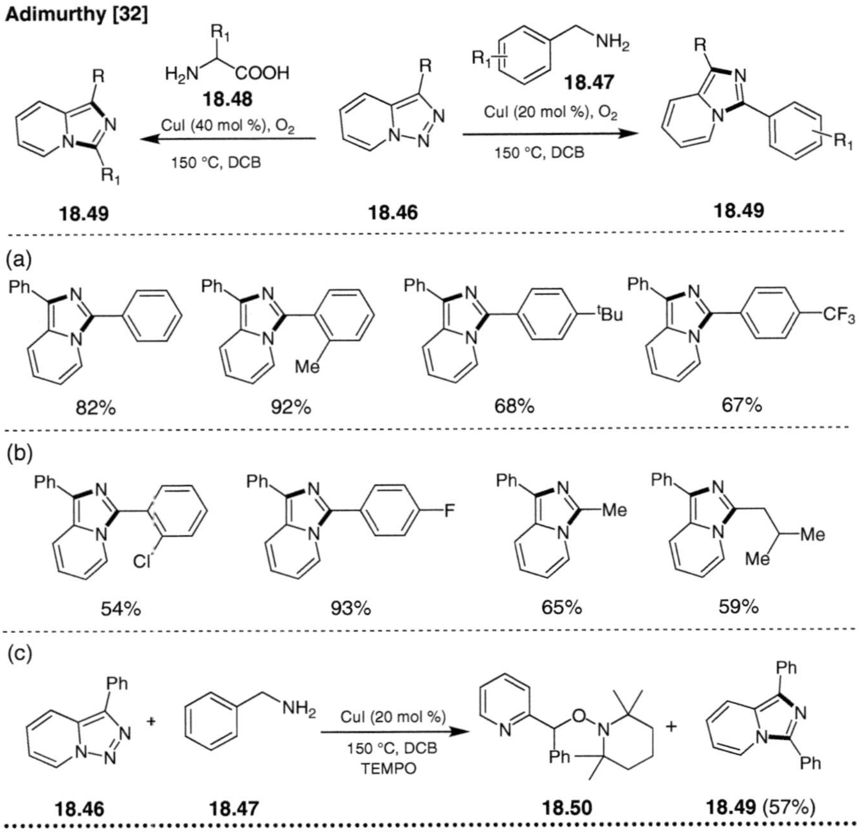

Scheme 18.22 Transannulations of pyridotriazole with amine derivatives and control experiments. (a) Transannulation from benzyl amine. (b) Transannulation from aminoacids. (c) Radical trapping experiment.

and represents a valuable complement to the synthesis of imidazo[1,5-*a*]pyridine derivatives (Scheme 18.22) [32]. A radical trapping reaction was performed to determine whether the reaction proceeds via a radical pathway or an ionic path; however, the TEMPO adduct (**18.50**) was not observed. Interestingly, the yield of the desired product (**18.49**) was significantly decreased. It is indicated the reaction might proceed through a carbene intermediate formation followed by oxidation to give imidazo[1,5-*a*]pyridines (Scheme 18.22c).

Based on the control experiments and literature reports, a possible mechanism was proposed, which is depicted in Scheme 18.23. Initially, the pyridotriazole (**18.46**) would undergo a ring opening to form a corresponding diazo intermediate **A**, which immediately reacts with copper metal and forms copper carbenoid **B** via nitrogen extrusion. The carbenoid intermediate **B** would react with benzylamine (**18.47**), followed by migratory insertion affords the intermediate **C**. The intermediate **C** would undergo an oxidative dehydrogenation followed by a single electron transfer to afford a benzyl radical intermediate **E**. Which further undergoes a

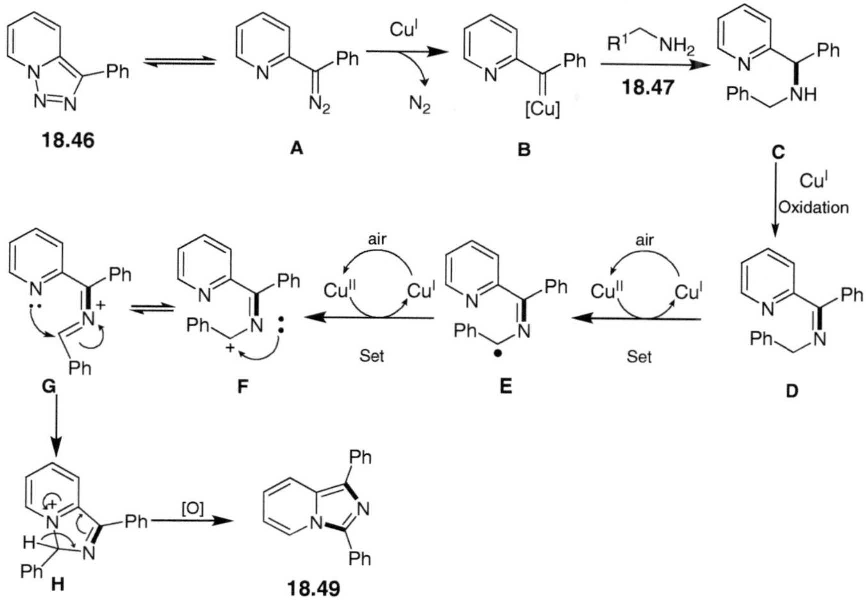

Scheme 18.23 Plausible mechanism.

single electron transfer process and generates the carbocation intermediate **F**, and subsequently **intermediate G**. Finally, the intermediate **G** is further involved in an intramolecular nucleophilic addition followed by cyclization and aromatization provides the desired imidazo[1,5-*a*]pyridine product **18.49**.

Chattopadhyay and co-workers recently disclosed a novel and efficient method for denitrogenative transannulation/cyclopropanations from the pyridotriazoles with alkyne or alkene via Co-catalyzed radical activation mechanism. Notably, the author developed an *in situ* formation of bicyclic pyridotriazole from the carbonyl compounds (**18.51**) and tosylhydrazine via the hydrazone intermediate, which further reacts with alkyne and provides the indolizine derivatives (**18.47**), exclusively (Scheme 18.24a). At the same time, alkyne replacement with alkene (**18.32**) obtains the cyclopropane scaffolds (**18.52**) with high diastereoselectivity and excellent yields, respectively (Scheme 18.24b). Moreover, the potential synthetic utility of the established methodology was demonstrated by the synthesis of unnatural racemic monomorine (**18.53**) in two steps (Scheme 18.24c) [33].

Further, various control experiments were performed to obtain information about the radical activation mechanism (Scheme 18.25). Initially, the TEMPO reagent was added to the reaction under standard developed conditions and observed transannulated product (**18.47**) with 98% yield. It assumed that due to the high reactivity of the Co-carbene radical, it may react rapidly with phenylacetylene to form the desired product (**18.47**) at high temperatures. Later, by reducing the reaction temperature from 80 to 60°C, the desired product was suppressed completely. These results revealed that the reaction proceeds through a radical-activation mechanism. Furthermore, utilizing the same conditions with a high equivalent of TEMPO

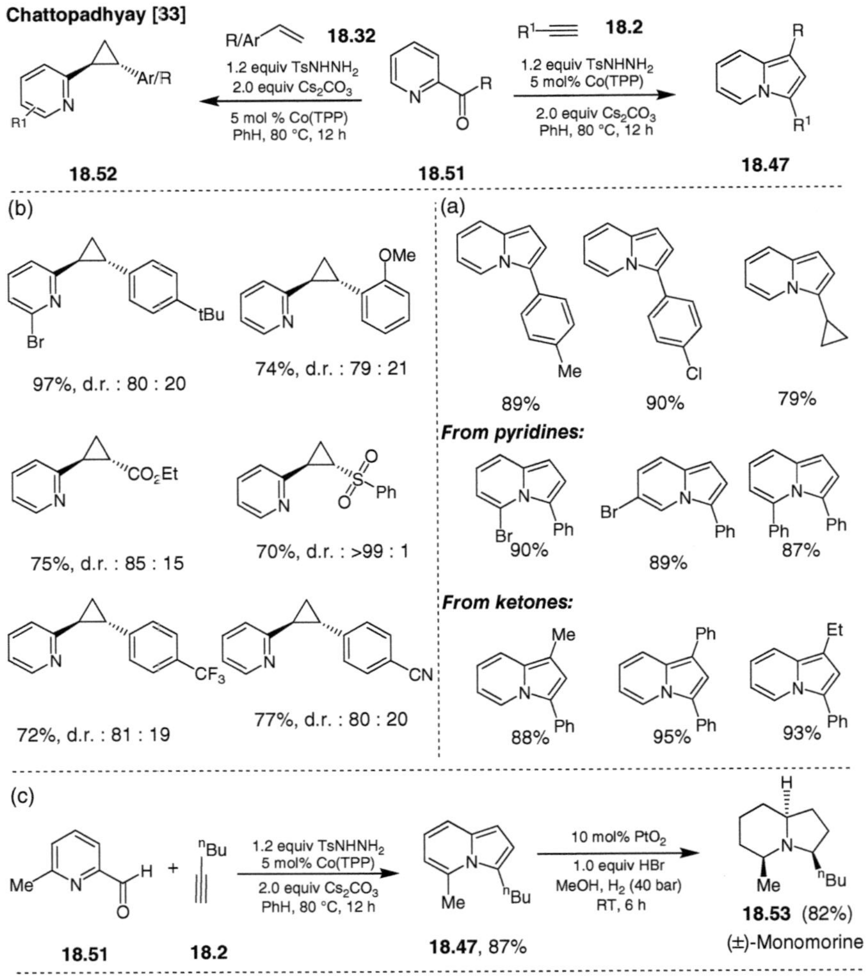

Scheme 18.24 Cobalt(II)-catalyzed transannulations and cyclopropanations. (a) Scope of transannulation. (b) Scope of cyclopropanation. (c) Synthesis of racemic monomorine. Source: Chattopadhyay [33]/John Wiley & Sons.

reagent delivered the radical trapping product (**18.55**), and it is characterized by various spectroscopic techniques (Scheme 18.25b). Based on the control experiments, a plausible mechanism is proposed, which is depicted in Scheme 18.25c. At first, the *in situ* formed pyridotriazole (**18.46**) would rapidly generate the 2-diazomethyl pyridine intermediate (**A**) via a ring-opening sequence, which further reacts with Co-catalyst (**B**) and affords an α-Co(III) pyridyl radical intermediate **C** via nitrogen extrusion. Then, the intermediate **C** reacts with an alkyne (**18.2**), giving the vinyl radical intermediate **D**. The intermediate **D** would undergo radical recombination that furnishes the desired indolizine product (**18.47**), along with Co-catalyst (**B**) regeneration to continue the catalytic cycle.

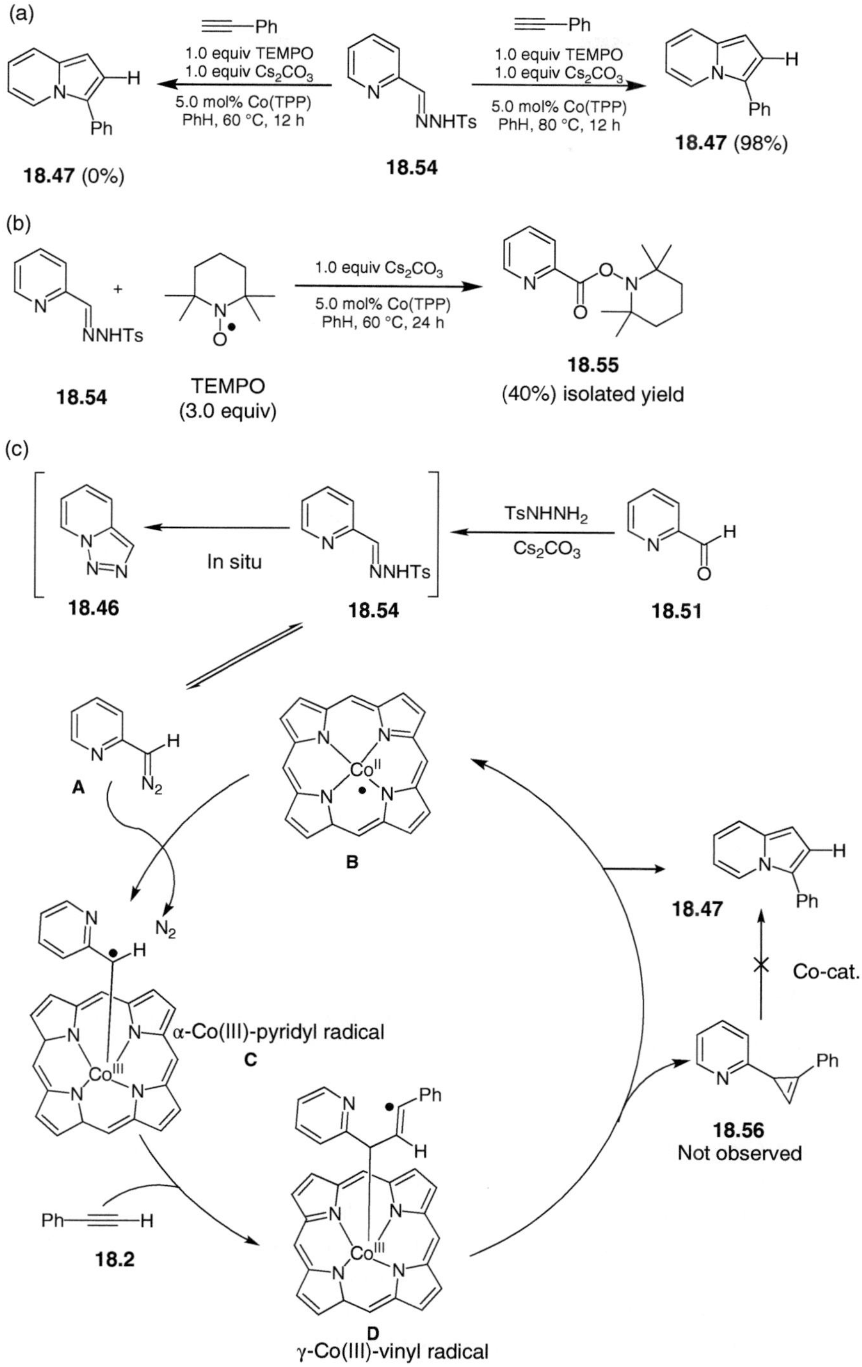

Scheme 18.25 Control experiments and proposed mechanism. (a) Control experiments. (b) Radical trapping reaction. (c) Proposed mechanism of transannulation.

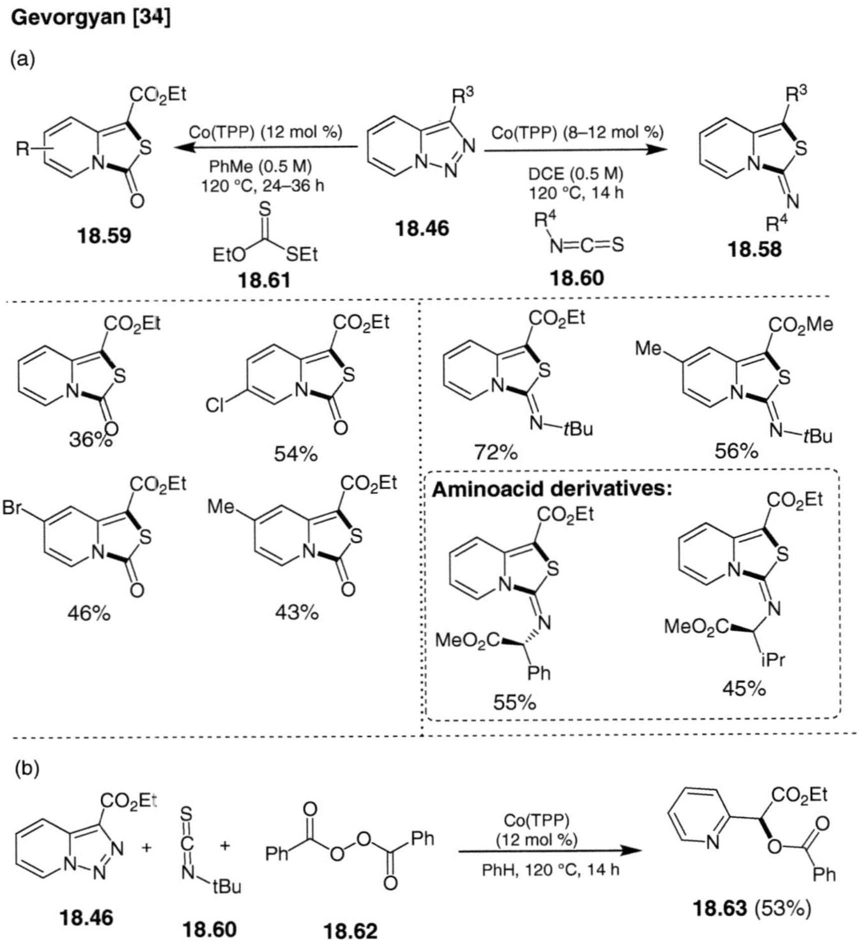

Scheme 18.26 Co-catalyzed transannulation of pyridotriazoles with isothiocyanates and xanthate esters. (a) Synthesis of imino-thiazolopyridines and oxo-thiazolopyridine derivatives. (b) Radical scavenging experiments. Source: Gevorgyan [34]/American Chemical Society.

In 2020, Gevorgyan group demonstrated an efficient transannulation strategy for the synthesis of imino-thiazolopyridines (**18.58**) and oxo-thiazolopyridine (**18.59**) derivatives via Co-catalyzed denitrogenative annulation reaction between pyridotriazoles (**18.46**) and isothiocyanates (**18.60**) or xanthate esters (**18.61**) (Scheme 18.26) [34]. The developed methodology proceeds under operationally simple conditions and exhibits wide functional-group tolerance with a broad substrate scope. Moreover, the synthesis of amino acid derivatives illustrated the potential synthetic utility of the following radical annulation reaction. The reaction mechanism follows the same as Scheme 18.25, and the mechanism is supported by the radical trapping experiments (**18.26b**).

18.5 Radical Denitrogenation of Substituted 3-Aminoindazoles

Indazole derivatives are an essential class of nitrogen-containing heterocycles that have been widely used in organic and medicinal chemistry because of their significant chemical and biological properties [35]. Particularly, 3-aminoindazoles (**18.64**) are readily available substrates and have recently arisen as versatile synthons for many denitrogenative transformations and offer a variety of nitrogen heterocycles. Moreover, 3-aminoindazoles have thus far proven to be effective arylating reagents via radical denitrogenative pathway.

In 2018, for the first time, Song et al. demonstrated an oxidative C—C bond formation between the 3-aminoindazoles (**18.64**) and aromatic/heteroaromatic scaffolds (**18.65**) via a copper-catalyzed denitrogenative pathway (Scheme 18.27).

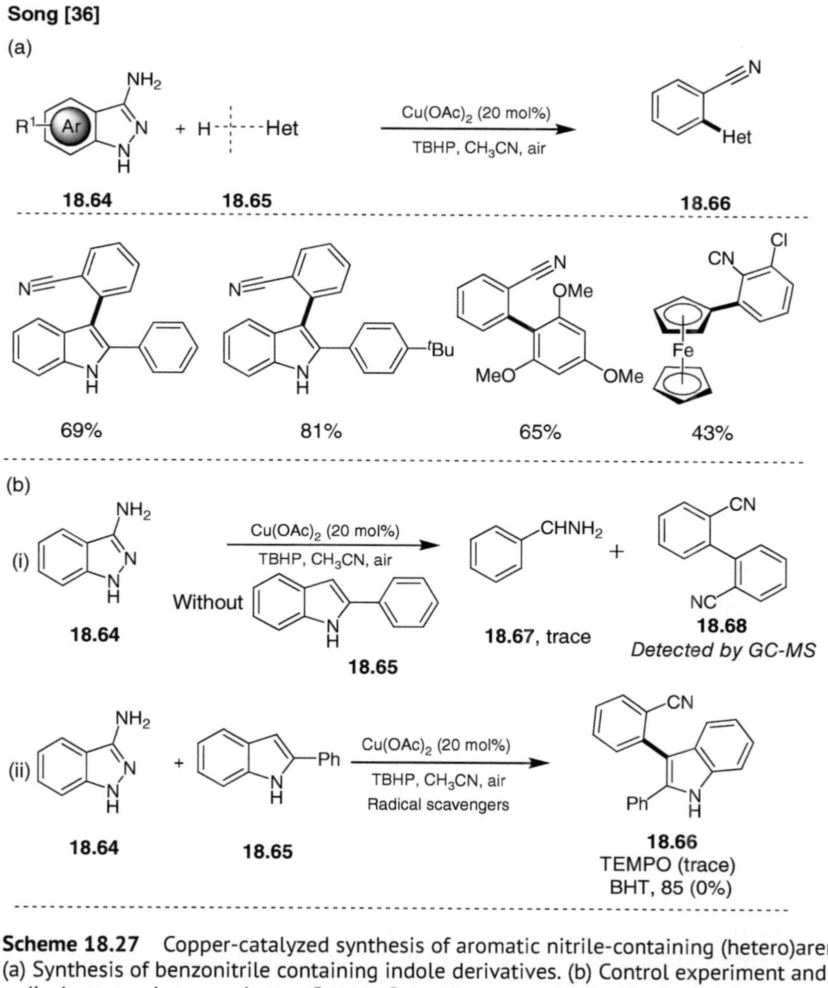

Scheme 18.27 Copper-catalyzed synthesis of aromatic nitrile-containing (hetero)arenes. (a) Synthesis of benzonitrile containing indole derivatives. (b) Control experiment and radical-scavenging experiment. Source: Song [36]/American Chemical Society.

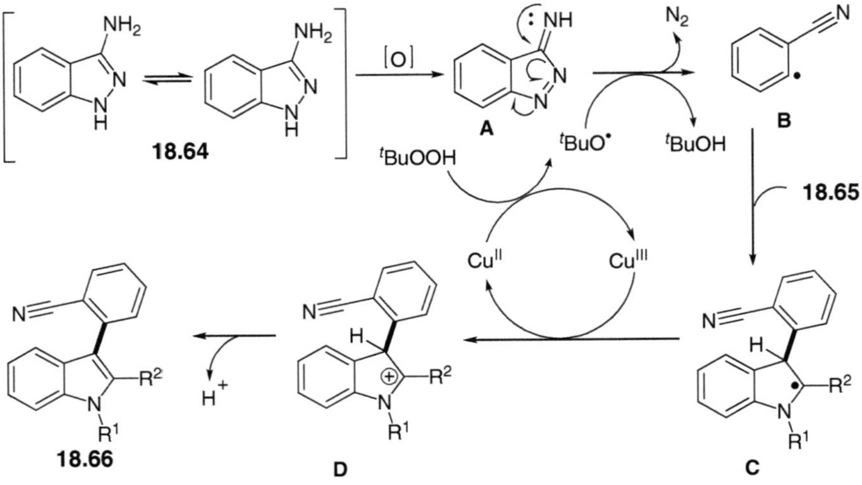

Scheme 18.28 Plausible mechanism.

A wide range of (hetero)arenes was well tolerated in this transformation to furnish the diversely functionalized aryl nitrile-derived (hetero)arenes (**18.66**) in good to excellent yields. The control experiments suggested that a radical-type reaction regime might be involved in this denitrogenative ring-opening of 3-aminoindazoles (Scheme 18.27b) [36].

The plausible mechanism is depicted in Scheme 18.28. Initially, oxidation of 3-aminoindazole (**18.64**) would give the intermediate **A**. Simultaneously, the Cu(II) undergoes a single electron transfer to peroxide to generate highly reactive tertbutoxyl radical (*t*BuO·) along with the Cu(III)-complex. Then, the active tertbutoxyl radical abstracts a proton from intermediate **A** and subsequent nitrogen extrusion to afford the aryl radical intermediate **B**. Concurrently, the intermediate **B** is attacked by the indole to give the intermediate **C**. Next, intermediate **C** further undergoes oxidation with Cu(III)-complex to deliver carbocation intermediate **D**. Finally, deprotonation of intermediate **D** provides the desired biaryl product (**18.66**).

Further, the same protocol extended the scope toward the formation of diverse C–S and C–Se bonds, and successfully achieved the corresponding 1,2-thiobenzonitriles (**18.69**) and 1,2-selanylbenzonitriles (**18.70**) derivatives (Scheme 18.29) [37].

Furthermore, the song group disclosed a novel Cu-catalyzed aromatic metamorphosis of 3-aminoindazoles (**18.71**) to the synthesis of diverse nitrile-containing triphenylenes (**18.72**) and fluorenes (**18.73**) via oxidative cleavage of two C—N bonds of aminoindazoles. Particularly, aryl-derived 3-aminoindazoles (**18.71**) were effectively converted underlying this Cu-catalyzed aromatic metamorphosis, providing the expected nitrile-containing triphenylenes (**18.72**) in decent yields. Moreover, alkenyl-derived 3-aminoindazoles (**18.74**) enable the assembly of the 9-benzylidene-9*H*-fluorene-1-carbonitriles (**18.73**) in moderate to good yields (Scheme 18.30a, b). The experimental and theoretical studies revealed that the reaction proceeds through the following steps in the reaction mechanism:

Song [37]

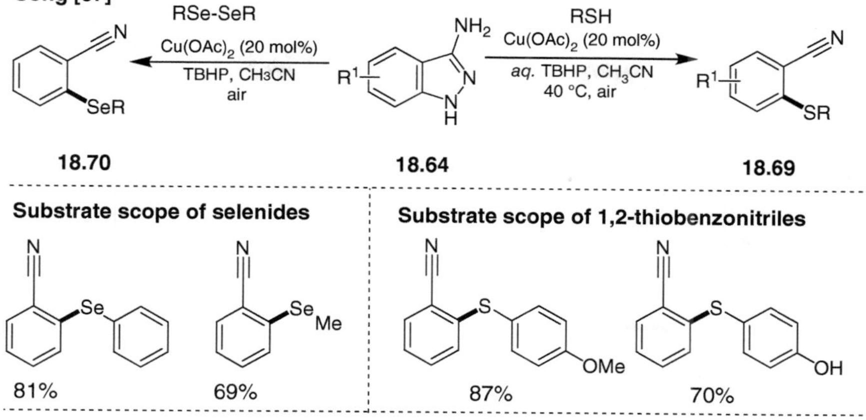

Scheme 18.29 Copper-catalyzed synthesis of aromatic nitrile-containing (hetero)arenes. Source: Song [37]/Royal Society of Chemistry.

Song [38]

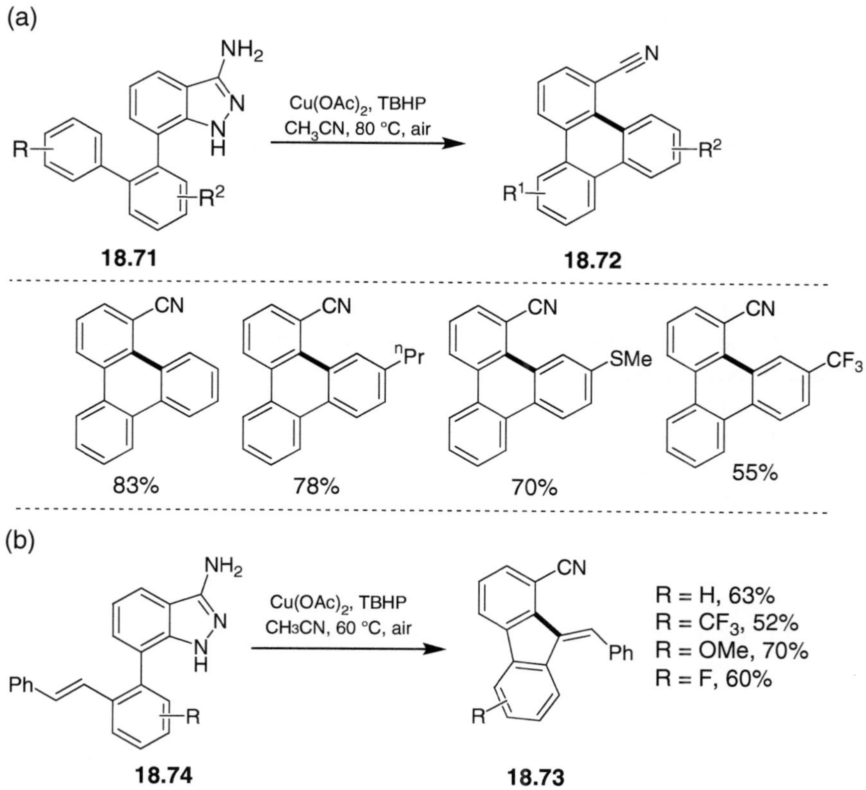

Scheme 18.30 Cu-catalyzed aromatic metamorphosis of 3-aminoindazoles. (a) Substrate scope for synthesis of triphenyleneses. (b) Substrate scope for synthesis of fluorenes. Source: Song [38]/American Chemical Society.

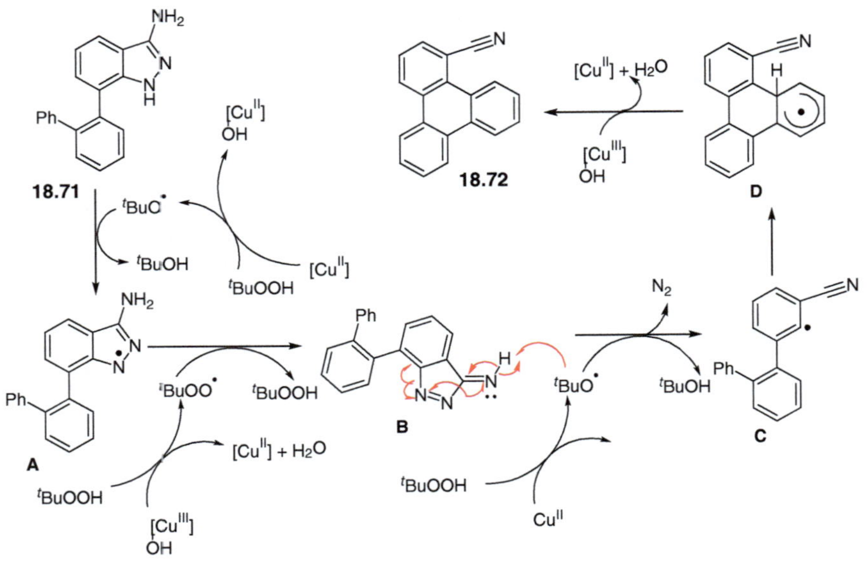

Scheme 18.31 Proposed mechanism.

oxidation/selective C–N bond activation/denitrogenation/radical cyclization (Scheme 18.31) [38].

In 2019, the same group developed a novel "extrude-and-sew" strategy for the construction of 1-amino isoquinolines (**18.75**) and 3-aminobenzothiophenes (**18.76**) via a Cu-catalyzed denitrogenative transannulation between 3-aminoindazoles (**18.64**) and enamine-esters (**18.77**) or mercapto-acetates (**18.78**). The substrate scope investigations revealed that the reaction was well tolerated for diversely functionalized 1-aminoisoquinoline and 3-aminobenzothiophene derivatives and obtained good yields (Scheme 18.32) [39]. The control experiments indicated that the reaction proceeds through a radical denitrogenative transannulation process to provide the arylated enamine intermediate, which undergoes cyclization with the help of the base. In the absence of the base, the reaction stops at the point of the C—H arylation step [40]. The proposed reaction mechanism is shown in Scheme 18.33, and it is similar to Scheme 18.31. The mechanism involves oxidation/selective C—N bond activation/denitrogenation/addition of the aryl radical to the enamine/oxidation followed by deprotonation/intramolecular nucleophilic addition. Finally, tautomerism delivers the desired annulated derivatives (Scheme 18.33).

In 2019, Cheng group developed a copper-catalyzed cascade denitrogenative transannulation/hydrolyzation strategy for the synthesis of 2,2-disubstituted indanones (**18.80**) from the 3-aminoindazoles (**18.64**) with *N*-(2-methylallyl)anilines (**18.79**). A broad range of indanones were conveniently obtained in moderate to excellent yields with good functional group tolerance under operationally simple reaction conditions (Scheme 18.34) [41]. To probe the mechanism, detailed mechanistic studies were performed. Notably, the reaction was evidently inhibited in

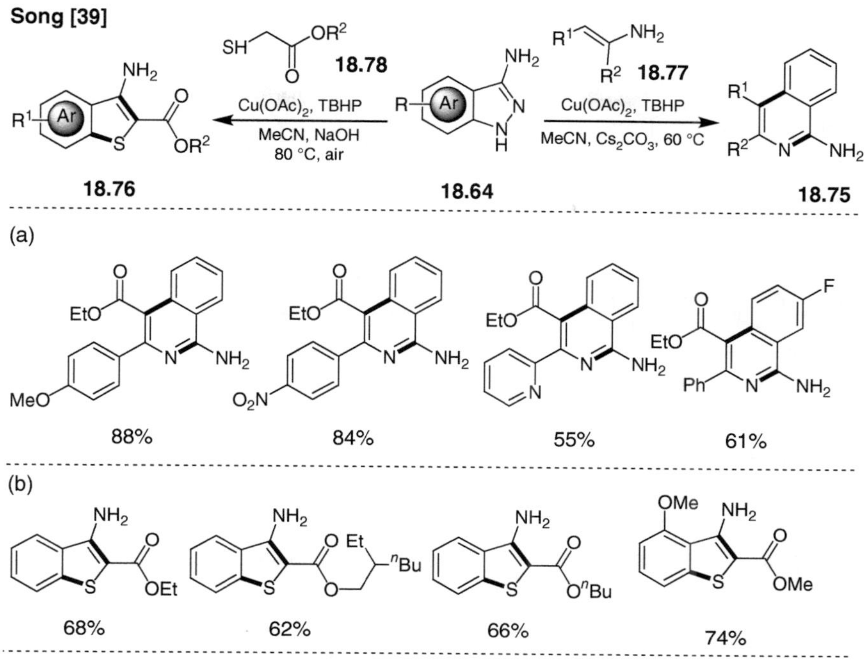

Scheme 18.32 Cu-catalyzed denitrogenative transannulation of 3-aminoindazoles. (a) Synthesis of 1-aminoisoquinolines. (b) Synthesis of 3-aminobenzothiophenes. Source: Song [39]/American Chemical Society.

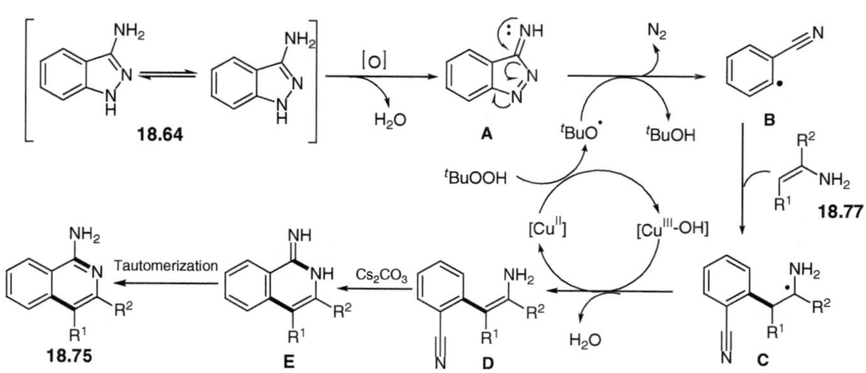

Scheme 18.33 Proposed mechanism for 1-amino isoquinolines.

the presence of radical scavengers such as TEMPO and BHT. In addition, product **18.81** was detected by GC-MS after the addition of 1,1-diphenylethylene. These results indicated that this transformation may proceed through a radical pathway. Moreover, the O^{18} isotopic labeling experiment revealed that the reaction proceeds through an imine formation followed by hydrolysis (Scheme 18.34b).

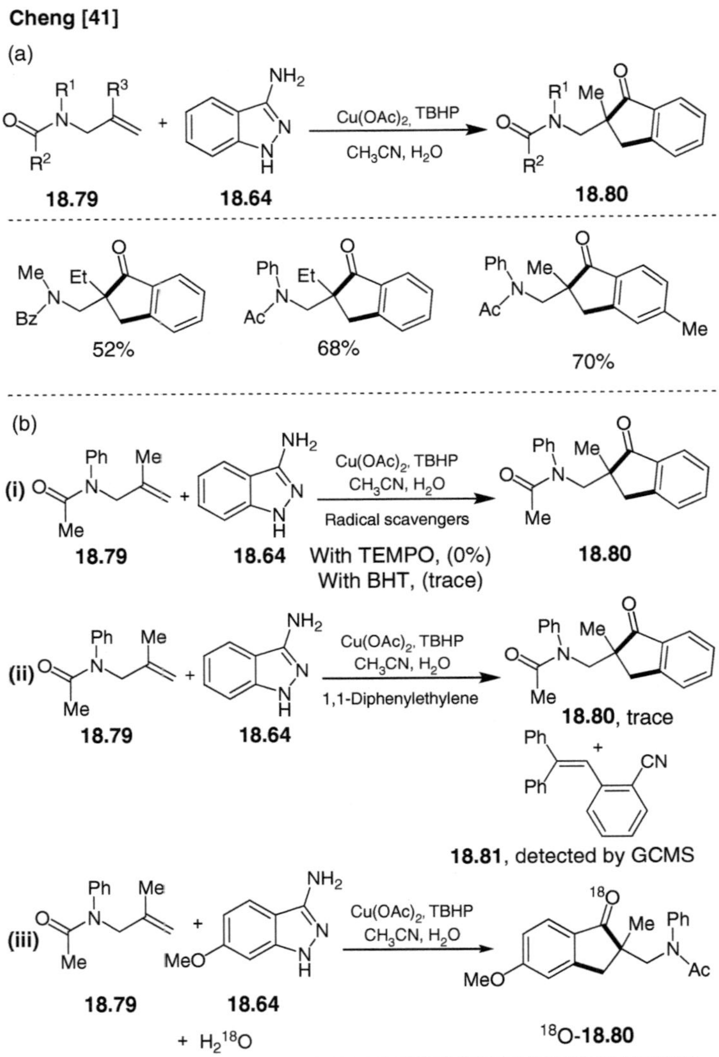

Scheme 18.34 Cu-catalyzed synthesis of disubstituted indanones. (a) 2,2-Desubstituted indanones. (b) Control experiments. Source: Cheng [41]/American Chemical Society.

Based on the control experiments, a plausible mechanism is proposed, which is depicted in Scheme 18.35. The mechanism follows the same steps of Scheme 18.31 for the formation of aryl radical intermediate-**B** from 3-aminoindazoles (**18.64**). The aryl radical intermediate **B** undergoes an intermolecular radical addition of the *N*-allyl aniline derivatives **18.79**, generating the intermediate **C**. Further, the intermediate **C** proceeds through the intramolecular radical addition to the nitrile moiety, giving an oxime intermediate **E**, along with the recovery of Cu(II) species. Finally, hydrolysis of oxime intermediate with water delivers the expected indanone derivatives (**18.80**).

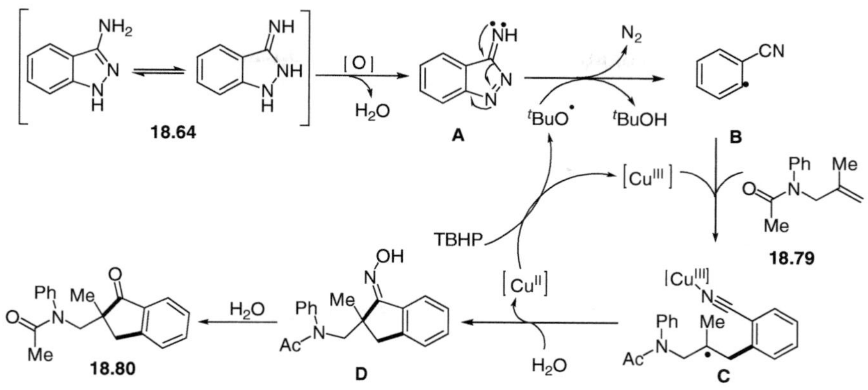

Scheme 18.35 Proposed mechanism.

18.6 Summary and Outlook

A series of efficient and remarkable strategies have been developed to synthesize valuable cyclic/acyclic molecules in a step-economical fashion via radical denitrogenative transformations of diverse poly-nitrogen heterocycles. Notably, most poly-nitrogen heterocycles are readily available through experimentally simple and environmentally friendly conditions. Moreover, it offers numerous opportunities and many breakthrough discoveries to construct a wide variety of heterocycle scaffolds, which are inaccessible from any traditional methods. The field of denitrogenative chemistry has continued to expand in the past decade as a growing number of light-triggered or transition-metal-catalyzed cascade reactions.

Despite significant progress in this rapidly developing research field, several limitations need to be addressed in the future: (i) the photocatalytic conditions utilize various peroxides as oxidants, which limits their utilization in large-scale reactions; hence to develop a catalytic system with ideal oxidants, such as molecular oxygen is more preferential. (ii) Most of the reactions followed the same catalytic or photolytic conditions, which could suppress the diversity. Thus, the goal is to develop new catalytic and operationally simple conditions that are more advantageous. (iii) Many nonenantioselective transformations have been discussed throughout the chapter; it would be a great invention to establish a catalytic enantioselective fashion of radical denitrogenative transformations of poly-nitrogen heterocycles soon. (iv) Even though various poly-nitrogen heterocycles are available, there is a need to introduce a wide variety of heterocycles as precursors, which could help to enhance the diversity in modern organic chemistry. We believe that more synthetic applications in this rapidly growing research area will be explored in the future.

References

1 (a) Kochi, J.K. (ed.) (1973). *Free Radicals, Vol. 1: Dynamics of Elementary Processes*. New York: Wiley Interscience. (b) Jasperse, C.P., Curran, D.P., and

Fevig, T.L. (1991). Radical reactions in natural product synthesis. *Chem. Rev.* 91: 1237–1286. c) Yan, M., Lo, J.C., Edwards, J.T., and Baran, P.S. (2016). Radicals: reactive intermediates with translational potential. *J. Am. Chem. Soc.* 138: 12692–12714.

2 (a) Xie, L.Y., Peng, S., Yang, L.H. et al. (2021). Aryl acyl peroxides for visible-light induced decarboxylative arylation of quinoxalin-2(1*H*)-ones under additive-, metal catalyst-, and external photosensitizer-free and ambient conditions. *Green Chem.* 23: 374–378. (b) Ding, G., Wang, Q., Liu, F. et al. (2021). A novel iron-chelating polyimide network as a visible-light-driven catalyst for photoinduced radical polymerization. *Chin. J. Catal.* 42: 141–151. (c) Narayanam, J.M. and Stephenson, C.R. (2011). Visible light photoredox catalysis: applications in organic synthesis. *Chem. Soc. Rev.* 40: 102–113. (d) Nicewicz, D.A. and MacMillan, D.W. (2008). Merging photoredox catalysis with organocatalysis: the direct asymmetric alkylation of aldehydes. *Science* 322: 77–80.

3 (a) Zollinger, H. (1994). *Diazo Chemistry I and II*. Weinheim: VCH. (b) Regitz, M. and Maas, G. (1986). *Diazo Compounds: Properties and Synthesis*. Orlando: Academic Press. (c) Doyle, M.P., McKervey, M.A., and Ye, T. (1998). *Modern Catalytic Methods for Organic Synthesis with Diazo Compounds*. New York: Wiley. (d) Padwa, A. and Hornbuckle, S.F. (1991). Ylide formation from the reaction of carbenes and carbenoids with heteroatom lone pairs. *Chem. Rev.* 91: 263–309.

4 (a) Yang, W.C., Chen, C.Y., Li, J.F., and Wang, Z.L. (2021). Radical denitrogenative transformations of polynitrogen heterocycles: building C—N bonds and beyond. *Chin. J. Catal.* 42: 1865–1875. (b) Rivas, M. and Gevorgyan, V. (2023). Advances in selected heterocyclization methods. *Synlett* 34: 1554–1562. (c) Roy, S., Das, S.K., Khatua, H. et al. (2021). Road map for the construction of high-valued N-heterocycles via denitrogenative annulation. *Acc. Chem. Res.* 54: 4395. (d) Das, S.K., Roy, S., and Chattopadhyay, B. (2023). Transition-metal-catalyzed denitrogenative annulation to access high-valued N-heterocycles. *Angew. Chem. Int. Ed.* 62: e202210912.

5 (a) von Fraunberg, K. and Huisgen, R. (1969). Kupferkatalysierte zerfallsreaktionen von imidaziden; reaktionen von kupfer-azenen. *Tetrahedron Lett.* 10: 2599–2602. (b) Huisgen, R., Fraunberg, K.V., and Sturm, H.J. (1969). Cycloadditionen von pyridyl-und pyrimidyl-aziden. *Tetrahedron Lett.* 10: 2589–2594. (c) Huisgen, R. and Fraunberg, K.V. (1969). Zur thermolyse des 2-azido-4.6-dimethyl-pyrimidins; reaktionen des azens. *Tetrahedron Lett.* 10: 2595–2598.

6 Wentrup, C. (2017). Flash vacuum pyrolysis of azides, triazoles, and tetrazoles. *Chem. Rev.* 117: 4562–4623.

7 (a) Chuprakov, S., Hwang, F.W., and Gevorgyan, V. (2007). Rh-catalyzed transannulation of pyridotriazoles with alkynes and nitriles. *Angew. Chem. Int. Ed.* 46: 4757–4759. (b) Horneff, T., Chuprakov, S., Chernyak, N. et al. (2008). Rhodium-catalyzed transannulation of 1,2,3-triazoles with nitriles. *J. Am. Chem. Soc.* 130: 14972–14974. (c) Gulevich, A.V. and Gevorgyan, V. (2013). Versatile reactivity of rhodium–iminocarbenes derived from N-sulfonyl triazoles. *Angew. Chem. Int. Ed.* 52: 1371–1373.

8 (a) Khalid, Z., Ahmad, H.A., Munawar, M.A. et al. (2017). 1,2,3-Benzotriazin-4 (3H)-ones: synthesis, reactions and applications. *Heterocycles* 94: 3–54. (b) Gyűjtő, I., Simig, G., Porcs-Makkay, M., and Volk, B. (2020). Synthesis and chemistry of 1,2,3-benzothiadiazine 1,1-dioxide derivatives: a comprehensive overview. *Chemistry* 2: 674–690.

9 (a) Clark, A.S., Deans, B., Stevens, M.F.G. et al. (1995). Antitumor imidazotetrazines. 32.1 synthesis of novel imidazotetrazinones and related bicyclic heterocycles to probe the mode of action of the antitumor drug temozolomide. *J. Med. Chem.* 38: 1493–1504. (b) Caliendo, G., Fiorino, F., Grieco, P. et al. (1999). Preparation and local anaesthetic activity of benzotriazinone and benzoyltriazole derivatives. *Eur. J. Med. Chem.* 34: 1043–1051.

10 (a) Miura, T., Yamauchi, M., and Murakami, M. (2008). Synthesis of 1(2H)-isoquinolones by the nickel-catalyzed denitrogenative alkyne insertion of 1,2,3-benzotriazin-4 (3*H*)-ones. *Org. Lett.* 10: 3085–3088. (b) Yamauchi, M., Morimoto, M., Miura, T., and Murakami, M. (2010). Enantioselective synthesis of 3, 4-dihydroisoquinolin-1 (2*H*)-ones by nickel-catalyzed denitrogenative annulation of 1, 2, 3-benzotriazin-4 (3*H*)-ones with allenes. *J. Am. Chem. Soc.* 132: 54–55.

11 Fang, Z.J., Zheng, S.C., Guo, Z. et al. (2015). Asymmetric synthesis of axially chiral isoquinolones: nickel-catalyzed denitrogenative transannulation. *Angew. Chem. Int. Ed.* 54: 9528–9532.

12 Nakamura, I., Nemoto, T., Shiraiwa, N., and Terada, M. (2009). Palladium-catalyzed indolization of *N*-aroylbenzotriazoles with disubstituted alkynes. *Org. Lett.* 11: 1055–1058.

13 (a) Bae, I., Han, H., and Chang, S. (2005). Highly efficient one-pot synthesis of *N*-sulfonylamidines by Cu-catalyzed three-component coupling of sulfonyl azide, alkyne, and amine. *J. Am. Chem. Soc.* 127: 2038–2039. (b) Cho, S.H., Yoo, E.J., Bae, I., and Chang, S. (2005). Copper-catalyzed hydrative amide synthesis with terminal alkyne, sulfonyl azide, and water. *J. Am. Chem. Soc.* 2005 (127): 16046–16047.

14 Wang, H. and Yu, S. (2015). Synthesis of isoquinolones using visible-light-promoted denitrogenative alkyne insertion of 1,2,3-benzotriazinones. *Org. Lett.* 17: 4272–4275.

15 (a) Xu, P., López-Rojas, P., and Ritter, T. (2021). Radical decarboxylative carbometalation of benzoic acids: a solution to aromatic decarboxylative fluorination. *J. Am. Chem. Soc.* 143: 5349–5354. (b) Chateauneuf, J., Lusztyk, J., and Ingold, K.U. (1988). Spectroscopic and kinetic characteristics of aroyloxyl radicals. 2. Benzoyloxyl and ring-substituted aroyloxyl radicals. *J. Am. Chem. Soc.* 110: 2886–2893. (c) Yi, J., Badir, S.O., Kammer, L.M. et al. (2019). Deaminative reductive arylation enabled by nickel/photoredox dual catalysis. *Org. Lett.* 21: 3346–3351. (d) Basch, C.H., Liao, J., Xu, J. et al. (2017). Harnessing alkyl amines as electrophiles for nickel-catalyzed cross couplings via C—N bond activation. *J. Am. Chem. Soc.* 139: 5313–5316.

16 Han, Y.Y., Wang, H., and Yu, S. (2016). Synthesis of biaryl sultams using visible-light-promoted denitrogenative cyclization of 1,2,3,4-benzothiatriazine-1,1-dioxides. *Org. Chem. Front.* 3: 953–956.

17 Thorat, V.H., Hsieh, J.C., and Cheng, C.H. (2020). Transition-metal-free tandem cyclization/N-arylation reaction: a method to access biaryl sultam derivatives via a diradical pathway. *Org. Lett.* 22: 6623–6627.

18 Chen, F., Hu, S., Li, S. et al. (2021). Visible-light-induced denitrogenative phosphorylation of benzotriazinones: a metal-and additive-free method for accessing ortho-phosphorylated benzamide derivatives. *Green Chem.* 23: 296–301.

19 Liu, B.X., Wang, F., Chen, Y. et al. (2022). Visible-light-promoted denitrogenative *ortho*-selenylation reaction of benzotriazinones: synthesis of ortho-selenylated benzamides and ebselen analogs. *Org. Chem. Front.* 9: 2418–2423.

20 (a) Matin, M.M., Matin, P., Rahman, M.R. et al. (2022). Triazoles and their derivatives: chemistry, synthesis, and therapeutic applications. *Front. Mol. Biosci.* 9: 864286. (b) Sahu, J.K., Ganguly, S., and Kaushik, A. (2013). Triazoles: a valuable insight into recent developments and biological activities. *Chin. J. Nat. Med.* 11: 456–465. (c) Celik, F., Unver, Y., Barut, B. et al. (2018). Synthesis, characterization and biological activities of new symmetric bis-1,2,3-triazoles with click chemistry. *Med. Chem.* 14: 230–241.

21 (a) Devaraj, N.K. and Finn, M.G. (2021). Introduction: click chemistry. *Chem. Rev.* 121: 6697–6698. (b) Hein, C.D., Liu, X.M., and Wang, D. (2008). Click chemistry, a powerful tool for pharmaceutical sciences. *Pharm. Res.* 25: 2216–2230. (c) Moses, J.E. and Moorhouse, A.D. (2007). The growing applications of click chemistry. *Chem. Soc. Rev.* 36: 1249–1262.

22 Aggarwal, S., Richards, W.J., and Fokin, V.V. (2023). Generation and aerobic oxidation of azavinyl captodative radicals. *J. Am. Chem. Soc.* 145: 24452–24458.

23 Kumar, D., Mishra, A., Mishra, B.B. et al. (2013). Synthesis of glycoconjugate benzothiazoles via cleavage of benzotriazole ring. *J. Org. Chem.* 78: 899–909.

24 Kumar, D., Mishra, B.B., and Tiwari, V.K. (2014). Synthesis of 2-N/S/C-substituted benzothiazoles via intramolecular cyclative cleavage of benzotriazole ring. *J. Org. Chem.* 79: 251–266.

25 Su, Y., Petersen, J.L., Gregg, T.L., and Shi, X. (2015). Ambient benzotriazole ring opening through intermolecular radical addition to vinyltriazole. *Org. Lett.* 17: 1208–1211.

26 Teders, M., Gómez-Suárez, A., Pitzer, L. et al. (2017). Diverse visible-light-promoted functionalizations of benzotriazoles inspired by mechanism-based luminescence screening. *Angew. Chem. Int. Ed.* 56: 902–906.

27 Teders, M., Pitzer, L., Buss, S., and Glorius, F. (2017). Regioselective synthesis of 2-substituted indoles from benzotriazoles and alkynes by photoinitiated denitrogenation. *ACS Catal.* 7: 4053–4056.

28 Jian, Y., Chen, M., Huang, B. et al. (2018). Visible-light-induced C (sp^2)—P bond formation by denitrogenative coupling of benzotriazoles with phosphites. *Org. Lett.* 20: 5370–5374.

29 Li, J., Wang, S., Zhao, J., and Li, P. (2022). Visible light-promoted radical-mediated ring-opening/cyclization of vinyl benzotriazoles: an alternative approach to phenanthridines. *Org. Lett.* 24: 5977–5981.

30 (a) Jones, G. (2002). The chemistry of the triazolopyridines: an update. *Adv. Heterocycl. Chem.* 83: 1–70. (b) Adam, R., Alom, S., Abarca, B., and Ballesteros, R. (2016). Reactivity of [1,2,3] Triazolo [1,5-*a*] pyridines as 1,3-dipoles. *Tetrahedron* 72: 8436–8441. (c) Abarca, B., Ballesteros, R., Ballesteros-Garrido, R. et al. (2008). Triazolopyridines. Part 26: The preparation of novel [1, 2, 3] triazolo [1,5-*a*] pyridine sulfoxides. *Tetrahedron* 64: 3794–3801. (d) Jones, G. and Abarca, B. (2010). The chemistry of the [1,2,3]-triazolo-[1,5-*a*] pyridines: an update. *Adv. Heterocycl. Chem.* 100: 195–252.

31 Chattopadhyay, B. and Gevorgyan, V. (2012). Transition-metal-catalyzed denitrogenative transannulation: converting triazoles into other heterocyclic systems. *Angew. Chem. Int. Ed.* 51: 862–872.

32 Joshi, A., Chandra Mohan, D., and Adimurthy, S. (2016). Copper-catalyzed denitrogenative transannulation reaction of pyridotriazoles: synthesis of imidazo [1,5-*a*] pyridines with amines and amino acids. *Org. Lett.* 18: 464–467.

33 Roy, S., Das, S.K., and Chattopadhyay, B. (2018). Cobalt(II)-based metalloradical activation of 2-(diazomethyl) pyridines for radical transannulation and cyclopropanation. *Angew. Chem. Int. Ed.* 57: 2238–2243.

34 Zhang, Z. and Gevorgyan, V. (2020). Co-catalyzed transannulation of pyridotriazoles with isothiocyanates and xanthate esters. *Org. Lett.* 22: 8500–8504.

35 (a) Zhang, S.G., Liang, C.G., and Zhang, W.H. (2018). Recent advances in indazole-containing derivatives: synthesis and biological perspectives. *Molecules* 23: 2783. (b) Wan, Y., He, S., Li, W., and Tang, Z. (2018). Indazole derivatives: promising anti-tumor agents. *Anti Cancer Agents Med. Chem.* 18: 1228–1234.

36 Zhou, Y., Deng, S., Mai, S., and Song, Q. (2018). Cu-catalyzed denitrogenative ring-opening of 3-aminoindazoles for the synthesis of aromatic nitrile-containing (hetero) arenes. *Org. Lett.* 20: 6161–6165.

37 Zhou, Y., Wang, Y., Lou, Y., and Song, Q. (2019). Facile synthesis of 1,2-thiobenzonitriles via Cu-catalyzed denitrogenative radical coupling reaction. *Chem. Commun.* 55: 10265–10268.

38 Cu-Catalyzed Zhou, Y., Lin, L., Wang, Y. et al. (2019). Cu-catalyzed aromatic metamorphosis of 3-aminoindazoles. *Org. Lett.* 21: 7630–7634.

39 Zhou, Y., Wang, Y., Lou, Y., and Song, Q. (2019). Cu-catalyzed denitrogenative transannulation of 3-aminoindazoles to assemble 1-aminoisoquinolines and 3-aminobenzothiophenes. *Org. Lett.* 21: 8869–8873.

40 Zhou, Y., Wang, Y., Song, Z. et al. (2020). Cu-catalyzed C—N bond cleavage of 3-aminoindazoles for the C—H arylation of enamines. *Org. Chem. Front.* 7: 25–29.

41 Teng, J., Sun, S., Yu, J.T., and Cheng, J. (2019). Copper-catalyzed cascade denitrogenative transannulation/hydrolyzation of 3-aminoindazoles toward 2,2-disubstituted indanones. *J. Org. Chem.* 84: 15669–15676.

19

Application of Denitrogenative Transformations in Synthesis of Natural Products

H. R. Lohitha and Vignesh Palani

Department of Organic Chemistry, Indian Institute of Science, Bangalore, Karnataka, India

19.1 Introduction

Owing to the widespread presence of heterocyclic scaffolds in natural products and medicinal drugs [1], the development of methodologies enabling elegant synthesis of such scaffolds has gained significant attention in recent years. As witnessed in the previous chapters, polynitrogen heterocycles are readily employed as versatile synthons for the formation of new C—C and C—*X* bonds and, in turn, a library of heterocyclic scaffolds [2, 3]. Over the years, denitrogenative transformations of polynitrogen heterocycles have emerged as a powerful strategy to assemble chemical structures that have previously been challenging to access [4–7]. The success of the general transformation hinges on the effective generation of highly reactive metal carbenoids or radical species from the corresponding polynitrogen heterocyclic precursors [8–10]. The developed methodologies illustrate the great synthetic potential of denitrogenative cascade reactions, which provide a direct and rapid approach to a myriad of heterocyclic scaffolds. This chapter highlights the versatility of employing denitrogenative transformations in various synthetic settings to access multiple natural products. The chosen total syntheses by several research groups stand as a testament to this rapidly evolving domain.

19.2 Enantioselective Synthesis of (*R*)-Cycloprodigiosin

In 2013, Schultz and Sarpong disclosed the first enantioselective synthesis of the natural product, (*R*)-cycloprodigiosin (Scheme 19.1b) [11]. Produced by bacterial strains including *Pseudoalteromonar (Altermonas) rubra*, *Pseudoalteromonas denitrificans*, and *Vibrio gazogenes*, (*R*)-cycloprodigiosin **19.1g** is identified to possess both proapoptotic anticancer and immunosuppressant activities [12]. Before this work, only one reported racemic synthesis of cycloprodigiosin by Wasserman and co-workers existed [13]. In this work, the authors envisioned assembling the pyrrole

Denitrogenative Transformation of Nitrogen Heterocycles: Synthesis, Reactions and Applications, First Edition. Edited by Pazhamalai Anbarasan.

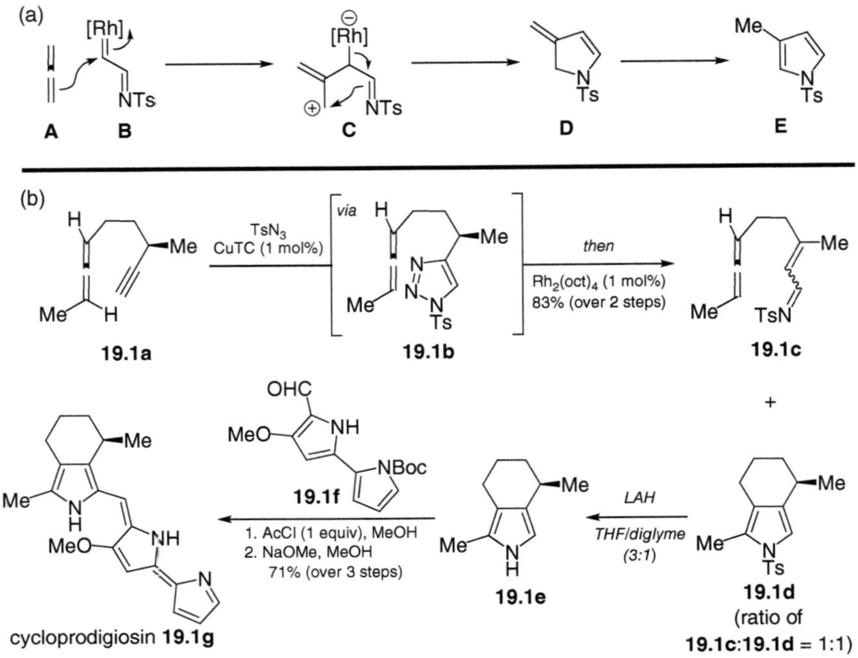

Scheme 19.1 (a) Proposed pyrrole annulation strategy. (b) Enantioselective synthesis of (*R*)-cycloprodigiosine.

moiety in (*R*)-cycloprodigiosin **19.1g** via an annulation strategy as outlined in Scheme 19.1a. Following the successful generation of the Rh-carbenoid **B**, the authors anticipated that the allene moiety **A** would react with **B** to generate intermediate **C**. A subsequent ring closure would furnish **D**, yielding pyrrole **E** upon aromatization.

The final route commenced with the synthesis of enantioenriched allenylalkyne **19.1a**, obtained in six steps from the known compound, as a mixture of diastereomers. Following the formation of *N*-sulfonyl-1,2,3-triazole **19.1b**, the reaction mixture was treated with $Rh_2(oct)_4$ in the same pot, which initiated the pyrrole annulation (*c.f.* **A** to **E**, Scheme 19.1a) to furnish a mixture of α,β-unsaturated imine **19.1c** and the desired pyrrole **19.1d** in a ratio of 1 : 1. The greater availability of the methine hydrogen toward migration in the intermediate Rh-carbenoid, resulted in competing formation of the undesired α,β-unsaturated imine **19.1c**. Efforts toward minimization of the formation of **19.1c** were met with unsatisfactory results. Nevertheless, the authors proceeded with the total synthesis of (*R*)-cycloprodigiosin **19.1g**, considering the overall satisfying isolated yield of pyrrole **19.1d**. Subsequent cleavage of the tosyl group under the lithium aluminum hydride (LAH) conditions furnished the (NH)-pyrrole **19.1e**. Finally, condensation of **19.1e** with aldehyde **19.1f** under Lindsley's conditions [14] yielded (*R*)-cycloprodigiosin **19.1g** in an overall 71% yield over the three steps from **19.d.**

19.3 Synthesis of Uhle's Ketone

3,4-Fused indole skeleton represents a key structural motif in various natural products possessing noteworthy biological activities (Scheme 19.2a). A few selected bioactive molecules include rugulovasine A **19.2a**, chanoclavine-I **19.2b**, lysergic acid **19.2c**, and hapalindole U **19.2d**. As such, methods for assembling 3,4-fused indole scaffolds have gained prominent attention. In 2014, Miura and Murakami reported a complementary method in this direction where they developed an intramolecular [3 + 2] annulation between α-imino Rh-carbenoid species and aryl rings to have straightforward access to 3,4-fused indole skeletons [15]. The synthetic utility of this method was then demonstrated in the synthesis of Uhle's ketone **19.2g**, which has often been employed in the synthesis of *Ergot* alkaloids [16] (Scheme 19.2b).

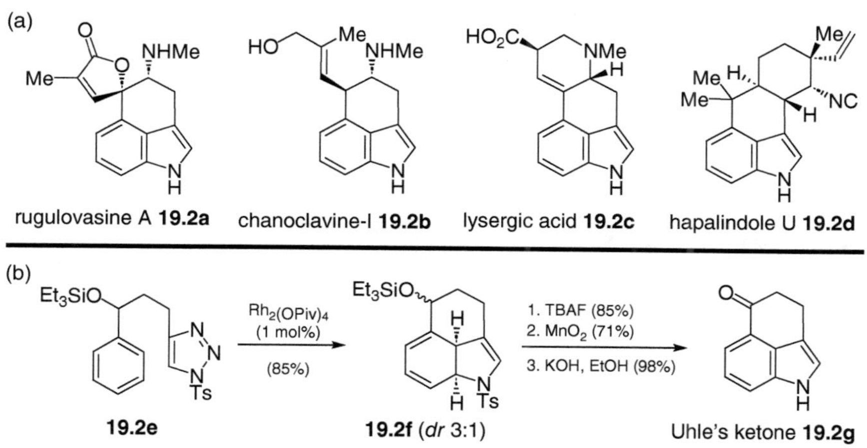

Scheme 19.2 (a) Natural products with 3,4-fused indole skeletons. (b) Synthesis of Uhle's ketone.

The authors synthesized *N*-sulfonyl-1,2,3-triazole **19.2e** from the corresponding 1-alkyne precursor via click reaction. Upon treatment with $Rh_2(OPiv)_4$, the generation of the α-imino Rh-carbenoid species was expected. Considering the high electrophilicity of the carbenoid carbon and the nucleophilic nature of the α-imino nitrogen, the α-imino Rh-carbenoid species was anticipated to behave as a 1,3-dipole equivalent to participate in a [3 + 2] annulation with unsaturated compounds. As envisioned, **19.2e** was directly transformed to the tricyclic intermediate **19.2f**, which was isolated as a mixture of diastereomers in a ratio of 3 : 1. The silyl group was then cleaved under tetrabutylammonium fluoride (TBAF) conditions. Subsequent oxidation in the presence of manganese dioxide yielded the desired 3,4-fused indole scaffold. A late-stage detosylation with potassium hydroxide completed the synthesis of Uhle's ketone **19.2g** in an overall short synthetic sequence.

19.4 Total Synthesis of (+)-Lysergol

In 2017, the utility of the intramolecular [3 + 2] annulation in constructing the 3,4-fused indole skeleton was subsequently showcased by Luo and co-workers in their synthesis of the ergot alkaloid, (+)-lysergol (Scheme 19.3) [17]. Following literature precedence [18], the authors synthesized the chiral monocyclic enamine **19.3a** from (*R*)-Garner's aldehyde in an overall seven-step sequence. The enamine **19.3a** was converted to the ring-enlarged intermediate **19.3c** in a two-step protocol. Toward this end, **19.3a** was treated with the *in situ* generated dibromocarbene and the resulting bicyclic intermediate was then ring-opened by employing silyl ketene acetal **19.3b** as the nucleophile to furnish **19.3c** as a mixture of diastereomers in a ratio of 2 : 1. The mixture then underwent Suzuki coupling with phenylboronic acid to provide the corresponding coupled product **19.3d** in a diastereomeric ratio of 2 : 1.

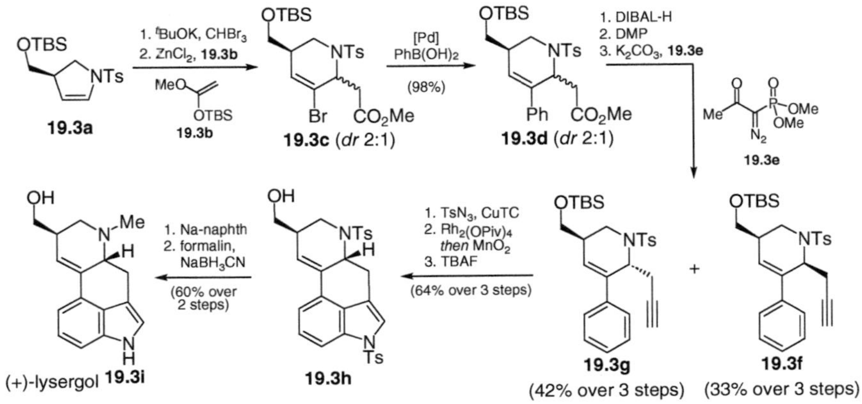

Scheme 19.3 Total synthesis of (+)-lysergol.

Following redox manipulation and Seyferth–Gilbert homologation by employing Bestmann–Ohira reagent **19.3e**, **19.3d** was readily translated to alkynes **19.3f** and **19.3g** in a straightforward three-step sequence. At this stage, the epimers were separable, and the authors proceeded with the diastereomer **19.3g** to construct the key 3,4-fused indole scaffold of (+)-lysergol **19.3i** via the envisioned [3 + 2] annulation strategy. Under standard protocol, a copper-catalyzed [3 + 2] cycloaddition was first effected between **19.3g** and tosyl azide to obtain the key *N*-sulfonyl-1,2,3-triazole intermediate. A one-pot rhodium-catalyzed intramolecular [3 + 2] annulated followed by oxidative aromatization then proceeded smoothly to assemble the 3,4-fused indole skeleton, which upon desilylation furnished the tetracyclic intermediate **19.3h** in 64% yield over three steps. A late-stage tosyl cleavage followed by *N*-methylation completed the total synthesis of (+)-lysergol **19.3i** in 11 steps starting from enantiopure monocyclic enamine **19.3a**. Notably, in a similar fashion, epimer **19.3f** was also translated to (−)-isolysergol, which is epimeric to **19.3i** at the fused position.

19.5 Total Synthesis of (–)-Chanoclavine I

The developed intramolecular [3 + 2] annulation strategy continued to find success in the synthesis of other ergot alkaloids as well. For instance, Cao and co-workers reported the synthesis of (–)-chanoclavine I **19.4e** in 2017 by employing the annulation strategy to construct the 3,4-fused indole moiety (Scheme 19.4) [19]. The authors accomplished an enantiospecific synthesis of **19.4e** by employing the (2*R*)-(+)-phenyloxirane **19.4a** derived building block **19.4b**. Having gained access to the alkyne **19.4b**, the stage was set to initiate the annulation protocol, which commenced with the click reaction between **19.4b** and tosyl azide to generate *N*-sulfonyl-1,2,3-triazole **19.4c**. Under the standard set of conditions, **19.4c** was transformed to the annulated tricyclic indole compound **19.4d** in an overall 77% yield over two steps. The compound **19.4d** now possesses the desired 3,4-fused indole skeleton present in the natural product (–)-chanoclavine I **19.4e**. Finally, a late-stage redox manipulation followed by tosyl cleavage completed the total synthesis of (–)-chanoclavine I **19.4e** in 13 steps starting from (2*R*)-(+)-phenyloxirane **19.4a** with an overall 17% yield.

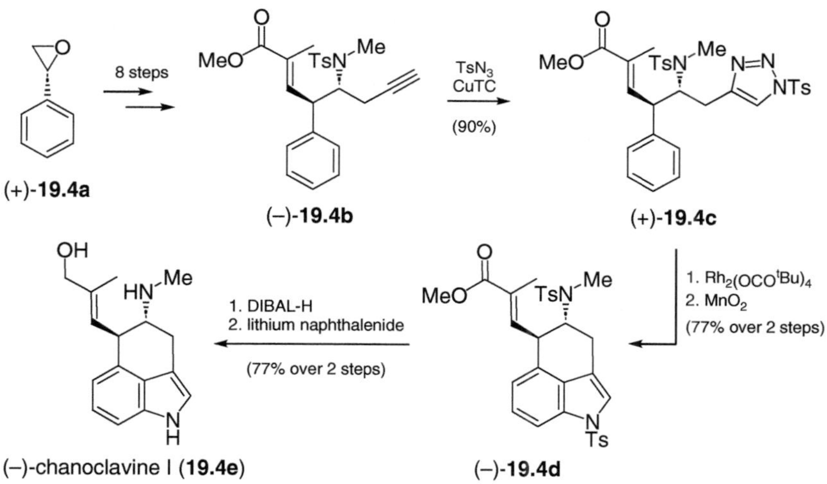

Scheme 19.4 Total synthesis of (–)-chanoclavine I.

Traditionally, synthesis of such ergot alkaloids commences with prefunctionalized indole starting building blocks. The annulation strategy enabled via successful generation of α-imino Rh-carbenoid species serves as a complementary approach to access the indole skeleton in a late-stage setting. For instance, before this work, there exist two independent syntheses of (–)-chanoclavine I **19.4e**, both of which commenced with 4-substituted indole derivatives. The first total synthesis of **19.4e** was disclosed by Genet and co-workers in 1994 in a 12-step synthetic sequence with a 1.6% overall yield [20]. After two years, the next synthesis of **19.4e** was reported by Yokoyama and Murakami in a total 16-step linear sequence with an overall 9%

yield [21]. Hence, this denitrogenative protocol offers an alternative platform to not only revise the retrosynthetic design but also provide an opportunity to improve the overall isolated yield.

19.6 Total Syntheses of Tuberostemospiroline and Stemona-Lactam R

The Rh-catalyzed denitrogenative transformation was extended to construct oxaspirocycles, a valuable structural motif in various biologically active natural products [22]. Despite its significant presence in bioactive molecules, syntheses of oxaspirocycle scaffolds have drawn considerable challenge owing to the difficulty in establishing the stereochemical configuration at the quaternary spiro center. Thus, developing new methods to construct spirocycles with well-defined stereochemistry is highly warranted. In 2014, Yang and co-workers reported an innovative method to build oxaspirocyles by employing α-imino Rh-carbenoid as a reactive species [23]. The synthetic utility of the method was further demonstrated in the stereoselective syntheses of tuberostemospiroline **19.5f** and stemona-lactam R **19.5g** (Scheme 19.5). Both **19.5f** and **19.5g** were isolated from Stemonaceae, primarily employed as a traditional medicine to treat respiratory disorders in China [24, 25]. The unique structure and intriguing bioactivity make these Stemona alkaloids attractive targets for total synthesis.

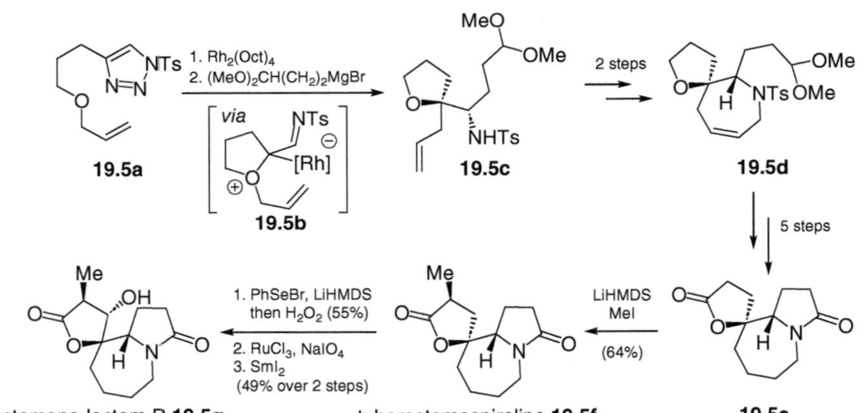

Scheme 19.5 Total synthesis of stemona-lactam R.

The authors initiated their study with the synthesis of *N*-sulfonyl-1,2,3-triazole **19.5a**. The authors envisioned that the generated α-imino Rh-carbene complex would intramolecularly engage with the ether oxygen to provide intermediate **19.5b**. A subsequent stereochemically defined [2,3]-sigmatropic rearrangement was anticipated to proceed in the oxonium ylide **19.5b** to migrate the allyl group. As predicted, the expected tetrahydrofuran-substituted imine was obtained, which was then reacted with (3,3-dimethoxypropyl) magnesium bromide to furnish amine **19.5c** as the significant diastereomer in a ratio of 20 : 1. Notably, when the Grignard addition was conducted at low temperature, the undesired stereoisomer was obtained as the major diastereomer. In a further two-step sequence, amine **19.5c** was elaborated to the spirocycle **19.5d**, which was then translated to the tricycle **19.5e** in a five-step linear sequence. A subsequent selective methylation completed the total synthesis of tuberostemospiroline **19.5f**. Skeletally, a β-hydroxy group had to be installed to complete the synthesis of stemona-lactam R **19.5g**. Toward this end, a three-step protocol was undertaken, which commenced with the α,β-unsaturation of the lactone moiety. A subsequent dihydroxylation with $NaIO_4$, followed by SmI_2-mediated dihydroxylation of the resultant diol, gave stemona-lactam R **19.5g** in an overall 27% yield over three steps starting from tuberostemospiroline **19.5f**.

19.7 Total Synthesis of (+)-Petromyroxol

Following the report by Yang and co-workers [23], the employment of denitrogenative transformations toward the assembly of oxaspirocycle framework found success in the total synthesis of other natural products. Toward this end, in 2015, Boyer explored the efficiency of this strategy in the synthesis of (+)-petromyroxol **19.6d** (Scheme 19.6) [26]. Petromyrxols are tetrahydrofuran-containing natural products isolated from water conditioned with larval sea lamprey. Due to the low natural abundance of **19.6d** (2.9 mg from over 1,00,000 l of water), extensive biological studies on **19.6d** were hampered [27]. The synthesis commenced with the formation of the requisite enantiopure alkyne **19.6a**. The asymmetric synthesis of **19.6a** proceeded in an overall four-step sequence from readily accessible starting material. Following the click reaction to form the *N*-sulfonyl-1,2,3-triazole **19.6b**, the generated Rh-carbenoid species interacted with the ether oxygen to assemble the required tetrahydrofuran core and the resulting tosyl imine was hydrolyzed to furnish the desired intermediate **19.6c** in an overall 60% yield starting from **19.6a**. The obtained 3-oxo-tetrahydrofuran intermediate **19.6c** was eventually

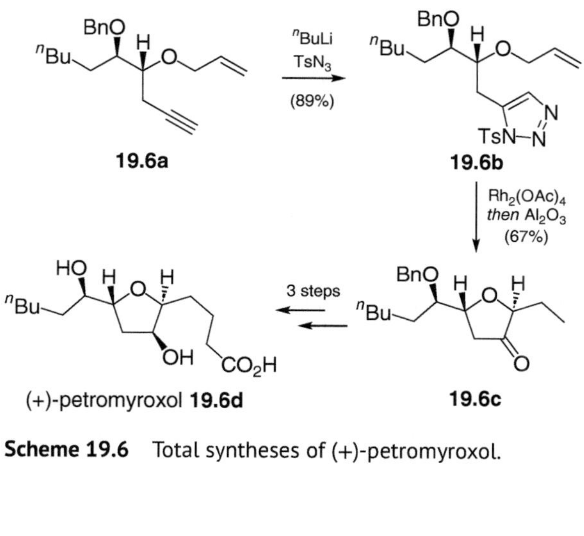

Scheme 19.6 Total syntheses of (+)-petromyroxol.

translated to the natural product, (+)-petromyroxol **19.6d**, in a further three-step sequence. In summary, the first enantioselective synthesis of (+)-petromyroxol **19.6d** was completed in only nine steps, starting from readily accessible building blocks with an overall yield of 20%.

In the two examples above, a denitrogenative protocol was employed to generate 2,2-disubstituted tetrahydrofuran compounds. In 2015, Sarpong and co-workers demonstrated the stereodivergent synthesis of saturated 2,3-disubstituted tetrahydrofuran compounds **19.7b** and **19.7c** through Rh(II) azavinyl carbene insertion into ethereal C(sp^3)—H bonds (Scheme 19.7a) [28]. Their study disclosed the possibility of altering the relative stereochemical configuration at the C2 and C3 positions by varying the Rh catalysts. It was observed that employing the sterically bulky $Rh_2(tpa)_4$ as the catalyst afforded the cis product with high diastereoselectivity, where $Rh_2(OAc)_4$ in the presence of 2-picoline favored the formation of the trans diastereomer. One of the isolated cis diastereomer **19.7d** was also showcased to be a valuable intermediate for the direct preparation of fused N-heterocycles **19.7f** and **19.7e** through different intramolecular C—N bond-forming methods such as C—H amination and the Pictet–Spengler reaction.

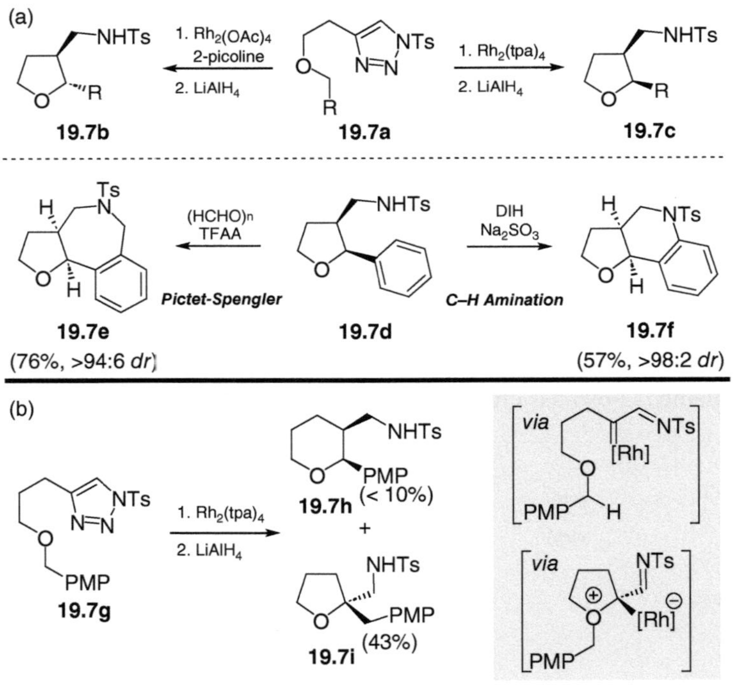

Scheme 19.7 (a) Stereodivergent synthesis of saturated 2,3-disubstituted tetrahydrofurans and the subsequent N-heterocycle synthesis via C—N bond forming reactions. (b) Substrate with extended tethers that favor the oxonium ylide formation.

In substrates such as **19.7a**, the Rh-carbenoid species prefer to undergo C—H insertion over forming the oxonium ylide to generate the corresponding oxetane-derived product. However, in homologated substrates such as

19.7g, the oxonium ylide generation overtook the C—H insertion pathway to favor the formation of the corresponding tetrahydrofuran-derived compound **19.7i** over the C—H inserted product **19.7h**. Considering the tunability of the denitrogenative methods toward forming varying heterocycles with different substitution patterns in high diastereoselectivity, it offers a unique opportunity to construct both polysubstituted and ring-fused tetrahydrofuran-derived compounds with high precision.

19.8 Synthesis of (±)-GSK1360707

Azabicyclo [n.1.0] frameworks have received considerable interest over the years as it is identified in numerous natural products and pharmaceutical agents. Rapidly constructing such bicyclic skeletons from linear precursors in one step is highly desirable over multistep synthesis. In 2015, Pan and Xu disclosed one such methodology where azabicyclo[4.1.0]/[5.1.0] carbaldehydes were synthesized via intramolecular Rh-catalyzed denitrogenative cyclopropanation of *N*-allyl- or *N*-homoallyl-tethered triazoles [29]. After careful optimization; the authors identified $Rh_2(OAc)_4$ to be the most suitable catalyst to effect this transformation. The method's utility was also demonstrated in synthesizing the antidepressant agent GSK1360707 (Scheme 19.8).

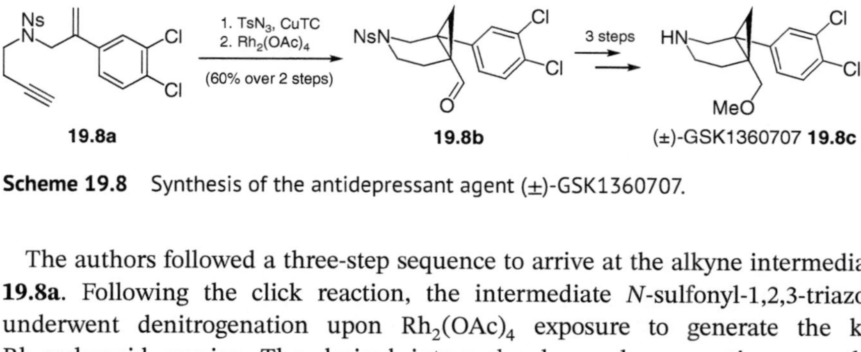

19.8a **19.8b** (±)-GSK1360707 **19.8c**

Scheme 19.8 Synthesis of the antidepressant agent (±)-GSK1360707.

The authors followed a three-step sequence to arrive at the alkyne intermediate **19.8a**. Following the click reaction, the intermediate *N*-sulfonyl-1,2,3-triazole underwent denitrogenation upon $Rh_2(OAc)_4$ exposure to generate the key Rh-carbenoid species. The desired intramolecular cyclopropanation smoothly assembled the azabicyclo[4.1.0] core. The immediately formed tosyl imine moiety was hydrolyzed in the same pot to obtain carbaldehyde **19.8b** as the only isolated product. After a few straightforward late-stage modifications, including reduction, methylation, and nosyl cleavage, the synthesis of GSK1360707 was completed.

19.9 Access to the Framework of Aspidosperma and Kopsia Indole Alkaloids

Indole alkaloids constitute one of the largest families of natural products, possessing unique chemical structures and prominent biological properties [30]. Selected few examples include vindoline **19.9a**, kopsifoline D **19.9b**, and aspidospermidine **19.9c** (Scheme 19.9a). A plethora of practical approaches continue to be developed toward constructing the indoline framework in a stereoselective fashion. Toward this end, Li and Zhai reported a complementary approach to assemble such scaffolds rapidly from relatively less complex starting materials by employing denitrogenative cascade tools. The authors envisioned a unique strategy of stitching the skeleton together by incorporating an intramolecular [3 + 2] cycloaddition strategy (Scheme 19.9b) [31].

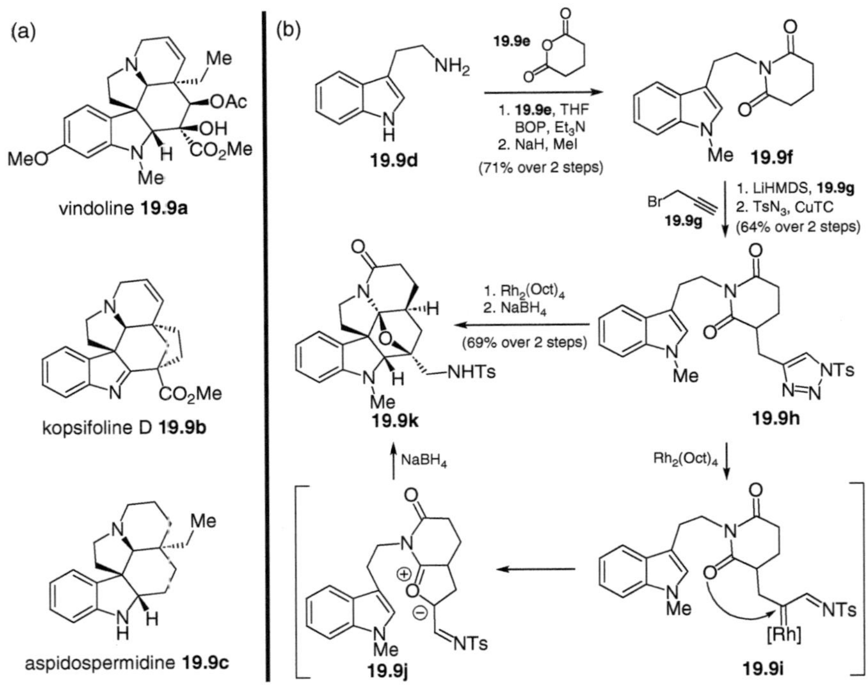

Scheme 19.9 (a) Selected examples of indole alkaloids. (b) Synthetic route to rapidly construct an indoline framework by employing the denitrogenative cascade tool.

To test the hypothesis, the authors executed a four-step straightforward synthesis to arrive at the essential *N*-sulfonyl-1,2,3-triazole intermediate **19.9h** starting from tryptamine **19.9d**. It was expected that upon successfully generating the electrophilic α-imino Rh-carbene complex **19.9i**, the intramolecularly positioned carbonyl oxygen would interact to furnish an oxonium ylide **19.9j**. The ylide would then serve as a C,C-1,3-dipole capable of undergoing a cycloaddition reaction with an olefin. In this example, the indole C2—C3 double bond behaves as the olefin to undergo [3 + 2] cycloaddition with the oxonium ylide to furnish the pentacyclic intermediate in a quick fashion. Upon subsequent reduction, pentacycle **19.9k** was synthesized as a single diastereomer by successfully implementing the denitrogenative intramolecular [3 + 2] annulation strategy. This annulation protocol could serve as a valuable entry to the existing toolbox of methods that construct polysubstituted indoline frameworks, owing to its operational simplicity and the mild conditions.

19.10 Total Synthesis of (±)-Aurantioclavine

Recently, Li and co-workers reported a denitrogenative mediated [4 + 3] cyclization reaction with several indole derivatives to rapidly construct azepino[5,4,3-*cd*]indole scaffolds. The broad utility of the method was demonstrated in the total synthesis of the azepinoindole alkaloid (±)-aurantioclavine **19.10f** (Scheme 19.10) [32]. This natural product represents the simplest member of the azepinoindole alkaloid family, isolated from *Penicillium aurantiovirens* by Kozlovskii and co-workers in 1981 [33]. Additionally, (±)-aurantioclavine **19.10f** is proposed as a biosynthetic intermediate in various communesins alkaloid synthesis [34]. Thus, devising a synthetic plan to access **19.10f** in a quick fashion will be an essential addition to the ongoing indole alkaloid-related studies.

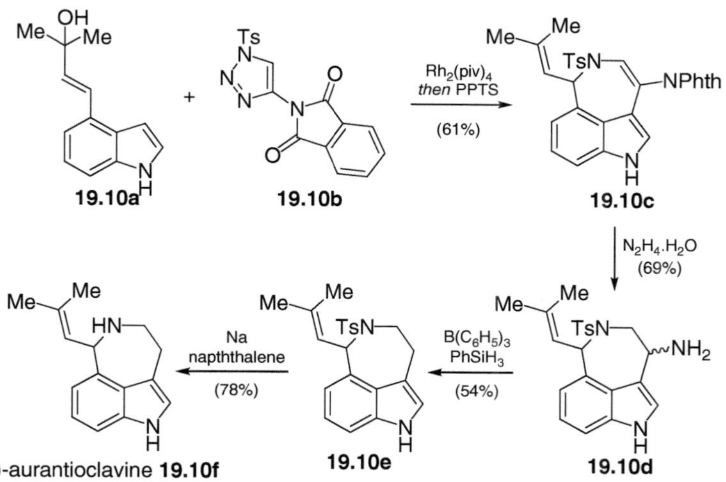

Scheme 19.10 Total synthesis of (±)-aurantioclavine.

Toward this end, the authors executed a [4 + 3] cyclization strategy between indole **19.10a** and the α-imino Rh-carbenoid species generated from *N*-sulfonyl-1,2,3-triazole **19.10b** to produce tricycle **19.10c** in 61% yield. Mechanistically, it was postulated that a nucleophilic addition from the C3 position of indole **19.10a** with the Rh-carbene complex would bind the two molecules together. A subsequent acid-promoted intramolecular SN2′ reaction would form the tricycle core, thereby constituting a net [4 + 3] cyclization reaction. Having gained access to **19.10c**, a straightforward reduction of the double bond followed by cleavage of the phthalimide group furnished intermediate **19.10d**. A subsequent deamination reaction followed by tosyl cleavage completed the total synthesis of (±)-aurantioclavine **19.10f**.

19.11 Total Synthesis of Nakafuran-8

In 2021, Li and co-workers disclosed an unusual type II intramolecular [3+2] dipolar cycloaddition strategy for synthesizing several bridged bicyclo[m.n.2] ring systems [35]. Key highlights of the strategy include the high regioselectivity and diastereoselectivity observed in the reaction outcome. This method facilitates synthesizing challenging bridged medium-sized polycyclic systems, which are common in natural products and potent pharmaceuticals. The developed method was further employed in the asymmetric synthesis of nakafuran-8 (Scheme 19.11). Nakafuran-8 **19.11h** is identified as an antifeedant against common reef fishes [36] and has been tested to possess inhibitory activity against human protein tyrosine phosphatase 1B, an enzyme involved in insulin signal regulation [37]. Structurally, **19.11h** possesses a unique bicyclo[4.2.2]decadiene framework decorated with a bridgehead all-carbon quaternary stereocenter and a fused furan ring. The authors thus undertook the daunting task of constructing the bicyclic core in a single-step operation by employing the devised [3+2] cycloaddition strategy.

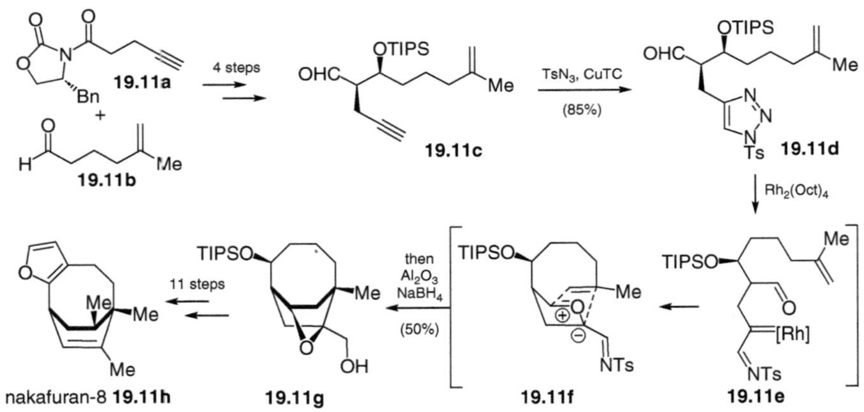

Scheme 19.11 Total synthesis of nakafuran-8.

Their synthetic route permitted a quick assembly of the key alkyne **19.11c** from readily available chiral compound **19.11a** and aldehyde **19.11b** in a short four-step sequence. The alkyne moiety in **19.11c** underwent the click reaction with TsN_3 to furnish *N*-sulfonyl-1,2,3-triazole **19.11d**. Under Rh-catalyzed conditions, denitrogenation was expected to generate α-imino Rh-carbene **19.11e**. With an aldehyde functionality in proximal, the formation of 1,3-dipole **19.11f** was anticipated. This sets the stage for the desired intramolecular [3+2] cycloaddition reaction between the oxonium ylide and the olefin component to assemble the polycyclic core. Subsequent hydrolysis and reduction in the same pot gave the desired polycyclic intermediate **19.11g** in an overall 50% yield. Having the core intact, a further 11-step sequence was employed to complete the synthesis of nakafuran-8 **19.11h**.

This strategy of constructing bridged bicyclo[m.n.2] ring systems from acyclic precursors in a single step offers a unique opportunity to rapidly arrive at such challenging scaffolds.

19.12 Photo-Induced Denitrogenative Transformation to Assemble an Indole Alkaloid Framework

The examples discussed thus far focused on transition metal-catalyzed denitrogenative transformations. Recently, efforts have been directed toward photo-induced denitrogenative transformations to generate radical intermediates and obtain varied reactivities. In this regard, Tang and co-workers explored the possibility of effecting a photo-induced denitrogenative annulation of a variety of 1-alkenylbenzotriazoles to construct polycyclic skeletons that are characteristic of monoterpene indole alkaloids (Scheme 19.12) [38]. 1,2,3-Benzotriazoles are an essential class of heterocycles with widespread applications in organic synthesis, materials science, and medicinal chemistry. In the past, 1,2,3-benzotriazole has been successfully identified as a precursor to undergo photo-mediated nitrogen extrusion to generate reactive N-centered radicals [39, 40]. Due to the appealing synthetic potential, this strategy has gained significant attention from the community to harness different methods that could innovatively leverage the generated N-centered radical intermediate.

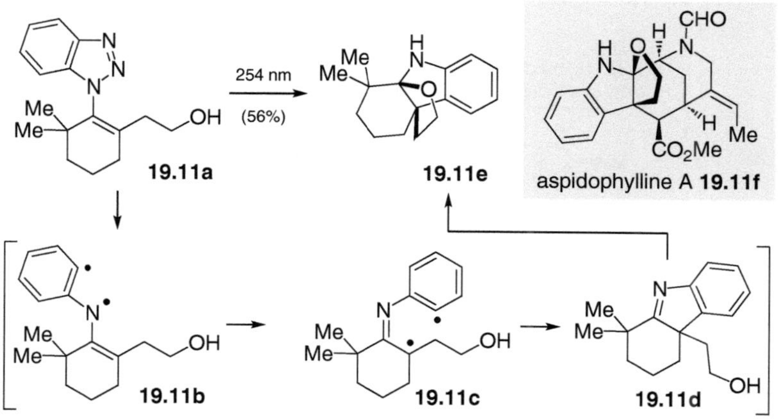

Scheme 19.12 Photo-induced denitrogenative annulation of 1-alkenylbenzotriazole with a tethered nucleophile.

In this work, the authors envisioned to effect the photo-induced denitrogenative annulation of 1-alkenylbenzotriazole with a tethered nucleophile to assemble selected indole alkaloid skeletons rapidly. The photoreaction was conducted with substrate **19.12a** with a 254 nm UV lamp as the light source. Interestingly, polycycle **19.12e** was isolated in 56% yield. Mechanistically, it is postulated that upon nitrogen extrusion, the generated diradical species **19.12b** would be under radical recombination via **19.12c** to furnish cyclic imine **19.12d**, which would then be trapped intramolecularly by the tethered hydroxy nucleophile to arise at **19.12e**. The assembled skeleton represents an integral portion of the indole alkaloid aspidophylline A **19.12f**. Considering the widespread synthetic potential of such transformations, it will be noteworthy to witness the growth of photo-mediated denitrogenative transformations to construct other challenging heterocycles.

19.13 Total Synthesis of Newbouldine and Withasomnine

Following pioneering work by Huisgen and co-workers demonstrating tetrazoles as a precursor to nitrile imine via denitrogenation [41], several groups have worked toward leveraging this strategy to achieve unique transformations. Toward this end, Gin and co-workers reported a photo-induced denitrogenation and subsequent intramolecular nitrile imine 1,3-dipolar cycloaddition by employing 2-alkyl-5-(methylthio)tetrazoles as precursors to afford a wide range of polycyclic pyrazoline products with high diastereoselectivity [42]. The synthetic utility of the method was further extended to the total synthesis of alkaloid natural products, newbouldine **19.13f** and withasomnine **19.13i** (Scheme 19.13). Both these natural products belong to the pyrrolopyrazole alkaloid family and are isolated from plant sources, which traditionally are known to possess rich medicinal properties against several ailments [43, 44].

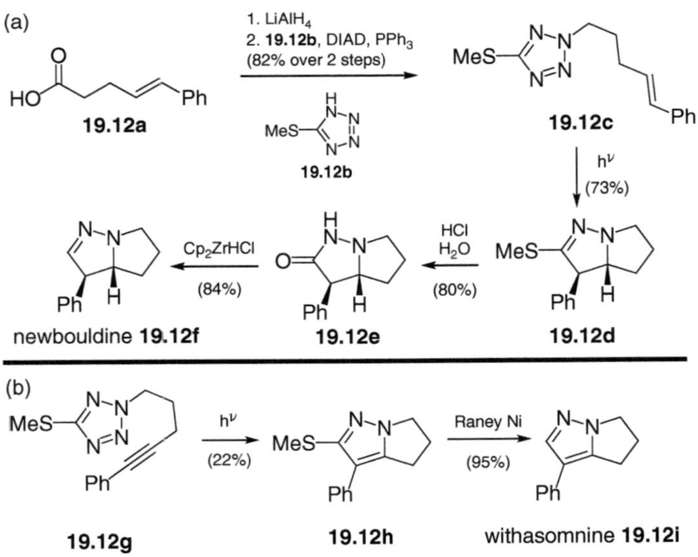

Scheme 19.13 Total synthesis of (a) newbouldine and (b) withasomnine.

The synthetic route to newbouldine **19.13f** commenced with the reduction of carboxylic acid **19.13a** followed by Mitsunobu reaction of the resulting alcohol with 5-(methylthio)tetrazole **19.13b** to afford the alkylated product **19.13c**. Upon photolysis, loss of molecular nitrogen from **19.13c** proceeded effectively to generate the nitrile imine intermediate, which underwent subsequent [3 + 2] cycloaddition *in situ* with the olefin group to furnish the syn-pyrrolopyrazoline **19.13d** in an overall 73% yield. The presence of the *C*-methylthio group is critical to effecting the initial photo-denitrogenation step as it is proposed to redshift the UV absorbance of the tetrazole component. Next, a two-step protocol involving initial hydrolysis of **19.13d** to pyrazolidinone **19.13e** followed by selective reduction of the hydrazide moiety with Schwartz's reagent gave rise to newbouldine **19.13f** in a five-step sequence with an overall 40% yield starting from **19.13a.** Analogously, alkynyl tetrazole **19.13g** underwent photo-induced cycloaddition to the corresponding pyrrolopyrazole **19.13h,** albeit in a low 22% yield. A subsequent Raney Ni-mediated desulfurization completed the synthesis of withasomnine **19.13i**.

19.14 Conclusion

In summary, developing synthetic strategies that leverage the denitrogenation protocol to effect novel and elegant transformations to construct heterocyclic building blocks is highly warranted. Notably, the reported total synthesis examples demonstrate the sheer potential of such transformation in elegantly synthesizing molecules with therapeutic potential. In several synthetic routes, it is evident that upon employing a denitrogenative cascade reaction, rapid complexity is built in a single-step operation from acyclic precursors. Additionally, many of the heterocycles can be assembled in a late-stage setting, bringing in a unique opportunity to redesign retrosynthetic analysis while leveraging methods involving denitrogenative protocols. It is thereby anticipated that this exciting development will broaden the scope of possible structural analogs that could be synthesized for several natural products and, in consequence, could streamline the field of drug discovery.

As witnessed, the poly-nitrogen heterocycles have been successfully explored as diazo surrogates to promote a myriad of denitrogenative transformations, including transannulations, ylide formations, insertions, rearrangements with several coupling partners such as alkenes, alkynes, carbocycles, C—*X* bonds and so on. Despite such advancements as independent methodologies, its application in the synthesis of bioactive molecules is still in its infancy. Although other transition metal catalysts have been examined, most of the transformations are limited to rhodium carboxylate-based catalysts to promote denitrogenation of *N*-sulfonyl-1,2,3-triazoles. With the exponential growth of photoredox catalysis, it will be interesting to witness several new combinations of transition metal catalysts to promote denitrogenation in other poly-nitrogen heterocycles to generate a library of reactive intermediates. With such developments, one can anticipate a profusion of denitrogenative transformations that enable bond constructions beyond those presented in this chapter and could offer new directions for innovative total syntheses.

References

1 Vitaku, E., Smith, D.T., and Njardarson, J.T. (2014). Analysis of the structural diversity, substitution patterns, and frequency of nitrogen heterocycles among U.S. FDA-approved pharmaceuticals. *J. Med. Chem.* 57: 10257–10274.

2 Nakamura, T. and Yamamoto, Y. (2004). Transition-metal-catalyzed reactions in heterocyclic synthesis. *Chem. Rev.* 104: 2127–2198.

3 Seregin, I.V. and Gevorgyan, V. (2007). Direct transition metal-catalyzed functionalization of heteroaromatic compounds. *Chem. Soc. Rev.* 36: 1173–1193.

4 Huisgen, R., Fraunberg, K.V., and Sturm, H.J. (1969). Cycloadditionen von pyridyl-und pyrimidyl-aziden. *Tetrahedron Lett.* 10: 2589–2594.

5 Chattopadhyay, B., Rivera, V.C.I., Chuprakov, S., and Gevorgyan, V. (2010). Fused tetrazoles as azide surrogates in click reaction: efficient synthesis of N-heterocycle-substituted 1,2,3-triazoles. *Org. lett.* 12: 2166–2169.

6 Das, S.K., Roy, S., and Chattopadhyay, B. (2023). Transition-metal-catalysed denitrogenative annulation to access high-valued N-heterocycles. *Angew. Chem. Int. Ed.* 62: e202210912.

7 Roy, S., Das, S.K., Khatua, H. et al. (2021). Road map for the construction of high-valued N-heterocycles via denitrogenative annulation. *Acc. Chem. Res.* 54: 4395–4409.

8 Desnoyer, A.N. and Love, J.A. (2017). Recent advances in well-defined, late transition metal complexes that make and/or break C—N, C—O and C—S bonds. *Chem. Soc. Rev.* 46: 197–238.

9 Katritzky, A.R., Lan, X., Yang, J.Z., and Denisko, O.V. (1998). Properties and synthetic utility of N-substituted benzotriazoles. *Chem. Rev.* 98: 409–548.

10 Yang, W.-C., Chen, C.-Y., Li, J.-F., and Wang, Z.-L. (2021). Radical denitrogenative transformations of polynitrogen heterocycles: building C—N bonds and beyond. *Chinese J. Catal.* 42: 1865–1875.

11 Schultz, E.E. and Sarpong, R. (2013). Application of in situ-generated Rh-bound trimethylenemethane variants to the synthesis of 3,4-fused pyrroles. *J. Am. Chem. Soc.* 135: 4696–4699.

12 Lee, J.S., Kim, Y.-S., Park, S. et al. (2011). Exceptional production of both prodigiosin and cycloprodigiosin as major metabolic constituents by a novel marine bacterium, Zooshikella rubidus S1-1. *Appl. Environ. Microbiol.* 77: 4967–4973.

13 Wasserman, H.H. and Fukuyama, J.M. (1984). The synthesis of (±)-cycloprodigiosin. *Tetrahedron Lett.* 25: 1387–1388.

14 Aldrich, L.N., Dawson, E.S., and Lindsley, C.W. (2010). Evaluation of the biosynthetic proposal for the synthesis of marineosins A and B. *Org. Lett.* 12: 1048–1051.

15 Miura, T., Funakoshi, Y., and Murakami, M. (2014). Intramolecular dearomatizing [3 + 2] annulation of a-imino carbenoids with aryl rings furnishing 3,4-fused indole skeletons. *J. Am. Chem. Soc.* 136: 1972–1975.

16 Berde, B. and Schild, H.O. (1978). Ergot alkaloids and related compounds. In: *Handbook of Experimental Pharmacology*. Berlin: Springer-Verlag.

17 Yuan, H., Guo, Z., and Luo, T. (2017). Synthesis of (+)-lysergol and its analogues to assess serotonin receptor activity. *Org. Lett.* 19: 624–627.

18 Inuki, S., Iwata, A., Oishi, S. et al. (2011). Enantioselective total synthesis of (+)-lysergic acid, (+)-lysergol, and (+)-isolysergol by palladium-catalyzed domino cyclization of allenes bearing amino and bromoindolyl groups. *J. Org. Chem.* 76: 2072–2083.

19 Lu, J.-T., Shi, Z.-F., and Cao, X.-P. (2017). Total synthesis of (−)-chanoclavine I and an oxygen-substituted ergoline derivatives. *J. Org. Chem.* 82: 7774–7782.

20 Kardos, N. and Genet, J.-P. (1994). Synthesis of (−)-chanoclavine I. *Tetrahedron: Asymmetry* 5: 1525–1533.

21 Yokoyama, Y., Kondo, K., Mitsuhashi, M., and Murakami, Y. (1996). Total synthesis of optically active chanoclavine I. *Tetrahedron Lett.* 37: 9309–9312.

22 Zhang, G., Wu, G., Zhu, T. et al. (2013). Meroterpenoids with diverse ring systems from the sponge-associated fungus *Alternaria* sp. JJY-32. *J. Nat. Prod.* 76: 1946–1957.

23 Fu, J., Shen, H., Chang, Y. et al. (2014). Concise stereoselective synthesis of oxaspirocycles with 1-tosyl-1,2,3-triazoles: application to the total syntheses of (±)-tuberostemospiroline and (±)-stemona-lactam R. *Chem. Eur. J.* 20: 12881–12888.

24 Hu, J.-P., Yang, D.-H., Lin, W.-H., and Cai, S.-Q. (2009). Alkaloids from the roots of *stemona tuberosa*. *Helv. Chim. Acta.* 92: 2125–2133.

25 Hitotsuyanagi, Y., Fukaya, H., Takeda, E. et al. (2013). Structures of stemona-amine B and stemona-lactams M–R. *Tetrahedron* 69: 6297–6304.

26 Boyer, A. (2015). Enantioselective synthesis of (+)-petromyroxol, enabled by rhodium-catalyzed denitrogenation and rearrangement of a 1-sulfonyl-1,2,3-triazole. *J. Org. Chem.* 80: 4771–4775.

27 Li, K., Huertas, M., Brant, C. et al. (2015). (+)- and (−)-petromyroxols: antipodal tetrahydrofurandiols from larval sea lamprey (*Petromyzon marinus* L.) that elicit enantioselective olfactory responses. *Org. Lett.* 17: 286–289.

28 Lindsay, V.N.G., Viart, H.M.-F., and Sarpong, R. (2015). Stereodivergent intramolecular $C(sp^3)$—H functionalization of azavinyl carbenes: synthesis of saturated heterocycles and fused N-heterocycles. *J. Am. Chem. Soc.* 137: 8368–8371.

29 Pan, X.-H., Jiang, P., Jia, Z.-H. et al. (2015). Expedient catalytic construction of azabicyclo[4.1.0]/[5.1.0] carbaldehydes via intramolecular cyclopropanation. *Tetrahedron* 71: 5124–5129.

30 Omar, F., Tareq, A.M., Alqahtani, A.M. et al. (2021). Plant-based indole alkaloids: a comprehensive overview from a pharmacological perspective. *Molecules* 26: 1997.

31 Li, Y., Zhang, Q., Du, Q., and Zhai, H. (2016). Rh-catalyzed [3 + 2] cycloaddition of 1-sulfonyl-1,2,3-triazoles: access to the framework of aspidosperma and kopsia indoles alkaloids. *Org. Lett.* 18: 4076–4079.

32 Duan, S., Xue, B., Meng, H. et al. (2021). Rhodium(II)-catalyzed [4+3] cyclization of triazoles with indole derivatives and its application in the total synthesis of (±)-aurantioclavine. *Chin. J. Chem.* 39: 1145–1152.

33 Kozlovskii, A.G., Solov'eva, T.F., Sakharovskii, V.G., and Adanin, V.M. (1981). Biosynthesis of "unusual" ergot alkaloids by the fungus penicillium aurantio-virens. *Dokl. Akad. Nauk SSSR* 260: 230–233.

34 Lin, H.C., Chiou, G., Chooi, Y.H. et al. (2015). Elucidation of the concise biosynthetic pathway of the communes in indole alkaloids. *Angew. Chem. Int. Ed.* 54: 3004–3007.

35 Hou, B.-L., Wong, J.J., Lv, N. et al. (2021). Facile generation of bridged medium-sized polycyclic systems by rhodium-catalysed intramolecular (3 + 2) dipolar cycloadditions. *Nat. Commun.* 12: 5239.

36 Schulte, G., Scheuer, P.J., and McConnell, O.J. (1980). Two furanosesquiterpene marine metabolites with antifeedant properties. *Helv. Chim. Acta.* 63: 2159–2167.

37 Huang, X.-C., Li, J., Li, Z.-Y. et al. (2008). Sesquiterpenes from the hainan sponge dysidea septosa. *J. Nat. Prod.* 71: 1399–1403.

38 Wang, Z., Chen, Y., Dong, Z., and Tang, Y. (2023). Natural product-oriented photo-induced denitrogenative annulations of 1-alkenylbenzotriazoles. *Molecules* 28: 363.

39 Katritzky, A.R. and Rachwal, S. (2010). Synthesis of heterocycles mediated by benzotriazole. 1. Monocyclic systems. *Chem. Rev.* 110: 1564–1610.

40 Katritzky, A.R. and Rachwal, S. (2011). Synthesis of heterocycles mediated by benzotriazole. 2. Bicyclic systems. *Chem. Rev.* 111: 7063–7120.

41 Huisgen, R., Seidel, M., Sauer, J. et al. (1959). The formation of nitrile imines in the thermal breakdown of 2,5-disubstituted tetrazoles. *J. Org. Chem.* 24: 892–893.

42 Pla, D., Tan, D.S., and Gin, D.Y. (2014). 5-(Methylthio)tetrazoles as versatile synthons in the stereoselective synthesis of polycyclic pyrazolines *via* photoinduced intramolecular nitrile imine–alkene 1,3-dipolar cycloaddition. *Chem. Sci.* 5: 2407–2415.

43 Schroter, H.-B., Neumann, D., Katritzky, A.R., and Swinbourne, F.J. (1966). Withasomnine. A pyrazole alkaloid from *Withania somnifera* Dun. *Tetrahedron* 19: 2895–2897.

44 Adesanya, S.A., Nia, R., Fontaine, C., and Pais, M. (1994). Pyrazole alkaloids from *Newbouldia laevis*. *Phytochemistry* 35: 1053–1055.

Index

Denitrogenative Transformation of Nitrogen Heterocycles: Synthesis, Reactions and Applications, First Edition. Edited by Pazhamalai Anbarasan.

b

n

q

r